CONCEPTS OF GENETICS

William S. Klug

Trenton State College at Hillwood Lakes

Michael R. Cummings

University of Illinois, Chicago

CONCEPTS OF GENETICS

Second Edition

Scott, Foresman and Company

Glenview, Illinois
London, England

Cover Computer Graphics: A, B, and Z DNA. Produced by
Nelson L. Max at the Lawrence Livermore National Laboratory.
Coordinates provided by Richard Dickerson of the University of
California at Los Angeles.

Part Opening Photos

Part One (pp. 10–11): Light micrograph of root tip cells in
various stages of division. (Courtesy of John McLeish.)

Part Two (166–67): Scanning electron micrograph of a
portion of a lampbrush chromosome. (Courtesy of Nicole
Angelier.)

Part Three (pp. 260–61): Electron microscopic visualization
of a bacteriophage.

Part Four (pp. 378–79): Electron microscopic visualization of
the process of transcription of DNA (RNA synthesis).
(Courtesy of Oscar Miller.)

Part Five (pp. 530–31): Golden poppy. (Grant Heilman
Photography.)

Library of Congress Catalog Card Number: 86-061996

ISBN: 0-673-18680-6

   4 5 6 7 8   VHJ   90 89 88 87

PREFACE
to first edition

The Title *Concepts of Genetics* conveys the basic pedagogic philosophy that has shaped this textbook. Throughout the text we have emphasized the concepts or general ideas that have been developed to explain genetic phenomena at all levels—the molecule, cell, organism, and population. The word *concept* implies that sense can be made of the abstract. Indeed, no discipline in biology better illustrates this transition in the acquisition of knowledge. We have attempted to minimize the presentation of highly technical and detailed information which for most students tends to blur the many important concepts and their logical development.

The information presented herein is a meld of the most significant findings in the science of heredity, including those in transmission genetics, cytogenetics, microbial genetics, molecular genetics, and population-evolutionary genetics. These include the original work of Mendel in the nineteenth century, the advances in genetic technology of the current and most recent decades, and those made between these periods.

Writing style has been carefully developed and edited solely with the student in mind. Our primary goals have been clear writing and meaningful figures, yet we have remained rigorous in our coverage of topics, believing that excess detail cannot be equated with rigor. This text is designed for use in all introductory genetics courses, whether offered in a semester, in a quarter, or in consecutive quarters.

For the instructor, we have created a flexible format and organization. The book is divided into five parts. The basic concepts of transmission genetics are discussed in Part 1. Within this section, meiosis is discussed before Mendelian genetics, which uses as its foundation the information about the behavior of chromosomes during gamete formation. Part 2 focuses on DNA—its structure, analysis, and replication. This section concludes with a chapter on the organization of DNA in chromosomes. Part 3 concentrates on genetic variability at both the chromosome and gene levels. The discussion of mutation and mutagenesis logically precedes the consideration of bacterial and viral genetics. These chapters provide a strong foundation for a more detailed consideration of molecular genetics in Part 4. Chapter 19 provides a modern synthesis of old and new findings concerning the organization of DNA into the functional unit of heredity, the gene. Part 5 includes discussions of developmental, somatic cell, and behavioral genetics as well as chapters on the genetics of populations and evolution. We believe that separate chapters on somatic cell genetics and behavior genetics are essential, since they represent two areas that are likely to expand our knowledge considerably over the next decade. In general, the various parts and the chapters within them may be used interchangeably. This is particularly true of the first two parts, which stand completely independently of each other.

This work has been in progress for over five years, allowing several revisions and reviews for content and clarity. In the final draft, we were able to avoid the organizational idiosyncrasies of our own courses. Instead, we have written a text that will be easily adaptable to a wide variety of course formats. We have never, however, lost sight of our primary goal—to write a text that will provide students with a rewarding learning experience and that conveys clearly the concepts of genetics.

William S. Klug
Michael R. Cummings
January, 1983

To those who mean the very most,
Donna, Cindy, Brad, and Dori
Lee Ann, Brendan, and Kerry

Captains and kings may rule the world,
but it is only the presence and memory of those we love
and with whom we share our intellectual and genetic heritage
which bring beauty to living and justify their good deeds

WILLIAM S. KLUG is presently serving as Professor of Biology and as Chairman of the Department of Biology at Trenton State College at Hillwood Lakes. He was previously a member of the faculty of Wabash College, Crawfordsville, Indiana. The focus of his research interest is on developmental genetics in *Drosophila*, with an emphasis on genes that control oogenesis. Dr. Klug is a member of the New Jersey State Panel of Science Advisors and has numerous research publications in scientific journals.

MICHAEL R. CUMMINGS, currently Associate Professor of Biological Sciences and Research Associate Professor at the Institute for the Study of Developmental Disabilities, University of Illinois, Chicago, has also served on the faculties of Northwestern University and Florida State University, Tallahassee. His research is concerned with the molecular consequences of aneuploidy, particularly in the molecular organization of chromosomes and the regulation of gene expression in Down syndrome. He has published many research papers in various scientific journals.

PREFACE

Creating the second edition of a textbook is both a source of great satisfaction and an opportunity. Satisfaction stems from the fact that the first edition has been successful and warrants the continued support of the publisher and the efforts of the authors. The opportunity presents itself as an invitation to build upon the strong points of the first edition, and to correct its shortcomings. If planned and executed carefully, the second edition of a text may not require substantial reorganization for many years.

We have been very gratified by the initial response to *Concepts of Genetics.* In general, it has been viewed as a well-organized, clearly written, and nicely illustrated text. Many faculty and students have indicated that the clarity and style of writing and the aesthetic and technical quality of the many figures are the best they have encountered in a science textbook. These responses have made the mammoth effort that went into the first edition worth every minute spent on it.

The second edition, which has required only minor reorganization and careful modernization, contains or reflects:

1 new chapters on quantitative genetics (Chapter 7) and recombinant DNA (Chapter 20), indicative of the importance these areas hold in teaching genetics.

2 the condensation of all sex-related genetic topics (sex linkage, sex determination, etc.) into Chapter 5.

3 modernization of all topics where new information has become available since 1983. In particular, information pertaining to genetic structure and function in eukaryotes has been extended. A large number of references that have appeared since 1983 have also been added to the Selected Readings at the end of each chapter.

4 the addition of new topics, including eukaryotic enhancers, oncogenes, fragile sites, splicing mechanisms, hybridomas, thalassemia, and restriction fragment length polymorphisms, with appropriate references.

5 expansion of the bacterial–viral genetics coverage in Chapter 14.

6 fine tuning of most existing figures and the creation of about 50 new ones.

7 about 75 new additions to the Problems and Discussion Questions. For the most part, the new problems are found near the end of each such section and are of the "more challenging" variety.

8 the creation of a new appendix (Appendix B) containing an expanded discussion of each human genetic disorder mentioned in the text. Additionally, an up-to-date human gene map of all chromosomes is included in that appendix.

More generally, we have made every effort to continue to emphasize concepts rather than excess detail. This approach seems to be even more important now than in 1983. As the information explosion continues to revolutionize genetics, it has become more and more difficult to do more than survey the most basic aspects of the vast field in a one-semester course. Our goal has been to create an improved edition that will continue to serve as a useful adjunct for faculty as they teach genetics, and as the basis of a rewarding learning experience for students.

William S. Klug
Michael R. Cummings

ACKNOWLEDGMENTS

Few, if any, textbooks are written without the support and professional contributions of many people. This book is no exception. First, we wish to thank Bob Lakemacher for his belief in and continued support of this project. Also at Merrill Publishing, we are extremely grateful to Linda Thornhill and Ben Shriver, who followed the manuscript of the second edition through the production process, and to Mary Lou Motl of Custom Editorial Productions, who ably served as the copy editor.

A textbook depends in large part on the input provided by reviewers. We were fortunate to have had the manuscript of the first edition assessed at several stages during its preparation. Among others, the reviewers included Glen C. Bewley, North Carolina State University; Charles C. Biggers, Memphis State University; H. E. Brockman, Illinois State University; Raymond P. Canham, Southern Methodist University; Lee Ehrman, State University of New York, Purchase; B. W. Greer, Knox College; Elliot S. Goldstein, Arizona State University; Victor J. Hoff, Austin State University; R. D. Jackson, Texas Tech University; Diana Johnson, George Washington University; Gary Kikudome, University of Missouri, Columbia; Robert M. Kitchin, University of Wyoming; Walter Rothenbuhler, Ohio State University; Eliot Spiess, University of Illinois, Chicago; and Dana L. Wrensch, Ohio State University.

The second edition has depended on the advice and expertise of still other individuals: Ted H. Emigh, North Carolina State University; Allan Stewart-Oaten, University of California, Santa Barbara; Thomas A. Cole, Wabash College; Kipp C. Kruse, Eastern Illinois University; Kenneth C. Jones, California State University, Northridge; Kenneth E. Michel, Slippery Rock University; Henry E. Schaffer, North Carolina State University; Richard Siegal, University of California at Los Angeles; Steven Weaver, University of Illinois, Chicago; Dennis Rio, Lewis University; and Barry Ganetzky, University of Wisconsin reviewed various portions of the manuscript. Additionally, we are extremely grateful to Barbara Hayhome, University of Nebraska at Omaha, and Kathleen Fleiszar, Kennesaw College, who examined the entire manuscript. Special thanks are also due to Elliott Goldstein, Arizona State University, for again sharing with us his wealth of knowledge in molecular genetics, and to Richard Davidson, University of Illinois at Chicago, for his many suggestions about somatic cell genetics.

The acquisition of photographs and permissions has placed us in contact with well over 100 geneticists. For both editions we are grateful to all those who dealt with us in a thoughtful and cooperative manner. Several have contributed above and beyond the call. In this regard, we thank Alexander Rich and those in his laboratory at the Massachusetts Institute of Technology who provided material for the cover of the first edition and many fine micrographs. Likewise, we are indebted to Oscar Miller, Jr., University of Virginia;

Jorge Yunis, University of Minnesota; Arthur E. Greene, Institute of Medical Research; and Irving Geis of New York City for their many contributions. Vickie Brewster, Iowa State University, created several valuable figures and provided innumerable other support services, including the preparation of the index.

On a day-to-day basis, we were dependent on a staff whose dedicated efforts were essential to meeting the many deadlines during the past several years. This staff included Irene Saproni and Saveria Symons at Trenton State College, and Helen Brown at the Institute for the Study of Developmental Disabilities. Their efficiency and pleasant manner aided immensely in the completion of this project.

The support of and discussions with colleagues have been no less important to us. We personally thank Tom Cole for his early participation in the conception of the project, Eliot Spiess for sharing his experiences as an author and his expertise in population genetics, Karl Gottesman for many discussions on writing style, and Al Eble and Joe Vena for their encouragement.

To all of these people we offer our sincere gratitude for their part in making this textbook a reality. Its success is in large part due to the high quality of their input.

William S. Klug
Michael R. Cummings
January, 1986

BRIEF CONTENTS

CONTENTS

1

An Introduction to Genetics

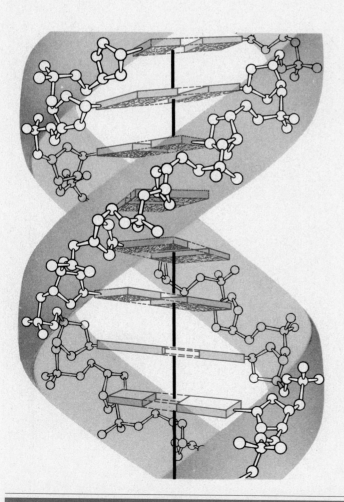

Welcome to the study of the discipline of genetics. You are about to embark on the exploration of a subject that many students before you have found to be the most interesting and fascinating in the field of biology. This is not surprising because an understanding of genetic processes is fundamental to the comprehension of life itself. Genetic information directs cellular function, determines an organism's external appearance, and serves as the link between generations in every species. Knowing how these processes occur is important to understanding the living world. Knowledge of genetic concepts also helps us to understand the other disciplines of biology. The topics studied in genetics overlap directly with molecular biology, cell biology, physiology, evolution, behavior, and ecology. Genetics is therefore said to unify biology and serve as its "core."

Interest in and fascination with this discipline further stem from the fact that, in genetics, countless initially vague and abstract concepts have been investigated in a logical fashion until they have become clearly and definitively understood. As a result, genetics has a rich history that exemplifies the nature of scientific discovery and the analytical approach used to acquire information. Scientific analysis, moving from the unknown to the known, is one of the major forces that attract students to biology.

But there is still another reason why the study of genetics is so appealing. Since its establishment, this field of study has expanded continuously. Every year large numbers of new findings are made. While it has been said that scientific knowledge doubles every ten years, one estimate holds that the doubling time in genetics is only two years. Certainly, over the past four decades, no two-year period has passed without some of the newly acquired information causing significant excitement for geneticists and for biologists in general. And each advance becomes part of an ever-expanding cornerstone upon which further progress is based. It is particularly stimulating to be in the midst of these developments, whether you are studying or teaching genetics.

BASIC CONCEPTS OF GENETICS

In this introductory chapter we would like to review some of the simple but basic concepts in genetics which you have undoubtedly already studied. By reviewing them at the outset, we can proceed through the text with a common foundation. We shall approach these basic concepts by asking and answering a series of questions. You may wish to write or think through an answer before reading the explanation of each question. Throughout the text, the answers to these questions will be expanded as more detailed information is presented.

What does genetics mean?

Genetics is the branch of biology concerned with the study of inherited variation. This discipline deals with the origin of genetic variation and the expression of traits resulting from this variation.

What is the center of heredity in a cell?

In eukaryotic organisms, the **nucleus** contains the genetic material. In prokaryotes such as bacteria, the genetic material exists in an unenclosed area of the cell called the **nucleoid region.** In viruses, which are not true cells, the genetic material is ensheathed in the protein coat referred to as the **viral head.**

What is the genetic material?

In eukaryotes and prokaryotes, **DNA** serves as the molecule storing genetic information. In viruses, either DNA or **RNA** serves this function.

What do DNA and RNA stand for?

DNA and RNA are abbreviations for **deoxyribonucleic acid** and **ribonucleic acid,** respectively. These are the two types of nucleic acids found in organisms. Nucleic acids, along with carbohydrates, lipids, and proteins, compose the four major classes of organic biomolecules found in living things.

How is DNA organized to serve as the genetic material?

DNA, while single-stranded in a few viruses, is usually a double-stranded molecule organized as a double helix. Contained within each DNA molecule are hereditary units called **genes,** which are part of a larger element, the **chromosome.**

What is a gene?

In simplest terms, the gene is the functional unit of heredity. In chemical terms, it is a linear array of nucleotides—the chemical building blocks of DNA and RNA. A more sophisticated approach is to consider it as an informational storage unit capable of undergoing replication, mutation, and expression. As investigations have progressed, the gene has been found to be a very complex genetic element.

What is a chromosome?

In viruses and bacteria, the chromosome is most simply visualized as a long, usually circular DNA molecule organized into genes. In eukaryotes, the chromosome is more complex. It is composed of a linear DNA molecule complexed with proteins. In addition to genes, the chromosome contains many regions that do not store genetic information. It is not yet clear what role, if any, is played by these components of chromosomes. Our knowledge of the chromosome, like the gene, is continually expanding.

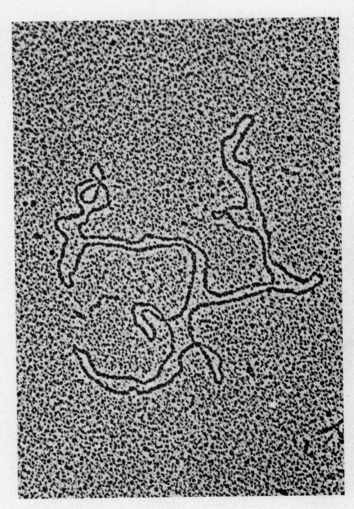

FIGURE 1.1

An electron micrograph of the DNA representing a viral chromosome. (Courtesy of Paul Godowski.)

When and how can chromosomes be visualized?

If the chromosomes are released from the viral head or the bacterial cell, they can be visualized under the electron microscope. In eukaryotes, chromosomes are most easily visualized under the light microscope when they are undergoing **mitosis** or **meiosis.** In these division processes, the chromosomes are tightly coiled and condensed. Following division, they uncoil and exist as **chromatin fibers** during interphase, when they can be studied under the electron microscope.

What is accomplished during the processes of mitosis and meiosis?

Mitosis is the process by which the genetic material of eukaryotic cells is duplicated and distributed during cell division. **Meiosis** is the process whereby cell division produces gametes in animals and plants. While mitosis occurs in somatic tissue and yields two progeny cells with an amount of genetic material identical to the progenitor cell, meiosis creates cells with precisely one-half of the genetic material. This accomplishment is essential if offspring arising from two gametes are to maintain an amount of genetic material characteristic of their parents and other members of the species.

What is the source of genetic variation?

Classically, there are two sources of genetic variation: **chromosomal** and **gene mutations.** The former, also called **chromosomal aberrations,** includes duplication, deletion, or rearrangement of chromosome segments. Gene mutations result from a change in the stored chemical information in DNA. In order to be studied, genetic variation must result in a detectable change in some characteristic of an organism.

How does DNA store genetic information, and how is it altered to produce variation?

The sequence of nucleotides in a segment of DNA constituting a gene is present in the form of a **genetic code.** This code specifies the chemical nature of proteins, which are the end product of genetic expression. Mutations are produced when a nucleotide is changed, which alters the genetic code.

How is the genetic code organized?

There are four types of nucleotides in DNA. The genetic code is a **triplet;** therefore, each combination of three nucleotides constitutes a code word. Almost all possible codes specify one of twenty **amino acids,** the chemical building blocks of proteins.

How is the genetic code expressed?

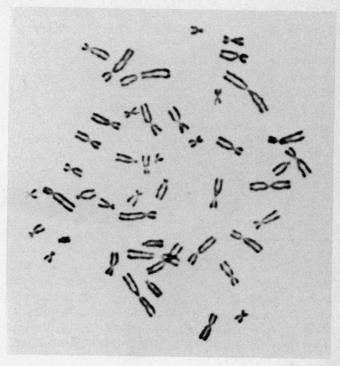

FIGURE 1.2

A photomicrograph of a standard mitotic metaphase preparation of human chromosomes. (Courtesy of Jorge J. Yunis.)

The coded information in DNA is first transferred during a process called **transcription** into a **messenger RNA (mRNA)** molecule. The mRNA subsequently associates with the cellular organelle, the **ribosome,** where it is **translated** into a protein molecule.

Are there exceptions where proteins are not the end product of a gene?

Yes. For example, genes coding for **ribosomal RNA (rRNA),** which is part of the ribosome, and for **transfer RNA (tRNA),** which is involved in the translation process, are transcribed but not translated. Therefore, RNA is sometimes the end product of stored genetic information.

Why are proteins so important to living organisms that they serve as the end product of the vast majority of genes?

Most proteins serve as highly specific biological catalysts, or **enzymes.** In this role, these proteins control cellular metabolism, determining which carbohydrates, lipids, nucleic acids, and other proteins are present in the cell. Many other proteins perform non-enzymatic roles. For example, hemoglobin, collagen, immunoglobulins, and some hormones are proteins.

Why are enzymes necessary in living organisms?
As biological catalysts, enzymes lower the activation energy required for most biochemical reactions and speed the attainment of equilibrium. Otherwise, these reactions would proceed so slowly as to be ineffectual in organisms living under the conditions on earth.

What key terms should be reviewed before proceeding through this text?
In addition to those appearing in bold type in the preceding discussion, there are several other particularly important terms. These are defined in the glossary at the end of the text: **alleles, complementarity, diploid** and **haploid, genome, homologous chromosomes,** and **locus.**

INVESTIGATIVE APPROACHES IN GENETICS

The scope of topics encompassed in the field of genetics is enormous. Studies have involved viruses, bacteria, and a wide variety of plants and animals and have spanned all levels of biological organization, from molecules to populations. With some overlap, most investigations have used one of four basic approaches.

The most classical investigative approach is the study of **transmission genetics,** in which the patterns of inheritance of traits are examined. Experiments are designed so that the transmission of traits from parents to offspring can be analyzed through several generations, and patterns of inheritance are sought that will provide insights into more general genetic principles. The first significant experimentation of this kind was performed by Gregor Mendel in the middle of the nineteenth century. Information derived from his work serves today as the foundation of transmission genetics. In human studies, where designed matings are neither possible nor desirable, **pedigree analysis** is used. In pedigree analysis, patterns of inheritance are traced through as many generations as possible, leading to predictions of the mode of inheritance of the trait under investigation.

The second approach involves **cytological investigations** of the genetic material. The earliest studies used the light microscope. The initial discovery in the early twentieth century of chromosome behavior during mitosis and meiosis was a critical event in the history of genetics. In addition to playing an important role in the rediscovery and acceptance of Mendelian principles, these observations served as the basis of the **chromosomal theory of inheritance.** This theory, which viewed the chromosome as the carrier of genes and the functional unit of transmission of ge-

FIGURE 1.3
Gregor Johann Mendel, who experimented with the garden pea *(Pisum sativum)* and discovered the fundamental principles of transmission genetics. (By permission of Bettmann Archives.)

netic information, was the foundation of both transmission and cytological studies throughout the first half of this century.

The light microscope continues to be an important research tool. It is useful in the investigation of chromosome structure and abnormalities and is instrumental in preparing **karyotypes,** which present the entire set of chromosomes characteristic of any species in an orderly pattern.

With the advent of electron microscopy, the repertoire of investigative approaches in genetics has grown. In high-resolution microscopy genetic molecules and their behavior during gene expression can be directly visualized.

The third general approach, **molecular and biochemical analysis,** has had the greatest impact on the growth of genetic information. Most often involving viruses and bacteria as experimental sources, molecular studies beginning in the early 1940s have consistently expanded our knowledge of the role of genetics in life processes. Detailed information is now available concerning the nature, expression, replication, and regulation of the genetic information. The precise nucleotide sequence of many genes has been determined and artificially propagated in the laboratory.

Recombinant DNA technology, where genes from any organism are literally spliced into bacterial or viral DNA and cloned *en masse,* is the most significant research tool yet developed. As a result, it is now possible to probe gene structure and function with a resolution heretofore impossible. This technology also has profound implications in medicine and agriculture.

The final approach involves the study of the **genetic structure of populations.** In these investigations scientists attempt to define how and why certain genetic variation is preserved in populations while other variation diminishes or is lost with time. Such information is critical to the understanding of the evolutionary process. Population genetics also allows us to predict gene frequencies in future generations.

Together these approaches have transformed a subject poorly understood in 1900 into one of the most advanced disciplines in biology today. As such, the impact of genetics on society has been immense. We shall discuss many examples of the applications of genetics in the following section and throughout the text.

FIGURE 1.4

Trofim Denisovich Lysenko, a Russian geneticist who embraced the Lamarckian ideas of inheritance of acquired characteristics.

THE SIGNIFICANCE OF GENETICS IN SOCIETY

In addition to acquiring information for the sake of extending knowledge in any discipline of science—an experimental approach called **basic research**—scientists conduct investigations to solve problems facing society or simply to improve the well-being of members of that society—an approach called **applied research.** The history of applied research shows that it is usually possible only as an extension of prior basic research. In order for the two types of research to be efficiently executed and ultimately to benefit society, they must be allowed to proceed in a reasonably uninhibited and rational manner. Scientists must be allowed to examine existing knowledge, determine the direction of their efforts, and freely exchange their ideas and findings with other scientists, as has been the case in most countries with respect to the field of genetics.

However, in one era in one country—the USSR under the government of Joseph Stalin—this was not the case. In order to illustrate the consequences of the loss of free scientific inquiry and the imposition of controlled research, we will recount this era. Then we will turn to a brief survey of some of the benefits society has enjoyed as the result of unrestrained research in genetics.

Soviet Genetics: Science and Politics

One of the most famous examples of an interaction between science and politics occurred in the Soviet Union. This instance happened to involve genetics. In particular, it had to do with the genetic and evolutionary theory of the inheritance of acquired characteristics. In the 1930s, Trofim Denisovich Lysenko, a young plant breeder from the Ukraine, espoused the idea that plant development and agricultural productivity could be improved by manipulating environmental conditions. Lysenko further stated that such improvements would be incorporated into the genetic material and thus passed on to future generations of plants.

Lysenko's ideas were derived from the Lamarckian theory of the inheritance of acquired characteristics put forth in 1809. Jean Lamarck believed that environmental factors could induce evolutionary change—that is, characteristics adaptable to a changing environment could be acquired and passed to future generations. Lamarck's theory had been popular until late in the nineteenth century, but by 1930 the experimental evidence firmly contradicted it.

Nevertheless, the political climate in the Soviet Union at that time provided a hospitable environment for the revival of Lamarck's ideas. First, the theory that environmental change could produce permanent genetic change was compatible with the Marxist political thesis that the proper social conditions would induce permanent changes in human behavior. Second, Ivan Pavlov, the Soviet scientist known for his work on the learned or conditioned reflex, emphasized the importance of environmental stimuli. Pavlov claimed, although he later retreated from this statement, to have found an example of a conditioned reflex that was inherited in mice.

Perhaps most significant in the acceptance of Lysenko's ideas was the poor condition of Soviet agriculture during this period. The Soviet government was anxious to improve agricultural production, and the traditional methods of selective breeding were not producing changes rapidly enough.

Facing a desperate situation, Stalin was impressed with Lysenko's ideas. Lysenko proposed that germination of the winter wheat crop could be speeded up by **vernalization,** the practice of subjecting plants to an artificial cold period to shorten the dormant period of the seeds, which are usually shed in autumn. The early shedding of the seeds permitted the planting of an additional crop that could be harvested before autumn. Lysenko claimed that because the changes induced by vernalization were permanent ones, this technique need be applied only once.

In fact, this practice did not lead to a permanent increase in agricultural output, but by the time this became apparent, Lysenko had become director of the Institute of Genetics of the USSR Academy of Science. Appointed to this position in 1940, he managed to suppress throughout the Soviet Union all work in genetics that ran contrary to his own views. He was responsible for destroying research records, laboratory supplies, and experimental material of those who opposed him, and in some cases, he had his opponents arrested and sentenced to prison.

One of Russia's leading geneticists, Nikolai Ivanovich Vavilov was a favorite target of Lysenko. Director of the Lenin Academy of Agriculture, Vavilov was one of the world's experts on wheat. He was imprisoned and sentenced to death for agricultural espionage in 1941. He died in prison from malnutrition in 1943.

During this time great advances were made in genetics, particularly in the United States. Scientifically based selective breeding programs had developed new varieties of hybrid grains whose yields were much greater than those of the old varieties. Yet none of this knowledge was made available to Soviet geneticists until 1964, when Lysenko finally fell from power and lost his stranglehold on Soviet genetics. One American botanist calculated that USSR corn production would have been increased by six million tons between 1947 and 1957 if only half the country's acreage had been planted with the new hybrid strains.

Unfortunately, Soviet genetics was held back for nearly a generation at a time when numerous significant advances were being made elsewhere in the world. Adherence to theories that were politically rather than scientifically acceptable was indeed costly to Soviet society.

Genetic Advances in Agriculture and Medicine

The major benefit to society as a result of genetic study has been in the areas of agriculture and medicine. Even though cultivation of plants and domestication of animals had begun long before, the rediscovery of Mendel's work in the early twentieth century spurred scientists to apply genetic principles to these processes. In both cases, the use of selective breeding and hybridization techniques has had the most significant impact.

In plants, four major categories of improvements have been possible: (1) more efficient energy utilization during photosynthesis, resulting in more vigorous growth and increased yields; (2) increased resistance to natural predators and pests, including insects and disease-causing microorganisms; (3) production of hybrids exhibiting a combination of superior traits derived from two different strains or even two different species; and (4) selection of genetic variants with increased protein value or an increased content of limiting amino acids that are essential in the human diet.

These improvements have resulted in a tremendous increase in yield and nutrient value in such crops as barley, beans, corn, oats, rice, rye, and wheat, among others. It is estimated that in the United States the use of improved genetic strains has led to a threefold increase in crop yield per acre. In Mexico, where corn is the staple crop, a significant increase in protein content and yield has occurred. A substantial effort has also been made to improve the growth of Mexican wheat. Led by Norman Borlaug, a team of researchers was able to develop a strain of wheat that incorporated favorable genes from wheat strains found in various parts of the world, creating a superior variety that is now grown in many underdeveloped countries besides Mexico. Because of this effort, which led to the well-publicized "Green Revolution," Borlaug received the Nobel Peace Prize in 1970. There is little question that this application of genetics has contributed to the well-being of our own species and to the maintenance of world peace.

Similarly, applied research in genetics has developed superior breeds of animals. Enormous increases in usable meat supplies produced per unit of food intake have occurred. For example, selective breeding has produced chickens that grow faster, produce

FIGURE 1.5
Norman Borlaug, recipient of the Nobel Peace Prize in 1970 for his research creating improved varieties of wheat. (Courtesy of the Rockefeller Foundation.)

FIGURE 1.6
A domesticated chicken produced from an extensive genetic breeding program. (Courtesy of DEKALB AgResearch, Inc.)

more high-quality meat per chicken, and lay greater numbers of larger eggs. In larger animals, including pigs and cows, the use of artificial insemination has been particularly important. Sperm samples derived from a single male with superior genetic traits may now be used to fertilize thousands of females located in all parts of the world.

Equivalent strides have been made in medicine as a result of advances in genetics, particularly since 1950. Numerous disorders in humans have been discovered to result from either a single mutation or a specific chromosomal abnormality. For example, the genetic basis of sickle-cell anemia, erythroblastosis fetalis, hemophilia, some forms of muscular dystrophy, Tay-Sachs disease, Down syndrome, and countless metabolic disorders is now well documented. It is estimated that over ten million children or adults in the United States suffer from some form of genetic affliction.

Recognition of the genetic basis of these disorders has provided direction for the development of treatment and preventive measures. For example, **genetic counseling** provides parents with objective information upon which they can base rational decisions. It is estimated that every childbearing couple stands an approximately three percent risk of having a child with some form of genetic anomaly.

Applied research in genetics has provided other medical benefits. Increased knowledge in **immunogenetics** has made possible compatible blood transfusions as well as organ transplants. The discovery of

FIGURE 1.7
The effects of breeding and selection in cattle. (Courtesy of Carolina Biological Supply Company.)

tissuebound antigens has led to the important concepts of **histocompatibility** and tissue typing. In conjunction with immunosuppressive drugs, transplant operations involving human organs, including the heart, liver, and kidney, are increasing annually.

Recombinant DNA technology is also an important part of applied genetics. By cloning human genes that code for such medically important molecules as insulin and interferon, bacteria can serve as the source of mass production of these gene products. Recombinant DNA techniques will also play an increasing and essential role in **human genetic engineering,** which involves the direct manipulation of the genetic material.

In the not-so-distant future, human genetic engineering will undoubtedly be used to alter the genetic constitution of individuals harboring genetic defects. While such a process presents ethical questions, it may correct serious genetic errors in members of our species.

In later chapters, these and other examples in agriculture and medicine will be discussed in greater detail. While other scientific disciplines are also expanding in knowledge, none have paralleled the growth of information that is occurring annually in genetics. By the end of this course, we are confident you will agree that the present truly represents the "Age of Genetics."

SELECTED READINGS

ANDERSON, W. F., and DIRCUMAKOS, E. G. 1981. Genetic engineering in mammalian cells. *Scient. Amer.* (July) 245: 106–21.

BAER, A. S., ed. 1973. *Heredity and society.* New York: Macmillian.

BRESLER, J. B., ed. 1973. *Genetics and society.* Reading, Mass.: Addison-Wesley.

COCKING, E. C., DAVEY, M. R., PENTAL, D., and POWER, J. B. 1981. Aspects of plant genetic manipulation. *Nature* 293: 265–70.

DAY, P. R. 1977. Plant genetics: Increasing crop yield. *Science* 197: 1334–39.

DUNN, L. C. 1965. *A short history of genetics.* New York: McGraw-Hill.

GARFIELD, E. 1981. Medical genetics: The new preventive medicine. *Current Contents—Life Sciences,* vol. 24, no. 36, pp. 5–20.

HAFEZ, E. S. E., and SEMM, K., eds. 1982. *In vitro fertilization and embryo transfer.* New York: Alan R. Liss.

HOPWOOD, D. A. 1981. The genetic programming of industrial microorganisms. *Scient. Amer.* (Sept.) 245: 91–102.

HUTTON, R. 1978. *Bio-revolution: DNA and the ethics of man-made life.* New York: New American Library.

JORAUSKY, D. 1970. *The Lysenko affair.* Cambridge, Mass.: Harvard University Press.

KING, R. C., and STANSFIELD, W. D. 1985. *A dictionary of genetics.* 3rd ed. New York: Oxford University Press.

LERNER, I. M., and LIBBY, W. J. 1976. *Heredity, evolution, and society.* 2nd ed. San Francisco: W. H. Freeman.

MEDVEDEV, Z. A. 1969. *The rise and fall of T. D. Lysenko.* New York: Columbia University Press.

NAGLE, J. J. 1979. *Heredity and human affairs.* 2nd ed. St. Louis: C. V. Mosby.

POPOVSKY, M. 1984. *The Vavilov affair.* Hamden, Conn.: Archon.

SILBERNER, J. 1982. Superchicken. *Science Digest* 90: 30–32.

STINE, G. J. 1977. *Biosocial genetics: Human heredity and social issues.* New York: Macmillan.

THOMPSON, J. S., and THOMPSON, M. W. 1980. *Genetics in medicine.* 3rd ed. Philadelphia: W. B. Saunders.

TORREY, J. G. 1985. The development of plant biotechnology. *Amer. Scient.* 73: 354–63.

WEINBERG, R. A. 1985. The molecules of life. *Scient. Amer.* (Oct.) 253: 48–57.

PART ONE
HEREDITY
AND
PHENOTYPE

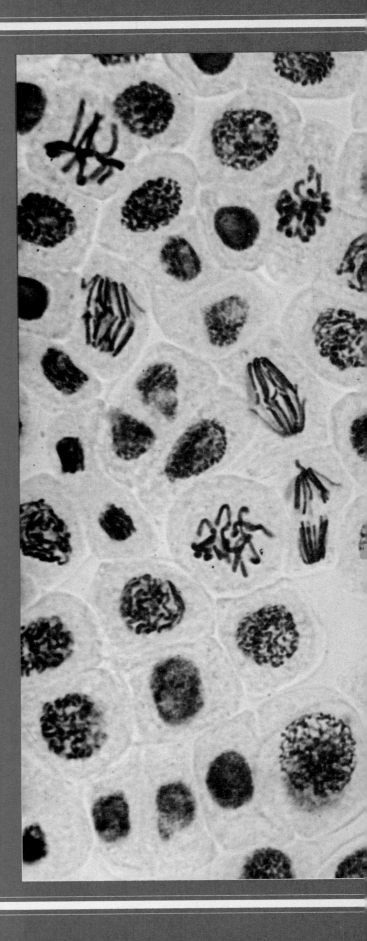

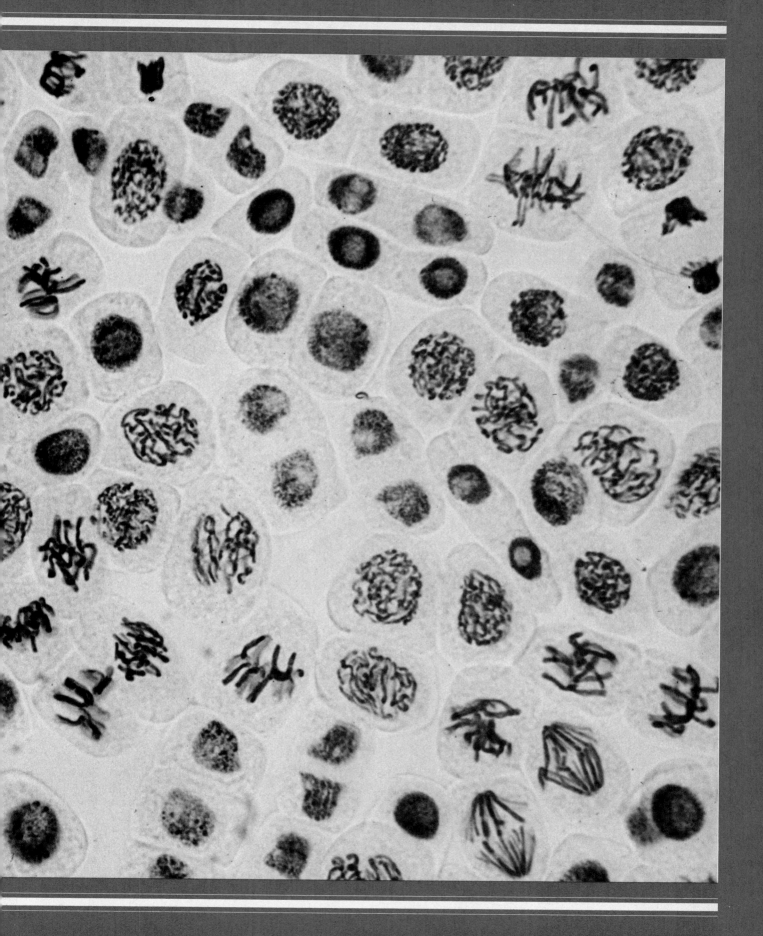

2

Mitosis and Meiosis: Transmission of Genetic Information

There exists in every living thing a substance referred to as the **genetic material.** Except in certain viruses, this material is composed of the nucleic acid, DNA. A molecule of DNA is organized into units called **genes** that direct all metabolic activities of cells. In this chapter, the topic of genetic continuity between cells and organisms is considered. The manner in which the genetic material is transmitted from one generation of cells to the next, and from organisms to their descendants, must be exceedingly precise.

Two major processes are involved in the transmission of the genetic material in eukaryotes: **mitosis** and **meiosis.** While the mechanisms of the two processes are similar in many ways, the outcomes are quite different. Mitosis leads to the production of two cells with an identical amount and type of genetic information. Meiosis, on the other hand, reduces the genetic information to precisely half. This reduction is essential if sexual reproduction is to occur without doubling the amount of genetic material at each generation. Strictly speaking, mitosis is that portion of cell division in which the hereditary components are precisely and equally divided into daughter cells. Meiosis is part of a special type of cell division leading to the production of sex cells: gametes and spores. Thus, meiosis is an essential step leading to the transmission of genetic information from an organism to its offspring.

CELL STRUCTURE

Before describing mitosis and meiosis, we will briefly review the structure of cells. A discussion of cell structure is important for three reasons:

1   During cell division most parts of cells must be distributed to the resulting division products.

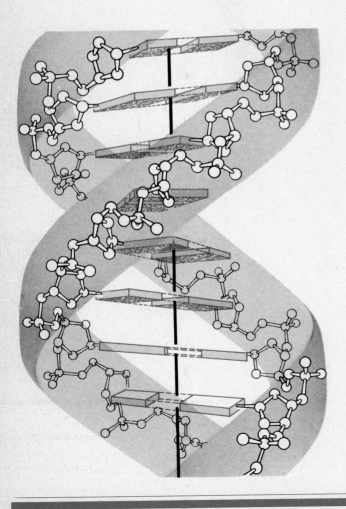

2   Certain structures, notably the **centrioles** and related **spindle fibers,** are essential to the mechanics of chromosome movement during mitosis and meiosis.

3   A basic principle in biology relates cell structure to genetics. This principle states that cell function is closely correlated with cell structure. Since the function of a cell is dependent on the expression of specific genetic information, we may conclude that the variation in structure noted in dissimilar cells is also dependent on specific genetic expression.

Before 1940, knowledge of cell structure was based on information obtained with the **light microscope.** Around 1940 the **transmission electron microscope** was in its early stages of development, and by 1950 many details of cell ultrastructure had been unveiled by electron microscopy. Under the electron microscope cells were seen as highly organized, precise structures. A new world of whorling membranes, miniature organelles, microtubules, granules, and filaments was revealed. These discoveries revolutionized thinking in the entire field of biology. We will be concerned with those aspects of cell structure relating to genetic study. As the parts of the cell are described, refer to Figure 2.1.

The entire cell is surrounded by a **plasma membrane,** an outer covering that defines the cell boundary and protects the cell from its immediate environment. This membrane is not lifeless and passive, but rather it actively controls the movement of materials such as gases, nutrients, and waste products into and out of the cell. In addition to this membrane, plant cells have an outer covering called the **cell wall,** a rigid structure primarily composed of a polysaccharide called cellulose. Bacterial cells also have a cell wall, but its chemical composition is quite different from that of the plant cell wall. The major component is a complex macromolecule called a **peptidoglycan.** As its name suggests, the molecule consists of peptide and sugar units. Long polysaccharide chains are cross-linked with short peptides, which impart great strength and rigidity to the bacterial cell. Some bacterial cells have still another covering, a **capsule.** It is a mucuslike polysaccharide that protects certain bacteria from phagocytic activity by the host during their pathogenic invasion of eukaryotic organisms.

Many, if not most, animal cells have a covering over the plasma membrane called a **cell coat.** Consisting of glycoproteins and polysaccharides, the chemical composition of the cell coat differs from comparable structures in either plants or bacteria. In addition to protecting the cell membrane, the cell coat provides biochemical identity at the surface of cells. Among other forms of molecular recognition, various antigenic determinants are part of the cell coat. For example, the **ABO and MN antigens** are found on the surface of red blood cells. In other cells, the **histocompatibility antigens,** which elicit an immune response during tissue and organ transplants, are part of the cell coat. All forms of biochemical identity at the cell surface are under genetic control, and many have been thoroughly investigated (see Chapter 4).

The presence of the **nucleus** and other membranous organelles characterizes eukaryotic cells and organisms. The nucleus houses the genetic material, DNA, which is found in association with large numbers of acidic and basic proteins. During nondivisional phases of the cell cycle this DNA/protein complex exists in an uncoiled, dispersed state called **chromatin.** As we will soon discuss, during mitosis and meiosis this material coils up and condenses into structures called **chromosomes.**

The lack of a nuclear envelope and membranous organelles is characteristic of the prokaryotes, which include bacteria and blue-green algae. The genetic material is, however, localized in the cell and during replication is found attached to the cell membrane. Prokaryotic DNA is not associated with many proteins, nor does it undergo the coiling or characteristic stages of mitosis found in eukaryotes.

An amorphous structure called the **nucleolus,** which is composed of RNA and protein, is also contained within the eukaryotic nucleus. This organelle is a processing center for the production of ribosomes. The nucleolus is generally associated with a specific chromosomal region called the **nucleolar organizer region (NOR).** This region contains the genetic information that codes for ribosomal RNA (rRNA). Following its synthesis, rRNA is processed and combined with many specific proteins to form the mature ribosome. Prokaryotic cells do not have a distinct nucleolus, but do contain genes that specify rRNA molecules.

The remainder of the cell enclosed by the plasma membrane, and excluding the nucleus, is composed of **cytoplasm.** Cytoplasm is a colloidal material containing numerous types of organelles. In eukaryotes it is highly compartmentalized by a membranous structure, the **endoplasmic reticulum.** The biochemical composition of the cytoplasm, as determined by the expression of the cell's genetic information, is directly related to the cell's function.

The endoplasmic reticulum (ER) greatly increases the cytoplasmic surface area available for synthetic activity and may be found studded with **ribosomes.** The amount of endoplasmic reticulum in any cell is correlated with the degree of synthetic activity exhibited by the cell. Ribosomes, which will be discussed in detail in Part 4, serve as nonspecific workbenches for the

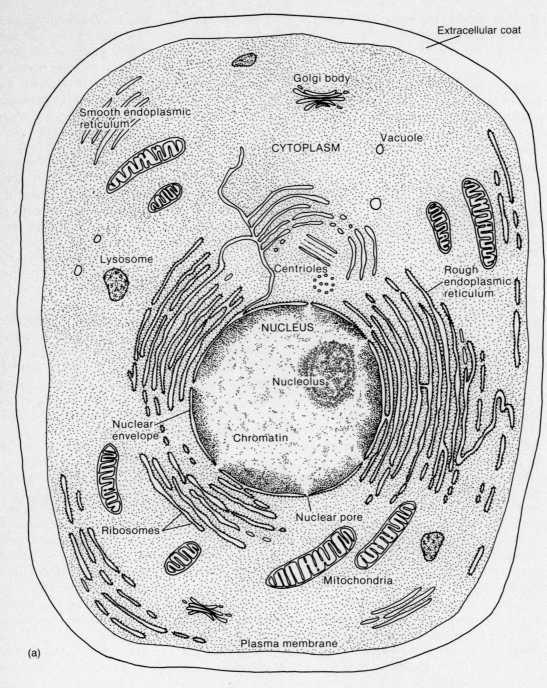

Extracular coat

Golgi body

Smooth endoplasmic reticulum

Vacuole

CYTOPLASM

Lysosome

Centrioles

Rough endoplasmic reticulum

NUCLEUS

Nucleolus

Nuclear envelope

Chromatin

Nuclear pore

Ribosomes

Mitochondria

Plasma membrane

(a)

FIGURE 2.1

(a) Drawing of a generalized animal cell as seen under the electron microscope. Emphasis has been placed on the cellular components discussed in the text. (b) Drawing of a generalized plant cell. Note the presence of the cell wall, large vacuoles, and chloroplasts, and the absence of centrioles and lysosomes in the plant cell compared with the animal cell.

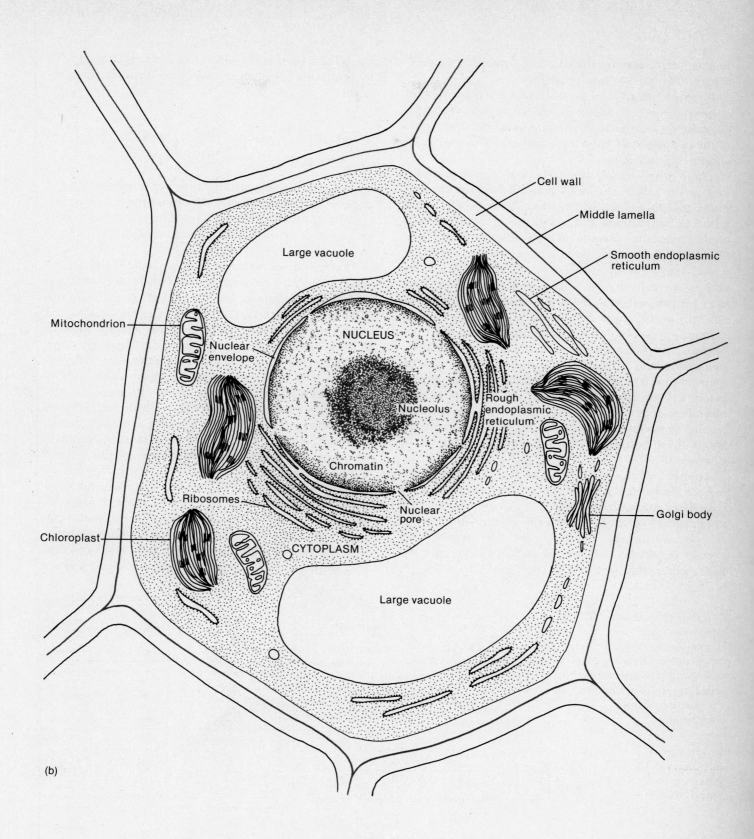

Cell wall

Middle lamella

Smooth endoplasmic reticulum

Large vacuole

Mitochondrion

Nuclear envelope

NUCLEUS

Nucleolus

Rough endoplasmic reticulum

Chromatin

Ribosomes

Nuclear pore

Golgi body

Chloroplast

CYTOPLASM

Large vacuole

(b)

translation of genetic information contained in messenger RNA (mRNA) into proteins.

Recent findings have revealed that the cytoplasm of eukaryotic cells contains an extensive **cytoskeleton.** Consisting of a matrix of intricate filaments, the lattice of the cytoskeleton provides a framework of structural support for the cell.

Three other cytoplasmic structures are very important in the eukaryotic cell's activities: **mitochondria, chloroplasts,** and **centrioles.** Mitochondria are found in both animal and plant cells and are the sites of the **oxidative phases of cell respiration.** These chemical reactions generate large amounts of ATP, an energy-rich molecule. Chloroplasts are found in plants and some protozoans. This organelle is associated with **photosynthesis,** the major energy-trapping process on earth. Both mitochondria and chloroplasts contain a type of DNA distinct from that found in the nucleus. Furthermore, these organelles can duplicate themselves and transcribe and translate their genetic information. It is interesting to note that the genetic machinery of mitochondria and chloroplasts closely resembles that of prokaryotic cells. This observation has led to speculation concerning the evolutionary origin of these organelles. This topic will be discussed in Chapter 15.

Animal cells also contain a pair of complex structures called the centrioles. These cytoplasmic bodies consist of microtubules and are associated with the organization of those spindle fibers that function in mitosis and meiosis. Centrioles are also associated in certain organisms with the formation of cilia and flagella, organelles involved in cell motility. Although there have been some reports of the presence of DNA in centrioles, most recent evidence disputes this proposal.

HOMOLOGOUS CHROMOSOMES, HAPLOIDY, AND DIPLOIDY

To describe mitosis and meiosis, we must employ the concept of **homologous chromosomes.** An understanding of this concept will also be of critical importance in the following chapters on transmission genetics.

Chromosomes are most easily visualized during mitosis; when they are examined carefully, they are seen to take on distinctive lengths and shapes. Each contains a condensed or constricted region called the **centromere,** which establishes the general appearance of each chromosome. Figure 2.2 illustrates chromosomes with centromere placements at different points along their axes. Extending from either side of the centomere are the arms of the chromosome. Depending on the position of the centromere, different arm ratios are produced. As Figure 2.2 illustrates, chromosomes are classified as **metacentric, submetacentric, acrocentric,** or **telocentric** on the basis of the centromere location.

When studying mitosis, we may make several other important observations. First, the number of chromosomes is species-specific. Generally, each somatic cell within members of the same species contains an identical number of chromosomes. This is called the **diploid number (2n).** When the lengths and centromere

FIGURE 2.2
Centromere locations and descriptions of chromosomes based on that location. Note that the shape of the chromosome during anaphase is determined by the position of the centromere.

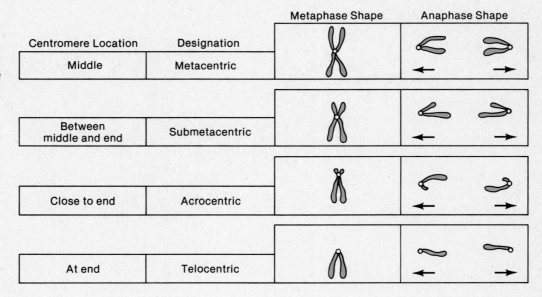

Centromere Location	Designation	Metaphase Shape	Anaphase Shape
Middle	Metacentric		
Between middle and end	Submetacentric		
Close to end	Acrocentric		
At end	Telocentric		

placements of all such chromosomes are examined, a second general feature is apparent. Nearly all of the chromosomes exist in pairs with regard to these two criteria. The members of each pair are called **homologous chromosomes.** For each chromosome exhibiting a specific length and centromere placement, another exists with identical features.

Figure 2.3 illustrates the identical physical appearance of members of homologous chromosome pairs. There, the mitotic chromosomes of a human male have been photographed, cut out of the print, and matched up, creating a **karyotype.** As you can see, human chromosomes have a 2n value of 46 and exhibit a diversity of sizes and centromere placements.

Collectively, the genes contained on one member of each homologous pair of chromosomes constitute the **haploid genome** of the species. The **haploid number** (**n**) of chromosomes is one half of the diploid number. Table 2.1 demonstrates the wide range of n values found in a variety of plants and animals.

Homologous pairs of chromosomes also have important genetic similarity. They contain identical gene sites, or **loci,** along their axes. Thus, they have identical genetic potential. In sexually reproducing organisms, one member of each pair is derived from the maternal parent (through the ovum) and one from the paternal parent (through the sperm). Thus, each diploid organism contains two copies of each gene as a consequence of biparental inheritance. As will be seen in the following chapters on transmission genetics, the members of each pair of genes, while influencing the same characteristic or trait, need not be identical. Alternate forms of the same gene are called **alleles.**

The concepts of haploid number, diploid number, and homologous chromosomes may be related to the process of meiosis. During the formation of gametes or spores, meiosis converts the diploid number of chromosomes to the haploid number. As a result, haploid gametes or spores contain precisely one member of each homologous pair of chromosomes. Following fu-

FIGURE 2.3
Mitotic chromosomes constituting the standard human male karyotype. With the exception of the X and Y chromosomes, all others are present in homologous pairs. On the basis of size and centromere placement, chromosomes are categorized into seven groups: (A:1-3; B:4-5; C:6-12; D:13-15; E:16-18; F:19-20; and G:21-22). These chromosomes have undergone a special staining procedure to produce their banded appearance. This procedure and the significance of bands is discussed in Chapter 11.

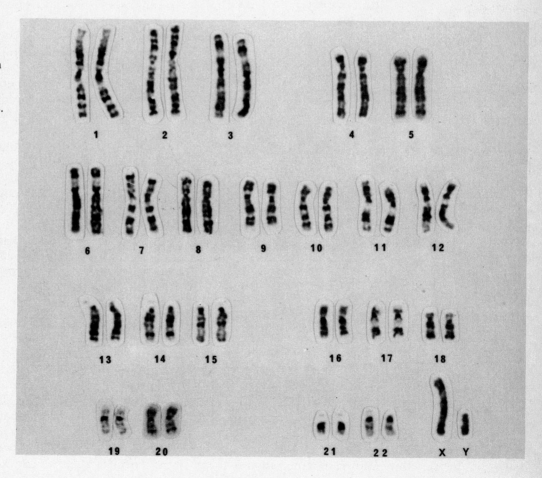

TABLE 2.1
The haploid number of chromosomes for representative organisms.

Common Name	Genus-Species	Haploid No.
Broad bean	*Vicia faba*	6
Cat	*Felis domesticus*	19
Cattle	*Bos taurus*	30
Chicken	*Gallus domesticus*	39
Corn	*Zea mays*	10
Cotton	*Gossypium hirsutum*	26
Dog	*Canis familiaris*	39
Evening primrose	*Oenothera biennis*	7
Frog	*Rana pipiens*	13
Fruit fly	*Drosophila melanogaster*	4
Garden onion	*Allium cepa*	8
Garden pea	*Pisum sativum*	7
Grasshopper	*Melanoplus differentialis*	12
Green alga	*Chlamydomonas reinhardi*	16
Horse	*Equus caballus*	32
House fly	*Musca domestica*	6
House mouse	*Mus musculus*	20
Human	*Homo sapiens*	23
Jimson weed	*Datura stramonium*	12
Mold	*Aspergillus nidulans*	8
Mosquito	*Culex pipiens*	3
Pink bread mold	*Neurospora crassa*	7
Rhesus monkey	*Macaca mulatta*	21
Slime mold	*Dictyostelium discoidium*	7
Snapdragon	*Antirrhinum majus*	8
Tobacco	*Nicotiana tabacum*	24
Tomato	*Lycopersicon esculentum*	12
Water lily	*Nymphaea alba*	80
Wheat	*Triticum aestivum*	21
Yeast	*Saccharomyces cerevisiae*	9

sion of two gametes in fertilization, the diploid number is reestablished, thus maintaining the constancy of genetic material from generation to generation.

There is one important exception to the concept of homologous pairs of chromosomes. In many species, one pair, the **sex-determining chromosomes,** may not be homologous in size, centromere placement,

arm ratio, or genetic potential. For example, in humans, males contain the Y chromosome in addition to one X chromosome (Figure 2.3), while females carry two homologous X chromosomes. The X and Y chromosomes are not strictly homologous. The Y is considerably smaller and lacks most of the gene sites contained on the X. Nevertheless, in meiosis they behave as homologues so that gametes produced by males receive either the X or Y chromosome.

MITOSIS

The process of mitosis is critical to all eukaryotic organisms. In single-celled organisms such as protozoans, fungi, and algae, mitosis is the basis of asexual reproduction. Multicellular diploid organisms begin life as single-celled fertilized eggs or **zygotes.** The mitotic activity of the zygote and the subsequent daughter cells is the foundation for development and growth of the organism. In adult organisms, mitotic activity is prominent in wound healing and other forms of cell replacement in certain tissues. For example, the epidermal skin cells of humans are continuously being sloughed off and replaced. Cell division also results in a continuous production of reticulocytes, which eventually shed their nuclei and replenish the supply of red blood cells in vertebrates. In abnormal situations, adult somatic cells may exhibit uncontrolled cell divisions, resulting in cancer.

It is generally observed that following cell division, the initial size of each new daughter cell is approximately one-half the size of its parent. However, the nucleus of each new cell is not appreciably smaller than the nucleus of the original cell. Quantitative measurements of DNA confirm that there are equivalent amounts of genetic material in the parent and daughter nuclei. Thus, during cell division, the mechanism for the division of the cytoplasm is different from the one by which the nucleus and its genetic material are divided.

The process of cytoplasmic division is called **cytokinesis.** The division of cytoplasm seems to require only a mechanism that results in a partitioning of the volume into two equal parts, followed by the enclosure of both new cells within a distinct plasma membrane. Cytoplasmic organelles either replicate themselves, arise from existing membrane structures, or are synthesized *de novo* (from the beginning) in each cell. The subsequent proliferation of these structures is a reasonable and adequate mechanism for replenishing the cytoplasm in daughter cells.

The division of the genetic material into daughter cells is more complex than cytokinesis and requires more precision. The chromosomes must first be exactly replicated and then accurately distributed into daughter cells. The end result is the production of two diploid cells from a single diploid cell.

Interphase and the Cell Cycle

Many cells undergo a continuous alternation of division and nondivision, the basis for the terms **mitosis** and **interphase,** respectively. It was once believed that interphase had no relation to mitosis and was merely the period of cell growth. Modern biochemical techniques have revealed, however, that a preparatory step essential to mitosis occurs during interphase: the replication of the genetic information.

The period of replication, designated the **S phase,** is when DNA synthesis occurs. It was originally detected using the technique of **autoradiography** (see Appendix A). When radioactive DNA precursors such as ^{3}H-thymidine are supplied to interphase cells, the initiation and termination of DNA synthesis can be determined by monitoring the uptake of the radioactive

substances. Investigations of this nature have demonstrated two periods during interphase, before and after S, when no DNA synthesis occurs. These are designated G_1 (gap I) and G_2 (gap II), respectively. During both of these periods, as well as during S, intensive metabolic activity, cell growth, and cell differentiation occur. By the end of G_2, the volume of the cell has roughly doubled, DNA has been replicated, and mitosis (M) is initiated. Continuously dividing cells then repeat this cycle (G_1, S, G_2, M) over and over. This concept of such a **cell cycle** is illustrated in Figure 2.4.

The activity of some cells is arrested, either permanently or temporarily, during the cell cycle. Such cells, in a sense, withdraw from the cycle and alter their metabolism and growth pattern. Thus, a separate designation, G_0, has been assigned to this phenomenon. As shown in Figure 2.4, this step occurs during the G_1 stage just prior to the initiation of DNA synthesis.

Once initiated, mitosis is a dynamic period of vigorous and continual activity. For discussion purposes, the entire process is subdivided into discrete phases, and specific events are assigned to each stage. These stages, in order of occurrence, are **prophase, metaphase, anaphase,** and **telophase** (Figure 2.5).

FIGURE 2.4
Diagrammatic view of the events and stages that comprise an arbitrary cell cycle.

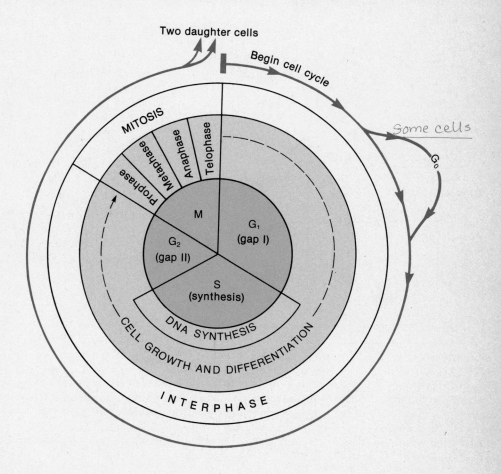

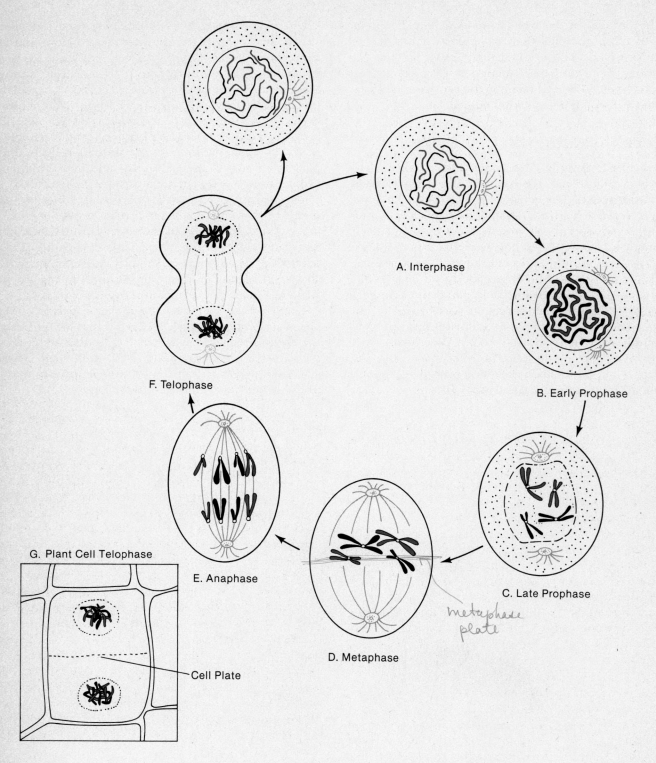

A. Interphase

B. Early Prophase

C. Late Prophase

D. Metaphase

metaphase plate

E. Anaphase

F. Telophase

G. Plant Cell Telophase

Cell Plate

Figure 2.5
Mitosis in an animal cell with a diploid number of 4. The events occurring in each stage are described in the text. Of the two homologous pairs of chromosomes, one contains longer, metacentric members and the other shorter, submetacentric members. For each pair, the maternal chromosome is colored and the paternal chromosome is black. The insert (G), showing the telophase stage in a plant cell, illustrates the formation of the cell plate and lack of centrioles.

Prophase

Almost one-third of mitosis is spent in **prophase,** and significant activity occurs during this stage. One of the early events in animal cell and lower plant prophase involves the migration of two pairs of centrioles to opposite ends of the cell. These structures are found just outside the nuclear envelope in an area of differentiated cytoplasm called the **centrosome.** It is thought that each pair of centrioles consists of one mature unit and a smaller, newly formed centriole.

The direction of migration of the centrioles is such that two poles are established at opposite ends of the cell. This creates an axis along which chromosomal separation occurs. Following the migration, the centrioles are responsible for the organization of cytoplasmic microtubules into a series of **spindle fibers.** Even though higher plant cells seem to lack centrioles, spindle fibers are nevertheless apparent during mitosis.

As the centrioles migrate, the nuclear envelope begins to break down and gradually disappears. In a similar fashion, the nucleolus disintegrates within the nucleus. While these events are taking place, the diffuse chromatin—the characteristic form of the genetic material during interphase—condenses until distinct threadlike structures, or chromosomes, are visible. Near the end of prophase, it is apparent that each chromosome is a double structure split longitudinally except at a single point of constriction, the centromere. Each member of the double structure is referred to as a **chromatid.** Each chromatid contains one DNA double helix running unbroken from end to end. Each pair represents **sister chromatids,** connected at the region of the centromere.* Thus in humans, who have a diploid chromosome number of 46, a cytological preparation of late prophase will reveal 46 chromosomal structures, each duplicated except at the centromeric region (see Figure 2.3). These are found randomly distributed in the area formerly occupied by the nucleus.

By the completion of prophase, spindle fibers have been laid down between the centrioles, which are now at opposite ends of the cell. The nucleolus and nuclear envelope are no longer visible, and sister chromatids are apparent. Parts (a) through (c) of Figure 2.5 illustrate interphase, early prophase, and late prophase.

---

*You may sometimes see the term *kinetochore* used synonymously with *centromere.* The kinetochore is actually a granule within the centromere that attaches to spindle fibers during mitosis. Two kinetochores form on opposite faces of the centromere, each attaching one or the other member of a pair of sister chromatids to the spindle fibers.

Metaphase

The distinguishing event of the metaphase stage is the migration of the centromeric region of each chromosome toward the equatorial plane. In some descriptions the term **metaphase** is applied strictly to the chromosome configuration following this movement. In such descriptions, **prometaphase** refers to the period of chromosome movement. The equatorial plane, also referred to as the **metaphase plate,** is the midline region of the cell, a line that runs perpendicular to the polarity established by the centrioles and spindle fibers.

Migration results from the binding of one or more spindle fibers to the centromere of each chromosome. This attachment (see footnote below) occurs prior to the movement of each chromosome, which is still a double structure. At the completion of metaphase, each centromere is now aligned at the plate with the chromosome arms extending outward in a random array. This configuration is shown in Figure 2.5(d).

Anaphase

Events critical to chromosome distribution during mitosis occur during its shortest stage, **anaphase.** It is during this phase that sister chromatids of each double chromosomal structure separate from each other and migrate to opposite ends of the cell. In order for complete separation to occur, each centromeric region must be divided into two. Once this has occurred, each chromatid is now referred to as a chromosome. Movement of the chromosomes to the opposite poles of the cell is dependent upon the centromere-spindle fiber attachment (Figure 2.6). The centromeres lead the way during migration, with the chromosome arms trailing behind. Depending on the location of the centromere along the chromosome, different shapes are assumed during separation.

At the completion of anaphase, the chromosomes have migrated to the opposite poles of the cell. The steps occurring during anaphase are critical in providing each subsequent daughter cell with an identical set of chromosomes. In human cells there would now be 46 chromosomes at each end, one from each original sister pair. Figure 2.5(e) illustrates anaphase prior to its completion.

Telophase

Telophase is the final stage of mitosis. The most significant event is cytokinesis, the dividing or partitioning of the cytoplasm. Cytokinesis is essential if two new cells are to be produced from one. The mechanism differs greatly in plant and animal cells. In plant

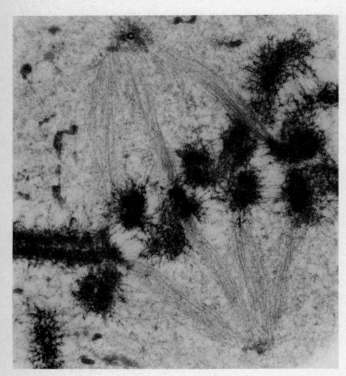
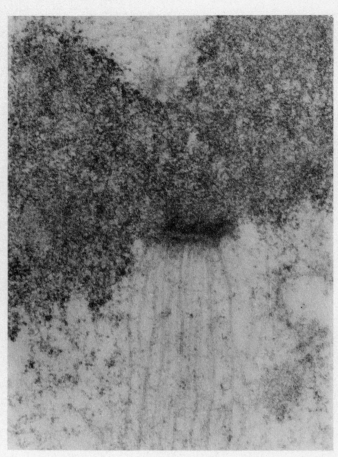

FIGURE 2.6
Electron micrographs revealing the attachment of spindle fibers to chromosomes during mitosis. In (a) spindle fibers emanating from both poles are apparent. In (b) spindle attachment to the kinetochore of the centromere can be seen. Photo (a) reproduced, with permission, from C. L. Rieder and S. S. Bowser. Correlative immunofluorescence and electron microscopy on the identical section of Eponembedded material. *J. Histochem. Cytochem.* 33 (in press). © 1985 by the Histochemical Society. Photo from C. L. Rieder, magnification = 12,000×. Photo (b) reproduced, with permission, from C. L. Rieder. Localization of ribonucleoprotein in the trilaminar kinetochore of P+K₁. *J. Ultrastr. Res.* 66: 109–119. © 1979 by Academic Press, Inc. Photo from C. L. Rieder, magnification = 55,000×.

cells, a **cell plate** is synthesized and laid down across the region of the metaphase plate. Animal cells, however, undergo a constriction of the cytoplasm, in much the same way a loop might be tightened around the middle of a balloon. The end result is the same: Two distinct cells are formed

It is not surprising that the process of cytokinesis varies among cells of different organisms. Plant cells, which are more regularly shaped and structurally rigid, require a mechanism for the deposition of new cell wall material around the plasma membrane. The cell plate, laid down during anaphase, becomes the **middle lamella.** Subsequently, the primary and secondary layers of the cell wall are deposited between the cell membrane and middle lamella on both sides

of the boundary between the two daughter cells. In animals, complete constriction of the cell membrane produces the **cell furrow** characteristic of newly divided cells.

Events necessary for the transition from mitosis to interphase are initiated during late telophase. These events represent a general reversal of those that occurred during prophase. In each new cell, the chromosomes begin to uncoil and become diffuse chromatin once again, while the nuclear envelope reforms around them. The nucleolus gradually reforms and is completely visible in the nucleus during early interphase. The spindle fibers also disappear. Telophase in animal and plant cells is illustrated in Figure 2.5(f) and 2.5(g).

Regulation of the Cell Cycle: Chemical and Genetic Control

Many cells alternate between mitosis and interphase, passing continuously from one cell cycle to another. For example, meristematic cells in plants and certain epithelial cells in animals exhibit this behavior. Figure 2.7 shows the portion of time spent in each stage of interphase and mitosis for a cultured cell that divides once every 16 hours.

As illustrated in Figure 2.4, other cells withdraw from the cell cycle, entering the G_0 stage. Some of these cells never divide again, while others can be induced to reenter the cycle. Different cell types, even in the same organism, vary widely in the lengths of their cell cycles. For example, animal cells exhibit cycles ranging from a few hours to several months. When examined carefully, it is apparent that most of this variation can be traced to the time spent in G_1. The total time necessary to complete S and G_2 is relatively constant in different cells.

Taken together, the above observations suggest that the cell cycle is tightly regulated. The discovery of the control mechanisms that regulate the cell cycle is significant for two reasons. First, such information extends our knowledge of the normal behavior of cells. Second, this information may be critical to our understanding of cancer, a disease process in which the regulation of the cell cycle is disrupted and cells divide continuously and more rapidly than is normal. We will briefly discuss several observations which suggest that the control mechanisms are at the molecular level and are genetically based.

Lester Goldstein and David Prescott used the protozoan *Amoeba proteus* to transplant nuclei from cells in one stage of interphase into the cytoplasm of cells in different stages of interphase, with the following results:

1 If a nucleus from a cell in the S phase is transplanted into the cytoplasm of a cell in the G_2 stage, the S nucleus slows down its rate of DNA synthesis.

2 If a G_2 nucleus, which has completed DNA replication, is transplanted into the cytoplasm of a cell in the S phase, the G_2 nucleus sometimes reinitiates DNA synthesis.

Additionally, studies in mammalian systems have shown that a G_1 nucleus placed in S phase cytoplasm initiates DNA synthesis sooner than expected.

These observations strongly suggest that some aspect of the cytoplasm influences the sequence of events in the cell cycle. It is probable that the cytoplasm contains molecular information that triggers DNA synthesis, thus initiating the S phase of the cycle. Once a cell enters the S phase and DNA replication is initiated, the normal sequence of events leading to mitosis will occur in the time sequence characteristic of that cell. Inhibitors of DNA synthesis will completely block the occurrence of these preliminary events and mitosis. These observations suggest that the most critical signal in the cell cycle is the one responsible for the initiation of DNA synthesis. Once synthesis is initiated, the cycle is self-perpetuating, leading to cell division.

Speculation exists that a division protein must be present in sufficient quantities to stimulate the cell to enter the S phase and thus complete the cell cycle. This protein is perhaps always present but may be normally unstable or inactive. However, upon conversion to an active state, its amount must then exceed a minimal threshold level or else the cell remains in the G_1 or G_0 stage. The observation that inhibitors that reduce protein synthesis also extend the G_1 stage is indirect evidence in support of this hypothesis. More direct evidence has been provided by the discovery of a protein called **statin**. Preliminary evidence has shown that this molecule is present *in vivo* in the nuclei of a variety of cells that have stopped dividing but absent

FIGURE 2.7
The time spent in each phase of one complete cell cycle of a human cell in culture.

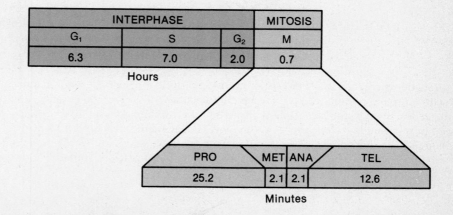

INTERPHASE			MITOSIS
G_1	S	G_2	M
6.3	7.0	2.0	0.7

Hours

PRO	MET	ANA	TEL
25.2	2.1	2.1	12.6

Minutes

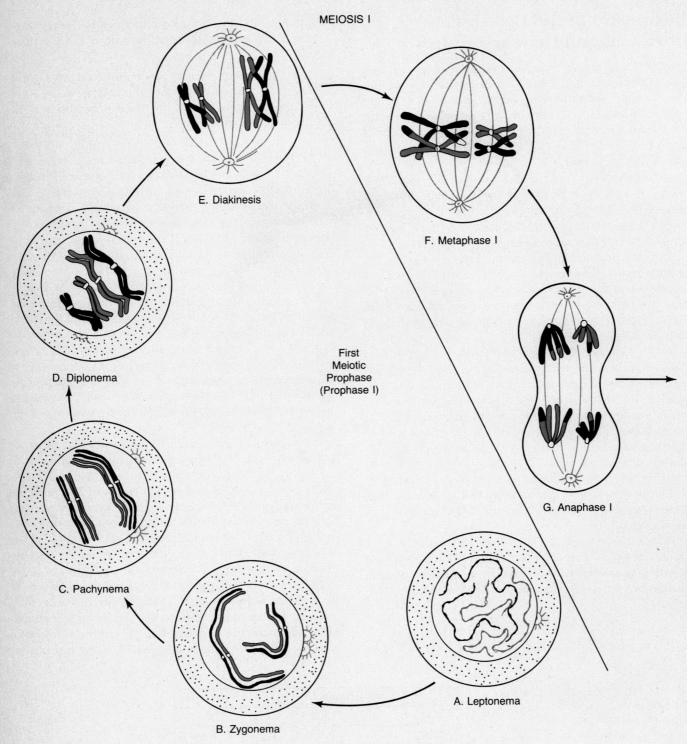

MEIOSIS I

E. Diakinesis

F. Metaphase I

D. Diplonema

First
Meiotic
Prophase
(Prophase I)

G. Anaphase I

C. Pachynema

A. Leptonema

B. Zygonema

FIGURE 2.8

A diagrammatic representation of the major events occurring during meiosis in a male animal with a diploid number of 4. The same chromosomes described in Figure 2.5 are followed, as described in the legend there. Note that the combination of chromosomes contained in spermatids is dependent on the random alignment of each tetrad and dyad on the equatorial plate during metaphase I and metaphase II. Numerous other combinations, which are not shown, can be formed. The events depicted in this figure should be correlated with the description in the text.

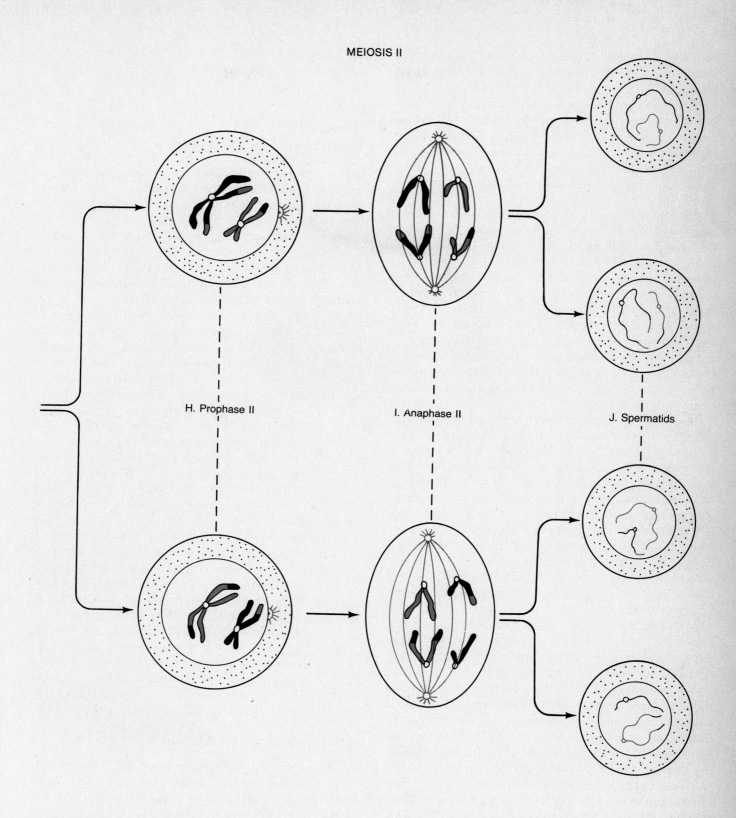

H. Prophase II

I. Anaphase II

J. Spermatids

in those undergoing division. *In vitro*, the protein is found in cultured senescent human fibroblasts that have arrested in the G_1 stage and fail to initiate any further DNA synthesis. It is also found in young fibroblasts that have arrested as a result of their culture conditions being altered. However, it is absent in actively dividing fibroblasts. While these observations do not prove that statin is the division protein, if one exists at all, they are interesting and merit future attention.

The division protein might also be **DNA polymerase,** the enzyme directly responsible for DNA replication. Another candidate is **thymidylate kinase,** an enzyme that converts thymidine to thymidine monophosphate, an essential step in the subsequent incorporation of this nucleotide into newly synthesized DNA. Both enzymes are found in relatively higher concentrations prior to the S phase. However, since the presence of both is predictable because they are needed for DNA replication, the evidence for either of them being the division protein is weak. Many other proteins, as well as other molecular species, must also be present at this time.

Other molecules that may function as division proteins are the **chalones,** which have been isolated from a variety of animal tissues. Most chalones demonstrate two properties: (1) they are endogenous mitotic inhibitors, and (2) their inhibitory influence is tissue-specific. Therefore, many animal cells produce chalones that usually arrest the cell cycle in the G_1 stage. This is consistent with our earlier conclusion that the transition from G_1 to S is the critical point of regulation during the cell cycle. Some chalones are proteins, while others appear to be glycoproteins. Further research involving chalones may clarify their importance in cell cycle regulation.

A second type of investigation has contributed evidence that the cell cycle is controlled by many different genes. Using the simple eukaryotic yeast *Saccharomyces cerevisiae*, Lee Hartwell was able to isolate about 150 independent mutations that interfere with the cell division cycle (**cdc mutations**). These were subsequently shown to represent 32 different genes. Among other problems, the mutant defects include the failure to initiate DNA synthesis and the inability to complete nuclear division and cytokinesis.

These observations allow us to conclude that the normal progress of the cell cycle is dependent upon a number of gene products found in the cytoplasm and probably the nucleus as well.

The studies of Hartwell and others suggest that regulation of the cell cycle is exceedingly complex. Once the control mechanisms in normal cells are defined, studies of the factors that disrupt the control mechanisms in proliferating cells of cancerous tissues will become more meaningful.

MEIOSIS AND SEXUAL REPRODUCTION

While mitosis produces identical daughter cells that contain equivalent genetic information, the process of meiosis, which results in gamete or spore formation, has a distinctly different outcome—the production of cells that contain only one-half the genetic information of somatic cells. During subsequent sexual reproduction, a gamete from each parent may combine during fertilization to reconstitute an amount of genetic material fully equivalent to that found in somatic parental cells. The events of meiosis must be highly specific. That is, it is insufficient to produce gametes with a random array of chromosomes equal to the haploid number. Instead, each gamete must contain an exact haploid genome, which includes precisely one member of each pair of homologous chromosomes. Fertilization then produces a diploid zygote with identical numbers and types of chromosomes, ensuring continuity of the species from generation to generation.

The process of sexual reproduction also ensures genetic variety among members of a species. Each offspring receives one copy of every gene on every chromosome from one parent as well as a second complete set of copies from the other parent. For any given gene site on a chromosome, called a **locus,** alternate forms of that gene may exist. These alternate forms are called **alleles** and have arisen from genetic changes or mutations. Sexual reproduction, therefore, reshuffles the combinations of alleles, producing an offspring that is never identical to either parent. This process constitutes one form of genetic recombination within species. A second type of genetic recombination—**crossing over,** or genetic exchange between homologous chromosomes during meiosis—may also occur. Crossing over produces an even greater potential for genetic variability among individuals.

An Overview of Meiosis

We have already established what must be accomplished during meiosis. Before systematically considering the stages of this process, we will briefly describe how diploid cells are converted to haploid gametes or spores. Unlike mitosis, in which each paternally and maternally derived member of any given homologous pair of chromosomes behaves autonomously during division, in meiosis homologous chromosomes pair together, or **synapse.** Each synapsed structure, called a

bivalent, gives rise to a unit, the **tetrad,** consisting of four chromatids; this demonstrates that both chromosomes have duplicated. In order to achieve haploidy, two divisions are necessary. In the first division, described as **reductional,** each tetrad separates and yields two **dyads,** each of which contains two sister chromatids joined at a common centromere. During the second division, described as **equational,** each dyad splits into two **monads** of one chromosome each. Thus, the two divisions may potentially produce four haploid cells.

The First Meiotic Division

During the first stage of meiosis, three significant events occur:

1 Homologous chromosomes pair up, or synapse, and take on the tetrad configuration, demonstrating that each chromosome has replicated.

2 The chromatid arms within a tetrad may overlap, forming **chiasmata.** This crosslike configuration is thought to result from genetic exchange or crossing over between homologues.

3 Following crossing over, the tetrads divide, yielding dyads.

We will now examine the actual stages leading to these events. As you read the following decriptions, you should identify each stage in Figure 2.8.

Prophase I

As in mitosis, DNA synthesis precedes meiosis, although the products of replication are not visible until the first prophase. The first meiotic prophase is a fairly lengthy one, and it has been further subdivided into five substages: leptonema,* zygonema,* pachynema,* diplonema,* and diakinesis.

Leptonema During the **leptotene stage,** the interphase chromatin material begins to condense, and the chromosomes, although still extended, become visible. Along each chromosome are **chromomeres,** localized condensations that resemble beads on a string.

Zygonema The chromosomes continue to shorten and thicken in the **zygotene stage.** Homologous chromosomes appear to be attracted to each other and begin to undergo zipperlike pairing or synapsis. Pairing is accompanied by the appearance of a unique ultra-

structural cell component, the **synaptonemal complex** (Figure 2.9). This structure looks like a tripartite ribbon, and as prophase continues, it is found in association with the synapsed homologues. Thus, it is speculated that the function of the synaptonemal complex is associated with chromosome pairing. At the completion of zygonema, the number of independent threadlike chromosomes is reduced by one-half because of synapsis. The paired structures are sometimes referred to as **bivalents.**

Pachynema The **pachytene stage** is the next one in the first prophase of meiosis. Coiling and shortening of the synapsed chromosomes continue. While such

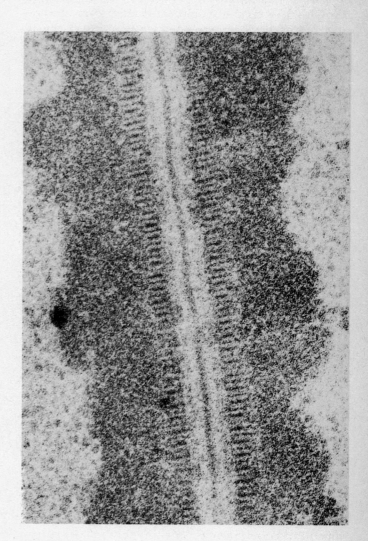

FIGURE 2.9
Electron micrograph of a portion of a synaptonemal complex derived from a pachytene bivalent of *Neotiella rutilans.* (Reproduced, with permission, from the *Annual Review of Genetics,* Volume 6. © 1972 by Annual Reviews Inc. Photo from D. von Wettstein.)

---

*These are the noun forms of these stages. The adjective forms (leptotene, zygotene, pachytene, and diplotene) are also used in the text of this chapter.

configurations are not emphasized in Figure 2.8(c), cytological preparations often reveal the chromosomes to be twisted around one another. During this stage, it first becomes apparent that each chromosome is really a double structure, thus providing visual evidence that chromosome replication has already occurred. As in mitosis, the two members of each chromosome are connected by a common centromere and are called **sister chromatids.** Thus, each bivalent now contains four members, (two pairs of sister chromatids) and is referred to as a **tetrad.** The number of tetrads is equivalent to the haploid number of the species.

During pachynema, a major meiotic event occurs that is of great signficance in genetics: **crossing over.** Crossing over is the reciprocal exchange of chromosome arms between nonsister chromatids of the tetrad. Within each tetrad, one or more such events occur. The points of exchange become apparent as crosslike structures called **chiasmata** in the next stage, diplonema. Biochemical studies confirm that a small amount of DNA synthesis accompanies the chromosome exchanges during the pachytene stage.

Diplonema During the ensuing **diplotene stage** the duplicated chromosomes of each tetrad begin to separate. However, one or more areas remain in contact where chromatids are intertwined. Each such area, called a **chiasma** (plural: chiasmata), is thought to represent a point where nonsister chromatids have undergone crossing over. Although the physical exchange between chromosome areas occurs during the pachytene stage, the evidence of crossing over is visible only when the duplicated chromosomes begin to separate.

Crossing over is an important source of genetic variability gained during gamete formation. New combinations of genetic material are formed during this process, mixing alleles from both members of a homologous chromosome pair.

Diakinesis Further shortening and condensation of the chromatids take place during the final stage of the first meiotic prophase, **diakinesis.** The chromosomes pull farther apart, but sister chromatids on either side of the chiasmata remain closely associated. As this separation proceeds the chiasmata move laterally toward the ends of the tetrad. This process, called **terminalization,** begins in late diplonema, and is completed during diakinesis. During this final period of prophase I, the nucleolus and nuclear envelope break down, and the two centromeres of each tetrad become attached to the recently formed spindle fibers.

Metaphase, Anaphase, and Telophase I

Following the first meiotic prophase stage, steps similar to those of mitosis occur. In the **metaphase stage of the first division**, the chromosomes have maximally shortened and thickened. The terminal chiasmata of each tetrad are visible and appear to be the only factor holding the nonsister chromosomes together. Each tetrad moves from a previous position along the nuclear envelope to the metaphase plate.

During the first division, a single centromere holds each pair of sister chromatids together. It does *not* divide. At the **first anaphase,** therefore, one-half of each tetrad (one pair of sister chromatids) is pulled toward each pole of the dividing cell. The products of the separation, or **disjunction,** of these homologous chromosomes are two **dyads.** At the completion of anaphase I, there is a series of dyads equal to the haploid number present at each pole.

If no crossing over had occurred in the first meiotic prophase, every dyad at each pole would consist solely of either the paternal or maternal chromatids. However, crossing over prevents this. The alignment of each tetrad prior to this first anaphase stage is random. The paternal half of each tetrad will be pulled to one or the other pole at random, and the maternal half will move to the opposite pole. This random **segregation** of dyads is the basis for the Mendelian principle of **independent assortment,** which we will discuss in Chapter 3. You may wish to return to this discussion when you study this principle.

In many organisms, **telophase of the first meiotic division** reveals a nuclear membrane forming around the dyads. Then, the nucleus enters into a short interphase period. In other cases, the cells go directly from the first anaphase into the second meiotic division. In any event, telophase is much shorter than the corresponding stage in mitosis.

The Second Meiotic Division

A second division of the sister chromatids is essential to achieve haploidy in the meiotic products. During **prophase II,** each dyad is composed of one pair of sister chromatids attached by a common centromere. During **metaphase II,** the centromeres are directed to the equatorial plate. Then, the centromere divides, and during **anaphase II** the sister chromatids of each dyad are pulled to opposite poles. Since the number of dyads is equal to the haploid number, **telophase II** reveals one member of each homologous chromosome pair present at each pole. Each chromosome is referred to as a **monad.** Not only has the haploid state been achieved, but, if crossing over has occurred, each monad is a combination of maternal and paternal genetic information. As a result, the offspring produced by any gamete will receive from it a mixture of genetic information originally present in his or her grandparents. Following cytokinesis in telophase II, potentially four haploid gametes may result from a single meiotic event.

SPERMATOGENESIS AND OOGENESIS

Although events that occur during the meiotic divisions are similar in all cells that participate in gametogenesis, there are certain differences between the production of a male gamete (spermatogenesis) and a female gamete (oogenesis) in most animal species.

Spermatogenesis takes place in the testes, the male reproductive organs. The process begins with the expanded growth of an undifferentiated diploid germ cell called a **spermatogonium.** This cell enlarges to become a **primary spermatocyte,** which undergoes the first meiotic division. The products of this division are called **secondary spermatocytes.** Each secondary spermatocyte contains a haploid number of dyads. The secondary spermatocytes then undergo the second meiotic division, and each of these cells produces two haploid **spermatids.** Spermatids go through a series of developmental changes, **spermiogenesis,** and become highly specialized, motile **spermatozoa** or **sperm.** All sperm cells produced during spermatogenesis receive equal amounts of genetic material and cytoplasm. Figure 2.10 summarizes these steps.

Spermatogenesis may be continuous or occur periodically in mature male animals, with its onset determined by the nature of the species' reproductive cycle. Animals that reproduce year-round produce sperm continuously, while those whose breeding period is confined to a particular season produce sperm only before and during that season.

In animal **oogenesis,** the formation of **ova** (singular: **ovum**), or eggs, occurs in the ovaries, the female reproductive organs. The daughter cells resulting from the two meiotic divisions receive equal amounts of genetic material, but they do *not* receive equal amounts of cytoplasm. Instead, during each division, almost all the cytoplasm of the **primary oocyte,** itself derived from the **oogonium,** is concentrated in one of the two daughter cells. The concentration of cytoplasm is necessary because the function of the mature ovum is to nourish the developing embryo following fertilization.

During the first meiotic anaphase in oogenesis, the tetrads of the primary oocyte separate, and the dyads move toward opposite poles. During the first telophase, the dyads present at one pole are pinched off with very little surrounding cytoplasm to form the **first polar body.** The other daughter cell produced by this first meiotic division contains most of the cytoplasm and is called the **secondary oocyte.** The first polar body may or may not divide again to produce two small haploid cells. The mature ovum will be produced from the secondary oocyte during the second meiotic division. During this division, the cytoplasm of the secondary oocyte again divides unequally, producing an **ootid** and a **second polar body.** The ootid then differentiates into the mature ovum. Figure 2.10 illustrates the steps leading to formation of the mature ovum and polar bodies.

FIGURE 2.10
Oogenesis and spermatogenesis in animal cells.

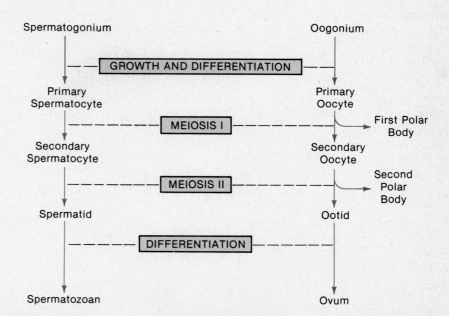

Unlike the divisions of spermatogenesis, the two meiotic divisions of oogenesis may not be continuous. In some animal species the two divisions may directly follow each other. In others, including the human species, the second division may not be complete until some time after sexual maturity is reached, even though the first division begins during the embryonic development of the individual. In humans, therefore, the two divisions are separated by 10 to 55 years, depending on when each ovulation event occurs during the life of the female!

THE SIGNIFICANCE OF MEIOSIS

The process of meiosis is critical to the successful sexual reproduction of all diploid organisms. It is the mechanism by which the diploid amount of genetic information is reduced to the haploid amount. In animals, meiosis leads to the formation of gametes, while in plants haploid spores are produced, which in turn lead to the formation of haploid gametes.

Each diploid organism contains its genetic information in the form of homologous pairs of chromosomes. Each pair consists of one member derived from the maternal parent and one from the paternal parent. Following meiosis, haploid cells contain a mixture of either the paternal or maternal representative of each homologous pair of chromosomes. Thus, meiosis results in a vast amount of genetic variation. Additionally, crossing over, which occurs in the first meiotic prophase stage, further reshuffles the genetic information. Crossing over occurs between the maternal and paternal members of each homologous pair, resulting in the production of even greater amounts of genetic variation in gametes.

Thus, the two most significant points about meiosis are that the process leads to:

1 the maintenance of a constant amount of genetic information between generations.

2 extensive genetic variation resulting from the mechanism of distribution of the genetic material.

Additionally, the products of meiosis play different roles in the life cycles of organisms, particularly in plants. In many single-celled algae and fungi, the predominant vegetative cells are haploid. They arise through meiosis and are proliferated by mitosis.

In multicellular plants, the life cycle alternates between the haploid **gametophyte** stage and the diploid **sporophyte** stage. One stage is usually predominant, depending on the species. For example, the predominant stage in the moss life cycle is the haploid gametophyte. Both male and female gametophytes may exist, and they lead to the production of haploid sperm and eggs. Following fertilization, the zygote is the initial stage of development of the diploid sporophyte. Within the sporophyte body, certain cells (the spore-mother cells) undergo meiosis, leading to the production of haploid spores. These spores germinate and lead to the production of the prominent male and female gametophytes. The cycle then repeats itself.

FIGURE 2.11
Diagram illustrating nondisjunction during the first and second meiotic divisions. In both cases, gametes are formed either containing two members of a specific chromosome or lacking it altogether. (Courtesy of Jorge J. Yunis, from Yunis and Chandler, 1979.)

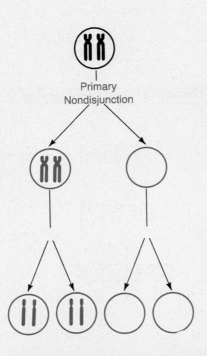

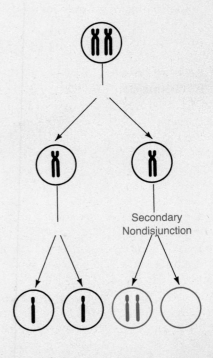

In the more advanced plants, meiosis results in the production of male microspores and female megaspores. In angiosperms, for example, these give rise to the haploid male and female parts of the gametophyte stage. The male microgametophyte results in pollen production, while the female megagametophyte is represented by the ovule. Fertilization yields a zygote that develops into the embryo housed within an embryo sac in the seed. Germination of the seed produces the diploid sporophyte, the predominant plant body of angiosperms.

Finally, it is important to know what happens when meiosis fails to achieve the expected outcome. Rarely, at either the first or second division, separation or disjunction of the chromatids of a tetrad or dyad does not occur. Instead, both members move to the same pole during anaphase. Such an event is called **nondisjunction,** because the two members fail to disjoin. If nondisjunction occurs at the first division stage, it is said to be a primary event; at the second division stage, it is called a secondary event.

The results of primary and secondary nondisjunction for just one chromosome of a diploid genome are shown in Figure 2.11. As can be seen, for the chromosome in question, gametes may be formed containing either two members or none at all. Following fertilization with a normal gamete, a zygote is produced with either three members (**trisomy**) or only one member (**monosomy**) of this chromosome. While these conditions are often tolerated in plants, they usually have severe or lethal effects in animals. Trisomy and monosomy will be described in greater detail in Chapter 12.

SUMMARY AND CONCLUSIONS

Mitosis and meiosis are essential processes that allow the transmission of genetic information from cells or organisms to their descendants. Mitosis is a form of cell division in which the hereditary components, which have earlier been replicated, are equally divided into daughter cells. Usually, the result is the conversion of one diploid cell into two diploid daughter cells. Meiosis, on the other hand, serves to reduce the diploid number of chromosomes to the haploid number. Such a conversion must be precise and is critical if organisms are not to double the amount of genetic material following sexual reproduction.

Cytoplasmic components, which must be divided during cell division, include the plasma membrane, nucleolus, endoplasmic reticulum, ribosomes, mitochondria, chloroplasts, and centrioles. The latter structures, found only in animal cells and some lower plants, are directly involved in organizing microtubules into spindle fibers during mitosis and meiosis. These fibers are intimately involved in the movement of chromosomes to opposite poles during mitosis and meiosis. The cytoplasm of eukaryotic cells contains a structural framework called the cytoskeleton.

To understand mitosis and meiosis, we must understand the concept of homologous chromosomes. Homologous pairs of chromosomes are derived from the maternal and paternal parents and, except for the sex chromosomes, share the same size, centromere placement, arm ratio, and potential genetic information. Following mitosis, daughter cells receive both a maternally and a paternally derived member of each homologous pair of chromosomes. During meiosis, however, daughter cells receive only the maternal or paternal copy of each pair, thus precisely reducing the diploid number to the haploid number.

Mitosis is just one part of the more extensive cell cycle. During interphase, the nondivisional stage, replication of the DNA of each chromosome occurs. This portion of interphase has been designated as the S phase. The G_1 and G_2 stages precede and follow the S phase, respectively. Some cells withdraw from the cell cycle prior to initiation of DNA synthesis. They are said to enter the G_0 stage. Studies suggest that the critical step in regulating cell division occurs before the S phase.

Mitosis is artificially divided into four discrete stages: prophase, metaphase, anaphase, and telophase. The condensation and coiling of chromatin into distinct chromosomal structures is characteristic of prophase, as is the gradual disappearance of the nucleolus and nuclear envelope. The pair of centrioles has also divided and begun to move to opposite ends of the cell. In metaphase, the chromosomes, each composed of two identical sister chromatids connected at a common centromere, have arranged themselves along the midline (or equatorial plane) of the cell. In anaphase, the centromeres divide and sister chromatids are pulled to opposite ends of the cell as a result of the action of spindle fibers. Telophase is largely a reversal of prophase and is characterized by cytokinesis. In animal cell mitosis, the cytoplasm is pinched into two cells. In plants, a cell plate is synthesized and laid down, dividing the original cell into two.

In meiosis, homologous chromosomes pair in synapsis. Since each member has previously replicated, meiotic prophase I is characterized by a haploid number of tetrad structures, each consisting of two pairs of sister chromatids. Two divisions are necessary to produce haploid cells. During the first meiotic prophase, nonsister chromatids exchange segments in the process called crossing over. Chiasmata represent this process. At the completion of meiosis, the division products contain a haploid content of genetic material.

Division of the cytoplasm is usually equal in spermatogenesis, but often unequal in oogenesis. This unequal distribution acts to conserve most of the cytoplasm in one egg and leads to the production of polar bodies.

The significance of meiosis is twofold: While maintaining a constant number of chromosomes from generation to generation in sexually reproducing organisms, the process also generates extensive individual genetic variation. Genetic variation is accomplished through the reshuffling of maternal and paternal homologues and through crossing over. In plants, meiosis produces cells with various critical functions during their life cycles.

PROBLEMS AND DISCUSSION QUESTIONS

1 What role do the following cellular components play in the storage, expression, or transmission of genetic information: (a) chromatin, (b) nucleolus, (c) ribosome, (d) mitochondrion, (e) centriole, (f) centromere?

2 Discuss the concepts of homologous chromosomes, diploidy, and haploidy. What characteristics are shared between two chromosomes considered to be homologous?

3 If two chromosomes of a species are the same length and have similar centromere placements yet are *not* homologous, what *is* different about them?

4 Describe the events that characterize each stage of mitosis.

5 If an organism has a diploid number of 16, how many chromatids are visible at the end of mitotic prophase? How many chromosomes are moving to each pole during anaphase of mitosis?

6 How are chromosomes generally named on the basis of centromere placement?

7 Contrast telophase in plant and animal mitosis.

8 Outline and discuss the events of the cell cycle. What experimental technique was used to demonstrate the existence of the S phase?

9 What is known about the regulation of the cell cycle?

10 Once DNA synthesis is initiated in the S phase of the cell cycle, the events leading to and including cell division continue automatically. What is the significance of this observation to investigations of cancer?

11 Contrast the end results of meiosis with those of mitosis.

12 Define and discuss the following terms: (a) synapsis, (b) bivalents, (c) chiasmata, (d) crossing over, (e) chromomeres, (f) sister chromatids, (g) tetrads, (h) dyads, (i) monads, and (j) synaptonemal complex.

13 If an organism has a diploid number of 16 in an oocyte,
    (a) How many tetrads are present in the first meiotic prophase?
    (b) How many dyads are present in the second meiotic prophase?
    (c) How many monads migrate to each pole during the second meiotic anaphase?

14 Contrast spermatogenesis and oogenesis. What is the significance of the formation of polar bodies?

15 Explain why meiosis leads to significant genetic variation while mitosis does not.

16 In Figure 2.8 the four spermatids formed as a result of meiosis each contain a unique combination of paternally and maternally derived chromosome segments (as represented by the different colors). These four different combinations of genetic information resulted from the random alignment of both sets of chromosomes on the equatorial plate during metaphase I and metaphase II. Had they become aligned differently, many other chromosome combinations might have been produced in the spermatids. With reference to Figure 2.8, how many other combinations are possible? Draw them.

17 During oogenesis in an animal species with a haploid number of 6, one dyad undergoes secondary nondisjunction. Following the second meiotic division, the involved dyad ends up intact in the ovum. How many chromosomes are present in: (a) the mature ovum, and (b) the second polar body? (c) Following fertilization by a normal sperm, what chromosome condition is created?

18 What are the only two stages during the meiotic divisions when tetrads are present?

19 During the first meiotic prophase, (a) when does crossing over occur?; (b) when does synapsis occur?; (c) during which stage are the chromosomes least condensed?; and (d) when are chiasmata first visible?

20 What is the role of meiosis in the life cycle of a higher plant such as an angiosperm?

SELECTED READINGS

ALBERTS, R., et al. 1983. *Molecular biology of the cell.* New York: Garland.

BAKER, B. A., et al. 1976. The genetic control of meiosis. *Ann. Rev. Genet.* 10: 53–134.

BASERGA, R., and KISIELESKI, W. 1963. Autobiographies of cells. *Scient. Amer.* (Aug.) 209: 103–110.

BRACHET, J., and MIRSKY, A. E. 1961. *The cell: Meiosis and mitosis.* Vol. 3. New York: Academic Press.

BROWN, W. 1972. *Textbook of cytogenetics.* St Louis: C. V. Mosby.

DeROBERTIS, E. D. P., and DeROBERTIS, E. M. F. 1981. *Essentials of cell and molecular biology.* Philadelphia: W. B. Saunders.

DUPRAW, E. 1968. *Cell and molecular biology.* New York: Academic Press.

GARBER, E. D. 1972. *Cytogenetics: An introduction.* New York: McGraw-Hill.

HARTWELL, L. H. 1971. Genetic control of the cell division cycle in yeast. IV. Genes controlling bud emergence and cytokinesis. *Exptl. Cell Res.* 69: 265–76.

————. 1978. Cell division from a genetic perspective. *J. Cell Biol.* 77: 627–37.

HARTWELL, L.H., et al. 1974. Genetic control of the cell division cycle in yeast. *Science* 183: 46–51.

KARP, G. 1984. *Cell biology.* 2nd ed. New York: McGraw-Hill.

MAZIA, D. 1961. How cells divide. *Scient. Amer.* (Jan.) 205: 101–120.

————. 1974. The cell cycle. *Scient. Amer.* (Jan.) 235: 54–64.

MOENS, P. B. 1973. Mechanisms of chromosome synapsis at meiotic prophase. *Int. Rev. Cytol.* 35: 117–134.

NOVIKOFF, A. B., and HOLTZMAN, E. 1970. *Cells and organelles.* New York: Holt, Rinehart and Winston.

PADILLA, G. M., and McCARTY, K. S., eds. 1982. *Genetic expression in the cell cycle.* New York: Academic Press.

PARDEE, A. B., et al. 1978. Animal cell cycle. *Ann. Rev. Biochem.* 47: 715-50.

PRESCOTT, D. M. 1973. *Cancer, the misguided cell.* New York: Pegasus–Bobbs-Merrill.

————. 1976. *Reproduction of eukaryotic cells.* New York: Academic Press.

PRESCOTT, D. M., and GOLDSTEIN, L. 1967. Nuclear-cytoplasmic interaction in DNA synthesis. *Science* 155: 469–70.

RAO, M. V. N. 1980. Nuclear proteins in programming cell cycles. *Int. Rev. Cytol.* 67: 291–315.

SIMCHEN, G. 1978. Cell cycle mutants. *Ann. Rev. Genet.* 12: 161–191.

SWANSON, C. P.; MERZ, T.; and YOUNG, W. J. 1981. *Cytogenetics, the chromosome in division, inheritance, and evolution.* 2nd ed. Englewood Cliffs, N. J.: Prentice-Hall.

WANG, E. 1985. A 57000-mol-wt protein uniquely present in nonproliferating and senescent human fibroblasts. *J. Cell Biol.* 100: 545–51.

WESTERGAARD, M., and vonWETTSTEIN, D. 1972. The synaptinemal complex. *Ann. Rev. Genet.* 6: 71–110.

WHEATLEY, D. N. 1982. *The centriole: A central enigma of cell biology.* New York: Elsevier/ North-Holland Biomedical.

WISSINGER, W. L., and WANG, R. J. 1983. Cell cycle mutants. *Int. Rev. Cytol.* 15 (supplement): 91–113.

YANISHEVSKY, R. M., and STEIN, G. M. 1981. Regulation of the cell cycle in eukaryotic cells. *Int. Rev. Cytol.* 69: 223–59.

YUNIS, J. J., and CHANDLER, M. E. 1979. Cytogenetics. In *Clinical diagnosis and management by laboratory methods*, ed. J. B. Henry, vol. 1. Philadelphia: W. B. Saunders.

ZIMMERMAN, A. M., and FORER, A., eds. 1981. *Mitosis/Cytokinesis.* New York: Academic Press.

3

Mendelian Genetics: Transmission of Genes Controlling Traits

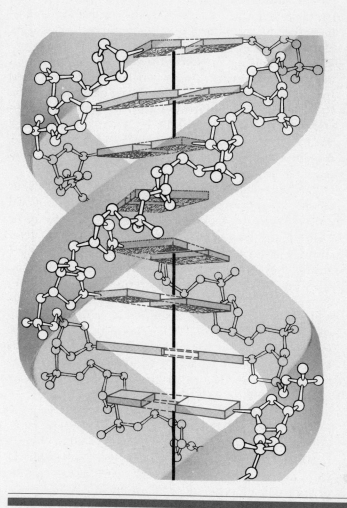

Even though inheritance has been recognized for thousands of years, the first significant insights into the mechanisms involved occurred only a little over a century ago. In 1866, Gregor Mendel published the results of a series of experiments that would lay the foundation for the formal discipline of genetics. In the ensuing years the concept of the gene as a distinct hereditary unit was established, and the ways in which genes are transmitted to offspring and control traits were clarified. Research in these areas was accelerated in the first half of the twentieth century, generating the interest so important to the acquisition of the knowledge derived in this field since 1950.

In this chapter we will focus on the development of the principles established by Mendel, now referred to as **Mendelian genetics.** These principles describe the transmission of genes from parents to offspring and were derived directly from his experimentation.

When Mendel began his studies of inheritance using *Pisum sativum*, the garden pea, there was little or no knowledge of chromosomes and the role and mechanism of meiosis. Nevertheless, he was able to determine that distinct **units of inheritance** exist. He also predicted their behavior during the formation of gametes. Subsequent investigators, with access to cytological data, were able to relate their observations of chromosome behavior during meiosis to Mendel's principles of inheritance. Once this correlation had been made, Mendel's postulates were accepted as the basis for the study of transmission genetics. Even today, they serve as the cornerstone of the study of inheritance.

GREGOR MENDEL

In 1822, Gregor Johann Mendel was born of a peasant family in the village of Heinzendorf, now part of Czechoslovakia. An excellent student in high school, Mendel studied philosophy for several years afterward and was admitted to the Augustinian Monastery in Brno in 1843, where he continued his studies. In 1849, he was relieved of pastoral duties and received a teaching appointment that lasted several years. From 1851 to 1853 he attended the University of Vienna, where he studied physics and botany. In 1854 he returned to Brno, where for the next sixteen years he taught physics and natural science.

In 1856 Mendel performed the first set of hybridization experiments with the garden pea. The research phase of his career lasted until 1868, when he was elected abbot. While his interest in genetics remained, his new responsibilities demanded most of his time. In 1884 Mendel died of a kidney disorder. The local newspaper paid him the following tribute: "His death deprives the poor of a benefactor, and mankind at large of a man of the noblest character, one who was a warm friend, a promoter of the natural sciences, and an exemplary priest. . . ."

Mendel's Experimental Approach

In 1865, Mendel first reported the results of some simple genetic crosses between certain strains of the garden pea. Although his was not the first attempt to provide experimental evidence pertaining to inheritance, Mendel's work is an elegant model of practical experimental design, insightful analysis, and data interpretation.

Mendel showed remarkable insight into the methodology necessary for good experimental biology. He chose an organism that was easy to grow and interbreed. The pea plant is self-fertilizing in nature, but is easily crossbred in designed experiments. It reproduces well and grows to maturity in a single season. Mendel worked with seven unit characters, visible features that were each represented by two contrasting forms or traits. For the character stem height, for example, he experimented with the traits *tall* and *dwarf*. He selected six other contrasting pairs of traits involving seed shape and color, pod shape and color, and pod and flower arrangement. True-breeding strains were available from local seed merchants. Each form appeared generation after generation in self-fertilizing plants; that is, the strains exhibiting these traits "bred true."

Mendel's success in an area where others had failed may be attributed to several factors in addition to the choice of a suitable organism. He restricted his examination to one or very few pairs of contrasting traits in each experiment. He also kept accurate quantitative records, a necessity in genetic experiments. From the analysis of his data, Mendel derived certain postulates that have become the principles of transmission genetics.

The significance of Mendel's experiments was not realized until the early twentieth century, well after his death. Once Mendel's publications were rediscovered by geneticists investigating the function and behavior of chromosomes, the implications of his postulates were immediately apparent. He had discovered the basis for transmission of hereditary traits!

THE MONOHYBRID CROSS

The simplest crosses performed by Mendel involved only one pair of contrasting traits. Each such breeding experiment is called a **monohybrid cross.** A monohybrid cross is made by mating individuals from two parent strains, each of which exhibits one of the two contrasting forms of the character under study. Initially we will examine the first generation of offspring of such a cross, and then we will consider the offspring of **selfing** or **self-fertilizing** individuals from this first generation. The original parents are called the P_1 or **parental generation,** their offspring are the F_1 or **first filial generation,** and the individuals resulting from the selfing of the F_1 generation are called the F_2 or **second filial generation.** We can, of course, continue to follow the F_3, F_4, F_5, and subsequent generations, if desirable.

The cross between peas with tall stems and dwarf stems is representative of Mendel's monohybrid crosses. *Tall* and *dwarf* represent contrasting forms or traits of the character of stem height. Unless tall or dwarf plants are crossed together or with another strain, they will undergo self-fertilization and breed true, producing their respective trait generation after generation. However, when Mendel crossed tall plants with dwarf plants, the resulting F_1 generation consisted only of tall plants. When members of the F_1 generation were selfed, Mendel observed that 787 of 1064 F_2 plants were tall, while 277 of 1064 were dwarf. Note that in this cross (Figure 3.1) the dwarf trait disappears in the F_1, only to reappear in the F_2 generation.

Genetic data are usually expressed and analyzed as ratios. In this particular example, many identical P_1 crosses were made and many F_1 plants—all tall—were produced. When these F_1 offspring were self-fertilized, 787 F_2 progeny were tall and 277 were dwarf—a ratio of approximately 2.8:1.0, or about 3:1.

Mendel made similar crosses between pea plants exhibiting each of the other pairs of contrasting traits.

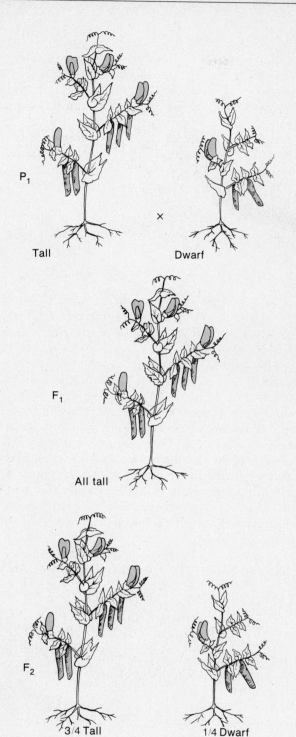

FIGURE 3.1
The F_1 and F_2 results of Mendel's monohybrid cross between tall and dwarf peas.

The results of these crosses are shown in Table 3.1. In every case, the outcome was similar to the tall/dwarf cross just described. All F_1 offspring were identical to one of the parents. In the F_2, an approximate ratio of

3:1 was obtained. Three-fourths appeared like the F_1 plants, while one-fourth exhibited the contrasting trait, which had disappeared in the F_1 generation.

It is appropriate to point out one further aspect of the monohybrid crosses. In each, the F_1 and F_2 patterns of inheritance were similar regardless of which P_1 plant served as the source of pollen, or sperm, and which served as the source of the ovum, or egg. The crosses could be made either way—that is, tall sperm fertilizing dwarf ova, or vice versa. These are called **reciprocal crosses.** Therefore, the results of Mendel's monohybrid crosses were not sex-dependent.

To explain these results, Mendel proposed the existence of particulate **unit factors** for each trait. He suggested that these factors serve as the basic units of heredity and are passed unchanged from generation to generation, determining various traits expressed by each individual plant. Using these general ideas, Mendel proceeded to hypothesize precisely how such factors could account for the results of the monohybrid crosses.

Mendel's First Three Postulates

Using the consistent pattern of results in the monohybrid crosses, reinforced seven times, Mendel derived the following three postulates or principles of inheritance.

1. UNIT FACTORS IN PAIRS
   Genetic characters are controlled by unit factors that exist in pairs in individual organisms.
   In the monohybrid cross involving tall and dwarf stems, a specific unit factor exists for each trait. Since the factors occur in pairs, three combinations are possible: two factors for tallness, two factors for dwarfness, or one of each factor. Every individual contains one of these three combinations, which determines stem height.

2. DOMINANCE/RECESSIVENESS
   When two unlike unit factors responsible for a single character are present in a single individual, one unit factor is dominant to the other, which is said to be recessive.
   In each monohybrid cross, the trait expressed in the F_1 generation is controlled by the dominant unit factor. The trait that is not expressed is controlled by the recessive unit factor. Note that this dominance/recessiveness relationship only pertains when unlike unit factors are present in pairs. The terms *dominant* and *recessive* are also used to designate the traits. In this case, tall stems are said to be dominant to the recessive dwarf stems.

TABLE 3.1
Mendel's monohybrid crosses.

Character	Phenotypes	F$_1$ Dominant	F$_2$ Results	F$_2$ Ratio
Stem	Tall/dwarf	Tall	787 tall 277 dwarf	2.84:1
Pods	Axial/terminal	Axial	651 axial 207 terminal	3.14:1
	Full/constricted	Full	882 full 299 constricted	2.95:1
	Yellow/green	Green	428 green 152 yellow	2.82:1
Seeds	Round/wrinkled	Round	5474 round 1850 wrinkled	2.96:1
	Yellow/green	Yellow	6022 yellow 2001 green	3.01:1
Flowers	Violet/white	Violet	705 violet 224 white	3.15:1

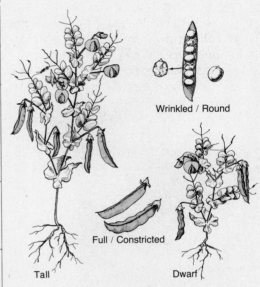

Wrinkled / Round

Full / Constricted

Tall

Dwarf

3. SEGREGATION

During the formation of gametes, the paired unit factors separate or segregate randomly so that each gamete receives one or the other.
If an individual contains a pair of like unit factors (e.g., both are specific for tall), then all gametes receive one tall unit factor. If an individual contains unlike unit factors (e.g., one for tall and one for dwarf), then each gamete has a 50 percent probability of receiving either the tall *or* dwarf unit factor.

These postulates provide a suitable explanation for the results of the monohybrid crosses. The tall/dwarf cross will be used to illustrate this explanation. Mendel reasoned that P$_1$ tall plants contained identical paired unit factors, as did the P$_1$ dwarf plants. During gamete formation the ova and pollen of tall plants all received one tall unit factor as a result of segregation. Likewise, ova and pollen of dwarf plants all received one dwarf unit factor. Following fertilization, all F$_1$ plants received one of each unit factor, reestablishing the paired relationship. Because tall is dominant to dwarf, all F$_1$ plants were tall.

When F$_1$ plants form gametes, the postulate of segregation demands that each gamete randomly receive *either* the tall *or* dwarf unit factor. Following random fertilization events during F$_1$ selfing, four F$_2$ combinations will result in equal frequency:

    (1) tall/tall
    (2) tall/dwarf
    (3) dwarf/tall
    (4) dwarf/dwarf

Combinations (1) and (4) will result in tall and dwarf plants, respectively. According to the postulate of dominance/recessiveness, combinations (2) and (3) will both yield tall plants. Therefore, the F$_2$ is predicted to consist of three-fourths tall and one-fourth dwarf, or a ratio of 3:1. This is approximately what Mendel observed in the cross between tall and dwarf plants. A similar pattern was observed in each of the other monohybrid crosses.

Modern Genetic Terminology

In order to illustrate the monohybrid cross and Mendel's first three postulates, we must introduce several new terms as well as a set of symbols for the unit factors. Traits such as tall or dwarf are visible expressions of the information contained in unit factors. The physical appearance of a trait is called the **phenotype** of the individual.

All unit factors represent units of inheritance called **genes** by modern geneticists. For any given character, such as plant height, the phenotype is determined by alternate forms of a single gene called **alleles.** For example, the unit factors representing tall and dwarf are alleles determining the height of the pea plant.

By one convention, the first letter of the recessive trait is chosen to symbolize the character in question. The lowercase letter designates that allele for the recessive trait, and the uppercase letter designates the allele for the dominant trait. Therefore, we let *d* stand for the dwarf allele and *D* represent the tall allele. When alleles are written in pairs to represent the two unit factors present in any individual (*DD*, *Dd*, or *dd*), these symbols are referred to as the **genotype.** This term reflects the genetic makeup of an individual whether it is haploid or diploid. By reading the genotype, it is possible to know the phenotype of the individual: *DD* and *Dd* are tall, while *dd* is dwarf. When both alleles are the same (*DD* or *dd*), the individual is said to be **homozygous** or a **homozygote;** when alleles are different (*Dd*), we use the term **heterozygous** or **heterozygote.** These symbols and terms are used in Figure 3.2 to illustrate the complete monohybrid cross. Given this information, we may explore the possible rationale used by Mendel to arrive at his postulates, keeping in mind that hindsight makes this task much easier.

What led Mendel to deduce unit factors in pairs? Since there were two contrasting traits for each character, it seemed logical that two distinct factors must exist. However, why does one of the two traits or phenotypes disappear in the F_1 generation? Observation of the F_2 generation helps to answer this question. The recessive trait and its unit factor do not actually disappear in the F_1; they are merely hidden or masked, only to reappear in one-fourth of the F_2 offspring. Therefore, Mendel concluded that one unit factor for tall and one for dwarf were transmitted to each F_1 individual; but because the tall factor or allele is dominant to the dwarf factor or allele, all F_1 plants are tall. Finally, how is the 3:1 F_2 ratio explained? As shown in Figure 3.2, if the tall and dwarf alleles of the

FIGURE 3.2
An explanation of the monohybrid cross shown in Figure 3.1. The symbols *D* and *d* are used to designate the tall and dwarf unit factors, respectively, in the genotypes of mature plants and gametes. All individuals are shown in unshaded rectangles. All gametes are shown in shaded circles.

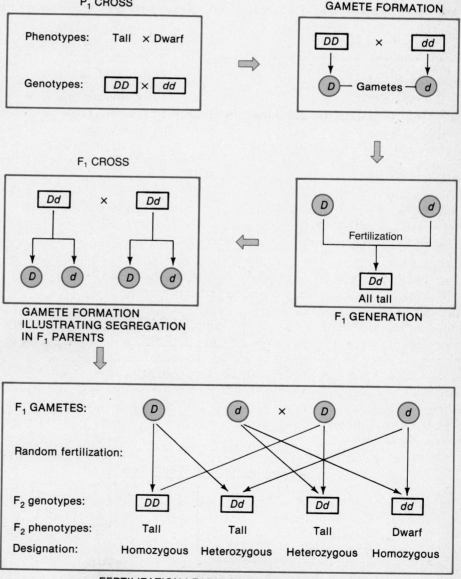

F_1 heterozygote segregate randomly into gametes, and if fertilization is random, this ratio is the natural outcome of the cross.

Because he operated without the hindsight that students of modern genetics enjoy, Mendel's analytical reasoning must be considered a truly outstanding scientific achievement. On the basis of rather simple, but precisely executed breeding experiments, he proposed that discrete particulate units of heredity exist, and he explained how they are transmitted from one generation to the next!

Of particular significance in Mendel's work was the evidence favoring **particulate units of inheritance.** This concept was unique in 1866. The most prevalent theory at that time suggested that offspring were a blend or mixture of their parents' traits. If this were true, then all of Mendel's F_1 plants would have exhibited an intermediate expression. In the tall/dwarf cross, the F_1 generation would all have been taller than the P_1 dwarf plants, but not so tall as the P_1 tall plants. The fact that Mendel's proposals for inheritance ran counter to the most widely accepted genetic theories has been cited as one reason why his work went "undiscovered" for over three decades.

Punnett Squares

The genotypes and phenotypes resulting from the recombination of gametes during fertilization can be easily visualized by constructing a **Punnett square.** Figure 3.3 illustrates this method of analysis for the $F_1 \times F_1$ monohybrid cross. All possible gametes are assigned to a column or a row, with the vertical column representing those of the male parent and the horizontal row those of the female parent. After entering the gametes in rows and columns, the new generation is predicted by combining the male and female gametic information for each combination and enter-

FIGURE 3.3
The use of a Punnett square in generating the F_2 ratio of the $F_1 \times F_1$ cross shown in Figures 3.1 and 3.2.

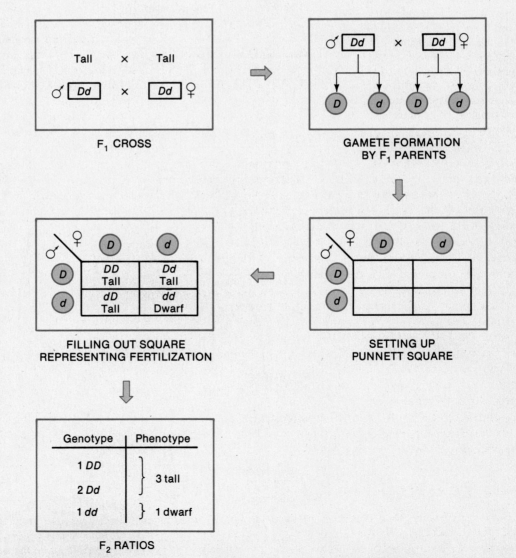

FIGURE 3.4

The test cross illustrated with a single character. In (a), the tall parent is homozygous. In (b), the tall parent is heterozygous. The genotypes of each tall parent are determined by examining the offspring when each is crossed to the homozygous recessive dwarf plant.

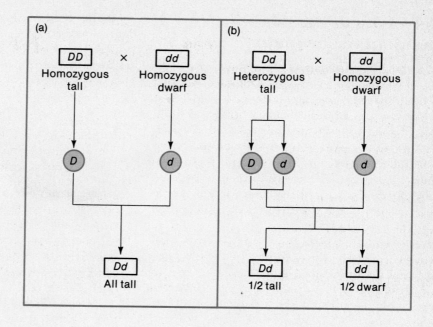

ing the resulting genotypes in the boxes. This process represents all possible random fertilization events. The genotypes and phenotypes of all potential offspring are ascertained by reading the entries in the boxes. The Punnett square method is particularly useful for more complex crosses.

The Test Cross: One Character

Tall plants produced in the F₂ generation are predicted to be of either the *DD* or *Dd* genotypes. We might ask whether there is a way to distinguish the genotype. Mendel devised a rather simple method that is still used today in breeding procedures of plants and animals: the **test cross.** The organism of the dominant phenotype but unknown genotype is crossed to a homozygous recessive individual. For example, if a tall plant of genotype *DD* is test-crossed to a dwarf plant, which must have the *dd* genotype, all offspring will be tall and *Dd* genotypically. However, if a tall plant is *Dd* and is crossed to a dwarf plant, one-half of the offspring will be tall (*Dd*) and the other half will be dwarf (*dd*). Therefore, a 1:1 tall/dwarf ratio demonstrates the heterozygous nature of the tall plant of unknown genotype. The basis for these conclusions is illustrated in Figure 3.4. The test cross reinforced Mendel's conclusion that separate unit factors control the tall and dwarf traits.

THE DIHYBRID CROSS

A natural extension of performing monohybrid crosses was for Mendel to design experiments where two characters were examined simultaneously. We will re-

fer to such a cross, involving two pairs of contrasting traits, as a **dihybrid cross.** For example, Mendel bred pea plants having *tall stems* and *round seeds* with those having *dwarf stems* and *wrinkled seeds*. The results of this cross are shown in Figure 3.5. The F₁ offspring were all tall and round. It is therefore apparent that tall is dominant to dwarf, as previously established, and that round is dominant to wrinkled. When the F₁ individuals were selfed, approximately 9/16 of the F₂ plants were tall and round, 3/16 were tall and wrinkled, 3/16 were dwarf and round, and 1/16 were dwarf and wrinkled.

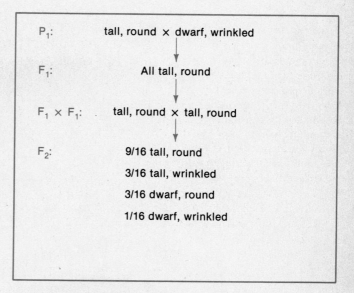

FIGURE 3.5

The F₁ and F₂ results of Mendel's dihybrid cross between tall, round and dwarf, wrinkled pea plants.

Mendel's Fourth Postulate: Independent Assortment

We can most easily understand the results of a dihybrid cross if we consider it theoretically as consisting of two monohybrid crosses conducted separately. Think of the two sets of traits as being inherited independently of each other. Thus, because tall is dominant to dwarf, all F_1 plants in the first theoretical cross would be tall-stemmed. In the second theoretical cross, all F_1 plants would have round seeds because round is dominant to wrinkled. When Mendel examined the F_1 plants of the dihybrid cross, all were tall and round, as predicted.

The predicted F_2 results of the first cross are 3/4 tall and 1/4 dwarf. Similarly, the second cross would yield 3/4 round and 1/4 wrinkled. Figure 3.5 shows that in the dihybrid cross, 12/16 F_2 plants are tall while 4/16 are dwarf, exhibiting the 3:1 ratio. Similarly, 12/16 F_2 plants have round seeds while 4/16 have wrinkled seeds, again revealing the 3:1 ratio.

Since it is clear that the two pairs of contrasting traits are inherited independently, we can predict the frequencies of all possible F_2 phenotypes based on the individual probabilities. **When two events occur independently but simultaneously, their combined probability is equal to the product of their individual probabilities of occurrence.** For example, the probability of an F_2 plant being tall *and* round is (3/4) · (3/4), or 9/16, since 3/4 of all F_2 plants should be tall and 3/4 of all F_2 plants should be round.

In a like way, the probabilities of the other three F_2 phenotypes may be calculated: tall (3/4) *and* wrinkled (1/4) are predicted to be present together 3/16 of the time; dwarf (1/4) *and* round (3/4) are predicted 3/16 of the time, and dwarf (1/4) *and* wrinkled (1/4) are predicted 1/16 of the time. These calculations are illustrated in Figure 3.6. They coincide closely with Mendel's observation of this cross.

On the basis of similar results in numerous dihybrid crosses, Mendel proposed a fourth postulate:

4. INDEPENDENT ASSORTMENT
During gamete formation, segregating pairs of unit factors assort independently of each other.

This postulate stipulates that segregation of any pair of unit factors occurs independently of all others. As a result of segregation, each gamete receives one member of every pair of unit factors. According to the postulate of independent assortment, all possible combinations of gametes will be formed in equal frequency.

Independent assortment is illustrated in the formation of the F_2 generation, shown in the Punnett square in Figure 3.7. Examine the formation of gametes by the F_1 plant. Segregation prescribes that every gamete receives either a *D* or *d* allele *and* a *W* or *w* allele. Independent assortment stipulates that all combinations (*DW*, *Dw*, *dW*, and *dw*) will be formed with equal probabilities.

In every $F_1 \times F_1$ fertilization event, each zygote has an equal probability of receiving one of the four combinations from each parent. If a large number of offspring is produced, 9/16 are tall and round, 3/16 are tall and wrinkled, 3/16 are dwarf and round, and 1/16 are dwarf and wrinkled, yielding what is designated as **Mendel's 9:3:3:1 dihybrid ratio.** This ratio is based on probability events involving segregation, indepen-

FIGURE 3.6
The determination of the combined probabilities of each F_2 phenotype for two independently inherited characters. The probability of each plant being tall *or* dwarf is independent of the probability of it being round or wrinkled.

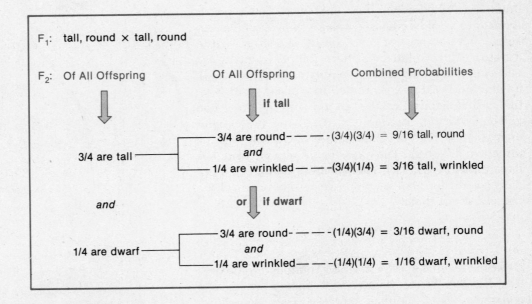

dent assortment, and random fertilization. Therefore, it is an ideal ratio. Because of chance deviation, particularly if small numbers of offspring are produced, the perfect ratio will seldom be approached.

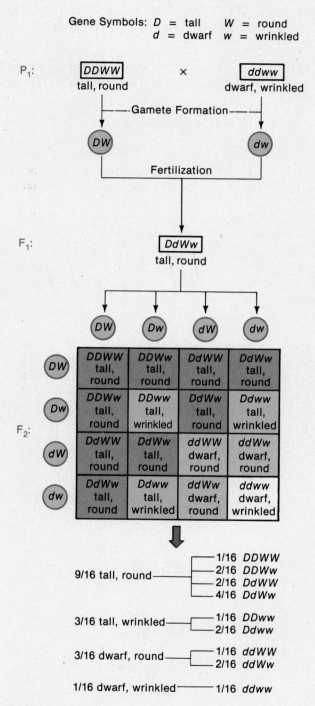

FIGURE 3.7

Diagram of a typical dihybrid cross. The F_1 heterozygous plants are self-fertilized to produce an F_2 generation, which is computed using a Punnett square. Both the phenotypic and genotypic F_2 ratios are shown.

The Test Cross: Two Characters

The test cross may also be applied to individuals that express two dominant traits, but whose genotypes are unknown. For example, the expression of the tall, round phenotype in the F_2 generation just described may result from the *DDWW, DDWw, DdWW,* and *DdWw* genotypes. If an F_2 tall, round plant is crossed with the homozygous recessive dwarf, wrinkled plant (*ddww*), analysis of the offspring will indicate the correct genotype of that tall, round plant. Figure 3.8 illustrates the use of the test cross with offspring resulting from a dihybrid cross.

THE TRIHYBRID CROSS

We have thus far considered inheritance by individuals of up to two pairs of contrasting traits. Mendel demonstrated that the identical processes of segregation and independent assortment apply to three pairs of contrasting traits in what is called a **trihybrid cross.**

While a trihybrid cross is somewhat more complex than a dihybrid cross, its results are easily calculated if the principles of segregation and independent assortment are followed. For example, consider the cross shown in Figure 3.9, where the gene pairs representing theoretical contrasting traits are symbolized *A/a, B/b,* and *C/c.* In the cross between *AABBCC* and *aabbcc* individuals, all F_1 individuals are heterozygous for all three gene pairs. Their genotype, *AaBbCc,* results in the phenotypic expression of the dominant *A, B,* and *C* traits. When F_1 individuals are parents, they all produce 8 different gametes in equal frequencies. At this point, we could construct a Punnett square with 64 separate boxes and read out the phenotypes. Because such a method is cumbersome in a cross involving so many factors, another method has been devised to calculate the predicted ratio.

The Forked-Line, or Branch Diagram, Method

It is much simpler to consider each contrasting pair of traits separately and then to combine these results using the **forked-line method,** which was illustrated in Figure 3.6. This method, also called a **branch diagram,** relies on the simple application of the laws of probability established for the dihybrid cross. Each gene pair is assumed to behave independently during gamete formation.

When the monohybrid cross AA × aa is made, we know that:

FIGURE 3.8

The test cross illustrated with two independent characters.

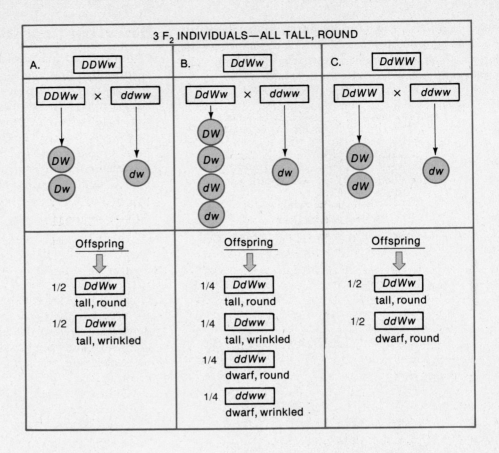

3 F₂ INDIVIDUALS—ALL TALL, ROUND		
A. `DDWw`	**B.** `DdWw`	**C.** `DdWW`

A. `DDWw` × `ddww`

Gametes: DW, Dw × dw

Offspring
- 1/2 `DdWw` tall, round
- 1/2 `Ddww` tall, wrinkled

B. `DdWw` × `ddww`

Gametes: DW, Dw, dW, dw × dw

Offspring
- 1/4 `DdWw` tall, round
- 1/4 `Ddww` tall, wrinkled
- 1/4 `ddWw` dwarf, round
- 1/4 `ddww` dwarf, wrinkled

C. `DdWW` × `ddww`

Gametes: DW, dW × dw

Offspring
- 1/2 `DdWw` tall, round
- 1/2 `ddWw` dwarf, round

1 All F₁ individuals have the genotype Aa and the phenotype represented by the *A* allele, which is called the *A* phenotype in the following discussion.

2 The F₂ generation consists of individuals with either the *A* phenotype or the *a* phenotype in the ratio of 3:1, respectively.

The same generalizations may be made for the *BB* × *bb* and *CC* × *cc* crosses. Thus, in the F₂ generation, 3/4 of all organisms will have phenotype *A*, 3/4 will have *B*, and 3/4 will have *C*. Similarly, 1/4 of all organisms will have phenotype *a*, 1/4 will have *b*, and 1/4 will have *c*. The proportions of organisms that express each phenotypic combination may be predicted by assuming that fertilization, following the independent assortment of these three gene pairs during gamete formation, is a random process.

The phenotypes of the F₂ generation calculated by the forked-line method are illustrated in Figure 3.10. They fall into the **trihybrid ratio of 27:9:9:9:3:3:3:1.** The same method may be applied when solving crosses involving any number of gene pairs, *provided* that all gene pairs assort independently from each other. We will see later that this is not always the case. However, it was apparently true for all of Mendel's characters.

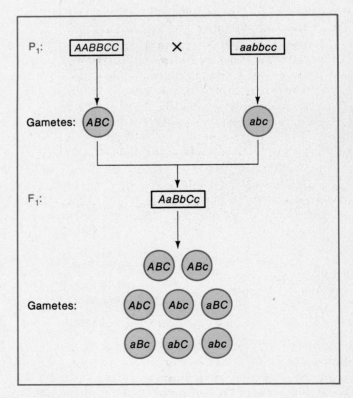

P₁: `AABBCC` × `aabbcc`

Gametes: ABC, abc

F₁: `AaBbCc`

Gametes: ABC, ABc, AbC, Abc, aBC, aBc, abC, abc

FIGURE 3.9

The formation of P₁ and F₁ gametes in a trihybrid cross.

FIGURE 3.10
The generation of the F_2 trihybrid phenotypic ratio using the forked-line, or branch diagram, method.

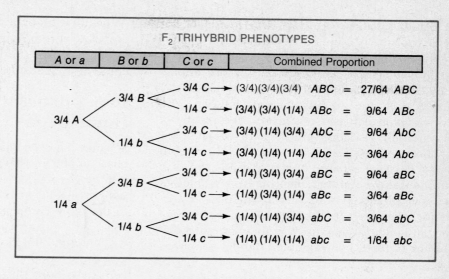

Note in Figure 3.10 that only phenotypic ratios of the F_2 generation have been derived. It is possible to generate genotypic ratios as well. To do so, we again consider the A/a, B/b, and C/c gene pairs separately. For example, for the A/a pair the F_1 cross is $Aa \times Aa$. Phenotypically, an F_2 ratio of 3/4 A:1/4 a is produced. Genotypically, however, the F_2 ratio is different; 1/4 AA: 1/2 Aa:1/4 aa will result. Using Figure 3.10 as a model, we would enter these genotypic frequencies on the left side of the calculation. Each would be connected by three bars to 1/4 BB, 1/2 Bb, and 1/4 bb, respectively. From each of these nine designations, three more bars would extend to the 1/4 CC, 1/2 Cc, and 1/4 cc genotypes. Thus, on the right side of the completed diagram, 27 genotypes and their frequencies of occurrence would appear. One of the problems at the end of this chapter asks you to use the forked-line or branch diagram method to determine the genotypic ratios generated in a trihybrid cross (see Problem 16).

It is obvious in crosses involving two or more gene pairs that the calculation of gametes and genotypic and phenotypic results is quite complex. There are several simple mathematical rules that will enable you to check the accuracy of various steps required in working genetics problems. First, you must determine the number of heterozygous gene pairs (n) involved in the cross. For example, where $AaBb \times AaBb$ represents the cross, $n = 2$; for $AaBbCc \times AaBbCc$, $n = 3$; for $AaBBCcDd \times AaBBCcDd$, $n = 3$ (because the B genes are not heterozygous). Once n is determined, 2^n is the number of different gametes that can be formed by each parent; 3^n is the number of different genotypes that result following fertilization; and 2^n is the number of different phenotypes that are produced from these genotypes. Table 3.2 summarizes these rules, which may be applied to crosses involving any number of gene pairs.

THE REDISCOVERY AND REINTERPRETATION OF MENDEL'S WORK

Mendel's work, initiated in 1856, was presented to the Brünn Society of Natural Science in 1865 and published one year later. However, his findings went largely unnoticed for about thirty-five years! Many reasons have been suggested to explain why the significance of his research was not immediately recognized.

TABLE 3.2
Simple mathematical rules useful in working genetics problems.

Number of Heterozygous Gene Pairs	Number of Different Types of Gametes Formed	Number of Different Genotypes Produced	Number of Different Phenotypes Produced
n	2^n	3^n	2^n
1	2	3	2
2	4	9	4
3	8	27	8
4	16	81	16

First of all, Mendel's adherence to mathematical analysis of probability events was quite an unusual approach in biological studies. Perhaps his approach seemed foreign to his contemporaries. More importantly, his conclusions drawn from such analyses did not fit well with the existing theories involving the cause of variation between organisms. The source of natural variation intrigued students of evolutionary theory. These individuals, stimulated by the proposal of Charles Darwin, believed that variation was of a **continuous nature** and that offspring were a blend of their parents' phenotypes. As we have mentioned earlier, Mendel theorized that variation was due to discrete or particulate units and was therefore of a **discontinuous nature.** For example, Mendel proposed that the F_2 offspring of a dihybrid cross were merely expressing traits produced by new combinations of previously existing unit factors. As a result, Mendel's theories did not follow the evolutionists' preconceptions about causes of variation.

Beyond the above speculations is still a further interpretation of why those who knew of Mendel's findings failed to grasp their significance. They did not realize that Mendel's postulates explained *how* inherited variation was transmitted, not *why* individuals expressing certain phenotypes survived preferentially over those expressing other phenotypes. It was this latter question that Charles Darwin had addressed in his **theory of natural selection.** Unfortunately, the collective vision of Mendel's colleagues was apparently obscured by the impact of Darwin's extraordinary theory of organic evolution.

The Rediscovery of Mendel's Work

In the very early twentieth century, two types of research led to the rediscovery of Mendel's work. Hybridization experiments similar to Mendel's were independently performed by three botanists, Hugo DeVries, Karl Correns, and Erich Tschermak. DeVries' work, for example, had focused on unit characters, and he demonstrated the principle of segregation in his experiments with several plant species. Apparently, he had searched the existing literature and found that Mendel's work had anticipated his own conclusions! Correns and Tschermak also drew conclusions similar to those of Mendel.

In 1902 two cytologists, Walter Sutton and Theodore Boveri, independently published papers, linking their discoveries of the behavior of chromosomes during meiosis to the Mendelian principles of segregation and independent assortment. They pointed out that the separation of chromosomes during meiosis could serve as the cytological basis of these two postulates.

While they thought that Mendel's unit factors were probably chromosomes rather than genes on chromosomes, their findings made Mendel's work the foundation of ensuing genetic investigations.

Unit Factors, Genes, and Homologous Chromosomes

Because the correlation between Sutton's and Boveri's observations and Mendelian principles is the foundation for the modern interpretation of transmission genetics, we will examine this correlation before moving to the more complex topics of the next several chapters.

As pointed out in Chapter 2, each species possesses a specific number of chromosomes in each somatic cell (body cell) nucleus. For diploid organisms, this number is called the **diploid number** ($2n$) and is characteristic of that species. During the formation of gametes, this number is precisely halved (n), and when two gametes combine during fertilization, the diploid number is reestablished. The chromosome number is not reduced in a random manner, however. It was apparent to early cytologists that the diploid number of chromosomes is composed of homologous pairs identifiable by their morphological appearance. The gametes contain one member of each pair. The chromosome complement of a gamete is thus quite specific, and the number of chromosomes in each gamete is equal to the haploid number.

With this basic information, we can see the correlation between the behavior of unit factors and chromosomes and genes. Figure 3.11 shows three of Mendel's postulates and the accepted explanation of each. Unit factors are really genes located on homologous pairs of chromosomes [Figure 3.11 (a)]. Members of each pair of homologues separate, or segregate, during gamete formation [Figure 3.11 (b)].

To illustrate the principle of independent assortment, it is important to distinguish between members of any given homologous pair of chromosomes. One member of each pair is derived from the **maternal parent,** while the other comes from the **paternal parent.** We represent their different origins by different colors. In Figure 3.11 (c) two pairs of segregating homologues are considered. They behave independently during gamete formation, with each gamete always receiving one of each pair. All possible combinations are shown. If we add the symbols used in Mendel's dihybrid cross (D, d and W, w) to the diagram, we see why equal numbers of the four types of gametes are formed.

From observations of the phenotypic diversity of living organisms, we see that it is logical to assume that

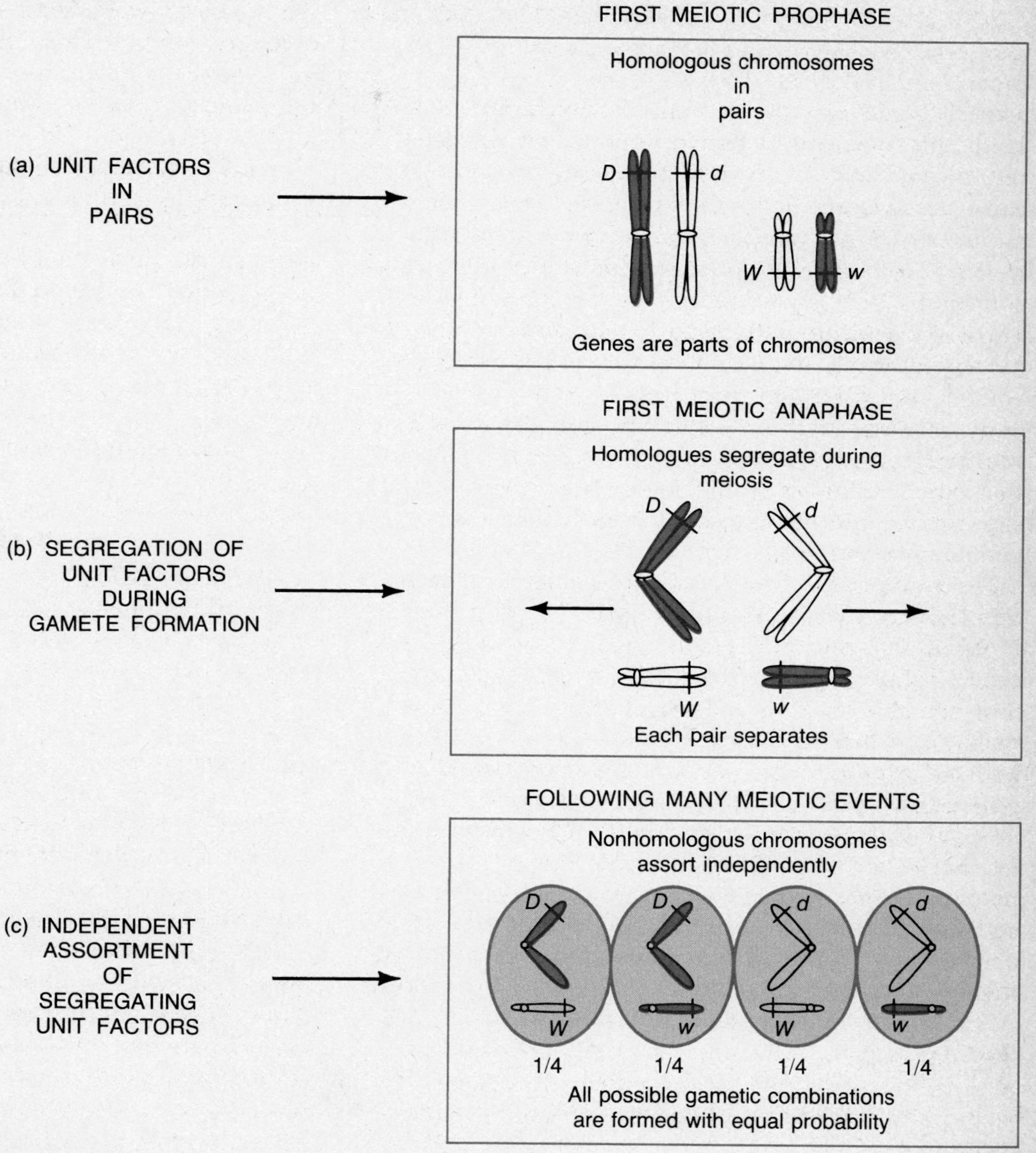

FIGURE 3.11
The correlation between the
Mendelian postulates of
(a) unit factors in pairs,
(b) segregation, and
(c) independent assortment,
and the presence of genes
located on homologous
chromosomes and their
behavior during meiosis.

(a) UNIT FACTORS
IN
PAIRS

FIRST MEIOTIC PROPHASE

Homologous chromosomes
in
pairs

D —— d

W —— w

Genes are parts of chromosomes

(b) SEGREGATION OF
UNIT FACTORS
DURING
GAMETE FORMATION

FIRST MEIOTIC ANAPHASE

Homologues segregate during
meiosis

D d

W w

Each pair separates

(c) INDEPENDENT
ASSORTMENT
OF
SEGREGATING
UNIT FACTORS

FOLLOWING MANY MEIOTIC EVENTS

Nonhomologous chromosomes
assort independently

D D d d

W w W w

1/4   1/4   1/4   1/4

All possible gametic combinations
are formed with equal probability

there are many more genes than chromosomes. Therefore, each homologue must carry genetic information for more than one trait. The currently accepted concept is that a chromosome is composed of a large number of linearly ordered, information-containing units called **genes.** Thus, Mendel's unit factors (which determine tall or dwarf stems, for example) actually constitute a pair of genes located on one pair of homologous chromosomes. The location on a given chromosome where any particular gene occurs is called its **locus.** The different forms taken by a given gene, called **alleles** (D or d), contain slightly different genetic information that determines the same character (stem length). Alleles are alternate forms of the same gene. Although we have only discussed genes with two alternative alleles, most genes have *more*

than two allelic forms. We will discuss the concept of **multiple alleles** in Chapter 4.

We conclude this section by reviewing the criteria necessary to classify two chromosomes as a homologous pair:

1   During mitosis and meiosis, when chromosomes are visible as distinct figures, both members of a homologous pair are the same size and exhibit identical centromere locations.

2   During early stages of meiosis, homologous chromosomes pair together, or synapse.

3   Although not generally microscopically visible, homologues contain identical, linearly ordered, gene loci.

INDEPENDENT ASSORTMENT AND GENETIC VARIATION

One of the major consequences of independent assortment is the production by individuals of genetically dissimilar gametes. Genetic variation results because the two members of any homologous pair of chromosomes are rarely, if ever, genetically identical. For them to be identical, homozygosity at every locus along both homologous chromosomes would be required. Various methods used to estimate the degree of heterozygosity show that on the average, 10 to 40 percent of the loci are occupied by different alleles. However, some geneticists feel that organisms that are not inbred display heterozygosity at most, if not all, of their loci. Therefore, since independent assortment leads to the production of all possible chromosome combinations, extensive genetic diversity results.

The number of possible gametes, each with different chromosome compositions, is 2^n, where n equals the haploid number. Thus, if a species has a haploid number of 4, then 2^4 or 16 different gamete combinations can be formed as a result of independent assortment. While this number is not great, consider the human species, where $n = 23$. If 2^{23} is calculated, we find that in excess of 8×10^6, or over 8 million, different types of gametes are represented. Since fertilization represents an event involving only one of approximately 8×10^6 possible gametes from each of two parents, each offspring represents only one of $(8 \times 10^6)^2$, or 64×10^{12}, potential genetic combinations! It is no wonder that, except for identical twins, each member of the human species demonstrates a distinctive appearance and individuality. This number of combinations is far greater than the number of humans who have ever lived on earth! Genetic variation resulting from independent assortment has been extremely important to the process of organic evolution in all organisms.

EVALUATING GENETIC DATA: CHI-SQUARE ANALYSIS

Mendel's 3:1 monohybrid and 9:3:3:1 dihybrid ratios are hypothetical predictions based on the following assumptions: (1) dominance/recessiveness; (2) segregation; (3) independent assortment; and (4) random fertilization. The last three processes are influenced by chance events and therefore are subject to normal deviation. This concept, **chance deviation,** is most easily illustrated by tossing a coin numerous times and recording the number of heads and tails observed. In each toss, there is a probability of 1/2 that a head will occur and a probability of 1/2 that a tail will occur.

Therefore, the expected ratio of many tosses is 1:1. If a coin were tossed 1000 times, usually *about* 500 heads and 500 tails would be observed. Any reasonable fluctuation from this hypothetical ratio (e.g., 486 heads and 514 tails) would be attributed to chance deviation.

As the total number of tosses is reduced, the impact of chance deviation increases. For example, if a coin were tossed only 4 times, the hypothetical ratio would predict 2 heads and 2 tails. However, you wouldn't be too surprised if all 4 tosses resulted in only heads or only tails. But, for 1000 tosses, 1000 heads or 1000 tails would be most unexpected. In fact, you might believe that such a result would be impossible. Actually, all heads or all tails in 1000 tosses would be predicted to occur with a probability of only $(1/2)^{1000}$. Since $(1/2)^{20}$ is equivalent to less than 1 in 1 million times, an event occurring with a probability of only $(1/2)^{1000}$ would be virtually impossible to achieve.

Two major points are significant here:

1 The outcomes of segregation, independent assortment, and fertilization, like coin tossing, are subject to random fluctuations from their predicted occurrences as a result of chance deviation.

2 As the sample size increases, the average deviation from the expected decreases. Therefore, a larger sample size diminishes the impact of chance deviation on the final outcome.

It is important in genetics to be able to evaluate observed deviation. When we assume that data will fit a given ratio such as 1:1, 3:1, or 9:3:3:1, we establish what is called the **null hypothesis.** It is so named because the hypothesis assumes that there is no real difference between the **measured values** (or ratio) and the **predicted values** (or ratio). Evaluation of the null hypothesis is accomplished by statistical analysis. On this basis, the null hypothesis may either: (1) be rejected, or (2) fail to be rejected. If it is rejected, any observed deviation from the expected cannot be attributed to chance alone. The null hypothesis and the underlying assumptions leading to it must be reexamined. If the null hypothesis fails to be rejected, any observed deviations are attributed to chance.

Thus, statistical analysis provides a mathematical basis for examining how well observed data fit or differ from predicted or expected occurrences, testing what is called the **goodness of fit.** Assuming that the data do not "fit" exactly, just how much deviation can be allowed before the null hypothesis is rejected? One of the simplest statistical tests devised to answer this question is **chi-square analysis (χ^2).** This test takes into account the observed deviation in each component of an expected ratio as well as the sample size and reduces them to a single numerical value. This

TABLE 3.3
Chi-square analysis.
(a) Hypothetical monohybrid cross.

Expected Ratio	Observed (o)	Expected (e)	Deviation (o − e)	Deviation² (d)²	Deviation²/Expected (d²/e)
3/4	740	3/4 (1000) = 750	740 − 750 = −10	$(-10)^2 = 100$	100/750 = 0.13
1/4	260	1/4 (1000) = 250	260 − 250 = +10	$(+10)^2 = 100$	100/250 = 0.40
TOTAL = 1000					$\Sigma\chi^2 = 0.53$
					p = 0.48

(b) Hypothetical dihybrid cross.

Expected Ratio	o	e	o − e	d^2	d^2/e
9/16	587	567	+20	400	0.71
3/16	197	189	+8	64	0.34
3/16	168	189	−21	441	2.33
1/16	56	63	−7	49	0.78
TOTAL = 1008				$\chi^2 = 4.16$	
				p = 0.26	

value (χ^2) is then used to estimate how frequently the observed deviation can be expected to occur strictly as a result of chance. The formula used in chi-square analysis is

$$\chi^2 = \Sigma\frac{(o - e)^2}{e}$$

where o is the observed value for a given category and e is the expected value for that category. Σ (sigma) represents the sum of the calculated values for each category of the ratio. Because (o − e) is the deviation in each case, the equation can be reduced to

$$\chi^2 = \Sigma\frac{d^2}{e}$$

Table 3.3(a) illustrates the step-by-step procedure necessary to make the χ^2 calculation for the F_2 results of a hypothetical monohybrid cross. Work from left to right, calculating and entering the appropriate numbers in each column. Regardless of whether the calculated deviation (o − e) is initially positive or negative, it becomes positive after the number is squared. Table 3.3(b) illustrates the analysis of the F_2 results of a hypothetical dihybrid cross. Based on your study of the calculations involved in the monohybrid cross, check to make certain that you understand how each number was calculated in the dihybrid example.

The final step in the chi-square analysis is to interpret the χ^2 value. To do so, you must initially determine the value of the **degrees of freedom (df)**, which is equal to n − 1 where n is the number of different categories into which each datum point may fall. For the 3:1 ratio, n = 2, so df = 2 − 1 = 1. For the 9:3:3:1 ratio, df = 3. Degrees of freedom must be taken into account because the greater the number of categories, the more deviation is expected as a result of chance.

With this accomplished, the χ^2 value must now be converted to the corresponding **probability value (p)**. Because this calculation is complex, the p value is usually located on a table or graph. Figure 3.12 shows the wide range of χ^2/p values for numerous degrees of freedom in both forms. We will use the graph to illustrate how to determine the p value. The caption for Figure 3.12(b) explains the use of the table.

These simple steps must be followed to determine p:

1 Locate the χ^2 value on the abscissa.

2 Draw a vertical line from this point to the line representing the appropriate df.

3 Extend a horizontal line from this point to the left until it intersects the ordinate.

4 Read off, interpolating if necessary, the corresponding p value.

For our two examples in Table 3.3, the p values of 0.48 and 0.26 may be determined in this way. These values are shown below the χ^2 values for both the monohybrid and dihybrid cross analyses in Table 3.3.

So far, we have been concerned only with the determination of p. The most important aspect of χ^2 analysis is understanding what the p value actually means. We will use the example of the dihybrid cross (p = 0.26) to illustrate. In these discussions, it is simplest to think of the p value as a percentage (e.g., 0.26 = 26 percent).

The p value indicates the probability of obtaining the observed deviation by chance alone. In our dihybrid example, the probability is 26 percent that this much deviation or more will occur owing to chance.

The larger the p value, the closer the data are to the predicted or ideal ratio.

Another description provides more specific statistical information. In our example, the p value indicates that, were the same experiment repeated many times, 26 percent of the trials would be expected to exhibit chance deviation as great or greater than that seen in the initial trial. Conversely, 74 percent of the repeats would show less deviation as a result of chance than initially observed.

These interpretations of the p value reveal that a hypothesis (a 9:3:3:1 ratio in this case) is never proved or disproved absolutely. Instead, a relative standard must be set to serve as the basis for either supporting or rejecting the hypothesis. This standard is generally a probability value of 0.05. In chi-square analysis, a value less than 0.05 makes it unlikely that the observed results could be obtained by chance alone. Instead,

FIGURE 3.12

(a) A graph used to convert χ^2 values to p values. The conversion of a χ^2 value of 0.53 with 1 degree of freedom to a probability value of 0.48 is illustrated. (b) A table showing χ^2 values for a variety of combinations of df and p. Any value greater than that shown at the p = 0.05 level for a particular df serves as the basis to reject the null hypothesis in question. In our example, a χ^2 value of 0.53 for a df of 1 is converted to a probability value between 0.20 and 0.50. From our graph in (a), the more precise value (p = 0.48) was estimated.

(a)

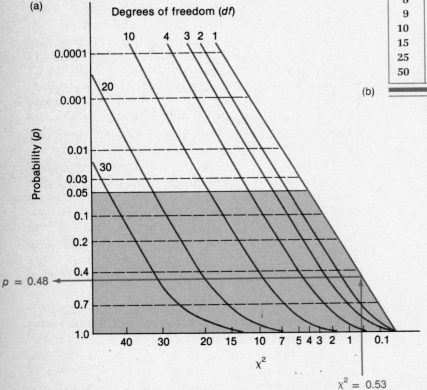

	Probabilities (p)				
df	0.90	0.50	0.20	0.05	0.01
1	0.02	0.46	1.64	3.84	6.64
2	0.21	1.39	3.22	5.99	9.21
3	0.58	2.37	4.64	7.82	11.35
4	1.06	3.36	5.99	9.49	13.28
5	1.61	4.35	7.29	11.07	15.09
6	2.20	5.35	8.56	12.59	16.81
7	2.83	6.35	9.80	14.07	18.48
8	3.49	7.34	11.03	15.51	20.09
9	4.17	8.34	12.24	16.92	21.67
10	4.87	9.34	13.44	18.31	23.21
15	8.55	14.34	19.31	25.00	30.58
25	16.47	24.34	30.68	37.65	44.31
50	37.69	49.34	58.16	67.51	76.15

(b)

FIGURE 3.13
(a) A representative pedigree for a single character through three generations.
(b) The most probable genotypes of each individual in the pedigree.

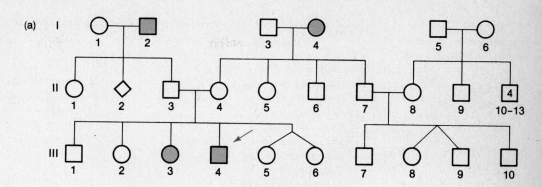

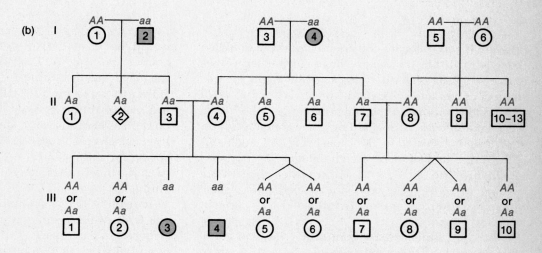

such a *p* value indicates substantial deviation from the predicted results and thus serves as the basis for rejecting the null hypothesis.

On the other hand, *p* values of 0.05 or greater (1.0 − 0.05) fail to reject the null hypothesis. In our example where *p* = 0.26, the hypothesis of independent assortment is supported by the experimental data. That is, the data do not provide any reason to reject the hypothesis. If we wished to be even more certain of our hypothesis, we could raise the relative standard, say to 0.10. In such a case, all *p* values less than 0.10 fail to support the hypothesis.

HUMAN PEDIGREES

In all crosses discussed so far, one of the two traits for each character has been dominant to the other. Based on this observation, two significant questions may be asked:

1 Does the expression of all genes occur in this fashion?

2 Is it possible to ascertain the mode of inheritance

of genes in organisms where designed crosses and the production of large numbers of offspring are impossible?

The answer to the first question is no. As we will see in Chapters 4 and 5, many modes of inheritance exist that modify the monohybrid and dihybrid ratios observed by Mendel.

The answer to the second question is yes. Even in humans the pattern of inheritance of a specific phenotype can be studied.

The simplest way to study this pattern is to construct a family tree indicating the phenotype of the trait in question for each member. Such a family tree is called a **pedigree.** By analyzing the pedigree, we may be able to determine how the gene controlling the trait is inherited.

Figure 3.13 shows the conventions used in constructing a pedigree. Circles represent females, and squares designate males. If the sex is unknown, a diamond is used. If a pedigree traces only a single trait, as Figure 3.13 does, the circles, squares, and diamonds are shaded if the phenotype being considered is ex-

TABLE 3.4
Representative recessive and dominant human traits.

Recessive Traits	Dominant Traits
Albinism	Brachydactyly
Alkaptonuria	Congenital stationary night blindness
Ataxia telangiectasia	Fascio-scapulo-humeral muscular dystrophy
Color blindness	Huntington's disease
Cystic fibrosis	Phenylthiocarbamide tasting (PTC)
Duchenne's muscular dystrophy	Pseudo-achondroplastic dwarfism
Galactosemia	
Hemophilia	
Lesch-Nyhan syndrome	
Phenylketonuria	
Sickle-cell anemia	

pressed. If two traits are considered, one way to follow them is to divide the square or circle into an upper and lower half. For example, a circle might be unshaded, completely shaded, or have either half shaded, depending on the phenotype.

The parents are connected by a horizontal line, and a vertical line leads to their offspring. All such offspring are called **sibs** and are connected by a horizontal **sibship line.** Sibs are placed from left to right according to birth order and are labeled with Arabic numerals. Each generation is indicated by a Roman numeral.

Twins are indicated by connected diagonal lines. **Monozygotic** or **identical twins** stem from a single line itself connected to the sibship line (see III-5,6 in Figure 3.13). **Dizygotic** or **fraternal twins** are connected directly to the sibship line (see III-8,9). A number within one of the symbols (II-10–13) represents numerous sibs of the same or unknown phenotypes. The individual whose phenotype drew the attention of a physician or geneticist is called the **propositus** (male) or **proposita** (female) and is indicated by an arrow (see III-4).

The pedigree shown in Figure 3.13 traces the theoretical pattern of inheritance of the human trait **albinism.** By analyzing the pedigree, we will see that albinism is inherited as a recessive trait.

Two of the parents of the first generation, I-2 and I-4, are affected. Because none of their offspring show the disorder, it is reasonable to conclude that the unaffected parents (I-1 and I-3) were homozygous normal individuals. Had they been heterozygous, one half of their offspring would be expected to exhibit albinism.

An unaffected second generation is characteristic of a rare recessive trait. If albinism were inherited as a dominant trait, one-half of the second generation would be expected to exhibit the disorder in the crosses involving the I-2 and I-4 parents. Inspection of the offspring constituting the third generation provides further support for the hypothesis that albinism is a recessive trait. Parents II-3 and II-4 are apparently both heterozygous, and approximately one-fourth of their offspring should be affected. Two of the six offspring do show albinism. This deviation from the expected ratio is characteristic of crosses with few offspring.

Individual II-7 is undoubtedly heterozygous, while II-8 is most likely homozygous normal, given the low frequency of the *a* allele in the population. If so, we would predict that none of their offspring (III-7,8,9, and 10) would be albino, and this is borne out.

Based on this pedigree analysis, and the conclusion that albinism is a recessive trait, the genotypes of all individuals can be predicted. For both the first and second generations, this can be done with some certainty. In the third generation, for normal individuals, we can only guess whether they are homozygous or heterozygous. These predictions are shown in Figure 3.13(b).

Pedigree analysis of many traits has been an extremely valuable research technique in human genetic studies. However, this approach does not usually provide the certainty in drawing conclusions that is afforded by designed crosses yielding large numbers of offspring. Nevertheless, when many independent pedigrees of the same trait or disorder are analyzed, consistent conclusions can often be drawn. Table 3.4 lists numerous human traits and classifies them according to their recessive or dominant expression. As we will see in Chapter 5, the genes controlling some of these traits are located on the sex-determining chromosomes.

SUMMARY AND CONCLUSIONS

More than a century has passed since Gregor Mendel performed breeding experiments with the garden pea—work that may be considered the cornerstone of the discipline of genetics. Mendel bred peas having one or more pairs of contrasting traits and kept accurate quantitative records of the observed results. Analysis of data derived from these crosses produced postulates that are now the principles of transmission genetics.

Mendel postulated that phenotypic characters are controlled by unit factors that exist in pairs and exhibit a dominance/recessiveness relationship in determining the phenotype. Unit factors are in fact alleles occupying identical loci on pairs of homologous chromosomes. Mendel also postulated that these unit factors segregate, or separate, during gamete formation. As a result, gametes randomly receive only one unit factor or the other. Succeeding geneticists discovered that the segregation occurs at the level of homologous pairs of chromosomes, separating alleles on one homologue from the corresponding alleles on the other homologue. These postulates were all derived from Mendel's analysis of monohybrid crosses.

From his data on dihybrid crosses, involving two pairs of contrasting traits, Mendel derived his final postulate, independent assortment. This postulate states that segregating pairs of unit factors assort independently of each other. As a result, all possible combinations of gametes are formed. Today, it is known that each pair of homologous chromosomes segregates independently of all other pairs.

Mendel's analysis of his experimentation was most remarkable because he knew nothing of chromosomes. While the relevance of his work was not immediately accepted by the scientific community, the discovery of chromosomes and their behavior in mitosis and meiosis led to the rediscovery and acceptance of his work in the early 1900s. His postulates apply regardless of the number of gene pairs involved in the cross, provided that each is located on a different pair of homologous chromosomes. In this chapter we have considered up to three gene pairs, constituting a trihybrid cross. To help in solving genetic crosses with this degree of complexity, the Punnett square and forked-line, or branch diagram, methods have been introduced.

Because all homologous pairs of chromosomes assort independently, there is extensive genetic variability in gametes and subsequent offspring. In one human, over eight million different chromosome combinations can be formed in gametes. Genetic variability is an important feature of the evolutionary process.

In all genetic experiments, variation from an expected ratio is anticipated owing to chance deviation. This variation can be analyzed statistically by using the chi-square test. Chi-square analysis predicts the probability of the variation arising as a result of chance and provides a basis for assessing the "fit" of the null hypothesis or ratio relative to the observed data. Based on this statistical test, the null hypothesis is either rejected or it fails to be rejected.

In the analysis of human mutant traits such as albinism, the pedigree method is a convenient means of studying and predicting the mode of inheritance. While less reliable than methods involving designed matings and large numbers of offspring, pedigree analysis has provided valuable insights into the inheritance of many traits and disorders occurring in humans.

PROBLEMS AND DISCUSSION QUESTIONS*

 1 In a cross between a black and a white guinea pig, all members of the F_1 generation are black offspring. The F_2 generation is made up of approximately 3/4 black and 1/4 white guinea pigs.

(a) Diagram this cross, showing the genotypes and phenotypes.

(b) What will the offspring be like if two F_2 white guinea pigs are mated?

(c) Two different matings were made between black members of the F_2 generation with the results shown below. Diagram each of the crosses.

Cross	Offspring
Cross 1	All black
Cross 2	3/4 black, 1/4 white

*When working genetics problems in this and succeeding chapters, always assume that members of the P_1 generation are homozygous, unless the information given indicates otherwise.

2   Albinism in humans is inherited as a simple recessive trait. For the following families, determine the genotypes of the parents and offspring. When two alternative genotypes are possible, list both.

(a) Two nonalbino (normal) parents have five children, four normal and one albino.

(b) A normal male and an albino female have six children, all normal.

(c) A normal male and an albino female have six children, three normal and three albinos

(d) Construct a pedigree of the families in (b) and (c). Assume that one of the normal children in (b) marries one of the albino children in (c) and that they have eight children.

3   Which of Mendel's postulates are illustrated by the pedigree in Problem 2? List and define these postulates.

4   Discuss the rationale used by Mendel in relating his monohybrid results to his postulates.

5   What advantages were provided by Mendel's choice of the garden pea in his experiments?

6   Pigeons may exhibit a checkered or plain pattern. In a series of controlled matings, the following data were obtained:

	F_1 Progeny	
P_1 Cross	**Checkered**	**Plain**
(a) checkered × checkered	36	0
(b) checkered × plain	38	0
(c) plain × plain	0	35

Then, F_1 offspring were selectively mated with the following results. The P_1 cross giving rise to each F_1 pigeon is indicated in parentheses.

	F_2 Progeny	
F_1 × F_1 Crosses	**Checkered**	**Plain**
(d) checkered (a) × plain (c)	34	0
(e) checkered (b) × plain (c)	17	14
(f) checkered (b) × checkered (b)	28	9
(g) checkered (a) × checkered (b)	39	0

How are the checkered and plain patterns inherited? Select and define symbols for the genes involved and determine the genotypes of the parents and offspring in each cross.

7   Mendel crossed peas having round seeds and yellow cotyledons with peas having wrinkled seeds and green cotyledons. All the F_1 plants had round seeds with yellow cotyledons. Diagram this cross through the F_2 generation using both the Punnett square and forked-line, or branch diagram, methods.

8   Determine the genotypes of the F_2 plants given here by analyzing the phenotypes of the offspring of these crosses.

F_2 Plants	Offspring
(a) round, yellow × round, yellow	3/4 round, yellow 1/4 wrinkled, yellow
(b) wrinkled, yellow × round, yellow	6/16 wrinkled, yellow 2/16 wrinkled, green 6/16 round, yellow 2/16 round, green
(c) round, yellow × round, yellow	9/16 round, yellow 3/16 round, green 3/16 wrinkled, yellow 1/16 wrinkled, green

(d) round, yellow × wrinkled, green   1/4 round, yellow
1/4 round, green
1/4 wrinkled, yellow
1/4 wrinkled, green

9 Which of the crosses in Problem 8 is a test cross?

10 Which of Mendel's postulates can only be demonstrated in crosses involving at least two pairs of traits? Define it.

11 Correlate Mendel's four postulates with what is now known about homologous chromosomes, genes, alleles, and the process of meiosis.

12 What is the basis for homology among chromosomes?

13 Distinguish between homozygosity and heterozygosity.

14 In *Drosophila*, *grey* body color is dominant to *ebony* body color, while *long* wings are dominant to *vestigial* wings. Work the following crosses through the F_2 generation and determine the genotypic and phenotypic ratios for each generation. Assume the P_1 individuals are homozygous.
(a) grey, long × ebony, vestigial
(b) grey, vestigial × ebony, long
(c) grey, long × grey, vestigial

15 How many different types of gametes can be formed by individuals of the following genotypes: (a) *AaBb*, (b) *AaBB*, (c) *AaBbCc*, (d) *AaBBcc*, (e) *AaBbcc*, and (f) *AaBbCcDdEe?* What are they in each case?

16 Using the forked-line, or branch diagram, method, determine the genotypic and phenotypic ratios of the trihybrid crosses (a) *AaBbCc* × *AaBBCC*, (b) *AaBBCc* × *aaBBCc*, and (c) *AaBbCc* × *AaBbCc*.

17 Mendel crossed peas with green seeds with those of yellow seeds. The F_1 generation produced only yellow seeds. In the F_2, the progeny consisted of 6022 plants with yellow seeds and 2001 plants with green seeds. Of the F_2 yellow-seeded plants, 519 were self-fertilized with the following results: 166 bred true for yellow and 353 produced a 3:1 ratio of yellow:green. Explain these results by diagraming the crosses.

18 In a study of black and white guinea pigs, 100 black animals were crossed individually to white animals and each cross was carried to an F_2 generation. In 94 of the cases, the F_1 individuals were all black, and an F_2 ratio of 3 black:1 white was obtained. In the other 6 cases, half of the F_1 animals were black and the other half were white. Why? Predict the results of crossing the black and white F_2 guinea pigs from the 6 exceptional cases.

19 Mendel crossed peas with round, green seeds to ones with wrinkled, yellow seeds. All F_1 plants had seeds that were round and yellow. Predict the results of test-crossing these F_1 plants.

20 Below are shown F_2 results of two of Mendel's monohybrid crosses. Calculate the χ^2 value and determine the *p* value for both. Which of the two shows a greater amount of deviation?

(a) Full pods	882
Constricted pods	299
(b) Violet flowers	705
White flowers	224

21 In one of Mendel's dihybrid crosses, he observed 315 smooth yellow, 108 smooth green, 101 wrinkled yellow, and 32 wrinkled green F_2 plants. Analyze these data using the chi-square test to see if
(a) It fits a 9:3:3:1 ratio
(b) The smooth:wrinkled traits fit a 3:1 ratio
(c) The yellow:green traits fit a 3:1 ratio

22 A geneticist, in assessing data that fell into two phenotypic classes, observed values of 250:150. She decided to perform chi-square analysis using two different null hypotheses: (a) the data fit a 3:1 ratio; and (b) the data fit a 1:1 ratio. Calculate the χ^2 values for each hypothesis. What can be concluded about each hypothesis?

23 For the following pedigree, predict the mode of inheritance and the resulting genotypes of each individual. Assume that the alleles A and a control the expression of the trait.

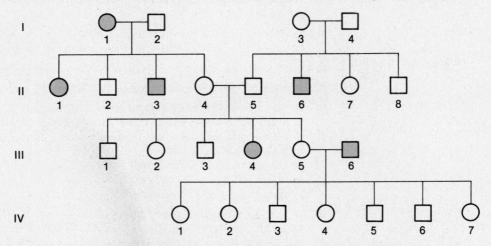

24 The following pedigree is for myopia in humans. Predict if the disorder is inherited as the result of a dominant or recessive allele. Determine the most probable genotype for each individual based on your prediction.

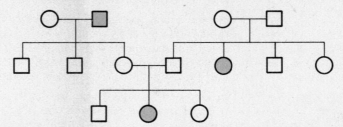

25 Thalassemia is an inherited anemic disorder in humans. Individuals can be completely normal, they can exhibit a "minor" anemia, or they can exhibit a "major" anemia. Assuming that only a single gene pair and two alleles are involved in the inheritance of these conditions, which phenotype is recessive?

SELECTED READINGS

DUNN, L. C. 1965. *A short history of genetics.* New York: McGraw-Hill.

OLBY, R. C. 1966. *Origins of Mendelism.* London: Constable.

PETERS, J., ed. 1959. *Classic papers in genetics.* Englewood Cliffs, N.J.: Prentice-Hall.

SNEDECOR, G. W., and COCHRAN, W. G. 1980. *Statistical methods.* 7th ed. Ames, Iowa: Iowa State University Press.

SOKAL, R. R., and ROHLF, F. J. 1981. *Biometry—The principles and practice of statistics in biological research.* 2nd ed. San Francisco: W. H. Freeman.

SOUDEK, D. 1984. Gregor Mendel and the people around him. *Am. J. Hum. Genet.* 36: 495–98.

STERN, C., 1950. *The birth of genetics.* (Supplement to *Genetics* 35.)

STERN, C., and SHERWOOD, E. 1966. *The origins of genetics: A Mendel source book.* San Francisco: W.H. Freeman.

STURTEVANT, A. H. 1965. *A history of genetics.* New York: Harper & Row.

VOELLER, B. R., ed. 1968. *The chromosome theory of inheritance: Classical papers in development and heredity.* New York: Appleton-Century-Crofts.

4

Modification of Mendelian Ratios: Allelic and Gene Interactions

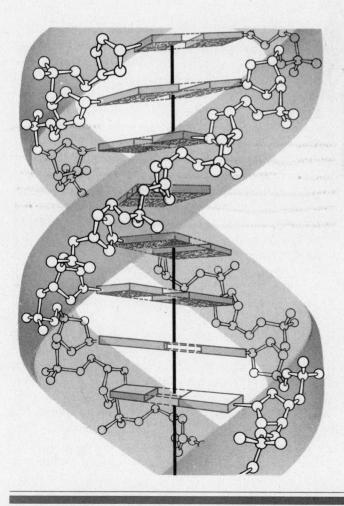

In Chapter 3, we discussed the simplest principles of transmission genetics. We saw that genes are present on homologous chromosomes and that these chromosomes **segregate** from each other and **independently assort** with other segregating chromosomes during gamete formation. These two postulates are the fundamental principles of gene transmission from parent to offspring. However, once an offspring has received the total set of genes, it is the expression of genes that determines the organism's phenotype. If gene expression does not adhere to a simple dominant/recessive mode, or if more than one pair of genes influences the expression of a single trait, the classic $3:1$ and $9:3:3:1$ F_2 ratios may be modified. Although in this and the next several chapters we consider more complex modes of inheritance, the fundamental principles set down by Mendel hold true in these situations as well.

In this chapter, our discussion will initially be restricted to the inheritance of traits that are under the control of only one gene. In diploid organisms, where homologous pairs of chromosomes exist, two copies of each gene influence such traits. The copies need not be identical since alternate forms of genes, or **alleles,** occur within populations. How alleles act to influence a given phenotype will be our major consideration. Then we will proceed to consider how a single phenotype may be controlled by more than one gene. This general phenomenon is referred to as **gene interaction,** indicating that phenotypes are frequently under the influence of more than one gene product. Numerous examples will be presented to illustrate a variety of heritable patterns observed in such situations.

POTENTIAL FUNCTION OF AN ALLELE

Following the rediscovery of Mendel's work in the early 1900s, research focused on how genetic information influences an individual's phenotype. This course of investigation, stemming from Mendel's findings, is called **neo-Mendelian genetics** (*neo* from the Greek word meaning since or new).

Each type of inheritance described in this chapter was investigated when observations of genetic data did not precisely conform to the expected Mendelian ratios. Hypotheses that modified and extended the Mendelian principles were proposed and tested with specifically designed crosses. The explanations for these observations were in accordance with the principle that a phenotype is under the control of one or more genes located at specific loci on one or more pairs of homologous chromosomes. If we adhere to the principles of segregation and independent assortment, we can predict accurately the transmission of any number of allele pairs.

To understand the various modes of inheritance, we must first examine the potential function of an **allele.** Alleles are alternate forms of the same gene; therefore, they contain modified genetic information and specify an alteration in the original gene product. For example, there are well over 100 alleles of the genes that specify the protein portion of human hemoglobin. Even though each specifies a modification of one of the components of hemoglobin, they all store information necessary for this molecule's chemical synthesis. Once manufactured, however, the product of an allele may or may not be functional. The allele that occurs most frequently in a population or the one that is arbitrarily designated as normal is called **wild type.** This common allele is usually dominant, such as the allele for tall plants in the garden pea, and its product is functional in the cell. Wild-type alleles thus serve as standards for comparison against all mutations occurring at a particular locus.

The process of **mutation** is a source of new alleles. Each allele may be recognized by a change in the phenotype. A new phenotype results from a change in functional activity of the cellular product controlled by that gene. Usually, the alteration or mutation is expressed as a loss of the specific wild-type function. For example, if a gene is responsible for the synthesis of a specific enzyme, a mutation in the gene may change the conformation of this enzyme, thus eliminating its affinity for the substrate. This mutation results in a total loss of function. On the other hand, another organism may have a different mutation in this gene, which results in an enzyme with a reduced or increased affinity for binding the substrate. This mutation may reduce or enhance rather than eliminate the functional capacity of the gene product. In either case, the phenotype may or may not be altered in a discernible way.

Although phenotypic traits may be affected by a single mutation, traits are often the result of many gene products. In the case of enzymatic reactions, most are part of complex metabolic pathways. Therefore, phenotypic traits may be under the control of more than one gene and the allelic forms of each gene involved. In the initial part of this chapter, examples will be restricted to only one gene pair and the alleles associated with it. Thus, in these cases we see a modification of the 3:1 monohybrid ratio. Then we will consider traits controlled by two genes and the accompanying modification of the 9:3:3:1 dihybrid ratio.

SYMBOLS FOR ALLELES

In Chapter 3, we learned to symbolize alleles for very simple Mendelian traits. We used the lowercase form of the initial letter of the name of a recessive trait to denote the recessive allele and the same letter in uppercase form to refer to the dominant allele. Thus, for tall and dwarf, where dwarf is recessive, *D* and *d* represent the alleles responsible for these respective traits.

As more complex inheritance patterns were investigated, a useful system was developed to discriminate between wild-type and mutant traits. In this system, the initial letter of the name of the mutant trait is selected. If the trait is recessive, the lowercase form is used; if it is dominant, the uppercase form is used. The contrasting wild-type trait is denoted by the same letter, but with a + as a superscript. For example, *white* is a recessive eye color mutation in the fruit fly, *Drosophila melanogaster*. The wild-type eye color is red. In this system *white* is denoted by *w* and red is denoted by w^+. *Notch* wing, a dominant mutation, is designated *N*, while the normal unnotched wing is designated N^+.

One advantage of this second system is that it may be further abbreviated whenever desired. In the abbreviated version, the wild-type trait is simply designated as +. This abbreviation is particularly useful when several genes located on the same chromosome are considered simultaneously. In this case, two horizontal lines are drawn representing the chromosomes, and the contrasting alleles are written directly above and below a point on the lines. For example, the *white* and *Notch* alleles are located on the same chromosome. A fly heterozygous for both genes can be designated

$$\frac{w\ N}{+\ +} \qquad \text{or} \qquad \frac{w\ +}{+\ N}$$

cis arrangement   trans arrangement
of   of
mutant alleles   mutant alleles

In the first case, designated the **cis** arrangement, both mutant alleles are on the same homologue, and the corresponding wild-type alleles are on the other. In the second example, designated as the **trans** arrangement, the mutant *white* allele is on one homologue, and the mutant *Notch* allele is on the other. The use of cis and trans in genetics has been "borrowed" from chemistry where the terminology describes the relative location of functional groups in geometric isomers.

Sometimes only a single line is used to designate both chromosomes:

$$\frac{w\ N}{+\ +} \quad \text{or} \quad \frac{w\ +}{+\ N}$$

Remember that, while each system is slightly different, the symbols used in each designate contrasting alleles located at single gene sites.

INCOMPLETE, OR PARTIAL, DOMINANCE

Incomplete, or **partial, dominance** in the offspring is based on the observation of intermediate phenotypes generated by a cross between parents with contrasting traits. For example, if plants such as four-o'clocks or carnations with red flowers are crossed with plants with white flowers, offspring may have pink flowers. It appears that neither red nor white flower color is dominant. Since some red pigment is produced in the F_1 intermediate-colored pink flowers, dominance appears to be incomplete or partial.

If this phenotype is under the control of a single pair of alleles where neither is dominant, the results of the F_1 (pink) $\times$ F_1 (pink) cross can be predicted. The resulting F_2 generation is shown in Figure 4.1, confirming the hypothesis that only one pair of alleles determines these phenotypes. The genotypic ratio (1:2:1) of the F_2 generation is identical to that of Mendel's monohybrid cross. Because there is no dominance, however, the phenotypic ratio is identical to the genotypic ratio. Note here that since neither of the alleles is recessive, either r or w can be used; in this case, r was chosen.

FIGURE 4.1
Incomplete dominance illustrated by flower color.

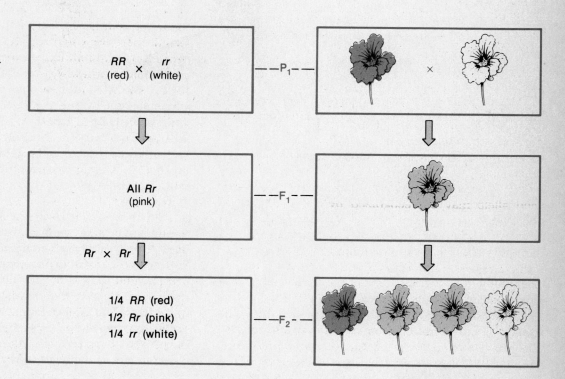

RR (red) $\times$ rr (white)	— P_1 —
All Rr (pink)	— F_1 —
$Rr \times Rr$	
1/4 RR (red) 1/2 Rr (pink) 1/4 rr (white)	— F_2 —

Incomplete dominance, which results in an intermediate expression of the overt phenotype, is relatively rare. However, even when complete dominance is evident, careful examination of the level of the gene product, rather than the phenotype, shows some intermediate gene expression. For example, in human biochemical disorders such as **Tay-Sachs disease,** homozygous recessive individuals are severely affected while heterozygotes are phenotypically normal. In affected individuals there is almost no activity of the enzyme **hexosaminidase.** Heterozygotes, on the other hand, express only about 50 percent of the enzyme activity found in homozygous normal individuals. This situation is not uncommon in enzyme disorders. It illustrates the somewhat arbitrary nature of the terms *dominance* and *recessiveness.*

CODOMINANCE

If one pair of alleles is responsible for the production of two distinct and detectable gene products, a situation unlike that of incomplete dominance or dominance/recessiveness arises. The distinct genetic expression of both alleles is called **codominance.** For example, the **MN blood groups** in humans are characterized by certain molecules called glycoproteins found on the surface of red blood cells. Discovered by Karl Landsteiner and Philip Levine, these molecules are **native antigens** that provide immunological identity to individuals. Each native antigen can elicit an antibody response if it is present on tissue transfused or transplanted to another individual lacking the antigen.

The MN system, under the control of a locus on chromosome 4, results in persons having blood group M, MN, or N. The MN phenotypes of offspring of parents with the various blood group combinations are as follows:

Parental Phenotypes	Offspring Phenotypes
M × M	All M
N × N	All N
M × N	All MN
M × MN	1/2 M : 1/2 MN
N × MN	1/2 N : 1/2 MN
MN × MN	1/4 M : 1/2 MN : 1/4 N

Examination of these results confirms that only a single pair of alleles is involved in MN inheritance. The results are consistent with the hypothesis that two alleles (L^M and L^N) determine the three phenotypes: $L^M L^M$ results in blood group M; $L^M L^N$ results in group MN; and $L^N L^N$ results in group N. The offspring of two heterozygous MN parents appear in a 1:2:1 ratio, the same as that found for the incomplete dominance mode of inheritance. However, while incomplete dominance produces an intermediate, blending effect in heterozygous individuals, codominant inheritance results in distinct evidence of the gene products of both alleles. Once again, the mode of inheritance can only be discerned by analyzing the specific gene products.

MULTIPLE ALLELES

Because the information stored in any gene is extensive, mutations may modify this information in many ways. Each change has the potential for producing a different allele. Therefore, at any given locus on the chromosome, the number of alleles within a population of individuals need not be restricted to only two. When three or more alleles are found for any particular gene, the mode of inheritance is called **multiple allelism.**

The concept of multiple alleles can only be studied in populations. Any individual diploid organism has, at most, two homologous gene loci, which may be occupied by different alleles. However, among members of a species, many alternative forms of the same gene may exist. The following examples illustrate the concept of multiple alleles. In several of the examples the relationship between genetics, immunology, and medicine is shown.

The ABO Antigens

The simplest possible case of multiple alleles is that in which there are three alleles of one gene. This situation exists in the inheritance of the **ABO blood types** in humans, discovered by Landsteiner in the early 1900s. The ABO system, like the MN blood types, is characterized by the presence of native antigens on the surface of red blood cells. The ABO antigens are distinct from the MN antigens and are under the control of a different gene, located on chromosome 9. As in the MN system, one combination of alleles in the ABO system exhibits a codominant mode of inheritance.

The ABO phenotype of any individual is ascertained by mixing a blood sample with antiserum containing type A or type B antibodies. If the antigen is present in the blood, it will **cross-react** with an antibody that was formed against it and cause clumping or agglutination of the red blood cells. When individuals are tested in this way, four phenotypes are revealed. Each individual has either the A antigen (A phenotype), the B antigen (B phenotype), the A and B antigens (AB phenotype), or neither antigen (O phenotype). In 1924, it was hypothesized that these phenotypes were inherited as the result of three alleles of a single gene. This hypothesis was based on studies of the blood types of many different families.

Although different designations may be used, we will use the symbols I^A, I^B, and I^O for the three alleles. The I designation stands for **isoagglutinogen,** another term for antigen. If we assume that the I^A and I^B alleles are responsible for the production of their respective A and B antigens and that I^O is an allele that does not produce any detectable A or B antigens, the various genotypic possibilities can be listed and the appropriate phenotype assigned to each:

Genotype	Antigen	Phenotype
$I^A I^A$	A	
$I^A I^O$	A	A
$I^B I^B$	B	
$I^B I^O$	B	B
$I^A I^B$	A, B	AB
$I^O I^O$	Neither	O

Note that in these assignments the I^A and I^B alleles behave dominantly to the I^O allele, but codominantly to each other. We can test the hypothesis that three alleles control ABO blood types by examining potential offspring from all possible matings, as shown in Table 4.1. If we assume heterozygosity wherever possible, we can predict which phenotypes can occur. These theoretical predictions have been upheld in numerous studies examining the blood types of children of parents with all possible phenotypic combinations. The hypothesis that three alleles control ABO blood types in the human population is now universally accepted.

Our knowledge of human blood types has several practical applications. Compatible blood transfusions can be achieved and decisions about disputed parentage more accurately made. The latter cases can occur when newborns are inadvertently mixed up in hospitals, but more commonly, an adult male is accused of fathering an illegitimate child. In both cases, an examination of the ABO phenotypes of the possible parents and the child may help to resolve the situation. Table 4.1 demonstrates numerous cases where it is impossible for a parent of a particular ABO phenotype to produce a child of a certain phenotype. In fact, the only mating that can result in offspring of all four phenotypes is between two heterozygous individuals, one showing the A phenotype and the other showing the B phenotype. On genetic grounds alone, a male or female may be unequivocally ruled out as the parent of a certain child. On the other hand, it should be obvious that this type of genetic evidence *never proves* parenthood.

TABLE 4.1

Potential phenotypes in the offspring of parents with all possible ABO blood type combinations, assuming heterozygosity whenever possible.

P₁ Generation		F₁ Phenotype			
Phenotypes	Genotypes	A	B	AB	O
A × A	$I^A I^O \times I^A I^O$	X	—	—	X
B × B	$I^B I^O \times I^B I^O$	—	X	—	X
O × O	$I^O I^O \times I^O I^O$	—	—	—	X
A × B	$I^A I^O \times I^B I^O$	X	X	X	X
A × AB	$I^A I^O \times I^A I^B$	X	X	X	—
A × O	$I^A I^O \times I^O I^O$	X	—	—	X
B × AB	$I^B I^O \times I^A I^B$	X	X	X	—
B × O	$I^B I^O \times I^O I^O$	—	X	—	X
AB × O	$I^A I^B \times I^O I^O$	X	X	—	—
AB × AB	$I^A I^B \times I^A I^B$	X	X	X	—

The Bombay Phenotype

The biochemical basis of the ABO blood type system has now been carefully worked out (Figure 4.2). The A and B antigens are actually carbohydrate groups that are bound to lipid molecules (fatty acids) protruding from the membrane of the red blood cell. The specificity of the A and B antigens is based on the terminal sugar of the carbohydrate group. Individuals possessing the I^A allele can add the modified monosaccharide, **N-acetyl-α-D-galactosamine,** to the carbohydrate chain. Individuals possessing the I^B allele can add the unmodified **α-D-galactose** molecule. Heterozygotes ($I^A I^B$) add either one or the other entity at the many sites available on each red blood cell. This latter phenomenon illustrates the biochemical basis of codominance in individuals of blood type AB.

The carbohydrate group that lacks either α-D-galactose or the modified N-acetyl-α-D-galactosamine at the terminal position is called the **H substance.** Type O individuals (of genotype $I^O I^O$) possess this on their red blood cell surface. In 1952 in Bombay, a woman was discovered who apparently lacked the H substance altogether. She demonstrated a most interesting genetic history and blood type. In need of a transfusion, she was found to lack both the A and B antigens and was thus typed as O. However, as shown

(a) COMPONENTS

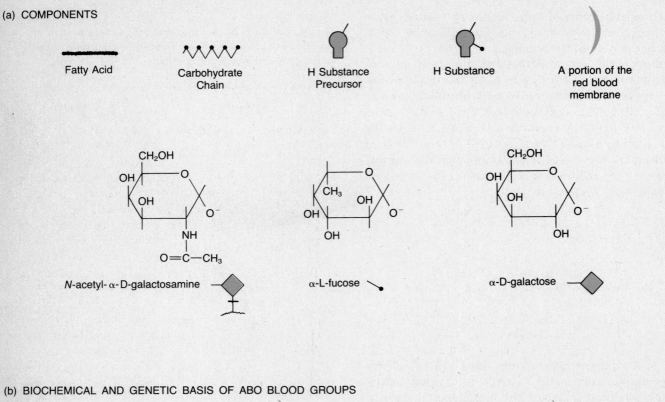

Fatty Acid

Carbohydrate Chain

H Substance Precursor

H Substance

A portion of the red blood membrane

N-acetyl- α - D-galactosamine

α-L-fucose

α-D-galactose

(b) BIOCHEMICAL AND GENETIC BASIS OF ABO BLOOD GROUPS

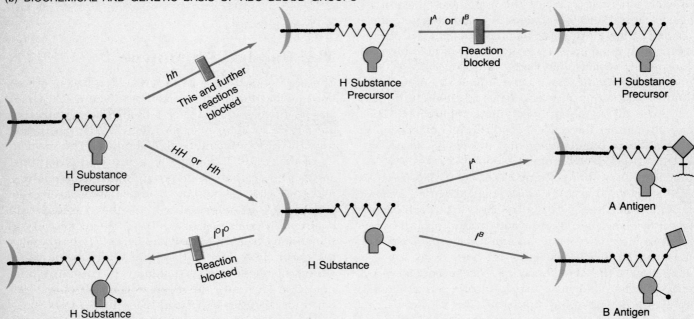

I^A or I^B

Reaction blocked

H Substance Precursor

H Substance Precursor

hh

This and further reactions blocked

H Substance Precursor

HH or *Hh*

I^A

A Antigen

$I^O I^O$

Reaction blocked

H Substance

I^B

B Antigen

H Substance

FIGURE 4.2

The genetic and biochemical basis of the ABO blood groups. In (a) several of the components are shown and identified diagrammatically. In (b) the biochemical reactions and the genetic control governing those reactions are illustrated.

in the partial pedigree in Figure 4.3, one of her parents was type AB and she was the obvious donor of an I^B allele to two of her offspring. Thus she was genetically type B but functionally type O!

It was subsequently shown that this woman was homozygous for a rare recessive mutation, *h*, which prevented her from synthesizing the complete H substance. The terminal portion of the carbohydrate chain protruding from the red cell membrane was shown to lack the monosaccharide, **α-L-fucose.** In the absence of fucose, the enzymes specified by the I^A and I^B alleles are apparently unable to recognize the incomplete H substance as a proper substrate. Thus, neither the terminal α-D-galactose or N-acetyl-α-D-galactosamine can be added, even though the enzymes capable of doing so are present and functional. As a result, the ABO system genotype cannot be expressed in individuals of genotype *hh*, and they appear as type O. To distinguish them from the rest of the population, they are said to demonstrate the **Bombay phenotype.** The frequency of the *h* allele is exceedingly low. Thus, the vast majority of the human population is of the *HH* genotype and can synthesize the H substance. The biochemical events that yield the various human blood types are illustrated in Figure 4.2.

The Secretor Locus

Still a third gene is known to affect the expression of the ABO blood type system: the **secretor locus.** In most people, the A and B antigens are present in various body secretions as well as on the membrane of red blood cells. The ability to secrete these antigens in body fluids such as saliva is under the influence of the dominant allele, *Se*, carried by more than three-fourths of the human population (*Se/Se* or *Se/se*). The minority of the population who do not secrete the antigens (*se/se*) lack an enzyme that normally modifies the H substance, rendering it water soluble. Thus, nonsecretors make the antigens as specified by the ABO loci but cannot secrete them.

The Rh Antigens

Another set of antigens thought by some geneticists to illustrate multiple allelism includes those designated Rh. Discovered by Landsteiner, Levine, and others around 1940, the Rh antigens have received a great deal of attention because of their direct involvement in the disorder **erythroblastosis fetalis.** The initial investigations led to the belief that in the human population only two alleles controlled the presence or absence of the antigen. It was thought that the Rh^+ allele determined the presence of the antigen and behaved as a dominant gene. The Rh^- allele seemed to result in the absence of the antigen.

Erythroblastosis fetalis, also referred to as **hemolytic disease of the newborn (HDN),** is a form of anemia. It occurs in an Rh-positive fetus whose mother is Rh-negative and whose father is Rh-positive, contributing that allele to the fetus. Such a genetic combination results in a potential immunological incompatibility between the mother and fetus. If fetal blood passes through the ruptured placenta at birth and enters the maternal circulation, the mother's immune system recognizes the Rh antigen as foreign and builds antibodies against it. During a second pregnancy, the antibody concentration becomes high enough that when maternal antibodies, which can pass across the placenta, enter the fetus's circulation, they begin to destroy the fetus's red blood cells. This causes the hemolytic anemia.

About 10 percent of all pregnancies demonstrate Rh incompatibility. However, for numerous reasons, less than 0.5 percent actually result in anemia. Currently, incompatible mothers are given anti-Rh antisera immediately after giving birth to an Rh-positive baby. This destroys any Rh-positive cells that have entered the mother's circulation so that she does not produce her own anti-Rh antibodies. Before this treatment was developed, many fetuses failed to survive to term. For those that did, complete blood transfusions were often necessary. Even then, many newborns failed to survive.

With the development of more refined antisera to test for the presence of the antigen, it became apparent that the genetic control of Rh antigens was much more complex than originally thought. For example, some presumed Rh-negative blood was found to contain the antigen, but in a different chemical form. Alexander Wiener has proposed the existence of at least eight alleles at a single Rh locus. Other workers, including Ronald A. Fisher, Robert R. Race, and Ruth

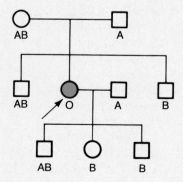

FIGURE 4.3

A partial pedigree of a woman displaying the Bombay phenotype. Functionally, her ABO blood group appears to be type O. Genetically, she should be type B.

Sanger, have proposed an alternative type of inheritance. These workers feel that there are three closely linked genes, each with two alleles involved in the inheritance of Rh factors. The term **linkage** is used to describe genes located on the same chromosome. We will discuss this concept in detail in Chapter 6.

The two systems of nomenclature designating the alleles are contrasted in Table 4.2. In the Wiener system, the presence of at least one of the four dominant alleles is sufficient to yield the Rh-positive blood type. Examination of the Fisher-Race system shows that the genetic locus bearing the set of D and d alleles is most critical. The presence of at least one dominant D allele results in the Rh-positive phenotype, while the dd genotype ensures the Rh-negative blood type. While the C, c, E, and e alleles specify distinguishable antigens, they are not immunologically significant. Because of the complexity of the antigenic patterns, it is difficult to favor one system over the other.

Histocompatibility Genes

The **histocompatibility complex** also illustrates multiple alleles. This immunological system in mammals determines the acceptability of tissue transplants between organisms. Unless two members of the same species are from highly inbred strains (or, in the case of humans, they are identical twins), tissue transplantation is generally unsuccessful. The host immune system recognizes native cell-surface antigens of donor tissue as foreign and produces antibodies that destroy the transplant. The genetic basis of histocompatibility antigens has been investigated extensively because of the potential lifesaving benefit of organ transplantation in humans.

TABLE 4.2

A comparison of the alleles involved in the Wiener system with those of the Fisher-Race system to explain the genetic basis of the Rh blood groups.

Wiener Nomenclature	Fisher-Race Nomenclature	Phenotype
R^1	CDE	
R^2	CDe	
R^0	cDE	Rh$^+$
R^z	cDe	
r	CdE	
r'	Cde	
r''	cdE	Rh$^-$
r^y	cde	

We shall focus our attention on what has been learned from the studies of mice and humans. Both groups contain one major histocompatibility complex (**MHC**) controlling the presence of many antigens that elicit a strong immunological response following tissue transplantation. Both species also contain numerous minor loci responsible for antigens that elicit a weaker immune response. Each major and minor locus is thought to have many alleles.

The major complex in the mouse, designated **H-2,** has been thoroughly studied. Located on chromosome 17, it is divided into four regions: K, I, S, and D. Each region controls several immunological traits, and each trait is presumed to be controlled by several genes. Each gene is thought to have a large number of alleles.

In humans, the **HLA** (human leukocyte A) complex on the short arm of chromosome 6 represents the MHC. It contains at least four genes, designated **HLA-A, HLA-B, HLA-C,** and **HLA-D.** A possible fifth locus, **DR,** is also recognized. It is not yet clear, however, whether it is a separate locus or a part of the D gene region. Each gene has up to 35 alleles, some of which have been well characterized, while others are only poorly understood. The combination of HLA genes located on a single chromosome is known as a **haplotype.** Each individual, being diploid, has two haplotypes, each with at least four genes.

An individual's HLA type can be determined by testing his or her white blood cells for the presence of specific HLA antigens. A significant finding is that haplotypes are almost always inherited as a unit. Figure 4.4 shows a theoretical distribution of haplotypes in a family with five children. Each child has only one haplotype in common with each parent.

The basis of inheritance of these "unit" haplotypes is the close proximity of the four loci along chromosome 6. The sequence of genes is known to be D–B–C–A, with locus A being closest to the end of the short arm and locus D being closest to the centromere. Individual loci are less than a single map unit apart. As we will see in Chapter 6, this close linkage minimizes the chance that crossing over will separate an existing set of alleles during gamete formation. In the family illustrated in Figure 4.4, for example, no separation by crossing over has occurred, and thus each haplotype has been transmitted intact.

A child has a one in four probability of matching up precisely with each brother or sister. Since there are other minor but important histocompatibility complexes, even if such a match occurs, it does not necessarily make siblings completely tolerant to the exchange of tissue during transplantation.

In recent years, transplantation of human kidneys, hearts, and other organs has become more frequent.

FIGURE 4.4

The inheritance of HLA haplotypes. Each haplotype consists of four loci (*A*, *B*, *C*, and *D*,) each specifying different antigenic determinants. Each diploid individual expresses two haplotypes, each inherited as a unit from one parent. The arrows indicate siblings with matching haplotypes. (From Thompson and Thompson, 1980, p. 191.)

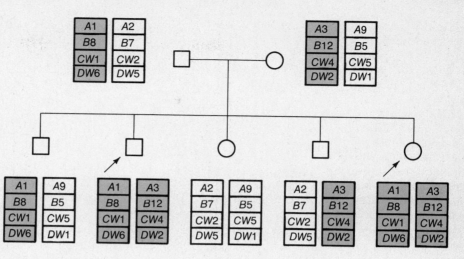

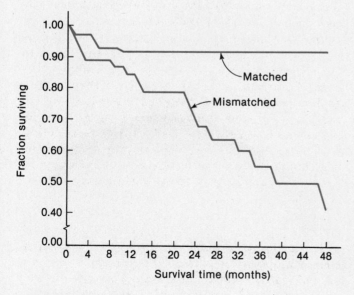

The success of a transplantation, as measured by the survival time of the donor organ in the recipient before it is immunologically rejected, depends on closely matching the HLA types. If the donor contains an allele that is absent in the recipient, the graft will be rejected. This allele will be expressed as a unique antigen and will be foreign to the recipient.

In the case of kidney transplants, a close relative is the most likely donor. Figure 4.5 vividly demonstrates the importance of matching HLA types, even between siblings. When HLA types were closely matched, over 90 percent of the grafts survived 48 months. In a more extensive study involving 25,000 kidney transplants over a 5-year period, there has been a comparable survival rate.

FIGURE 4.5

The survival of HLA-matched and unmatched siblings receiving kidney transplants. (From Singal et al., p. 246. © 1965 The Williams & Wilkins Co., Baltimore.)

A heart transplant obviously can be performed only with an organ from a deceased donor. Thus, an individual needing the transplant must wait, usually until an unrelated but close antigenic match is available. Even then, most transplants will eventually be rejected. However, immunosuppressive drugs, administered to the host, are effective in prolonging the acceptance of the transplant.

HLA and Disease Association

Certain HLA genes are now known to be associated with a predisposition to certain diseases. Figure 4.6 illustrates several cases where certain HLA antigens are more frequently found in diseased than in normal individuals. For example, over 90 percent of patients with the connective tissue disease **ankylosing spondylitis** have the B27 antigen, while its incidence in the general population is only about 7 percent. Table 4.3 further demonstrates the increased risk individuals show for other specific diseases when they express certain HLA alleles. For example, antigen B27, the culprit in ankylosing spondylitis, also predisposes one to **juvenile rheumatoid arthritis** and a disease of the eye, **anterior uveitis.**

These data indicate that knowledge of the HLA complex is important to our understanding of disease resistance in humans. Many theories have been put forward to explain such disease associations. One suggests that the foreign antigens (molecules or microorganisms) responsible for a specific disease may chemically mimic the HLA antigen that predisposes one to the disorder. Thus, the agent responsible for the disease is not recognized as foreign, and no immune response is mounted to combat its presence. As a result, the disease is much more likely to afflict an individual with a specific HLA antigen.

Another prevalent theory suggests that the HLA loci are not directly involved in the disease association.

FIGURE 4.6
Presence of certain HLA antigens in diseased and normal individuals. For example, 90 percent of individuals with the connective tissue disease, ankylosing spondylitis, contain the B27 antigen, while less than 10 percent of normal individuals contain this antigen. Psoriasis is a skin disorder; coeliac disease is a sensitivity to certain proteins in cereals; and multiple sclerosis is a progressive degenerative disorder of the nervous system caused by the destruction of the myelin sheath surrounding nerve fibers. (From McDevitt and Bodmer, 1974, p. 1267.)

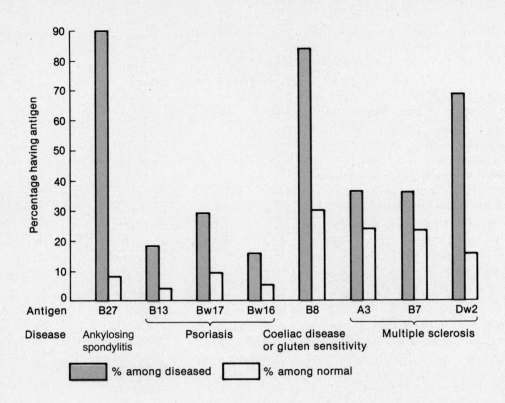

Rather, closely linked, but separate genes governing the expression of the immune response are considered responsible. Called **IR** or **Immune Response genes,** they either enhance or diminish the expression of other antibody-producing genes. Because they are so closely linked to the HLA loci, those that diminish the response are often closely associated with the expression of certain HLA antigens. While it is not clear which, if either, of these two theories is correct, the study of HLA disease associations may prove useful in the early diagnosis of many human disorders.

TABLE 4.3
The increased risk of developing certain disorders when a specific antigen is expressed, compared with a control population.

HLA Antigens	Disorder	Increased Risk
B5	Ulcerative colitis	9.3×
B27	Juvenile rheumatoid arthritis	4.7×
	Anterior uveitis (eye)	15.4×
Bw35	deQuervain's thyroiditis	22.2×
B8	Diabetes	2.1×
	Chronic active hepatitis	3.0×
Bw18	Hodgkin's disease	1.9×

Multiple Alleles in *Drosophila:* The *white* Locus

Many other phenotypes in plants and animals are known to be controlled by multiple allelic inheritance. In *Drosophila*, for example, where the induction of mutations has been used extensively as a method of investigation, many alleles are known at practically every locus. The *white* eye mutation, discovered by Thomas H. Morgan and Calvin Bridges in 1912, is only one of over 100 alleles that may occupy this locus. In this allelic series, eye colors range from complete absence of pigment in the *white* allele, to deep ruby in the *white-satsuma* allele, to orange in the *white-apricot* allele, to a buff color in the *white-buff* allele. These alleles are designated w, w^{sat}, w^a, and w^{bf}, respectively. In all cases studied, the total amount of pigment in these mutant eyes is reduced to less than 20 percent of that found in the brick red wild-type eye.

LETHAL ALLELES

Many gene products are essential to an organism's survival. Mutations resulting in the synthesis of a nonfunctional gene product can sometimes be tolerated in the heterozygous state; that is, one wild-type allele may be sufficient to produce enough of the essential product to allow survival. However, such a mutation behaves as a **recessive lethal,** and homozygous recessive individuals will not survive. The time of death will depend upon when the product is needed during development. In instances where one copy of the wild gene is not sufficient for normal development, even the heterozygote will not survive. In this case, the mutation is behaving as a **dominant lethal allele** because its presence somehow overrides the expression of the wild-type product.

In some cases, the surviving heterozygotes who possess one lethal allele may exhibit a phenotype different from the homozygous wild-type individual. Such an allele is behaving as a recessive lethal but is dominant with respect to the phenotype. For example, a mutation causing a yellow coat in mice was discovered in the early part of this century. The yellow coat varied from the normal agouti-coat phenotype. Crosses between the various combinations of the two strains yielded unusual results:

Cross A: agouti × agouti → all agouti

Cross B: yellow × yellow → 2/3 yellow: 1/3 agouti

Cross C: agouti × yellow → 1/2 yellow: 1/2 agouti

These results are explained on the basis of a single pair of alleles, one of which behaves as an autosomal recessive lethal. The mutant *yellow* allele (A^Y) is dominant to the wild-type *agouti* allele (A), so heterozygous mice will have yellow coats. However, the yellow allele also behaves as a recessive lethal; thus no homozygous yellow mice are ever recovered. These three crosses are diagramed in Figure 4.7.

Many genes are known to behave similarly in other organisms. In *Drosophila, Curly* wing (Cy), *Plum* eye (Pm), *Dichaete* wing (D), *Stubble* bristle (Sb), and *Lyra* wing (Ly) behave as recessive lethals but are dominant with respect to the expression of the mutant phenotype when heterozygous.

Other alleles are known to behave as dominant lethals. In humans, a disorder called **Huntington's disease** (also referred to as Huntington's chorea) is due to a dominant allele that behaves quite differently from the alleles just described. While individuals homozygous for the lethal gene apparently never survive through fetal development, heterozygotes develop normally to adulthood. Affected individuals then undergo gradual nervous and motor degeneration until they die, usually between the ages of 20 and 50 years. The disorder is particularly tragic because it has such a late onset. By that time, the affected individual may have produced a family, thereby passing this genetic disorder to one-half the offspring. The American folk singer and composer Woody Guthrie suffered from this crippling disease.

Dominant lethal alleles are rarely observed. In order for them to exist in a population, the affected individual must reproduce before the allele's lethality is expressed. If all affected individuals die before reaching the reproductive age, the mutant gene will not be passed to future generations. The mutation will thus disappear from the population unless it recurs as a result of mutation.

FIGURE 4.7
Inheritance patterns in three crosses involving the mutant *yellow* allele (A^Y) in the mouse. Note that the mutant allele behaves dominantly to the wild-type agouti (A) allele in controlling coat color, but it also behaves as a recessive lethal allele. The genotype A^YA^Y does not survive.

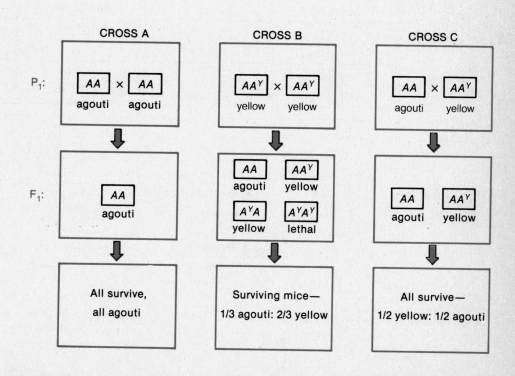

COMBINATIONS OF TWO GENE PAIRS

Each example discussed so far modifies Mendel's 3:1 F_2 monohybrid ratio. Therefore, combining any two of these modes of inheritance in a dihybrid cross will likewise modify the classical 9:3:3:1 ratio. Having established the foundation for the modes of inheritance of incomplete dominance, codominance, multiple alleles, and lethal genes, we can now deal with the situation of two modes of inheritance occurring simultaneously. Mendel's principle of independent as-

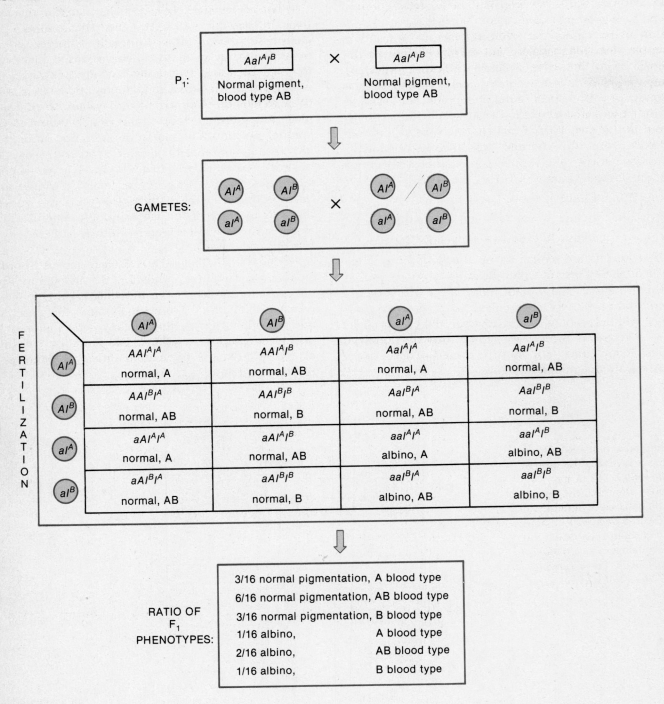

FIGURE 4.8
A theoretical modified dihybrid cross involving the ABO blood type and albinism in humans. The probabilities of each phenotype in the offspring of such a mating are calculated using a Punnett square.

sortment applies to these situations, provided that the genes controlling each character are not located on the same chromosome.

Suppose, for example, that a mating occurs between two humans who are both heterozygous for the autosomal recessive gene that causes albinism and who are both of blood type AB. What is the probability of any particular phenotypic combination occurring in each of their children? Albinism is inherited in the simple Mendelian fashion, and the blood types are determined by the series of three multiple alleles, I^A, I^B, and I^O. The solution to this problem is diagramed in Figure 4.8.

Instead of this dihybrid cross yielding the classical four phenotypes in a $9:3:3:1$ ratio, six phenotypes occur in a $3:6:3:1:2:1$ ratio, thus establishing the expected probability for each phenotype. While Figure 4.8 solves the problem using the conventional Punnett square, the forked-line method described in Chapter 3 simplifies this task (Figure 4.9). Recall that this method simply requires that the phenotypic ratios for each trait be computed individually (which can usually be done by inspection); all possible combinations can then be calculated.

We can deal in a similar way with any combination of two modes of inheritance. You will be asked to determine the phenotypes and their expected probabilities for many of these combinations when you solve the problems at the end of the chapter. In each case, the final phenotypic ratio is a modification of the $9:3:3:1$ dihybrid ratio.

FIGURE 4.9
The calculation of the probabilities in the mating illustrated in Figure 4.8 using the forked-line method.

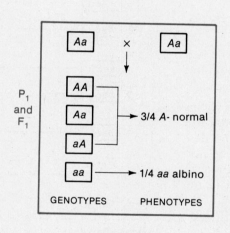

CONSIDERATION
OF PIGMENTATION
ALONE

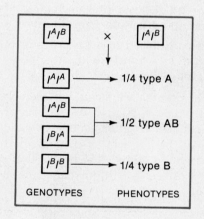

CONSIDERATION
OF BLOOD TYPES
ALONE

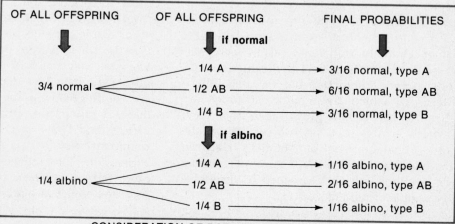

CONSIDERATION OF BOTH TRAITS TOGETHER

GENE INTERACTION

Soon after the rediscovery of Mendel's work, experimentation revealed that phenotypic characters were often under the control of more than one gene pair. This was a significant discovery because it revealed for the first time that genetic influence on the phenotype is much more sophisticated than envisioned by Mendel. Instead of single genes controlling the development of individual parts of the plant and animal body, it soon became clear that each phenotypic character is influenced by many gene products.

The concept of gene interaction does not mean that two or more genes, or their products, necessarily interact directly to influence a particular phenotype. Instead, this concept implies that the cellular function of numerous gene products is related to the development of a common phenotype. To clarify this point, we will present several examples that directly illustrate gene interaction at the biochemical level.

In the following discussion, we will make several assumptions and adopt certain conventions:

1   The genes considered in each cross are unlinked and therefore assort independently of one another during gamete formation.

2   If complete dominance exists between the alleles of any gene pair, such that AA and Aa or BB and Bb are equivalent in their genetic effects, the designations $A-$ or $B-$ will be used for both combinations. Therefore, the $(-)$ indicates that either allele may be present, without consequence to the phenotype.

3   All P_1 crosses will involve homozygous individuals (e.g., $AABB \times aabb$ or $AAbb \times aaBB$). Therefore, each F_1 generation will consist of only heterozygotes of genotype $AaBb$.

4   In each example, the F_2 generation produced from these heterozygous parents will be the main focus of analysis. When two genes are involved (Figure 4.10), the F_2 genotypes fall into four categories: $9/16\ A-B-$, $3/16\ A-bb$, $3/16\ aaB-$, and $1/16\ aabb$. Because of dominance, all genotypes in each category are equivalent in their effect on the phenotype.

The study of gene interaction has revealed inheritance patterns in which these four categories are grouped together in various ways, such that each grouping yields a different phenotype. Thus, the 9:3:3:1 ratio, characteristic of Mendel's dihybrid cross, is modified in several ways (Figure 4.11). In the next several sections, we shall proceed to discuss a number of examples from Figure 4.11.

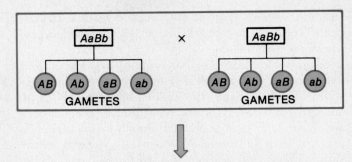

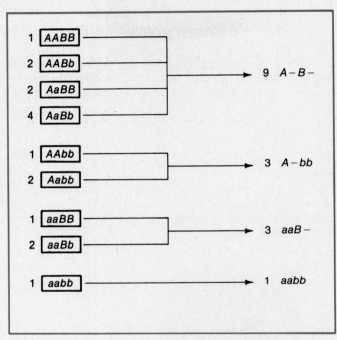

FIGURE 4.10

The generation of grouped genotypic classes resulting from a cross between two identical double heterozygotes.

Epistasis

Perhaps the best examples of gene interaction are those illustrating the phenomenon of **epistasis.** Derived from the Greek word meaning "stoppage," epistasis can in genetics be equated with the word *masking*. The phenomenon occurs when the expression of one gene pair masks or modifies the expression of another gene pair. This masking may occur under different conditions. First, the presence of two recessive alleles at one locus may prevent or override the expression of alleles at a second locus. Or, a single dominant allele at the first locus may influence the expression of the alleles at a second gene locus. Another form of epistasis occurs when the homozygous recessive alleles at one or the other of two genetic loci

Case	Organism	Character	F₂ Genotypes and Proportion Out of 16									Final Phenotypic Ratio
			AABB 1	*AABb* 2	*AaBB* 2	*AaBb* 4	*AAbb* 1	*Aabb* 2	*aaBB* 1	*aaBb* 2	*aabb* 1	
	Pea	Mendel's dihybrid	9				3		3		1	9:3:3:1
1	Mouse	Coat color	Agouti				Black		Albino			9:3:4
2	Squash	Color	White						Yellow		Green	12:3:1
3	Pea	Flower color	Purple				White					9:7
4	Squash	Fruit shape	Disc				Sphere				Long	9:6:1
5	Chicken	Color	White						Colored		White	13:3
6	Mouse	Color	White-spotted				White		Colored		White-spotted	10:3:3
7	Shepherd's purse	Seed capsule	Triangular								Ovoid	15:1
8	Flour beetle	Color	Red	Sooty	Red	Sooty	Black		Jet		Black	6:3:3:4

FIGURE 4.11

The basis of modified dihybrid F₂ phenotypic ratios resulting from crosses between doubly heterozygous F₁ individuals.

mask the expression of the dominant allele at the other locus. Thus, in this third case, two gene pairs complement one another such that at least one dominant allele at each locus is required to express a particular phenotype. Each of these three forms of epistasis will be examined in more detail.

An example of the homozygous recessive condition at one locus masking the expression of a second locus is seen in the inheritance of coat color in mice (Case 1 of Figure 4.11). Wild-type coat color is agouti, a grayish pattern formed by alternating bands of pigment on each hair. Agouti is dominant to nonagouti (black) hair caused by a recessive mutation, *b*. Thus, *B*− results in agouti, while *bb* yields nonagouti coat color. When homozygous, a recessive mutation, *a*, at a separate locus, eliminates pigmentation altogether, yielding albino mice.

In a cross between agouti (*AABB*) and albino (*aabb*), members of the F₁ are all *AaBb* and have agouti coat color. In the F₂ progeny of a cross between two F₁ double heterozygotes, the following genotypes and phenotypes are observed:

F₁: *AbBb* × *AaBb*

F₂ Ratio	Genotype	Phenotype	Final Phenotypic Ratio
9/16	*A−B−*	agouti	9/16 agouti
3/16	*A−bb*	black	3/16 black
3/16	*aaB−*	albino ⎫	4/16 albino
1/16	*aabb*	albino ⎭	

Gene interaction yielding the observed 9:3:4 F₂ ratio can be envisioned as a two-step process:

$$\text{Precursor Molecule (colorless)} \xrightarrow[A-]{\text{Gene } A} \text{Black Pigment} \xrightarrow[B-]{\text{Gene } B} \text{Agouti Pattern}$$

In the presence of an *A* allele, black pigment can be made from a colorless substance. In the presence of a *B* allele, the black pigment is deposited during the development of hair in a pattern producing the agouti phenotype. If the *bb* genotype occurs, all of the hair becomes black. If the *aa* genotype occurs, no black pigment is produced, regardless of the presence of the *B* or *b* alleles, and the mouse is albino. Therefore, the *aa* genotype masks or suppresses the expression of the *B* gene.

The second type of epistasis occurs when a dominant allele at one genetic locus masks the expression of the alleles of a second locus. For instance, Case 2 of Figure 4.11 deals with the inheritance of fruit color in summer squash. Here, the dominant allele A results in white fruit color regardless of the genotype at a second locus, B. In the absence of the dominant A allele (the aa genotype), BB or Bb results in yellow color while bb results in green color. Therefore, if two white-colored double heterozygotes ($AaBb$) are crossed together, an interesting genetic ratio occurs because of this type of epistasis. Of the offspring, 9/16 are $A-B-$ and are thus white. The 3/16 bearing the genotypes $A-bb$ are also white. Of the remaining squash, 3/16 are yellow ($aaB-$), while 1/16 are green ($aabb$). Thus, a modified ratio of 12:3:1 occurs.

Our third example (Case 3 of Figure 4.11) is demonstrated in a cross between two strains of white-flowered sweet peas. Unexpectedly, the F_1 plants were all purple, and the F_2 occurred in a ratio of 9/16 purple to 7/16 white. The proposed explanation for these results suggests that the presence of at least one dominant allele of each of two gene pairs is essential in order for flowers to be purple. All other genotype combinations yield white flowers. The two gene pairs appear to complement each other in producing the purple phenotype. On the other hand, the homozygous condition of either recessive allele masks the expression of the dominant allele at either locus.

The cross is shown as follows:

$$P_1: \frac{AAbb}{\text{white}} \times \frac{aaBB}{\text{white}}$$

$$F_1: \text{All } \frac{AaBb}{\text{purple}}$$

F_2 Ratio	Genotype	Phenotype	Final Phenotypic Ratio
9/16	$A-B-$	purple	9/16 purple
3/16	$A-bb-$	white	
3/16	$aaB-$	white	7/16 white
1/16	$aabb$	white	

We can envision the way in which two gene pairs might yield such results:

Gene A Gene B

Precursor Substance (colorless) $\xrightarrow{A-}$ Intermediate Product (colorless) $\xrightarrow{B-}$ Final Product (purple)

At least one dominant allele from each pair of genes is necessary to ensure both biochemical conversions to the final product, yielding purple flowers. In the cross above, this will occur in 9/16 of the F_2 offspring. All other plants have flowers that remain white.

These three examples illustrate in a simple way how two gene products interact to influence a common phenotype. In most, if not all, instances, many more than two genes interact to produce each phenotypic character. Provided that each such gene is homozygous for the wild-type alleles, these additional genes would not influence the outcome of the above crosses.

Novel Phenotypes

Other cases of gene interaction yield novel, or new, phenotypes in the F_2 generation, in addition to producing modified dihybrid ratios. Case 4 in Figure 4.11 depicts the inheritance of fruit shape in the summer squash *Cucurbita pepo*. When plants with *disc*-shaped fruit ($AABB$) are crossed to plants with *long* fruit ($aabb$), the F_1 generation all have *disc* fruit. However, in the F_2 progeny, fruit with a novel shape—*sphere*—appear as well as fruit exhibiting the parental phenotypes. The F_2 generation, with a modified 9:6:1 ratio, is as follows:

$$\frac{AaBb}{\text{disc}} \times \frac{AaBb}{\text{disc}}$$

F_2 Ratio	Genotype	Phenotype	Final Phenotypic Ratio
9/16	$A-B-$	disc	9/16 disc
3/16	$A-bb$	sphere	
3/16	$aaB-$	sphere	6/16 sphere
1/16	$aabb$	long	1/16 long

In this example of gene interaction, both gene pairs influence fruit shape equivalently. A dominant allele at either locus ensures a *sphere*-shaped fruit. In the absence of dominant alleles, the fruit is *long*. However, if both dominant alleles (A and B) are present, the fruit is flattened into a *disc* shape.

Another interesting example of an unexpected phenotype arising in the F_2 generation is the inheritance of eye color in *Drosophila melanogaster*. The wild-type eye color is brick red. When two autosomal recessive mutants, *brown* and *scarlet*, are crossed, the F_1 generation consists of flies with wild-type eye color. In the F_2 generation wild, scarlet, brown, and white-eyed flies are found in a 9:3:3:1 ratio. While this ratio is numerically the same as Mendel's dihybrid ratio, the *Drosophila* cross involves only one character, eye color. The diagram of this cross uses the gene symbols A to B to maintain consistency in this section. (The actual symbols for the two genes are *bw* and *st*.)

$$P_1: \frac{AAbb}{\text{brown}} \times \frac{aaBB}{\text{scarlet}}$$

$$F_1: \frac{AaBb}{\text{wild type}}$$

F_2 Ratio	Genotype	Phenotype	Final Phenotypic Ratio
9/16	$A-B-$	wild type	9/16 wild
3/16	$A-bb$	brown	3/16 brown
3/16	$aaB-$	scarlet	3/16 scarlet
1/16	$aabb$	white	1/16 white

This cross is an excellent example of gene interaction because the biochemical basis of eye color in this organism has been determined. *Drosophila*, as a typical arthropod, has compound eyes made up of individual visual units called **ommatidia.**

The wild-type eye color is due to the deposition and mixing of two separate pigments in each ommatidium. These include the bright red pigment **drosopterin** and the brown pigment **xanthommatin.** Each pigment is produced by separate biosynthetic path-

ways. Each step of each pathway is catalyzed by a separate enzyme and is thus under the control of a separate gene. As shown in Figure 4.12, the *brown* mutation, when homozygous, interrupts the synthesis of the bright red pigment. Thus, the eye contains only xanthommatin pigments and is brown. The recessive mutation *scarlet* located on a separate autosome, interrupts the synthesis of the brown xanthommatins and renders the eye color bright red in homozygous mutant flies. Each mutation apparently causes the

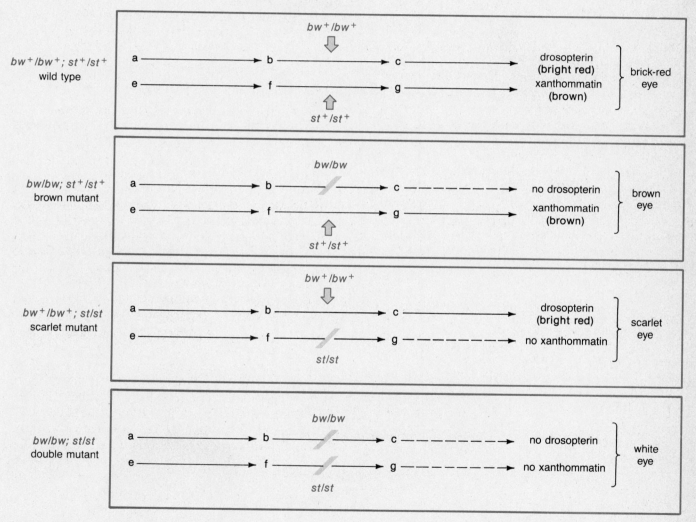

FIGURE 4.12

A theoretical explanation of the biochemical basis of the four eye-color phenotypes produced in a cross between flies with *brown* and *scarlet* eyes in *Drosophila melanogaster*. In the presence of at least one wild-type *bw*[+] allele, an enzyme is produced that converts substance b to c, and drosopterins are synthesized. In the presence of at least one wild-type *st*[+] allele, substance f is converted to g, and xanthommatins are synthesized. The homozygous presence of the recessive *bw* and *st* mutant alleles blocks the synthesis of these respective pigment molecules. Either none, one, or both of these pathways can be blocked, depending on the genotype.

INSIGHTS INTO SOLVING GENETICS PROBLEMS

Students demonstrate their knowledge of transmission genetics by solving genetics problems. Success at this task represents not only comprehension of theory but its application to more practical genetic situations. Most students find problem solving in genetics to be challenging but rewarding. This section is designed to provide some basic insights into the reasoning essential to this process.

Genetics problems are in many ways similar to algebraic word problems. The approach taken should be identical: (1) analyze the problem carefully; (2) translate words into symbols, defining each one first; and (3) solve the problem. The first two steps are most critical. The third step is largely mechanical.

The simplest problems are those that state all necessary information about the P_1 generation and ask you to find the expected ratios of the F_1 and F_2 genotypes and/or phenotypes. The following steps should always be followed when you encounter this type of problem:

1 Determine the genotypes of the P_1 generation.

2 Determine what gametes may be formed by the P_1 generation.

3 Recombine gametes either by the Punnett square method, the forked-line method, or, if the situation is very simple, by inspection. Read the F_1 phenotypes directly.

4 Repeat the process to obtain information about the F_2 generation.

Determining the genotypes from the given information requires an understanding of the basic theory of transmission genetics. For example, consider the following problem: *A recessive mutant allele*, black, *causes a very dark body in* Drosophila *when homozygous. The wild-type color is described as gray. What F_1 phenotypic ratio is predicted when a* black *female is crossed to a* gray *male whose father was* black?

To work this problem, you must understand dominance and recessiveness as well as the principle of segregation. Further, you must use the information about the male parent's father. You can work out the problem as follows:

1 Since the female parent is *black*, she must be homozygous for the mutant allele (*bb*).

2 The male parent is *gray* and thus, he must have at least one dominant allele (*B*). Since his father was *black* (*bb*), and since he received one of the chromosomes bearing those alleles, the male parent must be heterozygous (*Bb*).

From here, the problem is simple:

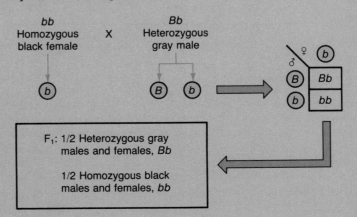

Problems may be more complex if they involve two independent traits. More complicated problems can also include cases involving multiple alleles, incomplete dominance, or epistasis. However, the most difficult types of problems are those that were faced by pioneering geneticists in the laboratory. In these problems they had to determine the mode of inheritance by working backwards from the observations of offspring to parents of unknown genotype. For example, consider the problem of comb shape inheritance in chickens, where walnut, rose, pea, and single are the observed distinct phenotypes: *How is comb shape inherited and what are the genotypes of the P_1 generation of each cross? Use the following data to answer the questions.*

Cross 1: single × single   ⟶ all single
Cross 2: walnut × walnut   ⟶ all walnut
Cross 3: rose × pea   ⟶ all walnut
Cross 4: Using F_1 walnut from Cross 3,
         walnut × walnut   ⟶ 93 walnut
                                   28 rose
                                   32 pea
                                   10 single

At first glance, this problem may appear quite difficult. The approach used in solving it must involve two steps. First, carefully analyze the data for any useful information. Once you determine something concrete, follow an empirical approach; that is, formulate a hypothesis and, in a sense, test it against the given data. Look for a pattern of inheritance that is consistent with all cases.

1 In this problem there are two immediately useful facts. First, P_1 singles breed true. Second, while P_1 walnut breeds true (Cross 2), a walnut phenotype is also produced in crosses between rose and pea (Cross 3). When these F_1 walnuts are crossed together (Cross 4), all four comb shapes are produced in a ratio that approximates a $9:3:3:1$. This observation should immediately suggest a cross involving two gene pairs because this ratio is identical to Mendel's dihybrid crosses. Since only one trait is involved (comb shape), epistasis must be occurring. This may serve as a working hypothesis, and we must now propose how the two gene pairs interact to produce each phenotype.

2 If we call the allele pairs *A, a* and *B, b*, we might predict that since walnut represents 9/16 in Cross 4, $A-B-$ will produce walnut. We might also hypothesize that in the case of Cross 2, the genotypes were $AABB \times AABB$ where walnut was seen to breed true. (Recall that $A-$ and $B-$ mean *AA* or *Aa* and *BB* or *Bb*, respectively.)

3 Since single is the phenotype representing 1/16 of the offspring of Cross 4, we could predict that this phenotype is the result of the *aabb* genotype. This is consistent with Cross 1.

4 Now we have only to determine the genotypes for rose and pea. The most logical prediction would be that at least one dominant *A* or *B* allele combined with the double recessive condition of the other allele pair can account for these phenotypes. For example,

$$A-bb \rightarrow \text{rose}$$
$$aaB- \rightarrow \text{pea}$$

If, in Cross 3, *AAbb* (rose) were crossed with *aaBB* (pea), all offspring would be *AaBb* (walnut). This is consistent with the data, and we must now only look at

Cross 4. We predict these walnut genotypes to be *AaBb* (as above), and from the cross

$$\frac{AaBb}{\text{(walnut)}} \times \frac{AaBb}{\text{(walnut)}}$$

we expect

9/16 $A - B -$ (walnut)
3/16 $A - bb$ (rose)
3/16 $aaB -$ (pea)
1/16 $aabb$ (single)

Our prediction is consistent with the information we were given. The initial hypothesis of the epistatic interaction of two gene pairs proves consistent throughout, and the problem has been solved.

This example illustrates the need to have a basic theoretical knowledge of transmission genetics. Then, you must search for the appropriate clues so that you can proceed in a stepwise fashion toward a solution. Mastering problem solving requires practice, but provides a great deal of satisfaction.

production of a nonfunctional enzyme. Flies that are double mutants and thus homozygous for both *brown* and *scarlet* lack both functional enzymes and can make neither of the pigments; they represent the novel white-eyed flies appearing in 1/16 of the F_2 generation.

Other Modified Dihybrid Ratios

The remaining cases (5–8) in Figure 4.11 illustrate additional modifications of the dihybrid ratio and provide still other examples of gene interactions. All cases (1–8) have two things in common. First in arriving at a suitable explanation of the inheritance pattern of each one, we have not violated the principles of segregation and independent assortment. Therefore, the added complexity of inheritance in these examples does not detract from the validity of Mendel's conclusions. Second, the F_2 phenotypic ratio in each example has been expressed in sixteenths. When a similar observation is made in crosses where the inheritance pattern is unknown, it suggests to geneticists that two gene pairs are controlling the observed phenotypes. You should make the same inference in the analysis of genetics problems. Other insights into solving genetics problems are provided in the boxed discussion on pages 74–76.

SUMMARY AND CONCLUSIONS

Since Mendel's postulates were accepted as fundamental laws of transmission genetics, many exceptions and extensions of modes of inheritance have been discovered.

In this chapter, we described the potential functions of alleles and outlined several systems of sym-

bols to denote them. Then we considered those modifications where single gene pairs influence the expression of single traits. For example, two alleles may exhibit incomplete dominance or codominance rather than the more classical dominant/recessive relationship. Or, many alleles may exist at a single locus. This concept of multiple alleles is a population phenomenon. Regardless of the number of alleles available at any single locus in a population, a diploid organism may have at most only two different ones. The ABO and Rh blood type systems and the major complex controlling histocompatibility during tissue transplantation are examples of multiple allelism in the human population. The *white* locus in *Drosophila* is also an excellent example of multiple allelism.

We have also considered lethal genes and crosses involving the inheritance of two different characters, each controlled by a modified mode of gene expression. When two different modes of inheritance are considered simultaneously, complex modifications of Mendel's classical 9:3:3:1 dihybrid ratio occur.

In the final part of this chapter, we considered the concept of gene interaction. Where two or more genes interact to influence a common phenotype, we found Mendel's classic dihybrid and trihybrid F_2 ratios to be modified. There are several general categories of gene interaction. In epistasis, alleles of one gene pair may mask or modify the expression of those of a second gene pair. In a second category, the combination of alleles from two different genes may result in novel phenotypes. Or, two genes may be part of two separate biochemical pathways, the products of which influence the same character.

We have concluded this chapter by examining approaches useful in arriving at solutions to genetics problems.

PROBLEMS AND DISCUSSION QUESTIONS

1 In shorthorn cattle, coat color may be red, white, or roan. Roan is an intermediate phenotype expressed as a mixture of red and white hairs. The following data were obtained from various crosses:

red × red   ⟶ all red
white × white ⟶ all white
red × white   ⟶ all roan
roan × roan   ⟶ 1/4 red: 1/2 roan: 1/4 white

How is coat color inherited? What are the genotypes of parents and offspring for each cross?

2 Contrast incomplete dominance and codominance.

3 In foxes, two alleles of a single gene, *P* and *p*, may result in lethality (*PP*), platinum coat (*Pp*), or silver coat (*pp*). What ratio is obtained when platinum foxes are interbred? Is the *P* allele behaving dominantly or recessively in causing lethality? in causing platinum coat color?

4 In mice, a short-tailed mutant was discovered. When it was crossed to a normal long-tailed mouse, 4 offspring were short-tailed and 3 were long-tailed. Two short-tailed mice from the F_1 generation were selected and crossed. They produced 6 short-tailed and 3 long-tailed mice. These genetic experiments were repeated three times with approximately the same results. What genetic ratios are illustrated? Hypothesize the mode of inheritance and diagram the crosses.

5 For the four major HLA loci in humans, assume that there are 10 *A* alleles, 35 *B* alleles, 20 *C* alleles, and 30 *D* alleles. How many different distinct haplotypes can be formed? If every individual can have any combination of two haplotypes, how many different diploid HLA genotypes are possible?

6 List all possible genotypes for the A, B, AB, and O phenotypes. Is the mode of inheritance of the ABO blood types representative of dominance? of recessiveness? of codominance?

7 For a gene coding for a protein consisting of 100 amino acids, what do you think is the upper theoretical limit for the number of alleles at a single locus in a very large population?

8 In a disputed parentage case, the child is blood type O while the mother is blood type A. What blood type would exclude a male from being the father? Would the other blood types prove that a particular male was the father?

9 The A and B antigens in humans may be found in water-soluble form in secretions, including saliva, of some individuals but not in others. The population thus contains "secretors" and "nonsecretors." The following inheritance patterns have established that this trait is inherited:

secretor × secretor	⟶ all secretors
nonsecretor × nonsecretor	⟶ all nonsecretors
secretor × nonsecretor	⟶ all secretors
F_1 secretors	⟶ F_2 = 3/4 secretors:
	1/4 nonsecretors

How is the trait inherited? Determine the genotypes of all parents and offspring.

10 In rabbits, a series of multiple alleles controls coat color in the following way: *C* is dominant to all other alleles and causes full color. The chinchilla phenotype is due to the c^{ch} allele, which is dominant to all alleles other than *C*. The c^h allele, dominant only to c^a (albino), results in the Himalayan coat color. Thus, the order of dominance is $C > c^{ch} > c^h > c^a$. For each of the three cases below, the phenotypes of the P_1 generations of two crosses are shown, as well as the phenotype of one member of the F_1 generation. For each case, determine the genotypes of the P_1 generation and the F_1 offspring and predict the results of making each cross between F_1 individuals as shown.

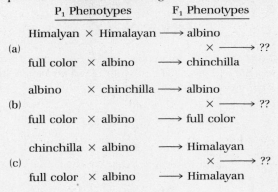

11 In the guinea pig, one locus involved in the control of coat color may be occupied by any of four alleles: *C* (black), c^k (sepia), c^d (cream), or c^a (albino). Like coat color in rabbits (Problem 10), an order of dominance exists: $C > c^k > c^d > c^a$. In the following crosses, write the parental genotypes and predict the phenotypic ratios that would result.
(a) sepia × cream, where both guinea pigs had an albino parent

    (b) sepia × cream, where the sepia guinea pig had an albino parent and the cream guinea pig had two sepia parents

    (c) sepia × cream, where the sepia guinea pig had two full color parents and the cream guinea pig had two sepia parents

    (d) sepia × cream, where the sepia guinea pig had a full color parent and an albino parent and the cream guinea pig had two full color parents.

12   Red flowers result from the genotype *WW*, white flowers from *ww*, and pink flowers from *Ww*. The flower shape will be straight if the genotype *CC* or *Cc* is present, while a genotype of *cc* causes the flowers to curl. What phenotypic ratio will result from a cross between two plants with pink flowers, both of which are heterozygously straight-shaped (*Cc*)?

13   In the snapdragon flower color may be *red, white,* or *pink.* Flower shape may be *personate* or *peloric.* From the data of the following hypothetical crosses, determine how flower color and shape are inherited.

    (a) red peloric × white personate ⟶ F_1 = all pink personate

    (b) red personate × white peloric ⟶ F_1 = all pink personate

    (c) pink personate × red peloric   ⟶ F_1 = 1/4 red personate

                                                   1/4 red peloric

                                                 1/4 pink personate

                                                 1/4 pink peloric

    (d) pink personate × white peloric ⟶ F_1 = 1/4 white personate

                                                 1/4 white peloric

                                               1/4 pink personate

                                                 1/4 pink peloric

    What phenotypic ratios would result from crossing the F_1 of (a) to the F_1 of (b)?

14   With reference to the eye color phenotypes produced by the recessive, autosomal, unlinked *brown* and *scarlet* loci in *Drosophila*, predict the F_1 and F_2 results of the following P_1 crosses. Recall that when both the *brown* and *scarlet* alleles are homozygous, no pigment is produced, and the eyes are white.

    (a) wild type × white

    (b) wild type × scarlet

    (c) brown × white

15   As discussed in the text, pigment in the mouse is only produced when the *C* allele is present. Individuals of the *cc* genotype have no color. If color is present, it may be determined by the *A, a* alleles. *AA* or *Aa* results in gray (agouti) color, while *aa* results in black coats.

    (a) What F_1 and F_2 genotypic and phenotypic ratios are obtained from a cross between *AACC* and *aacc* mice?

    (b) In three crosses between gray females whose genotypes were unknown and males of the *aacc* genotype, the following phenotypic ratios were obtained:

        (1) 8 gray   (2) 9 gray   (3) 4 gray

             8 colorless   10 black   5 black

                                                          10 colorless

    What are the genotypes of these female parents?

16   In some plants a red pigment, cyanidin, is synthesized from a colorless precursor. The addition of a hydroxyl group (OH$^-$) to the cyanidin molecule causes it to become purple. In a cross between two randomly selected purple plants, the following results were obtained:

                               94 purple

                               31 red

                               43 colorless

    How many genes are involved in the determination of these flower colors? Which genotypic combinations produce which phenotypes? Diagram the purple × purple cross.

17 In rats, the following genotypes of two independently assorting autosomal genes determine coat color:

$$A-B- \quad \text{(grey)}$$
$$A-bb \quad \text{(yellow)}$$
$$aaB- \quad \text{(black)}$$
$$aabb \quad \text{(cream)}$$

A third gene pair on a separate autosome determines whether any color will be produced. The CC and Cc genotypes allow color according to the expression of the A and B alleles. However, the cc genotype results in albino rats regardless of the A and B alleles present. Determine the F_1 phenotypic ratio of the following crosses:

(a) $AAbbCC \times aaBBcc$
(b) $AaBBCC \times AABbcc$
(c) $AaBbCc \times AaBbcc$
(d) $AaBBCc \times AaBBCc$
(e) $AABbCc \times AABbcc$

18 Given the inheritance pattern of coat color in rats, as described in Problem 17, predict the genotype and phenotype of the parents who produced the following F_1 offspring:

(a) 9/16 grey: 3/16 yellow: 3/16 black: 1/16 cream
(b) 9/16 grey: 3/16 yellow: 4/16 albino
(c) 27/64 grey: 16/64 albino: 9/64 yellow: 9/64 black: 3/64 cream
(d) 3/8 black: 3/8 cream: 2/8 albino
(e) 3/8 black: 4/8 albino: 1/8 cream

19 In dogs, true-breeding white strains produce all white in the F_1 when crossed to true-breeding brown strains. In numerous F_2 data, the following results were obtained: 37 white, 17 brown, 11 black. Explain how these coat colors are inherited.

20 As a result of your genetic analysis of Problem 19, consider the following question: Two dogs of different color, but neither of which was white, were mated. All of their offspring were white. What are the genotypes and phenotypes of the parents?

21 Body color in the flour beetle may be sooty, red, jet, or black. Can you devise a mode of inheritance such that a cross between two double heterozygous sooty individuals produces an F_2 in a ratio of 6/16 sooty: 3/16 red: 3/16 jet: 4/16 black?

22 Consider the genes and alleles involved in the Bombay phenotype and the ABO blood groups. How would you describe the genetic interaction between these alleles?

23 What types of offspring (and in what ratio) can be expected from two parents of blood type AB who are both heterozygous at the H locus?

24 Shown below are five maternal and paternal phenotypes (1–5), each designating the ABO, MN, and Rh blood group antigens. Each combination resulted in one of the five offspring shown to the right (a–e). Arrange the offspring with the correct parents such that all five cases are consistent. Is there more than one set of correct answers?

Parental Phenotypes	Offspring
(1) A, M, Rh^- $\times$ A, N, Rh^-	(a) A, N, Rh^-
(2) B, M, Rh^- $\times$ B, M, Rh^+	(b) O, N, Rh^+
(3) O, N, Rh^+ $\times$ B, N, Rh^+	(c) O, MN, Rh^-
(4) AB, M, Rh^- $\times$ O, N, Rh^+	(d) B, M, Rh^+
(5) AB, MN, Rh^- $\times$ AB, MN, Rh^-	(e) B, MN, Rh^+

25 An invading alien geneticist from a planet where genetic research was prohibited brought with him to earth two pure-breeding lines of pet frogs. One line croaked by uttering rib-it rib-it, and had purple eyes. The other line croaked more softly by muttering knee-deep knee-deep, and had green eyes. With a new-found sense of inquiry, he mated the two types of frogs. In the F_1, all frogs had blue eyes and uttered rib-it rib-it. He proceeded to make many $F_1 \times F_1$ crosses, and when he fully analyzed the F_2 data, he realized they could be reduced to the following ratio:

27/64 blue-eyed, rib-it utterer
12/64 green-eyed, rib-it utterer
9/64 blue-eyed, knee-deep mutterer
9/64 purple-eyed, rib-it utterer
4/64 green-eyed, knee-deep mutterer
3/64 purple-eyed, knee-deep mutterer

(a) How many total gene pairs are involved in the inheritance of both traits? Why?
(b) Of these, how many are controlling eye color? How can you tell? How many are controlling croaking?
(c) Assign gene symbols for all phenotypes and indicate the genotypes of the P_1 and F_1 frogs.
(d) Indicate the genotypes of the six F_2 phenotypes.
(e) After years of experiments, the geneticist isolated pure-breeding strains of all six F_2 phenotypes. Indicate the F_1 and F_2 phenotypic ratios of the following crosses using these pure breeding strains:

blue-eyed, knee-deep mutterer × purple-eyed, rib-it utterer

(f) One set of crosses with his true-breeding lines initially caused the geneticist some confusion. When he crossed true-breeding purple-eyed, knee-deep mutterers with true-breeding green-eyed, knee-deep mutterers, he often got different results. In some matings, all offspring were blue-eyed, knee-deep mutterers but in other matings, all offspring were purple-eyed, knee-deep mutterers. In still a third mating, 1/2 blue-eyed, knee-deep mutterers and 1/2 purple-eyed, knee-deep mutterers were observed. Explain why the results differed.
(g) In another experiment, the geneticist crossed two purple-eyed, rib-it utterers together with the results shown below. What were the genotypes of the two parents?

9/16 purple-eyed, rib-it utterer
3/16 purple-eyed, knee-deep mutterer
3/16 green-eyed, rib-it utterer
1/16 green-eyed, knee-deep mutterer

SELECTED READINGS

BACH, F. H., and VAN ROOD, J. J. 1976. The major histocompatibility complex—Genetics and biology. *New Eng. J. Med.* 295: 806–13.

BRINK, R. A., ed. 1967. *Heritage from Mendel.* Madison: University of Wisconsin Press.

CLARKE, C. A. 1968. The prevention of "Rhesus" babies. *Scient. Amer.* (Nov.) 219: 46–52.

CORWIN, H.O., and JENKINS, J. B. 1976. *Conceptual foundations of genetics: Selected readings.* Boston: Houghton-Mifflin.

CROW, J. F. 1966. *Genetics notes.* 6th ed. Minneapolis: Burgess.

DAUSSET, J. 1981. The major histocompatibility complex in man. *Science* 213: 1469–74.

DUNN, L. C. 1966. *A short history of genetics.* New York: McGraw-Hill.

EBRINGER, A. 1978. Genes and disease. *New Scient.*, 79: 865–67.

FAIRD, N. R., ed. 1981. *HLA in endocrine and metabolic disorders.* New York: Academic Press.

FOSTER, H. L., et al., eds. 1981. *The mouse in biomedical research. Vol. 1: History, genetics, and wild mice.* New York: Academic Press.

FOSTER, M. 1965. Mammalian pigment genetics. *Adv. in Genet.* 13: 311–39.

HOOD, L., STEINMETZ, M., and GOODENOW, R. 1982. Genes of the major histocompatibility complex. *Cell* 28: 685–87.

JENKINS, J. B. 1983. *Human genetics.* Menlo Park: Benjamin/Cummings.

KLEIN, J., et al. 1981. The traditional and a new version of the mouse H-2 complex. *Nature* 291: 455–60.

LEVITAN, M., and MONTAGU, A. 1977. *Textbook of human genetics.* 2nd ed. New York: Oxford University Press.

LINDSLEY, D. C., and GRELL, E. H. 1967. *Genetic variations of* Drosophila melanogaster. Washington, D.C.: Carnegie Institute of Washington.

McDEVITT, H. O., and BODMER, W. F. 1974. HLA, immune response genes and disease. *Lancet* 1: 1269–75.

NOLTE, D. J. 1959. The eye-pigmentary system of *Drosophila. Heredity* 13: 233–41.

PAWELEK, J. M., and KÖRNER, A. M. 1982. The biosynthesis of mammalian melanin. *Amer. Scient.* 70: 136–45.

PLOEGH, H. L., ORR, H. T., and STROMINGER, J. L. 1981. Major histocompatibility antigens: The human (HLA-A,-B,-C) and murine (H2-K, H-2D) class I molecules. *Cell* 24: 287–99.

RACE, R. R., and SANGER, R. 1975. *Blood groups in man.* 6th ed. Oxford, England: Blackwell Scientific Publishers.

RANSON, R., ed. 1982. *A handbook of* Drosophila *development.* New York: Elsevier Biomedical Press.

REILLY, P. 1981. HLA tests in the courts. *Amer. J. Hum. Genet.* 33: 1007–8.

RYDER, L. P., SVEJGAARD, A., and DAUSSET, J. 1981. Genetics of HLA disease association. *Ann. Rev. Genet.* 15: 169–88.

SINGAL, D. P., et al. 1965. Serotyping for homotransplantation. *Transplantation* 7: 246–58.

SNELL, G. D. 1981. Studies in histocompatibility. *Science* 213: 172–78.

STERN, C. 1973. *Principles of human genetics.* 3rd ed. San Francisco: W. H. Freeman.

THOMPSON, J. S., and THOMPSON, M. W. 1980. *Genetics and medicine.* 3rd ed. Philadelphia: W. B. Saunders.

VOELLER, B. R., ed. 1968. *The chromosome theory of inheritance—Classic papers in development and heredity.* New York: Appleton-Century-Crofts.

VOGEL, F., and MOTULSKY, A. G. 1979. *Human genetics: Problems and approaches.* New York: Springer-Verlag.

WATKINS, M. W. 1966. Blood group substances. *Science* 152: 172–81.

WIENER, A. S., ed. 1970. *Advances in blood groupings, III.* New York: Grune and Stratton.

YOSHIDA, A. 1982. Biochemical genetics of the human blood group ABO system. *Amer. J. Human Genet.* 34: 1–14.

ZIEGLER, I. 1961. Genetic aspects of ommochrome and pterin pigments. *Adv. in Genet.* 10: 349–403.

5

Sex Determination, Sex Differentiation, and Sex Linkage

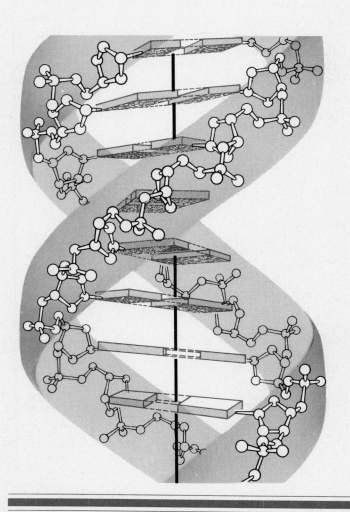

In the previous chapter, we emphasized two broad categories of genetic phenomena that are important in the production of phenotypic variability: allelic and gene interactions. The orderly transmission of the responsible genetic units from parents to offspring relies on the processes of segregation and independent assortment occurring during meiosis. The meiotic process also results in the production of haploid gametes so that following fertilization the resulting offspring have the diploid number of chromosomes characteristic of their species. Thus, meiosis ensures genetic constancy within members of the same species.

These events, which are involved in the perpetuation of all sexually reproducing organisms, depend ultimately on an efficient union of gametes during fertilization. Highly effective fertilization, in turn, depends on some form of sexual differentiation in organisms. This differentiation occurs as low on the evolutionary scale as bacteria and single-celled eukaryotic algae. In higher forms of life, the differentiation of the sexes is often apparent as phenotypic dimorphism in the male and female members of each species. The shield and spear (♂), the ancient symbols of iron and Mars, and the mirror (♀), the symbol of copper and Venus, represent the maleness and femaleness so acquired by individuals.

In this chapter we will first review several representative modes of **sexual differentiation** by examining the life cycle of three organisms often studied in genetics: the green alga, *Chlamydomonas;* the bread mold, *Neurospora;* and the corn plant, *Zea mays.* Then, we will investigate what is known about the genetic basis for the **determination of sexual differences.** While dissimilar, or heteromorphic, chromosomes such as the X-Y pair are often involved, they are not always the direct basis of sex determination. Following this discussion we will turn to the direct influence of sexual differentiation and sex chromosomes on patterns of inheritance and genetic expression. The topics of sex linkage, as well as sex-limited and sex-influenced inheritance, will be discussed. Sex linkage, the inheritance of traits under the control of genes located on the X chromosome, is another example of neo-Mendelian genetics. This section of the chapter thus extends the discussion of this topic initiated in Chapter 4. In the course of the chapter we will also consider numerous related topics involving humans specifically and mammals, more generally.

REPRESENTATIVE MODES OF SEXUAL DIFFERENTIATION

In the biological world, a wide range of reproductive modes is recognized. Asexual organisms are those for which no evidence of sexual reproduction is known. Other species alternate between short periods of sexual reproduction and prolonged periods of asexual reproduction. In most diploid eukaryotes, however, sexual reproduction is the only natural mechanism resulting in new members of a species.

In multicellular organisms, it is important to distinguish between **primary sexual differentiation,** which involves only the gonads where gametes are produced, and **secondary sexual differentiation,** which involves other organs, such as mammary glands and external genitalia. In plants and animals, the terms **unisexual, dioecious,** and **gonochoric** are equivalent; they all refer to an individual containing only male or female reproductive organs. Conversely, the terms **bisexual, monoecious,** and **hermaphroditic** refer to individuals containing both male and female reproductive organs, a common occurrence in both the plant and animal kingdoms. This latter group of organisms can produce fertile gametes of both sexes. The term **intersex** is usually reserved for individuals of intermediate sexual differentiation, who are most often sterile.

FIGURE 5.1
The life cycle of
Chlamydomonas.

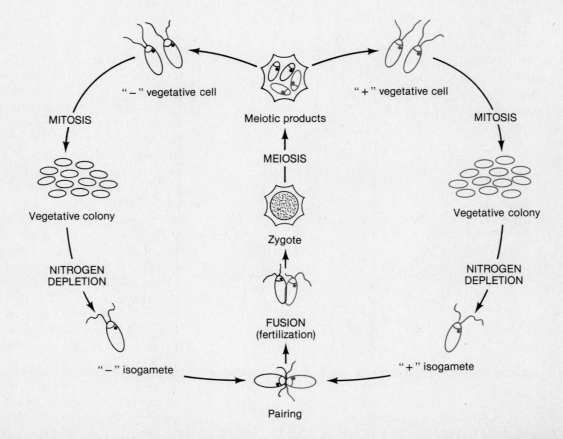

"–" vegetative cell   Meiotic products   "+" vegetative cell

MITOSIS   MEIOSIS   MITOSIS

Vegetative colony   Zygote   Vegetative colony

NITROGEN DEPLETION   FUSION (fertilization)   NITROGEN DEPLETION

"–" isogamete   Pairing   "+" isogamete

Chlamydomonas

The life cycle of the green alga *Chlamydomonas*, shown in Figure 5.1, is representative of organisms that exhibit only infrequent periods of sexual reproduction. Such organisms spend most of their life cycle in the haploid phase. Asexually reproducing organisms produce daughter cells by mitotic divisions. However, under unfavorable nutrient conditions, such as nitrogen depletion, certain daughter cells function as gametes. Following fertilization, a diploid zygote, which can withstand the unfavorable environment, is formed. When conditions become more suitable, meiosis ensues and haploid vegetative cells are produced.

In species such as these, there appears to be little visible difference between the haploid vegetative cells that reproduce asexually and the haploid gametes that are involved in sexual reproduction. The two gametes that fuse together during mating are not usually morphologically distinguishable. Such gametes are called **isogametes,** and species producing them are said to be **isogamous.**

In 1954, Ruth Sager and Sam Granik demonstrated that gametes in *Chlamydomonas* could be subdivided into two mating types. Producing clones derived from single haploid cells, they showed that cells from a given clone would mate with cells from some other clones but not all of them. When they tested mating abilities of large numbers of clones, all could be placed into either a **plus (+)** or a **minus (−)** mating category. Plus cells would only mate with minus cells, and vice versa, as represented in Figure 5.2. Following fertilization and meiosis, the four haploid cells (**zoospores**) produced were found to consist of two plus types and two minus types. Therefore, in this alga, a primitive means of sex differentiation exists even though there is no morphological indication that such differentiation has occurred.

If chemical extracts are prepared from cloned *Chlamydomonas* cells or their flagella, and these extracts are then added to cells of the opposite mating type, clumping or agglutination occurs. This observation suggests that despite the morphological similarity between isogametes, a chemical differentiation has occurred between them, influencing sex determination.

Neurospora

A more complex form of sexual differentiation occurs in Ascomycetes. One ascomycete, the bread mold *Neurospora crassa*, has been used in numerous biochemical and genetic studies. Its general life cycle is illustrated in Figure 5.3. The predominant vegetative phase is represented by a spongy mass, the **mycelium.** The mycelium is composed of intertwined branching filaments called **hyphae.** The nuclei found within the segmented hyphae are haploid. Most reproduction is asexual and occurs in one of two ways. First, hyphae may grow by mitosis and fragment, increasing the mass of the mycelium. Second, special haploid spores called **conidia** may be formed mitotically. These spores may become airborne, and when present on a suitable medium, such as bread, may originate new mycelia through mitosis.

Two mating types, designated *A* and *a*, have been recognized in *Neurospora*. Under proper nutrient conditions, mycelial masses of either type will form fruiting bodies, or **protoperithecia.** Within the fruiting body, a female structure (the **ascogonium**) develops. Extending out of the fruiting body are receptive hyphae called **trichogynes.** Mating occurs when a conidium of the opposite mating type contacts a trichogyne. When this occurs, multiple fertilization events soon follow. The conidial nucleus enters the trichogyne, dividing mitotically as it moves inward. These conidial nuclei fuse with the ascogonial cells. This female structure buds off numerous hyphae or proasci containing binucleate cells of opposite mating types. These nuclei then fuse to form a diploid zygote, and the cell containing it enlarges, forming a saclike structure called the **ascus.** Within one fruiting body, many such structures are formed.

The zygote immediately undergoes meiosis, and each of the four haploid products divides mitotically, yielding eight **ascospores.** The meiotic and mitotic products are retained in the ascus in the precise linear order of their formation. Thus, each half of the ascus contains separate products resulting from the first meiotic division. This observation has been very useful in examining crossover products because each asco-

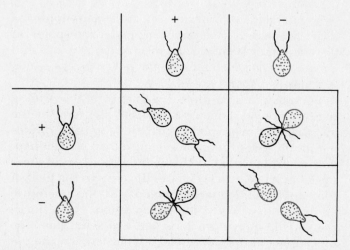

FIGURE 5.2
Mating types in *Chlamydomonas*.

FIGURE 5.3
Sexual differentiation
resulting in the production
of type *A* and type *a*
ascospores during the life
cycle of *Neurospora crassa*.

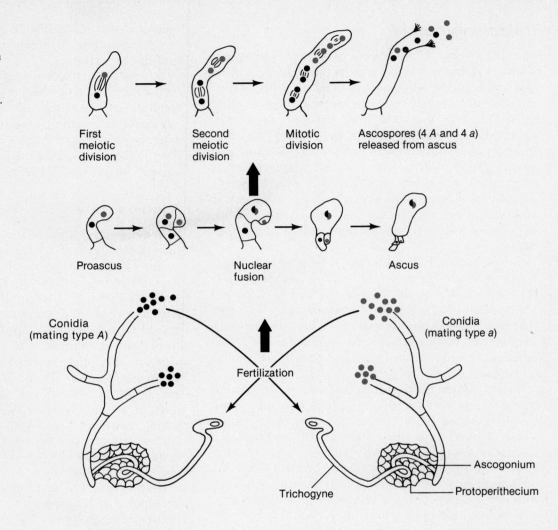

First meiotic division

Second meiotic division

Mitotic division

Ascospores (4 *A* and 4 *a*) released from ascus

Proascus

Nuclear fusion

Ascus

Conidia (mating type *A*)

Conidia (mating type *a*)

Fertilization

Trichogyne

Ascogonium

Protoperithecium

spore can be dissected from the ascus and germinated experimentally (see Chapter 6). The haploid ascospores are released from the ascus and may form new mycelia through mitosis and branching.

Like *Chlamydomonas*, the diploid phase of *Neurospora* is short lived. The existence of distinct mating types ensures that sexual reproduction does not occur between conidia derived from the same fruiting body or mycelium. This restriction promotes greater genetic variability through sexual reproduction.

Zea mays

In many plants, life cycles alternate between the haploid **gametophyte stage** and the diploid **sporophyte stage.** The processes of meiosis and fertilization link the two phases during the life cycle. The relative amount of time spent in the two phases varies between the major plant groups. In some nonseed plants, such as mosses, the haploid gametophyte phase and the morphological structures representing

this stage predominate. The reverse is true in all seed plants.

Maize (*Zea mays*) is an example of a monoecious seed plant that bears both male and female structures on a single diploid **sporophyte.** The life cycle of *Zea mays* is illustrated in Figure 5.4. **Microgametophytes** arise by meiosis from small haploid microspores in the **stamen** or **tassels.** Each microspore develops into a mature male gametophyte—the pollen grain—which contains two sperm nuclei.

Equivalent female diploid cells exist in the **pistil** of the sporophyte. Following meiosis, only one of the four haploid megaspores survives. It usually divides mitotically three times, producing a total of eight haploid nuclei enclosed in the embryo sac. Two of these nuclei unite near the center of the embryo sac to form the diploid **endosperm nucleus.** At the micropyle end of the sac where the sperm enters, three nuclei remain: the **oocyte nucleus** and two **synergids.** The other three **antipodal nuclei** are clustered at the opposite end of the embryo sac.

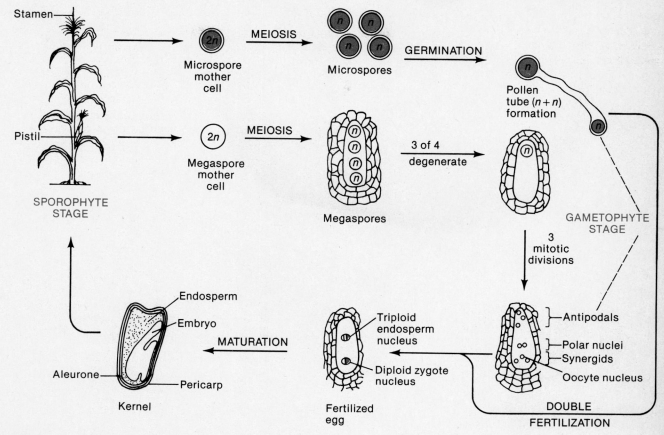

FIGURE 5.4
The life cycle of maize *(Zea mays).*

Pollination occurs when pollen grains make contact with the silks (or *stigma*) of the pistil and develop extensive pollen tubes that grow toward the embryo sac. When contact is made at the micropyle, the two sperm nuclei enter the embryo sac. **Double fertilization** occurs when one sperm nucleus unites with the haploid oocyte nucleus and the other sperm nucleus unites with two endosperm nuclei. This process results in a diploid zygote and a triploid endosperm, respectively. Each ear of corn may contain as many as 1000 such structures, each of which develops into a single kernel. Although the majority of kernels serve as nutrition for animal populations, including human beings, each structure, if allowed to germinate, will give rise to a new sporophyte.

The mechanism of sex determination and differentiation in a monoecious plant like *Zea mays*, where the tissues that form both male and female gametes are of the same genetic constitution, was difficult to comprehend at first. However, the discovery of a large number of mutant genes that disrupt normal tassel and pistil formation supports the concept that gene products

play an important role in sex determination and differentiation. Table 5.1 lists eleven mutations representing separate genes on a number of chromosomes of *Zea mays*. These mutant genes affect the differentiation of male or female tissue in several ways.

Those mutant genes that cause a sex reversal provide the most information. All mutations classified as *tassel seed* (*ts*) interfere significantly with tassel production when homozygous and induce the formation of female structures. Thus, it is possible for a single gene to cause a normally monoecious plant to become functionally female. On the other hand, the recessive mutations *silkless* (*sk*) and *barren stalk* (*ba*) interfere with the development of the pistil, resulting in plants with only functional male reproductive organs.

By manipulating these mutations in certain strains of corn, geneticists can produce dioecious, or single sex, strains. A plant of genotype *sk/sk; ts_2/ts_2* has no functional pistil along the stem, but has the tassels converted to pistils. This plant is female. The plant of genotype *sk/sk; ts_2^+/ts_2* also lacks pistils along the stem, but the tassels function normally to produce

TABLE 5.1
Mutants of *Zea mays* affecting the development of the tassel and pistil.

Chromosome	Locus	Mutant Symbol	Mutant Name	Mutant Phenotype
1	24	ts_2	Tassel seed	Tassel completely pistillate
1	119	ts_3	Tassel seed	Tassel with both pistillate and staminate flowers
1	158	ts_6	Tassel seed	Tassel completely pistillate
2	56	*sk*	Silkless	Pistils abort; no silks produced
3	55	ts_4	Silkless	Tassel with pistillate and staminate flowers
3	72	ba_1	Barren stalk	Ear shoots missing
4	56	ts_5	Tassel seed	Tassel with both silks and anthers
6	4	po_1	Polymitotic	Microspores divide rapidly without chromosome division; tassel sterile
6	17	ms_1	Male sterile	Anthers shriveled
8	14	ms_8	Male sterile	No anthers exserted; microsporophytes usually degenerate
9	67	ms_2	Male sterile	No anthers exserted

pollen. This is a male plant. If these two strains are interbred, both parental genotypes are produced, perpetuating the dioecious condition.

Data gathered from studies of these and other mutants suggest that the wild-type alleles of these genes interact in the sex determination of certain cells. They subsequently undergo sexual differentiation into either male or female structures, which in turn produce male or female gametes.

HETEROMORPHIC CHROMOSOMES AND SEX DETERMINATION

The earliest observations directly linking genetics and sex determination occurred around 1900. The research involved the cytological examination of sperm cells derived from insects. Variation in chromosome number between different sperm cells and between male and female somatic cells was observed and correlated with sex differences. From these observations geneticists derived a chromosomal basis for sex determination and the 1:1 sex ratio observed in unisexual species. Hypotheses concerning special **sex chromosomes** and their role in sexual differentiation conformed nicely with the rediscovery early in this century of Mendelian concepts.

Modes of Heterogametic Sex Determination

In 1891, H. Henking identified a nuclear structure, which he labeled the **X-body,** in the sperm of certain insects. Several years later, Clarance McClung showed that some grasshopper sperm contain a genetic structure, which he called a **heterochromosome,** but the remainder do not. He mistakenly associated the presence of the heterochromosome with the production of male progeny. In 1906, Edward B. Wilson clarified Henking and McClung's findings when he demonstrated that female somatic cells in the insect *Protenor* contain 14 chromosomes, including 2 X chromosomes. During oogenesis, an even reduction occurs, producing gametes with 7 chromosomes, including one X. Male somatic cells, on the other hand, contain only 13 chromosomes, including a single X chromosome. During spermatogenesis, gametes are produced containing either 6 or 7 chromosomes, without or with an X, respectively. Fertilization by X-bearing sperm results in female offspring, and fertilization by X-deficient sperm result in male offspring (Figure 5.5).

The presence or absence of the X chromosome in male gametes provides an efficient mechanism for sex determination in this species and also produces a 1:1 sex ratio in the resulting offspring. This mechanism, now called the **XX/XO** or **Protenor mode of sex de-**

FIGURE 5.5
The XX/XO or Protenor mode
of sex determination.

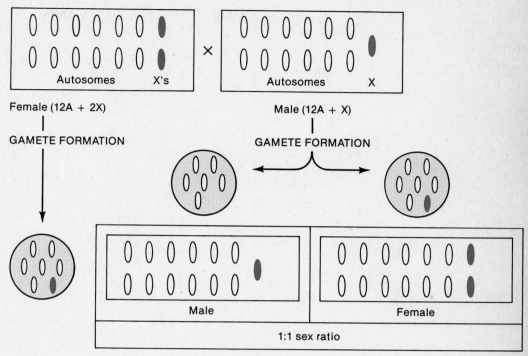

termination, depends on the random distribution of the X chromosome into one-half of the male gametes during segregation.

Wilson also experimented with the hemipteran insect *Lygaeus turicus*, in which both sexes have 14 chromosomes. Twelve of these are autosomes. Additionally, the females have 2 X chromosomes, and the males have only a single X and a smaller heterochromosome labeled the **Y chromosome.** Females in this species produce only gametes of the (6A + X) constitution, but males produce two types of gametes in equal proportions: (6A + X) and (6A + Y). Therefore, following fertilization, equal numbers of male and female progeny will be produced with distinct chromosome complements. This mode of sex determination is called the **Lygaeus** or **XX/XY type.**

The examples presented so far have implied that the male always has unlike chromosomes and produces two types of gametes. In these cases, the male is described as the **heterogametic sex;** the female, who has like chromosomes and produces a uniform gamete with regard to chromosome numbers, is called the **homogametic sex.** However, in many insects and other organisms, the female is the heterogametic sex and exhibits either the Protenor (XX/XO) or Lygaeus (XX/XY) mode of sex determination. To immediately

distinguish situations in which the female is the heterogametic sex, some workers use the notation **ZZ/ZW,** where ZW is the heterogamous female, instead of the XX/XY notation.

Studies have provided strong evidence that the female is the heterogametic sex for many organisms. For example, all moths and butterflies, most birds, some fish, reptiles, amphibians, and at least one species of plants (*Fragaria orientalis*) demonstrate such a pattern.

The situation with fowl illustrates the difficulty in establishing which sex is heterogametic and whether the Protenor or Lygaeus mode is operable. While genetic evidence supported the hypothesis that the female is the heterogametic sex, the cytological identification of the sex chromosomes was not accomplished until 1961. Their identification was difficult to achieve because of the large number of chromosomes (78), about half of which are very small microchromosomes. In each group of similar chromosomes, many pairs are homogenous in appearance. When the sex chromosomes were finally identified, the female was shown to contain an unlike chromosome pair, including a Y (or W) chromosome. Thus, in fowl, the female is indeed heterogametic and is characterized by the Lygaeus type of sex determination.

TABLE 5.2
A comparison of chromosome composition and sex in *Melandrium*.

Chromosome Composition		Sex
Number of Each Autosome	Sex Chromosomes	
2	XX	Normal female
2	XY	Normal male
2	XXX	Female
3	XX	Female
4	XXXX	Female
2	XYY	Male
2	XXY	Male
3	XY	Male
3	XXXY	Male
4	XXXY	Male

SOURCE: After Westergaard, 1953.

Sex Determination in *Melandrium*

Investigations with dioecious species of the angiosperm *Melandrium* have enhanced our knowledge of sex determination in heterogametic organisms. In 1923 it was discovered that the chromosome composition varied between male and female plants of these species. One set of chromosomes, designated X and Y, has been shown to determine maleness or femaleness. The male contains four pairs of autosomes plus an X and a Y chromosome. The Y chromosome in these plants is larger than the X chromosome. Female plants contain the same number of autosomes as the male but have two X chromosomes.

In 1953, Mogens Westergaard correlated abnormal chromosome compositions in plants with the sex of the individual, as shown in Table 5.2. He concluded that the Y chromosome has a strong masculinizing influence because a male plant is always produced when it is present. The X chromosome has a feminizing influence, but this influence is masked by the action of the Y chromosome. He based this conclusion on the observation that XXY and XXXY plants, for example, are male despite the presence of two or three X chromosomes.

Westergaard also investigated plants that contained the normal number of autosomes and one X chromosome, but only fragments of the Y chromosome. These studies allowed him to identify specific sex-determining functions for certain areas of the sex chromosome. Figure 5.6 shows these subdivisions for the sex chromosomes of a male *Melandrium*.

As shown in Figure 5.6, regions I and IV of the Y chromosome suppress female development, somehow counteracting the influence of region V of the X, which promotes female development. If either region I or IV is missing, bisexual development occurs. Region II promotes male development, and if male tissue develops, region III is essential for male fertility. In its absence, male tissue develops, but the plant is sterile. A region of the X chromosome is also designated IV because it has been identified as the only part of the chromosome that synapses with the Y chromosome during meiosis.

FIGURE 5.6
The subdivisions of the X and Y chromosomes of *Melandrium* and the proposed role of each chromosomal area.

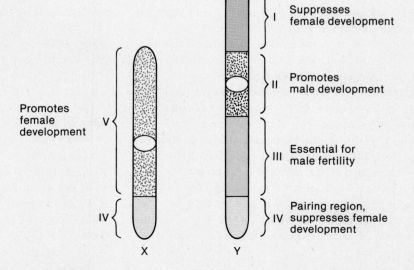

Promotes female development — V

IV

X

I — Suppresses female development

II — Promotes male development

III — Essential for male fertility

IV — Pairing region, suppresses female development

Y

In summary, genes localized on the Y chromosome initiate or trigger the development of male sex organs, and at the same time suppress the expression of the female-determining genetic information localized on the X chromosomes. However, the mechanisms involved in the interaction and expression of genetic information resulting in maleness or femaleness are unknown. It should be noted that these studies do not rule out the possible influence of autosomal genes in the process of sexual differentiation.

CHROMOSOME RATIOS IN *DROSOPHILA:* ANOTHER MODE OF SEX DETERMINATION

Studies of the fruit fly *Drosophila melanogaster* demonstrate yet another mode of sex determination. Cytological evidence reveals that the male *Drosophila* is the heterogametic (XY) sex (Figure 5.7). As in *Melandrium* we might assume that the Y chromosome determines maleness in this species were it not for the elegant work of Calvin Bridges. Bridges' work on sex determination in *Drosophila*, initiated in 1916, can be divided into two phases: a study of offspring resulting from nondisjunction of the X chromosomes in females, and subsequent work with progeny of triploid (3*n*) females.

Nondisjunction is the failure of paired chromosomes to segregate or separate during the anaphase stage of the first or second meiotic divisions. The result of nondisjunction is the production of two abnormal gametes, one of which contains an extra chromosome (*n* + 1) and the other one lacks a chromosome

(*n* − 1). Fertilization of such gametes produces (2*n* + 1) or (2*n* − 1) **aneuploid** zygotes. Bridges studied flies that were the product of gametes formed by nondisjunction of the X chromosome. In addition to the normal complement of six autosomes, these individuals had either an XXY or an XO chromosome composition. (The zero signifies that the second chromosome is absent.) The XXY flies were normal females, and the XO flies were sterile males. The presence of the Y chromosome in the XXY flies did not cause maleness, and its absence in the XO flies did not produce femaleness. From these data Bridges concluded that the Y chromosome in *Drosophila* lacks male-determining factors, but apparently contains genetic information essential to male fertility since the XO males were sterile.

Bridges was able to clarify the mode of sex determination in *Drosophila* by studying the progeny of triploid females (3*n*), which have *three* copies each of the haploid complement of chromosomes. These females apparently originate from unusual diploid eggs fertilized by normal haploid sperm. Triploid females have heavy-set bodies, coarse bristles, and coarse eyes, and may be fertile. During meiosis a wide range of chromosome complements is distributed into gametes, and these gametes give rise to offspring with a variety of abnormal genetic constitutions. A correlation between the sexual morphology, chromosome composition, and Bridges' interpretation is shown in Figure 5.8.

Bridges realized that the critical factor in determining sex is the **ratio of X chromosomes to the number of haploid sets of autosomes present.** Normal (2X:2A) and triploid (3X:3A) females each have a ratio equal to 1.0, and both are fertile. As the ratio exceeds

FIGURE 5.7
The chromosomal composition of male and female *Drosophila melanogaster.*

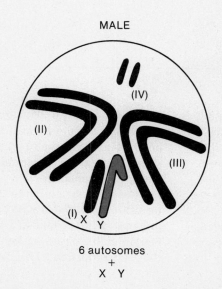

MALE

(IV)

(II)

(III)

(I) X Y

6 autosomes
+
X Y

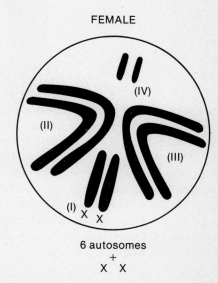

FEMALE

(IV)

(II)

(III)

(I) X X

6 autosomes
+
X X

FIGURE 5.8
Chromosome compositions, the ratios of X chromosomes to sets of autosomes, and the resultant sexes in *Drosophila melanogaster*.

CHROMOSOME COMPOSITION	CHROMO-SOME FORMU-LATION	RATIO OF X CHROMOSOMES TO AUTOSOME SETS	SEX
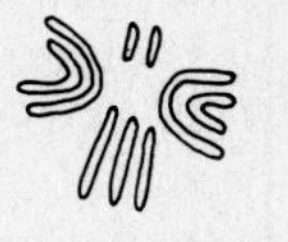	$\frac{3X}{2A}$	1.5	Metafemale
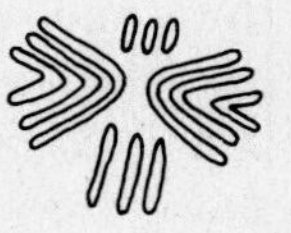	$\frac{3X}{3A}$	1.0	Female
	$\frac{2X}{2A}$	1.0	Female
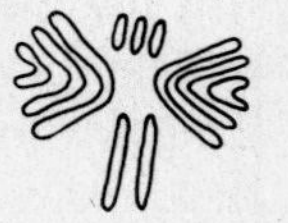	$\frac{2X}{3A}$	0.67	Intersex
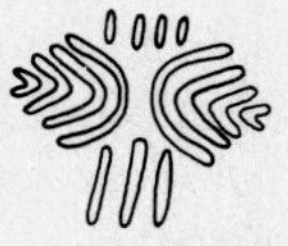	$\frac{3X}{4A}$	0.75	Intersex
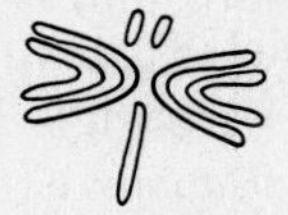	$\frac{X}{2A}$	0.50	Male
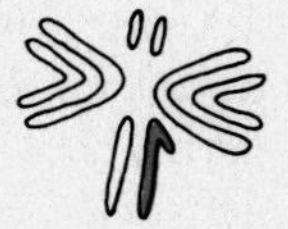	$\frac{XY}{2A}$	0.50	Male
	$\frac{XY}{3A}$	0.33	Metamale

unity (3X:2A, or 1.5, for example), what was originally called a **superfemale** is produced. Since this female is rather weak and infertile and has lowered viability, this type is now more appropriately called a **metafemale**.

Normal (XY:2A) and sterile (X:2A) males each have a ratio of 1:2, or 0.5. When the ratio decreases to 1:3, or 0.33, as in the case of an XY:3A male, infertile **metamales** result. Other flies recovered by Bridges in these studies contained an X:A ratio intermediate be-

tween 0.5 and 1.0. These flies were generally larger, and they exhibited a variety of morphological abnormalities and rudimentary bisexual gonads and genitalia. They were invariably sterile and were designated as **intersexes.**

These results indicate that in *Drosophila* male-determining factors are not localized on the sex chromosomes, but are instead found on the autosomes. Female-determining factors, however, are localized on the X chromosomes. Thus, with respect to primary sex determination, male gametes containing one of each autosome plus a Y chromosome result in male offspring *not* because of the presence of the Y chromosome but because of the lack of an X chromosome— quite a different situation from that previously described in *Melandrium*. This mode of sex determination is explained by the **genic balance theory.** Bridges proposed that a threshold for maleness is reached when the X:A ratio is 1:2 (X:2A), but that the presence of an additional X (XX:2A) alters this balance and produces female differentiation.

Bridges' conclusion received further support from the experiments of Jack Schultz and Theodosius Dobzhansky. They were able to obtain males, females, and intersexes that had gained or lost variously sized fragments of the X chromosome. Some of their results are shown in Table 5.3. The effects were proportional to the extent of the addition or subtraction of the X and were consistent with Bridges' genic balance interpretation of sex determination. This finding led to the conclusion that multiple factors critical to sex determination exist on the X chromosome.

Several mutant genes, which behave in a manner analogous to the sex-determining genes of *Zea mays*, have been located on the autosomes of *Drosophila*. The recessive autosomal gene *transformer* (*tra*), discovered by Alfred H. Sturtevant, is especially interesting. Homozygous mutant females are transformed into sterile males, but normal males are unaffected when homozygous for *tra*. Experiments by Robert C. King and Dietrich Bodenstein showed that the transplantation of normal but immature female gonadal tissue into the abdomens of *tra/tra* sterile males results in normal ovarian development. Thus, it appears that the abdominal environment can support the production of eggs, but that the *transformer* gene has interfered with primary sex determination in these flies.

TABLE 5.3

The effect of the addition or subtraction of variously sized fragments of the *Drosophila* X chromosome.

Chromosome Composition	Sex	Addition or Subtraction	Effect
2X/3A	Intersex	+	Toward female
2X/2A	Female	+	Toward metafemale
2X/3A	Intersex	−	Toward male

GENES ON THE X CHROMOSOME: SEX LINKAGE

By 1920, researchers had gained clear insights into the genetic basis of sex determination. The discovery of X and Y chromosomes is a particularly significant chapter in the story of neo-Mendelian genetics. With the role of these so-called sex chromosomes in sex determination established, research ensued involving the inheritance of traits controlled by genes located on the X and Y. Many questions were open to investigation. Two pertinent ones involved the genetic constitution of these chromosomes:

1 Are the X and Y chromosomes in a species homologous with regard to genetic loci?

2 Are there genes other than those involved in sex determination residing on the X and Y chromosomes?

As we will see in the remainder of this chapter, the answers to these and many other related questions were soon forthcoming.

Sex-Linked Genes in *Drosophila*

The first clue that genes reside on the X chromosome became evident during Thomas H. Morgan's genetic analysis of the *white* mutation in *Drosophila*. Recall from Chapter 4 that the *white* locus may be occupied by any member of a large series of multiple alleles. Morgan's work established that the inheritance pattern of the white-eye trait was clearly related to the sex of the parent carrying the mutant allele. Unlike the outcome of the typical monohybrid cross, Morgan's reciprocal crosses between white- and red-eyed flies did not yield identical results. In contrast, in all of Mendel's monohybrid crosses, F_1 and F_2 data were very similar in reciprocal crosses regardless of which P_1 parent exhibited the recessive mutant trait. Morgan's analysis led to the conclusion that the *white* locus is present on the X chromosome rather than one of the autosomes. As such, both the gene and the trait are said to be **sex linked.**[*]

Results of reciprocal crosses between white-eyed and red-eyed flies are shown in Figure 5.9. The obvious differences in phenotypic ratios in both the F_1 and F_2 generations are dependent on whether the P_1 white-eyed parent was male or female.

---

[*]Traditionally, the term *sex linked* has been used to describe genes located on the X chromosome. Thus, we will use this term. However, **X-linkage** is a more accurate description of this phenomenon.

FIGURE 5.9

The F_1 and F_2 results of T. H. Morgan's reciprocal crosses involving the sex-linked white-eye mutation in *Drosophila melanogaster.*

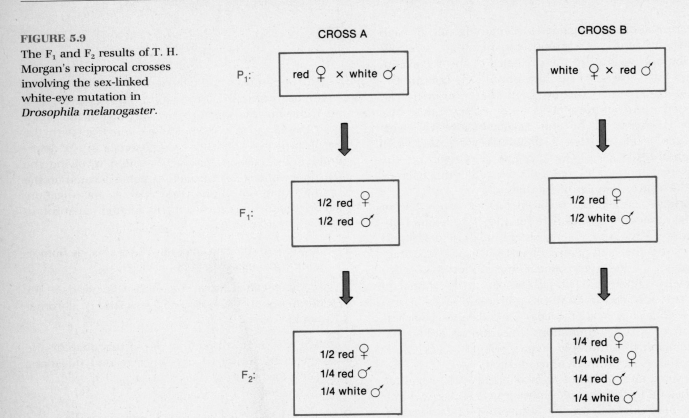

Morgan was able to correlate these observations with the difference found in chromosome composition between male and female *Drosophila*. As we have seen, one of the four pairs of chromosomes varies between sexes. This chromosome pair is involved in the sex determination mechanism and constitutes the **sex chromosomes.** All other chromosomes are called **autosomes.** Females possess two rod-shaped homologues called the X chromosomes and designated XX. Males possess a single X chromosome and a J-shaped Y chromosome, and they are designated XY. The complex chromosome composition of both sexes of *Drosophila* is shown in Figure 5.7.

On the basis of this correlation, the postulate was put forth that the gene for white eye is found on the X chromosome but is absent from the Y chromosome. Females thus have two available genes, one on each X chromosome, while males have only one available gene on their single X chromosome. This explanation supposes that the Y chromosome lacks loci homologous to those on the X chromosome. However, the X and Y chromosomes still behave as homologues in

that they do partially synapse with each other and segregate into gametes during meiosis.

Morgan's interpretation of sex-linked inheritance, shown in Figure 5.10, provides a suitable theoretical explanation for his results. Since the Y chromosome is essentially genetically blank or inert, whatever allele is present on the X chromosome of males will be expressed in the phenotype. Since males cannot be either homozygous or heterozygous for sex-linked genes, this condition is called **hemizygous.** In such cases, no alternative alleles exist, and the concept of dominance and recessiveness has no meaning.

One result of sex linkage is the so-called **crisscross pattern** of inheritance. Phenotypic traits controlled by recessive sex-linked genes are passed from the homogametic sex to the heterogametic sex in each generation. In *Drosophila* and other species where the male is the heterogametic sex, such traits are passed from mother to son. In order for a female to exhibit a recessive trait, both of her X chromosomes must contain the mutant allele. Since all male offspring receive one of their mother's two X chromosomes and are hemi-

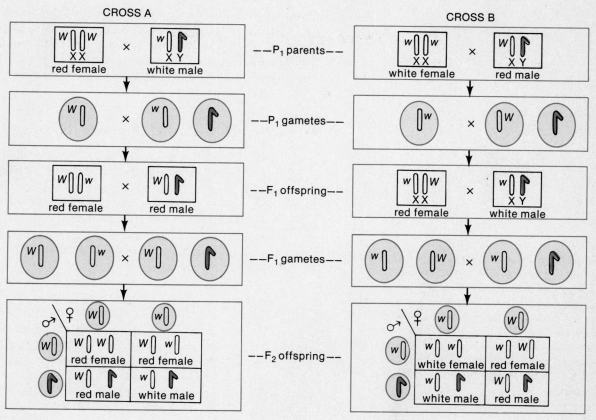

FIGURE 5.10

The chromosomal explanation of the results of the sex-linked crosses shown in Figure 5.9.

zygous for all alleles present on that X, they will express all the recessive sex-linked traits she expresses.

Drosophila Mosaics

The mode of sex determination and our knowledge of sex linkage in *Drosophila* help to explain the appearance of the extremely unusual fruit fly shown in Figure 5.11. This fly was recovered from a stock where all other females were heterozygous for the sex-linked genes *white* eye (*w*) and *miniature* wing (*m*). It is a **bilateral gynandromorph,** which means that one-half has developed as a male and the other as a female. If a zygote heterozygous for *white* eye and *miniature* wing were to lose one of the X chromosomes during the first mitotic division, the two cells would be of the XX and XO constitution, respectively. Thus, one cell would be female and the other would be male.

In the case of the bilateral gynandromorph, each of these cells is responsible for producing all progeny cells that make up either the right half *or* the left half of the body during embryogenesis. The cell of XO constitution apparently produced only identical progeny cells and gave rise to the left half of the fly, which was male. Since the male half demonstrated the *white, miniature* phenotype, the X chromosome bearing the *w+, m+* alleles was lost, while the *w, m*-bearing homologue was retained. All female cells of the right side of the body remained heterozygous for both mutant genes and therefore contained normal eye-wing phenotypes. Depending on the orientation of the spindle during the first mitotic division, gynandromorphs can be produced where the "line" demarcating male versus female development occurs at almost any place along or across the fly's body.

SEX DETERMINATION IN HUMANS

From the time dividing cells of humans were first observed, geneticists tried to determine accurately the chromosome number of our own species. The first significant attempt was made in 1912, when H. von Wini-

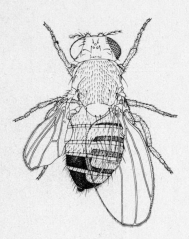

FIGURE 5.11

A bilateral gynandromorph of *Drosophila melanogaster* formed following the loss of one X chromosome in one of the two cells during the first mitotic division. The left side of the fly, composed of male cells containing a single X, expresses the mutant white-eye miniature-wing alleles. The right side is composed of female cells containing two X chromosomes heterozygous for the two recessive alleles. (From *Genetics*, Second Edition, by Robert C. King. Copyright © 1965 by Oxford University Press, Inc. Reprinted by permission.)

warter counted 47 chromosomes in a spermatogonial metaphase. At that time, geneticists believed that the sex-determining mechanism in humans was based on the presence of an extra chromosome in females; that is, females were thought to have 48 chromosomes. In the 1920s, Theophilus Painter observed between 45 and 48 chromosomes in testicular tissue and also discovered the small Y chromosome, which is now known to occur only in males. In his original paper, Painter favored 46 as the diploid number in humans, but he later concluded that 48 was the correct number in both males and females.

For thirty years, 48 was accepted as the correct number of human chromosomes. Then, in 1956, Joe Hin Tjio and Albert Levan introduced improvements in chromosome preparation techniques. These improved techniques led to a strikingly clear demonstration of metaphase cells showing that 46 was indeed the human diploid number. Later that same year, C. E. Ford and John L. Hamerton, also working with testicular tissue, confirmed this finding.

Within the normal 23 pairs of human chromosomes, one pair was shown to vary in configuration in males and females. These two chromosomes were designated the sex chromosomes. The human female has two X chromosomes, and the human male has one X and one Y chromosome. As with *Drosophila*, where XX and XY constitutions also yield females and males, respectively, these observations are insufficient to conclude that the Y chromosome determines maleness.

Klinefelter and Turner Syndromes

Around 1940 it was observed that two human abnormalities, the **Klinefelter** and **Turner syndromes,*** caused aberrant sexual development. The study of these genetic disorders has been instrumental in sex determination studies with humans. Individuals with Klinefelter syndrome [Figure 5.12(a)] have genitalia and internal ducts that are male, but their testes are underdeveloped and fail to produce sperm. Enlarged breast development may occur in some affected individuals, and mental retardation has also been noted. Although masculine development occurs, feminine sexual development is not entirely suppressed.

In Turner syndrome [Figure 5.12(b)], the affected individual has female external genitalia and internal ducts, but the ovaries are rudimentary. Other characteristic abnormalities include short stature (usually under five feet); a short, webbed neck; and a broad, shieldlike chest.

In 1959, the karyotypes of individuals with these syndromes were independently determined to be abnormal with respect to the sex chromosomes. Individuals with Klinefelter syndrome most often have an XXY complement in addition to 44 autosomes. People with this karyotype are designated **47,XXY.** Individuals with Turner syndrome have only 45 chromosomes, including just a single X chromosome, and are designated **45,X.** Both conditions result from nondisjunction of the sex chromosomes during meiosis.

These karyotypes and their corresponding sexual phenotypes allow us to conclude that, as in *Melandrium*, the Y chromosome determines maleness in humans. In its absence, the sex of the individual is female. The presence of the Y chromosome in the individual with Klinefelter syndrome is sufficient to determine maleness, even though its expression is not complete. Similarly, in the absence of a Y chromosome, as in the case of individuals with Turner syndrome, no masculinization occurs.

---

*Although the possessive form of the names of most syndromes (eponyms) is often used (e.g., Klinefelter's), the current preference is to use the nonpossessive form for syndromes, as recommended in McKusick, 1975.

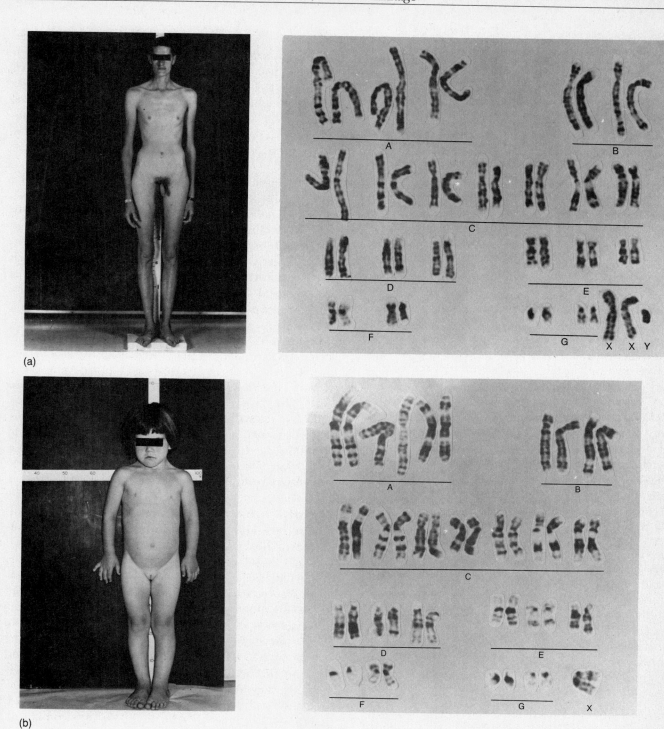

FIGURE 5.12

(a) The karyotype and a photograph of an individual with Klinefelter syndrome. In the karyotype, two X chromosomes and one Y chromosome are visible, creating the 47,XXY condition. (b) The karyotype and a photograph of an individual with Turner syndrome. In the karyotype, only one X chromosome is visible, creating the 45,X condition. [Chromosome preparations in (a) and (b) courtesy of Arthur E. Greene, Institute for Medical Research. Photo (a) by Dr. Richard C. Juberg, The Children's Medical Center, Dayton, Ohio. Photo (b) courtesy of Dr. Stella B. Kontras, Children's Hospital, Columbus, Ohio.]

Klinefelter syndrome occurs in 2 of 1000 male births. The karyotypes **48,XXXY; 48,XXYY; 49,XXXXY;** and **49,XXXYY** are similar phenotypically to **47,XXY.** Generally, more severe manifestations are seen in individuals with a greater number of X chromosomes.

Karyotypes other than 45,X also lead to Turner syndrome. These include mosaic individuals with two apparent cell lines, the most common chromosome combinations being **45,X/46,XY** and **45,X/46,XX.** Turner syndrome is observed in only 1 in 3000 female births, a frequency much lower than that for Klinefelter syndrome. One explanation for this difference is that a substantial majority of **45,X** fetuses die *in utero* and are aborted spontaneously.

47,XXX Syndrome

The presence of three X chromosomes (**47,XXX**) results in female differentiation. This syndrome, which is estimated to occur in 1 of 1200 female births, is highly variable in expression. Frequently, 47,XXX women are perfectly normal. In other cases, underdeveloped secondary sex characteristics, sterility, and mental retardation may occur. In rare instances, **48,XXXX** and **49,XXXXX** karyotypes have been reported. The syndromes associated with these karyotypes are similar to but more pronounced than the 47,XXX. Thus, the presence of additional X chromosomes appears to disrupt the delicate balance of genetic information essential to normal female development.

The XYY Condition

A third human trisomy involving the sex chromosomes has been discovered and intensively investigated. This case involves males with one X and two Y chromosomes in addition to the normal complement of 44 autosomes (**47,XYY**). Many studies have described a characteristic phenotype for these individuals.

In 1965, Patricia Jacobs discovered 9 of 315 males in a Scottish maximum security prison to have the 47,XYY karyotype. These males were significantly above average in height and were involved in criminal acts of serious social consequence. Of the 9, 7 were of subnormal intelligence, and all suffered personality disorders. In several other studies, similar findings were obtained.

Because of these investigations, the phenotype and frequency of the 47,XYY condition in criminal and noncriminal populations have been examined more extensively. Above average height and subnormal intelligence have been generally substantiated, and the frequency of males displaying this karyotype is indeed higher in penal and mental institutions compared with unincarcerated males. Table 5.4 compares the frequency of XYY individuals in newborns with the frequency found in various mental-penal settings. Each value represents data compiled from many individual studies. Note that when men are selected for above average height, the frequency of XYY cases increases dramatically. The precise extent of the association between the XYY karyotype and the increased frequency of such individuals in mental-penal settings is not yet clear.

The possible correlation between chromosome composition and behavior is of considerable interest. Is there a causal relationship between the presence of an extra Y chromosome and antisocial or criminal behavior? There is currently no strong supporting evidence for any such constant association. In fact, most 47,XYY individuals studied have no serious antisocial tendencies. Furthermore, scientists have seriously debated whether long-term study of this question should proceed.

In 1968 a long-term study was initiated to identify 47,XYY individuals at birth and to follow their behavioral patterns during preadult and adult development. By 1974 the two primary investigators, Stanley Walzer and Park Gerald, had identified about 20 XYY newborns in 15,000 births at Boston Hospital for Women. However, they soon came under great pressure to abandon the study. Led by a group called Science for the People, the opponents of the study argued that the investigation could not be justified and might cause great harm to those individuals who displayed this karyotype. They argued that (1) no association between the additional Y chromosome and abnormal behavior had been previously established, and (2) by "labeling" these individuals, a self-fulfilling prophecy might be created. That is, as a result of participation in the study, parents, relatives, and friends might treat those identified as 47,XYY differently, leading to the type of antisocial behavior expected of them. Despite the support of a government funding agency and the faculty at Harvard Medical School, Walzer and Gerald experienced great personal and professional stress. As a result, they abandoned the investigation in 1975.

Sex-Determining Function of the X and Y Chromosomes

The sex chromosome abnormalities just described demonstrate that both the X and Y chromosomes play a critical role in human sex determination. Indeed, the Y chromosome seems most decisive. When it is present, the sexual phenotype is male; and in its absence, female development occurs. The genetic function of the Y chromosome appears to be similar in all mam-

TABLE 5.4
Frequency of XYY individuals in various settings.

Setting	Restriction	Number Studied	Number XYY	Frequency XYY
Control population	Newborns	28,366	29	0.10%
Mental-penal	No height restriction	4,239	82	1.93%
Penal	No height restriction	5,805	26	0.44%
Mental	No height restriction	2,562	8	0.31%
Mental-penal	Height restriction	1,048	48	4.61%
Penal	Height restriction	1,683	31	1.84%
Mental	Height restriction	649	9	1.38%

SOURCE: Compiled from data presented in Hook, 1973, Tables 1–8. Copyright 1973 by the American Association for the Advancement of Science.

malian species. In the mouse, rabbit, rat, horse, cow, sheep, goat, pig, cat, and dog, all males have the XY karyotype and all females are XX.

Even though it is known that the presence of a Y chromosome leads to male development and that its absence leads to female development, just how the genetic information induces sexual differentiation is not yet clear. Some insights have been gained as a result of observations of partial loss of an X or Y chromosome in individuals with otherwise normal karyotypes. Partial deletion of one X in an apparent 46,XX female results in a variable expression of the Turner syndrome. If the short arm of the X is missing (**46,XXp-**), the complete syndrome results. If the long arm is missing (**46,XXq-**), incomplete ovarian development is usually the only consequence.

The significance of the Y chromosome is even more enigmatic because it is heterochromatic. As such, it remains condensed in interphase cells and in spermatozoa and can be identified by fluorescent staining techniques. Heterochromatic chromosomal material is thought to be genetically inert. Although the heterochromatic chromosome lacks information homologous to the X chromosome, the Y chromosome must contain genetic information essential for sex determination in humans.

Data have been collected on individuals who have one X chromosome and only part of a Y chromosome. When these deletions involve the short arm of the Y (**46,XYp-**), a sexual phenotype similar to that of individuals with Turner syndrome is seen. Thus, genetic information located on the short arm of the Y chromosome appears to be important for male development. The short arm of the Y chromosome also contains a region that synapses with the terminal portion of the short arm of the X chromosome during meiosis. This partial pairing region is essential to male spermatogenesis during segregation of the sex chromosomes.

Sex-Linked Inheritance in Humans

In humans, many genes and the respective traits controlled by them are recognized as being linked to the X chromosome. These sex-linked traits may be easily identified in pedigrees because of the crisscross pattern of inheritance. A pedigree for one form of human color blindness is shown in Figure 5.13. The mother in generation I passes the trait to all her sons but to none of her daughters. If the offspring in generation II marry normal individuals, the color-blind sons will produce all normal male and female offspring (III-1, 2, and 3); the normal-visioned daughters will produce normal-visioned female offspring (III-4, 6, and 7), as well as color-blind (III-8) and normal-visioned (III-5) male offspring.

Many sex-linked human genes have now been identified. For example, the genes controlling one form of hemophilia and one form of muscular dystrophy are located on the X chromosome. Additionally, numerous genes whose expression yields enzymes are sex-linked. Glucose-6-phosphate dehydrogenase and phosphoglycerate kinase are two examples. Further, the **Lesch-Nyhan syndrome** results from the mutant form of the X-linked gene product hypoxanthine-guanine-phosphoribosyl transferase (see Chapter 24).

Sexual Differentiation in Humans

It is important to distinguish between **sex determination** and **sexual differentiation** to comprehend how adult male and female humans arise. During early development, every human embryo undergoes a period when it is potentially hermaphroditic or bisexual. Gonadal primordia arise as a pair of ridges associated with each embryonic kidney. Primordial germ cells migrate to these ridges, where an outer cortex and inner medulla form. The **cortex** is capable of de-

FIGURE 5.13
(a) A human pedigree of the sex-linked color blindness trait. (b) The most probable genotypes of each individual in the pedigree.

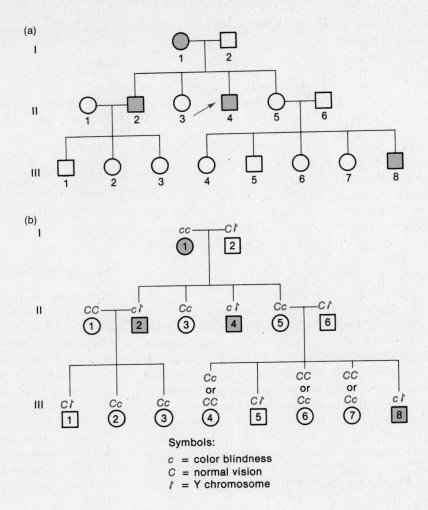

Symbols:
c = color blindness
C = normal vision
ʅ = Y chromosome

veloping into an ovary, while the inner **medulla** may develop into a testis. In addition, two sets of undifferentiated male (Wolffian) and female (Mullerian) ducts exist in each embryo.

The genital ridge is present after five weeks of gestation, and if these cells have the XY constitution, male development of the medullary region can be detected by the seventh week. If the cells have a normal XX constitution, no binding occurs, no male development is initiated, and the genital ridge is destined to form ovarian tissue later. Parallel development of the appropriate male or female duct system then occurs, and the other duct system degenerates. There is a substantial amount of evidence that once testes differentiation is initiated, the embryonic testicular tissue secretes a hormone that supports continued sexual differentiation. For example, in rabbits, if castration occurs after testes development has begun but before the duct system has developed, embryos develop female characteristics regardless of the presence of the Y chromosome.

As the twelfth week of human female development approaches, the oogonia within the ovaries begin meiosis and primary oocytes can be detected (see Figure 2.10). By the twenty-fifth week of gestation, all oocytes become arrested in meiosis, where all will remain dormant until puberty is reached, some ten to fifteen years later. Primary spermatocytes, on the other hand, are not formed in males until puberty is reached.

We can conclude that genetic information on the sex chromosomes is responsible for the primary sex determination event. Under normal conditions, once testicular or ovarian development has begun, subsequent early differentiation occurs under the influence of the male or female sex hormones. These hormones support, if not determine, expression of other secondary sexual characteristics during development.

The H-Y Antigen: A Holandric Gene Product

For many years, geneticists have sought to document the presence of genes on the human Y chromosome. In addition to identification of those genetic factors causing male sex determination, investigators have attempted to show that other genes reside on the Y chromosome. Such genes must exhibit a **holandric** or **Y-linked** inheritance pattern, with the trait in question being passed directly from the father and expressed in all his sons. While the evidence is questionable, one such trait, **hairy pinna** (hair developing on the cartilagenous projection of the external ear) has often been cited as one of the only known holandric traits in humans.

On the other hand, another gene that is responsible for the production of the **H-Y antigen** clearly fits the criteria for residence on the Y chromosome. The most specific insight into human sex determination has been gained from investigations of this male-specific gene product. Discovered in 1955 as a transplantation antigen in mice, it has been shown to be common to all mammalian species studied. In mice, the H-Y antigen is thought to be responsible for the females' rejection of male tissue transplants between members of highly inbred strains. In these strains, male-to-male, female-to-female, and female-to-male transplants are not rejected because of their genetic compatibility.

Recent studies, particularly by Stephen Wachtel, have suggested that the expression of this histocompatibility antigen is associated with a gene located on the Y chromosome. The strongest evidence has been provided by dose-related studies. If anti–H-Y antibodies are added to white blood cells, the H-Y antigen on the surface of the cells will bind the antibody. Human males with 47,XYY and 48,XXYY karyotypes bind about twice as much antibody as do males with only one Y chromosome (46,XY). Supporting evidence includes the nearly invariable association of testicular development, the H-Y antigen, and the presence of the short arm of the Y chromosome.

These studies suggest that a gene near the centromere on the short arm of the Y chromosome is involved in the expression of this antigen. It is not yet clear, however, whether this gene codes for the antigen, or is somehow involved in the regulation of the expression of an autosomally linked gene that codes for the antigen. Susumo Ohno has proposed that the embryonic gonad responds to the presence or absence of the H-Y antigen during primary sex determination events. In females, who do not produce the antigen because they lack the Y chromosome, no stimulation of the undifferentiated gonad to become testis occurs, and ovaries eventually develop. In males, Ohno has proposed that the H-Y antigen is involved in the stimulation of testicular development.

Still another observation suggests that the H-Y antigen may be involved in sex determination. In amphibians and birds, the female is the heterogametic sex. In such organisms, the female—not the male—expresses the H-Y antigen. It seems that the antigen has been preserved throughout evolution and may serve as the initial signal for primary sexual determination. While these and other observations do not prove this role, they have caused a great deal of excitement and have provided the impetus for more extensive investigations.

Sex Ratio in Humans

The presence of heteromorphic sex chromosomes in one sex of a species but not the other provides a potential mechanism for producing equal proportions of male and female offspring. The actual proportion of male to female offspring is called the **sex ratio.** It can be assessed in two ways. The **primary sex ratio** reflects the proportion of males and females *conceived* in a population. The **secondary sex ratio** reflects the proportion that are *born.* The secondary sex ratio is much easier to determine, but has the disadvantage of not accounting for disproportionate embryonic or fetal mortality, should it occur.

When the secondary sex ratio in the human population is determined using worldwide census data, it does not equal 1.0. For example, in the Caucasian population in the United States, the secondary ratio is 1.06, reflecting 106 males born for each 100 females. In the black population in the United States, the ratio is 1.025. In other countries the excess of male births is even higher. In Korea, the secondary sex ratio is 1.15.

To account for this discrepancy, it has been suggested that prenatal female mortality might be greater than prenatal male mortality. If so, it is possible that the primary sex ratio is 1.0 and that it is altered before birth. However, this hypothesis has been shown to be false. In fact, just the opposite occurs. In a Carnegie Institute study, reported in 1948, the sex of approximately 6000 embryos and fetuses recovered from miscarriages and abortions was determined. On the basis of the data derived from this study, the primary sex ratio was estimated to be 1.079. More recent data have placed this figure between 1.20 and 1.60! Therefore, many more males than females are conceived in the human population.

It is not clear why such a radical departure from the expected 1.0 primary sex ratio occurs. A suitable explanation can only be derived from examining the assumptions upon which the theoretical ratio is based:

1 Because of segregation, males produce equal numbers of X- and Y-bearing sperm.

2 Each type of sperm has equivalent viability and motility in the female reproductive tract.

3 The egg surface is equally receptive to both types of sperm.

There is no strong experimental evidence to suggest that any of these assumptions are invalid. However, it has been speculated that since the human Y chromosome is smaller than the X chromosome, Y-bearing sperm are of less mass and therefore more motile. If this is true, then the probability of a fertilization event leading to a male zygote is increased. Other explanations must also be considered.

THE X CHROMOSOME AND DOSAGE COMPENSATION

The presence of two X chromosomes in normal human females and only one X in normal human males is unique compared with the equal numbers of autosomes present in the cells of both sexes. On theoretical grounds alone, it is possible to speculate that this situation should create a genetic dosage problem between males and females for all X-linked genes. Females have two copies and males only one. The additional X chromosomes in both males and females exhibiting the various syndromes discussed earlier in this chapter should compound this dosage problem. In this section, we will describe certain research findings regarding X-linked gene expression which demonstrate that a **genetic dosage compensation mechanism** does indeed exist.

Barr Bodies and Dosage Compensation in Mammals

Murray L. Barr and Ewart G. Bertram's experiments with female cats, and Keith Moore and Barr's subsequent study with humans, demonstrate a genetic mechanism in mammals that compensates for X chromosome dosage disparities. Barr and Bertram observed a darkly staining body in interphase nerve cells of female cats. They found that this structure was absent in similar cells of males. In human females, this body can be easily demonstrated in cells derived from the buccal mucosa or in fibroblasts but not in similar male cells (Figure 5.14). This highly condensed structure, about 1 μm in diameter, lies against the nuclear envelope of interphase cells. It stains positively in the Feulgen reaction for DNA.

Current experimental evidence strongly suggests that this body, called a **sex chromatin body** or simply a **Barr body,** is an inactivated X chromosome. Ohno was the first to suggest that the Barr body arises from one of the two X chromosomes. This hypothesis is attractive because it provides a mechanism for dosage compensation. If one of the two X chromosomes is inactive in the cells of females, the dosage of genetic information that may be expressed in males and females is equivalent. Convincing but indirect evidence for this hypothesis comes from the study of the sex chromosome syndromes described earlier in this chapter. Regardless of how many X chromosomes exist, all but one of them appear to be inactivated and can be seen as Barr bodies. For example, none is seen in Turner 45,X females; one is seen in Klinefelter 47,XXY males; two in 47,XXX females; three in 48,XXXX females; and so on (Figure 5.15). Therefore, the number of Barr bodies follows an $N - 1$ rule, where N is the total number of X chromosomes present.

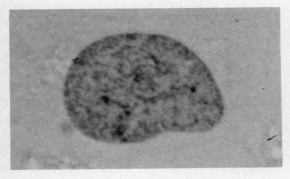 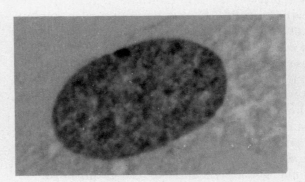

FIGURE 5.14
Photomicrographs of two human fibroblast nuclei. The cell on the left, derived from a male, lacks a Barr body. The cell on the right was derived from a female and demonstrates a densely staining Barr body on the periphery of the nucleus. (Courtesy of Ursula Mittwoch, from Yunis, 1974.)

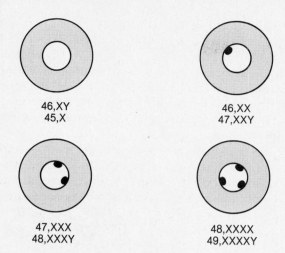

46,XY
45,X

46,XX
47,XXY

47,XXX
48,XXXY

48,XXXX
49,XXXXY

FIGURE 5.15
Diagrammatic representation of Barr body occurrence in various human karyotypes.

While this mechanism of inactivation of all but one X chromosome increases our understanding of dosage compensation, it further complicates our perception of other matters. Since one of the two X chromosomes is inactivated in normal human females, why then is the Turner 45,X individual not entirely normal? Why aren't females with the triplo-X and tetra-X karyotypes (47,XXX and 48,XXXX) normal? Further, in Klinefelter syndrome (47,XXY), X chromosome inactivation effectively renders such persons 46,XY. Why aren't they unaffected by the additional X chromosome in their nuclei?

One possible explanation is that chromosome inactivation does not normally occur in the very early stages of development of those cells destined to form gonadal tissues. Another explanation is that apparently not all of each X chromosome forming a Barr body is inactivated. As a result, excessive expression of certain X-linked genes still occurs despite apparent inactivation of additional X chromosomes. In spite of these possible explanations, the above observations remain enigmatic.

The Lyon Hypothesis

In mammalian females one X chromosome is of maternal origin, and the other is of paternal origin. Which one is inactivated? Is the inactivation random? Is the same chromosome inactive in all somatic cells? In 1961, Mary Lyon and Liane Russell independently proposed a hypothesis that answers these questions. They postulated that the inactivation of X chromosomes occurs randomly in somatic cells at a point early in embryonic development and that once inacti-

vation has occurred, all progeny cells have the same X chromosome inactivated.

This explanation, which has come to be called the **Lyon hypothesis,** was initially based on observations of female mice heterozygous for sex-linked coat color genes (Figure 5.16). The pigmentation of these heterozygous females was mottled, with large patches of skin expressing the color allele on one X and other patches expressing the allele on the other X. Indeed, if one or the other of the two X chromosomes was inactive in adjacent patches of cells, such a phenotypic pattern would result. Similar mottling occurs in the black and tan patches of female tortoise-shell cats. Such sex-linked coat color patterns do not occur in male cats since all cells are hemizygous for only one sex-linked coat color allele.

The most direct evidence in support of the Lyon hypothesis comes from studies of gene expression in clones of human fibroblast cells. Individual cells may be isolated following biopsy and cultured *in vitro*. If each culture is derived from a single cell, it is referred to as a **clone.** The synthesis of the enzyme **glucose-6-phosphate dehydrogenase (G-6-PD)** is controlled by a sex-linked gene. Numerous mutant alleles of this gene have been detected, and their gene products can

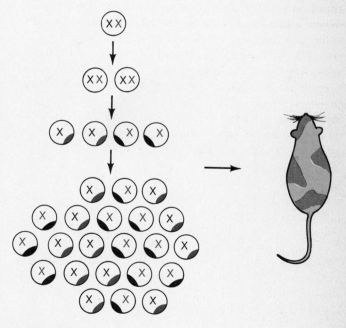

FIGURE 5.16
Diagrammatic representation of the Lyon hypothesis in the mouse. Following random inactivation of one of the two X chromosomes in female cells, all progeny cells inactivate the same chromosome. The adult is a mosaic, with some groups of cells expressing the alleles on one X chromosome and the other groups expressing the alleles present on the homologous X chromosome.

be differentiated from the wild-type enzyme by their migration pattern in an electrophoretic field. Fibroblasts have been taken from females heterozygous for different allelic forms of G-6-PD and studied.

The Lyon hypothesis predicts the results of the examination of numerous clones derived from a female heterozygous for two G-6-PD alleles. If inactivation of an X chromosome occurs randomly early in development and is permanent in all progeny cells, such a female should show two types of clones in approximately equal proportion. Each clone shows one electrophoretic variant of G-6-PD or the other, but never both. In 1963, Ronald G. Davidson, Harold M. Nitowski, and Barton Childs performed an experiment involving 14 clones from a single heterozygous female. Seven showed only one form of the enzyme, and seven showed only the other form. What was most important was that none of the 14 showed both forms of the enzyme. Studies of G-6-PD mutants thus provide strong support for the random permanent inactivation of either the maternal or paternal X chromosome.

Although not all of the existing evidence is as clear-cut as the preceding experiment, the Lyon hypothesis is generally accepted as valid. One extension of the hypothesis is that mammalian females are mosaics for all heterozygous sex-linked alleles. Some areas of the body express only the maternally derived alleles, and others express only the paternally derived alleles. The least understood aspect of the hypothesis concerns how inactivation occurs. While studies on the mechanisms that select and inactivate an X chromosome are just beginning, evidence suggests that DNA in inactivated chromosome regions has been chemically modified. One line of work has shown an association between the addition of methyl groups to cytosine residues and the condensation of chromosomes. Perhaps this process of methylation is part of the mechanism responsible for X chromosome inactivation.

SEX-LIMITED AND SEX-INFLUENCED INHERITANCE

There are numerous examples in different organisms where the sex of the individual plays a determining role in the expression of certain phenotypes. In some cases, the expression is absolutely limited to one sex; in others, the sex of an individual influences the expression of a phenotype that is not limited to one sex or the other. This distinction differentiates **sex-limited inheritance** from **sex-influenced inheritance.**

In domestic fowl, tail and neck plumage is often distinctly different in males and females (Figure 5.17), demonstrating sex-limited inheritance. Cock-feathering is longer, more curved, and pointed, while hen-feathering is shorter and more rounded. The inheritance of feather type is due to a single pair of autosomal alleles whose expression is modified by the individual's sex hormones.

As shown below, hen-feathering is due to a dominant allele, H; but regardless of the homozygous presence of the recessive h allele, all females remain hen-feathered. Only in males does the hh genotype result in cock-feathering.

	Phenotype	
Genotype	♀	♂
HH	Hen-feathered	Hen-feathered
Hh	Hen-feathered	Hen-feathered
hh	Hen-feathered	Cock-feathered

In the development of certain breeds of fowl, one allele or the other has become fixed in the population. In the Leghorn breed, all individuals are of the hh genotype, and thus all males and females show distinctive plumage. Sebright bantams are all HH, showing no sexual distinction in feathering.

The influence of sex hormones on the feathering phenotypes has been demonstrated in experiments where the gonads are removed from males and females of various genotypes. In all cases, gonadectomized birds become cock-feathered at the next molt.

Male   Female

FIGURE 5.17
Cock-feathering (left) versus hen-feathering (right) in domestic fowl. The feathers in the hen are shorter and less curved.

FIGURE 5.18
Pattern baldness, a sex-influenced trait in humans. (Courtesy of James F. Sunday.)

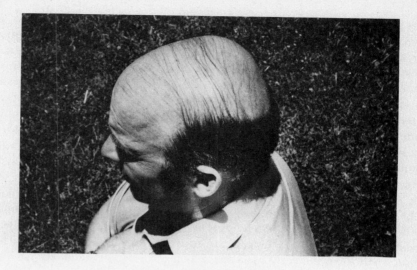

Cases of sex-influenced inheritence include pattern baldness in humans, horn formation in sheep, and certain coat patterns in cattle. In such cases, autosomal genes are responsible for the contrasting phenotypes displayed by both males and females. However, the heterozygous genotype may exhibit one phenotype in one sex and the contrasting one in the other. For example, **pattern baldness,** where the hair is very thin on the top of the head (Figure 5.18), is inherited in the following way:

	Phenotype	
Genotype	♀	♂
BB	Bald	Bald
Bb	Not bald	Bald
bb	Not bald	Not bald

Thus, while females may display pattern baldness, this phenotype is much more prevalent in males. When females do inherit the *BB* genotype, the phenotype is much less pronounced than in males.

SUMMARY AND CONCLUSIONS

Genetic variability produced during sexual reproduction is an integral part of the evolutionary process. In order for sexual reproduction to occur efficiently, sexual differentiation within members of a species is imperative. In this chapter, we reviewed several modes of sexual differentiation, including those in *Chlamydomonas*, *Neurospora*, and *Zea mays*, as well as in humans. In the more primitive organisms, the algae and fungi, sexual differentiation in the form of mating types occurs physiologically, even in the absence of

the more obvious morphological dimorphism characteristic of higher forms.

The genetic basis of sexual differentiation (i.e., the mechanisms of sex determination) has been a major area of research. Early studies showed that most often males and females have different chromosome compositions. One sex, called heterogametic, either lacks one chromosome or contains a Y chromosome. Despite the apparent presence of sex-determining chromosomes, careful investigation has revealed that genes located on the autosomes may be just as important. In *Drosophila*, a balance between information expressed by the X chromosomes and the autosomes allows for maleness, intersexual development, or femaleness.

When genes are located on the sex-determining chromosomes, reciprocal crosses often yield different results. This phenomenon, known as sex linkage, occurs because one sex or the other is heterogametic. Members of this sex lack a chromosome that is strictly homologous to the other member of the pair. As a result, in such individuals only one locus is present for genes on the sex chromosomes. The allele occupying that locus is directly expressed, even if it is recessive. Such expression is the basis of hemizygosity. Sex linkage leads to a modification of the normal 3:1 monohybrid ratio as well as to a crisscross pattern of inheritance. When males are the heterogametic sex, mothers homozygous for a recessive gene pass the trait to their sons but not necessarily to their daughters.

Modes of sex determination undoubtedly vary from organism to organism. In humans, the presence or absence of the Y chromosome appears to determine male or female development, respectively. The major evidence in support of this conclusion is based on the observations of sexual development of individuals with abnormal karyotypes, including those with Turner and Klinefelter syndrome. In this context, other human sex chromosome abnormalities have been discussed. However, once the chromosome signal serves to initiate male or female development of the bisexual gonad, further human sexual differentiation is enhanced by the influence of hormones.

In many vertebrates there is now strong evidence that a holandric gene product, the H-Y antigen is involved in the initial event leading to maleness. This Y-linked gene product is characteristic of the heterogametic sex. In humans, the gene responsible for the antigen is on the short arm of the Y chromosome.

The presence of one heterogametic sex in a species theoretically provides an efficient basis for a 1:1 sex ratio. Analysis of the human species reveals that the ratio at conception and at birth favors males at these stages. To account for this observation at birth, it has been proposed that Y-bearing sperm are more motile than X-bearing sperm because of the reduced mass of the smaller Y chromosome. From conception on, however, male mortality is higher than that in females.

We have also considered mammalian dosage compensation of sex-linked genetic information between males and females. Since the Y chromosome lacks most homologous information contained on the X chromosome, a disparity exists. In theory, females with two X's are capable of twice the expression of that in males with only one X. However, it has been found that mammalian females have one X chromosome inactivated early in development. This inactivated X chromosome is the sex chromatin or Barr body found adhering to the nuclear envelope of somatic cells. Inactivation of either the maternal or paternal X is random, but permanent in all progeny cells. Thus, researchers have looked for and found a mechanism for dosage compensation. As a result of random inactivation, mammalian females develop as genetic mosaics with respect to their expression of heterozygous sex-linked alleles.

Finally, we have discussed the concepts of sex-limited and sex-influenced inheritance. Hen-feathering in fowl is controlled by a dominant autosomal allele. However, this phenotype occurs in females regardless of the genotype, illustrating sex-limited inheritance. In humans, the phenotype of individuals heterozygous for the pattern baldness allele varies depending on their sex. Males express this phenotype when heterozygous, but females do not. Sex hormones are believed to modify gene expression, leading to sex-influenced inheritance.

PROBLEMS AND DISCUSSION QUESTIONS

1 Define the following terms: (a) heterogamy, (b) heteromorphic chromosome, (c) nondisjunction, (d) Barr body, and (e) Lyon hypothesis.

2 An insect species is discovered in which the heterogametic sex is unknown. A sex-linked recessive mutation for *reduced wing* size (*rw*) is found in this species. Contrast the F_1 and F_2 generations from a cross between a female with reduced wings and a normal male when:
   (a) the female is the heterogametic sex.
   (b) the male is the heterogametic sex.
   (c) Is it possible to distinguish between the Protenor and Lygaeus type of sex determination based on these crosses?

3 Contrast the Protenor and Lygaeus modes of sex determination.

4 Contrast the evidence leading to the explanation of the different modes of sex determination in *Drosophila* and humans. Does *Melandrium* show a mode of sex determination that parallels either of these species?

5 In *Drosophila*, females homozygous for the recessive sex-linked gene causing *singed* bristles (*sn*) are crossed to wild-type males. Both the males and females are heterozygous for the gene *transformer* (*tra*), which is located on chromosome 3. What phenotypic ratios in the male and female offspring will result from this cross?

6 Devise a method of nondisjunction in human female gametes that would give rise to Klinefelter and Turner syndrome offspring following fertilization by a normal male gamete.

7 A sex-linked dominant mutation in the mouse, *Testicular feminization (Tfm)*, eliminates the normal response to the testicular hormone testosterone during sexual differentiation. An XY animal bearing the *Tfm* allele on the X chromosome develops testes, but no further male differentiation occurs. The external genitalia of such an animal are female. From this information, what might you conclude about the role of the X and Y chromosomes in sex determination and differentiation in mammals?

8 Discuss the possible reasons why the primary sex ratio in humans is as high as 1.40.

9 It has been suggested that any male-determining genes contained on the Y chromosome in humans *are not* located on the area that synapses with the X chromosome during meiosis. What might be the outcome if such genes were located in this region?

10 What does the apparent need for dosage compensation mechanisms suggest about the expression of genetic information in normal diploid individuals?

11 Indicate the expected number of Barr bodies in interphase cells of the following individuals: Klinefelter syndrome; Turner syndrome; and karyotypes 47,XYY, 47,XXX, and 48,XXXX.

12 Cat breeders are aware that kittens with the calico coat pattern are almost invariably females. Why?

13 It is not uncommon to encounter self-sterile plant species. For example, in *Nicotiana*, pollen that falls on the stigma of the same plant will not develop in some cases, although the same pollen will fertilize another plant. Speculate on the evolutionary advantages of self-sterility.

14 When cows have twin calves of unlike sex (fraternal twins), the female twin is usually sterile and has masculinized reproductive organs. This calf is referred to as a **free martin.** In cows, twins may share a common placenta and fetal circulation. Based on the information in this chapter concerning sexual differentiation, discuss how a free martin comes to exist.

15 In certain *Drosophila* strains, the X chromosomes are attached to each other. Some flies contain attached-X chromosomes and a separate Y chromosome. What sex would this fly be? If a similar situation existed in the human species, what sex would the individual be? What syndrome would the individual exhibit?

16 An attached-X plus Y fly, as described in Problem 16, with *white* eyes is crossed to a fly of the opposite sex which has *miniature* wings (both are sex-linked, recessive genes). What proportion of phenotypes in male and female offspring is expected? Assume that flies with three or no X chromosomes do not develop to adulthood.

17 *Parthenogenesis* is the process in which unfertilized eggs initiate development and give rise to offspring. In birds such as the turkey, females are the heterogametic sex and usually of the Protenor type (X0). How could a parthenogenic male offspring arise?

18 In the wasp *Bracon hebetor*, a form of parthenogenesis resulting in haploid organisms is common. All haploids are males. When fertilization occurs, diploid individuals are almost always female. P. W. Whiting has shown that an X-linked gene with nine multiple alleles (*Xa, Xb, Xc,* etc.) controls sex determination. Any homozygous or hemizygous condition results in males, and any heterozygous condition results in females. If an *Xa/Xb* female mates with an *Xa* male and lays 50 percent fertilized and 50 percent unfertilized eggs, what proportion of male and female offspring will result?

19 The marine echiurid worm *Bonellia viridis* is an extreme example of the environment's influence on sex determination. Undifferentiated larvae either remain free-swimming and differentiate into females or they settle on the proboscis of an adult female and become males. If larvae that have been on a female proboscis for a short period are removed and placed in sea water, they develop as intersexes. If larvae are forced to develop in an aquarium where pieces of proboscises have been placed, they develop into males. Contrast this mode of sexual differentiation with that of mammals. Suggest further experimentation to elucidate the mechanism of sex determination.

20 In cats, yellow coat color is determined by the *b* allele and black coat color by the *B* allele. The heterozygous condition results in a color known as tortoise shell. These genes are sex-linked. What kinds of offspring would be expected from a cross of a black male and a tortoise-shell female? What are the chances of getting a tortoise-shell male?

21 A husband and wife have normal vision, although both of their fathers are red-green color-blind, which is inherited as a sex-linked recessive affliction. What is the probability that their first child will be
(a) a normal son?
(b) a normal daughter?
(c) a color-blind son?
(d) a color-blind daughter?

22 In humans, the presence of red blood cell antigen Xg^a has been shown to be a dominant sex-linked trait. Individuals are thus classified as positive or negative (i.e., $Xg^{(a+)}$ and $Xg^{(a-)}$). The ABO blood group system is determined by a set of autosomal multiple alleles. Suppose a woman of types A and $Xg^{(a-)}$ marries an AB and $Xg^{(a+)}$ man. With respect to these two loci, what types of offspring would be expected, and in what proportions?

23 In *Drosophila*, a sex-linked recessive mutation, *scalloped* (*sd*), causes irregular wing margins. Diagram the F_1 and F_2 results if:
(a) A *scalloped* female is crossed with a normal male.
(b) A *scalloped* male is crossed with a normal female.
Compare these results to those that would be obtained if *scalloped* were not sex-linked.

24 Another recessive mutation in *Drosophila*, *ebony* (*e*), is on an autosome (chromosome 3) and causes darkening of the body compared with wild-type flies. What phenotypic F_1 and F_2 male and female ratios will result if a *scalloped*-winged female with normal body color is crossed with a normal-winged *ebony* male? Work this problem by both the Punnett square method and the forked-line method.

25 In *Drosophila* the recessive gene for *white* eye (*w*) is sex linked, while the recessive gene for *dumpy* wing (*dp*) is located on chromosome 2. A female who was both red-eyed (homozygous) and dumpy-winged was crossed to a male whose mother was white-eyed and whose mother and father both had homozygously normal wings. Show the F_1 and F_2 phenotypic ratios of the sex, eye color, and wing shape.

26 In *Drosophila*, the sex-linked recessive mutation *vermilion* (*v*) causes bright red eyes, which is in contrast to brick red eyes of wild type. A separate autosomal recessive mutation, *suppressor of vermilion* (*su-v*), causes flies homozygous or hemizygous for *v* to have wild-type eyes. In the absence of vermilion alleles, *su-v* has no effect on eye color. Determine the F_1 and F_2 phenotypic ratios from a cross between a female with wild-type alleles at the *vermilion* locus, but who is homozygous for *su-v*, with a *vermilion* male who has wild-type alleles at the *su-v* locus.

27 While *vermilion* is sex linked and brightens the eye color, *brown* is an autosomal recessive mutation that darkens the eye. Flies carrying both mutations lose all pigmentation and are white eyed. Predict the F_1 and F_2 results of the following crosses:
(a) vermilion females × brown males
(b) brown females × vermilion males
(c) white females × wild males.

28 In spotted cattle, the colored regions may be mahogany or red. If a red female and a mahogany male, both derived from separate true-breeding lines, are mated and the cross carried to an F_2 generation, the following results are obtained:

F_1: 1/2 mahogany males
1/2 red females

F₂: 3/8 mahogany males
1/8 red males
1/8 mahogany females
3/8 red females

When the reciprocal of the initial cross is performed (mahogany female and red male), identical results are obtained. Explain these results by postulating how the color is genetically determined. Diagram the crosses.

29 Predict the F_1 and F_2 results of crossing a male fowl that is cock-feathered with a true-breeding hen-feathered female fowl. Recall that these traits are sex limited and were discussed in the text.

30 In fowl such as the Wyandotte breed, males and females may be *HH*, *Hh*, or *hh* for the sex limited genes determining plumage. What feather phenotypes will occur for each of these genotypes in male and female birds? In a cross between a hen-feathered male and a hen-feathered female, all female offspring were hen-feathered, but one-fourth of the male offspring were cock-feathered. What were the genotypes of the parents?

SELECTED READINGS

BACCI, G. 1965. *Sex determination.* New York: Academic Press.

BARR, M. L. 1966. The significance of sex chromatin. *Int. Rev. Cytol.* 19: 35–39.

BUHLER, E. M. 1980. A synopsis of the human Y chromosome. *Hum. Genet.* 55: 145–75.

BULL, J.T. 1983. *The evolution of sex-determining mechanisms.* Menlo Park: Benjamin Cummings.

CARR, D. H. 1971. Chromosomes and abortion. *Adv. Hum. Genet.* 2: 201–57.

COURT BROWN, W. M. 1968. Males with an XYY sex chromosome complement. *J. Med. Genet.* 5: 341–59.

DAVIDSON, R., NITOWSKI, H., and CHILDS, B. 1963. Demonstration of two populations of cells in human females heterozygous for glucose-6-phosphate dehydrogenase variants. *Proc. Natl. Acad. Sci.* 50: 481–85.

DRAYNA, D., and WHITE, R. 1985. The genetic linkage map of the human X chromosome. *Science* 230: 753–58.

ERICKSON, J.D. 1976. The secondary sex ratio of the United States 1969–71: Association with race, parental ages, birth order, paternal education and legitimacy. *Ann. Hum. Genet. Lond.* 40: 205–12.

FORD, E. H. R. 1973. *Human chromosomes.* New York: Academic Press.

GORDON, J. W., and RUDDLE, F. H. 1981. Mammalian gonadal determination and gametogenesis. *Science* 211: 1265–78.

HAMERTON, J. L. 1971. *Human cytogenetics.* Vol. 2. New York: Academic Press.

HASELTINE, F. P., and OHNO, S. 1981. Mechanisms of gonadal differentiation. *Science* 211: 1272–78.

HOOK, E. B. 1973. Behavioral implications of the human XYY genotype. *Science* 179: 139–50.

JACOBS, P. A., et al. 1974. A cytogenetic survey of 11,680 newborn infants. *Ann. Hum. Genet.* 37: 359–76.

LEWIS, K. R., and JOHN, B. 1968. The chromosomal basis of sex determination. *Int. Rev. Cytol.* 23: 277–379.

LYON, M. F. 1961. Gene action in the X-chromosome of the mouse (*Mus musculus* L.). *Nature* 190: 372–73.

————. 1962. Sex chromatin and gene action in the mammalian X chromosome. *Amer. J. Hum. Genet.* 14: 135–48.

————. 1972. X-chromosome inactivation and developmental patterns in mammals. *Biol. Rev.* 47: 1–35.

McKUSICK, V. A. 1962. On the X chromosome of man. *Quart. Rev. Biol.* 37: 69–175.

————. 1975. *Mendelian inheritance in man.* Baltimore: Johns Hopkins University Press.

McMILLEN, M. M. 1979. Differential mortality by sex in fetal and neonatal deaths. *Science* 204: 89–91.

MITTWOCH, U. 1967. *Sex chromosomes.* New York: Academic Press.

————. 1973. *Genetics of sex differentiation.* New York: Academic Press.

MORGAN, T. H. 1910. Sex limited inheritance in *Drosophila. Science* 32: 120–22.

NÖTHIGER, R., and STEINMANN-ZWICKY, M. 1985. Sex determination in *Drosophila. Trends in Genet.* 1: 209–14.

OHNO, S. 1967. *Sex chromosomes and sex-linked genes.* Berlin: Springer-Verlag.

————. 1978. *Major sex determining genes.* New York: Springer-Verlag.

POLANI, P. E. 1982. Pairing of X and Y chromosomes, non-inactivation of X-linked genes, and the maleness factor. *Hum. Genet.* 60: 207–11.

RUSSELL, L. B. 1961. Genetics of mammalian sex chromosomes. *Science* 133: 1795–1803.

SILVERS, W. K., and WACHTEL, S. S. 1977. H-Y antigen: Behavior and function. *Science* 195: 956–60.

SIMPSON, J. L. 1982. Abnormal sexual differentiation in humans. *Ann. Rev. Genet.* 16: 193–224.

VOELLER, B. R. ed. 1968. *The chromosome theory of inheritance—Classic papers in development and heredity.* New York: Appleton-Century-Crofts.

WACHTEL, S. S. 1977. H-Y antigen and the genetics of sex determination. *Science* 198: 797–99.

————. 1981. Conservation of the H-Y/H-W receptor. *Hum. Genet.* 58: 54–58.

————. 1983. *H-Y antigen and the biology of sex determination.* New York: Grune and Stratton.

WESTERGAARD, M. 1958. The mechanism of sex determination in dioecious flowering plants. *Adv. in Genet.* 9: 217–81.

WHITING, P. W. 1939. Multiple alleles in sex determination of *Habrobracon, J. Morph.* 66: 323–55.

6

Linkage, Crossing Over, and Chromosome Mapping

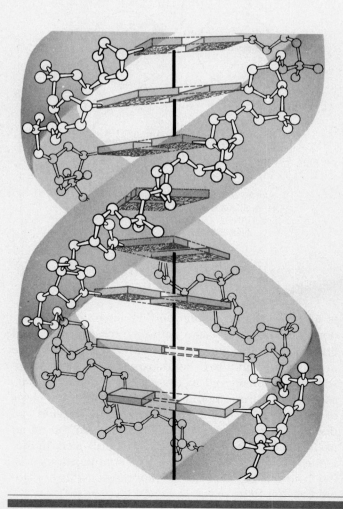

As early as 1903, Walter Sutton, one of the men who rediscovered Mendel's work, pointed out the likelihood that in organisms there are many more "unit factors" than chromosomes. Soon thereafter, genetic investigations with several organisms revealed that certain genes were not transmitted according to the law of independent assortment. When studied together in matings, these genes seemed to segregate as if they were somehow joined or linked together. Further investigations showed that such genes were part of the same chromosome, and they were indeed transmitted as a single unit.

We now know that most chromosomes consist of very large numbers of genes. Recent molecular studies have revealed that the average-sized chromosome contains sufficient DNA to encode thousands of genes. Genes that are part of the same chromosome are said to be **linked** and to demonstrate **linkage** in genetic crosses.

Since the chromosome, not the gene, is the unit of transmission during meiosis, linked genes are not free to undergo independent assortment. Instead, the alleles at all loci of one chromosome should, in theory, be transmitted as a unit during gamete formation. However, in many instances this does not occur. During the first meiotic prophase, when homologues are paired or synapsed, a reciprocal exchange of chromosome segments may take place. This event, called **crossing over,** results in the reshuffling or **recombination** of the alleles between homologues.

FIGURE 6.1

A comparison of the results of gamete formation where two heterozygous genes are (a) on two different pairs of chromosomes; (b) on the same pair of homologues, but with no exchange occurring between them; and (c) on the same pair of homologues, with an exchange between two nonsister chromatids.

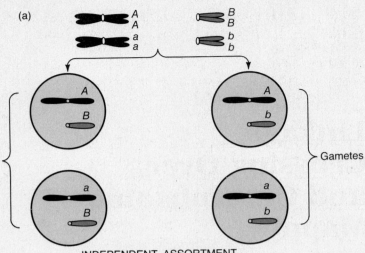

INDEPENDENT ASSORTMENT
OF TWO GENES ON TWO DIFFERENT
HOMOLOGOUS PAIRS OF CHROMOSOMES

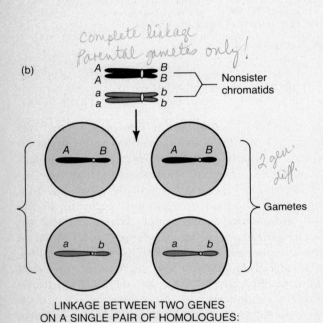

LINKAGE BETWEEN TWO GENES
ON A SINGLE PAIR OF HOMOLOGUES:
NO EXCHANGE OCCURS

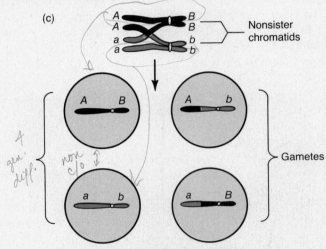

LINKAGE BETWEEN TWO GENES
ON A SINGLE PAIR OF HOMOLOGUES:
EXCHANGE OCCURS BETWEEN TWO
NONSISTER CHROMATIDS

The degree of crossing over between any two loci on a single chromosome is proportionate to the distance between them. Thus, the percentage of recombinant gametes varies, depending on which loci are being considered. This correlation serves as the basis for the construction of **chromosome maps.**

Crossing over is currently viewed as an actual physical breaking and rejoining process that occurs during meiosis. This exchange of chromosome segments provides for an enormous potential variation in the gametes formed by any individual. This type of variation, in combination with that resulting from independent assortment, ensures that all offspring will contain a diverse mixture of both maternal and paternal alleles. In addition to producing individual diversity, genetic variability is of paramount importance to the process of organic evolution.

LINKAGE VS. INDEPENDENT ASSORTMENT

In Figure 6.1, the meiotic consequences of independent assortment, linkage without crossing over, and linkage with crossing over are contrasted. For the sake of simplicity, only two homologous pairs of chromosomes are considered in the case of independent assortment, and one homologous pair in the two cases of linkage. Later in this chapter, we will provide evidence that crossing over actually occurs in the tetrad or four-strand stage of meiosis, where each homologue is in its duplicated state. The complications arising from this fact will also be more thoroughly considered at that time.

Figure 6.1(a) illustrates the results of independent assortment of two pairs of nonhomologous chromosomes, each containing one heterozygous gene pair. Four genetically different gametes are formed in equal proportions.

Figures 6.1(b) and (c) compare these results with those that occur if the same genes are instead linked on the same chromosome. If no crossing over occurs between the two genes [Figure 6.1(b)], only two genetically different gametes are formed. Each gamete receives the alleles present on one homologue or the other, which has been transmitted intact as the result of segregation. This case illustrates **complete linkage,** which is said to result in the production of only **parental,** or **noncrossover, gametes.** The two parental gametes are formed in equal proportions.

Figure 6.1(c) illustrates the results when crossing over occurs between two linked genes. As you will note, the crossover involves only two nonsister chromatids of the four chromatids present in the tetrad. This exchange generates two new allele combinations, called **recombinant** or **crossover gametes.** The two chromatids not involved in the exchange result in noncrossover gametes [as those in part (b) of this figure].

We know that the frequency with which crossing over occurs between any two linked genes is proportional to the distance separating the respective loci along the chromosome.

It is possible for two genes to be so close to each other that an insignificant number of crossover events occurs between them. This circumstance, complete linkage, results in the production of only parental gametes. If a distinct but small distance separates two genes, few recombinant and many parental gametes will be formed. As the distance between two genes increases, the proportion of recombinant gametes increases and that of the parental gametes decreases.

As will be explored later in this chapter, when two linked genes are considered, the number of recombinant gametes produced never exceeds 50 percent. Therefore, if the maximal number were produced, a 1:1:1:1 ratio of the four types would result. In such a case, transmission of two linked genes would be indistinguishable from that of two unlinked, independently assorting genes.

The Linkage Ratio

If complete linkage exists between two genes because of their proximity, and organisms with mutant alleles representing these genes are mated, a unique F_2 phenotypic ratio results that is characteristic of linkage. To illustrate this ratio, we will consider a cross involving the closely linked recessive mutant genes *brown* (*bw*) eye and *heavy* (*hv*) wing vein in *Drosophila melanogas-*

ter (Figure 6.2). The normal, wild-type alleles bw^+ and hv^+ are both dominant and result in red eyes and thin wing veins, respectively.

In this cross, flies with mutant brown eyes and normal thin veins are mated to flies with normal red eyes and mutant heavy veins. In more concise terms, brown-eyed flies are crossed with heavy-veined flies. Recall from the discussion on genetic symbols in Chapter 4 that linked genes may be represented by placing their allele designations above and below a single or double horizontal line. Those placed above the line are located on one homologue and those placed below on the other homologue. Thus, we may represent the P_1 generation as follows:

$$P_1: \quad \frac{bw \quad hv^+}{bw \quad hv^+} \quad \times \quad \frac{bw^+ \quad hv}{bw^+ \quad hv}$$

$$\text{brown, thin} \quad \text{red, heavy}$$

Since the genes are located on an autosome, no distinction for male and female is necessary.

In the F_1 generation, all flies are heterozygous for both gene pairs and exhibit the dominant traits of red eyes and thin veins:

$$F_1: \quad \frac{bw \quad hv^+}{bw^+ \quad hv}$$

$$\text{red, thin}$$

As shown in Figure 6.2, when the F_1 generation is interbred, because of complete linkage, each F_1 individual forms only parental gametes. Following fertilization, the F_2 generation will be produced in a 1:2:1 phenotypic and genotypic ratio. One-fourth of this generation will show brown eyes and thin veins; one-half will show both wild-type traits, red eyes and thin veins; and one-fourth will show red eyes and thick veins. In more concise terms, the ratio is 1 brown:2 wild:1 thick. Such a ratio is characteristic of complete linkage.

Figure 6.2 also demonstrates the results of a test cross with the F_1 flies. Such a cross produces a 1:1 ratio of brown, thin and red, thick flies. Had the genes controlling these traits been incompletely linked or located on separate autosomes, four phenotypes rather than two would have been produced.

When large numbers of mutant genes present in any given species are investigated in crosses similar to those just described, the genes located on the same chromosome will show evidence of linkage to one another. As a result, **linkage groups** can be established, one for each chromosome. Hence, the number of linkage groups should correspond to the haploid number of chromosomes. In organisms where large numbers of mutant genes are available for genetic study, this correlation has been upheld.

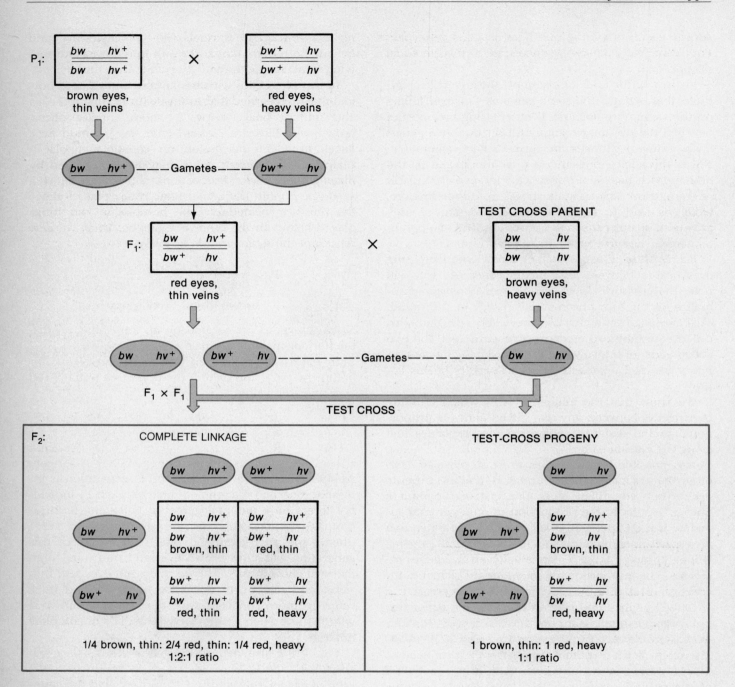

FIGURE 6.2

The results of a cross between two linked genes demonstrating complete linkage. The cross is carried to the F₂ generation. The results of a test cross with an F₁ individual are also illustrated.

INCOMPLETE LINKAGE, CROSSING OVER, AND CHROMOSOME MAPPING

If two genes linked on the same chromosome are selected randomly it is unlikely that their respective loci will be contiguous to each other. Therefore, because crossing over will occur between them, complete linkage is rarely achieved. Instead, crosses involving two linked genes usually produce a percentage of offspring resulting from recombinant gametes. This percentage is variable, depending upon which two genes are involved in the cross. This phenomenon was first studied and explained by two *Drosophila* geneticists, Thomas H. Morgan and his student, Alfred H. Sturtevant.

Morgan and Crossing Over

As you may recall from our earlier discussion of sex-linked genes in Chapter 5, Morgan first discovered the phenomenon of sex linkage. In his studies, he investigated numerous *Drosophila* mutations located on the X chromosome. When he analyzed crosses involving only one trait, he was able to deduce the mode of sex-linked inheritance. However, when he made crosses involving two sex-linked genes, his results were at first puzzling. For example, as shown in Figure 6.3(a), he crossed females with the mutant traits *yellow* body (*y*) and *white* eyes (*w*) with wild-type males (grey bodies and red eyes). The F_1 females were wild-type, while the F_1 males expressed both mutant traits. In the F_2, 98.7 percent of the offspring showed the parental phenotypes—*yellow, white*, and wild type. The remaining 1.3 percent of the flies were either yellow-bodied with red eyes or grey-bodied with white eyes. It was as if the genes had somehow separated from each other during gamete formation in the F_1 flies.

When he made crosses involving other sex-linked genes, the results were even more puzzling [Figure 6.3(b)]. The same basic pattern was observed, but the proportion of F_2 phenotypes differed; for example, in a cross involving white-eye, miniature-wing mutants, only 62.8 percent of the F_2 showed the parental phenotypes, while 37.2 percent of the offspring appeared as if the mutant genes had been separated during gamete formation.

In 1911, Morgan was faced with two questions: (1) What was the source of gene separation? and (2) Why did the frequency of the apparent separation vary depending on the genes being studied? The proposed answer to the first question was based on his knowledge of earlier cytological observations made by F. Janssens and others. Janssens had observed that synapsed homologous chromosomes in meiosis wrapped around each other, forming points of **chiasmata** [sing.: **chiasma**]. Morgan proposed that these chiasmata could represent points of genetic exchange, the crux of the so-called **chiasmatype theory.**

In the crosses shown in Figure 6.3, Morgan postulated that if an exchange occurred between the mutant genes on the two X chromosomes of the F_1 females, it would lead to the observed results. He suggested that such exchanges led to 1.3 percent recombinant gametes in the *yellow-white* cross and 37.2 percent in the *white-miniature* cross. On the basis of this and other experimentation, Morgan concluded that linked genes exist in a linear order along the chromosome, and that a variable amount of exchange occurs between any two genes.

As an answer to the second question, Morgan proposed that two genes located relatively close to each other along a chromosome are less likely to have a chiasma form between them than if the two genes are farther apart on the chromosome. Thus, the closer two genes are, the less likely it is that a genetic exchange will occur between them. Morgan proposed the term **crossing over** to describe the physical exchange leading to recombination. Although there still is some doubt concerning the validity of the chiasmatype theory, it did provide an adequate basis for Morgan's theory of genetic recombination and the subsequent work on crossing over and mapping.

Sturtevant and Mapping

Morgan's student, Alfred H. Sturtevant, was the first to realize that his mentor's proposal could be used to map the sequence of and distance between linked genes. Sturtevant argued that if the frequency of crossing over between genes is related to the distance between them, the recombination frequencies between a series of linked genes must be additive.

For example, Sturtevant compiled data on recombination between the genes represented by the *yellow, white*, and *miniature* mutants initially studied by Morgan. Frequencies of crossing over between each pair of these three genes were observed to be

(1)	*yellow, white*	0.5%
(2)	*white, miniature*	34.5%
(3)	*yellow, miniature*	35.4%

Since the sum of (1) and (2) is approximately equal to (3), Sturtevant suggested that the order of the genes on the X chromosome was *yellow-white-miniature*. The *yellow* and *white* genes are apparently close to each other because the recombination frequency is low. However, both of these genes are quite far apart from

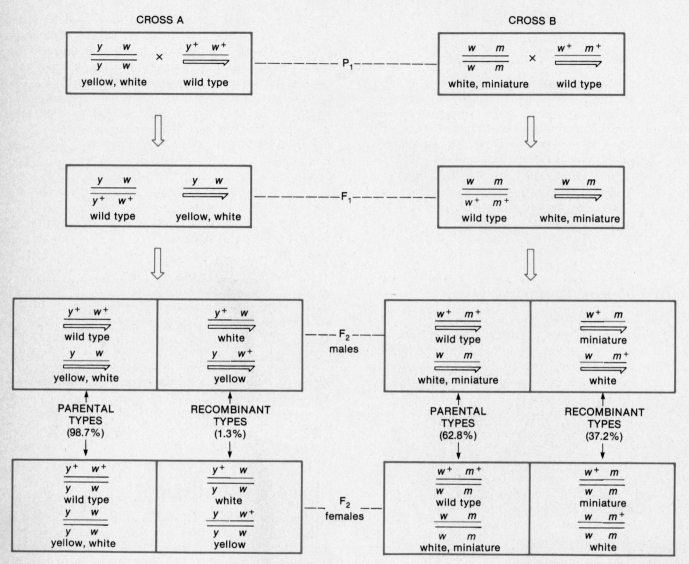

FIGURE 6.3
The F_1 and F_2 results of crosses involving the yellow-body, white-eye mutations and the white-eye, miniature-wing mutations. In the F_2 generation of Cross A, 1.3 percent of the flies demonstrate recombinant phenotypes, which are either *white* or *yellow*. In the F_2 generation of Cross B, 37.2 percent of the flies demonstrate recombinant phenotypes, which are either *miniature* or *white*.

miniature because the *white, miniature* and *yellow, miniature* combinations show large recombination frequencies. Since *miniature* shows more recombination with *yellow* than with *white* (35.4 vs. 34.5), it follows that *white* is between the other two genes.

Sturtevant also suggested that the frequency of exchange could be taken as a direct measure of the distance between two genes or loci along the chromosome. Using this information, he constructed a map of the three genes on the X chromosome. In the preceding example, the distance between *yellow* and *white*

would be 0.5 map unit, and between *yellow* and *miniature*, 35.4 map units. It follows that the distance between *white* and *miniature* is about 34.9 units, which is close to that observed. One map unit is directly equated with one percent recombination between two genes.* The simple map for these three genes is shown in Figure 6.4.

---

*In honor of Morgan's work, some geneticists have suggested that one map unit be called a **centimorgan (cM)**.

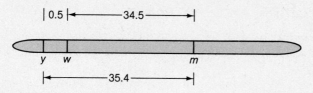

FIGURE 6.4
A simple map of the *yellow (y)*, *white (w)*, and *miniature (m)* genes on the X chromosome of *Drosophila melanogaster*. Each number represents the percentage of recombinant offspring produced in the three crosses, each involving two different genes.

In addition to these three genes, Sturtevant considered three other genes on the X chromosome and produced a more extensive map including all six genes. He and a colleague, Calvin Bridges, soon began a search for autosomal linkage in *Drosophila*. By 1923, they had clearly shown that linkage and crossing over were not restricted to sex-linked genes. Rather, they had discovered linked genes, on autosomes between which crossing over occurred.

During this work, they made another interesting observation. In *Drosophila*, crossing over was shown to occur only in females. The fact that no crossing over occurs in *Drosophila* males made the genetic analysis of autosomal inheritance and mapping much simpler to perform. However, crossing over does occur in males in most other organisms.

Although many refinements in chromosome mapping have developed since Sturtevant's initial work, his

basic principles are accepted as correct. They have been used to produce detailed chromosome maps of organisms for which large numbers of linked mutant genes are known. In addition to providing the basis for chromosome mapping, Sturtevant's findings were historically significant to the field of genetics. In 1910, the **chromosomal theory of inheritance** was still being widely disputed. Even Morgan was skeptical of the theory before conducting the bulk of his experimentation. Research has now firmly established that chromosomes contain genes in a linear order and that these genes are the equivalent of Mendel's unit factors.

Single Crossovers

Why should the relative distance between two loci influence the amount of recombination and crossing over observed between them? The basis for this variation is explained in the following analysis.

During meiosis, a limited number of crossover events occurs in each tetrad. These recombinant events occur randomly along the length of the tetrad. Therefore, the closer two loci reside along the axis of the chromosome, the less likely it is that a crossover event will occur between them. The same reasoning suggests that the farther apart two linked loci are, the more likely it is that a random crossover event will occur between them.

In Figure 6.5(a), a single crossover occurs between two nonsister chromatids, but not between the two loci; therefore, the crossover goes undetected because

FIGURE 6.5
Two cases of exchange between nonsister chromatids and the gametes subsequently produced. In (a) the exchange does not separate the alleles of the two genes, only parental gametes are formed, and the exchange goes undetected. In (b) the exchange separates the alleles, resulting in recombinant gametes.

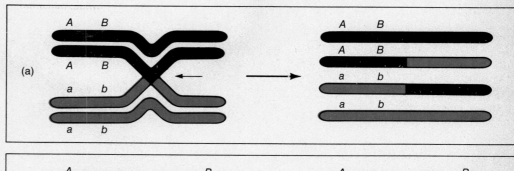

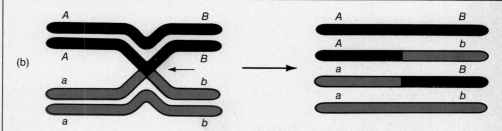

Exchange between
nonsister chromatids

Gametes

no recombinant gametes are produced. In (b), where two loci are quite far apart, the crossover occurs between them, yielding recombinant gametes.

The evidence supporting the concept that crossing over occurs in the four-strand tetrad stage will be discussed later in this chapter. However, we must point out certain consequences of this fact. When a single crossover occurs between two nonsister chromatids, the other two strands of the tetrad are not involved in this exchange and may enter a gamete unchanged. Thus, even if a crossover occurs 100 percent of the time between two linked genes, this recombinant event will subsequently be observed in only 50 percent of the gametes formed. This concept is diagrammed in Figure 6.6. For example, when 20 percent recombinant gametes are observed, crossing over actually occurred between these two loci in 40 percent of the tetrads. Therefore, the upper limit of observed crossing over is 50 percent. The general rule is that twice the percentage of recombinant gametes of tetrads are actually involved in a genetic exchange.

When two linked genes are greater than 50 map units apart, a crossover can theoretically be expected to occur between them in 100 percent of the tetrads. If this theoretical prediction is achieved, each tetrad yields equal proportions of the four gametes shown in Figure 6.6, just as if the genes were on different chromosomes and assorting independently. This occurrence was undoubtedly confusing to early geneticists and remained so until the processes of crossing over and linkage were sufficiently understood.

Multiple Crossovers

It is possible that in a single tetrad, two, three, or more exchanges will occur between nonsister chromatids as a result of several crossing over events. Double exchanges of genetic material result from **double crossovers,** as shown in Figure 6.7. For a double exchange to be studied, three genes must be investigated, each heterozygous for two alleles. Before we can determine the frequency of recombination between all three loci, we must review some simple probability calculations.

The probabilities of a single exchange occurring between the A and B or the B and C genes are directly related to the physical distance between each locus. The closer A is to B and B is to C, the less likely it is that a single exchange will occur between either of the two sets of loci. In the case of a double crossover, two separate and independent events or exchanges must occur simultaneously. **The mathematical probability of two independent events occurring simultaneously is equal to the product of the individual probabilities.**

Suppose that crossover gametes resulting from single exchanges between A and B are recovered 20 percent of the time and between B and C 30 percent of the time. The probability of recovering a double-crossover gamete arising from two exchanges, between A and B and B and C, is predicted to be $(0.2) \cdot (0.3) = 0.06$, or 6 percent. It is apparent from this calculation that the frequency of double-crossover gametes is always expected to be much lower than that of either single-crossover class of gametes.

FIGURE 6.6
The consequences of a single exchange between two nonsister chromatids occurring in the tetrad stage. Two noncrossover (parental) and two crossover (recombinant) gametes are produced.

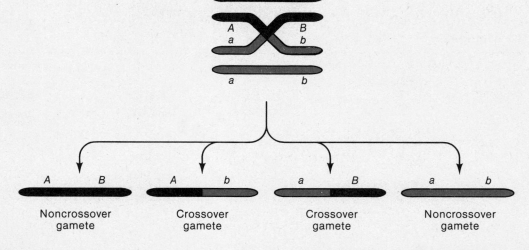

Noncrossover gamete   Crossover gamete   Crossover gamete   Noncrossover gamete

FIGURE 6.7
The results of a double exchange occurring between nonsister chromatids. The smaller arrows represent the points of exchange. Because the exchanges involve only two strands, two noncrossover gametes and two double crossover gametes are produced.

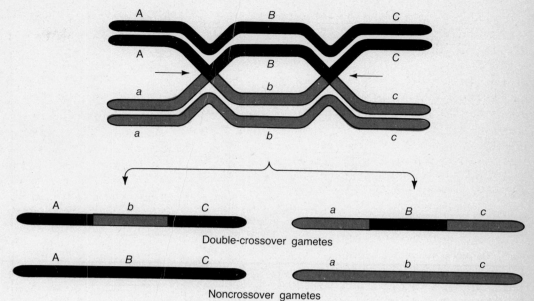

Double-crossover gametes

Noncrossover gametes

If three genes are relatively close together along one chromosome, the expected frequency of double-cross-over gametes is extremely low. For example, consider the A–B distance in Figure 6.7 to be 3 map units and the B–C distance in that figure to be 2 map units. The expected double-crossover frequency would be $(0.03) \cdot (0.02) = 0.0006$, or 0.06 percent. This translates to only 6 events in 10,000. Thus, in a mapping experiment involving closely linked genes, very large numbers of offspring are required in order to detect double-crossover events. In this example, it would be unlikely that a double crossover would be observed even if 1000 offspring were examined. If these probability considerations are extended, it is evident that even fewer triple and quadruple crossovers than double crossovers can be expected to occur.

Three-Point Mapping in *Drosophila*

The information presented in the previous section serves as the basis for simultaneously mapping three or more linked genes. To illustrate this, we will examine a situation involving three linked genes.

Three criteria must be met for a successful mapping cross:

1 The organism producing the crossover gametes must be heterozygous at all loci under consideration.

2 The cross must be constructed so that the genotypes of all gametes can be accurately determined by observing the phenotypes of the resulting off-

spring. This is necessary because the gametes and their genotypes can never be observed directly. Thus, each phenotypic class must reflect the genotype of the gametes of the parents producing it.

3 A sufficient number of offspring must be produced in the mapping experiment to recover a representative sample of all crossover classes.

These criteria are met in the three-point mapping cross from *Drosophila melanogaster* shown in Figure 6.8. In this cross, three sex-linked recessive mutant genes—*yellow* body color, *white* eye color, and *echinus* eye shape—are considered.

In the P_1 generation, males hemizygous for all three wild-type alleles are crossed to females that are homozygous for all three recessive mutant alleles. Therefore, the P_1 males are wild type with respect to body color, eye color, and eye shape. They are said to have a **wild-type** phenotype. The females, on the other hand, exhibit the three mutant phenotypes—*yellow* body color, *white* eyes, and *echinus* eye shape.

This cross produces an F_1 generation consisting of females heterozygous at all three loci and males that, because of the Y chromosome, are hemizygous for the three mutant alleles. Phenotypically, all F_1 females are wild type, while all F_1 males are *yellow*, *white*, and *echinus*. The genotype of the F_1 females fulfills the first criterion for constructing a map of the three linked genes; that is, it is heterozygous at the three loci and may serve as the source of recombinant gametes generated by crossing over. Note that because of the genotypes of the P_1 parents, all three mutant alleles are

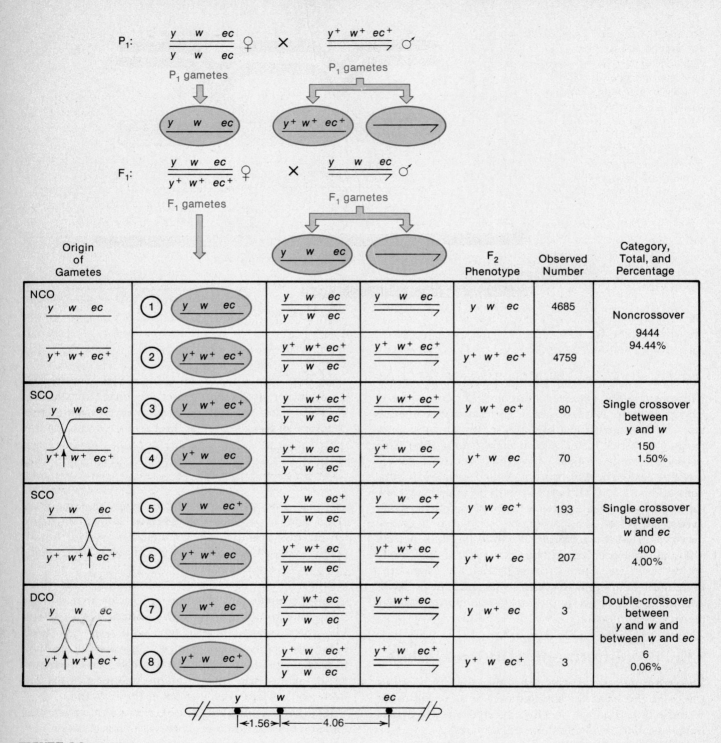

FIGURE 6.8

A three-point mapping cross involving the *yellow* (y or y⁺), *white* (w or w⁺), and *echinus* (ec or ec⁺) genes in *Drosophila melanogaster.*

on one homologue and all three wild-type alleles are on the other homologue. *Other arrangements are possible.* For example, the heterozygous F_1 female might have the *y* and *ec* mutant alleles on one homologue and the *w* allele on the other. This would occur if, in the P_1 cross, one parent was *yellow* and *echinus* and the other parent was *white.*

Now, returning to our original cross, the second criterion is met by virtue of the gametes formed by the F_1 males. Every gamete will contain either an X chromosome bearing the three mutant alleles or a Y chromosome, which is genetically inert. Whichever type participates in fertilization, the genotype of the F_1 female gamete will be expressed phenotypically in the F_2 male and female offspring. As a result, all F_1 noncrossover and crossover gametes can be detected by observing the F_2 phenotypes.

With these two criteria met, we can construct a chromosome map from the crosses illustrated in Figure 6.8. First, we must determine which F_2 phenotypes correspond to the various noncrossovers and crossover categories. Two of these can be determined immediately.

The first is the **noncrossover category** of F_2 phenotypes. These phenotypes are determined by the combination of alleles present in the parental gametes formed by the F_1 female. Each such gamete contains one or the other of the X chromosomes unaffected by crossing over. As a result of segregation, equal proportions of the two types of gametes, and subsequently the F_2 phenotypes, are produced. Because they are derived from a heterozygote, the genotypes of the two parental gametes and the phenotypes of the two F_2 phenotypes complement one another. For example, if one is wild type, the other is completely mutant. This is the case in the cross under consideration. In other situations, if one chromosome shows one mutant allele or trait, the other shows the other two mutant traits, and so on. They are therefore called **reciprocal classes** of gametes and phenotypes.

The two noncrossover phenotypes are most easily recognized because *they exist in the greatest proportion.* Figure 6.8 shows that classes (1) and (2) are present in the greatest numbers. Therefore, flies that are *yellow*, *white*, and *echinus* and those that are normal or wild type for all three characters constitute the noncrossover category and represent 94.44 percent of the F_2 offspring.

The second category that can be easily detected is represented by the double-crossover phenotypes. Because of their probability of occurrence, *they must be present in the least numbers.* Remember that this

group represents two independent but simultaneous single-crossover events. Two reciprocal phenotypes can be identified: class (7), which shows the mutant traits *yellow* and *echinus* but normal eye color; and class (8), which shows the mutant trait *white* but normal body color and eye shape. Together these double-crossover phenotypes constitute only 0.06 percent of the F_2 offspring.

The remaining four phenotypic classes represent two categories resulting from single crossovers. Classes (3) and (4), reciprocal phenotypes produced by single-crossover events occurring between the *yellow* and *white* loci, are equal to 1.50 percent of the F_2 offspring. Classes (5) and (6), constituting 4.00 percent of the F_2 offspring, represent the reciprocal phenotypes resulting from single-crossover events occurring between the *white* and *echinus* loci.

The map distances between the three loci can now be calculated. The distance between *y* and *w*, or between *w* and *ec*, is equal to the percentage of all detectable exchanges occurring between them. For any two genes under consideration, this will include all single crossovers as well as double crossovers. The latter are included because they represent two simultaneous single crossovers. For the *y* and *w* genes this includes classes (3), (4), (7), and (8), totaling 1.50% + 0.06%, or 1.56 map units. Similarly, the distance between *w* and *ec* is equal to the percentage of offspring resulting from an exchange between these two loci: classes (5), (6), (7), and (8), totaling 4.00% + 0.06%, or 4.06 map units. The map of these three loci on the X chromosome, based on these data, is shown at the bottom of Figure 6.8.

Determining the Gene Sequence

In the preceding example, the order or sequence of the three genes along the chromosome was assumed to be *y–w–ec.* In most mapping experiments, the gene sequence is not known, and this constitutes another variable in the analysis. Had the gene order been unknown in this example, it could have been easily determined. There are two methods by which to accomplish this. You should select one or the other method and adhere to its use in your own analysis.

Method I This method is based on the fact that there are only three possible orders, each containing one of the three genes in between the other two. Note that the right-to-left direction along the chromosome is meaningless. While we assumed *y–w–ec*, *ec–w–y* is fully equivalent because the data provide no basis for knowing which is correct. Therefore, only three possibilities exist, and one of them must be correct:

(I) *w–y–ec* (*y* is in the middle)

(II) *y–ec–w* (*ec* is in the middle)

(III) *y–w–ec* (*w* is in the middle)

If you use the following steps during your analysis, you will be able to determine gene order:

1 Assuming any of the three orders, first determine the **arrangement** of alleles along each homologue of the heterozygous parent giving rise to noncrossover and crossover gametes (the F_1 female in our example).

2 Determine whether a double-crossover event occurring within that arrangement will produce the **observed** double-crossover phenotypes. Remember that these phenotypes occur least frequently and can be easily identified.

3 If this order does not produce the correct phenotypes, try each of the other two orders. One must work!

These steps will be illustrated for the cross shown in Figure 6.8. The three possible orders are labeled (I), (II), and (III), as just shown. Either *y*, *ec*, or *w* must be in the middle.

1 Assuming *y* is between *w* and *ec*, the arrangement of alleles along the homologues of the F_1 heterozygote is

$$(I) \quad \frac{w \quad y \quad ec}{w^+ \ y^+ \ ec^+}$$

We know this because of the way in which the P_1 generation was crossed. The P_1 female contributed an X chromosome bearing the *w*, *y*, and *ec* alleles, while the P_1 male contributed an X chromosome bearing the w^+, y^+, and ec^+ alleles.

2 A double crossover within the above arrangement would yield the following gametes:

$$\underline{w \quad y^+ \ ec} \quad \text{and} \quad \underline{w^+ \ y \quad ec^+}$$

Following fertilization, the F_2 double-crossover phenotypes would correspond to these gametic genotypes, yielding offspring that are *white, echinus* and offspring that are *yellow*. Determination of the actual double crossovers reveals them to be *yellow, echinus* flies and *white* flies. Therefore, our assumed order is incorrect.

3 If we try the other two orders,

$$(II) \quad \frac{y \quad ec \quad w}{y^+ \ ec^+ \ w^+} \quad \text{and} \quad (III) \quad \frac{y \quad w \quad ec}{y^+ \ w^+ \ ec^+}$$

we see that arrangement II again provides **predicted** double-crossover phenotypes that do not correspond to the **actual** double-crossover phenotypes. The predicted phenotypes are *yellow, white* F_2 flies and *echinus* flies. Therefore, this order is also incorrect. However, arrangement III **will** produce the **observed** phenotypes, *yellow, echinus* flies and *white* flies. Therefore, this order, where the *w* gene is in the middle, is correct.

To summarize, utilizing this method is rather straightforward. Determine the arrangement of alleles on the homologues of the heterozygote yielding the crossover gametes. Then, test each of three possible orders to determine which one yields the observed double-crossover phenotypes. Whichever of the three does so represents the correct order. Testing the three possibilities in our example is summarized in Figure 6.9.

Method II   This method again requires that we know the arrangement of alleles along the homologues of the heterozygous parent. It also assumes that one of the three genes must be in the middle. It requires one further assumption:

Following a double-crossover event, the allele representing the middle gene will find itself present with the outside or flanking alleles present on the opposite parental homologue.

To illustrate, assume order (I), *w–y–ec*, in the arrangement

$$(I) \quad \frac{w \quad y \quad ec}{w^+ \ y^+ \ ec^+}$$

Following a double-crossover event, the *y* and y^+ alleles would find themselves switched to the arrangement

$$\frac{w \quad y^+ \ ec}{w^+ \ y \quad ec^+}$$

Following segregation, two gametes would be formed:

$$\underline{w \quad y^+ \ ec} \quad \text{and} \quad \underline{w^+ \ y \quad ec^+}$$

Since the genotype of the gamete will be expressed directly in the phenotype following fertilization, the double-crossover phenotypes will be

white, echinus flies   and   *yellow* flies

FIGURE 6.9

A summary of the three possible orders of the *white*, *yellow*, and *echinus* genes, the results of a double crossover in each case, and the resulting phenotypes produced. Note that this and subsequent figures use the conventions of w^+ or $+$, y^+ or $+$, etc., to denote the dominant allele, in contrast to the use of W or Y.

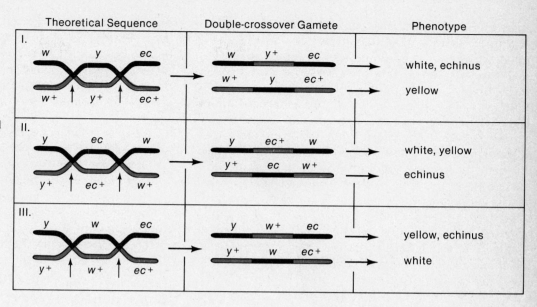

Note that the *yellow* allele, assumed to be in the middle, is now associated with the two outside markers of the other homologue, w^+ and ec^+. However, these predicted phenotypes do not coincide with the observed double-crossover phenotypes. Therefore, the *yellow* gene is not in the middle.

This same reasoning can be applied to the assumption that the *echinus* gene or the *white* gene is in the middle. In the former case, a negative conclusion will be reached. If we assume that the *white* gene is in the middle, the predicted and actual double crossovers coincide. Therefore, we conclude that the *white* gene is located between the *yellow* and *echinus* genes.

To summarize, this method is also straightforward. Determine the arrangement of alleles on the homologues of the heterozygote yielding crossover gametes. Then determine the actual double-crossover phenotypes. Simply select the single allele that has been switched so that it is now associated with two other alleles that have not been separated by crossing over.

In our example above, y, ec, and w are together in the F_1 heterozygote, as are y^+, ec^+, and w^+. In the F_2 double-crossover classes, it is w and w^+ that have been switched. The w allele is now associated with y^+ and ec^+, while the w^+ allele is now associated with the y and ec alleles. Therefore, the *white* gene is in the middle, and the *yellow* and *echinus* genes are the flanking markers.

A Mapping Problem in Maize

Having established the basic principles of chromosome mapping, we will now consider a related problem in maize where the gene sequence and interlocus distances are unknown.

This analysis differs from the preceding discussion in several ways and therefore will expand your knowledge of mapping procedures:

1 In the previous discussion the gene sequence was provided initially, and we used this information to explain how an unknown sequence is determined. In this analysis the sequence is initially unknown.

2 The previous mapping cross involved sex-linked genes. Here, autosomal genes are considered.

3 The previous cross was discussed in bits and pieces as each principle was established. Here, the analysis proceeds uninterrupted from beginning to end.

4 In the discussion of this cross we will make a transition in the use of symbols, as shown in Chapter 4. Instead of using the symbols bm^+, v^+, and pr^+, we will simply use $+$ to denote each wild-type allele. This symbol is less complex to manipulate, but requires a better understanding of mapping procedures.

When we consider three autosomally linked genes in maize, the experimental cross must still meet the same three criteria established for the X-linked genes in *Drosophila:* (1) one parent must be heterozygous for all traits under consideration; (2) the gametic genotypes produced by the heterozygote must be apparent from observing the phenotypes of the offspring; and (3) a sufficient sample size must be available for complete analysis.

In maize, the recessive mutant genes *bm* (brown midrib), *v* (virescent seedling), and *pr* (purple aleurone) are linked on chromosome 5. Assume that a female plant is known to be heterozygous for all three traits. Nothing is known about the arrangement of the mutant alleles on the maternal and paternal homologues, what sequence of genes exists, or the map distances between the genes. What genotype must the male plant have to allow successful mapping? In order to meet the second criterion, the male must be homozygous for all three recessive mutant alleles. Otherwise, offspring of this cross showing a given phenotype might represent more than one genotype, making accurate mapping impossible.

Figure 6.10 diagrams this cross. The offspring have been arranged in groups of two for each pair of reciprocal phenotypic classes. The two members of each

reciprocal class are derived from either no crossing over (**NCO**), one single-crossover event (**SCO**), or a double crossover (**DCO**).

To solve this problem, answer the following questions. It will be helpful to refer to Figures 6.10 and 6.11 while you consider them.

1 *What is the correct heterozygous arrangement of alleles in the female parent?*

Determine the two noncrossover classes, those that occur in the highest frequency. In this case, they are *+ v bm* and *pr + +*. Therefore, the arrangement of alleles on the homologues of the female parent must be as shown in Figure 6.11(a). These homologues segregate into gametes, unaffected by any recombination event. Any other arrangement of alleles could not yield the observed

FIGURE 6.10
The results of a three-point mapping cross in maize where the arrangement of alleles, the order of the genes, and the distance between them are all unknown.

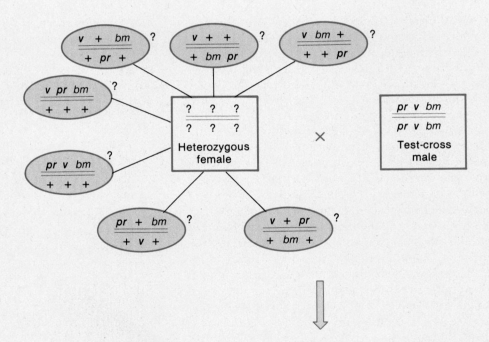

Offspring Phenotype			Number	Total Percentage	Exchange Classification
+	v	bm	230	467	Noncrossover
pr	+	+	237	42.1%	(NCO)
pr	v	+	82	161	Single-crossover
+	+	bm	79	14.5%	(SCO)
+	v	+	200	395	Single-crossover
pr	+	bm	195	35.6%	(SCO)
pr	v	bm	44	86	Double-crossover
+	+	+	42	7.8%	(DCO)

noncrossover classes. (Remember that _+ v bm_ is equivalent to _pr⁺ v bm,_ and that _pr + +_ is equivalent to _pr v⁺ bm⁺_.)

2 *What is the correct sequence of genes?*
The reasoning established in Method I will first be used to answer this question. We know that the arrangement of alleles is

$$\frac{+\ v\ bm}{pr\ +\ +}$$

But, is the assumed sequence correct? That is, will

a double-crossover event yield the observed double-crossover phenotypes following fertilization? Simple observation shows that it will not [Figure 6.11(b)]. Try the other two orders [Figure 6.11(c) and (d)], keeping the same arrangement:

$$\frac{+\ bm\ v}{pr\ +\ +}\ \text{and}\ \frac{v\ +\ bm}{+\ pr\ +}$$

Only the second case will yield the observed double-crossover classes [Figure 6.11(d)]. Therefore, the _pr_ gene is in the middle.

FIGURE 6.11
The steps used in producing a map of the three genes involved in the cross shown in Figure 6.10.

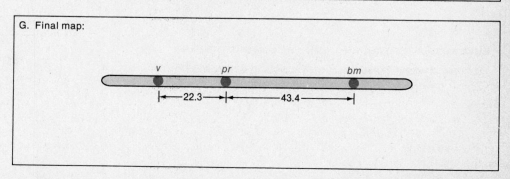

The same conclusion is reached if the problem is analyzed using Method II. In this case, no assumption of gene sequence is necessary. The arrangement of alleles in the heterozygous parent is

$$\frac{+ \ v \ bm}{pr \ + \ +}$$

The double-crossover gametes are also known:

$$pr \ v \ bm \quad \text{and} \quad + \ + \ +$$

It is obvious that the *pr* allele has shifted so as to be associated with *v* and *bm* following a double crossover. The latter two alleles were present together on one homologue, and they stayed together. Therefore, *pr* is the odd gene, so to speak, and it is in the middle.

3 *What is the distance between each pair of genes?* Having established the sequence of loci as *v pr bm*, we can now determine the distance between *v* and *pr* and between *pr* and *bm*. Remember that the map distance between two genes is calculated on the basis of all detectable recombinational events occurring between them. This includes both the single- and double-crossover events involving the two genes being considered.

Figure 6.11(e) shows that the phenotypes *v pr +*

and *+ + bm* result from single crossovers between *v* and *pr*, accounting for 14.5 percent of the offspring. By adding the percentage of double crossovers (7.8%) to the number obtained for single crossovers, the total distance between *v* and *pr* is calculated to be 22.3 map units.

Figure 6.11(f) shows that the phenotypes *v + +* and *+ pr bm* result from single crossovers between *pr* and *bm*, totaling 35.6 percent. With the addition of the double-crossover classes (7.8%) the distance between *pr* and *bm* is calculated to be 43.4 map units.

The final map for all three genes in this example is shown in Figure 6.11(g).

The Accuracy of Mapping Experiments

Until now, we have considered crossover frequencies to be directly proportional to the distance between any two loci along the chromosome. However, it is not always possible to detect all crossover events. A case in point is a double exchange that occurs between the two loci in question. As shown in Figure 6.12(a), if a double exchange occurs, the original arrangement of alleles on each nonsister homologue is recovered. Therefore, even though crossing over occurs, it is im-

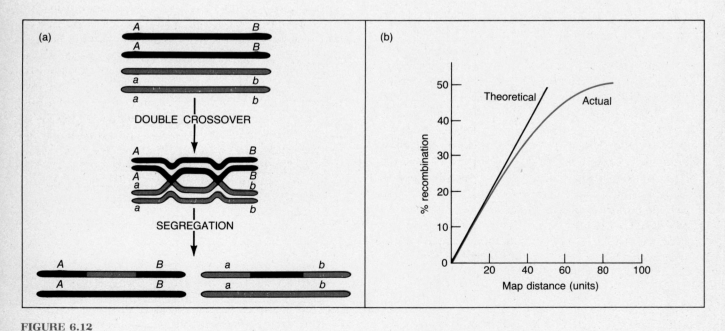

FIGURE 6.12
(a) An illustration of a double crossover that goes undetected because no rearrangement of alleles occurs. (b) A comparison of the theoretical and actual relationships between recombination and map distance, as studied in *Drosophila*, *Neurospora*, and maize. (After Perkins, 1962.)

possible to detect in these cases. This factor holds for all even-numbered exchanges between two loci.

As a result of this and other types of multiple exchanges, mapping determinations will usually slightly underestimate the actual distance between two genes. The farther apart two genes are, the greater the probability that undetected crossovers will occur. Therefore, the discrepancy is minimal for two genes that are relatively close together, but increases proportionately as the distance becomes greater. This fact is reflected in Figure 6.12(b), where the theoretical and actual relationships between recombination and map distance are graphed.

Accurate linkage distances between widely separated genes must therefore be determined by many different experiments. These experiments should utilize numerous genes in between the two in question. For example, if genes A and K are widely separated, mapping experiments using genes B, C, D, E, F, G, H, I, and J may be necessary, assuming these to be between A and K. If each distance is determined (e.g., A–B, B–C, C–D. . . J–K), then the distance between A and K is additive. That is, by adding the A–B, B–C, C–D. . . J–K distances together, we obtain a more accurate measurement of the distance between A and K.

Interference and the Coefficient of Coincidence

Still another factor tends to limit the accuracy of mapping data. This factor involves the actual reduction of the number of expected double crossovers when genes are reasonably close to one another along the chromosome. This reduction, called **interference**, will be illustrated for three-point mapping.

We have already considered the probability relationships between single- and double-crossover events. In theory, the percentage of **expected double crossovers** is predicted by multiplying the percentage of the total crossovers between each pair of genes. Remember that a double crossover (DCO) represents two single-crossover events. For example, the expected double-crossover frequency of the cross illustrated in Figures 6.10 and 6.11 may be calculated in the following manner:

$$DCO_{exp} = (0.223) \times (0.434) = 0.097 = 9.7\%$$

Frequently this predicted figure does not correspond precisely with the observed DCO frequency. Generally, there are fewer DCOs observed than predicted. In the maize cross, only 7.8 percent DCOs were observed. In some cases there are more DCOs than expected. These disparities are explained by the concept of interference, which is quantified by calculating the **coefficient of coincidence (C)**:

$$C = \frac{\text{Observed DCO}}{\text{Expected DCO}}$$

In the maize cross, we have

$$C = \frac{0.078}{0.097} = 0.804$$

Once C is calculated, interference (I) may be quantified using the simple equation

$$I = 1 - C$$

In the maize cross, we have

$$I = 1.000 - 0.804 = 0.196$$

If interference is complete and no double crossovers occur, $I = 1.0$. If fewer DCOs than expected occur, I is a positive number (as above) and **positive interference** has occurred. If more DCOs than expected occur, I is a negative number and **negative interference** has occurred.

In eukaryotic systems, positive interference is the rule. It appears that a crossover event in one region of a chromosome inhibits a second crossover in neighboring regions of the chromosome. In general, the closer genes are to one another along the chromosome, the more positive interference is observed and the lower the C value. In *Drosophila*, when three genes are clustered within 10 map units, interference is often complete and no double-crossover classes are recovered. This observation led to the proposal that interference may be explained by physical constraints that prohibit the formation of closely aligned chiasmata. A mechanical stress may be imposed on chromatids during crossing over such that one chiasma inhibits the formation of a second chiasma in the neighboring region. This interpretation is consistent with the finding that interference decreases as the genes in question are located farther apart. In the maize cross illustrated in Figure 6.11, the three genes are relatively far apart, and 80 percent of the expected double crossovers are observed.

Extensive Genetic Maps

In organisms such as *Drosophila*, maize, and the mouse, where large numbers of mutants have been discovered, extensive maps of each chromosome have been made. Partial maps of the four chromosomes of *Drosophila* are illustrated in Figure 6.13. Virtually every morphological feature of the fruit fly has been observed to be subject to mutation. The gene locus involved in determining an altered phenotype is first lo-

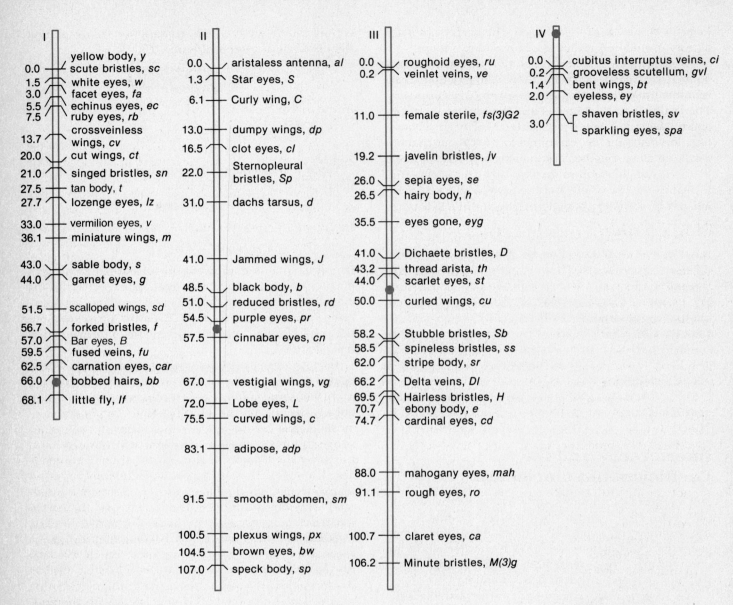

FIGURE 6.13
A partial genetic map of the four chromosomes of *Drosophila melanogaster*. The colored circle on each chromosome represents the position of the centromere.

calized to one of the four chromosomes, or linkage groups, and then mapped in relation to other linked genes of that group. As can be seen, the genetic map of the X chromosome is somewhat less extensive than autosome 2 or 3. In comparison to these three, autosome 4 is minuscule. Based on cytological evidence, the relative lengths of the genetic maps have been found to correlate with the relative physical lengths of these chromosomes.

In humans, many mutations are known. However,

matings do not produce sufficient numbers of offspring to allow mapping. Nevertheless, another technique—**somatic cell hybridization** (see Chapter 23)—has paved the way for establishing human linkage groups. In many cases, **chromosome banding techniques** (see Chapter 11) have been a useful adjunct to somatic cell hybridization studies in determining the position of numerous genes along each chromosome. Both techniques will be discussed in subsequent chapters.

OTHER ASPECTS OF GENETIC EXCHANGE

Up to this point in the chapter, many of the more complex topics involving crossing over have purposely been avoided so as not to complicate the difficult concepts involved in understanding chromosome mapping. We will now consider several topics that are essential to a comprehensive view of genetic exchange.

The Use of Haploid Organisms in Recombinational Studies

Many of the single-celled eukaryotes are haploid during the vegetative stages of their life cycle. The alga *Chlamydomonas* and the mold *Neurospora* illustrate this genetic condition. These organisms do form reproductive cells that fuse during fertilization, producing a diploid zygote. However, this structure soon undergoes meiosis, resulting in haploid vegetative cells that are then propagated by mitotic divisions. The life cycle of *Neurospora* was presented in Figure 5.3.

Haploid organisms have several important advantages in genetic studies compared with more advanced diploid eukaryotes. They can be cultured and manipulated in genetic crosses much more easily. In addition, a haploid organism contains only a single allele of each gene, which is expressed directly in the phenotype. This fact greatly simplifies genetic analysis. As a result, organisms such as *Chlamydomonas* and *Neurospora* have served as subjects of research investigations in many areas of genetics, including linkage and mapping studies.

In order to perform genetic experiments with such organisms, crosses are made, and following fertilization, the meiotic structures are isolated. Because in each of these structures all four meiotic products give rise to spores, each structure is called a **tetrad**. *This term has a different meaning here than when it was used earlier to describe a precise chromatid configuration in meiosis.* In any case, individual tetrads are isolated, and the resultant cells are grown and analyzed separately from those of other tetrads. In the results to be described, the data will reflect the proportion of tetrads that showed one combination of genotypes, the proportion that showed another combination, and so on. Thus, such experimentation is called **tetrad analysis.**

In the following sections, we will examine the use of such data to: (1) demonstrate that crossing over occurs in the four-strand stage; (2) determine the map distance between a gene and the centromere; (3) distinguish between independent assortment and linkage of two genes; and (4) map two linked genes.

Crossing Over in the Four-Strand Stage

The question of when crossing over occurs during meiosis is critical to understanding the process and consequence of genetic exchange in eukaryotes. There are two times at which crossing over can occur. First, exchange could take place before or early in meiosis in the **two-strand stage.** Alternatively, exchange could occur, as Janssen's theory suggests, at the **four-strand stage** after the chromosomes have duplicated into sister chromatids and formed tetrads.

If crossing over occurs at the two-strand stage, all four products of a single meiotic event will be recombinant gametes because each pair of sister chromatids in the tetrad is derived from one of the members of the two-strand stage. If, on the other hand, crossing over occurs between two nonsister chromatids in the four-strand stage, two parental (noncrossover) gametes and two recombinant gametes will be formed. These alternative possibilities are illustrated in Figure 6.14.

If crossing over occurs at the two-strand stage, then in a crossover experiment involving two distantly separated genes, a large percentage of offspring should be recovered that express the recombinant genotype. If two gene loci were situated so far apart that a crossover always occurred between them, 100 percent of the gametes would be of the recombinant type. In fact, as stated earlier, the maximum number of offspring recovered as a result of crossing over is 50 percent. This observation strongly favors, but does not directly prove, that exchange occurs at the four-strand stage.

To provide more conclusive evidence for the four-strand stage hypothesis, we can examine the results of an experiment with an organism from which all meiotic products may be recovered and observed. The organism used in this experiment is the ascomycete *Neurospora*, a haploid mold. Fertilization occurs, and meiosis results in the formation of a four-celled tetrad. Then, in *Neurospora*, a mitotic division of each product occurs, producing eight haploid cells called **ascospores.** Most important, the entire process takes place in a thin ascus sac that retains the haploid ascospore products in the order in which they are formed.

Carl L. Lindegren's observations of meiotic segregation in *Neurospora crassa*, the pink bread mold, strongly support the theory that crossing over occurs in the four-strand stage. He examined the various pos-

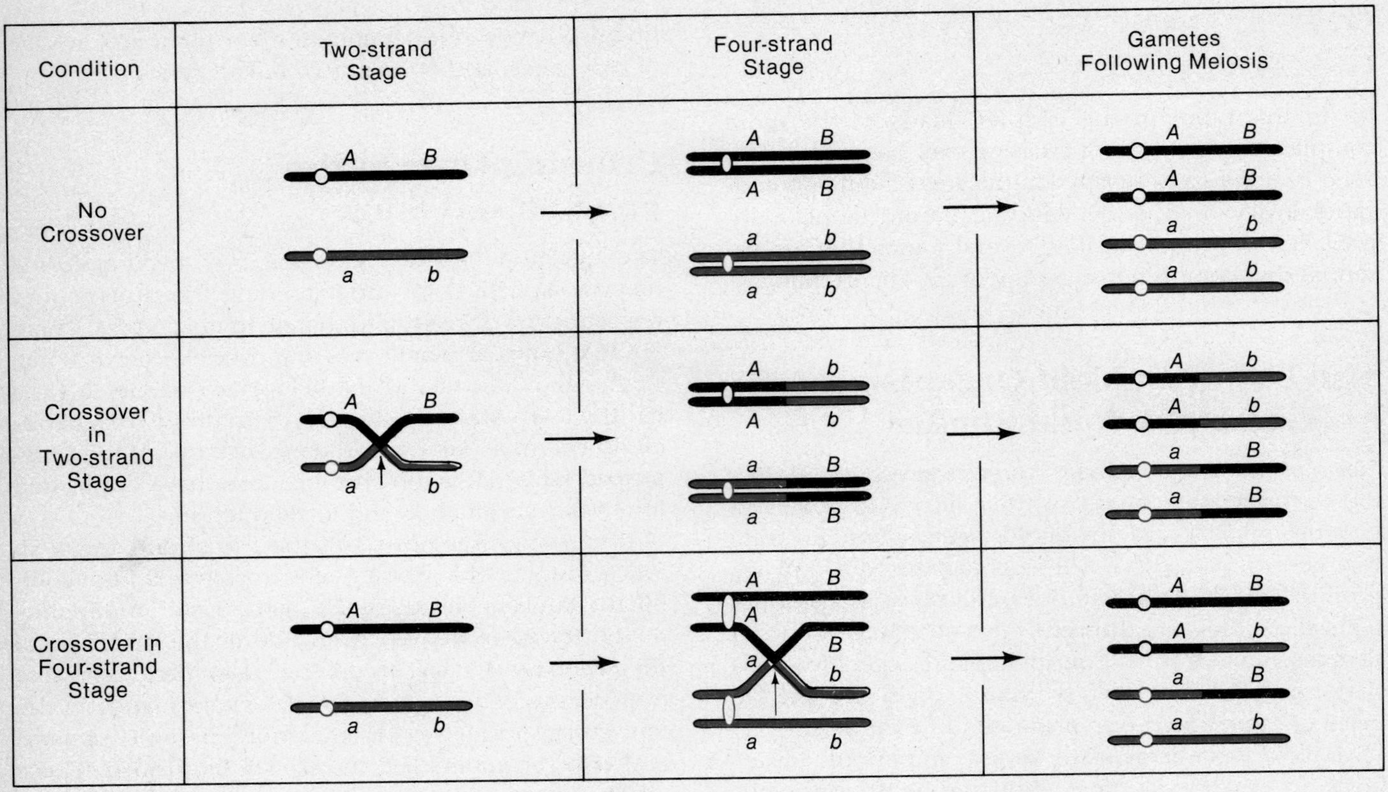

FIGURE 6.14
Comparison of the genotypes of gametes formed as a result of crossing over in the two-strand and four-strand stages with those when no crossing over occurs.

sibilities of ascospore formation resulting from a cross between an *albino* mutant strain (a) and one with normal pigmentation (+). His work focused on the results of a crossover that occurred in the region between the mutant *albino* locus and the centromere. Figure 6.15 illustrates the theoretical results for various alternatives of an exchange in this region for both the two- and four-strand stages. In case (a) no exchange occurs. In case (b) the results of crossing over in the two-strand stage are predicted. With or without a crossover event, the resulting ascus will always contain four pigmented ascospores and four unpigmented ascospores, in that sequence. The asci produced in cases (a) and (b) cannot be distinguished from each other.

If, on the other hand, a crossover occurs in this region during the four-strand stage, an alternate arrangement of ascospores in the ascus is predicted, as seen in case (c). Case (d) shows still another arrangement that occurs as a result of a slightly different exchange during the four-strand stage.

Lindegren observed asci with arrangements found in cases (a), (b), (c), and (d). His findings are consistent with the conclusion that crossing over indeed occurs in the four-strand stage and exclude crossing over in the two-strand stage, which cannot generate arrangements (c) and (d).

Similar findings have been drawn from studies of other organisms, including *Drosophila*. To date, no experiments have been reported that seriously dispute crossing over in the four-strand stage in eukaryotic organisms. The interpretation of these results is that crossing over occurs between any two nonsister chromatids of the four-strand stage. When such an event occurs, two of the four strands will be uninvolved in the exchange. Therefore, the theoretical upper limit of recombination between any two linked loci, however distantly located, is 50 percent. Crossing over also occurs between sister chromatids, but since each contains identical alleles, no new genotypic combinations are produced. Therefore, such crossing over is undetectable during genetic analysis.

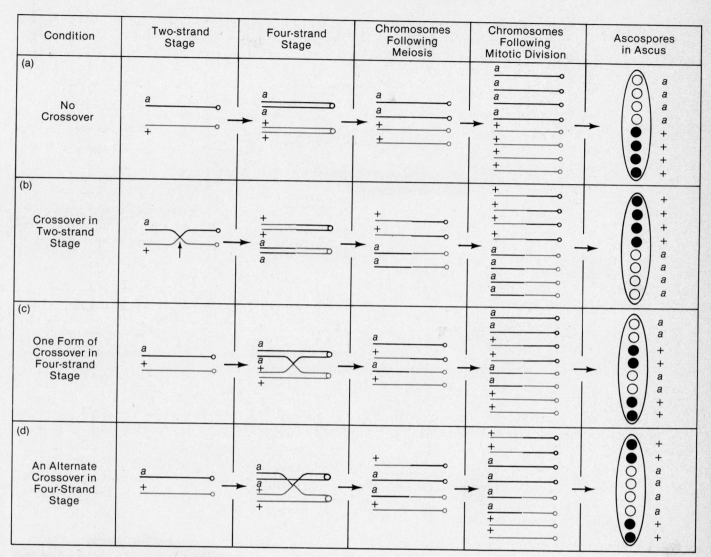

FIGURE 6.15

Four ways in which ascospore patterns can be generated in the asci of *Neurospora* as a result of genetic events. While the patterns produced in (a) and (b) cannot be distinguished from one another, these and the patterns produced in (c) and (d) were observed, leading to the conclusion that crossing over occurs in the four-strand stage.

Gene to Centromere Mapping

When a single gene is analyzed in *Neurospora*, as diagramed in Figure 6.15, the data can be used to calculate the map distance between the gene and the centromere. This process is sometimes referred to as **mapping the centromere.** It may be accomplished by experimentally determining the frequency of recombination using tetrad data.

If no crossover event between the gene under study and the centromere occurs, the pattern of ascospores in the ascus appears as shown in Figure 6.15(a) (*aaaa++++*). This pattern represents **first division segregation** because the two alleles are separated during the first meiotic division. However, a crossover event will alter this pattern, as shown in Figure 6.15(c) (*aa++aa++*) and 6.15(d) (*++aaaa++*). Actually, two other recombinant patterns may occur, depending on the chromatid orientation during the second meiotic division: *++aaaa++* and *aa++++aa*. All four of these patterns reflect **second division segregation** because the two alleles are not separated until the second meiotic division. Usually, the ordered tetrad data are condensed to reflect the genotypes of the identical ascospore pairs. Thus, six combinations are possible:

First Division Segregation

aa + +
+ + aa

Second Division Segregation

a + a +
+ a + a
+ a a +
a + + a

In order to calculate the distance between the gene and the centromere, a large number of asci resulting from a controlled cross must be scored. Using these data, the distance is calculated:

$$\frac{\frac{1}{2}(\text{second division segregant asci})}{\text{total asci scored}}$$

The recombination percentage is only one-half the number of second division segregants because crossing over in each of them has occurred in only two of the four strands during meiosis.

To illustrate, assume that *a* represents albino and + represents wild type in *Neurospora*. In crosses between the two genetic types, suppose the following data were observed:

65 first-division segregants
70 second-division segregants

The distance between *a* and the centromere is thus:

$$\frac{(\frac{1}{2})\,(70)}{135} = 0.259$$

or about 26 map units.

As the distance increases up to 50 units, in theory, all asci should reflect second division segregation. However, numerous factors prevent this. As in diploid organisms, accuracy is greatest when the gene and centromere are relatively close together.

Linkage and Gene Mapping in Haploid Organisms

Analysis similar to that described above can easily distinguish between linkage and independent assortment of two genes; it further allows mapping distances to be calculated between gene loci once linkage is established. In the following discussion, we will consider tetrad analysis in the alga, *Chlamydomonas*. With the exception that the four meiotic products are not or-

dered and *do not* undergo a mitotic division following the completion of meiosis, the general principles discussed for *Neurospora* also apply to *Chlamydomonas*.

To compare independent assortment and linkage, we will consider two theoretical mutant alleles, *a* and *b*, representing two distinct loci in *Chlamydomonas*. Suppose that 100 tetrads derived from the cross *ab* × + + yield the data shown in Table 6.1. As you can see, all tetrads produce one of three patterns. For example, all tetrads in category I produce two + + cells and two *ab* cells and are designated as **parental ditypes (P).** Category II tetrads produce two *a*+ cells and two *b*+ cells and are called **nonparental ditypes (NP).** Category III tetrads produce one cell each of the four possible genotypes and are thus termed **tetratypes (T).**

These data support the hypothesis that the genes represented by the *a* and *b* alleles are located on separate chromosomes. In order to understand why, you must refer to Figure 6.16. In parts (a) (category I) and (b) (category II) of this figure, the origin of parental and nonparental ditypes is demonstrated for two unlinked genes. If the Mendelian principle of independent assortment is in operation, approximately equal proportions of these tetrad types will be produced. That is, when the parental ditypes are equal to the nonparental ditypes, then the two genes are not linked. The data confirm this prediction. Since independent assortment has occurred, it can be concluded that the two genes are located on separate chromosomes.

The origin of category III, the **tetratypes,** is diagramed in Figure 6.16(c) and (d). The genotypes of tetrads in this category can be generated in two possible ways. Both involve a crossover event between one of the genes and the centromere. In Figure 6.16(c) the exchange occurs between gene *a* and the centromere, and in Figure 6.16(d), the exchange occurs between gene *b* and the centromere.

TABLE 6.1

Tetrad analysis in *Chlamydomonas*.

Category	I	II	III
Tetrad Type	Parental (P)	Nonparental (NP)	Tetratypes (T)
Genotypes Present	+ + + + a b a b	a + a + + b + b	+ + a + + b a b
Number of Tetrads	43	43	14

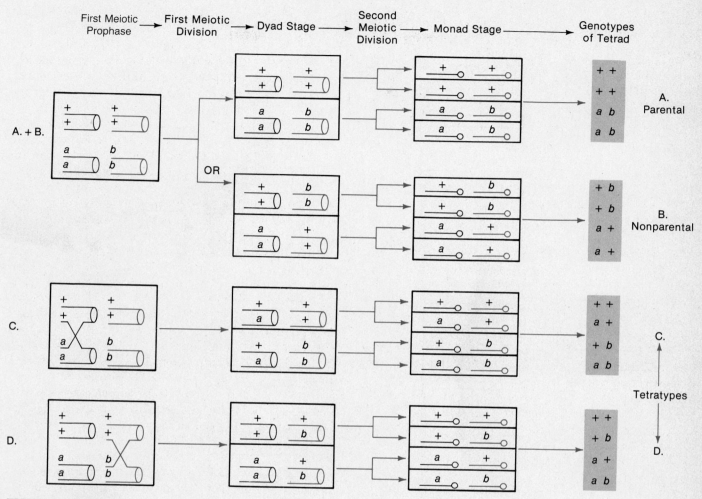

FIGURE 6.16
The origin of various genotypes found in tetrads in *Chlamydomonas* when two genes located on separate chromosomes are considered.

The production of tetratype tetrads does not alter the final ratio of the four genotypes present in the meiotic products. If the genotypes from all 100 tetrads (which yield 400 cells) are computed, 100 of each genotype are found. This 1:1:1:1 ratio is predicted according to the expectation of independent assortment.

Now consider the case where the genes *a* and *b* are linked. The same categories of tetrads will be produced. However, parental and nonparental ditypes will not necessarily occur in equal proportions; nor will the four genotypic combinations be found in equal numbers if the genotypes of all meiotic products are computed. For example, the following data might be encountered:

Category I (Parental)	Category II (Nonparental)	Category III (Tetratype)
64	6	30

Since the parental and nonparental categories are not produced in equal proportion, independent assortment is not in operation. Instead, we can conclude that the two genes are linked and proceed to determine the map distance between the two genes.

In the analysis of these data, we are concerned with the determination of which tetrad types represent genetic exchanges between the two genes. The **parental tetrads** arise only when no crossing over occurs between the two genes. The **nonparental tetrads** arise only when one type of double exchange occurs between two genes. The **tetratype tetrads** arise when either a single crossover occurs or when a second type of double exchange occurs between the two genes. The various types of exchanges described here are diagramed in Figure 6.17.

When the proportion of the three tetrad types has been determined, it is possible to calculate the map

FIGURE 6.17
The origin of various genotypes found in tetrads in *Chlamydomonas* when two genes located on the same chromosome are considered.

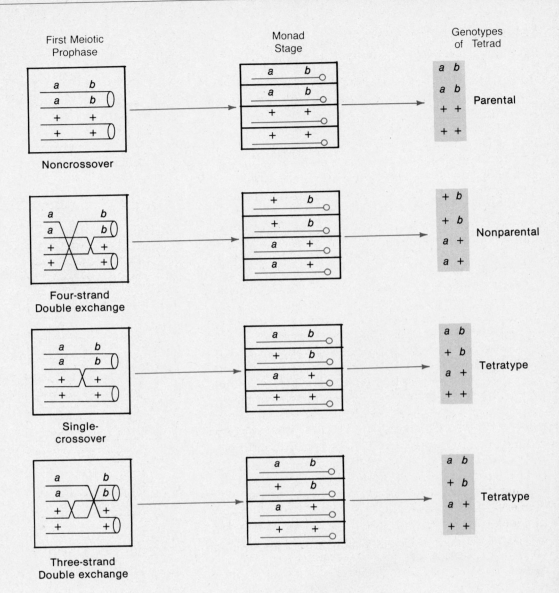

distance between the two linked genes. The following formula computes the exchange frequency, which is proportional to the map distance between the two genes:

$$\text{Exchange Frequency} = \frac{\text{NP} + \frac{1}{2}(\text{T})}{\text{Total Number of Tetrads}}$$

In this formula NP represents the nonparental tetrads; all meiotic products represent an exchange. The tetratype tetrads are represented by T; one-half of the meiotic products represents exchanges. The sum of the scored tetrads that fall into these categories is then divided by the total number of tetrads examined.

If this calculated number is multiplied by 100, it is converted to a percentage, which is directly equivalent to the map distance between the genes.

In our example, the calculation reveals that genes *a* and *b* are separated by 21 map units:

$$\frac{6 + \frac{1}{2}(30)}{100} = \frac{6 + 15}{100} = \frac{21}{100} = 0.21 \times 100 = 21\%$$

This discussion introduces linkage analysis and chromosome mapping in haploid organisms. We can extend such investigations to the consideration of three or more genes. In these cases, the gene sequence as well as map distances can be determined.

Cytological Evidence for Crossing Over

Visual proof that genetic crossing over in higher organisms is accompanied by an actual physical exchange between homologous chromosomes has been elegantly demonstrated independently by Curt Stern in *Drosophila* and by Harriet Creighton and Barbara McClintock in *Zea mays* (corn). Since the experiments are similar, we will consider only the work with corn. In Creighton and McClintock's work, two linked genes on chromosome 9 were studied. At one locus, the alleles *colorless* (c) and *colored* (C) control endosperm coloration. At the other locus, the alleles *starchy* (Wx) and *waxy* (wx) control the carbohydrate characteristics of the endosperm.

Corn plants were obtained that were heterozygous at both loci and contained unique cytological markers on one of the homologues. The markers consisted of a translocated part of chromosome 8, located at one end of chromosome 9, and a densely staining knob, located on the other end. The design of the experiment was straightforward: Obtain recombinant offspring from this heterozygous plant and examine their chromosomes to see if the cytological markers are now located on different homologues. If so, a physical exchange of chromosome segments has occurred during the crossover event.

The entire cross is illustrated in Figure 6.18. First examine the arrangement of alleles and cytological markers in the parents. Then study the genotypes and phenotypes of the offspring. The nonrecombinants are shown in the left column. In each case, the markers, alleles, and phenotypes are consistent with the prediction that no crossover occurred between the two loci. The right-hand column illustrates the arrangement of alleles and markers and the phenotypes of the four possible recombinant offspring. In each case, a crossover event has separated the two cytological markers as well as the C and wx alleles present together in the *colored, starchy* parent. Note that the *colorless, waxy* recombinant offspring (marked by a circled star) represents a unique phenotype compared with those of the nonrecombinant offspring. Observations of a variety of these rearrangements were made by Creighton and McClintock. Similar findings by Stern using *Drosophila* leave no doubt that an actual physical exchange between homologues occurs during genetic crossing over.

The Mechanism of Crossing Over

It has long been of interest to determine how and when during meiosis crossing over occurs. If we ac-

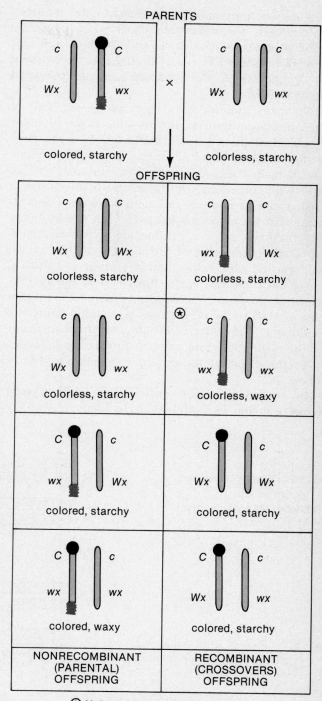

⊛ Unique recombinant phenotype

FIGURE 6.18

The phenotypes and chromosome composition observed in Creighton and McClintock's demonstration in maize that crossing over involves a breakage and rejoining process.

cept the evidence just discussed, as provided by Creighton, McClintock, and Stern, then crossing over must be considered to be the result of an actual physical exchange between homologous chromosomes. Researchers are particularly interested in the relationship between the chiasmata observed cytologically during prophase I of meiosis and the breakage and reunions presumed to occur during genetic crossing over. Are chiasmata the cytological manifestations of crossover events?

Two main theories have been proposed, both of which involve chiasmata, but in quite different ways. The **classical theory** holds that crossing over events are the result of physical strains imposed by chiasmata, which themselves occur fortuitously. While there is not necessarily a one-to-one relationship—a chiasma may or may not induce the breakage and rejoining that leads to crossing over—this theory holds that chiasmata are responsible for crossover events and clearly precede them. Since we know that chiasmata are first seen rather late in meiotic prophase I, during the diplotene stage, the classical theory predicts that crossing over occurs sometime after diplonema but before the chromosomes separate at meiotic anaphase I. The order of events presumed to occur and the resultant pairing relationships created according to the classical theory are illustrated in Figure 6.19(a).

The other proposal, called the **chiasmatype theory,** was first set forth by F. A. Janssens in the early twentieth century and later modified by John Belling and C. D. Darlington around 1930. It predicts that crossing over precedes chiasma formation and occurs early in meiotic prophase I, presumably in pachytene. Chiasmata are subsequently formed at points of genetic exchange. Thus, a chiasma present in diplotene is a cytological manifestation of one crossing over event. Chromosomes become wrapped around one another as a result of the previous exchange between homologues. These events are illustrated in Figure 6.19(b).

The chiasmatype theory has gained much support over the years, even though it does not account for all related observations. Several findings derived from the use of modern research techniques are pertinent to this subject. The first involves the discovery of DNA synthesis during meiosis. Yasuo Hotta and Herbert Stern have performed careful analysis using *Lilium* (lily) anthers, where the stages of meiotic prophase of microsporophytes can be easily identified and studied.

They have shown that a small but measurable amount of DNA replication occurs during the zygotene stage of meiotic prophase I. Recall from our earlier discussion of mitosis and meiosis in Chapter 2 that DNA replication occurs during the S phase of the cell cycle, well before mitosis and meiosis begin. Thus,

FIGURE 6.19
Two main theories proposed to account for the mechanism of crossing over.

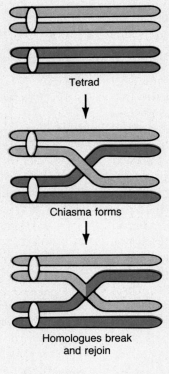

Tetrad

Chiasma forms

Homologues break and rejoin

(a) CLASSICAL THEORY

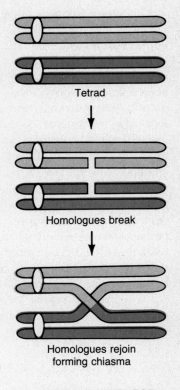

Tetrad

Homologues break

Homologues rejoin forming chiasma

(b) CHIASMATYPE THEORY

this finding has generated much interest. What is the role of this late-replicating DNA? It amounts to 0.3 percent of the total nuclear DNA; it is distributed generally among all chromosomes; and it is not strongly bonded covalently to the DNA already present. Importantly, if this DNA synthesis is inhibited, then chromosomal synapsis is also inhibited and meiosis arrests. These latter findings suggest that the newly synthesized DNA plays an important role in chromosome alignment during meiosis.

Coincidental with this DNA synthesis is the formation of the **synaptonemal complex (SC)** during the zygotene stage. The SC is found between synapsed homologues (see Figure 2.9) and is essential to chromosome alignment. Perhaps the DNA synthesis occurring during zygonema is directly related to SC formation between homologues. Interestingly, there is a possibility that the SC is also involved in crossing over. Several observations are particularly significant here. In male *Drosophila*, no crossing over occurs, and no SCs are found when spermatocytes are viewed under the electron microscope. In the silk moth, *Bombyx*, no crossing over occurs in females, and no synaptonemal complexes are observed in the oocytes. Because of the constant association between this organelle and crossing over, the participation of the SC in this event must be considered seriously.

Many geneticists feel that these events, characteristic of the zygotene stage, are essential to genetic crossing over which occurs in the subsequent pachytene stage. Just how crossing over occurs, however, is still a matter of debate and speculation. Some interesting biochemical information derived from Stern's studies of the lily bears on this issue.

It is generally held that the physical exchange occurring during crossing over involves breakage and rejoining of the DNA strands between homologues. Stern and coworkers have demonstrated the presence of an enzyme system during zygonema and pachynema capable of breaking and resealing DNA strands. For example, certain **endonucleases** are capable of "nicking" one of the two strands of the DNA double helix. Other enzymes called **ligases** are capable of rejoining broken ends of DNA strands. Additionally, a small but discernible amount of DNA synthesis occurs during the pachytene stage in *Lilium* and other organisms, such as wheat. This synthesis, unlike that occurring during zygonema, is of the repair type. It does not result in a net increase in DNA and thus does not represent replication of DNA. Inhibitors of premeiotic S-phase and zygonema DNA synthesis are not effective in preventing the DNA synthesis that occurs during pachynema.

With good reason, these discoveries have generated much enthusiasm and the hope that we will soon come to understand the precise molecular mechanisms underlying crossing over. Several popular models have been put forth. Because we have yet to discuss the specific chemistry of DNA structure, replication, and synthesis, we will postpone their presentation. Those already versed in DNA chemistry may wish to jump ahead to Figure 10.15 and the accompanying discussion. These models provide a suitable explanation for how crossing over might occur.

Mitotic Recombination

In 1939, Curt Stern demonstrated that exchanges similar to crossing over occur during mitosis in *Drosophila*. This finding was considered unusual because homologues do not normally pair up during mitosis in most organisms. However, such synapsis is apparently the rule in *Drosophila*. Since Stern's discovery, genetic exchange during mitosis has been shown to be a general event in certain fungi as well.

Stern observed small patches of mutant tissue in females heterozygous for the sex-linked recessive mutations *yellow* body and *singed* bristles. Under normal circumstances, a heterozygous female is completely wild type (gray-bodied with straight, long bristles). He explained these observations by postulating that, during mitosis in certain cells during development, homologous exchanges could occur between the loci for *yellow* and *singed* or between *singed* and the *centromere*, or that a double exchange could occur. These three possibilities are diagramed in Figure 6.20. After these three types of exchanges occur, tissues derived from the progeny cells are produced with *yellow* patches, adjacent *yellow* and *singed* patches (**twin spots**), and *singed* patches, respectively. The last type of tissue, which represents the double exchange, was found in the lowest frequency, similar to the less frequent double-crossover frequency seen in meiosis. The frequency of twin spots argues strongly against the appearance of *yellow* or *singed* tissue being due to two spontaneous but independent mutational events. If the twin spots arose in such a way, their frequency would occur in only one in many million flies.

Table 6.2 compares the relative frequency of occurrence for each of the three spotting types. These frequencies are correlated with the known genetic distances between the *yellow* and *singed* loci and the centromere, which effectively serves as an additional genetic marker. The data parallel the predicted frequencies if the exchanges occur according to the rules we established earlier for meiotic crossing over. This is

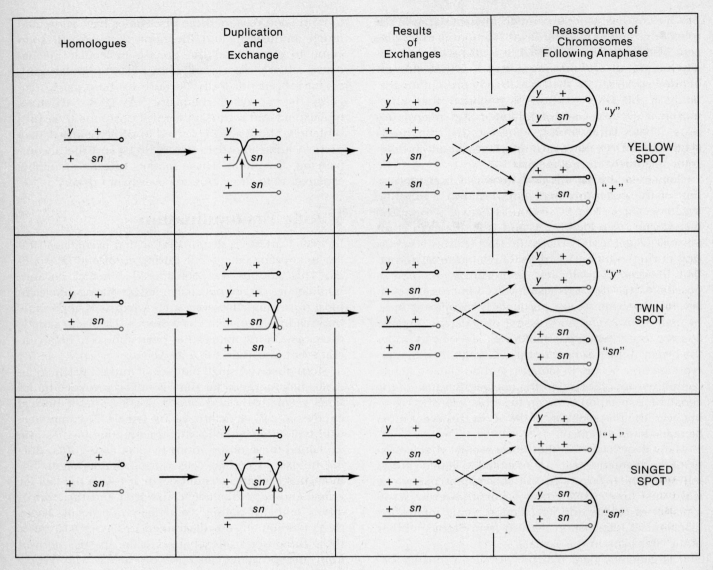

FIGURE 6.20
The production of mutant tissue in a female heterozygous for the recessive *yellow* and *singed*
alleles as a result of mitotic recombination in *Drosophila*.

TABLE 6.2

Comparison of the various
parameters related to mitotic
recombination studies in
Drosophila

Mutant Tissue Type	Region of Exchange	Relative Frequency of Exchange	Type of Exchange	Relative Distance
Yellow	*y–sn*	High	Single	21 map units
Yellow-singed (twin spot)	*sn*–centromere	Highest	Single	49 map units
Singed	*y–sn and* *sn*–centromere	Lowest	Double	————

additional evidence that the occurrence of mutant tissue spots is due to mitotic exchange in somatic cells.

In 1958 George Pontecorvo and others described a similar phenomenon in the fungus *Aspergillus*. Although the vegetative stage is normally haploid, some cells and their nuclei fuse, producing diploid cells that divide mitotically. Occasionally, crossing over occurs between linked genes during mitosis in this diploid stage so that resulting cells are recombinant. Pontecorvo referred to these events that produce genetic variability as the **parasexual cycle.** On the basis of such exchanges, genes can be mapped by estimating the frequency of recombinant classes.

As a rule, if mitotic recombination occurs at all in an organism, it does so at a much lower frequency than meiotic crossing over. While it is assumed that there is always at least one exchange per meiotic tetrad, mitotic exchange occurs in 1 percent or less of mitotic divisions in organisms that demonstrate it. Some researchers feel that the low frequency of exchange may be explained by a pairing of only some portions of homologous chromosomes.

Sister Chromatid Exchanges

In Chapter 10 we will discuss the mechanism of DNA replication. In one set of experiments, root tips of the broad bean *Vicia faba* were used as a source of dividing cells. In addition to establishing the mode of replication in eukaryotic cells, these experiments revealed that exchanges sometimes occur between sister chromatids of mitotic figures (see Figure 10.5).

Recall that in mitosis homologous chromosomes do not usually undergo synapsis. In the metaphase stage of mitosis, the number of distinct chromosome figures is equal to the diploid number. Each mitotic figure consists of two sister chromatids joined at the region of the centromere. These sister chromatids represent one chromosome that has duplicated itself, and they are therefore genetically identical. The chromatids will subsequently separate and move to opposite poles during anaphase.

Identification and study of sister chromatid exchanges (SCEs) are now facilitated by several modern fluorescent staining techniques. One experiment using fluorescent staining is illustrated in Figure 6.21. If cells are allowed to replicate for several generations in the presence of the thymidine analogue bromodeoxyuridine (BrdU), and the cells are then shifted to medium lacking the analogue, chromatids with or without BrdU are then distinguishable. Chromatids with both strands containing BrdU fluoresce less brightly than chromatids with only one strand of the double helix containing the analogue. BrdU is a molecule that

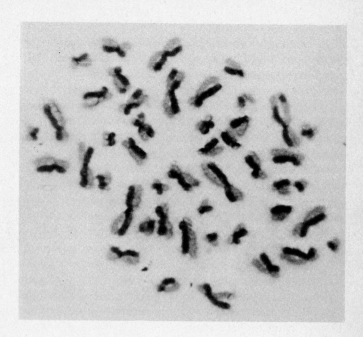

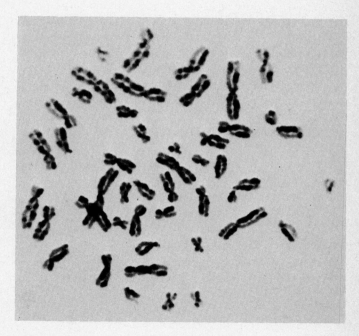

FIGURE 6.21

Demonstration of sister chromatid exchanges (SCEs) in mitotic chromosomes derived from the lymphocytes of a normal individual (a) and one with Bloom syndrome (b). Chromatids containing the thymidine analogue BrdU in one DNA strand fluoresce more brightly than those containing the analogue in both strands. (Micrographs courtesy of R. S. K. Chaganti from Chaganti, Schonberg, and German, 1974.)

quenches fluorescence; thus, the greater the amount of the analogue present in a chromatid, the less it fluoresces. In Figure 6.21 numerous instances of SCE events may be detected.

While the significance of sister chromatid exchanges is still uncertain, several observations have led to a renewed interest in this phenomenon. It is known, for example, that agents that induce chromosome damage (such as viruses, X-rays, ultraviolet light, and certain chemical mutagens) also increase the frequency of sister chromatid exchanges. This frequency of SCEs is also elevated in **Bloom syndrome,** an autosomal recessive disorder in humans. This rare genetic disease is characterized by retardation of growth, a great sensitivity of the facial skin to the sun, abnormal immunological function, and a predisposition to cancer. The chromosomes from cultured leucocytes and fibroblasts derived from homozygotes (*bl/bl*) are very fragile and unstable when compared with those derived from unaffected (+/+) and heterozygous (*bl/*+) individuals (Figure 6.21). Increased breaks and rearrangements between nonhomologous chromosomes are observed in addition to excessive amounts of sister chromatid exchanges.

The mechanisms of exchange between nonhomologous chromosomes and between sister chromatids may prove to be similar because the frequency of both events increases substantially in individuals with genetic disorders. These findings suggest that further study of sister chromatid exchange may contribute to an increased understanding of recombination mechanisms and the relative stability of normal and genetically abnormal chromosomes.

WHY DIDN'T MENDEL FIND LINKAGE?

We conclude this chapter by examining a modern-day interpretation of the experiments that serve as the cornerstone of transmission genetics—the crosses with garden peas performed by Mendel.

It has often been said that Mendel had extremely good fortune in his classical experiments with the garden pea. In none of his crosses did he encounter apparent linkage relationships between any of the seven mutant characters. Had Mendel obtained highly variable data characteristic of linkage and crossing over, these unorthodox ratios might have hindered his successful analysis and interpretation of data.

The simplest explanation for the absence of linkage is that each of the seven genes was located on a different linkage group or chromosome. Since *Pisum sativum* has a haploid number of 7, this speculation has been widely publicized. An article by Stig Blixt, reprinted on page 141 in its entirety from *Nature*, demonstrates the inadequacy of this hypothesis and explains why Mendel did not encounter ratios characteristic of linkage and crossing over.

SUMMARY AND CONCLUSIONS

This chapter examines linkage and the associated phenomenon of crossing over during meiosis. The concept of linkage implies that genes are coupled during gamete formation by virtue of their presence on the same chromosome. As a result, linked genes are not free to undergo independent assortment. However, genetic exchange as a result of crossing over offers a mechanism for reshuffling various alleles of heterozygous loci during gamete formation.

Early in this century, geneticists realized that crossing over could serve as the experimental basis for the construction of chromosome maps. We have described the criteria necessary for setting up crosses to detect crossing over. Further, we have discussed the rationale for analyzing the data of such crosses, leading to the determination of the sequences and interlocus distances between genes. For those organisms where large numbers of mutations are known, particularly *Drosophila*, corn, and the mouse, extensive genetic maps are now available. Such information is valuable because it refines our knowledge of the organization of the genetic material.

The concept of genetic recombination is of great interest in genetics. During gamete formation, genetic exchange leads to increased variability in offspring. Besides providing tremendous diversity within members of each species, this variability is an important part of organic evolution.

While the initial insights into linkage and chromosome mapping were gained using diploid organisms, these genetic phenomena have also been investigated in haploid eukaryotes, including *Chlamydomonas* and *Neurospora*. Studies with the latter organism provided strong evidence that the process of crossing over occurs during the four-strand tetrad stage of meiosis. While it is generally accepted that this recombinational event is the result of an actual breakage and rejoining between nonsister chromatids, other questions remain. For example, the precise relationships between chiasmata and crossing over and between synapsis, crossing over, and the synaptonemal complex

WHY DIDN'T GREGOR MENDEL FIND LINKAGE?

I t is quite often said that Mendel was very fortunate not to run into the complication of linkage during his experiments. He used seven genes and the pea has only seven chromosomes. Some have said that had he taken just one more, he would have had problems. This, however, is a gross oversimplification. The actual situation, most probably, is shown in Table 1. This shows that Mendel worked with three genes in chromosome 4, two genes in chromosome 1, and one gene in each of chromosome[s] 5 and 7. It seems at first glance that, out of the 21 dihybrid combinations Mendel theoretically could have studied, no less than four (that is, *a-i*, *v-fa*, *v-le*, *fa-le*) ought to have resulted in linkages. As found, however, in hundreds of crosses and shown by the genetic map of the pea[1], *a* and *i* in chromosome 1 are so distantly located on the chromosome that no linkage is normally detected. The same is true for *v* and *le* on the one hand, and *fa* on the other, in chromosome 4. This leaves *v-le*, which ought to have shown linkage.

TABLE 1

Relationship between modern genetic terminology and character pairs used by Mendel.

Character Pair Used by Mendel	Alleles in Modern Terminology	Located in Chromosome
Seed colour, yellow–green	*I–i*	1
Seed coat and flowers, coloured–white	*A–a*	1
Mature pods, smooth expanded–wrinkled indented	*V–v*	4
Inflorescences, from leaf axils– umbellate in top of plant	*Fa–fa*	4
Plant height, 1 m–around 0.5 m	*Le–le*	4
Unripe pods, green–yellow	*Gp–gp*	5
Mature seeds, smooth–wrinkled	*R–r*	7

Mendel, however, seems not to have published this particular combination and thus, presumably, never made the appropriate cross to obtain both genes segregating simultaneously. It is therefore not so astonishing that Mendel did not run into the complication of linkage, although he did not avoid it by choosing one gene from each chromosome.

Weibullsholm Plant Breeding Institute, Landskrona, Sweden and
Centro Energia Nucleare na Agricultura, Piracicaba, SP, Brazil

STIG BLIXT

Received March 5; accepted June 4, 1975.
[1]Blixt, S. 1974. In *Handbook of genetics*, ed. R. C. King. New York: Plenum Press.

are not yet resolved. Nevertheless, the mechanism of crossing over, and particularly its biochemical basis, is still of great interest to geneticists.

Recombination also occurs between mitotic chromosomes. Such exchanges occur between nonsister chromatids only in those organisms such as *Drosophila* where homologous chromosomes synapse in somatic cells. While of no direct genetic consequence, recombination also occurs between sister chromatids during mitosis. These are referred to as sister chromatid exchanges (SCEs). An elevated frequency of such

occurrences exists in the human disorder, Bloom syndrome.

We have concluded this chapter by reprinting an article entitled "Why didn't Gregor Mendel find linkage?" by Stig Blixt. While it has long been assumed that Mendel studied seven characters, each located on a separate chromosome, evidence now suggests that several loci were, in fact, linked. However, in each such case, the loci were sufficiently far apart to preclude the detection of linkage.

PROBLEMS AND DISCUSSION QUESTIONS

1 What is the significance of recombination to the process of evolution?

2 Describe the cytological events that suggest that crossing over occurs during the first meiotic prophase.

3 Why does more crossing over occur between two distantly linked genes than between two genes that are very close together on the same chromosome?

4 Why is a 50 percent recovery of single-crossover products the upper limit, even when crossing over *always* occurs between two linked genes?

5 Why are double-crossover events expected in lower frequency than single-crossover events?

6 What is the proposed basis for positive interference?

7 What essential criteria must be met in order to execute a successful mapping cross?

8 The genes *dumpy* wing (*dp*), *clot* eye (*cl*), and *apterous* wing (*ap*) are linked on chromosome 2 of *Drosophila*. In a series of two-point mapping crosses, the following genetic distances were determined:

$$
\begin{array}{ll}
dp\text{--}ap & 42 \\
dp\text{--}cl & 3 \\
ap\text{--}cl & 39
\end{array}
$$

What is the sequence of the three genes?

9 Consider two hypothetical recessive autosomal genes *a* and *b*. Where a heterozygote is test-crossed to a double homozygous mutant, predict the phenotypic ratios under the following conditions:
(a) *a* and *b* are located on separate autosomes.
(b) *a* and *b* are linked on the same autosome but are so far apart that a crossover always occurs between them.
(c) *a* and *b* are linked on the same autosome but are so close together that a crossover almost never occurs.
(d) *a* and *b* are linked on the same autosome about 10 map units apart.

10 In corn, colored aleurone (in the kernels) is due to the dominant allele *R*. The recessive allele *r*, when homozygous, produces colorless aleurone. The plant color (not the kernel color) is controlled by the gene pair *Y* and *y*. The dominant *Y* gene results in green color, while the homozygous presence of the recessive *y* gene causes the plant to appear yellow. In a test cross between a plant of unknown genotype and phenotype and a plant that is homozygous recessive for both traits, the following progeny were obtained:

Colored green	88
Colored yellow	12
Colorless green	8
Colorless yellow	92

Explain how these results were obtained by determining the exact genotype and phenotype of the unknown plant, including the precise association of the two genes on the homologues.

11 In the cross shown below involving two linked genes, *ebony* (*e*) and *claret* (*ca*), in *Drosophila*, where crossing over does not occur in males,

$$\frac{e \quad ca^+}{e^+ \quad ca} \times \frac{e \quad ca^+}{e^+ \quad ca}$$

offspring were produced in a 2 +: 1 *ca*: 1 *e* phenotypic ratio. These genes are 30 units apart on chromosome 3. What contribution did crossing over in the female make to these phenotypes?

12 With two pairs of genes involved (*P/p* and *Z/z*), a test cross (*ppzz*) with an organism of unknown genotype indicated that the gametes produced were in the following proportions:

PZ, 42.4%; *Pz*, 6.9%; *pZ*, 7.1%; and *pz*, 43.6%

Draw all possible conclusions from these data.

13 Two different female *Drosophila* were isolated, each heterozygous for the autosomally linked genes *b* (*black body*), *d* (*dachs tarsus*), and *c* (*curved wings*). These genes are in the order *d*—*b*—*c*, with *b* being closer to *d* than to *c*. Shown below is the genotypic arrangement for each female along with the various gametes formed by both. Identify which categories are noncrossovers (NCO), single crossovers (SCO), and double crossovers (DCO) in each case. Then, indicate the relative frequency in which each will be produced.

Female A

$$\frac{d\ b\ +}{+\ +\ c}$$

Female B

$$\frac{d\ +\ +}{+\ b\ c}$$

⇩ ————————Gametes———————— ⇩

(1) *d b c*		(5) *d + +*		(1) *d b +*		(5) *d b c*
(2) *+ + +*		(6) *+ b c*		(2) *+ + c*		(6) *+ + +*
(3) *+ + c*		(7) *d + c*		(3) *d + c*		(7) *d + +*
(4) *d b +*		(8) *+ b +*		(4) *+ b +*		(8) *+ b c*

14 In *Drosophila*, a cross was made between females expressing the three sex-linked recessive traits, *scute* (*sc*) bristles, *sable body* (*s*), and *vermilion eyes* (*v*), and wild-type males. In the F_1, all females were wild type, while all males expressed all three mutant traits. The cross was carried to the F_2 generation and 1000 offspring were counted with the results shown below. No determination of sex was made in the F_2 data.

Phenotype	Offspring
sc s v	314
+ + +	280
+ s v	150
sc + +	156
sc + v	46
+ s +	30
sc s +	10
+ + v	14

(a) Determine the genotypes of the P_1 and F_1 parents, using proper nomenclature.
(b) Determine the sequence of the three genes and the map distance between them.
(c) Are there more or fewer double crossovers than expected? Calculate the coefficient of coincidence. Does this represent positive or negative interference?

15 Another cross in *Drosophila* involved the recessive, sex-linked genes *yellow* (*y*), *white* (*w*), and *cut* (*ct*). A female that was yellow-bodied and white-eyed with normal wings was crossed to a male whose eyes and body were normal but whose wings were cut. The F_1 females were wild type for all three traits, while the F_1 males expressed the yellow-body, white-eye traits. The cross was carried to an F_2, and only male offspring were tallied. On the basis of the data shown below, a genetic map was constructed.

Phenotype	Male Offspring
$y + ct$	9
$+ w +$	6
$y\ w\ ct$	90
$+ + +$	95
$+ + ct$	424
$y\ w +$	376
$y + +$	0
$+ w\ ct$	0

(a) Diagram the genotypes of the F_1 parents.

(b) Construct a map, assuming that *white* is at locus 1.5 on the X chromosome.

(c) Were any double-crossover offspring expected?

(d) Could the F_2 female offspring be used to construct the map? Why or why not?

16 In *Drosophila*, *Dichaete* (*D*) is a chromosome 3 mutation with a dominant effect on wing shape. It is lethal when homozygous. The genes *ebony* (*e*) and *pink* (*p*) are chromosome 3 recessive mutations affecting the body and eye color, respectively. Flies from a *Dichaete* stock were crossed to homozygous *ebony*, *pink* flies, and the F_1 progeny, with a *Dichaete* phenotype, were back-crossed to the *ebony*, *pink* homozygotes. The results of this back cross were

Phenotype	Number
Dichaete	401
ebony, pink	389
Dichaete, ebony	84
pink	96
Dichaete, pink	2
ebony	3
Dichaete, ebony, pink	12
wild type	13

(a) Diagram this cross, showing the genotypes of the parents and offspring of both crosses.

(b) What is the sequence and interlocus distance between these three genes?

17 In *Drosophila*, these genes occur on chromosome 3: *fz*—*frizzled*; *h*—*hairy*; and *eg*—*eagle*. The cross

$$\frac{+\ +\ +}{fz\ h\ eg} \times \frac{fz\ h\ eg}{fz\ h\ eg}$$

yielded the following progeny phenotypes:

+	+	+	40
fz	+	eg	5
fz	h	eg	42
+	h	+	7
+	+	eg	2
fz	h	+	2
fz	+	+	1
+	h	eg	1

(a) Which parent was the male (remember that crossing over does not occur in male *Drosophila*)?

(b) Construct a map of these three genes, indicating the proper sequence and distances between the genes.

(c) Has there been positive or negative interference during this cross? Show your calculation.

18 In *Drosophila*, two mutations, *Stubble* (*Sb*) and *curled* (*cu*), are linked on chromosome 3. *Stubble* is a dominant gene that is lethal in a homozygous state, and *curled* is a recessive gene. If a female of the genotype

$$\frac{Sb\ cu}{+\ +}$$

is to be mated to detect recombinants among her offspring, what male genotype would you choose as a mate?

19 In *Drosophila*, a heterozygous female for the sex-linked recessive traits *a*, *b*, and *c* was crossed to a male that was phenotypically *a b c*. The offspring occurred in the following phenotypic ratios:

+	b	c	460
a	+	+	450
a	b	c	32
+	+	+	38
a	+	c	11
+	b	+	9

No other phenotypes were observed.

(a) What is the genotypic arrangement of the alleles of these genes on the X chromosomes of the female?

(b) Determine the correct sequence and construct a map of these genes on the X chromosome.

(c) What progeny phenotypes are missing? Why?

20 *Drosophila melanogaster* has one pair of sex chromosomes (XX or XY) and three autosomes, referred to as chromosomes 2, 3, and 4. A male fly with very short legs was discovered by a genetics student. Using this male, the student was able to establish a pure breeding stock of this mutant and found that it was recessive. This mutant was then incorporated into a stock containing the recessive gene *black* (body color located on chromosome 2) and the recessive gene *pink* (eye color located on chromosome 3). A female from the homozygous *black, pink, short* stock was then mated to a wild-type male. The F$_1$ males of this cross were all wild type and were then back-crossed to the homozygous *b p sh* females. The F$_2$ results appeared as follows:

	Wild	Pink*	Black, short	Black, pink, short
Females	63	58	55	69
Males	59	65	51	60

*Pink indicates that the other two traits are wild type, and so on.

(a) Based on these results, the student was able to assign *short* to a linkage group (a chromosome). Which one was it? Include a step-by-step reasoning.

(b) The experiment was subsequently repeated making the reciprocal cross, F$_1$ females back-crossed to homozygous *b p sh* males. It was observed that 85 percent of the offspring fell into the above classes, but that 15 percent of the offspring were equally divided among *b + p*, *b + +*, *+ sh p*, and *+ sh +* phenotypic males and females. How can these results be explained and what information can be derived from the data?

21 In a cross in *Neurospora* involving two alleles *B* and *b*, the following tetrad patterns were observed. Calculate the distance between the locus and the centromere.

Tetrad Pattern	Number
BBbb	36
bbBB	44
BbBb	4
bBbB	6
BbbB	3
bBBb	7

22 In *Neurospora*, the cross $a+ \times +b$ yielded only two types of ordered tetrads in approximately equal numbers:

	Spore Pair			
	1–2	3–4	5–6	7–8
Tetrad Type 1	$a+$	$a+$	$+b$	$+b$
Tetrad Type 2	$++$	$++$	ab	ab

What can be concluded?

23 Below are two sets of data derived from crosses in *Chlamydomonas*, involving three genes represented by the mutant alleles *a*, *b*, and *c*. Determine as much as you can concerning genetic arrangement of these three genes relative to one another. Describe the expected results of Cross 3.

	Genes	Tetrads		
Cross	Involved	P	NP	T
1	*a & b*	36	36	28
2	*b & c*	79	3	18
3	*a & c*			

24 Describe the rationale of the experiment which demonstrates that crossing over occurs in the four-strand stage in *Neurospora*.

25 In *Chlamydomonas*, a cross $ab \times ++$ yielded the following unordered tetrad data where *a* and *b* are linked.

(1)	$++$ $++$ ab ab 38	(4)	ab $a+$ $+b$ $++$ 17
(2)	$++$ ab $++$ ab 5	(5)	ab $++$ $+b$ $a+$ 2
(3)	$a+$ $a+$ $+b$ $+b$ 6	(6)	ab $+b$ $a+$ $++$ 3

(a) Identify the categories representing parental ditypes (P), nonparental ditypes (NP), and tetratypes (T).

(b) Explain the origin of category (2).

(c) Determine the map distance between *a* and *b*.

26 Why were there more "twin spots" observed by Stern than *singed* spots in his study of somatic crossing over? If he had been studying *tan* body color (locus 27.5) and *forked* bristles (locus 56.7) on the X chromosome of heterozygous females, what relative frequencies of tan spots, forked spots, and "twin spots" would you predict might occur?

27 Are mitotic recombinations and sister chromatid exchanges effective in producing genetic variability in an individual? In the offspring of individuals?

28 What possible conclusions can be drawn from the observations that no synaptonemal complexes are observed in male *Drosophila* and female *Bombyx* and that no crossing over occurs in these respective organisms?

29 A female of genotype

$$\frac{a \quad b \quad c}{+ \quad + \quad +}$$

produces 100 meiotic tetrads. Of these, 68 show no crossover events. Of the remaining 32, 20 show a crossover between *a* and *b*, 10 show a crossover betrween *b* and *c*, and 2 show a double-crossover between *a* and *b* and between *b* and *c*. Of the 400 gametes produced, how many of each of the 8 different genotypes will be produced? Assuming the order *a–b–c* and the allele arrangement shown above, what is the map distance between these loci?

30 If three linked autosomal genes, *a*, *b*, and *c*, in *Drosophila* were to be mapped, why would it be inadequate to cross a heterozygous female.

$$\frac{a \quad b \quad c}{+ \quad + \quad +}$$

to a male of identical genotype? What male should be selected?

SELECTED READINGS

ALLEN, G. E. 1978. *Thomas Hunt Morgan: The man and his science.* Princeton, N. J.: Princeton University Press.

BAKER, W. K. 1965. *Genetic analysis.* Boston: Houghton Mifflin.

BARRATT, R. W., et al. 1954. Map construction in *Neurospora crassa. Adv. in Genet.* 6: 1–93.

BLIXT, S. 1975. Why didn't Gregor Mendel find linkage? *Nature* 256: 206.

CATCHESIDE, D. G. 1977. *The genetics of recombination.* Baltimore: University Park Press.

CHAGANTI, R., SCHONBERG, S., and GERMAN, J. 1974. A manyfold increase in sister chromatid exchange in Bloom's syndrome lymphocytes. *Proc. Natl. Acad. Sci.* 71: 4508–12.

CREIGHTON, H. S., and McCLINTOCK, B. 1931. A correlation of cytological and genetical crossing over in *Zea mays. Proc. Natl. Acad. Sci.* 17: 492–97.

DRESSLER, D., and POTTER, H. 1982. Molecular mechanisms in genetic recombination. *Ann. Rev. Biochem.* 51: 727–61.

DUNN, L. C. 1965. *A short history of genetics.* New York: McGraw-Hill.

GARCIA-BELLIDO, A. 1972. Some parameters of mitotic recombination in *Drosophila melanogaster. Molec. Genet.* 115: 54–72.

GRANT, V. 1975. *Genetics of flowering plants.* New York: Columbia University Press.

GRELL, R. F., ed. 1974. *Mechanisms in recombination.* New York: Plenum Press.

HOTCHKISS, R. D. 1974. Molecular basis for genetic recombination. *Genetics* 78: 247–57.

HOTTA, Y., TABATA, S., and STERN, H. 1984. Replication and nicking of zygotene DNA sequences: Control by a meiosis-specific protein. *Chromosome* 90: 243–53.

HOWELL, S. H., and STERN, H. 1971. The appearance of DNA breakage and repair activities in the synchronous meiotic cycle of *Lilium. J. Mol. Biol.* 55: 357–78.

KING, R. C. 1970. The meiotic behavior of the *Drosophila* oocyte. *Int. Rev. Cytol.* 28: 125–68.

KUSHEV, V. F. 1974. *Mechanisms of genetic recombination.* New York: Plenum Press.

LATT, S. A. 1981. Sister chromatid exchange formation. *Ann. Rev. Genet.* 15: 11–56.

LINDSLEY, D. L., and GRELL, E. H. 1972. *Genetic variations of* Drosophila melanogaster. Washington, D.C.: Carnegie Institute of Washington.

MESELSON, M. S., and RADDING, C. M. 1975. A general model for genetic recombination. *Proc. Natl. Acad. Sci.* 72: 358–61.

MOENS, P. B. 1977. The onset of meiosis. In *Cell biology—A comprehensive treatise*, vol. 1, *Genetic mechanisms of cells*, ed. L. Goldstein and D. M. Prescott, pp. 93–109. New York: Academic Press.

MORGAN, T. H. 1911. An attempt to analyze the constitution of the chromosomes on the basis of sex-linked inheritance in *Drosophila. J. Exp. Zool.* 11: 365–414.

NEUFFER, M. G., JONES, L., and ZOBER, M. 1968. *The mutants of maize.* Madison, Wis.: Crop Science Society of America.

PERKINS, D. 1962. Crossing-over and interference in a multiply marked chromosome arm of *Neurospora. Genetics* 47: 1253–74.

PONTECORVO, G. 1958. *Trends in genetic analysis.* New York: Columbia University Press.

RADDING, C. M. 1978. Genetic recombination: Strand transfer and mismatch repair. *Ann. Rev. Biochem.* 47: 847–80.

STAHL, F. W. 1979. *Genetic recombination.* San Francisco: W. H. Freeman.

STERN, H., and HOTTA, Y. 1973. Biochemical controls in meiosis. *Ann. Rev. Genet.* 7: 37–66.

———. 1974. DNA metabolism during pachytene in relation to crossing over. *Genetics* 78: 227–35.

STURTEVANT, A. H. 1913. The linear arrangement of six sex-linked factors in *Drosophila,* as shown by their mode of association. *J. Exp. Zool.* 14: 43–59.

———. 1965. *A history of genetics.* New York: Harper & Row.

TAYLOR, J. H., ed. 1965, *Selected papers on molecular genetics.* New York: Academic Press.

VOELLER, B. R., ed. 1968. *The chromosome theory of inheritance: Classical papers in development and heredity.* New York: Appleton-Century-Crofts.

von WETTSTEIN, D., RASMUSSEN, S. W., and HOLM, P. B. 1984. The synaptonemal complex in genetic segregation. *Ann. Rev. Genet.* 18: 331–414.

WAKSVIK, H., MAGNUS, P., and BERG, K. 1981. Effects of age, sex and genes on sister chromatid exchange. *Clin. Genet.* 20: 449–54.

WHITEHOUSE, H. L. K. 1965. *Towards an understanding of the mechanisms of heredity.* New York: St. Martin's Press.

———. 1982. *Genetic recombination: Understanding the mechanisms.* New York: Wiley.

WOLFF, S., ed. 1982. *Sister chromatid exchange.* New York: Wiley-Interscience.

7

Quantitative Inheritance, Phenotypic Expression, and Heritability

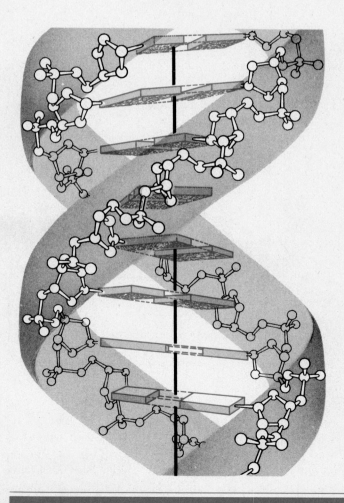

In preceding chapters we have dealt with phenotypic variation that is easily classified into distinct traits. Pea plants may be tall or dwarf; squash shape may be spherical, disk shaped, or elongated; and fruit-fly eye color red, brown, scarlet, or white. These phenotypes are examples of **discontinuous variation** where discrete phenotypic categories exist. There are many other traits in a population that demonstrate considerably more variation and are not easily categorized into distinct classes. Such phenotypes represent **continuous variation.** For example, in addition to his work with sweet peas, Mendel experimented with beans. In a cross between a purple- and a white-flowered strain, the F_1 was purple. However, the F_2 contained not only purple- and white-flowered bean plants, but ones with numerous intermediate shades. He was unable to explain these results satisfactorily, but recognized that they were inconsistent with most of his data derived from sweet peas. It was not until more than 50 years later that the inheritance of characters exhibiting continuous variation was explained.

It is now known that traits exhibiting continuous variation are often controlled by two or more genes and are termed **polygenic.** In cases where several genes make additive contributions to the phenotype, the trait is said to exhibit **quantitative or continuous variation.**

In this chapter, we will examine such patterns of inheritance in which a phenotypic trait is controlled by genes at two or more loci. We will also outline the statistical tools used by geneticists to study traits that exhibit continuous variation. In addition, we will consider the effects of nongenetic factors on gene expression. These include environmental effects and **heritability,** a concept used by geneticists to estimate the degree of genetic and environmental influence on the expression of traits controlled by genes at many loci.

QUANTITATIVE INHERITANCE: POLYGENES

In the late nineteenth century, Josef Gottlieb Kolreuter showed that when tall and dwarf tobacco plants were crossed, the F_1 generation was not all tall, as Mendel found with garden peas. Instead, the individual plants were all intermediate in height. When the F_2 generation was examined, individuals showed continuous variation in height, ranging from tall to dwarf. The majority of the F_2 plants were intermediate like the F_1, but a few were as tall or dwarf as the P_1 parents. These distributions are depicted in histograms in Figure 7.1. Note that the F_2 data have assumed a normal distribution, as evidenced by the bell-shaped curve drawn over the top of the bars in the histogram.

At the turn of the century, geneticists noted that many characters in different species had similar patterns of inheritance, such as height and stature in humans, seed size in the broad bean, grain color in wheat, and kernel number and ear length in corn. In each case, offspring in the succeeding generation seemed to be a blend of their parents' characteristics.

The issue of whether continuous variation could be accounted for in Mendelian terms caused considerable controversy in the early 1900s. Those who adhered to the Mendelian explanation of inheritance suggested that a large number of factors or genes were responsible for the observed patterns. This proposal, called the **multiple-factor** or **multiple-gene hypothesis,** implied that many factors or genes contributed to the phenotype in a cumulative or quantitative way. However, some geneticists argued that Mendel's unit factors could not account for the blending of parental phenotypes characteristic of these patterns of inheritance. As a result, geneticists became skeptical of Mendel's ideas and did not accept them for some time.

By 1920, the conclusions of several critical sets of experiments had been disclosed. These findings largely resolved the controversy and demonstrated that Mendelian factors could account for continuous variation. In one experiment, Edward M. East performed crosses between two strains of the tobacco plant *Nicotiana longiflora*. The fused inner petals of the flower, or **corollas,** of strain A were decidedly shorter than the corollas of strain B. With only minor variation, each strain was true breeding. Thus, the differences between them were clearly under genetic control.

When plants from the two strains were crossed, the F_1, F_2, and selected F_3 data (Figure 7.2) demonstrated a very distinct pattern. The F_1 generation contained corollas that were intermediate in length, compared with the P_1 varieties, and showed only minor variabil-

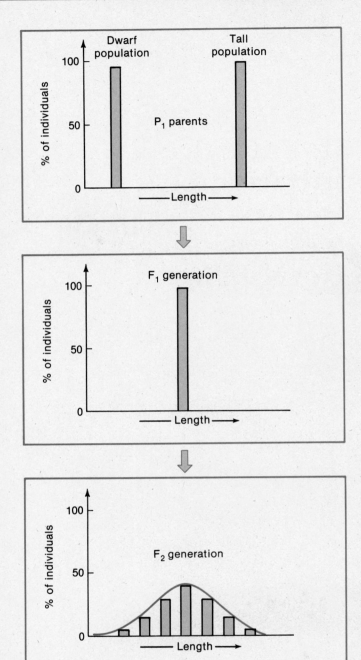

FIGURE 7.1
Histograms showing the percentage of individuals expressing various height phenotypes in a cross between dwarf and tall tobacco plants carried to an F_2 generation (Kolreuter's experiment).

ity among individuals. When the cross was carried to the F_2 generation, however, the range of corolla lengths increased considerably. While corolla lengths of the P_1 plants were about 40 mm and 94 mm, the F_1 generation contained plants with corollas that were all

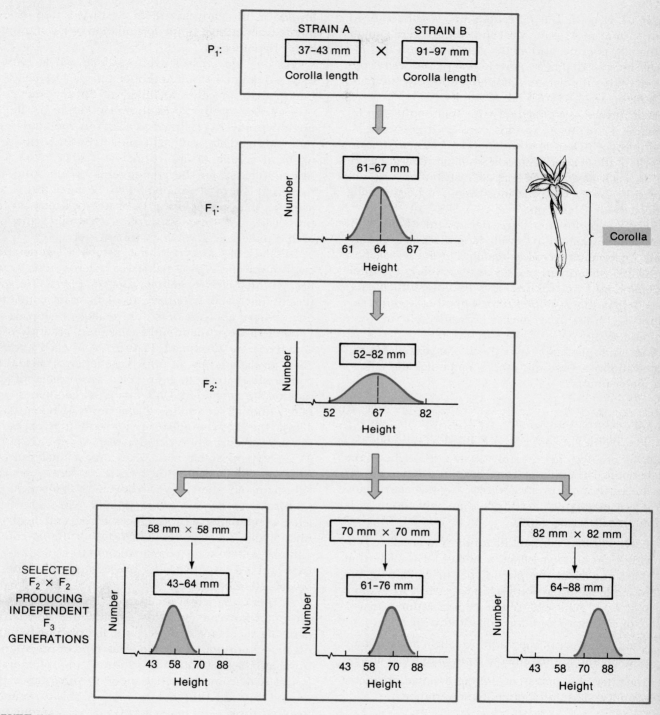

FIGURE 7.2

The F_1, F_2, and selected F_3 results of a cross between two strains of *Nicotiana* with different corolla lengths. Plants of strain A vary from 37 to 43 mm, while plants of strain B vary from 91 to 97 mm (experiments of E. M. East).

about 64 mm. In the F_2 generation, lengths ranged from 52 mm to 82 mm. Most individuals were similar to their F_1 parents, and as the deviation from this average increased, fewer and fewer plants were observed. When the data are plotted graphically (number vs. length), a bell-shaped curve results.

East further experimented with this population by selecting F_2 plants of various corolla lengths and allowing them to produce independent F_3 generations. Several of these F_3 generations are illustrated in Figure 7.2. In each case, a bell-shaped distribution was observed in the F_3, with the most frequent height being similar to the F_2 parents.

East's experiments thus demonstrated that although the variation in corolla length seemed continuous, experimental crosses resulted in the segregation of distinct phenotypic classes as observed in the three independent F_3 categories. This finding strongly suggested that the multiple-factor hypothesis could account for traits that deviate considerably in their expression.

The multiple-factor hypothesis, suggested by the observations of East and others, embodies the following major points:

1 Characters under such control can usually be quantified by measuring, weighing, counting, etc.

2 Two or more gene pairs, located throughout the genome, account for the hereditary influence on the phenotype in an additive way. Because many genes may be involved, inheritance of this type is often called **polygenic.**

3 Each pair *may* contain, in addition to an allele providing an additive effect, a nonadditive allele that does not affect the phenotype. In explaining the multiple-factor hypothesis, it is easiest to assume that each gene pair under consideration consists of one additive and one nonadditive allele.

4 The total effect of each additive allele at each gene is small and approximately equivalent to all others.

5 Together, the genes controlling a single character produce substantial phenotypic variation.

6 This genetic variation is enhanced by environmental factors. For example, despite their genetic basis, height in humans is influenced by individual diets, and plant characters are influenced by rainfall.

7 Analysis of polygenic traits requires the study of large numbers of progeny from a population of organisms.

These points can be illustrated by Herman Nilsson-Ehle's experiments involving grain color in wheat performed in the early twentieth century. These experiments were critical to the formulation of the multiple-factor hypothesis.

Nilsson-Ehle crossed white-grained wheat with a dark-red-grained strain and obtained an F_1 generation intermediate in color. Members of the F_1 generation were then crossed, and as shown in Figure 7.3, the F_2 phenotypes were classified as dark red, moderate red, red, intermediate red, light red, light-light red, and white in a ratio of approximately $1:6:15:20:15:6:1$, respectively. When the components of this ratio are totaled, the sum of 64 is obtained. Because this is the same as that found in the F_2 of trihybrid crosses, he proposed that three gene pairs controlled the observed pattern of grain-color inheritance.

Nilsson-Ehle assumed that in the F_1 generation each of the three gene pairs was heterozygous, being the product of two homozygous P_1 plants. He proposed that each heterozygous gene pair contained one additive and one nonadditive allele. Following independent assortment and fertilization, an array of F_2 genotypes are produced (Figure 7.3). If each additive allele, designated by an uppercase letter (A, B, and C), contributes equally to grain color, seven different gradations are generated. Only one of 64 genotypes contains either all six additive alleles or all six nonadditive alleles; these F_2 genotypes are identical to those of the P_1 dark red and white parents, respectively. Six of the 64 genotypes contain either five additive alleles or one additive allele, accounting for the moderate red and light-light red phenotypes, respectively. Fifteen of the 64 genotypes contain either four or two additive alleles, combinations producing the red and light red phenotypes, respectively. Finally, 20 of the 64 combinations contain some combination of three additive alleles and are responsible for the most prevalent phenotype, intermediate red.

This explanation fits nicely with the data. Multiple-factor inheritance as just described is now an acceptable mechanism to account for phenotypes displaying continuous variation. This type of inheritance, in which alleles contribute additively to a phenotype, is different from any other mode of inheritance discussed thus far. Although the Nilsson-Ehle experiment involved three gene pairs, there is no reason why two, four, or more gene pairs cannot function in controlling various phenotypes. If two gene pairs were involved in this type of cross, only five F_2 categories, in a $1:4:6:4:1$ ratio, would be expected. On the other hand, as four, five, six, or more gene pairs become involved, greater and greater numbers of classes would be expected to appear in more complex ratios. The number of phenotypes and the expected F_2 ratios of crosses involving up to five gene pairs are illustrated in Figure 7.4.

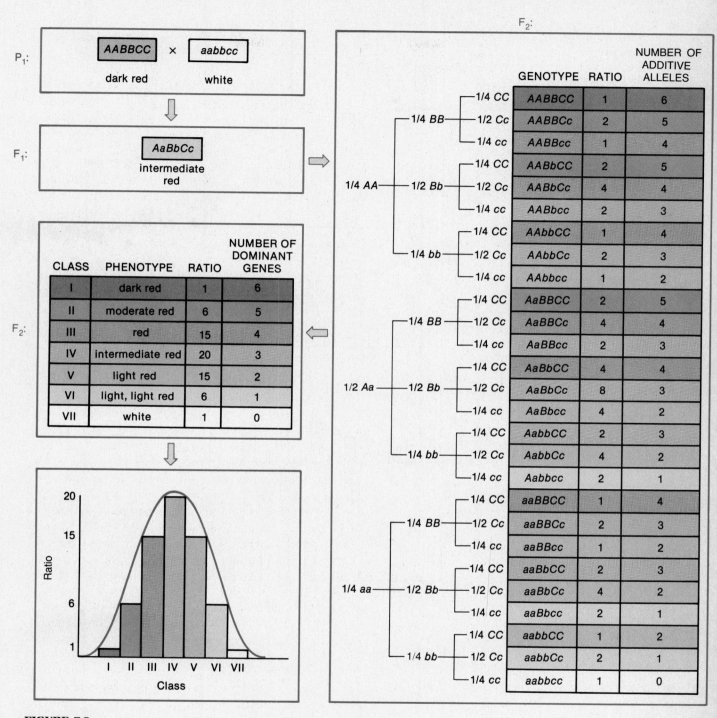

FIGURE 7.3

A detailed explanation of the F$_2$ results of a cross involving polygenic inheritance, where three genes are contributing to grain color in wheat.

FIGURE 7.4
The results of crossing two heterozygotes where polygenic inheritance is in operation with one to five gene pairs.

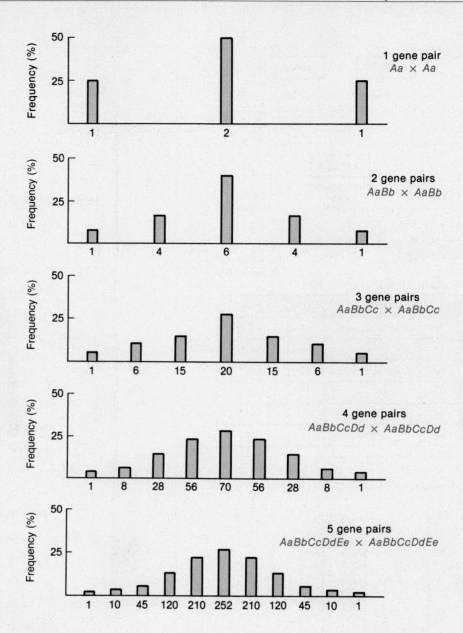

If the ratio of F_2 individuals resembling either of the P_1 parents can be determined, then the number of gene pairs involved (n) may be calculated using the following simple formula:

$$\frac{1}{4^n} = \text{ratio of } F_2 \text{ individuals resembling } \textit{either } P_1 \text{ class}$$

In Figure 7.3, 1/64 of the F_2 are either dark red *or* white like the P_1 classes; this ratio can be substituted on the right side of the equation prior to solving for *n*:

$$\frac{1}{4^n} = \frac{1}{64}$$

$$\frac{1}{4^3} = \frac{1}{64}$$

$$n = 3$$

Table 7.1 lists the ratio and the number of F_2 phenotypic classes produced in crosses involving up to five gene pairs.

Polygenic control is a significant concept because it is believed to be the mode of inheritance for a vast number of traits. For example, height, weight, and stature in all plants and animals, grain yield in crops, beef and milk production in cattle, egg production in chickens, as well as skin color and intelligence in humans are thought to be under polygenic control. Thus, knowledge of this mode of inheritance is of prime importance in animal breeding, agriculture, and sociological studies of humans. In each case of polygenic inheritance, the genotype establishes the potential range in which a particular phenotype may fall, while environmental factors determine exactly how much of the potential will be realized. In the crosses described in this section we have assumed an optimal environment, which minimizes variation resulting from that source.

ANALYSIS OF POLYGENIC TRAITS

Because traits inherited in a quantitative fashion cannot be described in the same way as Mendelian traits, it is necessary to provide a way of evaluating the inheritance of quantitative characters. In most cases, the outcome of quantitative genetic experiments is a set of measurements that can be expressed as a frequency diagram (Figure 7.5). These measurements, made from one or a series of crosses, are taken as a sample of all possible crosses that could have been made in the experimental population. If we had made an infinite number of such measurements, the frequency distribution would take on the shape of the curve in Figure 7.5, which is a normal, or bell-shaped, distribution

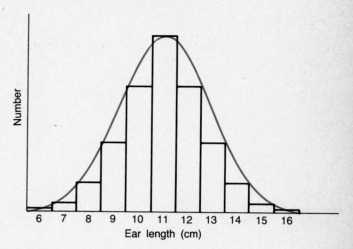

FIGURE 7.5
Frequency distribution of ear length in corn.

curve. Obviously, in many cases, it is inconvenient or impossible to count all the individuals in an experimental population or to collect exhaustive amounts of data. But gathering a smaller data sample that is unbiased, and yet representative of the measured group, is often a problem. In addition, individual variation is a universal attribute of living systems, and this variability is reflected in the data gathered from observations of organisms. How can we distinguish between normal variation that is strictly due to chance and that due to an experimental variable?

The solution to these and other problems with data analysis has been to employ the techniques of **statistics.** Statistical analysis serves three purposes:

1 Data can be mathematically reduced to provide a **descriptive summary** of the sample.

2 Data from a small but random sample can be utilized to infer information about groups larger than those from which the original data were obtained **(statistical inference).** The sample values are called **statistics** (symbolized by Roman letters) and are used as an estimate of the value for the population. These population values are called **parameters** (symbolized by Greek letters).

3 Two or more sets of experimental data may be compared to determine whether they represent significantly different populations of measurements.

Several statistical methods are useful in the analysis of traits that exhibit a normal distribution, including the mean, variance, standard deviation, and standard error of the mean.

TABLE 7.1
Determination of number of gene pairs from polygenic inheritance patterns.

n	Ratio	Number of Distinct F_2 Phenotypic Classes
1	1/4	3
2	1/16	5
3	1/64	7
4	1/256	9
5	1/1024	11

The Mean

The distribution of phenotypic values for ear length in Figure 7.5 tends to cluster around a central value of 11 cm. This clustering is called **central tendency,** the most common measurement of which is the mean ($\overline{X}$). The mean is simply the arithmetic average of a set of measurements or data and is calculated as

$$\overline{X} = \frac{\sum X_i}{n} \qquad (7.1)$$

where $\overline{X}$ is the mean, $\sum X_i$ represents the sum of all individual values in the sample, and n is the number of individual values.

Although the mean provides a descriptive summary of the sample, it is in itself of limited value. All values in the sample may be clustered near the mean, or they may be distributed widely around it. Figure 7.6 shows normal or symmetrical distributions with identical means, but a widely different range of values. These contrasting conditions represent a different type and amount of variation within the sample. Whether due to chance or to one or more experimental variables, such variation creates the need for methods to describe variation around the mean.

Variance

As seen in Figure 7.6(a), the range of values on either side of the mean will determine the width of the distribution curve. Measurement of the variation in a sample is called the variance (s^2 or V) and is used as an estimate of the variation present in an infinitely large population (σ^2). The variance for a sample is calculated as:

$$s^2 = V = \frac{\sum (X_i - \overline{X})^2}{n - 1} \qquad (7.2)$$

where the sum (Σ) of the squared differences between each measured value (X_i) and the mean ($\overline{X}$) is divided by one less than the total sample size ($n - 1$). To avoid the numerous subtraction functions necessary in calculating s^2 for a large sample, we can convert the equation to its algebraic equivalent:

$$s^2 = V = \frac{\sum X_i^2 - n\overline{X}^2}{n - 1} \qquad (7.3)$$

The variance is a valuable measure of sample variability. Two distributions may have identical means ($\overline{X}$), yet vary considerably in their concentration around the mean. The variance represents the average squared deviation of the measurements from the mean. Obviously, the larger the variance, the greater

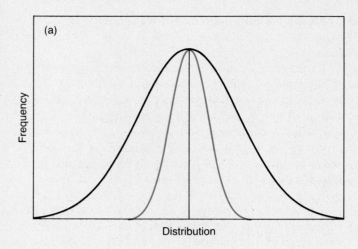

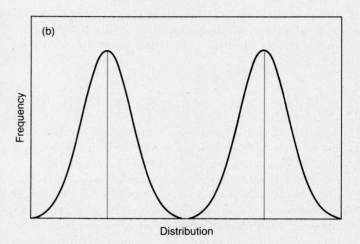

FIGURE 7.6

(a) Two normal frequency distributions with the same mean value, but different amounts of variation. (b) Two normal distributions with different mean values, but the same amounts of variation.

the range of values on either side of the mean. The estimation of variance has been particularly valuable in determining the degree of genetic control of traits where the immediate environment also influences the phenotype.

Standard Deviation

Since the variance is a squared value, its unit of measurement is also squared (cm^2, mg^2, etc.). To express variation around the mean in the original units of measurement, it is necessary to calculate the square root of the variance, a term called the **standard deviation (s):**

$$s = \sqrt{s^2} \qquad (7.4)$$

Table 7.2 shows what fraction of the individual values within a normal or bell-shaped distribution is included with different multiples of the standard deviation. The mean plus or minus one standard deviation $(\overline{X} \pm 1s)$ includes 68 percent of all values in the sample. Over 95 percent of all values are found within two standard deviations $(\overline{X} = 2s)$. As such, an important descriptive summary of a set of data is provided.

Standard Error of the Mean

To estimate how much the means of other similar samples drawn from the same population might vary, we can calculate the **standard error of the mean** $(S_{\overline{X}})$:

$$S_{\overline{X}} = \frac{s}{\sqrt{n}} \qquad (7.5)$$

where s is the standard deviation and $\sqrt{n}$ is the square root of the sample size. The standard error of the mean is a measure of the accuracy of the sample mean, that is, the variation of sample mean in replications of the experiment. Because it can be expected that the standard deviation of mean values will reflect less variance than the standard deviation of a set of individual measurements, the standard error is always less than the standard deviation. Since the standard error of the mean is computed by dividing s by $\sqrt{n}$, it is always a smaller value.

PHENOTYPIC EXPRESSION

Gene expression is often discussed as if the genes operate in a closed, "black box" system and as if the presence or absence of functional products directly determines the collective phenotype of an individual.

TABLE 7.2
Sample inclusion for various s values.

Multiples of s	% of Sample Included
$\overline{X} \pm 1s$	68.3
$\overline{X} \pm 1.96s$	95
$\overline{X} \pm 2s$	95.5
$\overline{X} \pm 3s$	99.7

The situation is actually much more complex. Gene products must function within the internal milieu of the cell; cells must interact with one another in various ways; and the organism must survive under diverse environmental influences. Thus, gene expression and the resultant phenotype are often modified through the interaction between an individual's particular genotype and the internal and external environment.

The degree of environmental influence may vary from inconsequential to subtle to very strong. Subtle interactions are the most difficult to detect and document, and have led to unresolvable "nature-nurture" conflicts in which scientists debate the relative importance of genes versus environment. How easily such conflicts are resolved depends on the characteristic being investigated, and even then it is sometimes impossible to provide definitive information. In this section we will deal with some of the variables known to modify gene expression.

Penetrance and Expressivity

Some mutant genotypes are always expressed by a distinct phenotype, while others produce a proportion of individuals whose phenotypes cannot be distinguished from wild type. The variable expression of a particular trait may be quantitatively studied by determining penetrance and expressivity. The percentage of individuals that show at least some degree of expression of a mutant genotype defines the **penetrance** of the mutation. For example, many mutant alleles in *Drosophila* are said to overlap with wild type. Flies homozygous for the recessive mutant gene *eyeless* yield phenotypes that range from the presence of normal eyes to the complete absence of one or both eyes (Figure 7.7). If 15 percent of mutant flies show the wild-type appearance, the mutant gene is said to have a penetrance of 85 percent. On the other hand, the range of expression of the genotype defines the **expressivity.** In the case of *eyeless*, although the average reduction of eye size is one-fourth to one-half, the range of expressivity is from complete loss of both eyes to completely normal eyes.

Examples such as the expression of the *eyeless* phenotype have provided the basis for experiments designed to determine the causes of extensive variation. If the laboratory environment is held constant and extensive variation is still observed, the genetic background may be investigated. It is possible that other genes are influencing or modifying the *eyeless* phenotype. On the other hand, if it is experimentally proven that the genetic background is not the cause for the phenotypic variation, environmental variables

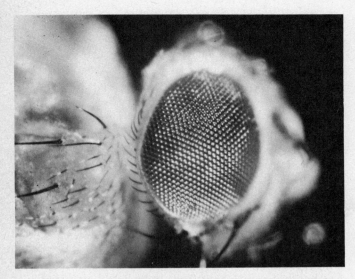

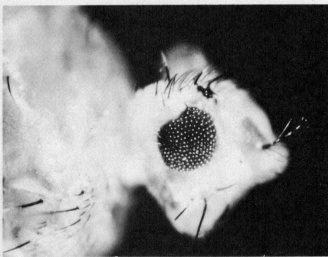

FIGURE 7.7
Photomicrographs of the wild-type and *eyeless* phenotypes in *Drosophila melanogaster*. (Courtesy of Carolina Biological Supply.)

such as temperature, humidity, and nutrition may be examined. In the case of the *eyeless* phenotype, it has been determined experimentally that both genetic background and environmental factors influence its expression.

Genetic Background: Suppression and Position Effects

With only certain exceptions, it is difficult to assess the specific relationship between the **genetic background** and the expression of a gene responsible for determining the potential phenotype.

The influence of genetic background may occur in two ways. First, the expression of other genes throughout the genome may have an effect on the phenotype produced by the gene in question. The phenomenon of **genetic suppression** is an example. Mutant genes such as *suppressor of vermilion (su-v)*, *suppressor of forked (su-f)*, and *suppressor of Hairy-wing (su-Hw)* in *Drosophila* completely or partially restore the normal phenotype to an organism that is homozygous (or hemizygous) for these recessive genes. For example, flies homozygous for both *vermilion* (a bright red eye color mutation) and *su-v* have eyes with wild-type color.

The phenomenon of suppression occurs in a wide variety of organisms. In microorganisms, where molecular studies are easier to perform, some suppressor gene products are molecules that function in the genetic translation process by correcting certain errors produced by the mutation of other genes. Transfer RNA in mutant form has been shown to have such an effect. It has also been hypothesized that a suppressor gene product might provide an alternate metabolic route to bypass a block in a biosynthetic pathway caused by the primary mutation. Thus, suppressor genes are excellent examples of the genetic background modifying primary gene effects.

Second, the physical location of a gene in relation to other genetic material may influence its expression. Such a situation is called a **position effect.** For example, if a gene is included in a **translocation** event (in which a region of a chromosome is relocated), the expression of the gene may be affected. This is particularly true if the gene is translocated to an area of the chromosome composed of **heterochromatin.** These regions are genetically inert and often appear condensed in interphase.

An example of a position effect involves female *Drosophila* heterozygous for the sex-linked recessive eye-color mutant *white (w).* If the region of the X chromosome containing the wild-type w^+ allele is translocated so that it is close to heterochromatin, expression of the w^+ allele is modified. Before translocation, the w^+/w genotype results in a wild-type brick red eye color. After translocation, the dominant effect of the w^+ allele is reduced. Instead of having a red color, the eyes are variegated, or mottled with red and white patches. A similar position effect is produced if a heterochromatic region is relocated next to the *white* locus on the X chromosome. Apparently, heterochromatic regions inhibit the expression of neighboring genes. Many other organisms exhibit a position effect, providing proof that alteration of the normal arrangement of genetic information can modify its expression.

Temperature Effects

Because chemical activity depends on the kinetic energy of the reacting substances, which in turn depends on the surrounding temperature, we can expect temperature to influence variation in phenotypes. For example, the evening primrose produces red flowers at 23°C and white flowers at 18°C. Siamese cats and Himalayan rabbits exhibit dark fur in regions of the nose, ears, and paws because the body temperatures of these extremities are slightly cooler. In these cases, it appears that an enzyme that is functional in pigment production at the lower temperatures of the extremities loses its catalytic function at the slightly higher temperatures throughout the rest of the body.

The most striking effect of temperature is the complete inhibition of mutant expression. Organisms carrying a certain mutant allele may show the mutant phenotype at one temperature, but if they are grown under a second temperature, they exhibit a wild-type phenotype. Mutations of this type are said to be **conditional** and are called **temperature sensitive;** they have been discovered in numerous organisms, including viruses, fungi, and *Drosophila.*

The effects of temperature are used in studying mutations that interrupt essential processes during development and are thus normally lethal to the organism. If viruses carrying a **temperature-sensitive lethal mutation** are allowed to infect bacteria cultured at 42°C, infection progresses until the essential gene product is needed and then arrests, because this is the **restrictive temperature.** If cultured at the **permissive temperature** of 25°C, infection occurs normally, new viruses are produced, and the bacteria are lysed. The use of temperature-sensitive mutations, which may be induced and isolated, has added immensely to the study of viral genetics.

David T. Suzuki and his colleagues discovered many temperature-sensitive mutations in *Drosphila.* Not only have dominant and recessive conditional lethal mutations been recovered in great numbers, but nonlethal temperature-sensitive mutations affecting morphological traits and behavior have been recovered. In most cases, temperatures of 22°C and 29°C are sufficient to demonstrate normal and abnormal development, respectively.

Nutritional Effects

Another example of conditional mutations involves nutrition. In microorganisms, mutations that prevent synthesis of nutrient molecules are quite common.

These **nutritional mutants** arise when an enzyme, essential to a biosynthetic pathway, becomes inactive. If the end product of a biochemical pathway can no longer be synthesized, and if that molecule is essential to normal growth and development, the mutation is lethal. If, for example, the bread mold *Neurospora* can no longer synthesize the amino acid leucine, proteins cannot be synthesized unless leucine is added to the growth medium. If leucine is added, the lethal effect is overcome. Nutritional mutants have been crucial to molecular genetic studies and also served as the basis for George Beadle and Edward Tatum's proposal, in the early 1940s, that one gene functions to produce one enzyme.

In humans, a slightly different set of circumstances is known. The presence or absence of certain dietary substances, which normal individuals may consume without harm, may adversely affect individuals with abnormal genetic constitutions. Often, a mutation may prevent an individual from metabolizing some substance commonly found in normal diets. For example, those afflicted with the genetic disorder **phenylketonuria** cannot metabolize the amino acid phenylalanine. Those with **galactosemia** cannot metabolize galactose. Other individuals are intolerant of the milk sugar lactose. Those with **diabetes** or **hypoglycemia** cannot adequately metabolize glucose. In each case, excessive amounts of the molecule accumulate in the body and become toxic, and a characteristic phenotype results. However, if the dietary intake of the molecule is drastically reduced or eliminated, the associated phenotype may be reversed.

The case of **lactose intolerance** illustrates the general principles involved. Lactose is a disaccharide consisting of a molecule of glucose and a molecule of galactose. It is present as 7 percent of human milk and 4 percent of cow's milk. To metabolize lactose, humans require the enzyme **lactase,** which cleaves the disaccharide. Adequate amounts of lactase are produced during the first few years after birth. However, in many races the levels of this enzyme soon drop drastically, and adults become intolerant of milk. The major phenotypic effect involves severe intestinal diarrhea, including flatulence and cramps. This condition is particularly prevalent in Eskimos, Africans, Asiatics, and Americans with these heritages. In these cultures, milk is usually used for food in different forms, including cheese, butter, and yogurt. In these forms, the amount of lactose is reduced significantly and adverse effects are largely eliminated. Thus, by altering the diet, phenotypic expression of a genetic trait can be modified. Other human genetic disorders will be discussed later in the text.

Onset of Genetic Expression

Not all genetic traits are expressed at the same time during an organism's life span. In most cases, the age at which a gene is expressed corresponds to the normal sequence of growth and development. In humans, the prenatal, infant, preadult, and adult phases require different kinds of genetic information. In a similar way, many genetic disorders can be expected to manifest themselves at different stages of life. Lethal genes account for many of the frequent spontaneous abortions and miscarriages occurring in the human population. These mutations are believed to alter genetic products essential to prenatal development. **Tay-Sachs disease** and the **Lesch-Nyhan syndrome** (see Chapter 24) are severe disorders of lipid and nucleic acid metabolism, respectively, but do not cause lethality until early in childhood. **Huntington's disease** (see Chapter 4) demonstrates a wide range of age of onset, but not usually before the age of 20, and most often after age 40. Observations of the development of these conditions support the concepts that the need for certain gene products may be more essential at certain times and that the internal environment of an organism changes with age. In the broadest sense, aging must be thought of as a process that begins following conception and continues until death.

Heredity vs. Environment

Many genetically determined traits are influenced by the environment and have considerable impact on the organism's phenotype. How do geneticists attempt to define the relative importance of heredity compared with that of the environment? Some traits demonstrate minimal variation under normal environmental circumstances, but other traits show distinct variability under the same environmental conditions. The latter case is particularly true for traits controlled by polygenic inheritance. Often, polygenic traits show such a large degree of variation in a population that geneticists can only speculate that a large but unknown number of genes provide the genetic potential, which is then modified by the environment.

Provided that a trait can be quantitatively measured, we can approach analytically the question of genetics versus environment. Experiments on plants and animals other than humans can test the causes of variation. Inbred strains containing individuals of a relatively homogeneous or constant genetic background can be generated, thus producing an **isogenic** population. Experiments are then designed to test the effects of the range of prevailing environmental conditions on phenotypic variability.

In studies with such isogenic lines, the **heritability index** or **ratio (H^2)** for a given trait can be calculated. Also called the **broad heritability** of the character being studied, H^2 measures the degree to which phenotypic variation (V_P) is due to genetic factors for a single population under the limits of environmental variability during the study. It is important to emphasize here that H^2 *does not* measure the proportion of the total phenotype attributed to genetic factors, but only the observed variation in the phenotype. Technically, the term H^2 is an expression of phenotypic variance, which is due to the sum of three components: environmental variance (V_E), genetic variance (V_G), and variance resulting from the interaction of genetics and environment (V_{GE}). Therefore, phenotypic variance (V_P) is theoretically expressed as

$$V_P = V_E + V_G + V_{GE} \qquad (7.6)$$

Since V_{GE} is nearly impossible to analyze, it is usually ignored. Therefore, the simpler equation is generally used:

$$V_P = V_E + V_G \qquad (7.7)$$

Broad heritability expresses that proportion of variance due to the genetic component:

$$H^2 = \frac{V_G}{V_P} \qquad (7.8)$$

A very high H^2 value indicates that the environmental conditions have had little impact on phenotypic variance in the population studied. A very low H^2 value indicates that the variation in the environment has been almost solely responsible for the observed phenotypic variation. Table 7.3 lists estimates of heritability for traits in a variety of organisms.

TABLE 7.3
Estimates of heritability for several traits.

Trait	Percent Heritability (H^2)
Mouse	
Tail length	60
Litter size	15
Drosophila	
Abdominal bristle number	52
Wing length	45
Egg production	18
Humans	
Asthma	80
Diabetes (late onset)	70
Ulcers	37

It is not possible to obtain an absolute H^2 value for any given character. If measured in a different population under a greater or lesser degree of environmental variability, H^2 might well change for that character.

Practically, heritability is most useful in animal and plant breeding as a measure of potential selection. Breeders are most interested in the improvement of economically important characters. From the standpoint of selection, V_G must be examined in a more complex way:

$$V_G = V_D + V_A + V_I \qquad (7.9)$$

where V_D is the genetic effect of dominant genes, V_A is due to the effect of additive genes, and V_I is the effect of interactive or epistatic genes. Of these components, V_A is most important in selection. This more limited estimate has been called **narrow heritability (h^2).** Therefore, heritability in breeding experiments becomes effectively

$$h^2 = \frac{V_A}{V_P} \qquad (7.10)$$

Narrow heritability estimates, while characterizing a particular population, have been very useful in agricultural programs. Based on these estimates, selection techniques have led to vast improvements in the quality and quantity of plant and animal products.

In humans, isogenic strains are obviously not available for study, nor can the environment be controlled for individuals or populations. However, twins are very useful subjects for studying the heredity versus environment question in humans. **Monozygotic** or **identical twins,** derived from the division and splitting of a single egg following fertilization, are identical in their genetic compositions. Although most identical twins are reared together and are exposed to very similar environments, some pairs are separated and raised in different settings. For any particular trait, average similarities or differences can be investigated. Such an analysis is particularly useful because characteristics that remain similar in different environments are believed to be inherited. These data can then be compared with a similar analysis of **dizygotic** or **fraternal twins,** who originate from two separate fertilization events. Dizygotic twins are thus no more genetically similar than any two siblings.

A form of quantitative analysis of characteristics of twins reared together may also be pursued. Twins are said to be **concordant** for a given trait if both express it or neither expresses it and **discordant** if one shows the trait and the other does not. Table 7.4 lists concordance values for various traits in both types of twins.

TABLE 7.4

A comparison of concordance for various traits between monozygotic (MZ) and dizygotic (DZ) twins.

Trait	Concordance	
	MZ	DZ
Blood types	100%	66%
Eye color	99	28
Mental retardation	97	37
Measles	95	87
Idiopathic epilepsy	72	15
Schizophrenia	69	10
Diabetes	65	18
Identical allergy	59	5
Tuberculosis	57	23
Cleft lip	42	5
Club foot	32	3
Mammary cancer	6	3

SOURCE: Derived from various sources.

These data must be examined very carefully before any conclusions are drawn. If the concordance value approaches 90 to 100 percent with monozygotic twins, we might be inclined to interpret this value as indicating a large genetic contribution to the expression of the trait. In some cases—blood types and eye color, for example—we know this is indeed true. In the case of measles, however, a high concordance value merely indicates that the trait is almost always induced by a factor in the environment—in this case, a virus.

Therefore, it is more meaningful to compare the difference between the concordance values of monozygotic and dizygotic twins. If these values are significantly higher for monozygotic twins than dizygotic twins, we suspect that there is a genetic component involved in the determination of the trait. We reach this conclusion because monozygotic twins, with identical genotypes, would be expected to show a greater concordance than genetically related, but not genetically identical, dizygotic twins. In the case of measles, where concordance is high in both types of twins, the environment is assumed to contribute significantly.

Even though a particular trait may be determined to be influenced substantially by genetic factors, it is often difficult to formulate a precise mode of inheritance based on available data. In many cases the trait is considered to be controlled by multiple-factor inheritance. However, when the environment is also exerting a partial influence, such a conclusion is particularly difficult to prove.

SUMMARY AND CONCLUSIONS

In this chapter, we have focused on genes with additive alleles that influence a character in a quantitative way. This type of inheritance, which controls characters that can be measured (height, weight, etc.), is called polygenic because many genes are usually involved.

We have also considered how polygenic characteristics can be analyzed using various statistical methods. Statistical analysis can be descriptive, can be used to make inferences about a population, or can be used to compare sets of data. The methods described include the mean, variance, standard deviation, and standard error of the mean.

In the second part of the chapter, we examined factors that modify phenotypic expression. Frequently, all members of a population that carry a mutant genotype do not express the corresponding mutant phenotype to the same extent. The percentage that expresses some evidence of the mutant phenotype is a measure of the penetrance of the mutant genotype. On the other hand, expressivity is a measure that reflects the range of expression.

Factors that modify phenotypic expression include genetic background, temperature, and nutrition, among others. The phenomena of genetic suppression and position effects have been used to illustrate the genetic background factors. Together, all such factors constitute the total environment in which genetic information is expressed. Because the need for certain gene products occurs at various periods in the development, growth, and aging processes, the time of onset of gene expression also varies. This observation is particularly evident in the study of inherited human disorders.

Often, it is difficult to ascertain the degree of impact of genetic and environmental factors in the establishment of a given phenotype. Nevertheless, for many characters, we can calculate heritability (H^2), which reflects the degree of genetic influence. Heritability is especially useful in assessing economically important characters in domestic animals and plants. From this information breeders can determine the degree of improvement to be expected as a result of selective breeding. Usually, this determination is based on the potential impact of additive alleles on the phenotype.

In humans, studies aimed at resolving the question of heredity versus environment focus on twins. The degree of concordance of a trait may be compared in monozygotic (identical) and dizygotic (fraternal) twins raised together or apart. Often, it can be shown that a trait has a genetic component, but the precise mode of inheritance is difficult or impossible to determine.

PROBLEMS AND DISCUSSION QUESTIONS

1   Assume that height in a plant is controlled by two gene pairs and that each additive allele contributes 5 cm to a base height of 20 cm (i.e., *aabb* is 20 cm).
    (a) What is the height of an *AABB* plant?
    (b) Predict the phenotypic ratios of F_1 and F_2 plants in a cross between *aabb* and *AABB*.
    (c) List all genotypes that give rise to plants which are 25 and 35 cm in height.

2   In a cross where three gene pairs determine weight in squash, what proportion of individuals from the cross *AaBbCC* × *AABbcc* will contain only two additive alleles? Which genotype or genotypes fall into this category?

3   An inbred strain of plants has a mean height of 24 cm. A second strain of the same species from a different geographical region also has a mean height of 24 cm. When plants from the two strains are crossed together, the F_1 plants are the same height as the parent plants. However, the F_2 generation shows a wide range of heights; the majority are like the P_1 and F_1 plants, but approximately 4 of 1000 are only 12 cm high, and about 4 of 1000 are 36 cm high.
    (a) What mode of inheritance is occurring here?
    (b) How many gene pairs are involved?
    (c) How much does each gene contribute to plant height?
    (d) Indicate one possible set of genotypes for the original P_1 parents and the F_1 plants that could account for these results.
    (e) Indicate three possible genotypes that could account for F_2 plants that are 18 cm high and F_2 plants that are 33 cm high.

4   Distinguish between epistasis and polygenic inheritance.

5 List as many human traits as you can that are likely to be under the control of a polygenic mode of inheritance.

6 Describe the difference between penetrance and expressivity.

7 Define and discuss the significance of the following terms: (a) position effect, (b) suppressor genes, (c) monozygotic and dizygotic twins, (d) concordance and discordance, and (e) heritability.

8 In the following table, average differences of height and weight between monozygotic twins (reared together and apart), dizygotic twins, and siblings are compared. Draw as many conclusions as you can concerning the effects of genetics and the environment in influencing these human traits.

Trait	MZ Reared Together	MZ Reared Apart	DZ Reared Together	Sibs Reared Together
Height (cm)	1.7	1.8	4.4	4.5
Weight (kg)	1.9	4.5	4.5	4.7

SOURCE: Newman, Freeman, and Holzinger, 1937.

9 In *Drosophila*, the sex-linked recessive mutation *vermilion* (*v*) causes bright red eyes, which is in contrast to brick red eyes of wild type. A separate autosomal recessive mutation, *suppressor of vermilion* (*su-v*), causes flies homozygous or hemizygous for *v* to have wild-type eyes. In the absence of vermilion alleles, *su-v* has no effect on eye color. Determine the F_1 and F_2 phenotypic ratios from a cross between a female with wild-type alleles at the *vermilion* locus, but who is homozygous for *su-v*, with a *vermilion* male who has wild-type alleles at the *su-v* locus.

10 In a series of crosses between plants of various heights to a 20-inch plant, the following results were obtained in the F_1 generations:
   (a) $4'' \times 20'' \rightarrow$ All $12''$
   (b) $8'' \times 20'' \rightarrow$ All $14''$
   (c) $12'' \times 20'' \rightarrow$ All $16''$
   (d) $16'' \times 20'' \rightarrow$ All $18''$

Propose an explanation for the inheritance of height in the above plant. Under the constraints of your explanation, predict the genotypes of plants of each height.

11 Corn plants from a test plot are measured, and the distribution of heights at 10-cm intervals is recorded below:

Height (cm)	100	110	120	130	140	150	160	170	180
Plants (no.)	20	60	90	130	180	120	70	50	40

Calculate (a) the mean height, (b) the variance, (c) the standard deviation, and (d) the standard error of the mean.

12 A dark red strain and a white strain of wheat are crossed to produce an intermediate or medium red F_1. The F_1 plants are self-crossed to produce an F_2 in a ratio of 1 dark red:4 medium-dark red:6 medium red:4 light red:1 white. Further crosses reveal that the dark red and white F_2 plants are true breeding.
   (a) Based on the ratio of offspring in the F_2, how many loci are involved in the production of color? How many alleles at each locus?
   (b) Are the allele effects additive?
   (c) How many units of color are produced by each allele?
   (d) How many genotypes are represented by each phenotypic class? What are they?

13 Height in humans depends on the additive action of genes. Assume that this trait is controlled by four loci, *R, S, T, U*, and that environmental effects are negligible. Dominant alleles contribute two units and recessive alleles contribute one unit to height.
   (a) Can two individuals of moderate height produce offspring that are much taller or shorter than either parent? How?
   (b) If an individual with the minimum height specified by these genes marries an individual of intermediate or moderate height, will any of their children be taller than the tall parent? Why?

14 The mean length and variance of corolla length in two true-breeding strains of *Nicotiana* and their progeny are as shown below.

Strain	Mean length (mm)	Variance
P_1 short	40.47	3.124
P_2 long	93.75	3.876
$F_1(P_1 \times P_2)$	63.90	4.743
$F_2(F_1 \times F_1)$	68.72	47.708

Calculate heritability of flower length in three strains.

SELECTED READINGS

BRINK, R. A., ed. 1967. *Heritage from Mendel.* Madison: University of Wisconsin Press.

CHAPMAN, A.B. 1985. *General and quantitative genetics.* Amsterdam: Elsevier.

CORWIN, H. O., and JENKINS, J. B. 1976. *Conceptual foundations of genetics: Selected readings.* Boston: Houghton-Mifflin.

CROW, J. F. 1966. *Genetics notes.* 6th ed. Minneapolis: Burgess.

DUNN, L. C. 1966. *A short history of genetics.* New York: McGraw-Hill.

FALCONER, D. S. 1981. *Introduction to quantitative genetics.* 2nd ed. New York: Longman.

FARBER, S. L. 1980. *Identical twins reared apart.* New York: Basic Books.

FELDMAN, M. W., and LEWONTIN, R. C. 1975. The heritability hang-up. *Science* 190: 1163–68.

FOSTER, H. L., et al., eds. 1981. *The mouse in biomedical research. Vol. 1: History, genetics, and wild mice.* New York: Academic Press.

FOSTER, M. 1965. Mammalian pigment genetics. *Adv. in Genet.* 13: 311–39.

LEVITAN, M., and MONTAGU, A. 1977. *Textbook of human genetics.* 2nd ed. New York: Oxford University Press.

LEWONTIN, R. C. 1974. The analysis of variance and the analysis of causes. *Amer. J. Hum. Genet.* 26: 400–11.

LINDSLEY, D. C., and GRELL, E. H. 1967. *Genetic variations of* Drosophila melanogaster. Washington, D.C.: Carnegie Institute of Washington.

MATHER, K. 1965. *Statistical analysis in biology.* London: Methuen.

NEWMAN, H. H., FREEMAN, F. N., and HOLZINGER, K. J. 1937. *Twins: A study of heredity and environment.* Chicago: The University of Chicago Press.

NOLTE, D. J. 1959. The eye-pigmentary system of *Drosophila. Heredity* 13: 233–41.

PAWELEK, J. M., and KÖRNER, A. M. 1982. The biosynthesis of mammalian melanin. *Amer. Scient.* 70: 136–45.

RANSON, R., ed. 1982. *A handbook of* Drosophila *development.* New York: Elsevier Biomedical Press.

STERN, C. 1973. *Principles of human genetics.* 3rd ed. San Franscisco: W. H. Freeman.

SUZUKI, D. T. 1970. Temperature-sensitive mutations in *Drosophila melanogaster. Science* 170: 695–706.

VOELLER, B. R., ed. 1968. *The chromosome theory of inheritance—Classic papers in development and heredity.* New York: Appleton-Century-Crofts.

ZIEGLER, I. 1961. Genetic aspects of ommochrome and pterin pigments. *Adv. in Genet.* 10: 349–403.

PART TWO DNA—THE CHEMICAL BASIS OF HEREDITY

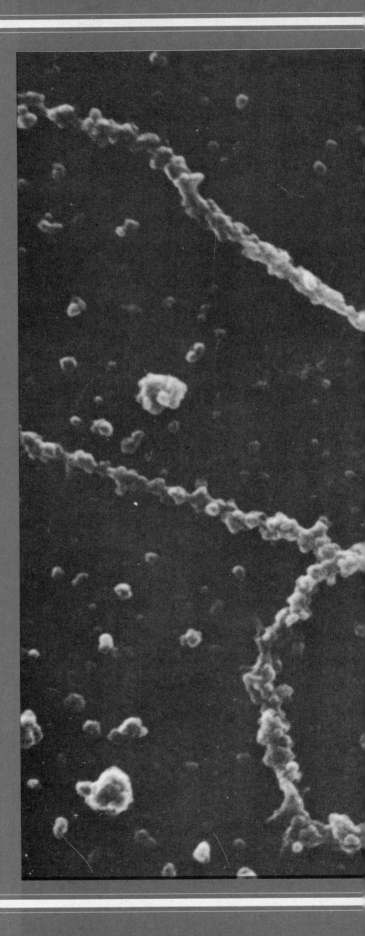

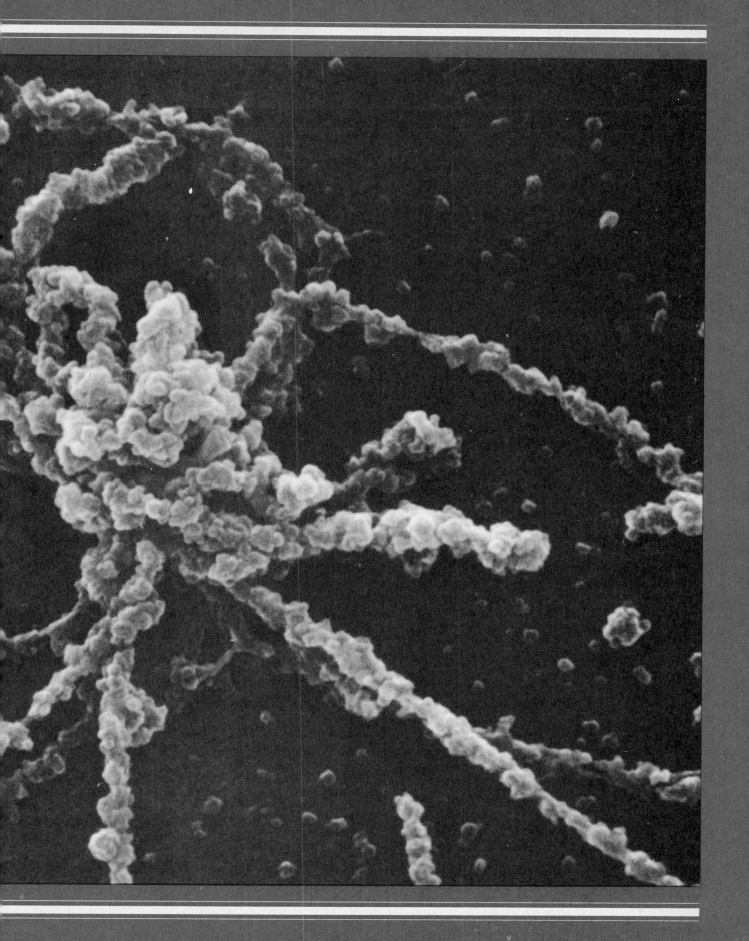

8

DNA—The Genetic Material

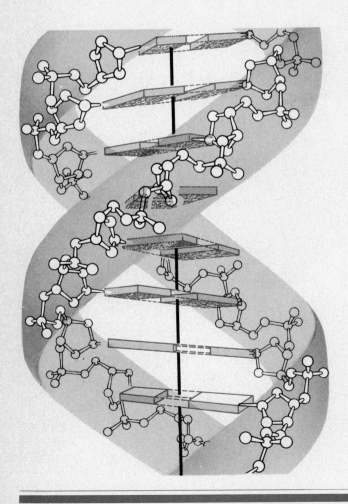

In Part 1 of this text, we discussed the presence of genes on chromosomes that control phenotypic traits and the way in which the chromosomes are transmitted through gametes to future offspring. Logically, there must be some form of information contained in genes which, when passed to a new generation, influences the form and characteristics of the offspring; this is called the genetic information. We might also conclude that this same information in some way directs the many complex processes leading to the adult form.

Until 1944 it was not clear what chemical component of the cell constitutes the genetic material, which in turn contains the genetic information. This component was believed to be associated with chromosomes and their individual units, the genes. Chromosomes were known to have both a nucleic acid and a protein component. In 1944, however, there emerged the first direct experimental evidence that the nucleic acid, DNA, serves as the informational basis for the process of heredity. In this chapter we will discuss the era in which DNA was confirmed as the genetic material for almost every living thing.

CHARACTERISTICS OF THE GENETIC MATERIAL

The genetic material has several characteristics or functions: **replication, storage of information, expression of that information,** and **variation by mutation.** Replication of the genetic material is one facet of cell division, a fundamental property of all living organisms. Once the genetic material of somatic cells has been replicated, it may be partitioned into daughter cells during the process of **mitosis.** During the formation of gametes, the genetic material is also replicated but is partitioned differently. This process is called **meiosis.** Although the products of mitosis and meiosis are different, these processes are both part of the more general phenomenon of cellular reproduction.

The concept of storage may be interpreted as unexpressed genetic information. When is genetic information not expressed? Information contained in sperm cells is a good example. Sperm cells contain a complete haploid set of genetic information, but during their formation many genes are not expressed. For example, they do not contain numerous gene products: hemoglobin, the oxygen-carrying molecule; trypsin and chymotrypsin, the digestive enzymes; or melanin, the major pigment molecule in mammals. However, these cells do exhibit highly specialized structures and contain numerous molecules related to the fertilization process and subsequent development. These structures and molecules are dependent on the expression of numerous genes. Unexpressed versus expressed genetic information is a characteristic of all cells. The study of this aspect is an area of active research in both molecular and developmental genetics.

Expression of the information stored in the genetic material is a complex process and is the basis for the concept of **information flow** within the cell. Figure 8.1 shows a simplified illustration of this concept. The initial event is the **transcription** of genetic information stored in DNA. Transcription results in the synthesis of three types of RNA molecules: **messenger RNA (mRNA), transfer RNA (tRNA),** and **ribosomal RNA (rRNA).** Of these, mRNAs are translated into proteins. Each type of mRNA is the product of a specific gene and leads to the synthesis of a different protein.

Translation, or protein synthesis, involves many molecular components, a supply of energy, and the cellular organelle, the **ribosome.** The ribosome consists of several types of rRNA plus a variety of individual proteins. The role of tRNA is to adapt the information present in mRNA to the correct amino acids during translation. Amino acids are the building blocks of proteins. In eukaryotic cells, transcription occurs in the nucleus and translation occurs in the cytoplasm.

The genetic material is also responsible for newly arising variability among organisms through the process of mutation. If a change in the chemical composition of DNA occurs, the alteration will be reflected during transcription and translation, perhaps affecting the specified protein. If a mutation is present in gametes, it will be passed to future generations and, with time, may become distributed in the population. Genetic variation, which also includes rearrangements

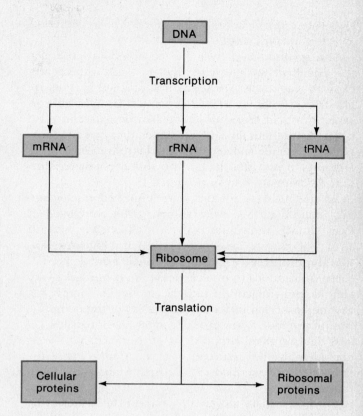

FIGURE 8.1

A simplified view of information flow involving DNA, RNA, and proteins within cells.

within and between chromosomes, provides the raw material for the process of evolution.

PROTEIN AS THE GENETIC MATERIAL

The idea that genetic material is physically transmitted from parent to offspring has been accepted for as long as the concept of inheritance has existed. Beginning in the late nineteenth century, research into the structure of biomolecules progressed considerably. Thus, the stage was set for the description of the genetic material in chemical terms. Although proteins and nucleic acid were both considered major candidates for the role of the genetic material, many geneticists, until the 1940s, favored proteins. Three factors contributed to this belief.

First, proteins are abundant in cells. Although the protein content may vary considerably, these molecules compose over 50 percent of the dry weight of cells. Since cells contain such a large amount and variety of proteins, it is not surprising that early geneticists believed that some of this protein could function as the genetic material.

The second factor was the accepted proposal for the chemical structure of nucleic acids during the early to mid-1900s. DNA was first studied in 1868 by Friedrick Miescher, a Swiss chemist. He was able to separate nuclei from the cytoplasm of cells and then isolate from them an acid substance that he called **nuclein.** Miescher showed that nuclein contained large amounts of phosphorus and no sulfur, characteristics that differentiate it from proteins.

As analytical techniques were improved, nucleic acids, including DNA, were shown to be composed of four similar molecules called nucleotides. Around 1910, Phoebus A. Levene proposed the **tetranucleotide hypothesis** to exlain the chemical arrangement of these nucleotides in nucleic acids. He proposed a very simple four-nucleotide unit as shown in Figure 8.2. Levene based his proposal on studies of the composition of the four types of nucleotides. Although his actual data revealed proportions of the four that varied considerably, he assumed a 1:1:1:1 ratio. The discrepancy was ascribed to inadequate analytical technique.

Since a single covalently bound tetranucleotide structure was relatively simple, geneticists believed nucleic acids could not provide the large amount of chemical variation expected for the genetic material. Thus, attention was directed away from nucleic acids as important genetic biomolecules, strengthening the speculation that proteins served as the genetic material.

The third contributing factor simply concerned the areas of most active research in genetics. Before 1940,

most geneticists were engaged in the study of transmission genetics and mutation. The excitement generated in these areas undoubtedly diluted the concern for finding the precise molecule that serves as the genetic material. Thus, proteins were the most promising candidate and were accepted rather passively.

Between 1910 and 1930, other proposals for the structure of nucleic acids were advanced, but they were generally overturned in favor of the tetranucleotide hypothesis. It was not until the 1940s that the work of Erwin Chargaff led to the realization that Levene's hypothesis was incorrect. Chargaff showed that, for most organisms, the 1:1:1:1 ratio was indeed inaccurate, thus discrediting Levene's hypothesis.

EXPERIMENTAL EVIDENCE FOR DNA: EXPERIMENTS WITH PROKARYOTES AND VIRUSES

The 1944 publication by Oswald Avery, Colin MacLeod, and MacLyn McCarty concerning the chemical nature of a "transforming factor" in bacteria marked the initial event leading to the acceptance of DNA as the genetic material. Along with the subsequent findings of other research teams, this work constituted direct experimental proof that, in the organisms studied, DNA, and not protein, is the biomolecule responsible for heredity. This period marked the beginning of an era of discovery unprecedented in the history of biology. In the ensuing years numerous discoveries have revolutionized our understanding of the molecular basis of life on earth. Their impact on biology parallels the work that followed the publication of Darwin's theory

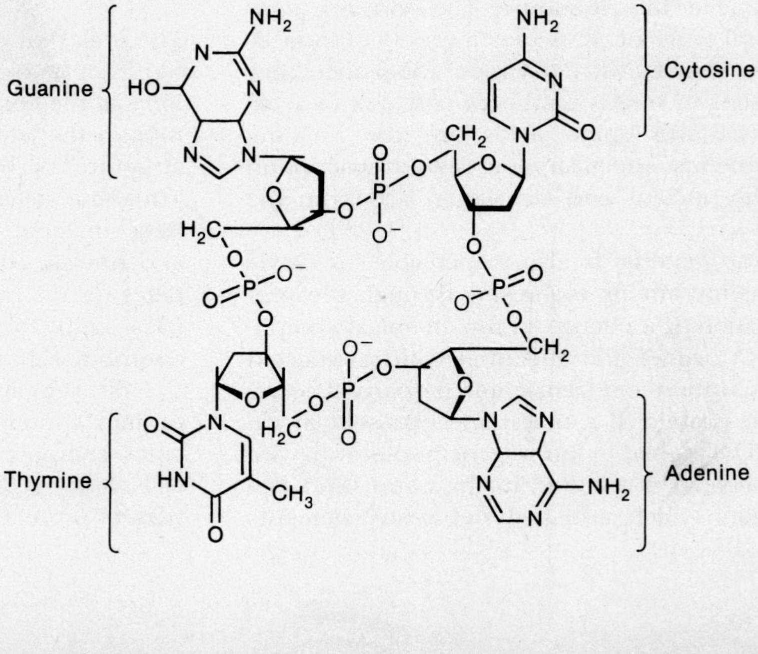

FIGURE 8.2

Levene's proposed structure of a DNA tetranucleotide containing one molecule each of the four nitrogenous bases: adenine, cytosine, guanine, and thymine.

of evolution and that following the rediscovery of Mendel's postulates of transmission genetics.

The initial evidence implicating DNA as the genetic material was derived from studies of prokaryotic bacteria and viruses that infect them. The reasons for their use will become apparent as the experiments are studied. Primarily, bacteria and viruses are capable of rapid growth because they complete life cycles in hours. They may also be experimentally manipulated and mutations may be easily induced and selected for. Thus, they are ideal for experimentation of this sort.

Transformation Studies

The research that provided the foundation for Avery, MacLeod, and McCarty's work was initiated in 1927 by Frederick Griffith, a medical officer in the British Ministry of Health. He performed experiments with several different strains of the bacterium *Diplococcus pneumoniae*. Some were **virulent** strains, which cause pneumonia in certain vertebrates (notably humans and mice), while some were **avirulent** strains, which do not cause illness.

The difference in virulence is related to the polysaccharide capsule of the bacterium. Virulent strains have this capsule, whereas avirulent strains do not. The nonencapsulated bacteria are readily engulfed and destroyed by phagocytic cells in the animal's circulatory system. Virulent bacteria, which possess the polysaccharide coat, are not easily engulfed; they multiply and cause pneumonia.

The presence or absence of the capsule is the basis for another characteristic difference between virulent and avirulent strains. Encapsulated bacteria form a **smooth,** shiny-surfaced colony (*S*) when grown on an agar culture plate; nonencapsulated strains produce **rough** colonies (*R*). Thus, virulent and avirulent strains may be distinguished easily by standard microbiological culture techniques. Further, the smooth colonies are large compared with the colonies of the rough strains.

Each strain of *Diplococcus* may be one of dozens of different types called **serotypes.** The specificity of the serotype is due to the detailed chemical structure of the polysaccharide constituent of the thick, slimy capsule. Serotypes are identified by immunological techniques and are usually designated by Roman numerals. In the United States, types I and II are most common in causing pneumonia. Griffith used types II and III in the critical experiments that led to new concepts about the genetic material. Table 8.1 summarizes the characteristics of Griffith's strains.

Griffith knew from the work of others that only living virulent cells would produce pneumonia in mice.

TABLE 8.1

Strains of *Diplococcus pneumoniae* used by Frederick Griffith in his original transformation experiments.

Serotype	Colony Morphology	Capsule	Virulence
II*R*	Rough	Absent	Avirulent
III*S*	Smooth	Present	Virulent

If heat-killed virulent bacteria are injected into mice, no pneumonia results, just as living avirulent bacteria fail to produce the disease. Griffith's critical experiment (Figure 8.3) involved a double injection into mice of living II*R* (avirulent) cells and heat-killed III*S* (virulent) cells. Neither cell type caused death in mice when injected alone. Griffith expected that the double injection would not kill the mice, but after five days all mice receiving double injections were dead. Analysis of the blood of the dead mice revealed large numbers of living type III*S* (virulent) bacteria! As far as could be determined, these III*S* bacteria were identical to the III*S* strain from which the heat-killed cell preparation had been made. The control mice, injected only with living avirulent II*R* bacteria for this set of experiments, did not develop pneumonia and remained healthy. This finding strongly suggested that the occurrence of the living III*S* bacteria in the dead mice was not caused by faulty technique or contamination, but that some interaction between the two types of injected bacteria had occurred.

Griffith suggested that the heat-killed III*S* bacteria were responsible for converting live avirulent II*R* cells into virulent III*S* ones. Calling the phenomenon **transformation,** he suggested that the **transforming principle** might be some part of the polysaccharide capsule *or* some compound required for capsule synthesis, although the capsule alone did not cause pneumonia. To use Griffith's term, the transforming principle from the dead III*S* cells served as a "pabulum" for the II*R* cells.

Griffith's work led other physicians and bacteriologists to research the phenomenon of transformation. By 1931, Henry Dawson, at the Rockefeller Institute, had confirmed Griffith's observations and extended his work one step further. Dawson and his coworkers showed that transformation could occur *in vitro* (in a test tube containing only bacterial cells); that is, injection into mice was not necessary for transformation to occur. By 1933, Lionel J. Alloway had refined the *in vitro* system by using crude extracts of *S* cells and living *R* cells. The soluble filtrate from the heat-killed *S*

CONTROLS

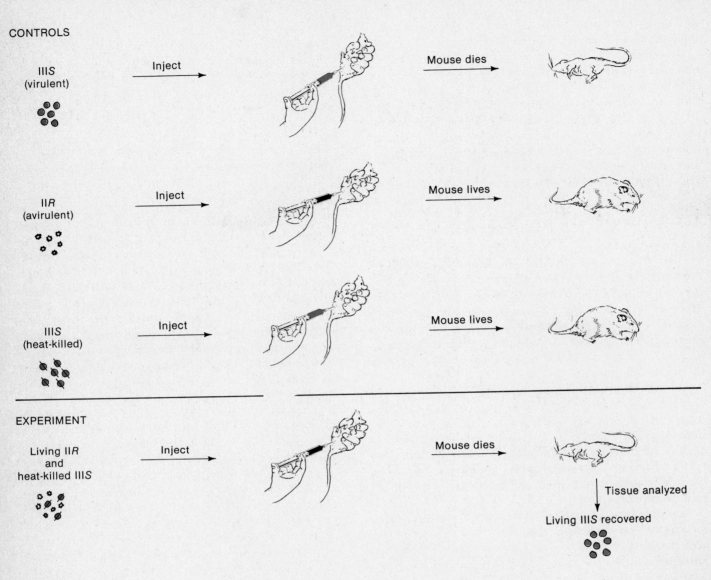

FIGURE 8.3
Summary of Griffith's transformation experiment.

cells was as effective in inducing transformation as were the intact cells! Alloway and others did not view transformation as a genetic event, but rather as a physiological modification of some sort. Nevertheless, the experimental evidence that a chemical substance was responsible for transformation was quite convincing. Many researchers during the 1930s—a time when very little genetic significance was attributed to nucleic acids—suspected that a protein or proteins would prove to be the transforming principle.

Then, in 1944, after ten years of work, Avery, MacLeod, and McCarty published their results in what is now regarded as a classic paper. They reported that they had obtained the transforming principle in a highly purified state, and that beyond reasonable doubt it was DNA. The details of their work are outlined in Figure 8.4.

These researchers began their isolation procedure with large quantities (50–75 liters) of liquid cultures of type IIIS virulent cells. The cells were centrifuged, collected, and heat-killed. Following washing and several extractions with the detergent deoxycholate (DOC), a soluble filtrate was obtained which, when tested, still contained the transforming principle. Protein was removed from the active filtrate by several chloroform extractions, and polysaccharides were enzymatically digested and removed. Finally, precipitation with ethanol yielded a fibrous nucleic acid that still retained the ability to induce transformation of type IIR avirulent cells.

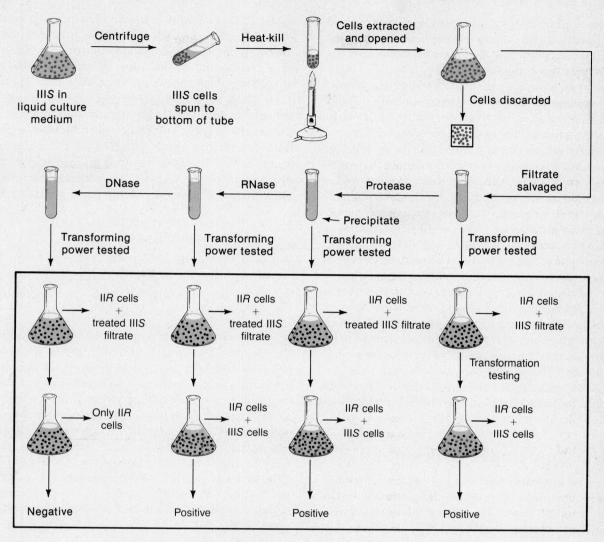

FIGURE 8.4
Summary of Avery, McCarty, and MacLeod's experiment demonstrating that DNA is the transforming principle.

Further testing established beyond doubt that the transforming principle was DNA. Treatment was performed with proteolytic (protein-digesting) enzymes and an RNA-digesting enzyme, **ribonuclease.** Such treatment destroyed the activity of any remaining protein and RNA. Nevertheless, transforming activity was not diminished. Chemical testing of the final product gave strong positive reactions for DNA. The final confirmation came with experiments using crude samples of the DNA-digesting enzyme **deoxyribonuclease,** which was isolated from dog and rabbit sera. This digestion was shown to destroy transforming activity. There could be no doubt that the active transforming principle was DNA!

The great amount of work, the confirmation and reconfirmation of the conclusions drawn, and the brilliant logic involved in the research of these three scientists are truly impressive. The conclusion to the 1944 publication was, however, very simply stated: "The evidence presented supports the belief that a nucleic acid of the desoxyribose type is the fundamental unit of the transforming principle of *Pneumococcus* type III."*

Avery and his coworkers recognized the genetic and biochemical implications of their work. They suggested that the transforming principle interacts with the *R* cell and gives rise to a coordinated series of enzymatic reactions that culminates in the synthesis of the type III capsular polysaccharide. They emphasized that, once transformation occurs, the capsular poly-

---

Desoxyribose is now spelled *deoxyribose*, and the genus *Pneumococcus* is now referred to as *Diplococcus*.

saccharide is produced in successive generations and that the transforming principle is replicated in daughter cells. Transformation is therefore heritable, and the process affects the genetic material.

Immediately after the publication of the report, several investigators turned to or intensified their studies of transformation in order to clarify the role of DNA in genetic mechanisms. In 1949, Harriet Taylor isolated an **extremely rough (ER)** mutant strain from a rough (R) strain. This ER strain produced colonies that were more irregular than R. DNA from R accomplished the transformation of ER to R. Thus, the R strain, which served as the recipient in the Avery experiments, was shown also to be able to serve as the DNA donor in transformation.

Transformation has now been shown to occur in *Hemophilus influenzae, Bacillus subtilis, Shigella paradysenteriae, Escherichia coli*, and several other microorganisms, including blue-green algae. One line of work pursued transformation of genetic traits other than colony morphology. Many traits were found to be transformable, particularly ones involving resistance to antibiotics and the ability to metabolize various nutrients. Thus, geneticists believed that most, if not all, traits could be transformed under suitable experimental conditions.

The Hershey-Chase Experiment

The second major piece of evidence supporting DNA as the genetic material was provided by the study of the bacterial virus T2. This virus, also called a **bacteriophage** or just a **phage,** has as its host the bacterium *Escherichia coli* and consists of a protein coat surrounding a core of DNA. Electron micrographs have revealed the phage's external structure to be composed of a hexagonal head plus a tail. The life cycle of bacteriophages such as T2 is shown in Figure 8.5.

In 1952, Alfred Hershey and Martha Chase published the results of experiments designed to clarify the events leading to phage reproduction. Several of the experiments clearly established the independent functions of phage protein and nucleic acid in the reproduction process associated with the bacterial cell. Hershey and Chase knew from existing data that

1   T2 phages consist of approximately 50 percent protein and 50 percent DNA.

2   Infection is initiated by adsorption of the phage by its tail fibers to the bacterial cell.

3   The production of new viruses occurs within the bacterial cell.

It would appear that some molecular component of the phage, DNA and/or protein, enters the bacterial cell and directs viral reproduction. Which was it?

Hershey and Chase used radioisotopes to follow the molecular components of phages during infection. Both ^{32}P and ^{35}S, radioactive forms of phosphorus and sulfur, were used. Since DNA contains phosphorus but not sulfur, ^{32}P effectively labels DNA. Since proteins contain sulfur but not phosphorus, ^{35}S labels protein. **This is a key point in the experiment.** If *E. coli* cells are first grown in the presence of ^{32}P or ^{35}S and then infected with T2 viruses, the progeny phage will have either a labeled DNA core or a labeled protein coat,

FIGURE 8.5
The life cycle of a typical bacteriophage.

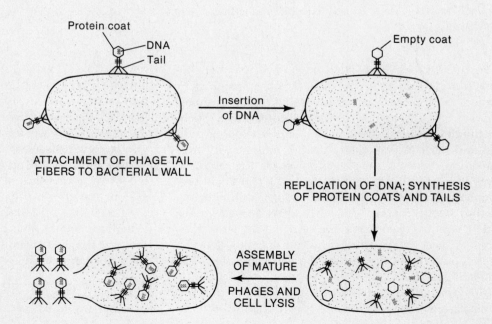

Protein coat
DNA
Tail
ATTACHMENT OF PHAGE TAIL
FIBERS TO BACTERIAL WALL

Insertion
of DNA

Empty coat

REPLICATION OF DNA; SYNTHESIS
OF PROTEIN COATS AND TAILS

ASSEMBLY
OF MATURE
PHAGES AND
CELL LYSIS

respectively. These radioactive phages may be isolated and used to infect unlabeled bacteria (Figure 8.6).

When labeled phage and unlabeled bacteria are mixed, an adsorption complex is formed as the phage attaches its tail fibers to the bacterial wall. These complexes were isolated and subjected to a high shear force by placing them in a blender. This force strips off the attached phages, which may then be analyzed separately (Figure 8.6). By tracing the radioisotopes, Hershey and Chase were able to demonstrate that most of the ^{32}P-labeled DNA had been transferred into the bacterial cell following adsorption; on the other hand, most of the ^{35}S-labeled protein remained out-side the bacterial cell and was recovered in the phage "ghosts" (empty phage coats) after the blender treatment. Following this separation, the bacterial cells, which now contained viral DNA, were eventually lysed as new phages were produced.

Hershey and Chase interpreted these results to indicate that the protein of the phage coat remains outside the host cell and is not involved in the production of new phage. On the other hand, and most importantly, phage DNA enters the host cell and directs phage multiplication. Thus, they had demonstrated that in phage T2, DNA, not protein, is the genetic material.

FIGURE 8.6
Summary of the Hershey-Chase experiment demonstrating that DNA and not protein is responsible for directing the reproduction of phage T2 during the infection of *E. coli.*

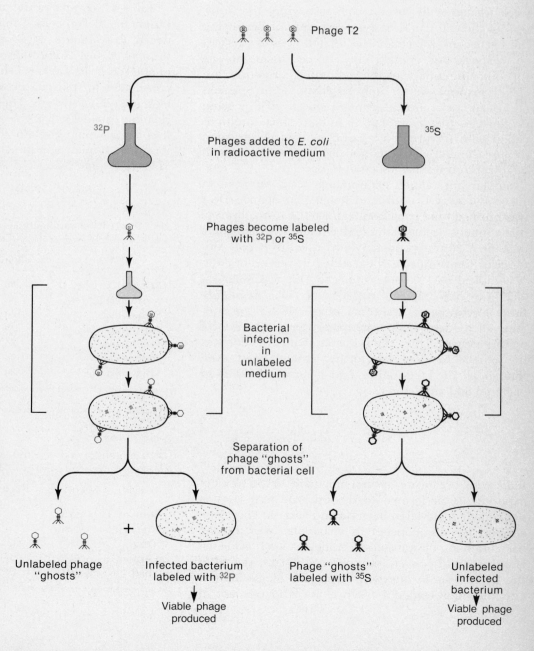

This experimental work, along with that of Avery and his colleagues, provided convincing evidence to most geneticists that DNA was the molecule responsible for heredity. Since then, many significant findings have been based on this supposition. These findings are discussed in detail in subsequent chapters.

Transfection Experiments

During the eight years following the publication of the Hershey-Chase experiment, additional research provided even more solid proof that DNA is the genetic material. These studies involved the same organisms used by Hershey and Chase.

In 1957, several reports demonstrated that if *E. coli* were treated with the enzyme **lysozyme,** the outer wall of the cell could be removed without destroying the bacterium. Enzymatically treated cells are naked, so to speak, and contain only the cell membrane as the outer boundary of the cell. Such structures are called **protoplasts,** or **spheroplasts.** John Spizizen and Dean Fraser independently reported that by using protoplasts, they were able to initiate phage multiplication with disrupted T2 particles. That is, provided protoplasts are used, it is not necessary for a virus to be intact in order for infection to occur.

Similar but refined experiments were reported in 1960 by George Guthrie and Robert Sinsheimer. DNA was purified from bacteriophage φX-174, a small phage that contains a single-stranded, circular DNA molecule of some 5386 nucleotides. When added to *E. coli* protoplasts, the purified DNA resulted in the production of complete φX-174 bacteriophages. This process of infection by only the viral nucleic acid, called **transfection,** proves conclusively that phage DNA alone contains all the necessary information for production of mature viruses. Thus, the evidence that DNA serves as the genetic material was further strengthened, even though all direct evidence had been obtained from bacterial and viral studies.

CIRCUMSTANTIAL EVIDENCE FOR DNA: EUKARYOTIC DATA

Eukaryotic organisms are not amenable to the types of experiments performed to demonstrate that DNA is the genetic material in bacteria and viruses. Therefore, support for this concept in eukaryotes was initially based only on circumstantial evidence. Such evidence is indirect and derived from observations incidental to the hypothesis in question. While any single item of circumstantial evidence taken alone is insufficient to

support the hypothesis, many different sorts of independent observations may lead to a common conclusion.

Distribution of DNA

Since it is generally accepted that chromosomes within the nucleus contain the genetic material, a correlation should exist between the total number of chromosomes and the molecule that functions as the genetic material. Meaningful comparisons may be made between gametes (sperm and eggs) and somatic or body cells. The latter are recognized as being **diploid** ($2n$) and containing twice the number of chromosomes as gametes, which are **haploid** (n).

Table 8.2 compares the amount of DNA found in a variety of chicken cells. There is a close correlation between the amount of DNA (in picograms) and the number of sets of chromosomes located in the nuclei of chicken cells. Table 8.3 shows that there is a similar correlation in the data obtained from five other species, including humans. The only exception to the correlation established is the data derived from human liver cells, which contain about three times the DNA found in human sperm. Although the reason is unknown, human liver cells are often found to vary in

TABLE 8.2

DNA in nuclei of various types of chicken cells.

Cell	Ploidy	DNA (picograms)
Sperm	n	1.26
Normablast*	$2n$	2.49
Spleen cells	$2n$	2.55
Pancreatic cells	$2n$	2.61
Heart cells	$2n$	2.45
Kidney cells	$2n$	2.20
Liver cells	$2n$	2.66

*A normablast is the nucleated precursor cell of an erythrocyte.

TABLE 8.3

DNA content of cells of various species (in picograms).

Species	Sperm (n)	Normablast ($2n$)	Liver cells ($2n$)
Human	3.25	7.30	10.36
Chicken	1.26	2.49	2.66
Trout	2.67	5.79	—
Carp	1.65	3.49	3.33
Shad	0.91	1.97	2.01

chromosome number (i.e., they can also be 3n, 4n, or 8n). Thus, the amount of DNA in these liver cells is expected to exceed twice that found in sperm.

No such correlation may be made between gametes and somatic cells for the other major classes of organic biomolecules, proteins, lipids, and carbohydrates. These data thus provide circumstantial or indirect evidence that DNA is the genetic material.

Mutagenesis

Ultraviolet light (**UV**) is one of a number of agents capable of inducing mutations in the genetic material. Bacteria and other simple organisms may be irradiated with various wavelengths of ultraviolet light, and the effectiveness of each wavelength measured by the number of mutations it induces. When the data are plotted, an action spectrum of ultraviolet light as a mutagenic agent is obtained. This **action spectrum** may then be compared with the **absorption spectrum** of any molecule suspected to be the genetic material (Figure 8.7). The molecule serving as the genetic material is expected to absorb at that wavelength found to be mutagenic.

UV light is most mutagenic at the wavelength (λ) of 260 nanometers (nm). Both DNA and RNA absorb UV light most strongly at 260 nm. On the other hand, protein absorbs most strongly at 280 nm. Thus, this indirect evidence supports both DNA and RNA as the genetic material and excludes protein.

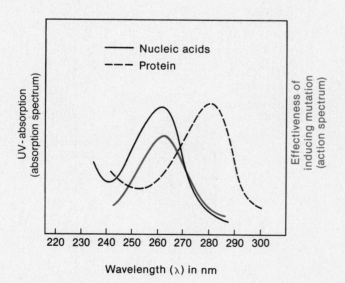

FIGURE 8.7
Ultraviolet absorption of nucleic acids and proteins compared with each wavelength's relative effectiveness in inducing mutations.

Metabolic Stability

If the genetic material is to provide continuity between generations and is to direct the metabolic activities of cells, the molecule is expected to remain stable. If this molecule is not stable but is continuously being broken down and resynthesized (the concept of **metabolic "turnover"**), it cannot easily perform its ascribed functions. Additionally, under unstable conditions the probability of errors or chemical mistakes arising during resynthesis is increased greatly.

In nondividing cells, DNA is metabolically stable compared with either RNA or protein, which have relatively high turnover rates. Metabolic stability can be assessed by using radioactive precursors specific to the class of molecules under study. For example, if radioactive precursors to DNA are administered to human leukemic patients, the isotope is incorporated rapidly into the dividing blood cells. As DNA synthesis occurs in these cells, it can be detected by the presence of the radioactive precursor. However, for periods as long as one year thereafter, no appreciable DNA synthesis can be detected in these patients' cartilage, brain, or muscle cells. Thus, when cells divide, active DNA synthesis occurs as expected. However, in nondividing cells DNA is stable compared with RNA and proteins, which are continuously being synthesized and broken down. Thus, DNA meets this criterion expected of the genetic material, while RNA and protein do not.

Association of DNA with Chromosomes

The genetic material should be found where it functions—in the nucleus as part of chromosomes. Both DNA and protein fit this criterion. However, protein is also abundant in the cytoplasm, while DNA is not. Both mitochondria and chloroplasts are known to perform genetic functions, and DNA is also present in these organelles. Thus, DNA is only found where primary genetic function is known to occur. Protein, however, is found in all parts of the cell. These observations favor DNA over protein as the genetic material.

DIRECT EVIDENCE FOR DNA: EUKARYOTIC DATA

While the circumstantial evidence just described does not constitute direct proof that DNA is the genetic material in eukaryotes, these observations spurred researchers to forge ahead under this assumption. Today, there is no doubt of the validity of this conclusion; nor was there any reason to suspect otherwise.

The strongest evidence has been provided by a current experimental procedure called **recombinant DNA research.** In this procedure, segments of eukaryotic DNA corresponding to specific genes are isolated and literally spliced into bacterial DNA. Such a complex can be inserted into a bacterial cell and its genetic expression monitored. If a eukaryotic gene is selected that is absent in bacterial genetic information, the presence of the corresponding eukaryotic protein product demonstrates directly that this DNA is functional in expressing genetic information. This has been shown to be the case in numerous instances. For example, the human genes specifying the hormone insulin and the immunologically important molecule interferon are produced by bacteria following recombinant DNA procedures.

As the bacterium divides, the eukaryotic DNA is replicated along with the host DNA and is distributed to daughter cells. As divisions continue, the eukaryotic genes are **cloned,** with each new bacterial cell containing an identical copy of the eukaryotic genetic information. As a result, large amounts of DNA containing specific eukaryotic genes may be isolated and studied in depth. This technique has paved the way for detailed analysis of the nucleotide sequence of specific genes, among other types of information.

The availability of vast amounts of DNA coding for specific genes has led to a second form of documentation that DNA serves as the genetic material. In the work of John Gurdon and others, billions of copies of specific genes were injected into oocytes of the frog *Xenopus laevis*. In these experiments, the egg served as a living test tube, and the transcription and translation of this genetic information were analyzed. Through such analysis, the products of these genes have been identified, confirming the informational role of DNA in genetic processes.

More recent work in the laboratory of Beatrice Mintz and others has strengthened this evidence. This research has demonstrated that DNA encoding the human β-globin gene, when microinjected into a fertilized mouse egg, is later found in the adult mouse tissue and can be transmitted to that mouse's prog-

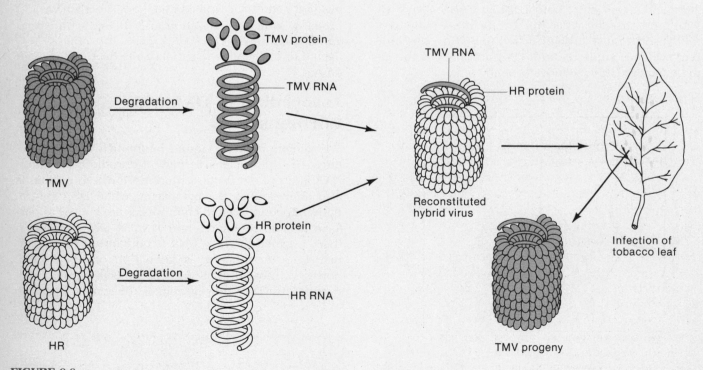

FIGURE 8.8
Reconstitution of hybrid tobacco mosaic viruses. In the hybrid, RNA is derived from the wild-type TMV virus, while the protein subunits are derived from the HR strain. Following infection, viruses are produced with protein subunits characteristic of the wild-type TMV strain and not those of the HR strain.

eny! Such mice are referred to as **transgenic.** More recent work has introduced DNA representing the growth hormone gene from rats into fertilized mouse eggs. About one-third of the resultant animals grew to twice the size of normal mice, indicating that the foreign DNA was present *and* functional. Subsequent generations receive the gene and display the trait that it governs.

We will pursue the topic of recombinant DNA again later in the text. The point to be made here is that in eukaryotes DNA has been shown directly to meet the requirement of expression of genetic information. Later we will see how DNA is stored, replicated, expressed, and mutated.

RNA AS THE GENETIC MATERIAL

Many plant viruses and some bacterial and animal viruses contain an RNA core rather than one composed of DNA. In these viruses, it would thus appear that RNA must serve as the genetic material—an exception to the general rule that DNA performs this function. In 1956, it was demonstrated than when purified RNA from **tobacco mosaic virus** (**TMV**) was spread on tobacco leaves, the characteristic lesions caused by TMV would appear later on the leaves. It was concluded that RNA is the genetic material of this virus.

Soon afterwards, another type of experiment with TMV was reported by Heinz Fraenkel-Conrat and B. Singer. These scientists discovered that the RNA core and the protein coat from wild-type TMV and other viral strains could be isolated separately. In their work, RNA and coat proteins were separated and isolated from TMV and a second viral strain, **Holmes ribgrass** (**HR**). Then, mixed viruses were reconstituted from the RNA of one strain and the protein of the other. When this "hybrid" virus was spread on tobacco leaves, the lesions that developed corresponded to the type of RNA in the reconstituted virus; that is, viruses with wild-type TMV RNA and HR protein coats produced TMV lesions (Figure 8.8) and vice versa. Again, it was concluded that RNA serves as the genetic material in these viruses.

In 1965 and 1966, Norman R. Pace and Sol Spiegelman further demonstrated that RNA from the phage Qβ could be isolated and replicated *in vitro*. Replication was dependent on an enzyme, **Qβ RNA replicase,** which was isolated from host *E. coli* cells following normal infection. When the RNA replicated *in vitro* was added to *E. coli* protoplasts, infection and viral multiplication occurred. Thus, RNA synthesized in a test tube can amply serve as the genetic material in these phages.

SUMMARY AND CONCLUSIONS

The existence of a genetic material is deducible from the observed patterns of inheritance in organisms. The functions of the genetic material are replication, storage, expression, and mutation of genetic information. Geneticists considered both proteins and nucleic acids as candidates for the genetic material. In the first part of the twentieth century, proteins were favored because protein chemistry had developed faster than nucleic acid chemistry. Proteins appeared to be more functionally diverse than nucleic acids, and Levene's tetranucleotide hypothesis resulted in an underestimation of the role of the nucleic acids.

By 1944, the tetranucleotide hypothesis was invalidated, and evidence establishing the importance of DNA in genetic processes began to accumulate. Except for those viruses containing only RNA, DNA was shown to be the genetic material through experimental and circumstantial evidence. The direct experimental evidence was derived from certain critical investigations designed to answer the specific question: Of what substance is the genetic material? These experiments included work on transformation of bacteria and infection of bacteria with bacteriophages or DNA from bacteriophages. Beyond reasonable doubt, these experiments established the genetic role of DNA in bacteria and most viruses.

Initially, only circumstantial evidence was available for eukaryotes. It included (1) the distribution of DNA in tissues; (2) the correspondence of the UV-absorption spectrum of DNA and the action spectrum for UV-induced mutagenesis; (3) the metabolic stability of DNA compared with other major classes of biomolecules; and (4) the localization of DNA in cellular organelles for which a genetic function has been established (nuclei, chloroplasts, mitochondria). Recent recombinant DNA techniques have now provided direct experimental evidence that DNA serves as the genetic material in all eukaryotes.

An important exception to the general rule involves a variety of viruses where RNA serves as the genetic material. These include certain bacteriophages as well as some animal and plant viruses.

The establishment of DNA as the genetic material paved the way for the expansion of knowledge in molecular genetics. In this regard, the research that went into this finding has served as the cornerstone for further important studies over the past three decades.

PROBLEMS AND DISCUSSION QUESTIONS

1 The functions ascribed to the genetic material are replication, expression, storage, and mutation. What does each of these terms mean?

2 Discuss the reasons why proteins were generally favored over DNA as the genetic material before 1940. What was the role of the tetranucleotide hypothesis in this controversy?

3 Contrast the various contributions made to an understanding of transformation by Griffith, Avery and his coworkers, and Taylor.

4 When Avery and his colleagues had obtained what was concluded to be purified DNA from the IIIS virulent cells, they treated the fraction with proteases, RNase, and DNase, followed by the assay for retention or loss of transforming ability. What were the purpose and results of these experiments? What conclusions were drawn?

5 Based on the information on transformation presented in this chapter, what other aspects of the process might you investigate to understand it more fully?

6 Why were ^{32}P and ^{35}S chosen for use in the Hershey-Chase experiment? Discuss the rationale and conclusions of this experiment.

7 Does the design of the Hershey-Chase experiment distinguish between DNA and RNA as the molecule serving as the genetic material? Why or why not?

8 Would an experiment similar to that performed by Hershey and Chase work if the basic design were applied to the phenomenon of transformation? Explain why or why not.

9 Why is the early evidence that DNA serves as the genetic material in eukaryotes called circumstantial? List and discuss these evidences.

10 What are the exceptions to the general rule that DNA is the genetic material in all organisms? What evidence supports these exceptions?

SELECTED READINGS

ALLOWAY, J. L. 1933. Further observations on the use of pneumococcus extracts in effecting transformation of type *in vitro. J. Exp. Med.* 57: 265–78.

AVERY, O. T., MacLEOD, C. M., and McCARTY, M. 1944. Studies on the chemical nature of the substance inducing transformation of pneumococcal types. Induction of transformation of pneumococcal types. Induction of transformation by a desoxyribonucleic acid fraction isolated from pneumococcus type III. *J. Exp. Med.* 79: 137–58. (Reprinted in Taylor, J. H. 1965. *Selected papers in molecular genetics.* New York: Academic Press.)

AYAD, S. R., and SHIMMIN, E. R. A. 1974. Properties of competence inducing factor of *Bacillus subtilis* 168I. *Biochem. Genet.* 11: 455–74.

CAIRNS, J., STENT, G. S., and WATSON, J. D. 1966. *Phage and the origins of molecular biology.* Cold Spring Harbor, N.Y.: Cold Spring Harbor Laboratory.

DAWSON, M. H. 1930. The transformation of pneumococcal types. I. The interconvertibility of type-specific S pneumococci. *J. Exp. Med.* 51: 123–47.

DeROBERTIS, E. M., and GURDON, J. B. 1979. Gene transplantation and the analysis of development. *Scient. Amer.* (Dec.) 241: 74–82.

DUBOS, R. J. 1976. *The professor, the Institute and DNA: Osward T. Avery, his life and scientific achievements.* New York: Rockefeller University Press.

FRAENKEL-CONRAT, H., and SINGER, B. 1957. Virus reconstruction. II, Combination of protein and nucleic acid from different strains. *Biochem. Biophys. Acta* 24: 530–48. (Reprinted in Taylor, J. H. 1965. *Selected papers in molecular genetics.* New York: Academic Press.)

GRIFFITH, F. 1928. The significance of pneumococcal types. *J. Hyg.* 27: 113–59.

GROBSTEIN, C. 1977. The recombinant DNA debate. *Scient. Amer.* (July) 237: 22–33.

GUTHRIE, G. D., and SINSHEIMER, R. L. 1960. Infection of protoplasts of *Escherichia coli* by subviral particles. *J. Mol. Biol.* 2: 297–305.

HAYES, W. 1968. *The genetics of bacteria and their viruses.* 2nd ed. New York: Wiley.

HERSHEY, A. D., and CHASE, M. 1952. Independent functions of viral protein and nucleic acid in growth of bacteriophage. *J. Gen. Phys.* 36: 39–56. (Reprinted in Taylor, J. H. 1965. *Selected papers in molecular genetics.* New York: Academic Press.)

HOTCHKISS, R. D. 1951. Transfer of penicillin resistance in pneumococci by the desoxyribonucleate derived from resistant cultures. *Cold Spr. Harb. Symp.* 16: 457–61. (Reprinted in Adelberg, E. A. 1960. *Papers on bacterial genetics.* Boston: Little, Brown.)

LEVENE, P. A., and SIMMS, H. S. 1926. Nucleic acid structure as determined by electrometric titration data. *J. Biol. Chem.* 70: 327–41.

McCARTY, M. 1980. Reminiscences of the early days of transformation. *Ann. Rev. Genet.* 14: 1–16.

———. 1985. *The transforming principle: Discovering that genes are made of DNA.* New York: W. W. Norton.

PACE, N. R., and SPIEGELMAN, S. 1966. *In vitro* synthesis of an infectious mutant RNA with a normal RNA replicase. *Science* 153: 64–67. (Reprinted in Subay, G. L. 1962. *Papers in biochemical genetics.* 2nd ed. New York: Holt, Rinehart and Winston.)

PALMITER, R. D., and BRINSTER, R. L. 1985. Transgenic mice. *Cell* 41: 343–45.

RAVIN, A. W. 1961. The genetics of transformation. *Adv. in Genet.* 10: 62–163.

SPIZIZEN J. 1957. Infection of protoplasts by disrupted T2 viruses. *Proc. Natl. Acad. Sci.* 43: 694–701.

STENT, G. S., and CALENDAR, R. 1978. *Molecular genetics: An introductive narrative.* 2nd ed. San Francisco: W. H. Freeman.

STEWART, T. A., WAGNER, E. F., and MINTZ, B. 1982. Human β-globin gene sequences injected into mouse eggs, retained in adults, and transmitted to progeny. *Science* 217: 1046–48.

TOMASZ, A. 1969. Cellular factors in genetic transformation. *Scient. Amer.* (Jan.) 220: 38–44.

WEINBERG, R. A. 1985. The molecules of life. *Scient. Amer.* (Oct.) 253: 48–57.

9

Nucleic Acids: Structure and Analysis

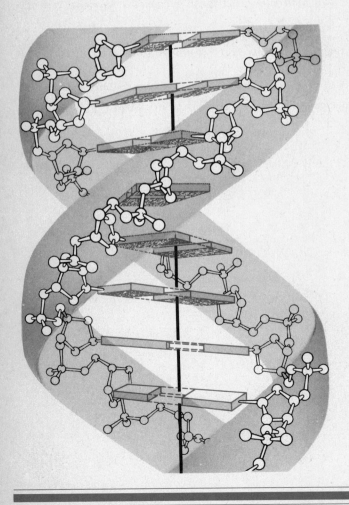

Chapter 8 provided evidence supporting the hypothesis that DNA is the genetic material in all organisms except certain viruses, where RNA serves this function. While this evidence was accumulating, other research was being performed to determine the chemical structure of nucleic acids. When the importance of DNA in genetic processes was realized, this work was intensified, with the hope of discerning not only the structural basis of this molecule but the relationship of its structure to the functional characteristics of the genetic material (e.g., replication, storage, expression, and mutation).

From 1940 to 1953, significant work concerning DNA structure was performed by Erwin Chargaff, Maurice Wilkins, Rosalind Franklin, Linus Pauling, Francis Crick, and James Watson, among others. These scientists sought information that might answer the most significant and intriguing question in the history of biology: How does DNA serve as the basis of life? The answer was believed to depend strongly on the chemical structure of the DNA molecule, given the complex but orderly functions ascribed to it.

These efforts were rewarded in 1953 when Watson and Crick set forth their hypothesis for the double-helical nature of DNA. The assumption that the molecule's functions would be more easily clarified once its general structure was determined proved to be correct. In this chapter we are concerned with nucleic acid structure and its analysis. In subsequent chapters, we will consider in greater detail how DNA functions in the direction of life processes.

THE NUCLEOTIDE: THE BASIC UNIT

Nucleotides are the building blocks of all nucleic acid molecules. Sometimes called mononucleotides, these structural units consist of three essential components: a **nitrogenous base**, a **pentose sugar** (5 carbons), and **phosphoric acid** (a phosphate group). There are two kinds of nitrogenous bases: the nine-membered double-ringed **purines** and the six-membered single-ringed **pyrimidines.** Two types of purines and three types of pyrimidines are found commonly in nucleic acids. The two purines are **adenine** and **guanine,** abbreviated **A** and **G.** The three pyrimidines are **cytosine, thymine,** and **uracil,** abbreviated **C, T,** and **U.** The chemical structures of A, G, C, T, and U are shown in Figure 9.1. Both DNA and RNA contain A, C, and G; only DNA contains the base T, whereas only RNA contains the base U. Each nitrogen or carbon atom of the ring structures of purines and pyrimidines is designated by a number without a prime sign. Note that corresponding atoms in the two rings are numbered differently in most cases.

The pentose sugars found in nucleic acids give them their names. Ribonucleic acids (RNA) contain **ribose,** while deoxyribonucleic acids (DNA) contain **deoxyribose.** Figure 9.2 shows the straight-chain (Fisher projection formula) and ring (Haworth formula) structures for these two pentose sugars. Each carbon atom is distinguished by a number and a prime sign (e.g., C-1′, C-2′, etc.). As you can see, deoxyribose is missing one hydroxyl group at the C-2′ position compared with ribose. This is the only difference between the two sugars.

If a molecule is composed of a purine or pyrimidine base and a ribose or deoxyribose sugar, the chemical unit is called a **nucleoside.** If a phosphate group is added to the nucleoside, the molecule is now called a

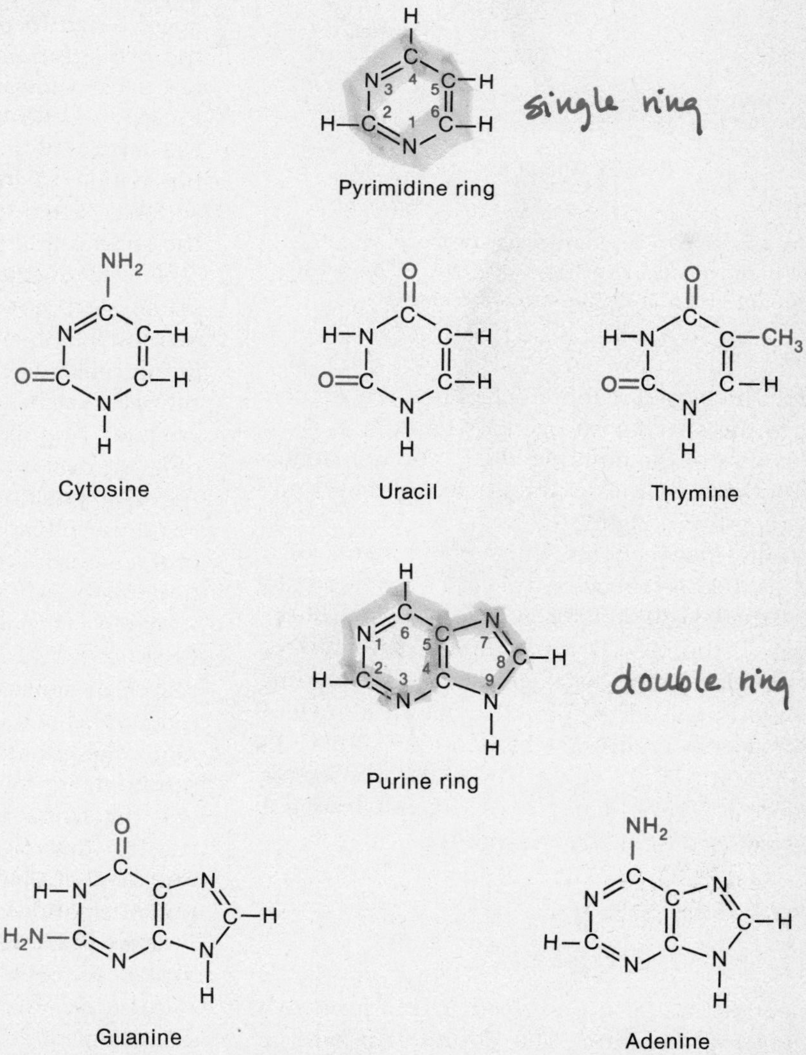

FIGURE 9.1
Chemical structures of the pyrimidines and purines that serve as the nitrogenous bases in RNA and DNA.

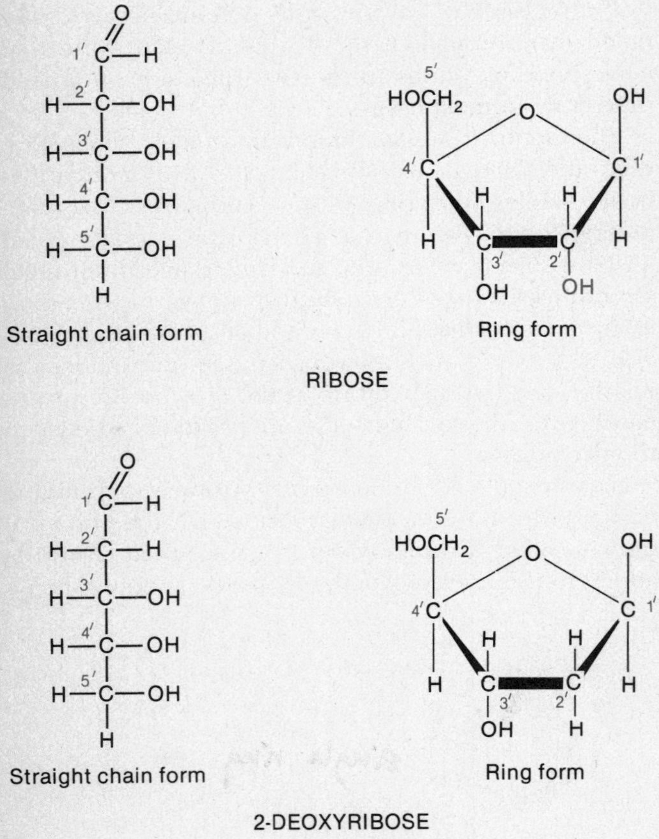

Straight chain form   Ring form

RIBOSE

Straight chain form   Ring form

2-DEOXYRIBOSE

FIGURE 9.2
Chemical structures of ribose and 2-deoxyribose, which serve as the pentose sugars in RNA and DNA, respectively. Both the straight-chain and ring forms are shown.

nucleotide. Nucleosides and nucleotides are named according to the specific nitrogenous base (A, T, G, C, or U) that is part of the building block. The nomenclature and general structure of the nucleosides and nucleotides are given in Figure 9.3.

The bonding between the three components of a nucleotide is highly specific. The C-1′ atom of the sugar is involved in the chemical linkage to the nitrogenous base. If the base is a purine, the N-9 atom is covalently bonded to the sugar. If the base is a pyrimidine, the bonding involves the N-1 atom. In a nucleotide, the phosphate group may be bonded to the C-2′, C-3′, or C-5′ atom of the sugar. The C-5′—phosphate configuration is shown in Figure 9.3. It is by far the most prevalent one in biological systems.

NUCLEOSIDE DIPHOSPHATES AND TRIPHOSPHATES

Mononucleotides are also described by the term **nucleoside monophosphate.** The addition of one or two phosphate groups result in **nucleoside diphos-**

phates and **triphosphates,** as illustrated in Figure 9.4. The triphosphate form is significant because it serves as the precursor molecule during nucleic acid synthesis within the cell. Additionally, the triphosphates **adenosine triphosphate (ATP)** and **guanosine triphosphate (GTP)** are important in the cell's bioenergetics because of the large amount of energy involved in the addition or removal of the terminal phosphate group. The hydrolysis of ATP or GTP to ADP or GDP and inorganic phosphate (P_i) is accompanied by the release of a large amount of energy in the cell. When these chemical conversions are coupled to energy-requiring reactions, the energy produced may be used to drive them to equilibrium. As a result, ATP and GTP are involved in most cellular activities.

POLYNUCLEOTIDES

The linkage between two mononucleotides consists of a phosphate group linked to two sugars. A **phosphodiester** bond is formed, because phosphoric acid has been joined to two alcohols (the hydroxyl groups on the two sugars) by an ester linkage on both sides. Figure 9.5(a) shows the resultant phosphodiester bonds in DNA and RNA. Each structure has a C-5′ end and a C-3′ end. The joining of two nucleotides forms a dinucleotide; of three nucleotides, a trinucleotide; and so forth. When long chains of nucleotides are formed, the structure is called a **polynucleotide.**

Since drawing the structures in Figure 9.5(a) is time consuming and complex, a schematic shorthand method has been devised [Figure 9.5(b)]. The vertical line represents the carbons of the pentose sugar; the nitrogenous base is attached at the top, or the C-1′ position. The diagonal line, with the Ⓟ in the middle of it, is attached to the C-3′ position of one sugar and the C-5′ position of the neighboring sugar; it represents the phosphodiester bond. Several modifications of this shorthand method are in use, and they can be understood in terms of these guidelines.

While Levene's tetranucleotide hypothesis (see Chapter 8) was generally accepted before 1940, research in subsequent decades showed it to be incorrect. It was shown that DNA does not necessarily contain equimolar quantities of the four bases. Additionally, the molecular weight of DNA molecules was determined to be in the range of 10^6 to 10^9 daltons, far in excess of that of a tetranucleotide. The current view of DNA is that it consists of exceedingly long polynucleotide chains.

Long polynucleotide chains would account for the observed molecular weight and would explain the most important property of DNA—genetic variation. If each nucleotide position in this long chain may be occupied by any one of four nucleotides, extraordinary

FIGURE 9.3
The structure and names of the nucleosides and nucleotides of RNA and DNA.

NUCLEOSIDES

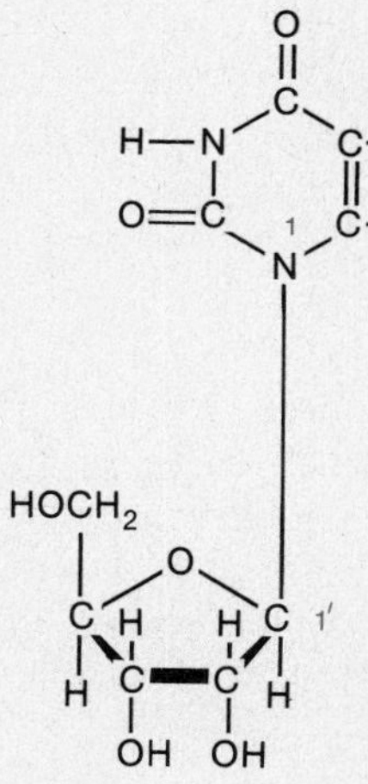

NUCLEOTIDES

Uridine

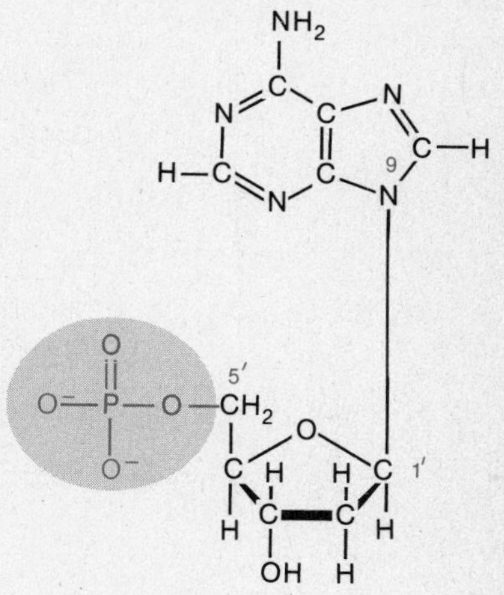

Deoxyadenylic acid

Ribonucleosides

Ribonucleotides

Adenosine

Adenylic acid

Cytidine

Cytidylic acid

Guanosine

Guanylic acid

Uridine

Uridylic acid

Deoxyribonucleosides

Deoxyribonucleotides

Deoxyadenosine

Deoxyadenylic acid

Deoxycytidine

Deoxycytidylic acid

Deoxyguanosine

Deoxyguanylic acid

Deoxythymidine

Deoxythymidylic acid

variation is possible. For example, a polynucleotide that is 1000 nucleotides in length may be arranged 4^{1000} different ways, each one different from all other possible sequences. This potential variation in molecular structure is essential if DNA is to serve the function of storing the vast amounts of chemical information necessary to direct cellular activities.

THE STRUCTURE OF DNA

In 1953, James Watson and Francis Crick proposed that the structure of DNA is in the form of a **double helix.** Their proposal was published in a short paper in *Nature*, which is reproduced in its entirety on pages 190 and 191. In a sense, this publication constituted the finish line in a highly competitive scientific race to

obtain what some consider to be the most significant finding in the history of biology. This "race," as recounted in Watson's book *The Double Helix*, demonstrates the human interaction, genius, frailty, and intensity involved in the scientific effort that eventually led to the elucidation of DNA structure.

The available data, crucial to the development of the proposal, came primarily from two sources: base composition analysis of hydrolyzed samples of DNA, and X-ray diffraction studies of DNA. The analytical success of Watson and Crick may be attributed to model building that conformed to the above types of existing data. If the structure of DNA may be analogized by a puzzle, Watson and Crick, working in the Cavendish Laboratory in Cambridge, England, were the first to put together all of the pieces successfully.

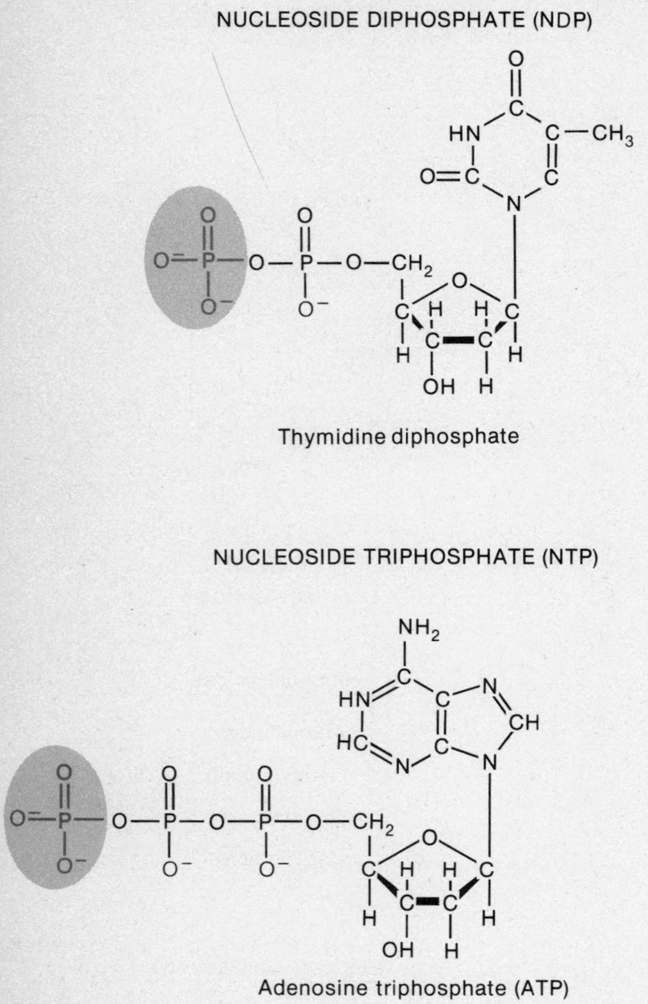

NUCLEOSIDE DIPHOSPHATE (NDP)

Thymidine diphosphate

NUCLEOSIDE TRIPHOSPHATE (NTP)

Adenosine triphosphate (ATP)

FIGURE 9.4
The basic structures of nucleoside diphosphates and
triphosphates, as illustrated by thymidine diphosphate and
adenosine triphosphate.

(a)

3′ to 5′
phosphodiester
bonds

(b)

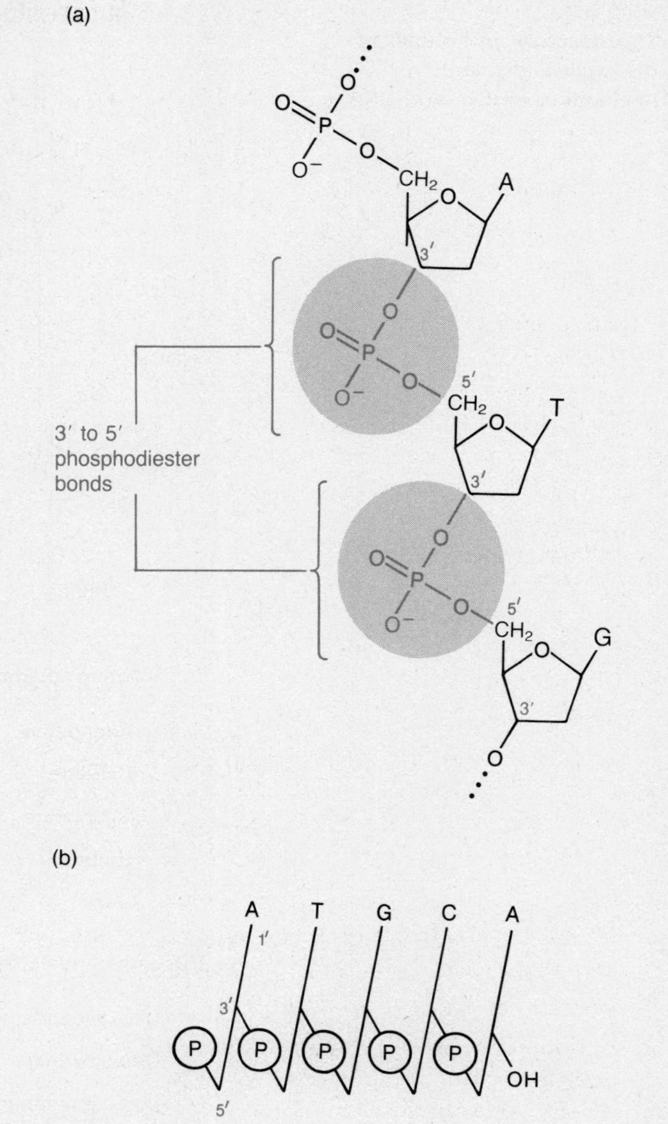

FIGURE 9.5
(a) The linkage of nucleotides by the formation of C-3′–C-5′
phosphodiester bonds, producing a polynucleotide chain.
(b) A shorthand notation for a polynucleotide chain.

Base Composition Studies

Between 1949 and 1953, Erwin Chargaff and his colleagues used chromatographic methods to separate
the four bases in DNA samples from various organisms. Quantitative methods were then used to determine the amounts of the four bases from each source.
Table 9.1 provides some of Chargaff's original data as
well as a representative sample of the analysis of more
recent base composition data derived from other organisms.

The analysis of these data led to the following conclusions:

1   The number of adenine residues equals the number of thymine residues in the DNA of any species
(columns 1, 2, and 5). Also, the number of guanine
residues is equivalent to the number of cytosine
residues (columns 3, 4, and 6).

2   The sum of the purines (A + G) equals the sum of
the pyrimidines (C + T), as shown in column 7.

TABLE 9.1

DNA base composition data.

(a) Chargaff's data.

Source	Approximate Percent*			
	A	T	G	C
Ox thymus	26	25	21	16
Ox spleen	25	24	20	15
Yeast	24	25	14	13
Avian tubercle bacilli	12	11	28	26
Human sperm	29	31	18	18

SOURCE: From Chargaff, 1950.

*Moles of nitrogenous constituent per mole of P (often, the recovery was less than 100 percent).

(b) Base compositions of DNAs from various sources.

Source	Base Composition				Base Ratio			Asymmetry Ratio
	1	2	3	4	5	6	7	8
	A	T	G	C	A/T	G/C	(A + G)/(C + T)	(A + T)/(C + G)
Human	30.9	29.4	19.9	19.8	1.05	1.00	1.04	1.52
Sea urchin	32.8	32.1	17.7	17.3	1.02	1.02	1.02	1.58
E. coli	24.7	23.6	26.0	25.7	1.04	1.01	1.03	0.93
Sarcina lutea	13.4	12.4	37.1	37.1	1.08	1.00	1.04	0.35
T7 bacteriophage	26.0	26.0	24.0	24.0	1.00	1.00	1.00	1.08

3 The ratio of (A + T)/(C + G) does not necessarily equal one; further, this ratio varies greatly among species, as shown in column 8.

4 Although not shown in Table 9.1, for any given species the base composition of DNA is identical, within experimental error, in any given somatic tissue.

These conclusions indicate a definite regularity or pattern of base composition of DNA molecules. These data served as the initial clue to "the puzzle." Additionally, they directly refute the tetranucleotide hypothesis, which stated that all four bases are present in equal amounts.

X-Ray Diffraction Analysis

When fibers of a DNA molecule are subjected to X-ray bombardment, these rays are scattered according to the molecule's atomic structure. The pattern of scatter may be captured as spots on photographic film and analyzed, particularly for the overall shape of and regularities within the molecule. This process, X-ray dif-

fraction analysis, was successfully applied to protein structure analysis by Linus Pauling. The technique had been attempted on DNA as early as 1938, but was unsuccessful because the available samples were impure.

Between 1950 and 1953, Pauling in the United States and Rosalind Franklin in England (working in Wilkins' laboratory) obtained data from highly purified DNA samples. Pauling, on the basis of his observations, incorrectly postulated a **triple helix** model for DNA. Franklin's work also supported a helical structure. Most significant, her data demonstrated two distinctive regularities (or periodicities) of 0.34 nm (3.4 Å) and 3.4 nm (34 Å) along the axis of DNA. However, she did not propose a definitive model of DNA. One of the X-ray photographs of DNA is shown in Figure 9.6.

The Watson-Crick Model

Watson and Crick published their analysis of DNA structure in 1953 (see pp. 190–191). By building models under the constraints of the information just dis-

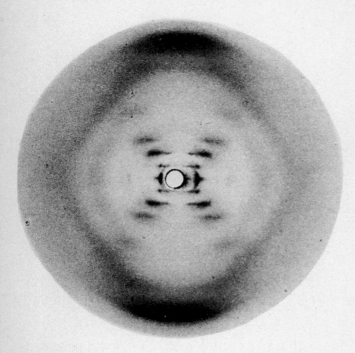

FIGURE 9.6

An X-ray diffraction photograph of the B form of crystallized DNA. The dark patterns at the top and bottom provide an estimate of the periodicity of nitrogenous bases, which are 3.4 Å apart. The central pattern is indicative of the molecule's helical structure (From Franklin and Gosling, 1953. Reprinted by permission from *Nature*, Vol. 171, pp. 738. Copyright © 1953 Macmillan Journals Limited. Photo courtesy of M. H. F. Wilkins, Biophysics Department, King's College, London.)

cussed, they proposed the double-helical form of DNA as shown in Figure 9.7. This model has the following major features:

1 Two right-handed helical polynucleotide chains are coiled around a central axis; the coiling is **plectonic,** meaning that the two coils can only be separated by completely unwinding them.

2 The two chains are **antiparallel;** that is, one is upside down with respect to the other; their C-5'-to-C-3' orientations are in opposite directions.

3 The bases of both chains are flat structures, lying perpendicular to the axis; they are "stacked" on one another, 0.34 nm (3.4 Å) apart.

4 Each complete turn of the helix is 3.4 nm long; thus, 10 bases exist in each chain per turn.

5 The nitrogenous bases of opposite chains are electrostatically attracted to one another as the result of the formation of **hydrogen bonds** (described in the following discussion): specifically, only A-T and G-C pairs are allowed.

6 In any segment of the molecule, alternating larger **major grooves** and smaller **minor grooves** are apparent along the axis.

7 The double helix measures 2.0 nm (20 Å) in diameter.

A more recent and accurate analysis of the form of DNA that served as the basis for the Watson-Crick model has revealed a minor structural difference. A precise measurement of the number of base pairs (bp) per turn has demonstrated a value of 10.4 rather than the 10.0 predicted by Watson and Crick. Where, in the classical model, each base pair is rotated around the helical axis 36°, relative to the adjacent base pair, the new finding requires a rotation of 34.6°. Thus, there are slightly more than 10 base pairs per 360° turn.

Point 2 in the model requires special emphasis. Antiparallelity means that from one end of the helix one chain is in the 5'-to-3' orientation while the other chain is in the 3'-to-5' orientation. Given the constraints of bond angles, a double helix of the nature described by Watson and Crick could not be constructed if both chains were in the same orientation.

One of the most significant features of the structure proposed by Watson and Crick is the specificity of base pairing. Chargaff's data had suggested that the amounts of A equaled T and that G equaled C. Watson and Crick realized that when placed opposite each other in the model, the members of each such base pair formed hydrogen bonds, providing the chemical stability necessary to hold the two chains together. Arranged in this way, both major and minor grooves became apparent along the axis. Further, with one purine (A or G) opposite one pyrimidine (T or C) as each "rung of the spiral staircase" of the proposed helix, the Watson-Crick model conformed to the 2.0-nm diameter suggested by X-ray diffraction studies.

The specific A-T and G-C base pairing is the basis for the concept of **complementarity.** This term is used to describe the chemical affinity provided by the hydrogen bonds between the bases. As we will see, this concept is very important in the processes of DNA replication and gene expression.

Two questions are particularly worthy of discussion. First, why aren't other base pairs possible? Watson and Crick discounted the A-G and C-T pairs because these represent purine-purine and pyrimidine-pyrimidine pairings, respectively. These pairings would lead to alternating diameters of more than and less than 20 nm because of the respective sizes of the purine and pyrimidine rings; additionally, the three-dimensional configurations formed by such pairings do not produce the proper alignment leading to sufficient hydrogen bond formations. It is for this latter

FIGURE 9.7
(a) A schematic representation of the DNA double helix as proposed by Watson and Crick. The ribbonlike strands constitute the sugar-phosphate backbones, and the horizontal rungs constitute the nitrogenous base pairs, of which there are 10 per complete turn. The major and minor grooves are apparent. The solid vertical rod has been placed through the center of the helix.
(b) A representation of the antiparallel nature of the two strands of the helix.
(c) The hydrogen bonds formed between cytosine and guanine and between thymine and adenine.

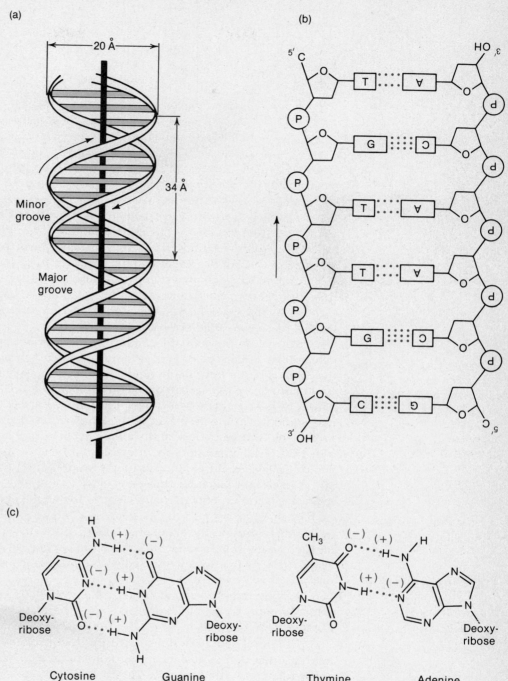

reason that A-C and G-T pairings were also discounted, even though these pairs each consist of one purine and one pyrimidine.

The second question concerns hydrogen bonds. Just what is the nature of such a bond, and is it strong enough to stabilize the helix? A **hydrogen bond** is a very weak electrostatic attraction between a covalently bonded hydrogen atom and an atom with an unshared electron pair. The hydrogen atom assumes a partial positive charge, while the unshared electron pair—characteristic of covalently bonded oxygen and nitrogen atoms—assumes a partial negative charge. These opposite charges are responsible for the weak chemical attraction. As oriented in the double helix, adenine forms two hydrogen bonds with thymine, and guanine forms three hydrogen bonds with cytosine. Although two or three hydrogen bonds taken alone are very weak, two or three thousand bonds in tandem

MOLECULAR STRUCTURE OF NUCLEIC ACIDS: A STRUCTURE FOR DEOXYRIBOSE NUCLEIC ACID

We wish to suggest a structure for the salt of deoxyribose nucleic acid (D. N. A.). This structure has novel features which are of considerable biological interest. A structure for nucleic acid has already been proposed by Pauling and Corey.[1] They kindly made their manuscript available to us in advance of publication. Their model consists of three intertwined chains, with the phosphates near the fibre axis, and the bases on the outside. In our opinion, this structure is unsatisfactory for two reasons: (1) We believe that the material which gives the X-ray diagrams is the salt, not the free acid. Without the acidic hydrogen atoms it is not clear what forces would hold the structure together, especially as the negatively charged phosphates near the axis will repel each other. (2) Some of the van der Waals distances appear to be too small.

Another three-chain structure has also been suggested by Fraser (in the press). In his model the phosphates are on the outside and the bases on the inside, linked together by hydrogen bonds. This structure as described is rather ill-defined, and for this reason we shall not comment on it.

We wish to put forward a radically different structure for the salt of deoxyribose nucleic acid. This structure has two helical chains each coiled round the same axis. We have made the usual chemical assumptions, namely, that each chain consists of phosphate diester groups joining β-D-deoxyribofuranose residues with 3′,5′ linkages. The two chains (but not their bases) are related by a dyad perpendicular to the fibre axis. Both chains follow right-handed helices, but owing to the dyad the sequences of the atoms in the two chains run in opposite directions. Each chain loosely resembles Furberg's[2] model No. 1; that is, the bases are on the inside of the helix and the phosphates on the outside. The configuration of the sugar and the atoms near it is close to Furberg's "standard configuration," the sugar being roughly perpendicular to the attached base. There is a residue on each chain every 3.4 Å in the z-direction. We have assumed an angle of 36° between adjacent residues in the same chain, so that the structure repeats after 10 residues on each chain, that is, after 34 Å. The distance of a phosphorus atom from the fibre axis is 10 Å. As the phosphates are on the outside, cations have easy access to them.

The structure is an open one, and its water content is rather high. At lower water contents we would expect the bases to tilt so that the structure could become more compact.

The novel feature of the structure is the manner in which the two chains are held together by the purine and pyrimidine bases. The planes of the bases are perpendicular to the fibre axis. They are joined together in pairs, a single base from one chain being hydrogen-bonded to a single base from the other chain, so that the two lie side by side with identical z-co-ordinates. One of the pair must be a purine and the other a pyrimidine for bonding to occur. The hydrogen bonds are made as follows: purine position 1 to pyrimidine position 1; purine position 6 to pyrimidine position 6.

If it is assumed that the bases only occur in the structure in the most plausible tautomeric forms (that is, with the keto rather than the enol configurations) it is found that only specific pairs of bases can bond together. These pairs are: adenine (purine) with thymine (pyrimidine), and guanine (purine) with cytosine (pyrimidine).

In other words, if an adenine forms one member of a pair, on either chain, then on these assumptions the other member must be thymine; similarly for guanine and cytosine. The sequence of bases on a single chain does not appear to be restricted in any way. However, if only specific pairs of bases can be formed, it follows that if the sequence of bases on one chain is given, then the sequence on the other chain is automatically determined.

It has been found experimentally[3,4] that the ratio of the amounts of adenine to thymine, and the ratio of guanine to cytosine, are always very close to unity for deoxyribose nucleic acid.

It is probably impossible to build this structure with a ribose sugar in place of the deoxyribose, as the extra oxygen atom would make too close a van der Waals contact.

The previously published X-ray data[5,6] on deoxyribose nucleic acid are insufficient for a rigorous test of our structure. So far as we can tell, it is roughly compatible with the experimental data, but it must be regarded as unproved until it has been checked against more exact results. Some of these are given in the following communications. We were not aware of the details of the results presented there when we devised our structure, which rests mainly though not entirely on published experimental data and stereochemical arguments.

It has not escaped our notice that the specific pairing we have postulated immediately suggests a possible copying mechanism for the genetic material.

Full details of the structure, including the conditions assumed in building it, together with a set of co-ordinates for the atoms, will be published elsewhere.

We are much indebted to Dr. Jerry Donohue for constant advice and criticism, especially on interatomic distances. We have also been stimulated by a knowledge of the general nature of the unpublished experimental results and ideas of Dr. M. H. F. Wilkins, Dr. R. E. Franklin and their coworkers at King's College, London. One of us (J. D. W.) has been aided by a fellowship from the National Foundation for Infantile Paralysis.

J. D. Watson
F. H. C. Crick
Medical Research Council Unit for the Study of the
Molecular Structure of Biological Systems
Cavendish Laboratory, Cambridge.
April 2.

[1]Pauling, L., and Corey, R. B., *Nature*, 171, 346 (1953); *Proc. U.S. Nat. Acad. Sci.*, 39, 84 (1953).
[2]Furberg, S., *Acta Chem. Scand.*, 6, 634 (1952).
[3]Chargaff, E., for references see Zamenhof, S., Brawerman, G., and Chargaff, E., *Biochim. et Biophys. Acta*, 9, 402 (1952).
[4]Wyatt, G. R., *J. Gen. Physiol*, 36, 201 (1952).
[5]Astbury, W. T., Symp. Soc. Exp. Biol. 1, Nucleic Acid, 66 (Camb. Univ. Press, 1947).
[6]Wilkins, M. H. F., and Randall, J. T., *Biochim. et Biophys. Acta*, 10, 192 (1953).

(which would be found in two long polynucleotide chains) are capable of providing great stability to the helix.

Still another stabilizing factor is the arrangement of sugars and bases along the axis. In the Watson-Crick model, the hydrophobic or "water-fearing" nitrogenous bases are stacked almost horizontally on the interior of the axis, thus shielded from water. The hydrophilic sugar-phosphate backbone is on the outside of the axis, where both components may interact with water. Taken together, these molecular arrangements provide significant stabilization to the helix.

The Watson-Crick model had an immediate effect on the emerging discipline of molecular biology. Even in their 1953 article, the authors noted, "It has not escaped our notice that the specific pairing we have postulated immediately suggests a possible copying mechanism for the genetic material." Two months later, in a second article in *Nature*, Watson and Crick pursued this idea, suggesting a specific mode of replication of DNA—the semiconservative model. The second article also alluded to two new concepts: the storage of genetic information in the sequence of the bases, and the mutation or genetic change that would result from alteration of the bases. These ideas have received vast amounts of experimental support since 1953 and are now universally accepted. Thus, the "synthesis" of ideas by Watson and Crick was a remarkable feat and highly significant in the history of genetics and biology.

A-, B-, C-, D-, AND E-DNA

Under different conditions of isolation, purification, and crystallization, several forms of DNA have been recognized. At the time Watson and Crick performed their analysis, two forms—**A-DNA** and **B-DNA**—were known. Watson and Crick's analysis was based on X-ray studies of the B form, which is present under a more hydrated set of conditions and is believed to be the biologically significant conformation.

While DNA studies around 1950 relied on the use of diffraction of fibers, more recent investigations have been performed using **single crystal X-ray analysis.** The earlier studies achieved limited resolution of about 5 Å, but single crystals diffract X-rays at about 1 Å, near atomic resolution. As a result, every atom is "visible" and much greater structural detail is available during analysis.

Using these modern techniques, the A form of DNA,

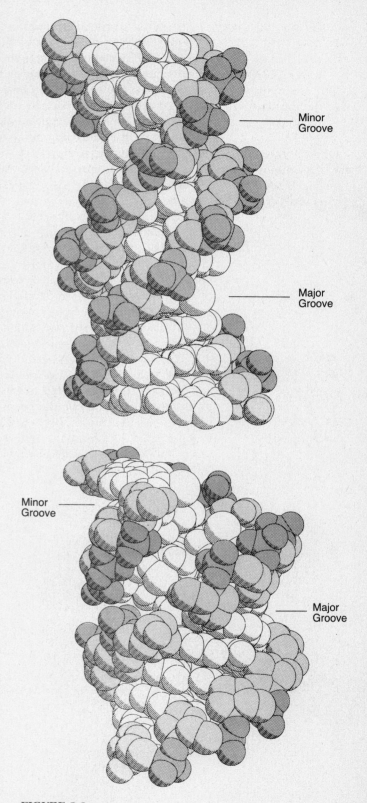

FIGURE 9.8
Uniform-scale drawings of DNA helices of B-DNA (top) and A-DNA (bottom). Phosphates are darkly shaded, sugars are lightly shaded, and bases are unshaded. (From Dickerson et al., 1982, Fig. 1. Copyright 1982 by the American Association for the Advancement of Science.)

which received minimal attention for over 25 years, has now been scrutinized (Figure 9.8). Like B-DNA, it assumes a right-handed double-helical configuration. However, A-DNA is slightly more compact, with 11 base pairs in each complete turn of the helix, which is 23 Å in diameter. The orientation of the bases is also somewhat different. They are tilted and displaced laterally in relation to the axis of the helix (Figure 9.9). As a result of these differences, the appearance of the major and minor grooves is modified compared with those in B-DNA.

It is not yet clear to what extent A-DNA occurs under biological conditions. However, under certain conditions, short synthetic double-stranded fragments (e.g., CCGG, GGCCGGCC, and GGTATACC along each strand) show a preference for assuming the A configuration. A-T–rich polymers show a preference for the B form. Since A-DNA occurs under conditions of decreased hydration, it has been suggested that it may be formed physiologically as a result of interaction with hydrophobic molecules or changing cellular conditions. If so, it is possible that DNA structure is a dynamic one, being altered under different conditions. This is a particularly attractive idea because of the

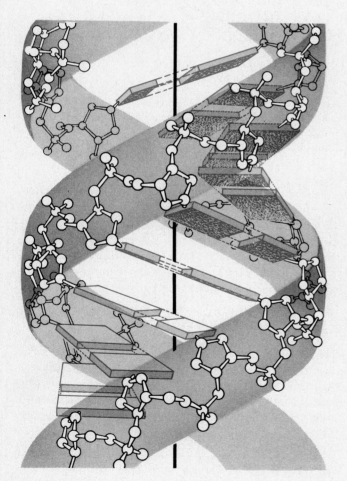
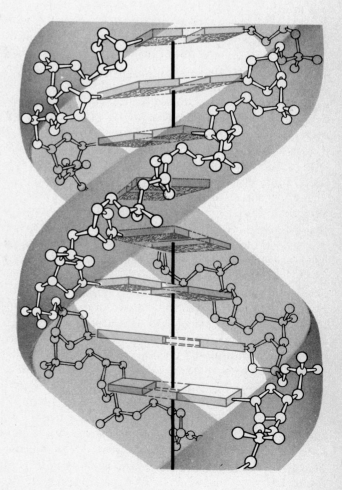

FIGURE 9.9

A comparison of the orientation of base pairs within the right-handed double helix of A-DNA (shown on the left) and B-DNA (shown on the right). In A-DNA, base pairs are tilted and pulled away from the helical axis. In B-DNA, base pairs are perpendicular to the helical axis. (Courtesy of Irving Geis.)

need for recognition sites along the helix during the various stages of gene activity.

Three other right-handed forms of DNA helices have been discovered. These have been designated C-, D-, and E-DNA. **C-DNA** is found to occur under even greater dehydration conditions than those observed during the isolation of A-, and B-DNA. It has only 9.3 base pairs per turn and is, thus, less compact. Its helical diameter is 19 Å. Like A-DNA, C-DNA does not have its base pairs lying flat; instead they are tilted relative to the axis of the helix. Two other forms, **D-DNA** and **E-DNA** occur in helices lacking guanine in their base composition. They have even fewer base pairs per turn: 8 and 7½, respectively.

Z-DNA

Still another form of DNA was discovered by Andrew Wang, Alexander Rich, and their colleagues in 1979 when they examined a DNA fragment of the hexanucleoside pentaphosphate d(CpGpCpGpCpG). Called **Z-DNA**, it takes on the rather remarkable configuration of a **left-handed double helix.** It had been known as early as 1972 that a dramatic shift in symmetry of the helix occurred under the conditions of high salt concentration. It has now been confirmed that this shift is the result of the conversion from a right-handed to a left-handed helix. The two configurations are illustrated and compared in Figures 9.10 and 9.11.

Z-DNA, like A- and B-DNA, consists of two antiparallel chains held together by Watson-Crick base pairs. Beyond these characteristics, Z-DNA is quite different. The left-handed helix is 18 Å in diameter, contains 12 base pairs per turn, and assumes a zigzag conformation (hence its name). The major groove present in B-DNA is nearly eliminated in Z-DNA.

Z-DNA has now been the subject of extensive analysis with increasing evidence accumulating that suggests it may be present in equilibrium with B-DNA under cellular conditions. Because of its more compact form, the phosphate groups on opposite strands of Z-DNA are closer together, creating a greater electrostatic repulsion than occurs in B-DNA. As a result, Z-DNA is energetically less stable and is not usually favored in an equilibrium existing between the two forms. However, DNA sequences consisting of alternating purines and pyrimidines have an affinity for forming Z-DNA. In addition to the hexanucleoside sequence described above, the following have been shown to form Z-DNA: poly(dG-dT), and the hexamer

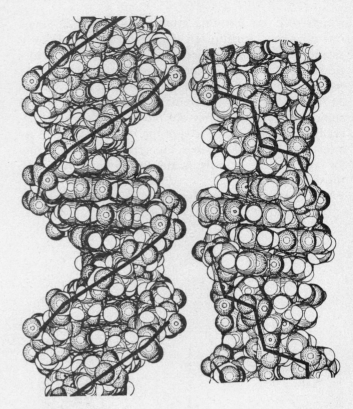

FIGURE 9.10

A comparison of the B and Z forms of DNA. The B form is a right-handed helix, and the Z form is a left-handed helix. (Computer-generated by Gary J. Quigley and courtesy of Alexander Rich, Department of Biology, MIT.)

sequence d(CpGpTpApCpG),* where the C-5 carbon atoms of the cytosine residues are brominated. Addition of a methyl group to the C-5 atom of cytosine facilitates and stabilizes the salt-induced formation of Z-DNA by poly(dA-m^{5}dC). The relative affinity of alternating purine-pyrimidine dinucleotide sequences to form Z-DNA has been summarized:

$$m^5CG > CG > TG = CA > TA$$

Of the various types of DNA structure that differ from B-DNA, the discovery of Z form has been the most exciting to geneticists. In equilibrium with B-DNA, the double helix must now be viewed as a dynamic structure with a single sequence of nucleotides exhibiting different conformations at different times. However, before this concept is fully accepted, Z-DNA

---

*Note that this and all other oligomers (short polynucleotide fragments) thus far discussed are of a sequence that is self-complementary. That is, the sequence in one direction is complementary to the sequence in the opposite direction. Thus, any two chains may form a double-helical molecule.

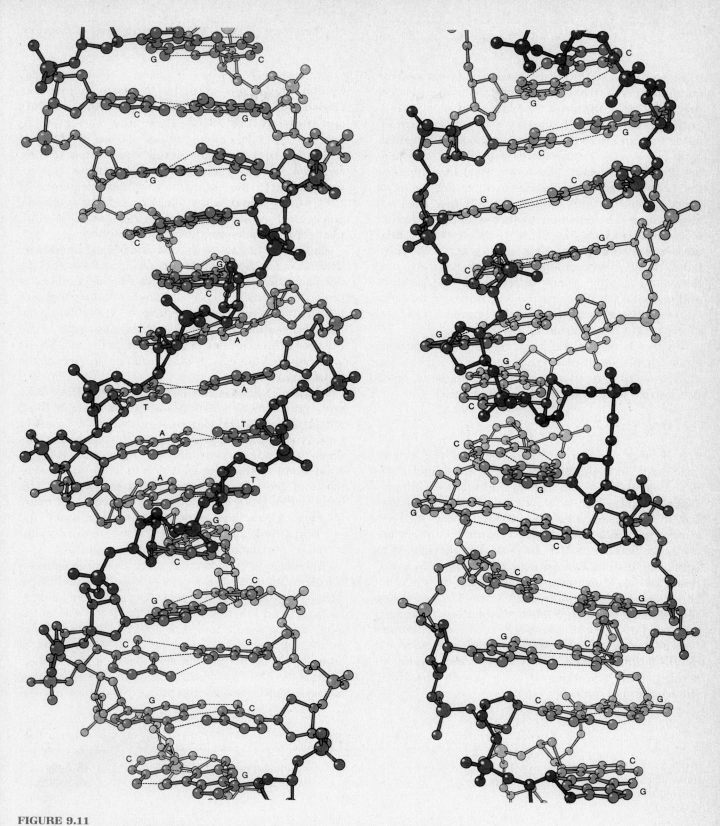

FIGURE 9.11
A comparison of the right-handed helix of B-DNA (shown on the left) with the left-handed helix of Z-DNA (shown on the right). The B-DNA helix was generated from the central 10 base pairs of the dodecamer CGCGAATTCGCG. The Z-DNA helix was generated from the central base pairs of an alternating CGCGCG sequence. Note that the major groove has nearly disappeared in Z-DNA. (Courtesy of Irving Geis.)

must be demonstrated directly to exist under cellular conditions. Additionally, the physico-chemical factors that stabilize Z-DNA must be established.

As we will see in Chapter 11, evidence is accumulating in support of the presence of Z-DNA in genetically active cells. Furthermore, factors that can stabilize Z-DNA are being discovered. Most convincing is the presence in *Drosophila* and other organisms of proteins that not only bind specifically to Z-DNA but also induce the transition to that structure. Another factor known to stabilize Z-DNA is referred to as **negative supercoiling.** Prevalent in circular DNA molecules, negative supercoiling occurs where DNA is underwound slightly. There are proteins that produce and maintain negative supercoiling. However, this generates an energy strain that can be relaxed if a portion of the B-DNA is converted to Z-DNA.

While we cannot yet draw absolute conclusions about Z-DNA, its discovery has generated a great deal of excitement. We can expect this to increase when its precise role in biological systems is clarified.

THE STRUCTURE OF RNA

The second category of nucleic acids is the ribonucleic acids, or RNA. The structure of these molecules is similar to DNA, with several important exceptions. While RNA also has as its building blocks nucleotides linked into polynucleotide chains, ribose replaces deoxyribose and uracil replaces thymine. Another important difference is that most RNA is thought of as being single stranded. However, RNA molecules sometimes fold back on themselves following their synthesis; such a configuration results when regions of complementarity occur in positions that allow base pairs to form. Furthermore, some animal viruses that have RNA as their genetic material contain it in the form of double-stranded helices. Thus, there are several instances where RNA does not exist strictly as a linear, single-stranded molecule.

At least three classes of cellular RNA molecules function during the expression of genetic information: **ribosomal RNA (rRNA), messenger RNA (mRNA),** and **transfer RNA (tRNA).** These molecules all originate as complementary copies of one of the two strands of DNA segments during the process of transcription. That is, their nucleotide sequence is complementary to the deoxyribonucleotide sequence of DNA, which served as the template for their synthesis. Since uracil replaces thymine in RNA, uracil is complementary to adenine during transcription.

Each class of RNA may be characterized by its size, sedimentation behavior in a centrifugal field, and genetic function. Sedimentation behavior depends on a molecule's density, mass, and shape, and its measure is called the **Svedberg coefficient (S).** Table 9.2 relates the S values, molecular weights, and approximate number of nucleotides of the major forms of RNA. As you can see, there is a wide variation in size of the three classes of RNA.

Ribosomal RNA is the largest of these RNA molecules and usually constitutes about 80 percent of all RNA in the cell. The various forms of rRNA found in prokaryotes and eukaryotes differ distinctly in size. These molecules constitute an important chemical component of **ribosomes,** which function in the synthesis of proteins, a process called **translation.** The large size of rRNA is undoubtedly related to its many chemical interactions with (1) proteins, forming the functional ribosome, and (2) the other components involved in translation, including mRNA and tRNA.

Messenger RNA molecules carry chemical information from the DNA of the gene to the ribosome, where translation into protein occurs. They vary considerably in length and function according to (1) the size of the gene serving as the template for transcription of any mRNA species, and (2) the size of, or the number of amino acids in, the protein encoded by a given gene.

Transfer RNA, the smallest class of RNA molecules, carries amino acids to the ribosome during transla-

TABLE 9.2
Sedimentation coefficients, molecular weights, and number of nucleotides for various RNAs.

RNA Type	Abbreviation	Sedimentation Coefficient	Molecular Weight	Number of Nucleotides
Ribosomal RNA	rRNA	5S	35,000	105
		16S (*E. coli*)	550,000–600,000	1541
		23S (*E. coli*)	1,100,000	3000
		18S (mammalian)	700,000	1740
		28S (mammalian)	1,800,000	4850
Transfer RNA	tRNA	4S	23,000–30,000	75–90
Messenger RNA	mRNA	6–25S	25,000–1,000,000	75–3000

tion. Since more than one tRNA molecule interacts simultaneously with the ribosome, the molecule's smaller size may facilitate these interactions. More is known about the secondary structure of this molecule than any of the other RNAs. All tRNAs conform to a two-dimensional cloverleaf model, as shown in Figure 9.12.

We will discuss the functions of the three classes of RNA in much greater detail later in the text (see Chapter 18). Our purpose in this section has been to contrast the structure of DNA, which stores genetic information, with that of RNA, which functions in the expression of that information.

ANALYSIS OF NUCLEIC ACIDS

Since 1953, the role of DNA as the genetic material and the role of RNA in transcription and translation have been clarified through detailed analysis of nucleic acids. We will consider several methods of analysis of these molecules in this chapter. Some of these, as well as other research procedures, are presented in greater detail in Appendix A.

Absorption of Ultraviolet Light (UV)

Nucleic acids strongly absorb ultraviolet light at wavelengths of 254 to 260 nm. This is the direct result of

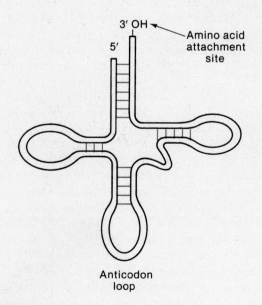

FIGURE 9.12
Two-dimensional cloverleaf model of transfer RNA (tRNA). Parallel regions of the chain are held together by hydrogen bonds between strings of three to seven pairs of complementary nitrogenous base pairs. The amino acid attachment site and the anticodon loop are identified.

the interaction between UV light at these wavelengths and the ring systems of the purines and pyrimidines. Thus, any molecule containing nitrogenous bases (i.e., nucleosides, nucleotides, and polynucleotides) can be analyzed using UV light. This technique is especially important in the localization, isolation, and characterization of nucleic acids.

UV analysis is used in conjunction with many standard procedures that separate molecules. For example, compounds containing a nitrogenous base can be separated by a technique called **paper chromatography.** As illustrated in Figure 9.13, a mixture of nucleotides or short polynucleotides can be separated on the basis of the solubility coefficient of each component in various solvents. The molecules are first allowed to migrate up the paper in a specific solvent, effecting the primary separation of the mixture. Then, the paper is dried, turned 90°, and placed in a second solvent, which further separates the molecules. The paper may then be exposed to UV light. The compounds containing a nitrogenous base will appear as dark spots against the fluorescing paper. Each spot may then be cut out, eluted from the paper, and investigated further.

Another way that molecules of nucleic acid mixtures can be separated is by subjecting them to one of several possible **gradient centrifugation** procedures, as shown in Figure 9.14. The mixture may be layered on top of a solution prepared so that a concentration gradient has been formed from top to bottom. Then the entire mixture is centrifuged at high speeds in an ultracentrifuge. The mixture of molecules will migrate downward, with each component moving at a different rate. Centrifugation is stopped, and the gradient is then eluted from the tube. Each fraction can then be measured spectrophotometrically for absorption at 260 nm. In this way, the position of a nucleic acid fraction can be located along the gradient and the fraction isolated.

Ultraviolet spectrometry is also used to quantitatively measure nucleic acids, since absorption is dependent on the concentration of the solute being studied. This principle also applies to other molecules such as proteins, which absorb UV light more intensely at 280 nm than at 260 nm. Other applications of the general use of UV absorption are discussed throughout the text.

Sedimentation Behavior

The gradient centrifugations just described rely on the sedimentation behavior of molecules in solution. There are two major types of sedimentation techniques employed in the analysis of nucleic acids: **sedi-**

FIGURE 9.13
Separation of a mixture of nucleotides using two-dimensional chromatography and analysis under ultraviolet light.

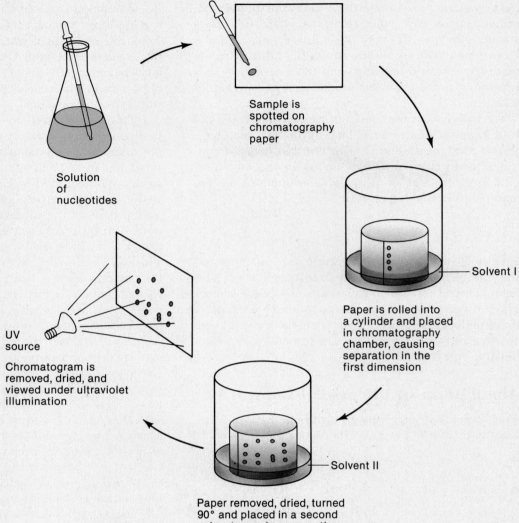

Solution of nucleotides

Sample is spotted on chromatography paper

Paper is rolled into a cylinder and placed in chromatography chamber, causing separation in the first dimension

Solvent I

Paper removed, dried, turned 90° and placed in a second solvent, causing separation in the second direction

Solvent II

UV source

Chromatogram is removed, dried, and viewed under ultraviolet illumination

mentation velocity and **sedimentation equilibrium.** Both require the use of high-speed centrifugation to create large centrifugal forces upon molecules in a gradient solution.

Sedimentation velocity centrifugation is usually performed in a gradient of sucrose and may employ an analytical centrifuge, which enables the migration of the molecules during centrifugation to be monitored with ultraviolet absorption optics. Thus, the "velocity of sedimentation" may be determined. This velocity has been standardized in units called Svedberg coefficients (*S*), as mentioned earlier.

In this technique, the molecules are layered on top of the gradient, and the gravitational forces created by

centrifugation drive them toward the bottom of the tube. Two forces work against this downward movement: (1) the viscosity of the solution creates a frictional resistance, and (2) part of the force of diffusion is directed upward. Under these conditions, the key variables are the mass and shape of the molecules being examined. In general, the greater the mass, the greater is the sedimentation velocity. However, the molecule's shape affects the frictional resistance. Therefore, two molecules of equal mass but different shape will sediment differently.

One use of this technique is the determination of **molecular weight (MW).** If certain physical chemical properties of a molecule under study are also known,

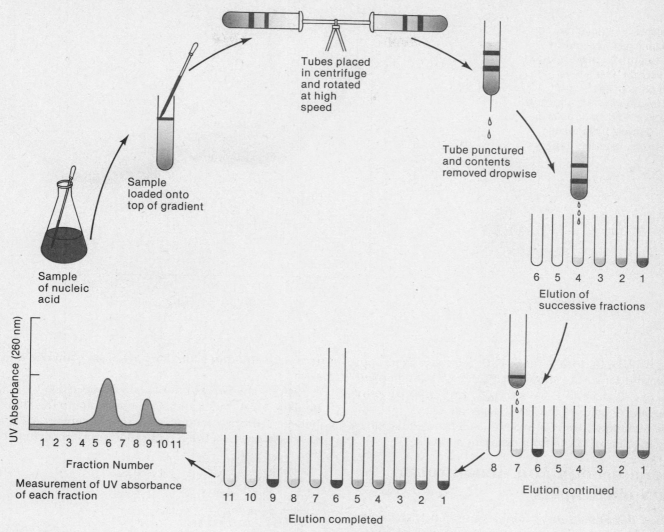

Sample loaded onto top of gradient

Tubes placed in centrifuge and rotated at high speed

Tube punctured and contents removed dropwise

Sample of nucleic acid

Elution of successive fractions

6 5 4 3 2 1

Elution continued

8 7 6 5 4 3 2 1

Measurement of UV absorbance of each fraction

UV Absorbance (260 nm)

1 2 3 4 5 6 7 8 9 10 11
Fraction Number

Elution completed

11 10 9 8 7 6 5 4 3 2 1

FIGURE 9.14

Separation of a mixture of two types of nucleic acid using gradient centrifugation. Following separation, successive samples are eluted from the bottom of the tube. Each is measured for absorbance of ultraviolet light at 260 nm, producing a profile of the sample in graphic form.

the MW can be calculated based on the sedimentation velocity. *S* values increase with molecular weight, but are not directly proportional to it.

In the second technique, sedimentation equilibrium centrifugation (sometimes called **density gradient centrifugation**), a density gradient is created that overlaps the densities of the individual components of a mixture of molecules. Usually, the gradient is made of a heavy metal salt such as cesium chloride (CsCl). During centrifugation the molecules migrate until they reach a point of neutral buoyant density. At this point, the gravitational force on them is equal and opposite

to the upward diffusion force and no further migration occurs. If DNAs of different densities are used, they will separate as the molecules of each density reach equilibrium with the corresponding density of CsCl. The gradient may be eluted and the components isolated (Figure 9.14). When properly executed, this technique provides high resolution in separating mixtures of molecules varying only slightly in density.

Sedimentation equilibrium centrifugation studies may also be used to generate data on the base composition of double-stranded DNA. Because G-C base pairs contain three hydrogen bonds, compared with

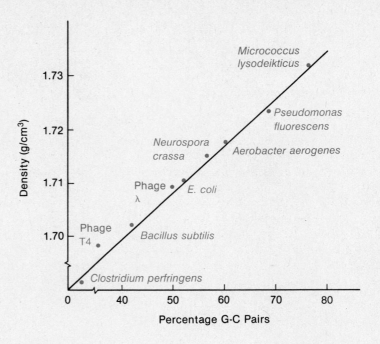

FIGURE 9.15
Percentage of guanine-cytosine (G-C) base pairs in DNA plotted against buoyant density for a variety of microorganisms. [With permission from Schildkraut, Marmur, and Doty, 1962, p. 430. Copyright: Academic Press Inc. (London) Ltd.]

the two bonds in A-T pairs, G-C pairs are more compact and dense. As shown in Figure 9.15, the percentage of G-C pairs in DNA is directly proportional to the buoyant density of the DNA molecule. Thus, by using this technique, we can make a useful molecular characterization of DNAs from different sources.

Denaturation and Renaturation of Nucleic Acids

When **denaturation** of double-stranded DNA occurs, the hydrogen bonds of the duplex structure break, the duplex unwinds, and the strands separate. However, no covalent bonds break. During strand separation, which can be induced by heat or chemical treatment, the viscosity of DNA decreases, and both the UV absorption and the buoyant density increase. The increase in UV absorption of heated DNA in solution, called the **hyperchromic effect,** is easiest to measure. Denaturation as a result of heating is sometimes referred to as **melting.**

Since G-C base pairs have one more hydrogen bond than A-T pairs, they are more stable to heat treatment. Thus, DNA with a greater proportion of G-C pairs than A-T pairs requires more heat to denature completely. When absorption at 260 nm is monitored and plotted against temperature during heating, a **melting profile** of DNA is obtained. The midpoint of this profile, or curve, is called the **melting point (T_m)** and represents the point at which 50 percent of the strands are unwound or denatured (Figure 9.16). Analysis of melting profiles provides a characterization of DNA and an al-

ternate method of estimating the base composition of DNA.

Thermal denaturation of DNA is reversible by slow cooling. This renaturation process is due to the affinity between the long sequence of complementary A-T and G-C base pairs. As the renaturation process occurs, the

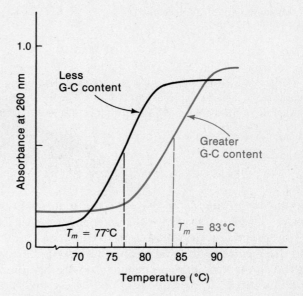

FIGURE 9.16
Comparison of the increase in UV absorbance with an increase in temperature for two DNA molecules with differing G-C contents. The molecule with a melting point (T_m) of 83°C has a greater G-C content than the molecule with a T_m of 77°C.

hyperchromic effect is reversed and can be monitored as the UV absorption decreases.

Molecular Hybridization Techniques

The property of denaturation/renaturation of nucleic acids is the basis for one of the most powerful and useful techniques in molecular genetics—**molecular hybridization.**

Denaturation occurs when double-stranded nucleic acid molecules are separated into single strands. Renaturation occurs when they return to their original duplex state. Provided that a reasonable degree of base complementarity exists, two nucleic acid strands from different sources will undergo molecular hybridization under the proper conditions. Thus, hybridizaton is possible not only between DNA strands from different species, but between DNA and RNA strands. For example, an RNA molecule will hybridize with the segment of DNA from which it was transcribed.

Figure 9.17 illustrates how the process of DNA-RNA hybridization occurs. In this example, the DNA strands are heated, causing strand separation, and then slowly cooled in the presence of single-stranded RNA. If the RNA has been transcribed on the DNA used in the experiment, and is therefore complementary to it, hybridization will occur. Several methods are available for monitoring the amount of double-stranded molecules produced following strand separation.

FIGURE 9.17
Diagrammatic representation of the process of molecular hybridization between DNA fragments and RNA that has been transcribed on one of the fragments.

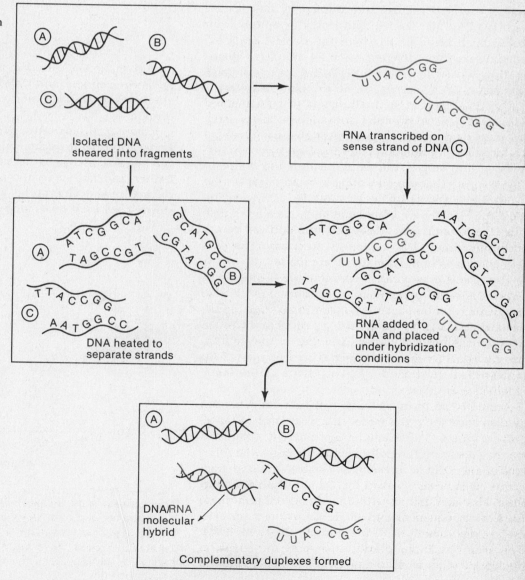

In the 1960s molecular hybridization techniques contributed to the increase in our understanding of transcriptional events occurring at the gene level. Refinements of this process have occurred continually and have been the forerunners of work in evolutionary homology and gene isolation studies. The technique can even be performed using cytological preparations, which is called ***in situ* molecular hybridization** (see Appendix A). We will refer to the process of molecular hybridization throughout the text.

Reassociation Kinetics and Repetitive DNA

One extension of molecular hybridization procedures is the technique that measures the **rate** of reassociation of complementary strands of DNA derived from a single source. This technique, called **reassociation kinetics,** was first refined and studied by Roy Britten and David Kohne. To distinguish this technique from molecular hybridization, where the nucleic acids examined are from different sources, the term **reannealing** is used to describe the hybridization of complementary DNA fragments from the same organism.

The DNA used in such studies is first fragmented into small pieces several hundred base pairs long. Fragments are produced as a result of shear forces introduced during isolation. After the DNA has been dissociated into single strands by heating, the temperature is lowered and reannealing is monitored. During reannealing, pieces of single-stranded DNA collide randomly. If they are complementary, a stable double strand is formed; if not, they separate and are free to encounter other DNA fragments randomly. The process continues until all matches are made.

The results of such an experiment are presented in Figure 9.18. The percentage of reassociation of DNA fragments is plotted against a logarithmic scale of normalized time, C_0t. In this term, C_0 is equal to the initial DNA concentration in moles per liter of nucleotides, and t is equal to time in seconds. The derivation of C_0t is described in Appendix A. The shape of the curve obtained is an ideal second-order reaction.

A great deal of information can be obtained from studies comparing the reassociation of DNA of different organisms. For example, for different organisms we may compare the point in the reaction when one-half of the DNA is present as double-stranded fragments. This point is called the $C_0t_{1/2}$, or **half reaction time.** Provided that all DNA fragments contain unique nucleotide sequences and all are about the same size, $C_0t_{1/2}$ varies directly with the **complexity** of the DNA. Designated as ***X,*** complexity represents the length in nucleotide pairs of all unique DNA fragments laid end

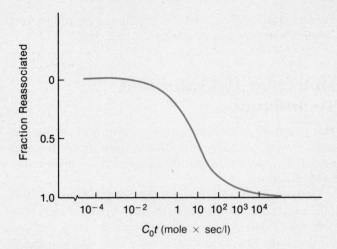

FIGURE 9.18
The time course of DNA reannealing when all fragments are unique. The curve is an ideal second-order reaction. Note that the abscissa is a logarithmic scale of C_0t.

to end. If the DNA used in an experiment represents the entire genome, and if all DNA sequences are different from one another, then X is equal to the size of the genome.

Figure 9.19 illustrates what is found when DNAs from various sources are compared. As can be seen, as genome size increases, the curves obtained are shifted farther and farther to the right, indicative of an increased reassociation time. Reassociation occurs more slowly because it takes longer for matches to be made

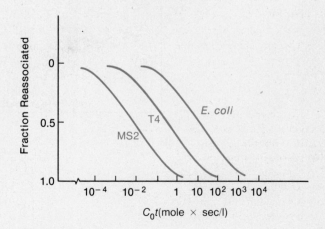

FIGURE 9.19
Comparison of the reannealing rate of DNA derived from phage MS-2, phage T4, and *E. coli*. The genome of T4 is larger than MS-2, and that of *E. coli* is larger than T4. (From Britten and Kohne, 1968, Fig. 2. Copyright © 1968 by the American Association for the Advancement of Science.)

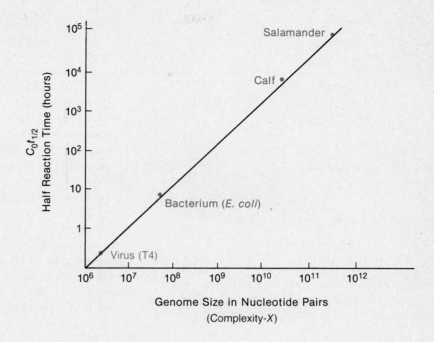

if there are initially greater numbers of unique DNA fragments. This is so because collisions are random; more sequences present will result in greater numbers of mismatches before the correct match is made.

As shown in Figure 9.20, $C_0t_{1/2}$ is indeed directly proportional to the size of the genome. If the entire genome consists of unique DNA sequences, reassociation experiments can be used to determine the genome size of organisms. This method has been particularly useful in studying viruses and bacteria.

However, when reassociation kinetics of DNA from eukaryotic organisms were first studied, a surprising observation was made. The data showed that some of the DNA segments reassociated even more rapidly than those derived from *E. coli*! The remainder, as ex-

pected because of its greater complexity, took longer to reassociate. For example, Britten and Kohne examined DNA derived from calf thymus tissue (Figure 9.21). Based on these observations, they hypothesized that the rapidly reassociating fraction must represent sequences present many times in the calf genome. This interpretation would explain why these DNA segments reassociate so rapidly. On the other hand, they hypothesized that the remaining DNA segments consist of unique nucleotide sequences present only once in the genome; because there are more of these unique sequences, increasing the DNA complexity in calf thymus (compared with *E. coli*), reassociation takes longer. The *E. coli* curve has been added to Figure 9.21 for the sake of comparison.

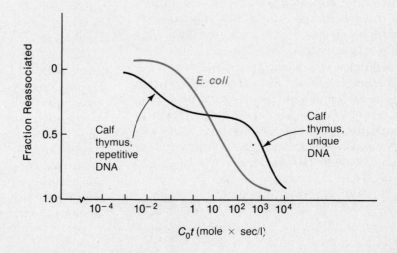

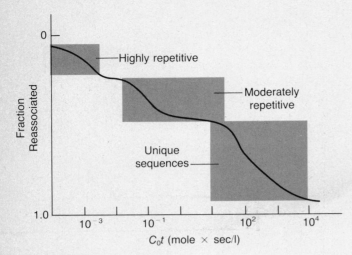

FIGURE 9.22
Reassociation analysis of eukaryotic DNA containing highly repetitive, moderately repetitive, and unique sequences. This pattern is similar to that found in mammals, including humans.

The copies present many times in the genome are collectively referred to as **repetitive DNA.** Repetitive DNA is characteristic of eukaryotic genomes and is important to our understanding of how genetic information is organized in chromosomes. Careful study has shown that, while there is a continuum of reassociation characteristics displayed by eukaryotic DNA, there are two major categories of repetitive DNA, as illustrated in Figure 9.22. The first reanneals very rapidly (much more rapidly than *E. coli* DNA) and is called **highly repetitive DNA.** The sequences present in this fraction occur between 100,000 and 1,000,000 times in the genome and are between 5 and 300 base pairs (bp) in length. Most frequently the sequences range from 5 to 15 bp long.

The second category is **moderately repetitive DNA.**[*] As the name implies, fewer copies are present per genome than is the case for highly repetitive sequences. Because of this, moderately repetitive DNA reanneals more slowly than the highly repetitive fraction. Each sequence is present 10 to 3000 times per genome, but most frequently, from 10 to 100 times. On the average, these sequences consist of about 300 bp. The remainder of DNA is nonrepetitive and said to consist of **unique sequence DNA,** which is usually present only once in the genome. These sequences are most frequently 1000 to 1500 bp long.

Although we will explore the significance of repetitive DNA sequences in Chapter 11, a short discussion is appropriate here. Highly repetitive sequences are usually much too short to serve as genes. Found clustered in several regions of the chromosomes, including the centromere, the role they play is unknown. Much of moderately repetitive DNA is also nongenic and is interspersed throughout the genome. However, some of this fraction consists of sequences that are transcribed and thus constitute repeat or duplicate copies of various genes. Included in this fraction are genes coding for ribosomal RNA, ribosomal proteins, and histones (positively charged proteins bound to DNA in eukaryotes). On the other hand, most genes are present as only a single copy within the genome and are part of the nonrepetitive or unique sequence fraction. However, it is important to point out that the majority of this fraction is not represented by structural genes encoding polypeptide chains.

Using a variety of modern techniques, it has been established that a pattern of interspersion of moderately repetitive and unique sequences often exists. Most commonly, the moderately repetitive sequences are about 300 bp and they alternate with unique sequences of 800 to 1500 bp. This pattern is called **short-period interspersion,** and it predominates in the DNA of such organisms as mammals, sea urchins, and toads. In humans, this type of interspersion pattern is primarily due to the presence of large numbers of copies of a short transposable DNA element called **alu** (see Chapter 11). **Long-period interspersion** also exists, where both the moderately repetitive sequences and unique sequences are much longer, e.g., 14,000 bp and 5600 bp, respectively. This pattern predominates in *Drosophila* and birds, where transposable elements longer than alu are part of the moderately repetitive fraction.

The proportion of the three fractions present in the genome varies in different organisms. In lower eukaryotes almost all DNA consists of nonrepetitive sequences with only 10 to 20 percent existing as highly or moderately repetitive. In the cells of most animals about half of the DNA is nonrepetitive and the rest moderately and highly repetitive. In plants and certain amphibians, up to 80 percent falls into the repetitive categories.

The discovery of repetitive sequences has extended our knowledge of and generated increased interest in the organization of DNA in eukaryotes. While the genetic role these repeated sequences play is still the subject of speculation, future research will undoubtedly clarify this issue. We will discuss repetitive DNA again in Chapters 11 and 19.

---

[*]Moderately repetitive DNA is also referred to as middle-repetitive DNA.

SUMMARY AND CONCLUSIONS

This chapter recounts some of the important findings made during one of the most significant periods in the history of biology. As information became available on the chemical structure of nucleotides, and evidence was accumulating that DNA is the genetic material, research focused on the structure of complex nucleic acids. With the knowledge of DNA base composition supplied by Chargaff and the X-ray crystallographic data of Wilkins and Franklin, Watson and Crick constructed a detailed model of DNA structure. Their model proposed the double-helical structure of two long, antiparallel polynucleotide chains. The most critical aspect of this model was the complementarity between the purine and pyrimidine nitrogenous bases, adenine to thymine and guanine to cytosine. As pointed out by Watson and Crick in their 1953 paper, not only is the complementarity integral to the double-helical structure, but base pairing provides the basis for the mechanism of DNA replication. Transcription and translation also depend on base complementarity.

Several forms of DNA exist and have been studied. DNA studied by Wilkins and Franklin assumed the B configuration, which serves as the basis for the Watson-Crick model. The A, C, D, and E forms, like the B form, are right-handed helices but exhibit slightly altered base-stacking properties. One of the most exciting recent discoveries in genetics is that of Z-DNA by Wang and Rich. This molecule assumes a left-handed helix and an overall conformation substantially different from either the A or B forms. The physiological and genetic significance of Z-DNA is currently being investigated.

As background for ensuing chapters, we have briefly discussed the structure of RNA molecules. Differences in the sugar molecule, one nitrogenous base, and the single-stranded conformation of RNA are in contrast to DNA. Three types of RNA are transcribed as molecules complementary to the nucleotide sequence in DNA: ribosomal, messenger, and transfer RNAs. These molecules all participate in the translation process, serving as the basis for information flow within the cell.

The structure of DNA lends itself to various forms of analysis, which have in turn led to studies of the functional aspects of the genetic machinery. Absorption of ultraviolet light, sedimentation properties, and denaturation/reassociation procedures are among the most important analytical tools for the study of nucleic acids. Ultraviolet absorption has been instrumental in identifying the presence of nucleic acids in mixtures and gradients. Through the techniques of sedimentation velocity and equilibrium centrifugation, nucleic acids have been separated and characterized. The property of heat denaturation/reassociation is the basis of the molecular hybridization technique, which is the one most often used to analyze nucleic acid structure and function. When reassociation kinetics were first studied, the class of repetitive DNA was discovered in the genome of higher organisms. Careful analysis has revealed two categories of sequences: highly repetitive and moderately repetitive. These are in addition to unique sequences, which include all single copy genes. Almost all repetitive DNAs, as well as the majority of the unique sequences, are nongenic. The role of repetitive DNA is not yet fully understood, but as we will see in Chapter 11, this discovery provided important insights into our understanding of chromosome structure.

PROBLEMS AND DISCUSSION QUESTIONS

1 Draw the chemical structure of the three components of a nucleotide and then link the three together. What atoms are removed from the structures when the linkages are formed?

2 How are the carbon and nitrogen atoms of the sugars, purines, and pyrimidines numbered?

3 Adenine may also be named 6-amino purine. How would you name the other four nitrogenous bases using this alternate system? ($=O$ is oxy, and $—CH_3$ is methyl.)

4 Draw the chemical structure of a dinucleotide composed of A and G. Opposite this structure, draw the dinucleotide TC in an antiparallel (or upside down) fashion. Form the possible hydrogen bonds.

 5 Describe the various characteristics of the Watson-Crick double-helix model for DNA.

 6 What evidence did Watson and Crick have at their disposal in 1953? What was their approach in arriving at the structure of DNA?

 7 Had Chargaff's data from a single source indicated the following, what might Watson and Crick have concluded?

A	T	C	G
%	29	19	21	31

   Why would this conclusion be contradictory to Wilkins and Franklin's data?

 8 How do covalent bonds differ from hydrogen bonds? Define base complementarity.

 9 List three main differences between DNA and RNA.

10 What are the three types of RNA molecules? How is each related to the concept of information flow?

11 What component of the nucleotide is responsible for the absorption of ultraviolet light? How is this technique important in the analysis of nucleic acids?

12 Distinguish between sedimentation velocity and sedimentation equilibrium centrifugation.

13 What chemical characteristics determine a molecule's Svedberg coefficient?

14 What is the basis for determining base composition using density gradient centrifugation?

15 What is the state of DNA following denaturation?

16 What is the hyperchromic effect? How is it measured? What does T_m imply?

17 Why is T_m related to base composition?

18 Which of the following terms mean approximately the same thing: denaturation, renaturation, annealing, melting, hybridization?

19 What is the chemical basis of molecular hybridization?

20 What did the Watson-Crick model suggest about the replication of DNA?

21 Compare the following curves representing reassociation kinetics. What can be said about the DNAs represented by each set of data compared with *E. coli?*

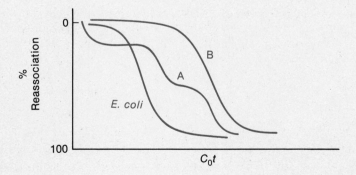

22 Draw the predicted C_0t curve if the following mixture of nucleic acid molecules were analyzed in a single experiment: 10^6 copies of a short, identical sequence; 10^3 copies of a much longer DNA sequence; and 10^2 unique sequences of DNA.

23 A genetics student was asked to draw the chemical structure of an adenine- and thymine-containing dinucleotide derived from DNA. The student made six major errors. His answer is shown below. One of them is circled, numbered ①, and explained. Find

the other five. Circle them, number them ②—⑥, and briefly explain each, following the example below.

EXPLANATIONS

Extra phosphate should not be present.

24 *Newsdate: March 1, 1995.* A unique creature has been discovered during exploration of outer space. Recently, its genetic material has been isolated and analyzed. This material is similar in some ways to DNA in its chemical makeup. It contains in abundance the 4-carbon sugar erythrose and a molar equivalent of phosphate groups. Additionally, it contains six nitrogenous bases: adenine (A), guanine (G), thymine (T), cytosine (C), hypoxanthine (H), and xanthine (X). These bases exist in the following relative proportion or ratio:

$$A/T/H = 1; C/G/X = 1 (A = T = H \text{ and } C = G = X)$$

X-ray diffraction has established a regularity to the molecule and a constant diameter of about 30 Å.

Together, these data have suggested a model for the structure of this molecule.
(a) Propose a general model of this molecule. Describe it briefly.
(b) What base-pairing properties must exist for H and for X in the model?
(c) Given the constant diameter of 30 Å, do you think that H and X are *either* (1) both purines or both pyrimidines, *or* (2) one is a purine and one is a pyrimidine?

25 The DNA of the bacterial virus T4 produces a $C_0t_{1/2}$ of about 0.5 and contains 10^5 nucleotide pairs in its genome. How many nucleotide pairs are present in the genome of the virus MS-2 and the bacterium *E. coli*, whose respective DNAs produce $C_0t_{1/2}$ values of 0.001 and 10.0?

SELECTED READINGS

AZORIN, F., and RICH, A. 1985. Isolation of Z-DNA binding proteins from SV40 minichromosomes: Evidence for binding to the viral control region. *Cell* 41: 365–74.

BRITTEN, R. J., and KOHNE, D. E. 1968. Repeated sequences in DNA. *Science* 161: 529–40.

——. 1970. Repeated segments of DNA. *Scient. Amer.* (April) 222: 24–31.

CANTOR, C. R. 1981. DNA choreography. *Cell* 25: 293–95.

CHARGAFF, E. 1950. Chemical specificity of nucleic acids and mechanism for their enzymatic degradation. *Experientia* 6: 201–9.

CRICK, F. H. C., WANG, J. C., and BAUER, W. R. 1979. Is DNA really a double helix? *J. Mol. Biol.* 129: 449–61.

DAVIDSON, J. N. 1976. *The biochemistry of the nucleic acids.* 8th ed. New York: Academic Press.

DICKERSON, R. E., et al. 1982. The anatomy of A-, B-, and Z-DNA. *Science* 216: 475–85.

DICKERSON, R. E. 1983. The DNA helix and how it is read. *Scient. Amer.* (June) 249: 94–111.

FRANKLIN, R. E., and GOSLING, R. G. 1953. Molecular configuration in sodium thymonucleate. *Nature* 171: 740–41.

GEIS, I. 1983. Visualizing the anatomy of A, B and Z-DNAs. *J. Biomol. Struc. Dynam.* 1: 581–91.

HALL, B. D., and SPIEGLEMAN, S. 1961. Sequence complementarity of T2-DNA and T2-specific RNA. *Proc. Natl. Acad. Sci.* 47: 137–46.

LEHNINGER, A. L. 1973. *Short course in biochemistry.* New York: Worth.

LENG, M. 1985. Left-handed Z-DNA. *Biochim. Biophys. Acta* 825:339–44.

OLBY, R. 1974. *The path to the double helix.* Seattle: University of Washington Press.

PARKER, G. E., and MERTENS, T. R. 1973. *Life's basis: Biomolecules.* New York: Wiley.

PAULING, L., and COREY, R. B. 1953. A proposed structure for the nucleic acids. *Proc. Natl. Acad. Sci.* 39: 84–97.

PECK, L. J., NORDHEIM, A., RICH, A., and WANG, J. C. 1982. Flipping of cloned d(pCpG)$_n$·d(pCpG)$_n$ DNA sequences from right-handed to left-handed helical structure by salt, Co(III) or negative supercoiling. *Proc. Natl. Acad. Sci.* 15: 4560–64.

RICH, A., NORDHEIM, A., and WANG, A. H.-J. 1984. The chemistry and biology of left-handed Z-DNA. *Ann. Rev. Biochem.* 53: 791–846.

SCHILDKRAUT, C. L., MARMUR, J., and DOTY, P. 1962. Determination of the base composition of deoxyribonucleic acid from its buoyant density in CsCl. *J. Mol. Biol.* 4: 430–43.

SCHMIDTKE, J., and EPPLEN, J. T. 1980. Sequence organization of animal nuclear DNA. *Hum. Genet.* 55: 1–18.

STENT, G. S., ed. 1981. *The double helix: Text, commentary, review, and original papers.* New York: W. W. Norton.

STRYER, L. 1981. *Biochemistry.* 2nd ed. San Francisco: W. H. Freeman.

WANG, H. J., et al. 1979. Molecular structure of a left-handed double helical DNA fragment at atomic resolution. *Nature* 282: 680–82.

WATSON, J. D. 1968. *The double helix.* New York: Atheneum.

_____. 1976. *Molecular biology of the gene.* 3rd ed. Menlo Park, Calif.: W. A. Benjamin.

WATSON, J. D., and CRICK, F. C. 1953a. Molecular structure of nucleic acids. A structure for deoxyribose nucleic acids. *Nature* 171: 737–38.

_____. 1953b. Genetic implications of the structure of deoxyribose nucleic acid. *Nature* 171: 964.

WHITE, E. H. 1971. *Chemical background for the biological sciences.* 2nd ed. Englewood Cliffs, N. J.: Prentice-Hall.

WILKINS, M. H. F., STOKES, A. R., and WILSON, H. R. 1953. Molecular structure of desoxypentose nucleic acids. *Nature* 171: 738–40.

WOLF, G. 1964. *Isotopes in biology.* New York: Academic Press.

VAN NORMAN, R. W. 1971. *Experimental biology.* 2nd ed. Englewood Cliffs, N. J.: Prentice-Hall.

ZIMMERMAN, B. 1982. The three-dimensional structure of DNA. *Ann. Rev. Biochem.* 51: 395–428.

10

Replication and Synthesis of DNA

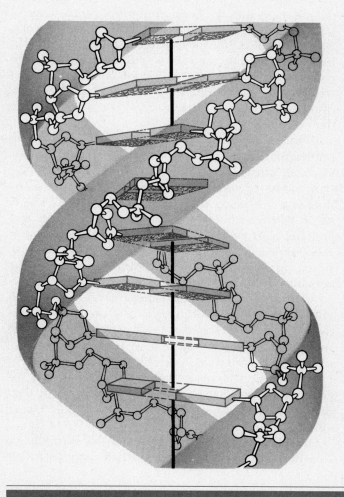

Following Watson and Crick's proposal for the structure of DNA, scientists focused their attention on how this molecule is replicated or duplicated. This process is an essential function of the genetic material and must be executed precisely if genetic continuity between cells is to be maintained following cell division. As discussed in Watson and Crick's early papers in 1953, the model of the double helix gave them the initial insight into how replication could occur. This mode, called **semiconservative replication,** has since received strong experimental support from studies of viruses, prokaryotes, and eukaryotes.

Once the general mode of replication was made clear, research was intensified to determine the precise details of DNA synthesis. What has since been discovered is that numerous enzymes and other proteins are needed to copy a DNA helix. Because of the complexity of the chemical events during synthesis, this subject remains an extremely active area of research.

In this chapter, we will discuss the mode of replication as well as the chemical synthesis of DNA. The research leading to this knowledge is still another link in our understanding of life processes at the molecular level.

THE MODE OF DNA REPLICATION

It was apparent to Watson and Crick that because of the arrangement and nature of the nitrogenous bases, each strand of a DNA double helix could serve as a template for the synthesis of its complement. They proposed that if the helix were unwound, each nucleotide along the two parent strands would have an affinity for its complementary nucleotide. As we learned in Chapter 9, the complementarity is due to the potential hydrogen bonds that can be formed. If thymidylic acid were present, it would "attract" adenylic acid; if guanidylic acid were present, it would "attract" cytidylic acid; and so on. If these nucleotides were then covalently linked into polynucleotide chains along both templates, the result would be the production of two new but identical double strands of DNA. This concept is illustrated in Figure 10.1. Each replicated DNA molecule would consist of an "old" and a "new" strand. Therefore, this mechanism is called **semiconservative replication.**

There are two other possible modes of replication that also rely on the parental strands as a template. In **conservative replication** [Figure 10.2(a)], synthesis of complementary polynucleotide chains occurs as described above; following synthesis, however, the two newly created strands are brought back together, and the parental strands reassociate. The original helix is thus "conserved."

In the second alternative mode, called **dispersive replication,** the parental strands are seen to be dispersed into two new double helices following replication [Figure 10.2(c)]. This mode would involve cleavage of the parental strands during replication. Therefore, it is the most complex of the three possibilities and, as such, is least likely. It could not, however, be ruled out as an experimental model. The theoretical results of two generations of replication by the conservative and dispersive modes are compared with those of the semiconservative mode in Figure 10.2.

The Meselson-Stahl Experiment

In 1958, Matthew Meselson and Franklin Stahl published the results of an experiment providing strong evidence that semiconservative replication is the mode used by cells to produce new DNA molecules. *E. coli* cells were grown for many generations in a medium where $^{15}NH_4Cl$ (ammonium chloride) was the only nitrogen source. A "heavy" isotope of nitrogen, ^{15}N, contains one more neutron than the naturally occurring ^{14}N isotope. After many generations, all nitrogen-containing molecules, including the nitrogenous bases of DNA, contained the heavy isotope in the *E. coli* cells. DNA containing ^{15}N may be distinguished from ^{14}N-containing DNA by the use of **sedimentation equilibrium centrifugation** in a cesium chloride gradient (see Chapter 9). The dense ^{15}N-DNA will reach equilibrium in the gradient at a point closer to the bottom than does ^{14}N-DNA.

In this experiment, uniformly labeled ^{15}N cells were transferred to a medium containing only $^{14}NH_4Cl$. Thus, all subsequent synthesis of DNA during replication contained the "lighter" isotope of nitrogen. The time of transfer was taken as time zero ($t = 0$). The *E. coli* cells were allowed to replicate during several generations with cell samples removed at various intervals. From each sample, DNA was isolated and subjected to sedimentation equilibrium analysis. The actual results are depicted in Figure 10.3.

After one generation, the isolated DNA was all present in a single band of intermediate density—the expected result for semiconservative replication. Each replicated molecule would have been composed of one new ^{14}N-strand and one old ^{15}N-strand, as seen in Figure 10.4(a). This result was not consistent with the conservative replication mode, in which two distinct bands would have been predicted to occur [Figure 10.4(b)].

After two cell divisions, DNA samples showed two density bands: one was intermediate and the other

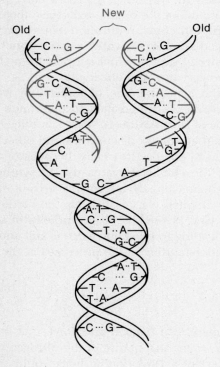

FIGURE 10.1
General model of semiconservative replication of DNA.

FIGURE 10.2
The results of two rounds of replication of DNA for each of the three possible modes of replication. All new synthesis is shown in color.

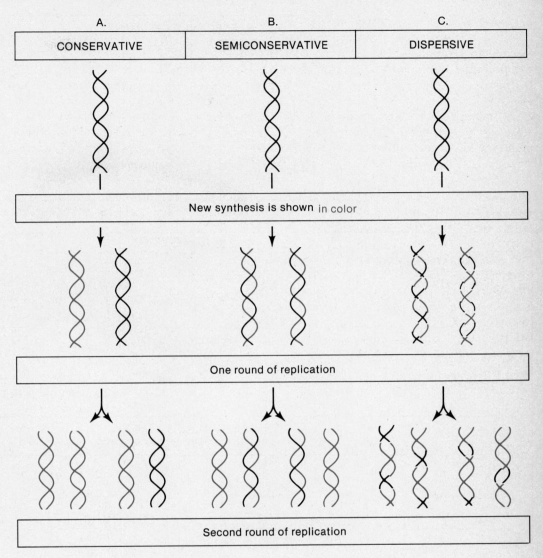

was lighter, corresponding to the ^{14}N position in the gradient. Similar results occurred after a third generation, except that the proportion of the ^{14}N-band increased. Figure 10.4(c) shows that if replication were dispersive, all subsequent generations after $t = 0$ would demonstrate DNA of an intermediate density. In each subsequent generation the ratio $^{14}N/^{15}N$ would increase, and the hybrid band would become lighter and lighter, eventually approaching the ^{14}N-band. The results of the Meselson-Stahl experiment thus provided strong support for the semiconservative mode of DNA replication, as postulated by Watson and Crick.

Semiconservative Replication in Eukaryotes

In 1957, the year before the work of Meselson and his colleagues was published, evidence was also presented that supported semiconservative replication in an eukaryotic organism. J. Herbert Taylor, Philip Woods, and Walter Hughes experimented with root tips of the broad bean *Vicia faba*. Root tips are an excellent source of dividing cells. These researchers examined the chromosomes of these cells following replication of DNA. They were able to monitor the process

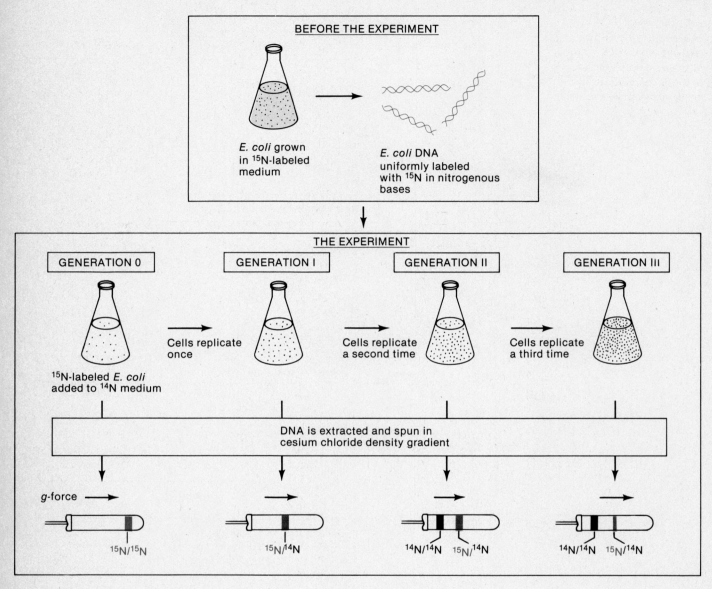

FIGURE 10.3
The Meselson-Stahl experiment. Note that in contrast to Figures 10.1 and 10.2, this figure shows new synthesis in black and the original DNA in color.

of replication by labeling DNA with ^{3}H-thymidine, a radioactive precursor of DNA, and performing autoradiography. In this experiment, labeled thymidine is found only in association with chromosomes that contain newly synthesized DNA.

The technique of autoradiography, discussed in detail in Appendix A, is a cytological procedure that allows the isotope to be localized within the cell. In this procedure, a photographic emulsion is placed over a section of cellular material (root tips in this experiment), and the preparation is stored in the dark. The slide is then developed, much as photographic film is

processed. Since the radioisotope emits energy, the emulsion turns black at the approximate point of emission following development. The end result is the presence of dark spots or grains on the surface of the section, localizing newly synthesized DNA within the cell.

Root tips were grown for approximately one generation in the presence of the radioisotope and then placed in unlabeled medium, where cell division continued. At the conclusion of each generation, cultures were arrested at metaphase by the addition of colchicine, and chromosomes were examined by autora-

FIGURE 10.4
The expected results of two
generations of replication in
the Meselson-Stahl
experiment for each of the
three possible modes of
replication.

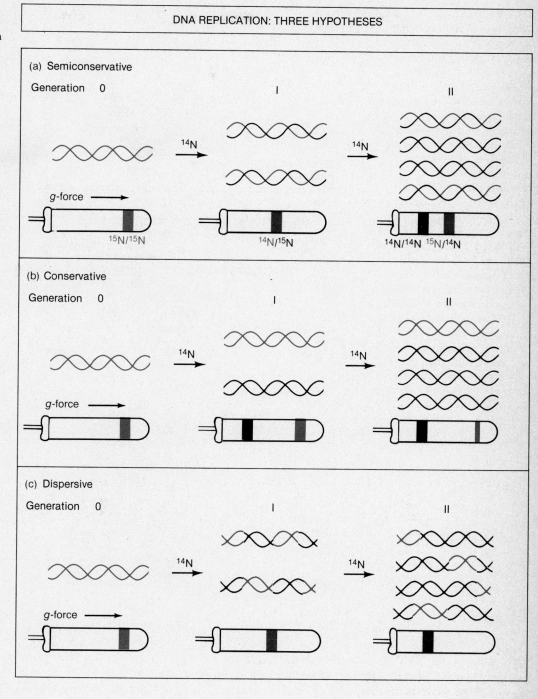

FIGURE 10.4
The expected results of two generations of replication in the Meselson-Stahl experiment for each of the three possible modes of replication.

diography. Figure 10.5 illustrates replication of a single chromosome over two division cycles as well as the distribution of grains. The results are compatible with the semiconservative mode of replication. After the first division, radioactivity is detected over both sister chromatids. This finding is expected because each chromatid will contain one radioactive DNA strand and an unlabeled strand. After the second division, which takes place in an unlabeled medium, only one

of the two new sister chromatids should be radioactive. With only the minor exception of the demonstration of **sister chromatid exchanges** (see Chapter 6), this result was observed.

Together, the Meselson-Stahl experiment and the experiment by Taylor, Woods, and Hughes soon led to the general acceptance of the semiconservative mode of replication. The same conclusion has been reached in studies with other organisms. Since this mode is

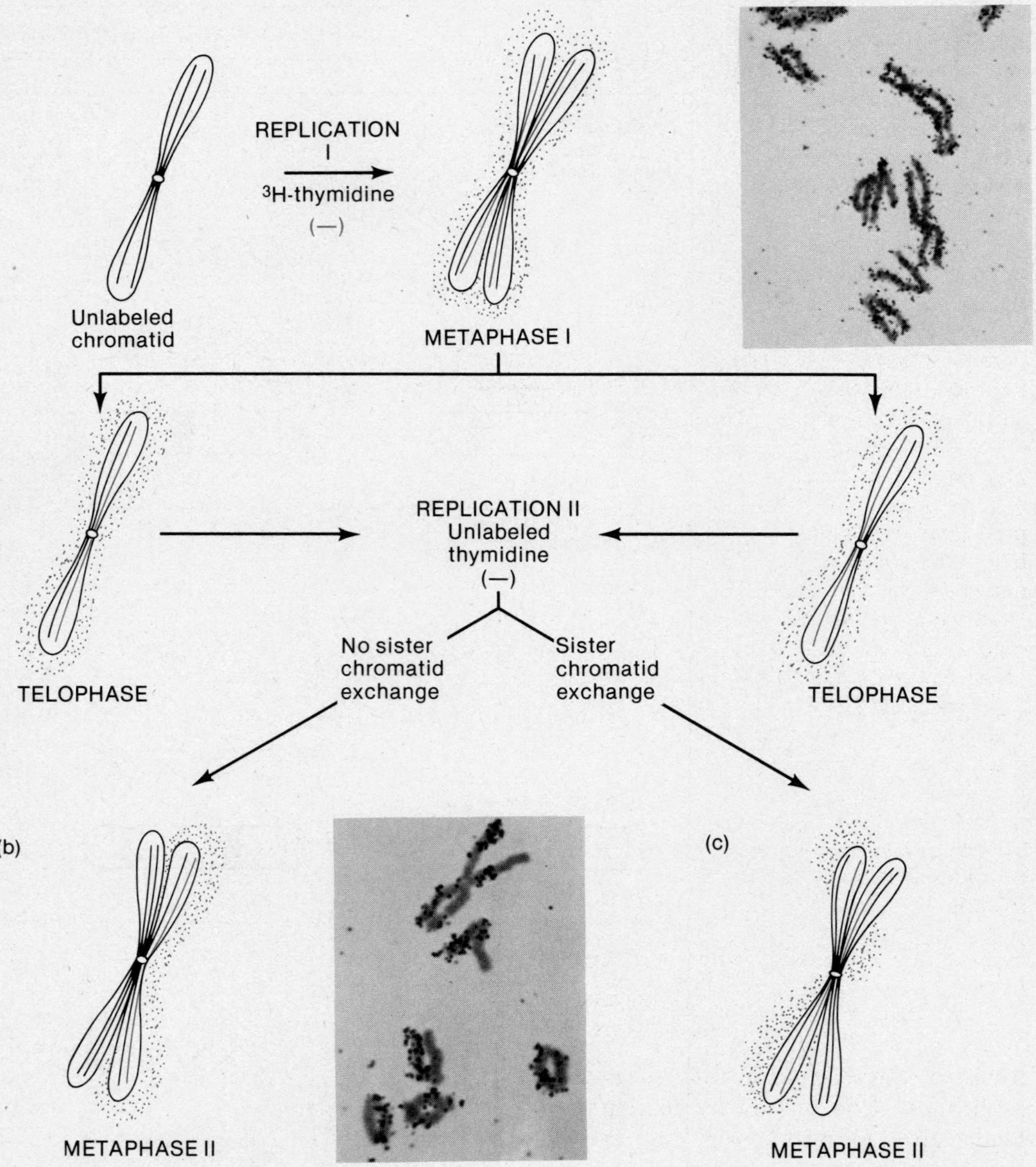

FIGURE 10.5
Depiction of the experiment by Taylor, Woods, and Hughes demonstrating the semiconservative mode of replication of DNA in root tips of *Vicia faba*. In (a) an unlabeled chromatid proceeds through the cell cycle in the presence of ^{3}H-thymidine. As it enters mitosis, both sister chromatids of each chromosome are labeled, as shown by autoradiography. After a second generation of replication in the absence of ^{3}H-thymidine, only one chromatid of each chromosome is expected to be surrounded by grains (b). In all cases, except where a reciprocal exchange has occurred between sister chromatids (c), the expectation was upheld. (Photos courtesy of J. Herbert Taylor.)

suggested by the double-helix model of DNA, these experiments also strongly supported Watson and Crick's proposal for DNA structure.

REPLICATION AND UNWINDING

In order for semiconservative replication to occur, the double helix, with thousands and thousands of turns, must be unwound to make the template surfaces available. Given the speed of DNA replication, scientists wondered how unwinding could occur. Is the entire helix first unwound, or do the processes of unwinding and replication occur simultaneously?

Watson and Crick suggested that the processes occurred together.

By 1963, evidence proved that they were correct. Using autoradiography, John Cairns examined DNA replication in *E. coli*. The cells were grown in a medium containing radioactive thymidine. At predetermined time intervals, a sample of cells was collected and the DNA extracted with great care. The DNA molecules were spread on grids and covered with a photographic emulsion. After being stored in the dark for more than a month, the emulsion was developed and the grids examined under the electron microscope. The electron micrographs showed the radioactive

FIGURE 10.6
Autoradiographic demonstration of DNA replication of the circular chromosome of *E. coli*, as viewed under the electron microscope. Shown diagrammatically in (a) is a parent molecule, partially radioactive from prior growth in ^{3}H-thymidine. It is allowed to begin replication in the continued presence of the same radioisotope. In (b), an interpretive drawing of the autoradiograph is presented. Replication from a single origin (o) has proceeded a little over one-half of the length of the molecule. New synthesis of DNA occurs at the advancing replication fork (rf). In the autoradiograph, the density of grains coincides with whether one or two strands are radioactive. Note that this model assumes unidirectional replication starting at a fixed point and proceeding along both strands.

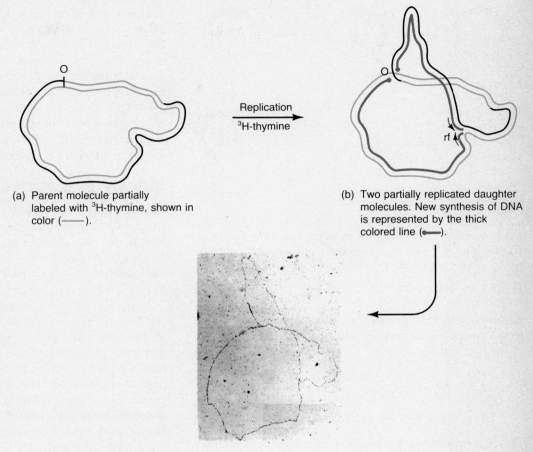

(a) Parent molecule partially labeled with ^{3}H-thymine, shown in color (———).

Replication
^{3}H-thymine

(b) Two partially replicated daughter molecules. New synthesis of DNA is represented by the thick colored line (●——).

tracks of the circular chromosome in the process of replication. One of Cairns' micrographs is shown in Figure 10.6 along with two interpretive diagrams. The first [Figure 10.6(a)] reveals the extent of radioactive labeling of the DNA strands of the parent molecule. The second [Figure 10.6(b)] suggests how replication of more than half of the parent molecule has proceeded in the continued presence of ^{3}H-thymidine. This interpretation can be compared to the autoradiograph. All of the new DNA synthesis results in a greater density of grains, thereby revealing the pattern of replication. Cairns interpreted his results to indicate that replication begins at a single position along the circular chromosome and proceeds in a **unidirectional** mode (counterclockwise in Figure 10.6) along both strands. In this model, at any given time, there is a single point of growth along both strands, creating what is called the **replication fork** [see rf in Figure 10.6(b)]. This fork progresses along the chromosome as unwinding and replication proceed.

While Cairns' experiments were initially interpreted to demonstrate that replication proceeds in only one direction, more recent investigations have revealed that replication is in fact **bidirectional** in almost all organisms studied. The consequences of bidirectional synthesis on a circular chromosome such as that in *E. coli* are illustrated in Figure 10.7. Later in this chapter, we will see exactly how unwinding and synthesis of DNA occur at each of the two replication forks. Since the two parent strands run antiparallel, an enzyme that synthesizes DNA in either a 3′ to 5′ or a 5′ to 3′ direction cannot immediately use both strands at a replication fork as unwinding occurs.

Each point of origin and the area of DNA that is subsequently replicated from it is called a **replicon.** In bacteria, which contain a single circular chromosome, this entire unit represents a replicon, since there is only one point of origin of replication. In eukaryotic organisms, where there is much more DNA than in bacteria, there are many replicons on each

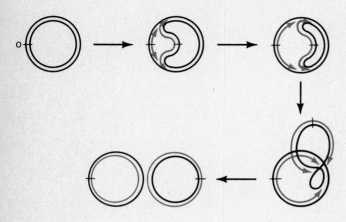

FIGURE 10.7
Interpretation of how the DNA of a circular chromosome such as that of *E. coli* replicates bidirectionally from a fixed point, o. New synthesis on each strand is shown in color.

chromosome. Having many points of origin and localized unwinding allows an entire chromosome to be replicated rapidly.

SYNTHESIS OF DNA IN MICROORGANISMS

The determination that replication is semiconservative indicates only how the finished strands associate with one another once synthesis is completed. A much more complex issue is how the **synthesis** of long complementary polynucleotide chains occurs on a DNA template. As in most studies of molecular biology, this question was first approached by using microorganisms. Research began about the same time as the Meselson-Stahl work, and even today this topic is an active area of investigation. What is most apparent in this research is the tremendous chemical complexity of the biological synthesis of DNA.

DNA Polymerase I

Studies of the enzymology of DNA replication were first reported by Arthur Kornberg and colleagues in 1957. They isolated an enzyme from *E. coli* that was able to direct DNA synthesis in a cell-free *(in vitro)* system. The enzyme is now called **DNA polymerase I,** since it was the first of several to be isolated. Kornberg determined the following requirements for *in vitro* DNA synthesis under the direction of the enzyme:

1 All four deoxyribonucleoside triphosphates (dATP, dCTP, dGTP, dTTP = dNTP*)

2 Mg^{++} ions

3 Template DNA

If any one of the four deoxyribonucleoside triphosphates was omitted from the reaction, no synthesis occurred. If derivatives of these precursor molecules other than the nucleoside triphosphate were used (nucleotides or nucleoside diphosphates), synthesis did not occur. If no template DNA was added, synthesis of DNA occurred, but only after prolonged reaction times. Thus, Kornberg's enzyme appeared capable of performing exactly the type of DNA synthesis required for semiconservative replication.

Fidelity of Synthesis

Having shown how DNA was synthesized, Kornberg sought to demonstrate the accuracy, or fidelity, with which the enzyme had replicated the DNA template. Since the nucleotide sequences of the template and the product could not be determined in 1957, he had to rely initially on several indirect methods.

One of Kornberg's approaches was to compare the nitrogenous base compositions of the template added to the reaction mixture with those of the recovered product. Table 10.1 shows Kornberg's base composition analysis of three different DNA templates. These may be compared with the product synthesized in each case. Within experimental error, the base composition of each product agreed with the template DNAs used. These data suggested that the templates were replicated faithfully.

Kornberg also used the **nearest-neighbor frequency test** to test the fidelity of copying. With this technique, he determined the frequency with which any two bases occur adjacent to each other along the polynucleotide chain. As illustrated in Figure 10.8, this

TABLE 10.1
Base composition of DNA template and product in Kornberg's early work.

Organism	Template/Product	%A	%T	%G	%C
T2	Template	32.7	33.0	16.8	17.5
	Product	33.2	32.1	17.2	17.5
E. coli	Template	25.0	24.3	24.5	26.2
	Product	26.1	25.1	24.3	24.5
Calf	Template	28.9	26.7	22.8	21.6
	Product	28.7	27.7	21.8	21.8

SOURCE: From Kornberg, 1960. Copyright 1960 by the American Association for the Advancement of Science.

*dNTP designates the deoxyribose forms of the four nucleoside triphosphates; in a similar way, dNMP refers to the monophosphate forms.

FIGURE 10.8
The theory of nearest-neighbor analysis. Initially, one of the four nucleotides (C) contains ^{32}P in the 5′-phosphate group. Following synthesis of a polynucleotide chain and enzymatic treatment with phosphodiesterase, the radioactive phosphate group is transferred to the nearest neighbors (G and T). Subsequent analysis reveals the percentage of time each of the four nucleotides (A, T, C, and G) have become radioactive.

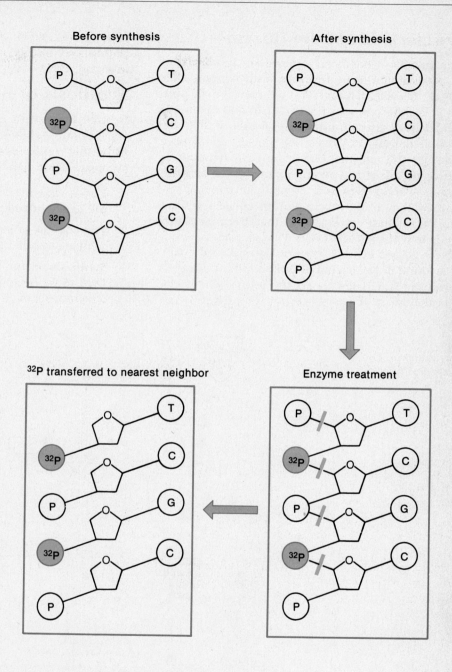

test relies on the enzyme **spleen phosphodiesterase,** which cleaves the polynucleotide chain differently from the way in which the chain was assembled. During synthesis of DNA, 5′-nucleotides are inserted; that is, each nucleotide is added with the phosphate on the C-5′ of deoxyribose. The enzyme cleaves between the phosphate and the C-5′ atom, thereby producing 3′-nucleotides. If the phosphates on only one of the four nucleotides (cytidylic acid, for example) are radioactive during DNA synthesis, then after enzymatic cleavage a radioactive phosphate will be attached to the base that is the "nearest neighbor" of all cytidylic acid nucleotides. Following four separate experiments, in which only one of the four nucleotide types is made

radioactive, the frequency of all sixteen possible nearest neighbors can be calculated. An example of such data is presented in Problem 10.24 at the end of this chapter.

When this technique was applied to the DNA template and resultant product from a variety of experiments, Kornberg found general agreement between the nearest-neighbor frequencies of the two. This type of analysis is a more stringent measure of the fidelity of copying than the base composition analysis. Thus, DNA polymerase I seemed the likely candidate for the synthesis of DNA during replication within the cell, even though Kornberg's experiments were, by necessity, performed *in vitro.*

Characteristics of the Enzyme

DNA polymerase I has been determined to consist of about 1000 amino acid residues with a molecular weight of 109,000 daltons. Three-dimensionally, the molecule is spheroid, and *in vitro* it can add about 650 mononucleotides per minute per molecule at 37°C. Intact double-helical DNA does not provide an effective template for the enzyme; but if the DNA template is denatured or "nicked," opening up one of the strands, then each strand serves as a template for copying.

In other analyses, Kornberg and his colleagues were able to derive indirect evidence that DNA synthesis proceeds in a 5'-to-3' direction. That is, each new nucleotide is added at its 5'-phosphate end to the C-3' of the growing polynucleotide. Each addition creates a new exposed 3'-OH group on the sugar, which can then participate in the next reaction. Since Kornberg's

work, more direct evidence has confirmed this hypothesis. Synthesis occurring in this way is illustrated in Figure 10.9.

Synthesis of Biologically Active DNA

Despite Kornberg's extensive work, researchers were still not sure whether DNA polymerase I was the enzyme that replicates DNA *in vivo*. Several of their concerns are listed here:

1   The *in vitro* rate of synthesis, 650 nucleotides/minute, is about 100 times less than the *in vivo* rate.

2   The enzyme is itself incapable of replicating intact double-stranded DNA.

3   Synthesis occurs only in the 5'-to-3' direction. For DNA to replicate *in vivo*, it was believed that 3'-to-5' synthesis might also be required.

FIGURE 10.9
Demonstration of 5'-to-3'
synthesis of DNA.

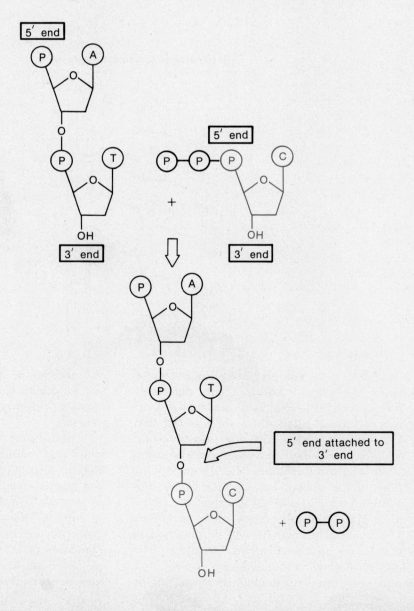

4 In addition to its synthetic capabilities, the enzyme can function as an **exonuclease.** That is, the enzyme can also *cleave* nucleotides one at a time from the end of a polynucleotide chain. This property suggested the possibility that the enzyme's main function *in vivo* might be related to a repair function for the removal and correction of errors in DNA.

Faced with the uncertainty of the true cellular function of DNA polymerase I, Kornberg pursued another approach. He reasoned that if the enzyme could be used to synthesize **biologically active DNA** *in vitro*, then DNA polymerase I must be the major catalyzing force for DNA synthesis within the cell. The term *biological activity* implies that the DNA synthesized is capable of supporting metabolic activities and directing reproduction of the organism from which it was originally derived.

In 1967, Mehran Goulian, Kornberg, and Robert Sinsheimer showed that the DNA of the small bacteriophage φX174 could be completely copied by DNA polymerase I *in vitro*, the new product isolated, and used to successfully transfect *E. coli* protoplasts. This process produced mature phages from the synthetic DNA, thus demonstrating biological activity! The ingenious experimental design is outlined in Figure 10.10.

The phage φX174 provided an ideal experimental system because it contains a very small (5386 nucleotides), circular, single-stranded DNA molecule as its genetic material. Since the molecule is a closed circle, the experiment depended on the isolation of a second enzyme, **DNA ligase** (also called the **polynucleotide joining enzyme**), which joins the two ends of the linear molecule following replication.

In the normal course of φX174 infection, the circular, single-stranded DNA referred to as the **(+) strand** enters the *E. coli* cell and serves as a template for the synthesis of the complementary **(−) strand.** The two strands (+ and −) remain together in a circular double helix called the **replicative form (RF).** The RF serves as the template for replication of itself, and subsequently only (+) strands are produced. These strands are then packaged into viral coat proteins to form mature virus particles.

As shown in Figure 10.10, Goulian, Kornberg, and Sinsheimer's experiment began with ^{3}H-labeled (+) strands. Replication using DNA polymerase I and DNA ligase occurred; the strands were denatured; the (−) strand was isolated; and the process repeated with the (−) strand serving as the template. Isolation of the (−) strand depended on the use of the base analogue **5-bromodeoxyuridine,** which may be successfully incorporated in place of thymidine. In the chemical structure of this analogue, a bromine atom is substituted for a carbon atom in the methyl group at the C-5′ position of the pyrimidine ring, increasing the mass and therefore the density. As a result, the (−) strands can be isolated from the (+) strands using sedimentation equilibrium centrifugation. Eventually, newly synthesized infectious (+) strands were synthesized, isolated in a similar way, and added to the bacterial protoplasts. Transfection occurred, resulting in the production of mature phages. Therefore, the synthetic DNA had successfully directed reproduction!

This demonstration of biological activity was viewed as a precise assessment of faithful copying. If even a single error in one of the 5386 nucleotides had occurred, the change would most likely have constituted a lethal mutation, precluding the production of viable phages.

DNA Polymerase II and III

Although the synthesis of DNA under the direction of polymerase I did demonstrate biological activity, a more serious reservation about the enzyme's true biological role was raised in 1969. Peter DeLucia and John Cairns reported the discovery of a mutant strain of *E. coli* that was deficient in polymerase I activity. The mutation was designated *polA1*. In the absence of the functional enzyme, this mutant strain of *E. coli* replicated its DNA and successfully reproduced! Other properties of the mutation led DeLucia and Cairns to conclude that in the absence of polymerase I, these cells were highly deficient in their ability to "repair" DNA. For example, the mutant strain was highly sensitive to ultraviolet light and radiation, both of which damage DNA and are mutagenic. Nonmutant bacteria are able to repair a great deal of UV-induced damage.

These observations led to two conclusions:

1 There must be another enzyme present in *E. coli* cells that is responsible for replicating DNA *in vivo*.

2 DNA polymerase I may serve only a repair function *in vivo*. This function is now believed by Kornberg and others to occur during the normal course of replication and to be critical to the fidelity of DNA synthesis.

To date, two unique DNA polymerases have been isolated from polymerase I-deficient cells. These two enzymes have also been isolated from normal cells that also contain polymerase I.

Purified **DNA polymerase II** has a molecular weight of about 120,000 daltons. This enzyme cannot totally replicate DNA, but can repair gaps of less than 100 nucleotides in the polynucleotide strands. Its biological function may be related to repair. It synthesizes DNA in a 5′-to-3′ mode, similar to polymerase I.

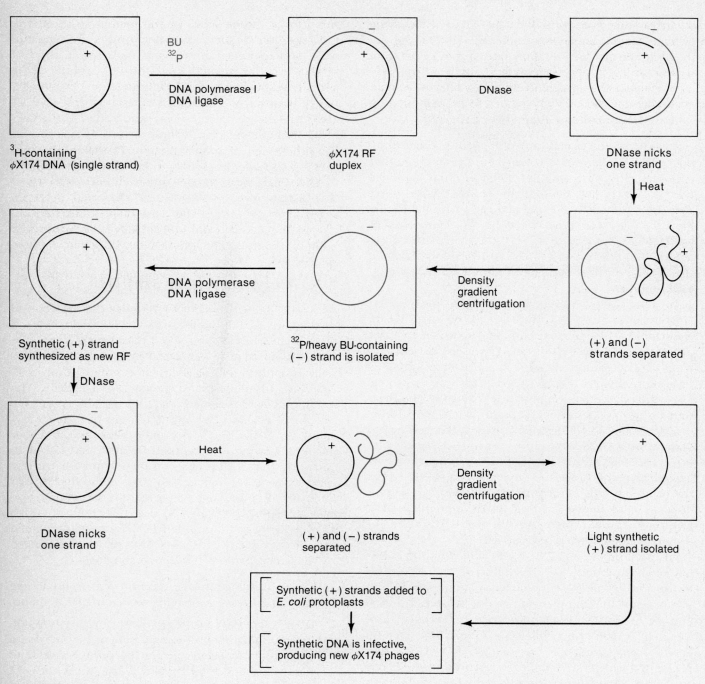

FIGURE 10.10

Schematic representation of Goulian, Kornberg, and Sinsheimer's *in vitro* replication and isolation of synthetic (+) DNA strands of phage φX174. The DNA successfully transfected *E. coli* protoplasts and thus demonstrated biological activity.

DNA polymerase III, discovered in 1972, is difficult to detect because of its instability during its isolation. This property probably explains why Kornberg did not detect polymerase III in his early studies. The core enzyme has a molecular weight of 175,000 daltons and has a 15-fold increase in catalytic activity compared with polymerase I.

Like polymerase I, polymerase III requires a 3'-OH group in order to add a nucleotide. Synthesis is therefore 5'-to-3'. Although scientists currently agree that polymerase III is the major enzyme responsible for cellular DNA replication in *E. coli*, they do not yet know all of the precise details pertaining to enzyme function in the cell. The physiologically active form of the enzyme is believed to be very complicated. The **holoenzyme** contains 11 to 13 polypeptides including the α, ε, and θ core subunits. Of these, the **α polypeptide** is responsible for the synthesis of new DNA strands. This complex is ATP-dependent and can add 700 nucleotides/second at 37°C using a single-stranded template. Its exonuclease activity also allows the enzyme to proofread its own product and then correct any errors.

OTHER ASPECTS OF DNA SYNTHESIS

Additional aspects of DNA synthesis rely primarily on data obtained from studies with *E. coli* and infecting bacteriophages. These findings are the foundation of a tentative model for DNA synthesis, presented following this discussion.

The Rolling-Circle Model of Replication

The way certain viruses achieve the replication of circular DNA structures has been studied extensively. In the bacteriophages λ and φX174, this is achieved by a mechanism referred to as the **rolling-circle model.** In the general model [Figure 10.11(a)], a double-stranded circular DNA molecule is first "nicked" by a site-specific endonuclease, opening up one of the two strands of the double helix. The open strand contains a free 3'-OH end and a free 5'-PO4 end. The complementary closed strand serves as a template, and DNA polymerase adds nucleotides to the free 3' end. As the circular structure "rolls" while synthesis continues, the 5' end

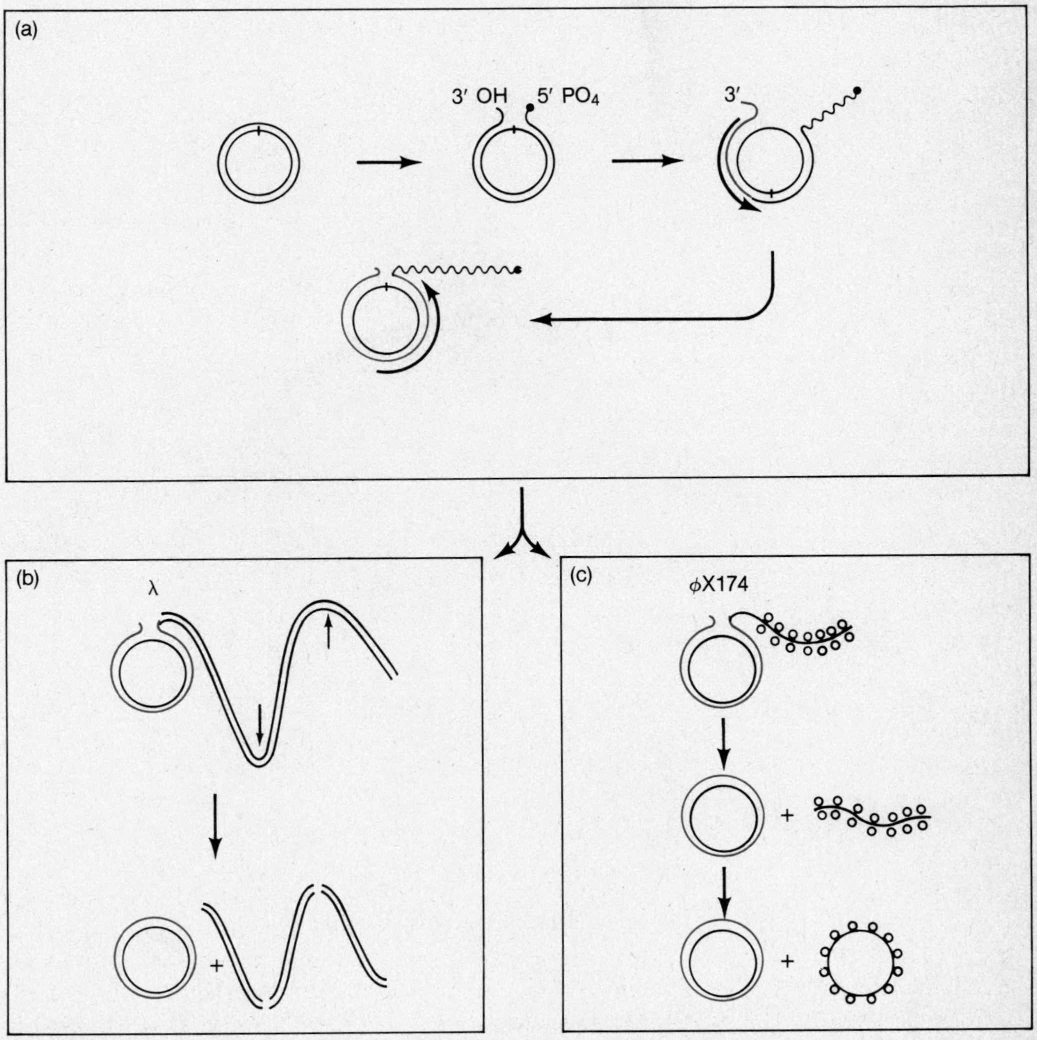

FIGURE 10.11
The rolling-circle model of replication. Part (a) shows the general sequence of events, and parts (b) and (c) show the differences between phages λ and φX174 following the replication by several turns (λ) or one turn (φX174) of the circular DNA.

of the open strand is displaced, creating a tail. Since the molecular shape created resembles the Greek letter σ (sigma), rolling-circle replication has also been called the σ mode.

When replication has occurred within the entire circle, the tail consists of one complete DNA strand. In phage λ [Figure 10.11(b)], this process continues until several complete lengths of the strand are produced. In the meantime, the tail has served as the template for the synthesis of its complement, creating a double-stranded molecule [Figure 10.11(b)]. Such a structure

is called a **concatemer;** it is subsequently cut by a nuclease into complete λ chromosomes, which are packaged into the viral heads.

In φX174 [Figure 10.11(c)], the mature viruses contain only single-stranded (+) DNA molecules. Thus, as the 5′ end is displaced during rolling-circle replication, the tail is the (+) strand. As it is displaced, it is covered with phage proteins. Once the tail is cut from the circular structure, DNA ligase seals it covalently and it becomes part of the newly assembled phage particle.

FIGURE 10.12
Helical unwinding of DNA during replication as accomplished by unwinding proteins, HDPs (helix-destabilizing proteins), and DNA gyrase.

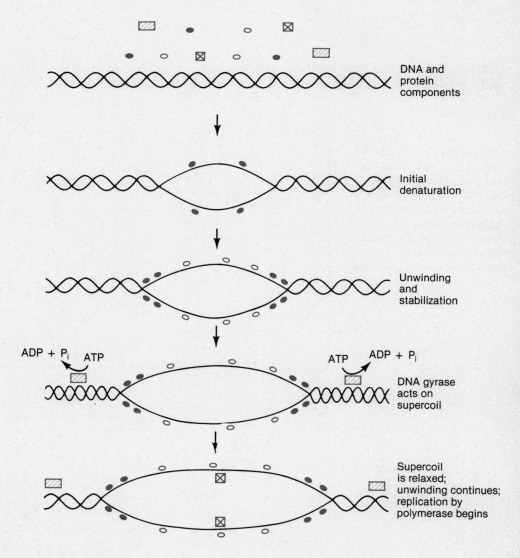

DNA and protein components

Initial denaturation

Unwinding and stabilization

ADP + P$_i$ ATP   ATP   ADP + P$_i$

DNA gyrase acts on supercoil

Supercoil is relaxed; unwinding continues; replication by polymerase begins

● Unwinding proteins

○ Helix-destabilizing proteins (HDP)

▨ DNA gyrase

⊠ DNA polymerase III*

Unwinding the DNA Helix

The presence of molecules that open or unwind the helix is undoubtedly essential to the observed rate of DNA synthesis *in vivo*. The manner in which the two strands unwind (or unzip) prior to DNA synthesis is intriguing. A number of proteins capable of localized denaturation of DNA well below its melting temperature (T_m) have been discovered. They range in molecular weight from 10,000 to 75,000 daltons and bind cooperatively along the helix, causing unwinding at the points ahead of the replication forks (Figure 10.12).

Some of these molecules are called **unwinding proteins;** others are referred to as **helix-destabilizing proteins,** or **HDPs;** and one protein is called **DNA gyrase.** The unwinding proteins, also called **helicases,** initiate localized denaturing of the helix, causing it to unwind. This reaction requires energy derived from the hydrolysis of ATP. The HDPs, also called **single-strand binding proteins (SSBPs),** bind to the single strands that are created, stabilizing the template that is to be replicated. As unwinding continues, supercoiling is created farther down the helix. DNA gyrase, a member of a larger group of enzymes called **DNA topoisomerases,** can function to reduce the tension produced by inducing **negative supercoils.** The enzyme accomplishes this by causing a break in both strands and a topological manipulation of the DNA. Then, in a more relaxed conformation, the gaps created in both strands of the DNA are resealed by DNA ligase. These reactions also require the hydrolysis of ATP to drive them. DNA gyrase can cause about 100 negative supercoils per minute and is inhibited by antibiotics that also inhibit the entire process of DNA replication in *E. coli*. Therefore, the action of this enzyme is indispensable.

In combination with the polymerase complex, these proteins create an array of molecules that participate in DNA synthesis. Kornberg, who has continued his investigations into DNA synthesis, has referred to the entire package as a **replisome.**

Initiation of Synthesis

Once a small portion of the helix is unwound, initiation of synthesis may occur. As previously mentioned, DNA polymerase III requires a free 3′ end in order to synthesize a polynucleotide chain. This prompted researchers to investigate how the first nucleotide can be added, since no free hydroxyl group is present. There is now evidence that RNA is involved in initiating DNA synthesis. It is thought that a short segment of RNA, complementary to DNA, is first synthesized on the DNA template. The RNA is made under the direction of a complex of proteins called the **primosome.** This includes some of the proteins illustrated in Figure 10.12 as well as a form of the enzyme RNA polymerase called **primase.** The short segment of RNA that is synthesized serves as a **primer,** and it is to this primer that DNA polymerase III begins to add deoxyribonucleotides. Obviously, the RNA polymerase does not require a free 3′ end to initiate synthesis. After DNA synthesis is completed at an area adjacent to the RNA primer, the RNA segment is clipped off by an exonuclease and the gap is then filled, perhaps by the action of DNA polymerase I. Recall that this enzyme also functions as an exonuclease, so it could be responsible for removing the segment of RNA as well.

RNA priming has been recognized in viruses, bacteria, and several eukaryotic organisms and is thought to be a universal phenomenon. Initiation of DNA synthesis as described above occurs at a single position, the **origin,** in viral and bacterial chromosomes. The origin and the unit of DNA that is subsequently replicated is called a **replicon.** In viruses and bacteria, this includes the entire chromosome. In eukaryotes, multiple replicons are present along each chromosome. Central to the concept of the replicon is the inclusion of control elements that regulate the initiation of replication once during each cell cycle.

Continuous and Discontinuous DNA Synthesis

We must now reconsider the fact that the two strands of a double helix are antiparallel to each other. One runs in the 5′-to-3′ direction, while the other has the opposite polarity. This structural characteristic is responsible for two features of DNA synthesis, which are illustrated in Figure 10.13. First, since DNA polymerase III synthesizes DNA in a 5′-to-3′ direction only, simultaneous synthesis of antiparallel strands must occur in one direction along one strand and in the opposite direction on the other.

Second, as the strands unwind and the replication fork progresses down the helix, only one strand can serve as a template for **continuous synthesis.** This strand is called the **leading strand.** As the fork progresses, many points of initiation are necessary on the opposite, or **lagging, strand** resulting in **discontinuous synthesis.**

Evidence in support of discontinuous synthesis was first provided by Reiji and Tuneko Okazaki and their colleagues. They discovered that when bacteriophage DNA is replicated in *E. coli*, some of the newly formed DNA is found as small fragments containing 1000 to 2000 nucleotides. These pieces, called **Okazaki frag-**

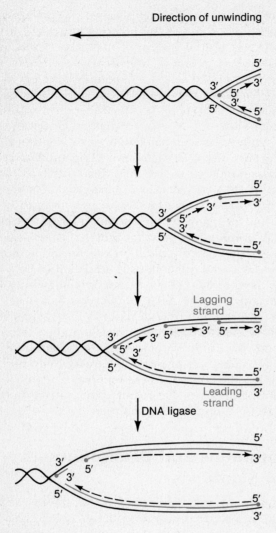

Direction of unwinding

DNA ligase

FIGURE 10.13
Illustration of the opposite polarity of synthesis along the two strands of DNA necessitated by the requirement of 5'-to-3' synthesis of DNA polymerase III. On the lagging strand, synthesis must be discontinuous, resulting in the production of Okazaki fragments, which are sealed by DNA ligase.

ments, must then be enzymatically joined to create longer DNA strands. As infection proceeds, the Okazaki fragments, with low molecular weight, are indeed replaced by longer and longer DNA strands of higher molecular weight.

Discontinuous synthesis of DNA requires an enzyme that can unite the smaller products into the longer continuous molecule. **DNA ligase** has been shown to be capable of performing this function. The evidence that DNA ligase does perform this function during DNA synthesis is strengthened by the observation of a **ligase-deficient mutant** strain of *E. coli.* In this strain, Okazaki fragments accumulate in particu-

larly large amounts. Apparently, they are not joined adequately. There is also other evidence that DNA polymerase I may be involved in joining discontinuously synthesized fragments. Other experimentation indicates that both strands may be synthesized discontinuously and that this form of synthesis is the rule rather than the exception. This conflicting evidence will undoubtedly be resolved soon.

DNA SYNTHESIS—A TENTATIVE MODEL

The model presented in Figure 10.14 summarizes the preceding discussions of DNA synthesis. It takes into account the following points:

1 Synthesis is initiated at one or more specific points along the DNA helix.

2 Unwinding proteins denature, or open up, the helix; HDPs stabilize the denatured region.

3 A unique RNA polymerase synthesizes a short RNA primer on both DNA strands.

4 DNA polymerase III synthesizes the complementary strands of DNA in the 5'-to-3' direction.

5 Synthesis is bidirectional.

6 On at least one DNA strand, synthesis is discontinuous, producing Okazaki fragments.

7 The Okazaki fragments are sealed by DNA ligase.

8 As unwinding proceeds, DNA gyrase reduces the tension created farther down the helix.

9 Primer RNA is removed. This is thought to be accomplished by DNA polymerase I, which then directs the synthesis of the short DNA complement in the area previously occupied by the primer.

10 As this process proceeds along the length of the replicon, semiconservative replication is achieved.

Since the investigation of DNA synthesis is still an extremely active area of research, this model will no doubt be extended in the future. The main question is whether the model is universal in nature. It does, however, provide a summary of DNA synthesis against which genetic phenomena may be interpreted.

GENETIC CONTROL OF REPLICATION

Much of what we know and have outlined in the previous section concerning the details of DNA replication in viruses and bacteria has been based on the genetic analysis of the process. For example, we have

FIGURE 10.14

A model of DNA synthesis incorporating helical unwinding, RNA priming, and bidirectional, discontinuous synthesis.

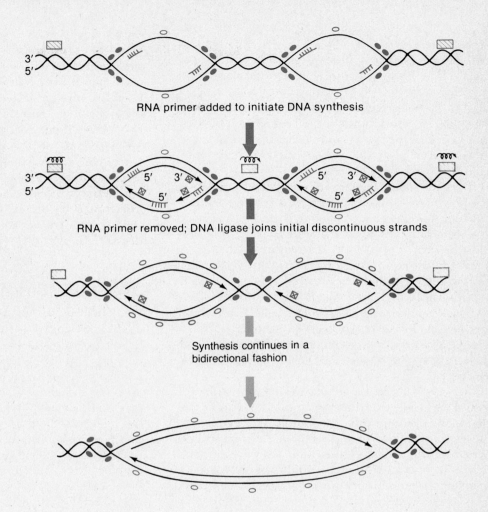

RNA primer added to initiate DNA synthesis

RNA primer removed; DNA ligase joins initial discontinuous strands

Synthesis continues in a bidirectional fashion

already discussed the *polA1* mutation, the study of which revealed that DNA polymerase I is not the major enzyme responsible for replication. Many other mutations have been isolated that interrupt or seriously impair some aspect of replication. These can be studied if they are **conditional mutations** that express the mutant condition at a restrictive temperature but function normally at a permissive temperature. Investigation of such mutants provides insights into the product and the associated function of the normal, nonmutant gene.

As shown in Table 10.2, the enzyme product or its general role in replication has been ascertained for a variety of genes in *E. coli.* For example, numerous mutations specifying the subunits of polymerases II and III have been isolated. Genes have also been identified that encode products involved in specification of the origin of synthesis, helix-unwinding and stabilization, initiation and priming, relaxation of supercoiling, repair, and ligation. The discovery of such a large group of genes attests to the complexity of the process of replication, even in the relatively simple prokaryote. This complexity is not unexpected, given the enor-

mous quantity of DNA that must be unerringly replicated in a very brief time. As we will see, the process is even more difficult to investigate in eukaryotes.

TABLE 10.2

A list of various *E. coli* mutant genes and their products or functions.

Mutant Gene	Enzyme or Gene Function
polA	DNA polymerase I
polB	DNA polymerase II
polC	DNA polymerase III
dnaX,Z	DNA polymerase III subunits
dnaA,I,P	Initiation
dnaB,C	Priming
oriC	Origin of replication
gyrA,B	Gyrase subunits
lig	Ligase
rep	Helicase
ssb	Single-stranded binding proteins
rpoB	RNA polymerase subunit

EUKARYOTIC DNA SYNTHESIS

Most features of DNA synthesis found in microorganisms are now thought to apply also to eukaryotic systems. However, the DNA of eukaryotes is complexed with a variety of proteins, some of which maintain the molecule's structural integrity. The presence of these proteins and the general complexity of the eukaryotic cell compared with the prokaryotic cell have made analysis much more difficult.

Eukaryotic cells contain three different kinds of DNA polymerases, corresponding to polymerase I, II, or III in *E. coli.* These are called **α, β,** and **γ.** It appears the α form is the major enzyme involved in the replication of DNA. The β form may be involved in DNA repair, while the γ form is the only DNA polymerase found in mitochondria. Presumably, it is unique in its function within that organelle. It appears that the DNA polymerases of eukaryotes have the same fundamental requirements for DNA synthesis as bacterial and viral systems: four deoxyribonucleoside triphosphates, a divalent cation (Mg^{++} or Mn^{++}), a template, and a primer.

Data and observations derived from autoradiographic and electron microscopic studies have provided many other insights. Most often, eukaryotic DNA synthesis appears to be bidirectional, moving in both directions and creating two replication forks from each point of origin. As one would expect because of the increased amount of DNA in eukaryotes compared to prokaryotes, many more points of origin are present. In mammals, there are about 25,000 replicons present in the genome, each consisting of an average of 100,000 to 200,000 base pairs (100–200 kb). In *Drosophila*, there are about 3500 replicons per genome of an average size of 40 kb. The many points of origin are thought not to be random along each chromosome, but instead to represent specific points of initiation of DNA synthesis. Not all origins are activated at the same time during the S phase of the cell cycle. Most DNA regions that represent genes are replicated preferentially early in the S phase, while the nongenic DNA areas are replicated later during that phase. These areas, called **euchromatin** and **heterochromatin,** respectively, will be discussed in greater detail in Chapter 11.

The *in vivo* synthesis of eukaryotic DNA is five to ten times slower than that of *E. coli.* However, most general aspects of chain elongation are thought to be similar. The actions of a variety of proteins—DNA helicase, HDP, and topoisomerase—are believed to modulate strand separation that precedes RNA priming by a primase enzyme. On the lagging strand, synthesis is discontinuous, resulting in Okazaki fragments that are linked together by DNA ligase. These fragments are about ten times smaller (100–150 nucleotides) than in prokaryotes. This size happens to coincide roughly with that of the basic repeating structural unit of eukaryotic chromatin, the **nucleosome,** but it is not yet clear whether there is a relationship between these two observations. Nucleosomes, consisting of about 150 base pairs of DNA and an octamer of histone proteins, will be discussed extensively in Chapter 11. While synthesis may occur continuously on the leading strand, it is not clear whether it can proceed uninterrupted for the full length of a replicon. Thus, synthesis on the leading strand may be semidiscontinuous.

DNA RECOMBINATION

We conclude this chapter by returning to a topic discussed in Chapter 6—**genetic recombination.** There, it was pointed out that the process of crossing over depends on breakage and rejoining of the DNA strands between homologues. Now that we have introduced the chemistry and replication of DNA, it is appropriate to consider how recombination occurs at the molecular level. In general, the following information pertains to genetic exchange between any two homologous double-stranded DNA molecules, whether they be viral or bacterial chromosomes or eukaryotic homologues during meiosis.

While there are several models available to explain crossing over, they all share certain common features. First, all are based on the initial proposals put forth independently by Robin Holliday and Harold L. K. Whitehouse in 1964. They also depend on the complementarity between DNA strands for their precision of exchange. Finally, each model relies on a series of enzymatic processes in order to accomplish genetic recombination.

One such model is illustrated in Figure 10.15. It begins (a) with two paired DNA duplexes or homologues, each of which has a single-stranded nick introduced (b) at an identical position by an endonuclease. The ends of the strands produced by these cuts are then displaced and subsequently pair with their complements on the opposite duplex (c). A ligase then seals the loose ends (d), creating hybrid duplexes called **heteroduplex DNA molecules.** The exchange creates a cross-bridged or **Holliday structure.** The position of this cross-bridge can then move down the chromosomes by branch migration (e). This occurs as a result of a zipperlike action as hydrogen bonds are broken and then reformed between complementary bases of the displaced strands of each duplex. This migration

FIGURE 10.15

A possible molecular sequence depicting how genetic recombination occurs as a result of the breakage and rejoining of heterologous DNA strands. The electron micrograph shows DNA in a chi-form structure similar to the diagram in (g). The DNA is from the Col. E1 plasmid of *E. coli.* (Photograph courtesy of David Dressler.)

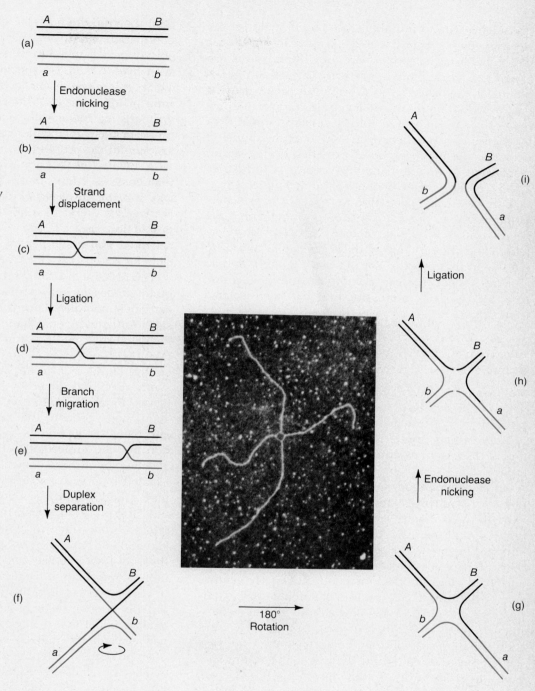

yields an increased length of heteroduplex DNA on both homologues.

If the duplexes now separate (f), and the bottom portions rotate 180° (g), an intermediate planar structure called a **chi form** is created. If the two strands on opposite homologues previously uninvolved in the exchange are now nicked by an endonuclease (h), and ligation occurs (i), recombinant duplexes are created.

Note that the arrangement of alleles has been altered as a result of the crossover that has occurred.

If the original nicks (b) do not occur at the same point on each duplex, the same end result can still be achieved. The overlapping single strands must first be clipped by an exonuclease and the gaps filled in by DNA polymerase. This tailoring of the duplexes must be accomplished before ligation (d) can occur.

Evidence supporting the above model includes the electron microscopic visualization of chi-form planar molecules from bacteria where four duplex arms are joined at a single point of exchange (Figure 10.15). Additionally, the discovery in *E. coli* of the **recA protein** has provided important evidence. This molecule promotes the exchange of reciprocal single-stranded DNA molecules as it must occur in step (c) of the model. It enhances the hydrogen bond formation during strand displacement, thus initiating heteroduplex formation. Finally, all enzymes essential to the nicking and ligation process have been discovered and investigated. Mutations that prevent genetic recombination have been found in a number of genes in viruses and bacteria. Thus, the products of these genes play an essential role in that process. Recall from Chapter 6 that an enzyme system that can accomplish the nicking and resealing essential to crossing over has been isolated from *Lilium*.

SUMMARY AND CONCLUSIONS

The means by which DNA is replicated, as well as the details of its synthesis, have constituted a fundamental problem in the study of molecular genetics. The entire process, which results in the duplication or replication of DNA, is an essential function of the genetic material.

Three general modes of replication are theoretically possible: semiconservative, conservative, and dispersive. All three modes rely on the complementarity of bases along the axis of the polynucleotide chain. Strong evidence favors semiconservative replication in both prokaryotes and eukaryotes. In 1958, Meselson and Stahl elegantly demonstrated that newly synthesized DNA from *E. coli* consists of a double helix composed of one old and one new strand. In their experiments, they used the heavy isotope ^{15}N and sedimentation equilibrium centrifugation. One year earlier, Taylor, Woods, and Hughes, using autoradiography to study dividing root tips of the broad bean, drew the same conclusion for replication of eukaryotic DNA.

The enzymatic synthesis of long polynucleotides on a DNA template is an involved and detailed process. Most information about this process has been derived from the study of bacteria and their phages. Relevant information about enzymatic synthesis was first reported in 1957, when Kornberg isolated DNA polymerase I from *E. coli*. He specified the conditions under which this enzyme can synthesize DNA *in vitro*. In spite of his clear 1967 demonstration of the *in vitro* synthesis of biologically active φX174 DNA, several properties of the enzyme cast doubt on its replicative function in the cell *(in vivo)*.

In 1969, a mutant strain of *E. coli* was discovered that possessed little or no polymerase I activity but was still able to replicate its DNA. Subsequently, DNA polymerase II and III were discovered. The true function of polymerase II remains unresolved, but it is now accepted that a complex of polymerase III is responsible for replication of DNA *in vivo*.

The process of DNA synthesis, which appears to be quite similar in both prokaryotes and eukaryotes, is extremely complex. Recent data indicate that synthesis is initiated by enzymatic unwinding of the double helix, followed by RNA primer formation. Synthesis of DNA is discontinuous and bidirectional under the control of polymerase III. Short Okazaki fragments are synthesized and joined by still another enzyme, DNA ligase. In addition to the enzymes performing the above steps, proteins that stabilize the unwound helix and relax the supercoils created ahead of the replication fork are important to the process. Mutant genes affecting all of these enzymes and proteins have been isolated in viruses and bacteria, attesting to the complex genetic control of the entire process. While the unit of replication, called the replicon, consists of the entire chromosome in viruses and bacteria, each eukaryotic chromosome contains many such units. Synthesis of DNA in eukaryotes proceeds five to ten times more slowly than in bacteria.

We have concluded this chapter by returning to the topic of genetic recombination and presenting a molecular model that may explain how exchange occurs between genetic molecules. As with all other metabolic processes, recombination relies on a series of enzymes. In this case, they cut, snip, redirect, and reseal DNA strands.

The study of DNA replication is particularly fascinating because the process must occur faithfully and is indispensable for reproduction of cells and organisms.

PROBLEMS AND DISCUSSION QUESTIONS

1. Compare conservative, semiconservative, and dispersive modes of DNA replication.

2. Review the experimental design, results, and conclusions of the Meselson-Stahl experiment.

3. In the Meselson-Stahl experiment, which of the three modes of replication could be ruled out after one round of replication? After two rounds?

4. Predict the results of the experiment by Taylor, Woods, and Hughes if replication were (a) conservative and (b) dispersive.

5. Reconsider Problem 24 in Chapter 9. In the model you proposed, could the molecule be replicated semiconservatively? Why? Would other modes of replication work?

6. What are the requirements for the *in vitro* synthesis of DNA under the direction of DNA polymerase I?

7. In Kornberg's initial experiments, he actually grew *E. coli* in Anheuser-Busch beer vats. Why do you think this was necessary?

8. How did Kornberg test the fidelity of copying DNA by polymerase I?

9. Which of Kornberg's tests is the more stringent assay? Why?

10. Which characteristics of DNA polymerase I led to doubts that its *in vivo* function is the synthesis of DNA leading to complete replication?

11. Explain the theory of nearest-neighbor frequency.

12. What is meant by "biologically active" DNA?

13. Why was the phage φX174 chosen for the experiment demonstrating biological activity?

14. Outline the experimental design of Kornberg's biological activity demonstration.

15. What was the significance of the *polA1* mutation?

16. Summarize the properties of polymerase II and III.

17. Distinguish between (a) unidirectional and bidirectional synthesis and (b) continuous and discontinuous synthesis of DNA.

18. List the proteins that unwind DNA during *in vivo* DNA synthesis. How do they function?

19. Define and indicate the significance of (a) Okazaki fragments, (b) DNA ligase, and (c) primer RNA.

20. Outline the current model for DNA synthesis.

21. Why should DNA synthesis be more complex in eukaryotes than in bacteria? How is DNA synthesis similar in the two types of organisms?

22. If the analysis of DNA from two different microorganisms demonstrated very similar nearest-neighbor frequencies, is the DNA of the two organisms identical in: (a) amount, (b) base composition, and (c) nucleotide sequences?

23. Analysis of nearest-neighbor data led Josse, Kaiser, and Kornberg in 1961 (*J. Biol. Chem.* 236: 864–75) to conclude that the two strands of the double helix are in opposite polarity to one another. Demonstrate your understanding of the nearest-neighbor technique by determining the outcome of such an analysis if the following molecule is in the antiparallel versus the parallel strand mode:

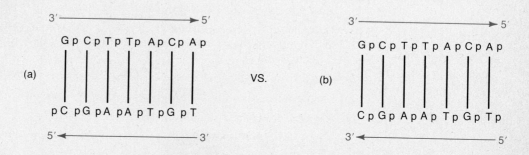

24 Shown below is the frequency with which each of the 16 possible nearest neighbors occurs in a bacterial DNA.

(a) If you assume that this DNA is double stranded and complementary, but it is not known whether the strands are parallel or antiparallel, which nearest neighbors should be equal in either case? Are they?

(b) Now consider the case where T and A exist as the nearest neighbors of C, e.g., C→T and C→A. If the strands are parallel, what nearest-neighbor combinations should be equal to C→T and to C→A? If the strands are antiparallel, which combinations should be equal to C→T and to C→A?

(c) What other combinations should be equal to one another, assuming a complementary, antiparallel double helix? Which ones are actually equal in the data?

Nearest Neighbor

		G	C	A	T
Initially Labeled Nucleotide	T	0.068*	0.045	0.035	0.025
	A	0.063	0.060	0.025	0.022
	G	0.088	0.133	0.035	0.060
	C	0.110	0.088	0.068	0.063

*The data indicate the frequency of each of the 16 combinations, and in this case, T→G.

25 Refer to the model presented in Figure 10.15. What polarity relationship must exist between the two inner DNA strands in parts (a) and (b)?

SELECTED READINGS

ALBERTS, B., and STERNGLANZ, R. 1979. Recent excitement in the DNA replication problem. *Nature* 269: 655–60.

CAIRNS, J. 1963. The bacterial chromosome and its manner of replication as seen by autoradiography. *J. Mol. Biol.* 6: 208–13.

CATCHESIDE, D. G. 1977. *The genetics of recombination.* Baltimore: University Park Press.

CHALBERG, M. D., and KELLEY, T. J. 1982. Eukaryotic DNA replication: Viral and plasmid model systems. *Ann. Rev. Biochem.* 51: 901–34.

COZZARELLI, N. R. 1980. DNA gyrase and the supercoiling of DNA. *Science* 207: 953–60.

DAVIDSON, J. N. 1976. *The biochemistry of nucleic acids.* 8th ed. New York: Academic Press.

DeLUCIA, P., and CAIRNS, J. 1969. Isolation of an *E. coli* strain with a mutation affecting DNA polymerase. *Nature* 224: 1164–66.

DENHARDT, D. T., and FAUST, E. A. 1985. Eukaryotic DNA replication. *BioEssays* 2: 148–53.

DRESLER, D., and POTTER, H. 1982. Molecular mechanisms in genetic recombination. *Ann. Rev. Biochem.* 51: 727–61.

GEFTER, M. L. 1975. DNA replication. *Ann. Rev. Biochem.* 44: 45–78.

GELLERT, M. 1981. DNA topoisomerases. *Ann. Rev. Biochem.* 50: 879–910.

HOLLIDAY, R. 1964. A mechanism for gene conversion in fungi. *Genet. Res.* 5: 282–304.

HOOD, L. E., WILSON, J. H., and WOOD, W. B. 1975. *Molecular biology of eukaryotic cells—A problems approach.* Menlo Park, Calif.: W. A. Benjamin.

JOSSE, J., KAISER, A. D., and KORNBERG, A. 1961. Enzymatic synthesis of deoxyribonucleic acid. VIII. Frequencies of nearest neighbor base sequences in deoxyribonucleic acid. *J. Biol. Chem.* 236: 864–75.

KORNBERG, A. 1960. Biological synthesis of DNA. *Science* 131: 1503–8.

———. 1969. Active center of DNA polymerase. *Science* 163: 1410.

———. 1974. *DNA synthesis.* San Francisco: W. H. Freeman.

———. 1979. Aspects of DNA replication. *Cold Spr. Harb. Symp.* 43: 1–10.

————. 1980. *DNA replication.* San Francisco: W. H. Freeman.

————. 1982. *DNA replication: 1982 Supplement.* San Francisco: W. H. Freeman.

LEHMAN, I. R. 1974. DNA ligase: Structure, mechanism, and function. *Science* 186: 790–97.

LEHNINGER, A. L. 1982. *Principles of biochemistry.* New York: Worth.

MESELSON, M. S., and RADDING, C. M. 1975. A general model for genetic recombination. *Proc. Natl. Acad. Sci.* 72: 358–61.

MESELSON, M., and STAHL, F. W. 1958. The replication of DNA in *Escherichia coli. Proc. Natl. Acad. Sci.* 44: 671–82.

OGAWA, T., and OKAZAKI, T. 1980. Discontinuous DNA synthesis. *Ann. Rev. Biochem.* 49: 421–57.

OKAZAKI, T., et al. 1979. Structure and metabolism of the RNA primer in the discontinuous replication of prokaryotic DNA. *Cold Spr. Harb. Symp.* 43: 203–22.

SCHEKMAN, R., WEINER, A., and KORNBERG, A. 1974. Multienzymes systems of DNA replication. *Science* 186: 987.

SHEININ, R., HUMBERT, J., and PEARLMAN, R. E. 1978. Some aspects of eukaryotic DNA replication. *Ann. Rev. Biochem.* 47: 277–316.

SZOSTAK, J. W., ORR-WEAVER, T. L., and ROTHSTEIN, R. J. 1983. The double-strand-break repair model for recombination. *Cell* 33: 25–35.

TAYLOR, J. H., WOODS, P. S., and HUGHES, W. C. 1957. The organization and duplication of chromosomes revealed by autoradiographic studies using tritium-labeled thymidine. *Proc. Natl. Acad. Sci.* 48: 122–28.

TOMIZAWA, J., and SELZER, G. 1979. Initiation of DNA synthesis in *E. coli. Ann. Rev. Biochem.* 48: 999–1034.

WANG, J. C. 1982. DNA topoisomerases. *Scient. Amer.* (July) 247: 94–108.

WATSON, J. D. 1976. *Molecular biology of the gene.* 3rd ed. Menlo Park, Calif.: W. A. Benjamin.

WHITEHOUSE, H. L. K. 1982. *Genetic recombination: Understanding the mechanisms.* New York: Wiley.

WICKNER, S. H. 1978. DNA replication proteins of *E. coli. Ann. Rev. Biochem.* 47: 1163–91.

WOOD, W. B., WILSON, J. H., BENBOW, R. M., and HOOD, L. E. 1975. *Biochemistry—A problems approach.* Menlo Park, Calif.: W. A. Benjamin.

ZUBAY, G. L., and MARMUR, J., eds. 1973. *Papers in biochemical genetics.* 2nd ed. New York: Holt, Rinehart and Winston.

11

The Organization of DNA in Chromosomes

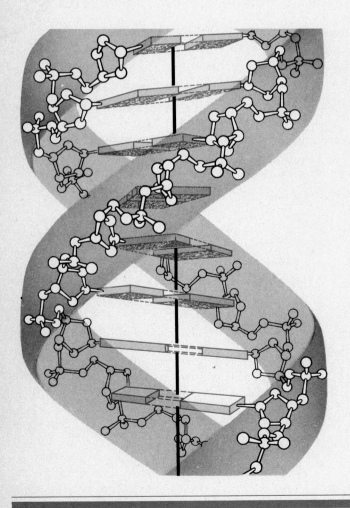

The elucidation of how DNA is organized in chromosomes is currently an extremely active area of research. There has been much interest in this topic because it is believed that the determination of the spatial arrangement of the genetic material and associated molecules will provide valuable insights into other aspects of genetics. For example, how the genetic information is stored, expressed, and regulated must be related to its organization. In eukaryotes, how the chromatin fibers characteristic of interphase are condensed into chromosome structures visible during mitosis and meiosis is also of great interest.

Investigations have involved viruses, bacteria, and a variety of eukaryotes. The genetic material has been studied using numerous approaches, including molecular analysis and direct visualization by light and electron microscopy. Despite these sophisticated approaches, we do not yet completely understand all molecular components directly associated with genetic material.

In this chapter, we provide a survey of the various forms of chromosome organization. Wherever possible, the physical arrangement of the genetic material will be related to its function. In a later chapter, we will focus directly on the organization of individual genes.

VIRAL AND BACTERIAL CHROMOSOMES

In comparison with higher forms, the chromosomes of viruses and bacteria are much less complicated. They consist of a single nucleic acid molecule which is often, but not always, circular in form. Such chromosomes are largely devoid of associated proteins, which are characteristic of eukaryotes. Because there is but a single chromosome, much less genetic information is stored in viruses and bacteria than in the multiple units of higher forms. These characteristics have greatly simplified analysis, which has now provided a fairly comprehensive view of viral and bacterial chromosomes.

Organization of Viral DNA and RNA

The chromosomes of viruses consist of a single nucleic acid molecule—either DNA or RNA—which is either single or double stranded. Viruses cannot reproduce in the absence of their host cell, because they depend on the host's biosynthetic machinery to duplicate their genetic material and to synthesize their protein coat. The DNA or RNA contained within the protein coat, or **capsid,** becomes functional only after the injection into a host cell.

Viral nucleic acid molecules may exist as ring structures, or they may take the form of linear molecules. The single-stranded DNA of the **φX174 bacteriophage** and the double-stranded DNA of the **polyoma virus** are ring-shaped nucleic acid molecules both within the protein coat of the mature virus and within the host cell. The **bacteriophage lambda (λ),** on the other hand, possesses a linear double-stranded DNA molecule prior to infection, which closes to form a ring upon infection of the host cell. The two free ends of the DNA molecule are described as being "sticky"; that is, because they are single stranded and composed of complementary base sequences, they have a chemical affinity for forming a ring. Following infection, an enzymatic reaction involving **polynucleotide ligase** occurs, forming a covalent bond and effecting closure of the ring. This process is illustrated in Figure 11.1. Circularization of the nucleic acid after invasion of the host is thought to be related to the mode of replication of the molecule, the so-called **rolling-circle** model (see Chapter 10). Still other viruses, such as the **T-even series of bacteriophages,** have linear, double-stranded chromosomes of DNA, which do not form circles inside the bacterial host.

Viral nucleic acid molecules have been visualized under the electron microscope. Figure 11.2 shows a mature bacteriophage lambda with its double-

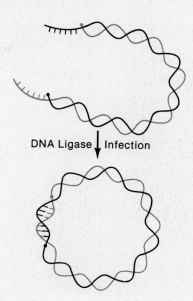

FIGURE 11.1

Schematic diagram of the complementary single-stranded tails at the ends of the chromosome of bacteriophage λ. Under the action of DNA ligase, the linear molecule is converted to a covalently sealed circular structure.

stranded DNA molecule in the circular configuration. One constant feature shared by viruses, bacteria, and eukaryotic cells is the ability to package an exceedingly long DNA molecule into a relatively small volume. In lambda, the DNA is 17 μm long and must fit into the phage head, which is less than 0.1 μm on any side. In Figure 11.2, the phage head is enlarged six times more than the DNA molecule.

In Table 11.1, the measured lengths of the genetic molecules are compared with the size of the head in several other viruses. In each case a similar packaging feat must be accomplished. The dimensions given for phage T2 may be compared with the micrograph of both the DNA and viral particle shown in Figure 11.3.

When the diameter as well as the length of the nucleic acid molecule is taken into account, the minimum volume of the entire chromosome can be estimated and compared to the volume of the viral head. Seldom does the space available in the head exceed the chromosome volume by more than a factor of two. In many cases, almost all space is filled, indicating nearly perfect packing. Just how the chromosome is folded, coiled, and condensed is not yet known. In single-stranded RNA viruses like the tobacco mosaic virus (TMV), the capsid is assembled around the chromosome (see Figure 9.8). The protein subunits of the

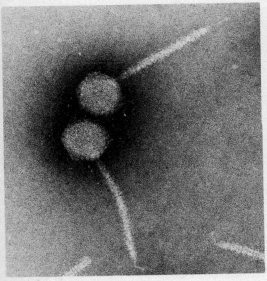

(a)

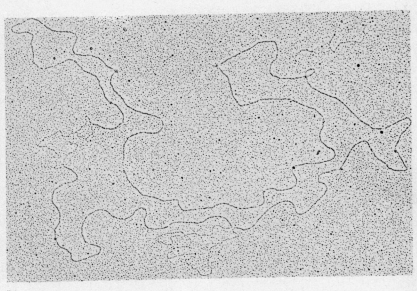

(b)

FIGURE 11.2
(a) Electron micrograph of phage λ and (b) the DNA isolated from it. Note that the
enlargement of (a) is six times greater than (b). [Photo (a) courtesy of Robley C. Williams.
Photo (b) by Hans Ris from Westmoreland, 1969. Copyright 1969 by the American Association
for the Advancement of Science.]

capsid are thought to interact with the RNA molecule and thus play a role in the packing phenomena. On the other hand, DNA viruses with nearly spherical heads, such as the T-even group, first assemble the capsid. Then, the chromosome is inserted into the head which, when filled, is sealed by addition of a tail. Once packed within the head, the genetic material is functionally inert until released into a host cell.

Organization of Bacterial DNA

Bacterial chromosomes are also relatively simple in form compared with those of eukaryotic cells. They always consist of a double-stranded DNA molecule that is compartmentalized in a region of the cell called the **nucleoid.** The nucleoid, which may occupy more than half the volume of the cell, is found near the pe-

TABLE 11.1
The genetic material of
representative viruses and
bacteria.

Organism	Nucleic Acid			Overall Size of Virus or Bacteria (μm)
	Type	SS or DS	Length (μm)	
VIRUSES				
φX174	DNA	SS	2.0	0.025 × 0.025
Tobacco mosaic virus	RNA	SS	3.3	0.30 × 0.02
Lambda phage	DNA	DS	17.0	0.07 × 0.07
T2 phage	DNA	DS	52.0	0.07 × 0.10
BACTERIA				
Hemophilus influenzae	DNA	DS	832.0	1.00 × 0.30
Escherichia coli	DNA	DS	1200.0	2.00 × 0.50

NOTE: SS = single stranded; DS = double stranded.

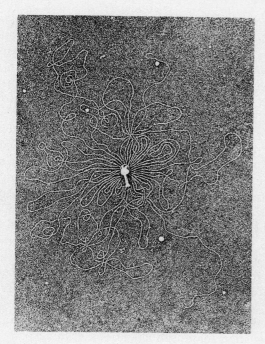

FIGURE 11.3

Electron micrograph of bacteriophage T2, which has had its DNA released by osmotic shock. Note that the molecule is linear. (From Kleinschmidt, 1962, p. 857. By permission of Elsevier Biomedical Press.)

riphery, where part of the chromosome is associated with the cell membrane. When the tightly compacted material is released from the cell during experimentation, it can be photographed under the electron microscope and measured. Figure 11.4 shows the DNA molecule released from the bacterium *Hemophilus influenzae*. The molecule is 832 μm long, while the bacterial cell measures only 0.3 by 1.0 μm. It is of interest to note that the diameter of a bacterial DNA strand measured from electron micrographs corresponds approximately to the prediction of the Watson-Crick double-helix model—about 20 Å.

Escherichia coli, the most extensively studied bacterium, has a still larger circular chromosome, measuring approximately 1200 μm (1.2 mm) in length. When the cell is lysed and the chromosome released, the DNA is thought to be associated with several types of proteins, including those called HU and H. They are small but abundant in the cell and contain a high percentage of positively charged amino acids that can bond ionically to the negative charges of the phosphate groups in DNA. As we will soon see, these proteins resemble structurally similar molecules called **histones** that are found associated with eukaryotic DNA. Whether the bacterial counterparts serve precisely the same role remains to be seen. It seems likely

FIGURE 11.4

Electron micrograph of the bacterium *Hemophilus influenzae*, which has had its DNA released by osmotic shock. The chromosome is approximately 830 μm long. [With permission from MacHattie, 1965, p. 648. Copyright: Academic Press Inc. (London) Ltd.]

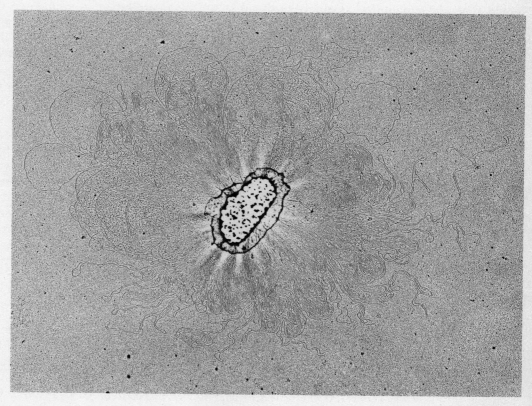

that they do play a role in packaging the chromosome within the nucleoid region. Unlike the tightly packed chromosome of a virus, the bacterial chromosome is not functionally inert. In spite of the compacted condition of the bacterial chromosome, transcription readily occurs.

EUKARYOTIC CHROMOSOMES: GROSS STRUCTURE

The structure and organization of the genetic material in **eukaryotic cells** is much more intricate than in viruses or bacteria. This complexity is due to the greater amount of DNA per chromosome and the presence of large numbers of proteins associated with DNA in eukaryotes. For example, while DNA in the *E. coli* chromosome is 1200 μm long, the DNA in human chromosomes ranges from 14,000 to 73,000 μm in length as a linear duplex. In a single human nucleus, all 46 chromosomes contain sufficient DNA to extend almost 2 meters! This genetic material, along with its associated proteins, is contained within a nucleus that usually measures about 5 μm in diameter.

For many reasons this intricacy is to be expected. It parallels the structural and biochemical diversity of the many types of eukaryotic cells making up a single organism. In a single multicellular organism, cells assume specific functions by "division of labor." It is assumed that any functional capability is based upon biochemical activity. Since this activity is under the direction of the genetic information, a highly ordered regulatory system governing the readout of this information must exist if dissimilar cells are to perform a variety of different functions. Such a system must in some way be imposed on or related to the molecular structure of the genetic material.

While bacteria can reproduce themselves, they never exhibit a detailed process similar to mitosis. As was pointed out in Chapter 2, eukaryotic cells exhibit a highly organized cell cycle. During interphase, the genetic material is finely dispersed throughout the nucleus in chromatin form. As mitosis begins, the chromatin condenses greatly, and during prophase it is compressed into recognizable chromosomes. This condensation represents a contraction in length of some 10,000 times for each chromatin fiber. This highly regular condensation-uncoiling cycle poses special organizational problems in eukaryotic genetic material.

Mitotic Chromosomes

Initially, biologists knew of eukaryotic chromosome structure only from observations made with the light microscope. Most information came from studies of mitotic chromosomes. By observing preparations of mitotic chromosomes from different species, geneticists and cytologists are able to determine the diploid number characteristic of any species and to note the varying sizes and gross morphology of chromosome pairs.

By the metaphase stage of mitosis it becomes apparent that each chromosome is really a double structure consisting of two **sister chromatids.** Sister chromatids are held together at a single point, the **centromere,** which is the area of attachment to the spindle fibers. Chromosomes for any given species are classified by the location of the centromere and the overall size of the chromosome. The **karyotype** consists of a micrograph of the chromosome pairs in metaphase arranged by size and centromere location. The number of chromosome pairs in the karyotype is equal to the haploid number. In species with a low haploid number, each pair of chromosomes may be distinct in gross morphology from all other pairs.

In humans, however, there is considerable morphological similarity between some pairs. In an initial classification scheme, human mitotic chromosome pairs were divided into seven categories (A–G), determined by size and by centromere position. The centromere position establishes the arm lengths on either side of the centromere (see Figure 2.3). This scheme is called the **Denver classification system** because it was established at a conference held in that city in 1960. Since then, revised procedures for staining mitotic chromosomes have been devised to produce **chromosome banding.** The various patterns observed allow all 23 human chromosomes to be distinguished from one another. We will discuss this topic later in this chapter.

The study of mitotic chromosomes and karyotypes has been important in several genetics disciplines. The way in which chromatin folds up into mitotic chromosomes is of interest to students of structure. **Chromosome mutations** or **aberrations** (abnormal numbers or arrangements of chromosomes) may be detected in karyotypes and linked to human disorders. The evolutionary relationships between different species may be gauged by similarities between chromosome number and morphology.

Polytene Chromosomes

Studies of other unique chromosomes with the light microscope have yielded additional information about gross chromosome morphology. Interphase cells from a variety of organisms contain giant **polytene chromosomes.** They are found in various dipteran larval cells (salivary, midgut, rectal, and malpighian excretory tubules) and in several species of protozoans and plants. The large amount of information obtained from studies of these chromosomes has provided a model system for more recent investigations. Such structures were first observed by Balbiani in 1881 and are illustrated in Figure 11.5.

The large size and distinctiveness of such chromosomes result from the many DNA strands that compose them. Many replication cycles occur without strand separation or cytoplasmic division. As replication proceeds, the homologues remain paired, ultimately creating chromosomes having 1000 to 5000 DNA strands remaining in parallel register with one another. In conjunction with associated chromosomal proteins, giant chromosomes are created. Their parts may be more easily studied than those comprising a haploid interphase chromosome because of their dramatic visibility.

When polytene chromosomes are observed under the light microscope, each is seen to be composed of a linear series of alternating **bands** and **interbands** (see insert in Figure 11.5). The banding pattern is distinctive for each homologous pair of chromosomes in any given species. Individual bands are sometimes called **chromomeres,** a more generalized term describing lateral condensations of material along the axis of a chromosome. Each polytene chromosome is 200 to 600 μm long. Each of the 1000 or more fibers is a continuous DNA double helix.

The genetic information, arranged as a linear series of bands, is contained in these fibers. In order for any particular gene, represented by all or part of a band, to become active in transcription, a localized uncoiling usually occurs. As the DNA unreels, the enzymes essential to transcription have access to it. A micrograph and a diagrammatic interpretation of this description are shown in Figure 11.6. Such localized sites of genetic activity are referred to as **puffs** because of their appearance.

That puffs are visible manifestations of transcription is evidenced by their high rate of incorporation of radioactively labeled uridine into RNA as assayed by autoradiography. Bands that are not extended into puffs incorporate much less uridine or none at all. During development in insects such as *Drosophila* and the midge fly *Chironomus*, the study of bands reveals differential gene activity. A characteristic pattern of band

FIGURE 11.5

Giant-polytene chromosomes derived from salivary gland cells of *Drosophila melanogaster*. The X chromosome, the right and left arms of chromosomes 2 and 3, and the small chromosome 4 are seen projecting from the chromocenter, where the centromeres of all chromosomes are aggregated. The insert depicts bands and interband regions along the axis of each chromosome. [With permission from Lefevre, 1976. Copyright: Academic Press Inc. (London) Ltd.]

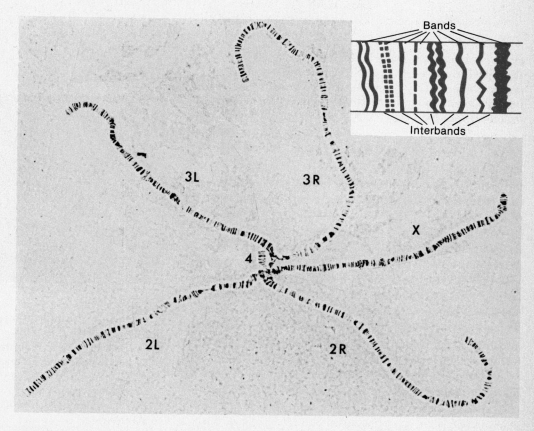

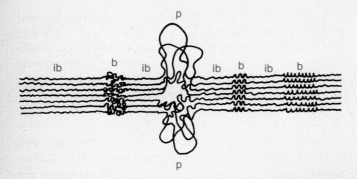

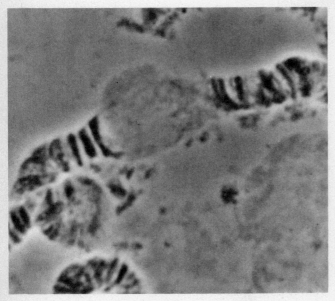

FIGURE 11.6

Photograph of a puff within a polytene chromosome from *Drosophila melanogaster.* The schematic representation depicts the uncoiling of strands within a band (b) region to produce a puff (p) in polytene chromosomes. Interband regions are labeled ib.

formation, equated with gene activation, occurs during development. This topic is pursued in more detail in Chapter 21.

Until recently, it has been unclear what, if anything, is different about the DNA composing bands compared with interbands. The bands are densely staining and vary in their axial length from 0.05 to 0.5 μm. They appear to be the obvious source of genetic expression since puffs emanate from them. The interband regions, on the other hand, are genetically quiescent and have been thought by some geneticists to contain fewer strands of DNA, lowering their polytene level compared with adjacent bands. A number of interesting observations have somewhat clarified the matter.

Ann and Pierre Spierer have examined closely the DNA content of a contiguous 315-kb (315,000 base

pairs) region of the *Drosophila* genome. This region spans 15 bands and adjoining interbands on chromosome 3. Their data support the concept that the degree of replication and level of polyteny do not vary in bands of different size or in interband regions. Thus, in spite of a quite varied degree of compaction, the number of DNA strands is believed to be constant in different regions of the polytene chromosome.

Another observation of interest occurred in 1981 and involved studies of Z-DNA by Alfred Nordheim, Alexander Rich, and their colleagues. They prepared highly specific fluorescent antibodies against Z-DNA and exposed acid-fixed *Drosophila* polytene chromosomes to such a preparation. Observations revealed that most antibody binding occurred in interband regions (Figure 11.7) although several bands and puffs also appeared to contain Z-DNA. Subsequent studies in 1983 by Ronald Hill and David Stollar involved unfixed "native" chromosomes and revealed no binding of anti-Z-DNA antibodies to any part of the chromosome. However, when acetic acid, the normal fixative, was added to the chromosome preparation, Z-DNA antibody binding initially occurred in the interband regions. After prolonged exposure to acid, binding also occurred in the band regions. The acid fixative is known to remove chromosomal proteins and would appear to be unmasking Z-DNA or inducing the formation of this conformation. However, the precise interpretation of the presence and role, if any, of Z-DNA in polytene chromosomes remains unclear.

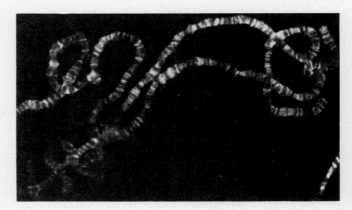

FIGURE 11.7

Fluorescent micrograph showing how antibodies prepared against Z-DNA bind to polytene chromosomes of *Drosophila melanogaster.* Specific binding has occurred only in the areas that appear white. These areas are exclusively interband regions. (From Nordheim et al., 1981. Reprinted by permission from *Nature*, Vol. 294, pp. 417-22. Copyright © 1981 Macmillan Journals Limited. Photo supplied by Alexander Rich.)

Cytological Mapping

By the early 1930s **genetic** or **linkage maps,** which gave the sequence and distances between many genes on the chromosomes of *Drosophila*, had been worked out (see Chapter 6). In that decade, these linkage maps were successfully correlated with the order of bands seen on the polytene chromosomes of *Drosophila*. This was accomplished by studying the altered banding patterns produced by flies with **chromosomal aberrations.** Aberrations (see Chapter 12) include duplications, deletions, and other rearrangements of the chromosomal material. In flies that demonstrated both a mutant phenotype and an aberration detectable in polytene chromosomes, the mutation was assigned to a region or band along the chromosome. In this way, a **cytological map** of many mutations on the *Drosophila* polytene chromosomes was compiled. A part of the *Drosophila* cytological map and its correlation with genetic loci identified by linkage studies are shown in Figure 11.8. Both the linkage and cytological maps are in agreement on the sequence of genes. However, the relative distance between any two loci as determined by linkage mapping and cytological techniques differs considerably in many instances.

For many years, it was postulated that each band on the *Drosophila* chromosomes represented one gene. Geneticists counted approximately 5000 bands on the four chromosomes of this species, and this number seemed to agree quite well with the estimated number of genes in *Drosophila*. More current information, however, has led geneticists to question whether there are not, in fact, more than 5000 genes. For example, there is sufficient DNA on the X chromosome alone to represent approximately 30,000 genes if each is composed of 1000 nucleotide pairs. Each of the 1000 chromomeres or bands on the X chromosome could represent as many as 30 genes.

Burke Judd and colleagues have extensively examined a small region of the *Drosophila* X chromosome consisting of about 15 bands. They induced and analyzed over 100 mutations mapped in this general region. Careful analysis revealed that the mutations fell into **complementation groups,** the number of which was equivalent to the number of bands in the region studied. A complementation group (see Chapter 19) is the equivalent of one functional genetic unit, or gene. When each complementation group was assigned to a particular band, it was concluded that a single band probably contains only a single gene. If this is true, and if each band contains sufficient DNA to code for 20 to 30 genes, what role is played by the remaining DNA?

Judd's work in the early 1970s has been supported more recently by the findings of Pierre Spierer, Welcome Bender, and David Hogness, who have investigated regions on chromosome 3 of *Drosophila*. Their work again suggests that the number of complementation groups is approximately equal to the number of bands. There are at least two possible explanations for the excess DNA present in each band. They are not mutually exclusive. First, perhaps the eukaryotic gene contains much more DNA than predicted by the size of an average protein. If so, what is the role of noncoding DNA contained within a gene, and what is its fate during transcription and translation? On the other hand, perhaps much of eukaryotic DNA is noncoding and interspersed between classical genes. If so, what is its role in genetic processes? As we will see, both explanations have been shown to be correct. In the former case, genes have been shown to contain nucleotide sequences that are transcribed but spliced out of mRNA before translation. (This topic is discussed in detail in Chapters 18 and 19.) In the second case, as we will see later in this chapter, noncoding repetitive DNA is interspersed between genes, occupying a variable portion of eukaryotic genomes.

Studies of polytene chromosomes have provided valuable insights into other areas of genetics. For example, polytene chromosomes are used in studies of chromosomal aberrations (see Chapter 12), developmental genetics (see Chapter 22), and evolution-population genetics (see Chapters 25 and 26).

Lampbrush Chromosomes

Another type of chromosome that has provided insights into chromosomal structure is the **lampbrush chromosome.** It was given this name because it is similar in appearance to the brushes used to clean lamp chimneys in centuries past. Lampbrush chromosomes were first discovered in 1892 in sharks and are now known to be characteristic of vertebrate oocytes as well as spermatocytes of some insects. Most experimental work has been done with material taken from amphibian oocytes.

These unique chromosomes are easily isolated from oocytes in the diplotene stage of the first prophase of meiosis, where they are active in directing the metabolic activities of the developing cell. The homologues are seen as synapsed pairs held together by chiasmata, but instead of condensing, as do most meiotic chromosomes, the lampbrush chromosomes are often extended to lengths of 500 to 800 μm. Later in meiosis they revert to their normal length of 15 to 20 μm. Thus, lampbrush chromosomes are interpreted as uncoiled and unfolded versions of the normal meiotic chromosomes.

Figure 11.9 shows a diagrammatic interpretation of these structures as well as a variety of micrographs

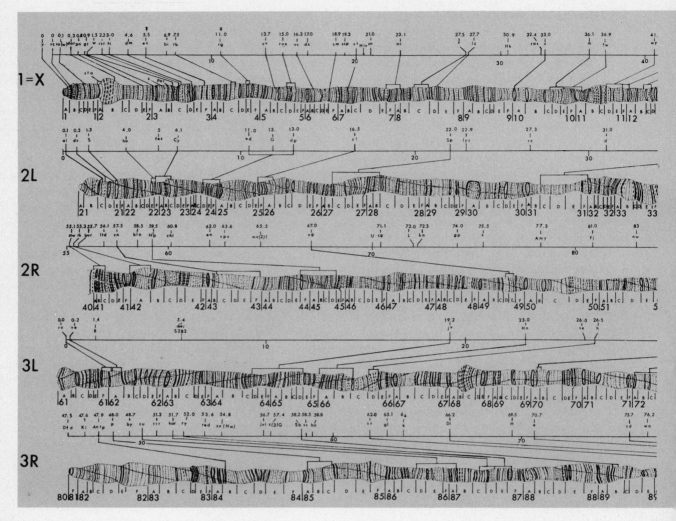

FIGURE 11.8
Comparison of the cytological map and the genetic map of the chromosomes of *Drosophila melanogaster*. The large designations at the left indicate the chromosome number and the right or left arm. The cytological map has been devised from the polytene chromosomes of larval salivary gland cells. It is divided into 102 major divisions, each of which is subdivided into six lettered subdivisions (A–F). The genetic map and many gene sites are shown above each cytological map. (From King, 1975, Fig. 1.)

obtained in different ways. Collectively, they provide significant insights into the morphology of these chromosomes. In part (a), the meiotic configuration is depicted. The linear axis of each member of the bivalent structure contains large numbers of repeating condensations. As in polytene chromosomes, the more general term, chromomere, describes each condensation. Each chromomere appears to support a pair of **lateral loops,** which give the chromosome its distinctive appearance. In part (b), a light micrograph from the oocyte nucleus of an amphibian is seen under phase contrast optics. In part (c), an extended axis and ad-

jacent loops are seen using the scanning electron microscope (SEM).

The structure and the functional activities of lampbrush chromosomes have been studied with enzymatic digestion and radioactive tracer techniques. Although considerable amounts of RNA and protein are associated with the chromosome, neither RNase nor protease enzymes disrupt the linear integrity of either the axis or the loops. It has been concluded, therefore, that DNA provides the skeletal structure of the chromosome, which is associated with protein and RNA.

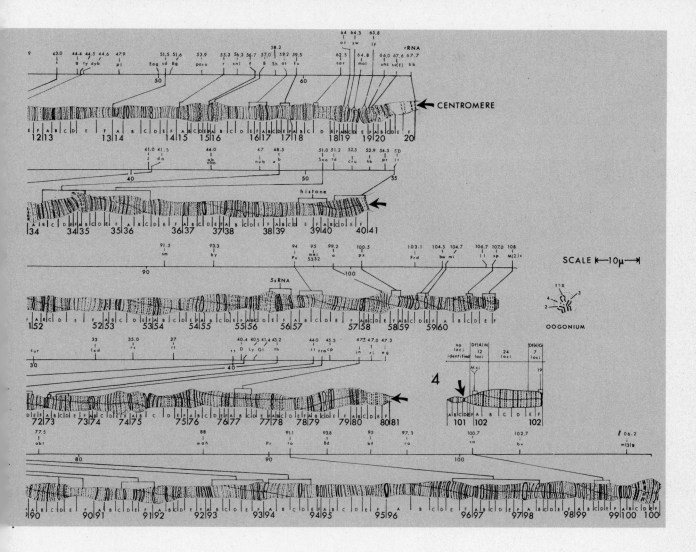

Part (d) of Figure 11.9 presents another SEM, but at a much higher magnification. This micrograph provides a detailed view of one chromomere and the loop emanating from it. Clearly, the chromomere is a coiled up chromosome fiber containing primarily a DNA molecule and associated protein, referred to as DNP (for deoxyribonucleoprotein). Finally, in part (e), a transmission electron micrograph (TEM) is presented, showing one complete loop and the matrix surrounding it. Each loop is thought to be composed of one double helix, and the central axis is made up of two DNA helices. This hypothesis is consistent with the belief that each chromosome is composed of a pair of sister chromatids. When oocytes are incubated in the presence of radioactive RNA precursors, the sites of transcription may be determined by autoradiographic analysis. Such studies reveal that the loops are active in the synthesis of RNA. The lampbrush loops, similar in a way to puffs in polytene chromosomes, represent DNA that has been reeled out from the central chromomere axis during transcription. As with polytene chromosomes, the study of lampbrush chromosomes has provided many insights into the arrangement and function of the genetic information.

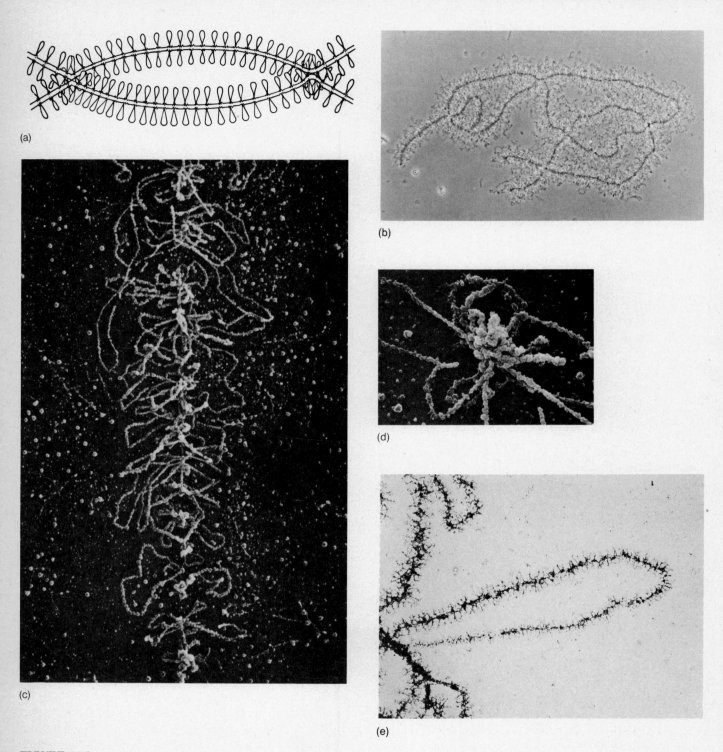

(a)

(b)

(c)

(d)

(e)

FIGURE 11.9

Lampbrush chromosomes. (a) Schematic drawing of a single lampbrush chromosome.
(b) Photomicrograph from the oocyte nucleus of the newt *Notophthalmus viridescens* viewed
under phase contrast. (Courtesy of Joe Gall.) (c) Scanning electron micrograph of a lampbrush
chromosome of the newt, *Pleurodeles.* (Courtesy of Nicole Angelier.) (d) Scanning electron
micrograph of a chromomere and associated loops of a lampbrush chromosome of
Pleurodeles. (Courtesy of Nicole Angelier.) (e) Transmission electron micrograph of a loop
extending from the axis of a lampbrush chromosome and showing the RNA and protein
matrix covering each loop. (From O. L. Miller, 1965, p. 79.)

EUKARYOTIC CHROMOSOMES: MOLECULAR ORGANIZATION

Early studies of the structure of eukaryotic genetic material concentrated on intact chromosomes, preferably large ones, because of the limitation of light microscopy. The development of techniques for biochemical analysis, as well as the examination of relatively intact eukaryotic chromatin and mitotic figures under the electron microscope, have greatly enhanced our understanding of chromosome structure. Recently, a combination of these techniques has been used to elucidate the organization of genetic material contained in the mitochondrion and the chloroplast.

Chromatin Composition and Structure

As established earlier, the genetic material of viruses consists of naked strands of DNA or RNA. In bacteria, the chromosome also consists of naked DNA, even though DNA-binding proteins have been isolated and studied. Their precise role remains unclear. In eukaryotes, the structure of the chromosome is much more complex. A substantial amount of protein is associated with the chromosomal DNA in all phases of the eukaryotic cell cycle. Thus, the eukaryotic genetic material is composed of **nucleoprotein;** such material is generally referred to as **chromatin,** particularly during interphase, when it is uncoiled. The associated proteins are divided into basic, positively charged **histones** and less positively charged **nonhistones.** The histones seem to be intimately associated with chromatin structure, while the nonhistone proteins are thought to play other roles, including genetic regulation. Despite the presence of protein in the chromatin, the DNA component is universally believed to be the part that stores genetic information.

Research in the past several years has made it possible to develop a general model for chromatin structure (Figure 11.10). This model is based on the assumption that chromatin fibers, composed of DNA

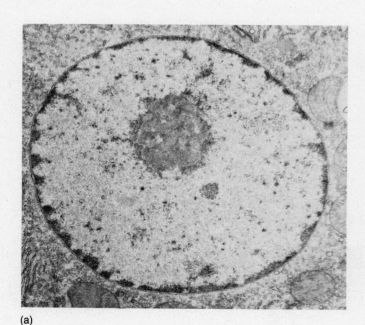

(a)

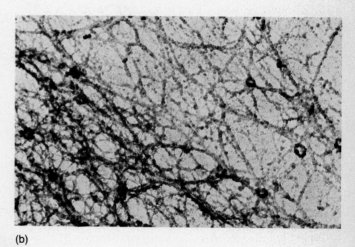

(b)

(c)

FIGURE 11.10

Eukaryotic chromatin seen in different ways. (a) Electron micrograph of a thin section cut through a mouse liver nucleus. Because the fibers have been sectioned, they appear as the lighter specks. The dense circular structure is the nucleolus. (b) Whole-mount electron micrograph of chromatin fibers derived from a mouse liver nucleus. The circular structures are nuclear pores. (c) Model depicting the supercoiling that is thought to occur and cause condensation of the chromatin fiber. (Micrographs courtesy of David E. Comings and Tadashi A. Okada.)

and protein, must undergo extensive coiling and folding in order to fit into the cell nucleus.

Of the proteins associated with DNA, the histones are now believed to be essential to the structural integrity of chromatin. Histones contain large amounts of the positively charged amino acids lysine and arginine. Thus, histones can bond electrostatically to the negatively charged phosphate groups of nucleotides. Recall that a similar interaction has been proposed for several bacterial proteins, including those designated HU and H. There are five main types of histones (Table 11.2); together in chromatin, they exist in a 1:1 mass ratio with DNA.

X-ray diffraction studies show that histones play an important role in chromatin structure. Under such investigation, chromatin produces regularly spaced diffraction rings, suggesting that repeating structural units occur along the chromatin axis. If the histone molecules are chemically removed from chromatin, the regularity of the diffraction rings disappears. In 1970, John Pardon and his colleagues proposed that the DNA-histone complex would conform to the observed X-ray diffraction pattern if it were twisted into a supercoiled helix, an idea suggested earlier by other workers. Such a supercoiled complex is illustrated in Figure 11.10(c). In such a model, the basic DNA-histone fiber is approximately 100 Å in diameter. Supercoiling can increase this dimension to 300 Å.

This basic model for chromatin has been refined recently. Between 1973 and 1977, observations from a large number of workers in many laboratories served as the basis for the development of a general model for chromatin. Together these observations may well be regarded as a significant breakthrough in this discipline:

1 Digestion of chromatin by certain endonuclease enzymes, such as **micrococcal nuclease,** yields DNA fragments that are approximately 200 base pairs in length or multiples thereof. Such work was reported by Dean Hewish and Leigh Burgoyne in 1973 ånd subsequently by others. These data suggest that enzymatic digestion is not random, for if it were, we would expect a wide range of fragment sizes. They also suggest that chromatin consists of repeating units, each of which is protected from enzymatic cleavage except where any two units join each other. Thus, the area between all units is attacked and cleaved by the nuclease. Such information agrees with the results of X-ray diffraction studies, which suggest regular spacing arrangements in chromatin.

2 In 1974, Ada and Donald Olins reported on electron microscopic observations of chromatin prepared by methods different from those of most earlier work. Two such micrographs appear in Figure 11.11; they show chromatin fibers composed of linear arrays of spherical particles. The particles occur regularly along the axis of a chromatin strand and resemble beads on a string. These particles are now referred to as ***v*-bodies** or **nucleosomes** (*v* is the Greek letter *nu*). These findings conform nicely to the earlier proposal, which suggested the existence of repeating units and which is supported by both the X-ray diffraction studies and the analysis of enzymatic digestion of chromatin.

3 In 1975, Roger Kornberg published the results of his study of the precise interactions of histone molecules and DNA in chromatin. His work showed that histones H2A, H2B, H3, and H4 occur as two types of tetramers: $(H2A)_2 \cdot (H2B)_2$, and $(H3)_2 \cdot (H4)_2$. He suggested that each repeating nucleosome unit consists of one of each tetramer, which together interact with about 200 base pairs of DNA. These data correlate well with the two previous observations and provide the basis for a model that explains the interaction of histone and DNA in chromatin.

4 Between 1975 and 1977, a more accurate understanding of the nucleosome emerged. Nuclease digestion of chromatin from diverse tissues and species was shown to yield wide variation in the number of DNA base pairs in the nucleosome unit. However, when digestion time is extended, DNA is removed from both the entering and exiting strands, creating a **nucleosome core particle** consisting of 146 base pairs. This number is consistent in all organisms studied.

. The DNA lost in this prolonged digestion—that is, the difference in length between the nucleosome and the core particle—is responsible for link-

TABLE 11.2
Categories and properties of histone proteins.

Histone Type	Lysine-Arginine Content	Molecular Weight (daltons)
H1	Lysine-rich	21,000
H2A	Slightly lysine-rich	14,500
H2B	Slightly lysine-rich	13,700
H3	Arginine-rich	15,300
H4	Arginine-rich	11,300

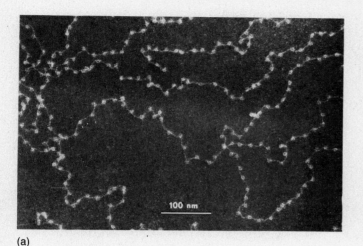

(a)

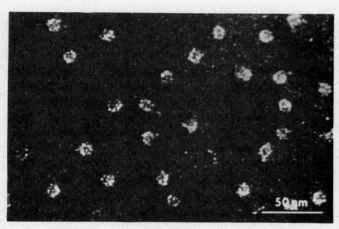

(b)

FIGURE 11.11
(a) Dark-field electron micrograph of nucleosomes present in chromatin derived from a chicken erythrocyte nucleus.
(b) Dark-field electron micrograph of nucleosomes produced by micrococcal nuclease digestion. (From Olins and Olins, 1978, Figs. 1 and 4.)

ing nucleosomes together. This **linker DNA** is associated with histone H1. Extended nuclease digestion of the nucleosome first creates an intermediate structure containing 166 base pairs. Then, following further digestion, during which the conversion of this intermediate form to the 146-base-pair core particle occurs, histone H1 falls off. This observation provides evidence that histone H1 is located very near the initial portion of the DNA of the near core particle but not as an integral part of it.

5 In 1977 John T. Finch, Aaron Klug, and other investigators, who had performed X-ray and neutron scattering analysis of crystallized core particles,

proposed a model of this core particle (Figure 11.12). In this model, the 146-base-pair length of DNA is a secondary helix surrounding the core of eight histone molecules. The superhelical DNA does not quite complete two full turns, yielding a flattened wedge shape of the dimensions 57 Å × 110 Å × 110 Å. Because of this shape, the core particle of the nucleosome was called a **platysome.** Although this model does not establish the precise arrangement of histones within the helix, it provides a close approximation of the basic unit making up chromatin.

6 The extensive investigation of nucleosomes now provides the basis for predicting how the initial stages of packaging of chromatin occur (Figure 11.13). In its most extended state under the electron microscope, the chromatin fiber is about 100 Å in diameter, a size consistent with the longer dimension of the ellipsoidal nucleosome. The nucleosome itself condenses the DNA of chromatin at least tenfold. This **packing ratio** is calculated by dividing the length of DNA in 200 base pairs (3.4 Å × 200 = 6800 Å) by the height of the nucleosome (60 Å).

The 100-Å fiber of chromatin is apparently further condensed into a fiber 300 Å in diameter, as viewed under the electron microscope. This transition has been studied carefully and can be achieved *in vitro* by increasing the ionic strength of the surrounding medium. The H1 histone is essential to this condensation since, in its absence, the transition cannot occur. The 300-Å fiber appears to consist of five or six nucleosomes coiled closely together. This structure has been called a **solenoid**

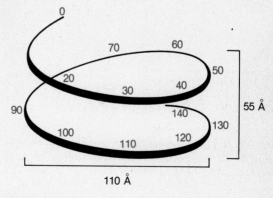

FIGURE 11.12
Representation of the DNA core of the nucleosome, called a platysome. The numbers indicate the relative positions of the 146 nucleotides contained in the core particle. (After Finch et al., 1977, Fig. 8.)

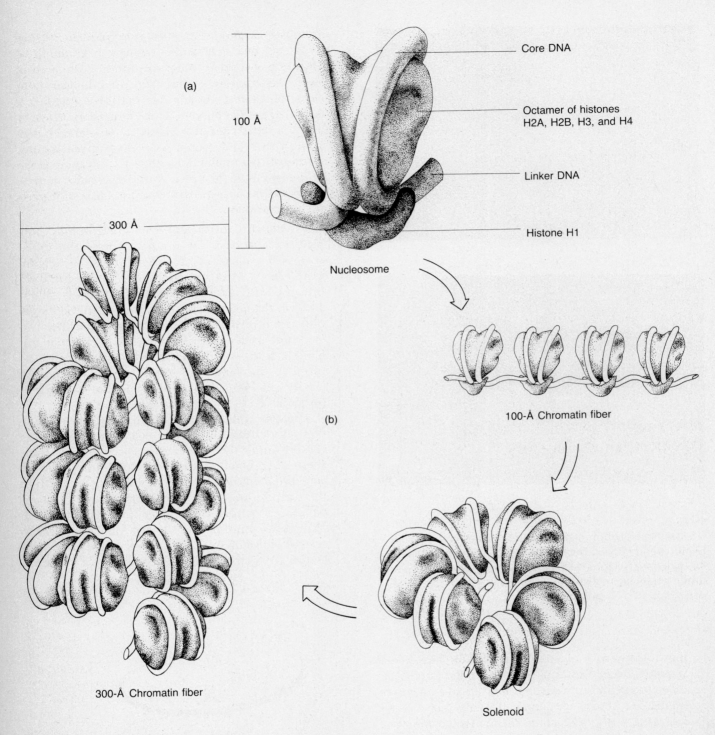

(a)

100 Å

300 Å

Core DNA

Octamer of histones
H2A, H2B, H3, and H4

Linker DNA

Histone H1

Nucleosome

(b)

100-Å Chromatin fiber

Solenoid

300-Å Chromatin fiber

FIGURE 11.13
(a) General model of the association of histones and DNA in the nucleosome. (b) Schematic
illustration of the way in which the chromatin fiber may be coiled into a more condensed
structure. (Courtesy of Vickie Brewster.)

and its formation increases the initial condensation by another factor of five, yielding an overall packing ratio of about 50.

These 100- and 300-Å fibers are probably characteristic of interphase chromatin. The larger fiber can undoubtedly be further bent, coiled, and packed as even greater condensation occurs during the formation of mitotic chromosomes. There, a packing ratio of approximately 5000 must be achieved.

EUKARYOTIC CHROMOSOMES: ELECTRON MICROSCOPIC OBSERVATIONS

We have so far considered what has been learned about the eukaryotic chromosome from studies involving light microscopy and from the molecular analysis of the chromatin fiber. In this section, we will examine three aspects of the eukaryotic chromosome that have been initially derived from electron microscopic observations.

The Folded-Fiber Model for Mitotic Chromosomes

As we have just discussed, during the transition from interphase to mitosis, chromatin must undergo considerable compression and condensation. It has been estimated that a 5000-fold contraction in the length of DNA occurs. This process must be extremely precise, given the highly ordered nature and consistent appearance of the mitotic chromosomes in all organisms. Electron microscopic observations of mitotic chromosomes have provided an excellent overview of the intact chromosome following condensation. A whole-mount electron micrograph of a human mitotic chromosome is shown in Figure 11.14(a). In areas of greatest spreading, individual fibers similar to those seen in interphase chromatin are apparent. Very few fiber ends seem to be present. In some cases, none can be seen. Instead, individual fibers always seem to loop back into the interior. Such fibers are obviously twisted and coiled around one another, forming the regular pattern of the mitotic chromosome. The fibers are so tightly packed that, unless uncoiled slightly, the chromosome has a uniform or homogenous appearance throughout much of its length. You should compare this illustration with the light microscopic view of human chromosomes shown in Figure 2.3.

Observations of mitotic chromosomes in varying states of coiling led Ernest DuPraw to postulate the **folded-fiber model,** illustrated in Figure 11.14(c). During metaphase, each chromosome consists of two sister chromatids joined at the centromeric region. Each arm of the chromatid appears to consist of a single fiber wound up much like a skein of yarn. The fiber is composed of double-stranded DNA and protein tightly coiled together. When these intact chromosomes are viewed under the scanning electron microscope, a pattern very similar to the model emerges [Figure 11.14(b)]. An orderly coiling-twisting-condensing process appears to be involved in the transition of the interphase chromatin to the more condensed, genetically inert mitotic chromosomes. Certainly, the more recent findings involving nucleosome and solenoid structures provide a basis for understanding the initial coiling and packing. While it is not yet clear exactly how further packing is accomplished, the model provides an excellent illustration of what must be achieved.

Synaptonemal Complex Organization

The electron microscope has also been used to visualize an additional ultrastructural component of the chromosome seen only in cells undergoing meiosis. This structure, mentioned in Chapter 2 (Figure 2.9), is intimately associated with synapsed homologues during the pachytene stage of the first meiotic prophase and is called the **synaptonemal complex.*** In 1956, Montrose Moses observed this complex in spermatocytes of crayfish, and Don Fawcett saw it in pigeon and human spermatocytes. Because there was not yet any satisfactory explanation of the mechanism of synapsis or of crossing over and chiasmata formation, many researchers became interested in this structure. With few exceptions, the ensuing studies revealed the synaptonemal complex to be present in most plant and animal cells visualized during meiosis.

Figure 2.9 is an electron micrograph of the synaptonemal complex. It is composed primarily of a triplet set of parallel strands. The **central element** of this tripartite structure is usually less dense and thinner (100–150 Å) than the two identical outer elements (500 Å). The outer structures, called **axial** or **lateral elements,** are distinctly and intimately associated

---

*An alternate spelling of this term is *synaptinemal complex.*

FIGURE 11.14
(a) Whole-mount transmission electron micrograph of a human mitotic D-group chromosome. (Courtesy of Walter F. Engler.)
(b) Scanning electron micrograph of a human mitotic chromosome. (From Golomb and Bahr, 1971, p. 1025. Copyright 1971 by the American Association for the Advancement of Science.)
(c) Depiction of the coiling and folding of the chromatin fiber into the structure characteristic of a mitotic chromosome. (Redrawn from DuPraw, 1970.)

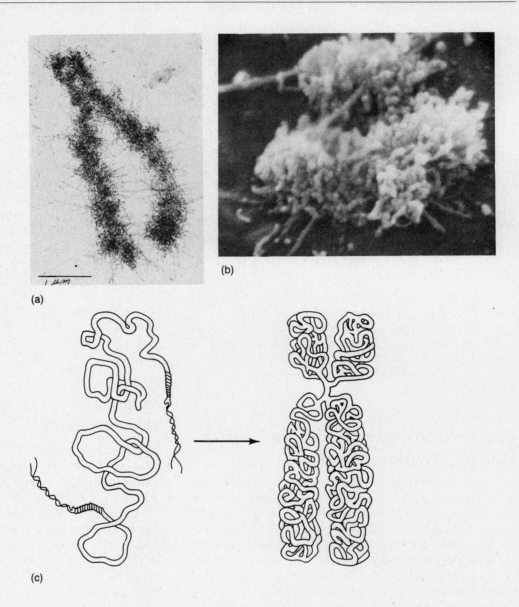

1 μm

(a)

(b)

(c)

with the chromatin of the synapsed homologues on either side. Selective staining has revealed that these axial elements contain protein, DNA and perhaps RNA. Some DNA fibrils traverse the axial elements, making connections with the central element, which is composed primarily of protein. Figure 11.15 provides a diagrammatic interpretation of the electron micrograph and the above description.

The formation of the complex is initiated prior to the pachytene stage. As early as leptonema of the first meiotic prophase, lateral elements are seen in association with sister chromatids. Homologues have yet to associate with one another and are randomly dispersed in the nucleus. By the next stage, zygonema, homologous chromosomes begin to align with one another, but remain distinctly apart by some 300 nm.

Then, during pachynema, an intimate association between homologues occurs as formation of the complex is completed. In some diploid organisms, this occurs in a zipperlike fashion beginning with the telomeres, which may be attached to the nuclear envelope.

It is now agreed that the synaptonemal complex is the vehicle for the pairing of homologues and their subsequent segregation during meiosis. However, synapsis can occur in certain cases where no synaptonemal complexes are formed. Thus, it is possible that the function of this structure may be more extensive than just its involvement in the formation of bivalents.

In certain instances where no synaptonemal complexes are formed during meiosis, synapsis is not so complete and crossing over is reduced or eliminated.

FIGURE 11.15
(a) Drawing of the synaptonemal complex (sc) observed during meiosis. Parts (b) and (c) show a schematic model of the interaction of theoretical components of this organelle with the chromatin fiber of the bivalents. (ce, central element; le, lateral element; c, chromatin.) (From King, 1970, pp. 136 and 138.)

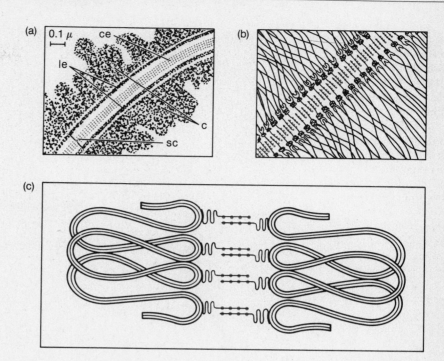

For example, in *Drosophila melanogaster* meiotic crossing over rarely, if ever, occurs in males, and it is not observed in females homozygous for the third chromosomal mutant gene *c(3)G*. Synaptonemal complexes are not usually observed in either case. This information suggests that the synaptonemal complex must be present in order for chiasmata to form and crossing over to occur. In addition, it has been hypothesized that the genetic product coded for at the *c(3)G* locus may in some way be involved in the biosynthesis of the complex.

There may be other, yet undiscovered molecular components that are essential to the function of the synaptonemal complex. Even so, the discovery of the complex is significant to the field of genetics. If the complex is critical to synapsis during meiosis, then all sexual reproduction is dependent on it. If the complex is essential to crossing over, then it is largely responsible for eukaryotic genetic recombination, enhancing the role of that process in evolution.

DNA of Mitochondria and Chloroplasts

The results of a number of experiments have demonstrated that both **mitochondria** and **chloroplasts** contain their own genetic information. It was observed that certain mutations in yeast, other fungi, and plants correlated with altered functioning of mitochondria or the chloroplasts. These mutations were not always inherited in the ratios expected by Mendelian principles. Often these traits seemed to follow a **maternal mode of inheritance;** that is, the mutation would be passed on only if it were possessed by the female parent. When it became clear that both mitochondria and chloroplasts are self-replicating, it seemed logical that these organelles might contain their own genetic information.

Later autoradiographic studies showed that chloroplasts and mitochondria could incorporate radioactive DNA precursors. Thus, geneticists set out to look for DNA in these organelles with the electron microscope. Electron microscopists not only documented the presence of DNA in both organelles, but they also saw DNA in a form quite unlike that seen in the chromosomes of eukaryotic cells.

Mitochondria are smaller than chloroplasts and contain less DNA. **Mitochondrial DNA (mtDNA)** has a form almost identical to the DNA seen in some viruses. It is a circular fiber with a diameter roughly equivalent to that predicted for a double helix. Figure 11.16(a) shows a molecule of mtDNA.

In vertebrates there are 5 to 10 mtDNA molecules per organelle, each about 5 μm in length. The DNA consists of some 16,000 base pairs (16 kb), enough to encode 15 to 20 gene products. In protists such as *Euglena* and higher plants such as the sweet pea, mtDNA is 30 to 50 μm long and contains 100 to 150 kb. As we will learn in Chapter 15, many of these products, in-

FIGURE 11.16
(a) Electron micrograph of mitochondrial DNA (mtDNA). [From Nass, 1970, Plate Ia. Copyright: Academic Press Inc. (London) Ltd.]
(b) Electron micrograph of chloroplast DNA derived from corn. (Courtesy of Richard Kolodnar.)

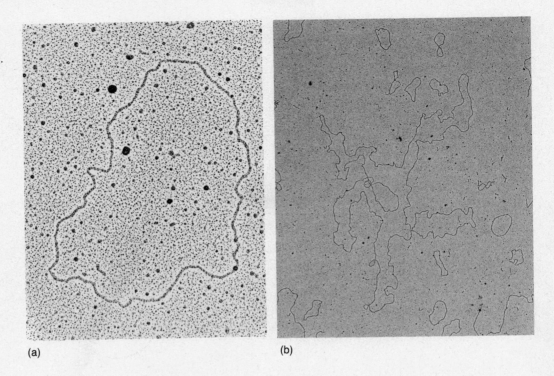

(a)   (b)

cluding respiratory enzymes, ribosomal proteins, transfer RNAs, and ribosomal RNA, have been identified.

DNA released from chloroplasts (**cpDNA**) is consistently longer than most mtDNA, being 40 to 60 μm and containing 130 to 200 kb. There are typically 30 or more such molecules per organelle. Like mtDNA, cpDNA is circular in most, if not all, plants. Figure 11.16(b) shows an electron micrograph of DNA released from a chloroplast. Its increased size compared with mtDNA is readily apparent.

Because mitochondrial and chloroplast DNA are similar in appearance to bacterial DNA, a theory of the origin of these organelles has been proposed. This hypothesis, called the **endosymbiont theory,** states that mitochondria and chloroplasts may have originated as bacterial-like particles that became incorporated into eukaryotic cells. In the evolution of this symbiotic relationship, the particles lost their ability to function independently. We will discuss the genetics of mitochondria and chloroplasts in more detail in Chapter 15.

HETEROCHROMATIN

All recent evidence supports the concept that the DNA of each eukaryotic chromosome consists of one continuous double-helical fiber along its entire length. A continuous fiber is the basis of the **unineme model** of DNA within a chromosome and is in keeping with what we have seen in viral and bacterial chromosomes. This finding might lead one to suspect that the chromatin fiber of each unit would demonstrate structural uniformity along its length. However, in the early part of this century, it was observed that some parts of the fiber remain condensed and stain deeply during interphase, but most do not. In 1928, E. Heitz coined the terms **heterochromatin** and **euchromatin** to describe the parts of chromosomes that remain condensed and those that are uncoiled, respectively.

Subsequent investigation has revealed a number of characteristics of heterochromatin that distinguish it from euchromatin. Heterochromatin areas are genetically inactive either because they lack genes or contain genes that are repressed. Also, heterochromatin replicates later during the S phase of the cell cycle than does euchromatin. This characteristic has been established using autoradiography and is thought to be a secondary feature of the extreme condensation. The discovery of heterochromatin provided the first clues that parts of eukaryotic chromosomes may not be related to storage of information. Instead, it is believed that some chromosome regions may be involved in the maintenance of the chromosome's structural integrity.

Heterochromatin is found in the genetic material of every species. Early cytological studies showed that areas closely adjacent to the centromeres were composed of heterochromatin. The ends of chromosomes, called **telomeres,** are also considered to be heterochromatic. In certain cells, notably the polytene-chromosome-containing salivary gland cells of *Drosophila*, the heterochromatin of centromeres may clump together to form a **chromocenter** (see Figure 11.5). This observation suggests that some molecular affinity involved in heterochromatin is responsible for this adherence.

Heterochromatin has been subdivided into two types. **Constitutive heterochromatin** is present at homologous sites on pairs of chromosomes and always exists in a genetically inert state. This form has been shown to contain DNA sequences that are not organized into genes. Telomeres and centromeric regions are examples of constitutive heterochromatin.

The second type of heterochromatin, called **facultative,** is chromatin with the faculty, or potential, to become heterochromatic. For example, one of the two X chromosomes in mammalian females exhibits this property when a **Barr body** is formed (see Chapter 5). In certain species, such as the mealy bug, one entire haploid set of chromosomes becomes heterochromatic. In both of these examples, only one member of a homologous pair or set of chromosomes becomes heterochromatic, in contrast with constitutive heterochromatin. The Y chromosome of many species, lacking a true homologue, is classified as facultative heterochromatin. It remains condensed during interphase, and the majority of its length is genetically inert.

Facultative heterochromatin may contain genetic information, but it is not used by the organism once it becomes condensed. It is believed that the extreme condensation and coiling of heterochromatic material physically precludes genetic activity in the same way that the condensation of mitotic and meiotic chromosomes renders them inactive.

When certain heterochromatic areas from one chromosome are translocated to a new site on the same or another nonhomologous chromosome, genetically active areas sometimes become genetically inert if they now lie adjacent to the translocated heterochromatin. This influence on existing euchromatin is one example of what is more generally referred to as a **position effect.** That is, the position of a gene or groups of genes relative to all other genetic material may affect their expression. This topic is discussed more extensively in Chapter 12.

Heterochromatin, Satellite DNA, and Repetitive DNA

Chapter 9 and Appendix A describe techniques useful in the analysis of DNA. The use of two of these—sedimentation equilibrium centrifugation and reassocia-

tion kinetics—has provided important information about the nature of DNA in chromatin. Specifically, the DNA of heterochromatin and euchromatin differs in molecular composition. With other techniques, functional differences between the two forms of chromatin have been determined. The nucleotide composition of the DNA of a particular species is reflected in its density, which can be measured with sedimentation equilibrium centrifugation. When the DNA of any eukaryotic species is analyzed in this way, the majority of it shows up as a single main peak or band of uniform density. However, one or more satellite peaks represent DNA that differs slightly in nucleotide composition and density from main-band DNA. This minor component, called **satellite DNA,** seldom represents more than 10 percent of the total DNA. A profile of main-band and satellite DNA from the mouse is shown in Figure 11.17. It should be noted that bacteria, as representative prokaryotes, contain only main-band DNA.

The significance of satellite DNA remained an enigma until the mid 1960s, when the techniques for measuring the reassociation kinetics of DNA became available. These techniques, developed by Roy Britten and David Kohne, allowed geneticists to determine the rate of reassociation of fragmented DNA. Britten and Kohne showed that when complementary strands of DNA fragments were separated by heating and then reassociated by cooling, certain portions were capable of reannealing more rapidly than others. They concluded that rapid reannealing was characteristic of multiple DNA fragments composed of identical or

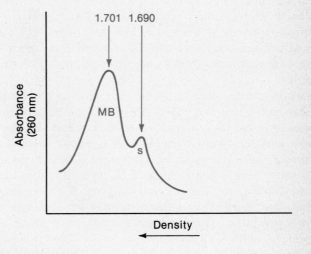

Figure 11.17
Separation of main-band (MB) and satellite (s) DNA from the mouse using sedimentation equilibrium centrifugation of a CsCl gradient.

nearly identical nucleotide sequences—the basis of **repetitive DNA.**

As discussed in Chapter 9, repeating nucleotide sequences are classified as either highly or moderately repetitive DNA. Evidence has now accumulated linking satellite DNA to repetitive DNA. Further, this specific DNA fraction has been shown to be present in heterochromatic regions of chromosomes. The following paragraphs review information supporting and clarifying these conclusions.

When satellite DNA is subjected to analysis by reassociation kinetics, it falls into the category of highly repetitive DNA. It therefore consists of short sequences repeated a large number of times. The available evidence further suggests that these sequences are present as tandem repeats clustered at various positions in the genome.

Satellite DNA is found in very specific chromosome areas known to be heterochromatic—the **centromeric regions.** This was discovered in 1969 when several workers, including Mary Lou Pardue and Joe Gall, applied the technique of **in situ hybridization** to the study of satellite DNA. This technique (see Appendix A) involves the molecular hybridization between an isolated fraction of radioactive DNA or RNA and the DNA contained in the chromosomes of a cytological preparation. Following the hybridization procedure, autoradiography is performed to locate the chromosomal areas complementary to the fraction of DNA or RNA.

In their work, Pardue and Gall demonstrated that RNA transcribed by mouse satellite DNA hybridizes with DNA of centromeric regions of mouse mitotic chromosomes (Figure 11.18). Thus, several conclusions can be drawn. Satellite DNA differs from main-band DNA in its molecular composition, as established by buoyant density studies. It is also composed of short repetitive sequences. Finally, satellite DNA is found in the heterochromatic centromeric regions of chromosomes.

These conclusions establish the basis of the structural and functional differences between heterochromatin and euchromatin. They further reflect the complexity of the genetic material in eukaryotes. Instead of viewing the chromosome as a linear series of genes, we must now see it as quite variable in its molecular organization and functional capacity. For example, there are nucleotide sequences that do not specify gene products. Are these genetically inert DNA sequences of heterochromatin related to the maintenance of the structural integrity of chromatin? Or, might they be involved in a regulatory role of transcription? Such noninformational sequences may play more than one role. There is no doubt that more extensive investigation will continually expand and modify our concept of eukaryotic genetic organization.

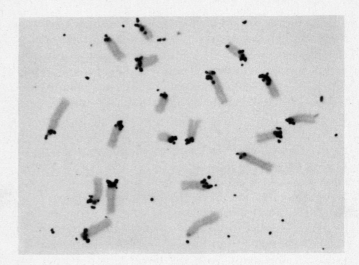

FIGURE 11.18

In situ hybridization between RNA transcribed by mouse satellite DNA and mitotic chromosomes. The grains in the autoradiograph localize the chromosome regions containing satellite DNA sequences. (From Pardue and Gall, 1972. Copyright Keter Publishing House Jerusalem Ltd.)

Heterochromatin Identification: Chromosome Banding

Until about 1970, mitotic chromosomes viewed under the light microscope could be distinguished only by their relative sizes and the positions of their centromeres. Even in organisms with a low haploid number, two or more chromosomes are often indistinguishable from one another. However, around 1970, differential staining along the longitudinal axis of mitotic chromosomes was made possible by new cytological techniques. These methods are now called **chromosome banding techniques** because the staining patterns are similar to the bands of polytene chromosomes.

One of the first chromosome banding techniques was discovered by Pardue and Gall during the development of the *in situ* hybridization procedure. They recognized that if chromosome preparations were heat denatured and then treated with Giemsa stain, a unique staining pattern emerged. The centromeric regions of mitotic chromosomes preferentially took up the stain! Thus, this cytological technique stained a specific area of the chromosome composed of constitutive heterochromatin. A diagram of the mouse karyotype treated in this way is shown in Figure 11.19. The staining pattern is referred to as **C-banding.**

FIGURE 11.19
Karyotypes of a male (a) and female (b) mouse where chromosome preparations were processed and stained to demonstrate C-bands. (From Chen and Ruddle, 1971, p. 54.)

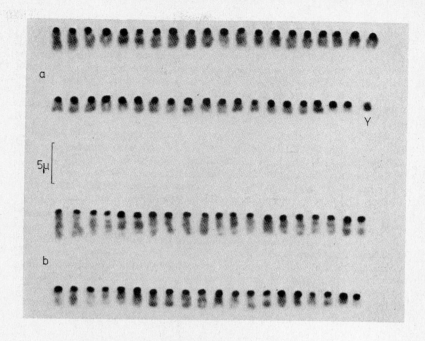

Still other chromosome banding techniques were developed about the same time. A group of Swedish researchers, led by Tobjorn Caspersson, used a technique that provided even greater staining differentiation of metaphase chromosomes. They used fluorescent dyes that bind to nucleoprotein complexes and produce unique banding patterns. When the chromosomes are treated with the fluorochrome **quinacrine mustard** and viewed under a fluorescent microscope, precise patterns of differential brightness are seen. Each of the 23 human chromosome pairs can be distinguished by this technique. The bands produced by this method are called **Q-bands.**

Another banding technique, which produces a staining pattern nearly identical to the Q-bands, has been developed by Tau-Chiuh Hsu and Frances E. Arrighi. This method, producing **G-bands** (Figures 11.20 and 11.21), involves the digestion of the mitotic chromosomes in the cytological preparations with the proteolytic enzyme **trypsin** followed by Giemsa staining. Another technique results in the reverse G-band staining pattern, called an **R-band** pattern. In 1971, a meeting was held in Paris to establish the nomenclature for these various patterns. Still other banding techniques are now available.

Intense efforts are currently underway to elucidate the molecular mechanisms involved in producing these banding patterns. The variety of staining reactions under different conditions reflects the heterogeneity and complexity of chromosome composition. The variations in nucleic acid composition that exist along the longitudinal axis of any given chromosome, with its alternating sections of heterochromatin and euchromatin, suggest functional diversity. For example, the following section describes the molecular organization of two chromosome components, the centromere and the telomere. As additional information becomes available, a clearer and more unified concept of genetic organization is expected.

Centromeres, Telomeres, and DNA Sequences

Linear chromosomes of most eukaryotes contain two regions that serve rather remarkable roles. The first is the **centromere** which, of course, is the region of the chromosome that attaches to one or more spindle fibers and is pulled to one of the poles during the anaphase stages of mitosis and meiosis. In sister chromatids during mitosis, the common centromere must divide and the daughter structures must always move to opposite poles. In the first meiotic division, on the other hand, the common centromere of each pair of sister chromatids must remain a single structure. But, the two structures from each pair of homologous chromosomes must align during synapsis, attach properly to spindle fibers, and then move in opposite directions during anaphase I. During the second meiotic division, the centromeres of sister chromatids must behave as they do subsequently in mitosis.

Maintenance of these patterns is essential to the fidelity of chromosome distribution during mitosis and meiosis. Most estimates of infidelity during the transmission of chromosomes are exceedingly low: 10^{-5} to

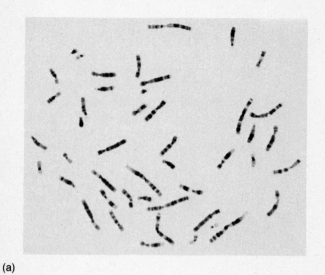

(a)

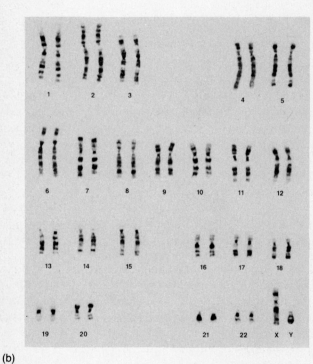

(b)

FIGURE 11.20
(a) G-banded human metaphase preparation. (b) G-banded
karyotype of a normal male showing approximately 400
bands per haploid set of chromosomes. Chromosomes were
derived from cells in midmetaphase. (From Yunis, 1981.)

10^{-6}, or one error per 100,000 to 1,000,000 cell divisions. As a result, it is generally assumed that an unusual or specific nucleotide sequence exists in the
DNA making up the centromere.

Preliminary information derived from the study of
yeast is now providing support for the above assump-

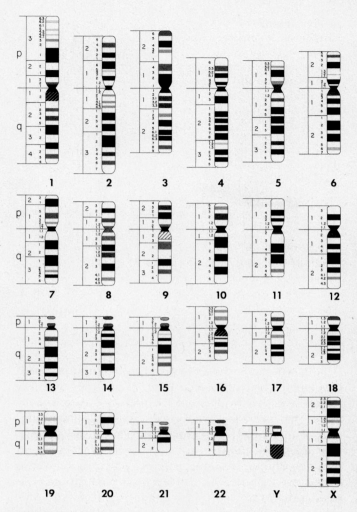

FIGURE 11.21
Schematic representation and nomenclature of human
chromosomes at the 400-band stage using G-banding.
Different shades represent varying intensities of bands.
(From Yunis, 1981.)

tion. Because yeast chromosomes are about 100 times
smaller than those in higher eukaryotes, and their
centromere region is also proportionately smaller
(yeast chromosomes attach to only one spindle fiber),
it has been relatively easy to isolate and examine centromeric DNA.

Thus far, the centromeric regions of four chromosomes (CEN 3, 4, 6 and 11) have been studied in great
detail. Several common features have been revealed.
All contain a central region consisting of 82 to 89 bp
containing greater than 90 percent AT residues. Flanking either side are short, conserved regions of 11 to 14
bp whose sequences are remarkably similar in all four
centromeres. Table 11.3 compares the three regions of
the four chromosomes.

While it is not yet clear how these sequences relate to the function of the centromere, several interesting observations have been made. A part of region I can be deleted without loss of function, but deletion of only parts of region II or III interrupts the normal function of the centromere. Changes in the nucleotide sequence by induced, site-specific mutations also disrupt the centromere's normal activity. Even though these studies have not centered on the higher eukaryotes, we can expect parallel, albeit somewhat more complex, findings from the study of more advanced organisms.

The second important structure is that portion of the linear chromosome constituting each end, called the **telomere.** Its major function is to provide stability, aiding in the maintenance of individual integrity of the chromosome. This role is accomplished by virtue of rendering chromosome ends generally inert in interactions with other chromosome ends. In contrast to broken chromosomes, whose ends are "sticky" and readily rejoin other such ends, telomere regions do not fuse with one another or with broken ends. Thus, it has been thought that some aspect of the DNA sequence of chromosome ends must be unique compared with most other chromosome regions.

As with centromeres, the problem has been approached by investigating the smaller chromosomes of less advanced eukaryotes, such as protozoans and yeast. The idea that all telomeres in a given species might share a common nucleotide sequence has now been borne out.

Two types of sequences have been discovered. The first type, called simply **telomeric DNA sequences,** consists of short tandem repeats. It is this group that contributes to the stability and integrity of the chromosome. In a variety of lower eukaryotes, these take the form of

$$5'-[C_{1-8}{}^{A}_{T}1_{-4}]-3'$$

For example, in the ciliate *Tetrahymena*, over fifty tandem repeats of the hexanucleotide sequence CCCCAA occur. In another ciliate, *Stylonchia*, CCCCAAAA is re-

peated many times. In the slime mold *Physarum*, the tandem repeat CCCTA occurs.

It is not yet clear how such sequences might relate to the creation of inert chromosome ends. Several interesting observations may eventually provide clues to the explanation. Within the tandem repeats of telomeric DNA sequences, several single-stranded nucleotide gaps are present. The structure created is apparently resistant to the ligase action that would normally seal the gaps. A second observation is that the number of tandem repeats at the end of each chromosome may increase during each cell generation. For example, in the flagellate parasite *Trypanosoma*, seven to ten base pairs are added during each generation of a cell clone. However, there is apparently some mechanism limiting the absolute length of the terminus of the chromosome. Although we do not yet comprehend the significance of these observations, they are intriguing.

In addition to the above satellitelike sequences made up of repetitive DNA, a number of examples of what are called **telomere-associated sequences** are known. These have been identified by the *in situ* molecular hybridization technique described earlier in this chapter (see Figure 11.18). For example, in the amphibian *Xenopus laevis*, the gene encoding a small ribosomal RNA component (5S RNA) is tandemly repeated a varying number of times at the end of the long arms of all chromosomes. In *Drosophila*, a tandem repeat of a 3-kb sequence has been identified at the telomere of all polytene chromosomes. In yeast, a 6.7-kb repeat has been found. The precise contribution of these and other similar sequences found in other organisms to chromosome stability remains a mystery. However, it is safe to conclude that the telomere is a very special part of all chromosomes and is required as part of the normal structure of these genetic units.

Direction of Future Studies

Recent investigations have transformed the study of euchromatin and heterochromatin from an interesting but poorly understood topic into one of the most ex-

TABLE 11.3

Nucleotide sequence information derived from DNA of the centromere regions of chromosomes 3, 4, 6, and 11 of yeast.

		Centromere Regions	
	I	II	III
CEN3	ATAAGTCACATGAT	--------------88 bp, 93% AT --------------	TGATTTCCGAA
CEN4	AAAGGTCACATGCT	--------------82 bp, 93% AT --------------	TGATTACCGAA
CEN6	ATCACGTG.C.TAT	--------------87 bp, 94% AT --------------	TGTTTTCCGAA
CEN11	ATAAGTCACATGAT	--------------89 bp, 94% AT --------------	TGATTTCCGAA

citing areas of genetics. The banding techniques that differentiate regions on metaphase chromosomes allow cytogeneticists to detect chromosomal aberrations with greater precision. This knowledge can also be applied to the study of evolution. It is now possible to observe differences in certain chromosome regions from organisms at different points in the phylogenetic scale and to determine precisely similarities of karyotypes between closely related species.

In Chapter 19, we will probe further into the structure of "classical" genes, which store transcribable genetic information. These genes constitute the euchromatic segments of chromosomes, which contain unique nucleotide sequences.

SUMMARY AND CONCLUSIONS

The organization of the molecular components of chromosomes has been of great interest from the time of their discovery. Knowing how these components are organized will help us to understand how the genetic material functions. Visualization of viral and bacterial chromosomes under the electron microscope has revealed rather simple, usually circular DNA helices closely corresponding to the Watson-Crick model of this molecule. The more classical studies of the gross structure of mitotic figures, polytene chromosomes, and lampbrush chromosomes relied on light microscopic observations. These studies have provided valuable insights into the morphology of eukaryotic chromosomes. Mitotic figures may be arranged into the karyotype of any organism. The study of karyotypes has provided important clues concerning the genetic basis of certain human disorders and has been important to evolutionary studies. The appearance of bands and puffs in polytene chromosomes

and loops in lampbrush chromosomes has been correlated with gene activity. Recently, Z-DNA has been found to occur in polytene chromosomes.

Biochemical analysis of the eukaryotic chromatin fiber has revealed that it is a nucleoprotein. Histone and nonhistone proteins are intimately bound to DNA. Histones, positively charged proteins, are part of a repeating unit along the chromosome axis, the nucleosome. As such, histones are important to the structural integrity of the fiber. The nucleosome may be important in facilitating the conversion of the extensive chromatin fibers characteristic of interphase into the highly condensed chromosomes seen in mitosis. The folded-fiber model, based on electron microscopic observations, suggests how such a transition takes place. Other ultrastructural observations have revealed the presence of the synaptonemal complex during meiosis in all organisms where synapsis and crossing over occur. Through the use of electron microscopy, DNA has been found in mitochondria and chloroplasts. It is similar in appearance to that of viruses and bacteria.

Biochemical and cytological investigations have extended our knowledge of functional and structural heterogeneity along the axis of chromosomes. Genetically inert heterochromatin has been clearly distinguished from potentially active euchromatin, and new insights into the structure of heterochromatin have been gained. In some instances, heterochromatin is composed of repeating or repetitive nucleotide sequences and can be identified cytologically. Chromosome banding techniques have confirmed that the axis of the mitotic chromosome is not chemically homogeneous. DNA analysis has revealed highly conserved, unique nucleotide sequences in both the centromere and telomere regions, distinguishing them from other parts of the chromosome. The findings presented are significant to many other areas of genetics.

PROBLEMS AND DISCUSSION QUESTIONS

1 Compare and contrast the chemical nature, size, and form assumed by the genetic material of viruses and bacteria.

2 Why might it be predicted that the organization of eukaryotic genetic material would be more complex than that of viruses or bacteria?

3 Describe the formation of polytene chromosomes. What do the terms *band, interband, chromomere,* and *puff* signify?

4 Correlate the observed structure of the lampbrush chromosome complex with what you have learned about the first meiotic prophase in Chapter 2. How do the enzyme digestion experiments support this model?

5 Describe the sequence of research findings leading to the most recent model of chromatin structure. What is the molecular composition and arrangement of the nucleosome? What is the solenoid?

6 Speculate as to why a structure such as the synaptonemal complex is necessary for precisely accomplishing synapsis and crossing over.

7 When chloroplasts and mitochondria are isolated, they are found to contain ribosomes that are similar to prokaryotic cells. What does this suggest about the function of DNA contained in these organelles? Does this observation support the endosymbiont theory?

8 Provide a comprehensive definition of heterochromatin.

9 List examples of constitutive and facultative heterochromatin.

10 Why are the cytological banding techniques considered to be very significant to the study of genetics?

11 Mammals contain a diploid genome consisting of at least 10^9 base pairs. If this amount of DNA is present as chromatin fibers where each group of 200 base pairs of DNA is combined with 9 histones into a nucleosome, and each group of 5 nucleosomes is combined into a solenoid, achieving a final packing ratio of 50, determine:
(a) the total number of nucleosomes in all fibers.
(b) the total number of solenoids in all fibers.
(c) the total number of histone molecules combined with DNA in the diploid genome.
(d) the combined length of all fibers.

12 Assume that a viral DNA molecule is in the form of a 50 μm long circular rod of a uniform 20-Å diameter. If this molecule is contained within a viral head that is a sphere with a diameter of 0.08 μm, will the DNA molecule fit into the viral head, assuming complete flexibility of the molecule? Justify your answer mathematically.

SELECTED READINGS

ANGELIER, N., et al. 1984. Scanning electron microscopy of amphibian lampbrush chromosomes. *Chromosoma* 89: 243–53.

ARRIGHI, F. E., and HSU, T. C. 1971. Localization of heterochromatin in human chromosomes. *Cytogenetics* 10: 81–86.

BEERMAN, W., and CLEVER, U. 1964. Chromosome puffs. *Scient. Amer.* (April) 210: 50–58.

BLACKBURN, E. H. 1984. Telomeres: Do the ends justify the means? *Cell* 37: 7–8.

BLACKBURN, E. H., and SZOSTAK, J. W. 1984. The molecular structure of centromeres and telomeres. *Ann. Rev. Biochem.* 53: 163–94.

BRADBURY, E. M., MACLEAN, N., and MATHEWS, H. R. 1981. *DNA, chromatin and chromosomes.* New York: Wiley.

BRITTEN, R. J., and KOHNE, D. 1968. Repeated sequences in DNA. *Science* 161: 529–40.

BRITTEN, R. J., GRAHAM, D. E., and NEUFELD, B. R. 1974. Analysis of repeating DNA sequences by reassociation. In *Methods of enzymology*, vol. 29, eds. L. Grossman and K. Moldave, pp. 363–418. New York: Academic Press.

BROWN, S. W. 1966. Heterochromatin. *Science* 151: 417–25.

BURLINGAME, R. W., et al. 1985. Crystallographic structure of the octameric histone core of the nucleosome at a resolution of 3.3 Å. *Science* 228: 546–53.

BUTLER, P. J. G. 1983. The folding of chromatin. *CRC Critical Rev. Biochem.* 15: 57–91.

CALLAN, H. G. 1982. Lampbrush chromosomes. *Proc. Roy. Soc. London*, Series B, 214: 417–88.

CARBON, J. 1984. Yeast centromeres: Structure and function. *Cell* 37: 351–53.

CHEN, T. R., and RUDDLE, F. H. 1971. Karyotype analysis utilizing differentially stained constitutive heterochromatin of human and murine chromosomes. *Chromosoma* 34: 51–72.

COLD SPRING HARBOR SYMPOSIUM ON QUANTITATIVE BIOLOGY. 1974. *Chromosome structure and function.* Vol. 38. Cold Spring Harbor, New York.

COMINGS, D. E. 1972. The structure and function of chromatin. *Adv. Hum. Genet.* 3: 237–431.

———. 1974. The structure of human chromosomes. In *The cell nucleus*, ed. H. Busch, vol. 1, pp. 538–64. New York: Academic Press.

CORNEO, G., et al. 1968. Isolation and characterization of mouse and guinea pig satellite DNA. *Biochemistry* 7: 4373–79.

DuPRAW, E. J. 1970. *DNA and chromosomes.* New York: Holt, Rinehart and Winston.

DUTRILLAUX, B., and LEJEUNE, J. 1975. New techniques in the study of human chromosomes: Methods and applications. *Adv. Hum. Genet.* 5: 119–56.

FELSENFELD, G. 1978. Chromatin. *Nature* 271: 115–22.

FINCH, J. T., et al. 1977. Structure of the nucleosome core particles of chromatin. *Nature* 269: 29–36.

FRENSTER, J. H. 1974. Ultrastructure and function of heterochromatin and euchromatin. In *The cell nucleus,* ed. H. Busch, vol. 1, pp. 565–80. New York: Academic Press.

GALL, J. G. 1963. Kinetics of deoxyribonuclease on chromosomes. *Nature* 198: 36–38.

GOLOMB, H. M., and BAHR, G. F. 1971. Scanning electron microscopic observations of surface structure of isolated human chromosomes. *Science* 171: 1024–26.

GREEN, B. R., and BURTON, H. 1970. *Acetabularia* chloroplast DNA: Electron microscopic visualization. *Science* 168: 981–82.

HEWISH, D. R., and BURGOYNE, L. 1973. Chromatin sub-structure. The digestion of chromatin DNA at regularly spaced sites by a nuclear deoxyribonuclease. *Biochem. Biophys. Res. Comm.* 52: 504–10.

HILL, R. J., and STOLLAR, D. 1983. Dependence of Z-DNA antibody binding to polytene chromosomes on acid fixation and DNA torsional strain. *Nature* 305: 338–40.

HOOD, L., WILSON, J., and WOOD, W. 1975. *Molecular biology of eukaryotic cells—A problems approach.* Menlo Park, Calif.: W. A. Benjamin.

HSU, T. C. 1973. Longitudinal differentiation of chromosomes. *Ann. Rev. Genet.* 7: 153–77.

HSU, T. C., and ARRIGHI, F. E. 1971. Distribution of constitutive heterochromatin in mammalian chromosomes. *Chromosoma* 34: 243–53.

ISENBERG, I. 1979. Histones. *Ann. Rev. Biochem.* 48: 159–92.

JUDD, B. H., SHEN, M., and KAUFMAN, T. 1972. The anatomy and function of a segment of the X chromosome of *Drosophila melanogaster. Genetics* 71: 139–56.

KING, R. C. 1970. The meiotic behavior of the *Drosophila* oocyte. *Int. Rev. Cytol.* 28: 125–68.

———, ed. 1975. *Handbook of genetics.* Vol. 3. New York: Plenum Press.

KLEINSCHMIDT, A. K. 1962. Darstellung und langenmessungen des gesamten desoxyribsenucleinsäure—inhaltes von T2-bakteriophagen. *Biochimica Biophysica Acta* 61: 857–64.

KORNBERG, R. D. 1975. Chromatin structure: A repeating unit of histones and DNA. *Science* 184: 868–71.

———. 1977. Structure of chromatin. *Ann. Rev. Biochem.* 46: 931–54.

KORNBERG, R. D., and KLUG, A. 1981. The nucleosome. *Scient. Amer.* (Feb.) 244: 52–64.

LEFEVRE, G. 1976. A photographic representation and interpretation of the polytene chromosomes of *Drosophila melanogaster* salivary glands. In *The genetics and biology of* Drosophila, eds. M. Ashburner and E. Novitski, vol. 1A. New York: Academic Press.

LEWIN, B. 1980. *Gene expression 2, Eukaryotic chromosomes.* 2nd ed. New York: Wiley.

LILLEY, D. M. J., and PARDON, J. F. 1979. Structure and function of chromatin. *Ann. Rev. Genet.* 13: 197–233.

MacHATTIE, L. A., et al. 1965. Electron microscopy of DNA from *Hemophilus influenzae. J. Mol. Biol.* 11: 648–49.

MILLER, O. L. 1965. Fine structure of lampbrush chromosomes. *Natl. Canc. Inst. Monogr.* 18: 79–99.

MOSES, M. J. 1968. Synaptinemal complex. *Ann. Rev. Genet.* 2: 363–412.

NASS, M. M. K., and BUCK, C. A. 1970. Studies on mitochondrial tRNA from animal cells, II. Hybridization of aminoacyl-tRNA from rat liver mitochondria with heavy and light complementary strands of mitochondrial DNA. *J. Mol. Biol.* 54: 187–98.

NORDHEIM, A., et. al. 1981. Antibodies to left handed Z-DNA bind to interband regions of *Drosophila* polytene chromosomes. *Nature* 294: 417–22.

OLINS, A. L., and OLINS, D. E. 1974. Spheroid chromatin units (*v* bodies). *Science* 183: 330–32.

———. 1978. Nucleosomes: The structural quantum in chromosomes. *Amer. Scient.* 66: 704–11.

PARDUE, M. L., and GALL, J. G. 1969. Molecular hybridization of radioactive DNA to the DNA of cytological preparations. *Proc. Natl. Acad. Sci.* 64: 600–604.

———. 1970. Chromosomal localization of mouse satellite DNA. *Science* 168: 1356–58.

———. 1972. Chromosome structure studied by nucleic acid hybridization in cytological preparations. *Chromosomes Today* 3: 47–52.

RICHMOND, T. J., et al. 1984. Structure of the nucleosome core particle at 7Å resolution. *Nature* 311: 532–37.

SCHNEDL, W. 1978. Structure and variability of human chromosomes analyzed by recent techniques. *Hum. Genet.* 41: 1–9.

SPIERER, P. 1984. A molecular approach to chromosome organization. *Devel. Genet.* 4: 333–39.

SPIERER, A., and SPIERER, P. 1984. Similar level of polyteny in bands and interbands of *Drosophila* giant chromosomes. *Nature* 307: 176–78.

SUMNER, A. T. 1982. The nature and mechanism of chromosome banding. *Cancer Genet. and Cytogenet.* 6: 59–88.

THERMAN, E. 1980. *Human chromosomes, structure, behavior, effects.* New York: Springer-Verlag.

vonHOLT, C. 1985. Histones in perspective. *Bioessays* 3:120–24.

vonWETTSTEIN, D., RASMUSSEN, S. W., and HOLM, P. B. 1984. The synaptonemal complex in genetic segregation. *Ann Rev. Genet.* 18: 331–413.

WARING, M., and BRITTEN, R. J. 1966. Nucleotide sequence repetition: A rapidly reassociating fraction from mouse DNA. *Science* 154: 791–804.

WESTERGAARD, M., and vonWETTSTEIN, D. 1972. The synaptinemal complex. *Ann. Rev. Genet.* 6: 71–110.

WESTMORELAND, B., et al. 1969. Mapping of deletions and substitutions in heteroduplex DNA molecules of bacteriophage lambda by electron microscopy. *Science* 163: 1343–46.

WISCHNITZER, S. 1976. The lampbrush chromosomes: Their morphology and physiological importance. *Endeavour* 35: 27–31.

YUNIS, J. J. 1981. Chromosomes and cancer: New nomenclature and future directions. *Hum. Pathol.* 12: 494–503.

YUNIS, J. J., and YASMINEH, W. G. 1971. Heterochromatin, satellite DNA, and cell function. *Science* 174: 1200–1209.

PART THREE
GENETIC
VARIATION

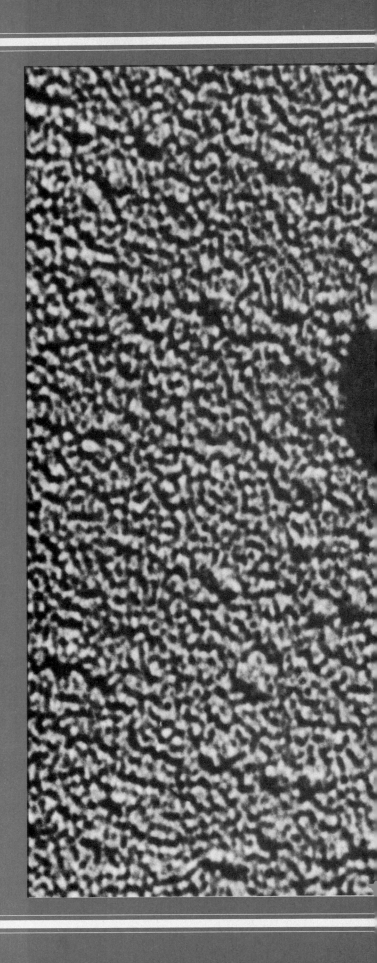

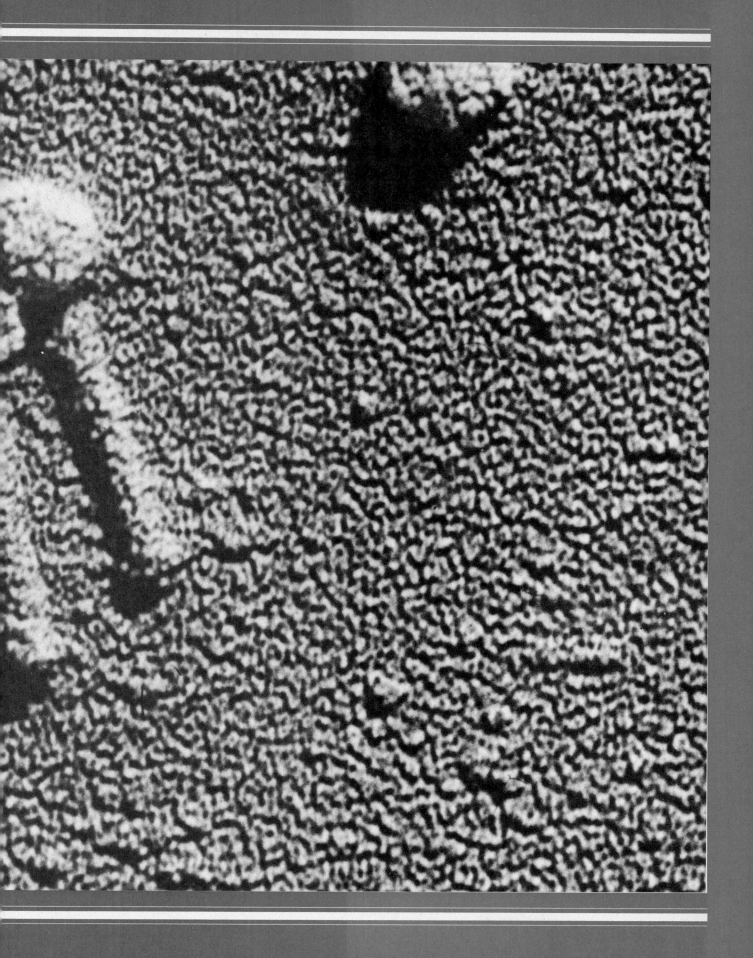

12

Variations in Chromosome Number and Arrangement

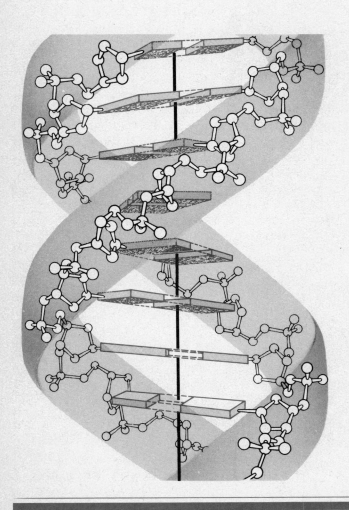

Genetic variation is indispensable to the process of evolution and to individual identity among members of any species. Alleles, one source of such variation, have already been discussed in Chapters 3 and 4. Alleles arise through mutation and result in alternative forms of a gene. We have also examined various mechanisms of recombination, including independent assortment and crossing over, which shuffle the genetic material during gamete formation. These processes have the effect of compounding the possibilities for genetic variation in offspring.

Some changes in the genetic material are more substantial than alterations of individual genes. These involve modifications at the level of the chromosome. Members of each species contain a characteristic haploid number (*n*) of different chromosomes. Together, these are sometimes referred to as "one set" of chromosomes, and the genetic information contained within them is called the **haploid genome.** Diploid organisms, of course, have two sets of these chromosomes. Modifications include variations in chromosome numbers as well as rearrangements of the genetic material either within or between chromosomes. Taken together, all such changes may be referred to as **chromosome mutations** or **chromosome aberrations.** They have classically been thought of as mutations because of the distinctive phenotypes that sometimes result. The more modern term *aberrations* distinguishes the change in the number of chromosomes or arrangement of genes from the simpler point mutations, which usually involve only a change in a nucleotide within a gene (see Chapter 13).

While individuals containing a chromosome aberration often are abnormal, in many instances no functional impairment results. This is particularly true when the aberration involves rearrangement of a chromosome segment but not the gain or loss of genetic material. As we shall see, even the presence of additional genetic material in an otherwise diploid genome, particularly in plants, may be tolerated. In some cases, the addition has provided important variation during the evolution of plants. In this chapter we will consider the many types of aberrations, their consequences, and their role in the evolutionary process.

VARIATION IN CHROMOSOME NUMBER

Variation in chromosome number ranges from the addition or loss of one or more chromosomes to the addition or loss of one or more haploid sets of chromosomes. When an organism gains or loses one or more chromosomes, but not a complete set, the condition of **aneuploidy** is created. This is contrasted with the condition of **euploidy,** where one or more complete haploid sets of chromosomes are found. Those with one set are **monoploid;** those with two sets are **diploid;** and when more than two sets exist, an organism is said to be **polyploid.** We shall discuss each of these categories and subsets within them in turn. Table 12.1 provides an organizational framework for the categories and subsets as well as a brief indication of what each term means.

ANEUPLOIDY

The most common examples of **aneuploidy,** where an organism has a chromosome number other than an exact multiple of the haploid set, are cases in which a single chromosome is either added to or lost from a normal diploid set. Such circumstances can arise as a result of primary or secondary nondisjunction (see Figure 2.11). The loss of one chromosome produces a $2n - 1$ complement and is called **monosomy;** the gain of one chromosome produces a $2n + 1$ complement and is described as **trisomy.** The $2n + 2$ and $2n + 3$ conditions are called **tetrasomy** and **pentasomy.**

Monosomy

Monosomy for one of the sex chromosomes is fairly common. In fact, it characterizes the **Protenor mode** of sex determination **(XX/XO).** When sex is determined in the **Lygaeus (XX/XY) mode,** XO individuals may also occur, provided that such individuals have a normal complement of autosomes. In *Drosophila*, the XO chromosome complement results in a normal appearing, but sterile male. In humans, this monosomy results in females with Turner syndrome (see Chapter 5).

TABLE 12.1
Terminology for variation in chromosome numbers.

Term	Explanation
Aneuploidy	$2n$ plus or minus chromosomes
Monosomy	$2n - 1$
Trisomy	$2n + 1$
Tetrasomy, pentasomy, etc.	$2n + 2, 2n + 3$, etc.
Euploidy	Multiples of n
Monoploidy	n
Diploidy	$2n$
Polyploidy	$3n, 4n, 5n, \ldots$
Triploidy	$3n$
Tetraploidy, pentaploidy, etc.	$4n, 5n$, etc.
Autopolyploidy	Multiples of the same genome
Allopolyploidy	Multiples of different genomes

Monosomy for one of the autosomes is not so easily tolerated, particularly in animals. In *Drosophila*, flies monosomic for the very small chromosome 4—a condition referred to as **Haplo-IV**—develop more slowly, exhibit a reduced body size, and have impaired viability. Monosomy for the larger chromosomes 2 and 3 is apparently lethal because such flies have never been recovered.

The failure of monosomic individuals to survive in many animal species is at first quite puzzling, since at least a single copy of every gene is present in the remaining homologue. However, if just one of those genes is represented by a lethal allele, the unpaired chromosome condition leads to the death of the organism. This occurs because monosomy unmasks recessive lethals that are tolerated in heterozygotes carrying the corresponding wild-type alleles.

Another explanation is that the expression of genetic information during early development is very delicately regulated so that a sensitive equilibrium of gene products is required to ensure normal development. This requirement does not appear to be so stringent in the plant kingdom. Monosomy for autosomal chromosomes has been observed in maize, tobacco, the evening primrose *Oenothera*, and the jimson weed *Datura*, among other plants. Nevertheless, such monosomic plants are usually less viable than their diploid derivatives. Haploid pollen grains, which undergo extensive development before participating in fertilization, are particularly sensitive to the lack of one chromosome.

Partial Monosomy: Cri-du-Chat Syndrome

In humans, autosomal monosomy has not been reported beyond birth. Individuals with such chromosome complements are undoubtedly conceived, but none apparently survive embryonic and fetal development. There are, however, examples of survivors with **partial monosomy,** where only part of one chromosome is lost. These cases are also referred to as **segmental deletions.** One such case was first reported by Jérôme LeJeune in 1963 when he described the clinical symptoms of the **cri-du-chat syndrome.** This syndrome is associated with the loss of about one-half the short arm of chromosome 5 (Figure 12.1). Infants with this sydrome exhibit multiple anatomic malformations, including gastrointestinal and cardiac complications, and are severely mentally retarded. Abnormal development of the glottis and larynx is characteristic of individuals with this syndrome. An infant afflicted with this syndrome has a cry similar to that of the meowing of a cat, thus giving the syndrome its name.

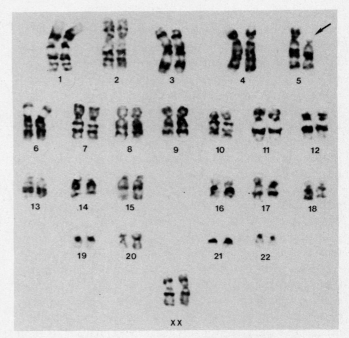

FIGURE 12.1

A representative karyotype and two children exhibiting cri-du-chat syndrome (46,5p-). In the karyotype, the arrow identifies the nearly complete absence of the short arm of one member of the chromosome 5 homologues. (Courtesy of Louise E. Wilkins.)

Since 1963, over 300 cases of cri-du-chat syndrome have been reported worldwide. An incidence of 1 in 50,000 live births has been estimated. The size of the deletion appears to influence the physical, psychomotor, and mental skill levels of those children who survive. While the effects of the syndrome are severe, many individuals achieve a level of social development in the trainable range. Those who receive home care and early special schooling may be ambulatory, develop self-care skills, and learn to communicate verbally.

Trisomy

In general, the effects of trisomy parallel those of monosomy. However, the addition of an extra chromosome produces somewhat more viable individuals in both animal and plant species than does the loss of a chromosome.

As in monosomy, the sex chromosome variation of the trisomic type has a less dramatic effect on the phenotype than autosomal variation. *Drosophila* females with three X chromosomes and a normal complement of autosomes (3X:2A) may be fertile but less viable than normal 2X:2A females. In humans, the 47,XXY, 47,XYY, and 47,XXX conditions lead to viable individuals. However, the addition of a large autosome to the diploid complement in both *Drosophila* and humans has severe effects and is usually lethal during development.

In plants, trisomic individuals are viable, but their phenotype may be altered. A classical example involves the jimson weed *Datura*, whose diploid number is 24. Twelve different primary trisomic conditions are possible, and examples of each one have been recovered. Each trisomy alters the phenotype of the capsule sufficiently (Figure 12.2) to produce a unique phenotype. These capsule phenotypes were first thought to be caused by point mutations.

In plants as well as animals, trisomy may be detected during cytological observations of meiotic divisions. Since three copies of one of the chromosomes are present, pairing configurations are irregular. At any particular region along the chromosome length, only two of the three homologues may synapse. At various regions, however, different members of the trio may be paired. One such example, called a **trivalent,** is diagramed in Figure 12.3. In some cases, one bivalent and

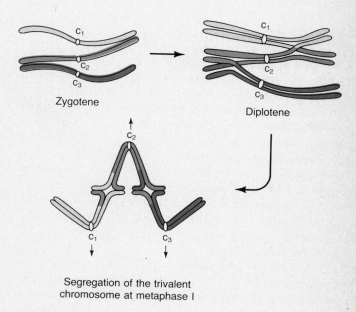

FIGURE 12.3

One possible pairing arrangement and segregation pattern of members of a trivalent chromosome during the first meiotic division. Two chiasmata have formed in the diplotene stage.

FIGURE 12.2

Capsule phenotypes of the fruits of *Datura stramonium*. In comparison with wild type, each phenotype is the result of trisomy of one of the twelve chromosomes characteristic of the haploid genome. (After Blakeslee, 1934.)

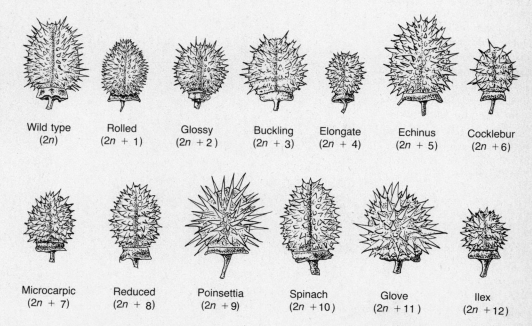

| Wild type ($2n$) | Rolled ($2n + 1$) | Glossy ($2n + 2$) | Buckling ($2n + 3$) | Elongate ($2n + 4$) | Echinus ($2n + 5$) | Cocklebur ($2n + 6$) |

| Microcarpic ($2n + 7$) | Reduced ($2n + 8$) | Poinsettia ($2n + 9$) | Spinach ($2n + 10$) | Glove ($2n + 11$) | Ilex ($2n + 12$) |

one univalent (an unpaired chromosome) maybe present instead of a trivalent during the first metaphase stage. The trivalent is usually arranged on the spindle so that during anaphase one member moves to one pole, and two go to the opposite pole. Meiosis thus produces gametes that can perpetuate the trisomic condition.

Down Syndrome

The only human autosomal trisomy in which a significant number of individuals survive longer than a year past birth was discovered in 1866 by Langdon Down and was originally called **mongoloid idiocy** or **mongolism.** Down chose this name because of the prominence of the epicanthic fold in the eyelid, a phenotype characteristic of members of the Mongoloid race. The condition is now known to result from trisomy of chromosome 21, one of the G group (Figure 12.4), and is more appropriately called **Down syndrome** or simply **trisomy 21.** Its official designation is **47,21 +.** This trisomy is found in approximately 3 infants in every 2000 live births.

The external phenotype of these individuals is so similar that many of them might appear to be siblings at first glance. They are short and may have small, round heads; protruding, furrowed tongues, which cause the mouth to remain partially open; and short, broad hands with fingers with characteristic dermatoglyphic palm and fingerprint patterns. Physical, psy-chomotor, and mental development is retarded and the IQ is seldom above 70. Their life expectancy is shortened, averaging about 20 years. A significant number of Down infants do not survive the first year after birth. They are prone to respiratory disease and heart malformations and show an incidence of leukemia approximately 15 times higher than that of the normal population. However, close medical scrutiny throughout the lives of Down individuals has extended their survival significantly.

One way in which this trisomic condition may originate is through **nondisjunction** of chromosome 21 during meiosis. Failure of paired homologues to disjoin during anaphase I or II can result in male or female gametes with the $n + 1$ chromosome composition. Following fertilization with a normal gamete, the trisomic condition is created. Chromosome banding studies have shown that while the additional chromosome may be derived from either the mother or father, the ovum is most often the source.

Before the development of banding techniques, other indirect evidence, derived from studies of the age of mothers giving birth to Down infants, supported this conclusion. Figure 12.5 shows the analysis of the distribution of maternal age and the incidence of Down syndrome newborns. The frequency of Down births dramatically increases as the age of the mother increases. While the frequency is about 1 in 1000 at maternal age 30, a tenfold increase to a frequency of 1 in 100 is noted at age 40. The frequency increases still

FIGURE 12.4
The karyotype and a photograph of a child with Down syndrome. In the karyotype, three members of the G-group chromosome 21 are present, creating the 47,21 + condition. (Chromosome preparation courtesy of Arthur E. Greene, Institute for Medical Research. Photo courtesy of Tom Hutchinson.)

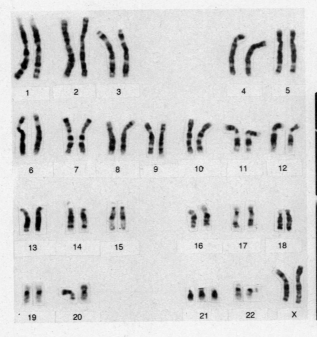

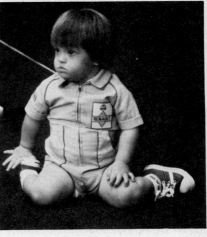

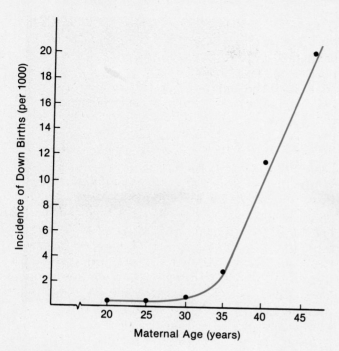

FIGURE 12.5

Relative incidence of Down syndrome births contrasted with maternal age. (From Hamerton, 1971.)

further to about 1 in 50 at age 45. Recently, a relationship has also been shown between a paternal age over 55 and an increased frequency of afflicted infants. However, a strong correlation has yet to be fully established.

While the nondisjunctional event that produces Down syndrome seems more likely to occur during oogenesis in older women, we do not know with certainty why this is so. In human females, all primary oocytes have been formed by birth. Therefore, once ovulation begins, each succeeding ovum has been arrested in meiosis for about a month longer than the one preceding it. Thus, women 30 or 40 years old produce ova that are significantly older and arrested longer than those they ovulated 10 or 20 years previously. However, it is not yet known whether ovum age is the cause of the increased incidence of nondisjunction leading to Down syndrome.

These statistics pose a serious problem for the woman who becomes pregnant late in her reproductive years. Genetic counseling early in the pregnancy serves two purposes. First, it informs the parents about the probability that their child will be affected and educates them about Down syndrome. Although some individuals with Down syndrome must be institutionalized, others are trainable and may be cared for

at home. Further, they are noted as being affectionate, loving children. Second, a genetic counselor may recommend the prenatal diagnostic technique of **amniocentesis.** This technique requires the removal and culture of fetal cells from the amniotic fluid. The karyotype of the fetus may then be determined by cytogenetic analysis. If the fetus is diagnosed as having Down syndrome, a therapeutic abortion is one option the parents may consider. The process of amniocentesis is more completely discussed in Chapter 20.

Since Down syndrome appears to be caused by a random error—nondisjunction of chromosome 21 during maternal or paternal meiosis—the disorder is not expected to be inherited. Nevertheless, a similar condition called **familial Down syndrome** does run in families and has been reported many times. Familial Down syndrome involves a **translocation** of chromosome 21, another type of chromosomal aberration which we will discuss later in this chapter.

Patau Syndrome

In 1960, Klaus Patau and his associates observed an infant with gross developmental malformations with a karyotype of 47 chromosomes (Figure 12.6). The additional chromosome was medium-sized, one of the acrocentric D group. It is now designated as chromosome 13. The trisomy 13 condition has since been described in many newborns and is called **Patau syndrome (47,13 +).** Affected infants are severely mentally retarded, thought to be deaf, and characteristically have a harelip, cleft palate, and polydactyly. Autopsy has revealed congenital malformation of most organ systems, a condition indicative of abnormal developmental events occurring as early as five to six weeks of gestation. The mean survival of these infants is less than six months.

The average maternal and paternal ages of parents of affected infants are higher than the ages of parents of normal children, but parental ages are not as high as the average maternal age in cases of Down syndrome. Both male and female parents average about 32 years of age when the affected child is born. Because the condition is so rare, occurring as infrequently as 1 in 10,000 live births, it is not known if the origin of the extra chromosome is more often maternal, paternal, or whether it arises equally from either parent.

Edwards Syndrome

In 1960, John H. Edwards and his colleagues reported on an infant trisomic for a chromosome in the E group, now known to be chromosome 18 (Figure 12.7). This aberration has been named **Edwards syndrome**

FIGURE 12.6

The karyotype and a photograph of an infant with Patau syndrome. In the karyotype, three members of the D-group chromosome 13 are present, creating the 47,13+ condition. (Chromosome preparation courtesy of Arthur E. Greene, Institute for Medical Research. Photo courtesy of Dr. David D. Weaver, Indiana University School of Medicine.)

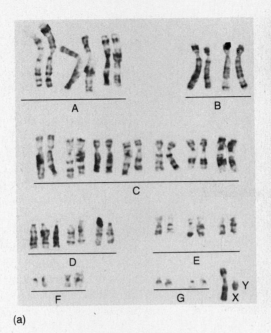

(a)

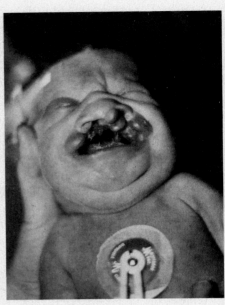

(b)

(**47,18+**). These infants are smaller than the average newborn. Their skulls are elongated in an anterior-posterior direction and their ears set low and malformed. A webbed neck, congenital dislocation of the hips, and a receding chin are often characteristic of such individuals. Although the frequency of trisomy 18 is somewhat greater than that of trisomy 13, the survival time is about the same, less than four months on the average. Death is usually caused by pneumonia or heart failure. The phenotype of this child, like that of individuals with Down and Patau syndromes, illustrates that the presence of an extra autosome produces congenital malformations and reduced life expectancy.

Again, the average maternal age is high—34.7 years by one calculation. One unusual aspect of the syndrome is the preponderance of afflicted female infants. In one set of observations based on 143 cases, 80 percent were female.

Viability in Human Aneuploid Conditions

The reduced viability of individuals with recognized monosomic and trisomic conditions leads us to believe that many other aneupolid conditions may arise but that the affected fetuses do not survive to term. This observation has been confirmed by karyotypic analysis of spontaneously aborted fetuses. These studies have revealed some striking statistics. About 15 to 20 percent of all conceptions are terminated in spontaneous abortion. About 30 percent of all spontaneous abortuses demonstrate some form of chromosomal anomaly, and approximately 90 percent of all chromosomal anomalies are terminated prior to birth as a result of spontaneous abortion.

Approximately 65 percent of spontaneous abortuses demonstrating chromosomal abnormalities are aneuploids. The aneuploid with highest incidence among abortuses is the 45,X condition, which produces an infant with Turner syndrome if the fetus survives to term. After compiling many sets of data, David H. Carr found that 23.8 percent of all abnormal karyotypes recovered from abortuses were of the 45,X variety. It has been estimated that liveborn Turner syndrome infants represent only 2 percent of 45,X conceptuses.

Carr's review of this subject also revealed that 42 percent of the cases were trisomic for one of the chromosome groups. Trisomies for every human chromosome have been recovered. Monosomies were almost never found, however, even though nondisjunction should produce $n - 1$ gametes with a frequency equal to $n + 1$ gametes. This finding suggests that gametes lacking a single chromosome are functionally impaired to a serious degree or that the embryo dies very early in its development. Various forms of polyploidy and other miscellaneous chromosomal anomalies were also found in Carr's study.

These observations support the hypothesis that normal embryonic development requires a normal diploid chromosome complement to maintain a deli-

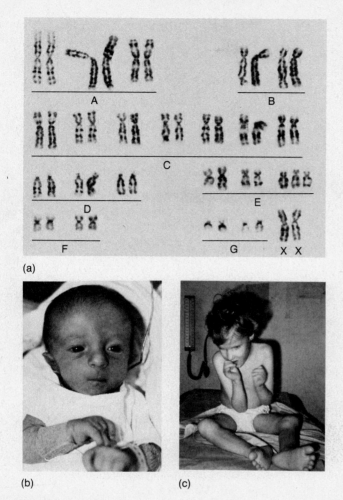

(a)

(b)   (c)

FIGURE 12.7
The karyotype and photographs of children with Edwards
syndrome. In the karyotype, three members of the E-group
chromosome 18 are present, creating the 47,18+ condition.
(Chromosome preparation courtesy of Arthur E. Greene,
Institute for Medical Research. Photos courtesy of Dr. David
D. Weaver, Indiana University School of Medicine.)

cate equilibrium in the expression of genetic information. The prenatal mortality of most aneuploids is an
efficient barrier against the introduction of a general
form of genetic anomaly into the human population.

POLYPLOIDY AND ITS ORIGINS

The term **polyploidy** describes instances where more
than two multiples of the haploid chromosome set are
found. The naming of polyploids is based on the number of sets of chromosomes found: a **triploid** has 3n
chromosomes; a **tetraploid** has 4n; a **pentaploid,** 5n;

and so forth. Several general statements may be made
about polyploidy. This condition is relatively infrequent in most animal species, but is well known in
lizards, amphibians, and fish. It is much more common in plant species. Odd numbers of chromosome
sets are not usually transmitted reliably from generation to generation because a polyploid organism with
an uneven number of homologues usually does not
produce genetically balanced gametes. For this reason,
triploids, pentaploids, etc., are not usually found in
species that depend solely upon sexual reproduction
for propagation.

Polyploidy can originate in two ways: (1) the addition of one or more extra sets of chromosomes, identical to the normal haploid complement of the same
species, results in **autopolyploidy,** and (2) the combination of chromosome sets from different species may
occur as a consequence of hybridization and results
in **allopolyploidy.** Thus, the distinction between auto-
and allopolyploidy is based on the genetic origin of
the extra chromosome sets.

In our discussion of polyploidy, we will use the following symbols to clarify the origin of additional chromosome sets. For example, if A represents the haploid
set of chromosomes of any organism, then

$$A = a_1 + a_2 + a_3 + a_4 \ldots a_n$$

where a_1, a_2, and so on represent individual chromosomes, and where n is the haploid number. Thus, a
normal diploid organism might be represented simply
as AA.

Autopolyploidy

In **autopolyploidy,** each additional set of chromosomes is identical to the parent species. Therefore, **triploids** are represented as AAA, **tetraploids** are $AAAA$,
and so forth.

Autotriploids may arise in several ways. A failure of
all chromosomes to segregate during meiotic divisions
may produce a diploid gamete. If such a gamete is fertilized by a haploid gamete, a zygote with three sets of
chromosomes is produced. Or, occasionally two
sperm may fertilize an ovum, resulting in a triploid zygote. Triploids may also be produced under experimental conditions by crossing diploids with tetraploids. Diploid organisms produce gametes with n
chromosomes while tetraploids produce $2n$ gametes.
Upon fertilization, the desired triploid is produced.

Because they have an even number of chromosomes, **autotetraploids** (4n) are theoretically more
likely to be found in nature than autotriploids. Unlike
triploids, which often produce genetically unbalanced
gametes, tetraploids are more likely to produce bal-

anced gametes, when involved in sexual reproduction. Tetraploid cells may be produced experimentally from diploid cells by applying cold or heat shock during meiosis or by applying colchicine to somatic cells undergoing mitosis. **Colchicine,** an alkyloid derived from the autumn crocus, interferes with spindle formation, and thus replicated chromosomes cannot migrate to the poles. If chromosomes are not lost, when the effect of colchicine is diminished and such a cell reenters interphase, it will be converted to 4n.

In general, autopolyploids are larger than their diploid relatives. Often, the flower, ovum, and fruit of plants are increased in size. This increase seems to be due to larger cell size rather than greater cell number. Although autopolyploids do not contain new or unique information compared with the diploid relative, such varieties may be of greater commercial value. Economically important triploid plants include several potato species of the genus *Solanum,* the popular Winesap apples, commercial bananas, the seedless watermelons, and the cultivated tiger lily *Lilium tigrinum.* These plants are propagated asexually. Diploid bananas contain hard seeds, but the commercial, triploid, "seedless" variety has edible seeds. Tetraploid alfalfa, coffee, peanuts, and McIntosh apples are also of economic value because they are either larger or grow more vigorously than their diploid or triploid counterparts. The commercial strawberry is an octoploid. These observations attest to the importance of autopolyploidy in domesticated plants.

Allopolyploidy

Polyploidy may also result from hybridization of two closely related species. Thus, if a haploid ovum from a species with chromosome sets *AA* is fertilized by sperm from a species with sets *BB,* the resulting hybrid is *AB,* where $A = a_1, a_2, a_3, \ldots, a_n$ and $B = b_1, b_2, b_3, \ldots, b_n$. The hybrid will be sterile because of its inability to produce viable gametes. If the new *AB* genetic combination undergoes a natural or induced chromosomal doubling, a fertile *AABB* tetraploid is produced. These events are illustrated in Figure 12.8. Since this polyploid contains the equivalent of four haploid genomes and since the hybrid contains unique genetic information compared with either parent, such an organism is called an **allotetraploid** (from the Greek word *allo,* meaning other or different).

Allopolyploidy is the most common natural form of polyploidy in plants because the chance of forming balanced gametes is much greater than in other types of polyploidy. Since two homologues of each specific chromosome are present, meiosis may occur normally

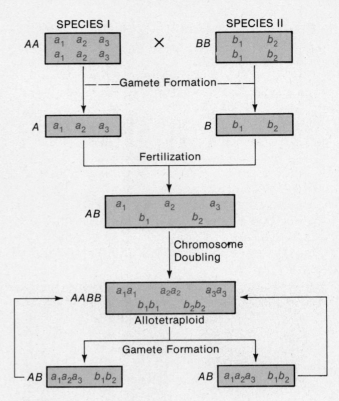

FIGURE 12.8

The origin and propagation of an allotetraploid. Species I contains a haploid genome *(A)* consisting of three chromosomes, a_1, a_2, and a_3. Species II contains a haploid genome *(B)* consisting of two chromosomes, b_1 and b_2. Following fertilization between members of the two species and chromosome doubling, a fertile allotetraploid containing four complete haploid genomes *(AABB)* is formed.

(Figure 12.8), and fertilization may successfully propagate the tetraploid sexually. This discussion assumes the simplest situation, where none of the chromosomes in set *A* are homologous to those in set *B*. In hybrids of very closely related species, some homology between *a* and *b* chromosomes is likely. In this case, meiotic pairing is more complex. Multivalents, which are more complex than the trivalents shown in Figure 12.3, may be formed, resulting in the production of unbalanced gametes. In such cases, aneuploid varieties of allotetraploids may arise. Allopolyploids are rare in most animals because mating behavior is most often species-specific, and thus the initial step in hybridization is unlikely to occur.

A classical example of allotetraploidy in plants is the cultivated species of American cotton, *Gossypium*. This species has 26 pairs of chromosomes: 13 are large and 13 are much smaller. When it was discovered that Old World cotton had only 13 pairs of large chromosomes, allopolyploidy was suspected. After an examination of wild American cotton revealed 13 pairs of small chromosomes, this speculation was strengthened. J. O. Beasley was able to reconstruct the origin of cultivated cotton experimentally. He crossed the Old World strain with the wild American strain and then treated the hybrid with colchicine to double the chromosome number. The result of these treatments was a fertile allotetraploid variety of cotton. It contained 26 pairs of chromosomes and characteristics similar to the cultivated variety.

Often allotetraploids exhibit characteristics of both parental species. An interesting example, but one with no practical economic importance, is that of the hybrid formed between the radish *Raphanus sativus* and the cabbage *Brassica oleracea*. Both species have a haploid number of 9. The hybrid consists of 9 *Raphanus* and 9 *Brassica* chromosomes (9R + 9B). While hybrids are almost always sterile, some fertile allotetraploids (18R + 18B) have been produced. Unfortunately, the root of this allotetraploid is more like the cabbage and its shoot resembles the radish. Had the converse occurred, the hybrid might have been of economic importance.

A much more successful commercial hybridization has been performed using the grasses wheat and rye. Wheat (genus *Triticum*) has a basic genome of 7 chromosomes. The diploid species have 14 chromosomes. Other cultivated autopolyploids exist, including tetraploid ($4n = 28$) and hexaploid ($6n = 42$) varieties. Rye (genus *Secale*) also has a genome consisting of 7 chromosomes. The only cultivated species is diploid ($2n = 14$).

Using the technique outlined in Figure 12.8, including colchicine treatment of the F_1 hybrid seedlings, geneticists have produced various allopolyploids. When tetraploid wheat is crossed with diploid rye, and the F_1 treated with colchicine, a hexaploid hybrid ($6n = 42$) is derived. The hybrid, designated *Triticale*, represents a new genus. It contains 14 pairs of wheat chromosomes and 7 pairs of rye chromosomes. Octoploids may also be produced from crosses between hexaploid wheat and diploid rye. The hexaploid hybrid, initially described, is more stable. Fertile hybrid varieties derived from various wheat and rye species may be crossed together or back-crossed to parental strains. These crosses have created a wide variety of members of the genus *Triticale*.

The hybrid plants demonstrate characteristics of both wheat and rye. For example, certain hybrids combine the high protein content of wheat with the high content of the amino acid lysine of rye. The lysine content is low in wheat and thus is a limiting nutritional factor. Wheat is considered a high-yielding grain, while rye is noted for its versatility of growth in unfavorable environments. *Triticale* species combining both traits have the potential of significantly increasing grain production. Programs designed to improve crops through hybridization have long been underway in several underdeveloped countries of the world, as discussed in Chapter 1.

Endopolyploidy

Certain cells in an otherwise diploid organism have been observed to be polyploid. The term **endopolyploidy** describes this general phenomenon. Although endopolyploidy is technically different from the polyteny described in Chapter 11, both conditions result in an increased nuclear amount of genetic information. In polytene nuclei, even though the number of copies of each chromosome may be 1000 to 2000 times greater than the haploid number, only *n* chromosomes remain visible. This occurs because the chromosomes replicate but remain paired. In the case of endopolyploidy, replication and separation of chromosomes occur without nuclear division. The process leading to endopolyploidy is called **endomitosis.**

Vertebrate liver cell nuclei, including human ones, often contain $4n$, $8n$, or $16n$ chromosome sets. The stem and parenchyma tissue of apical regions of flowering plants are also often endopolyploid. Cells lining the gut of mosquito larvae attain a $16n$ ploidy, but during the pupal stages such cells undergo very quick reduction divisions, giving rise to smaller diploid cells. In the water strider *Gerris*, wide variations in chromosome numbers are found in different tissues, with as many as 1024 to 2048 copies of each chromosome in the salivary gland cells. Since the diploid number in this organism is 22, the nuclei of these cells may contain over 40,000 chromosomes!

Although the role of endopolyploidy is not clear, the proliferation of chromosome copies often occurs in cells where multiples of certain gene products are required. In fact, it is well established that certain genes, whose product is in high demand in *every* cell, exist naturally in multiple copies in the genome. Ribosomal and transfer RNA genes are examples of multiple-copy genes. In certain cells of organisms, it may be necessary to replicate the entire genome, allowing an even greater rate of expression of various genes.

SIGNIFICANCE OF VARIATION IN CHROMOSOME NUMBER

The study of variation in chromosome number has enhanced our understanding of genetic expression as well as our knowledge of evolution. Examples of monosomy and trisomy in both plant and animal species have made it clear that the diploid genome expresses genetic information in a delicately balanced way. In humans, this equilibrium of genetic expression is apparently carefully regulated, and the loss or addition of even a single chromosome is sufficiently disruptive to be lethal or severely deleterious during early development. Plant aneuploids survive better than animal aneuploids, but plant aneuploids are not particularly successful either. Aneuploidy has not played a significant role in plant evolution.

With some exceptions, polyploidy is relatively uncommon in the animal kingdom, probably because of the disruption of the sex chromosome balance. However, the process has had a pronounced impact on plant evolution. Naturally existing autopolyploids are quite adaptive, and those with even numbers of haploid genomes are quite successful in sexual reproduction. The majority of the plant genera contain one or more successful polyploid species. In fact, it has been suggested that most flowering plants with a haploid chromosome number greater than 12 have arisen as a result of polyploidy. Related species often contain multiples of a common number of chromosomes. For example, potatoes have species with 24, 36, 48, 60, 72, 96, 108, 120, and 144 chromosomes. We have already seen that the genus *Triticum* exists in diploid, tetraploid, or hexaploid varieties. These are examples of autopolyploidy.

By providing a mechanism for rapid evolution, allopolyploidy has also been significant in the formation of new plant varieties. If the chromosomes of the two parent species do not share any homology, allotetraploids are likely to exhibit a more extensive phenotypic variation, as compared with the two parent species. Such plants reproduce successfully. Experimental production of these varieties has played an important role in agriculture.

VARIATION IN CHROMOSOME STRUCTURE AND ARRANGEMENT

The second general class of chromosome aberrations includes structural changes that delete, add, or rearrange substantial portions of one or more chromosomes. Included in this category are **deletions** of part of a chromosome, **duplications** of genes or part of a chromosome, and **rearrangements** of genetic material in which a chromosome segment is either inverted, exchanged with a segment of a nonhomologous chromosome, or merely transferred to one. Before discussing these aberrations, we will present several general statements pertaining to them.

In most instances, these changes are due to one or more breaks along the axis of a chromosome, followed by either the loss or rearrangement of some genetic material. Chromosomes can break spontaneously, but the rate of breakage can be increased by the use of chemicals or radiation. Although the ends of chromosomes, the **telomeres,** do not readily fuse with ends of "broken" chromosomes or with other telomeres, the ends produced at points of breakage are "sticky" and can rejoin other such ends. If breakage-rejoining does not merely reestablish the original relationship, and if the alteration occurs in germ plasm, the gametes will contain the structural rearrangement, which will be heritable.

If the aberration is found in one homologue but not the other, unusual but characteristic pairing configurations are formed during meiotic synapsis. These patterns are useful in identifying the type of change that has occurred.

DELETIONS OR DEFICIENCIES

When a portion of a chromosome is lost, the missing piece is referred to as a **deletion** or a **deficiency.** The deletion may occur either at one end or from the interior of the chromosome. These are called **terminal** or **intercalary deletions,** respectively. Both result from one or more breaks in the chromosome. The portion retaining the centromere region will usually be maintained and that segment without the centromere will eventually be lost in progeny cells following mitosis or meiosis. During synapsis between a chromosome with a large intercalary deficiency and a normal complete homologue, the unpaired region of the normal homologue "buckles out." Such a configuration is called a **deficiency loop** or **compensation loop.** The origins of both types of deletion and the formation of such a loop are diagrammed in Figure 12.9.

If too much genetic information is lost as a result of a deletion, the aberration may be lethal and never becomes available for study. As seen in the cri-du-chat syndrome, where only part of the short arm of chromosome 5 is lost, a deficiency need not be very great before the effects become severe.

A final consequence of deletions may be noted in organisms heterozygous for a deficiency. Consider the mutant *Notch* phenotype in *Drosophila*. In these flies,

FIGURE 12.9
Origins of a terminal (a) and intercalary (b) deletion. In part (c), pairing occurs between a normal chromosome and one with an intercalary deletion by looping out the undeleted portion.

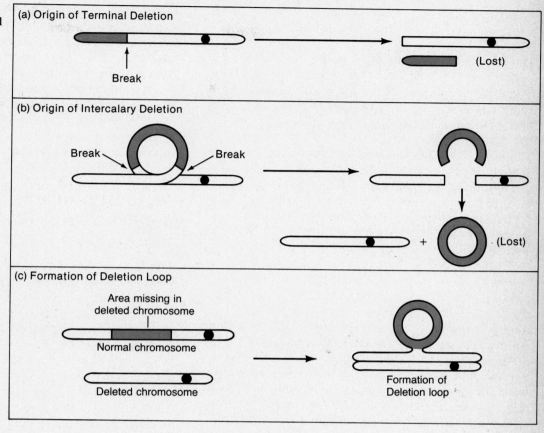

the wings are notched out on the posterior and lateral margins. Data from breeding studies seem to indicate that the phenotype is controlled by a sex-linked dominant gene since heterozygous females have notched wings and transmit this trait to one-half of their male progeny. The gene also appears to behave as a recessive lethal because homozygous *N/N* females and hemizygous *N/ ⁄* males are never recovered. It has also been noted that if notched-winged females are also heterozygous for the closely linked recessive *white*-eye, *facet*-eye, or *split*-bristle mutations, they express these mutant phenotypes as well as *Notch*. Because these mutations are recessive, heterozygotes should express the normal, wild-type phenotypes. These genotypes and phenotypes are summarized in Table 12.2.

These observations, which identified *Notch* as a deficiency rather than a mutation, were explained through a cytological examination of the polytene X chromosome of heterozygous *Notch* females. A deficiency loop was found along the X chromosome from band 3C2 through band 3C11, as shown in Figure 12.10. These bands had previously been shown to include the *white*, *facet*, and *split* loci, among others. On the genetic map, this region corresponds to loci 1.5 to 3.0. Thus, this region's deficiency in one of the two

homologous X chromosomes has two distinct effects. First, it results in the *Notch* phenotype. Second, by deleting the loci for genes whose mutant alleles are present on the other X chromosome, the deficiency creates a hemizygous condition so that the recessive *white*, *facet*, or *split* alleles are expressed. This type of phe-

TABLE 12.2

Notch genotypes and phenotypes.

Genotype	Phenotype
$\dfrac{N^+}{N}$	*Notch* female
$\dfrac{N}{N}$	*Lethal*
$\dfrac{N}{\Longrightarrow}$	
$\dfrac{N^+\ w}{N\ \ w^+}$	*Notch, white* female
$\dfrac{N^+\ fa\ \ spl}{N\ \ fa^+\ spl^+}$	*Notch, facet, split* female

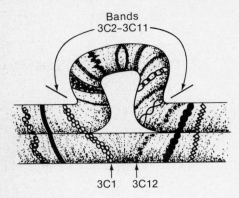

Bands
3C2–3C11

3C1   3C12

FIGURE 12.10
Deletion loop formed in salivary chromosomes
heterozygous for the *Notch* deletion on the X chromosome
of *Drosophila melanogaster*. The deficiency encompasses
bands 3C2 through 3C11.

notypic expression of recessive genes in association
with a deletion is called **pseudodominance.**

Many independently arising *Notch* phenotypes
have been investigated. The common deficient band
for all *Notch* phenotypes has now been designated as
3C7. In every case that *white* was also expressed pseu-
dodominantly, the common deficient band was 3C2.
The bands that cytologically distinguish the *white* locus from the *Notch* locus have been confirmed in this
manner.

DUPLICATIONS

When any part of the genetic material—a single locus
or a large piece of a chromosome—is present more
than once in the genome, it is called a **duplication.** As
in deletions, pairing in heterozygotes may produce a

compensation loop. Duplications may arise as the result of unequal crossing over between synapsed chromosomes during meiosis (Figure 12.11) or through a
replication error prior to meiosis. In the former case,
both a duplication and a deficiency are produced.

Three interesting aspects of duplications will be
considered. First, they may result in gene redundancy.
Second, as with deletions, duplications may produce
phenotypic variation. Third, according to one convincing theory, duplications have also been an important
source of genetic variability during evolution.

Gene Redundancy and Amplification—Ribosomal RNA Genes

Although many gene products are not needed in every
cell of an organism, other gene products are known to
be essential components of all cells. For example, ribosomal RNA must be present in abundance in order
for protein synthesis to occur. The more metabolically
active a cell is, the higher is the demand for this molecule. In theory, a single copy of the gene for the precursor rRNA molecule would be inadequate in most
cells. Studies using the technique of molecular hybridization, which allows the determination of the percentage of the genome coding for specific RNA sequences, have shown that there are multiple copies of
the genes coding for rRNA. Such DNA is called **rDNA,**
and the general phenomenon is called **gene redundancy.** For example, in *E. coli*, about 0.4 percent of the
haploid genome consists of rDNA. This is equivalent to
5 to 10 copies of the gene. In *Drosophila melanogaster*,
0.3 percent of the haploid genome, equivalent to 130
copies, consists of rDNA. While the presence of multiple copies of the same gene is not restricted to those
coding for rRNA, we will focus on them in this section.

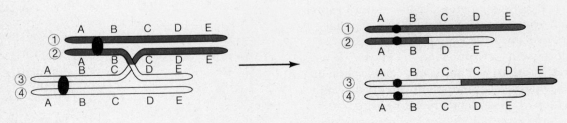

FIGURE 12.11
The origin of duplicated and deficient chromosomes as a result of unequal crossing over. The
tetrad at the left is mispaired during synapsis. A single crossover between strands 2 and 3
results in a deficient and a duplicated chromosome. The two chromatids uninvolved in the
crossover event remain normal in their gene sequence.

Studies of *Drosophila* have documented the need for the extensive amounts of rRNA and ribosomes made possible by multiple copies of these genes. In this organism, the apparent X-linked mutation *bobbed* has, in fact, been shown to be due to a deletion of a variable number of genes coding for rRNA. Mutant flies have low viability, are underdeveloped, and have bristles reduced in size. General development as well as bristle formation, which occurs very rapidly during normal pupal development, apparently depend on a great number of ribosomes to support translation. Many *bobbed* alleles have been studied, and each has been shown to involve a deletion, often of a unique size. The extent to which both viability and bristle length are decreased correlates well with the relative number of RNA genes deleted. Thus, we may conclude that the normal rDNA redundancy observed in wild-type *Drosophila* is the minimum required for adequate ribosome production during normal development.

In some cells, particularly oocytes, even the normal redundancy of rDNA may be insufficient to provide sufficient amounts of rRNA and ribosomes. Oocytes store abundant nutrients in the ooplasm for use by the embryo during early development (see Chapter 2). Probably more ribosomes are included in the oocytes than in any other cell type. By considering how the amphibian *Xenopus laevis* acquires this abundance of ribosomes, we will see a second way in which the amount of rRNA is increased. This phenomenon is referred to as **gene amplification.**

The genes that code for rRNA are located in an area of the chromosome known as the **nucleolar organizer region (NOR).** The NOR is intimately associated with the nucleolus, which is a processing center for ribosome production. Molecular hybridization analysis has shown that each NOR in *Xenopus* contains the equivalent of 400 redundant gene copies coding for the rRNA precursor molecule. Even this number of genes is apparently inadequate to synthesize the vast amount of ribosomes which must accumulate in the amphibian oocyte to support development following fertilization. To further amplify the number of rRNA genes, the rDNA is selectively replicated, and each new set of genes is released from its template. Since each new copy is equivalent to the NOR, multiple small nucleoli are formed around each NOR in the oocyte. As many as 1500 of these "micronucleoli" have been observed in a single oocyte. If we multiply the number of micronucleoli (1500) by the number of gene copies in each NOR (400), we see that amplification in *Xenopus* oocytes can result in over half a million gene copies! In each copy is transcribed only 20 times during the maturation of the oocyte, in theory, sufficient

copies of rRNA may be produced to result in over one billion ribosomes.

The *Bar* Eye Mutation in *Drosophila*

Duplications may cause phenotypic variation that at first might appear to be caused by a simple point mutation. The *Bar*-eye phenotype in *Drosophila* is a classical example. Instead of the normal oval eye shape, *Bar*-eyed flies have narrow, slitlike eyes. This phenotype appears to be inherited as a sex-linked mutation. Since both heterozygous females and hemizygous males exhibit the trait, but homozygous females show a more pronounced phenotype than either of these two cases, the inheritance is more accurately described as **semidominant.**

In the early 1920s Alfred H. Sturtevant and Thomas H. Morgan discovered and investigated this "mutation." As illustrated in Figure 12.12, normal wild-type females (B^+/B^+) have an average of 779 facets in each eye. Heterozygous females (B/B^+) have an average of 358 facets, while homozygous females (B/B) average only 68 facets. Females are occasionally recovered with even fewer facets and are designated as *double Bar* (B^+/B^D).

About ten years later, Calvin Bridges and Herman J. Muller compared the polytene X chromosome banding pattern of the *Bar* fly with that of the wild-type fly. Their studies revealed that one copy of region 16A of the X chromosome was present in wild-type flies and that this region was duplicated in *Bar* flies and triplicated in *double Bar* flies. These observations provided evidence that the *Bar* phenotype is not the result of a conventional mutation but is instead a duplication. The *double Bar* condition originates as a result of unequal crossing over, which produces the triplicated 16A region [Figure 12.12(b)].

Figure 12.12 also illustrates what we have previously referred to as a **position effect.** When the eye facet phenotypes of *B/B* and B^+/B^D flies are compared, an average of 68 and 45 facets are found, respectively. In both cases, there are two extra 16A regions. When the repeated segments are distributed on the same homologue instead of on two homologues, the phenotype is more pronounced. Thus, the same amount of genetic information produces a slightly different phenotype depending on the position of the genes.

The Role of Gene Duplication in Evolution

One of the most intriguing aspects of the study of evolution is the consideration of the mechanisms for genetic variation. The origin of unique gene products

(a)

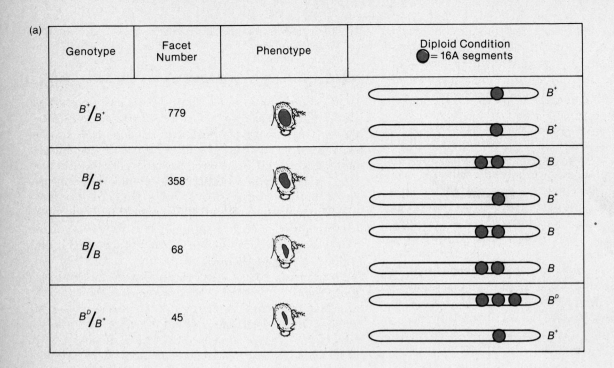

Genotype	Facet Number	Phenotype	Diploid Condition ⬤ = 16A segments
B^+/B^+	779		B^+ B^+
B/B^+	358		B B^+
B/B	68		B B
B^D/B^+	45		B^D B^+

(b) Origin of B^D allele as a result of unequal crossing over

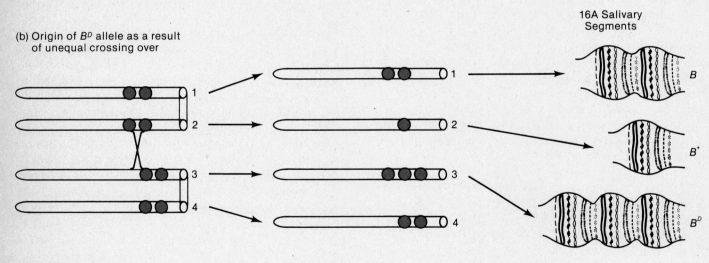

FIGURE 12.12
Summary of the *Bar* duplication in *Drosophila melanogaster.*

present in phylogenetically advanced organisms but absent in less advanced, ancestral forms is a topic of particular interest. In other words, how do "new" genes arise?

In 1970, Susumo Ohno published a provocative monograph entitled *Evolution by Gene Duplication* in which he suggests that gene duplication has been essential to the origin of new genes during evolution. Ohno's thesis is based on the supposition that the

gene products of all unique genes, present as only a single copy in the genome, are indispensable to the survival of any species during evolution. Therefore, unique genes are not free to accumulate mutations sufficient to alter their primary function and give rise to new genes.

However, if an essential gene were to become duplicated, the new copy would be inherited in all future generations. Since it is an extra copy, so to speak, major mutational changes in it are not expected to jeopardize the species because the original gene still provides the genetic information for its essential function. The duplicated copy is now free to undergo large numbers of mutational changes within the organisms of any species over long periods of time. Over short intervals, the new genetic information may be of no practical advantage. However, over long periods of time, the gene may change sufficiently so that its product assumes a divergent role in the cell. The new function may impart an adaptability and survival advantage to organisms carrying such unique genetic information. Thus, Ohno has outlined a mechanism through which sustained genetic variability may have originated.

Ohno's thesis is based on the discovery of proteins with divergent function but whose amino acid sequence is very similar. According to current knowledge of the genetic code, amino acid sequence homology suggests nucleotide sequence homology. Similarity at the gene level suggests a common origin. So far, many pairs of molecules with related but distinct functions that share amino acid homology have been discovered. These include trypsin and chymotrypsin, myoglobin and hemoglobin, hemoglobin subunits, and the light and heavy immunoglobulin chains, to name but a few. As more and more protein molecules are sequenced, the list continues to grow.

As more data have accumulated, Ohno's thesis has gained support. As a result, gene duplication must be considered an important mechanism in the production of genetic variability and in the process of evolution.

INVERSIONS

The **inversion,** another class of structural variation, is a type of chromosomal aberration in which a segment of a chromosome is turned around 180° and reinserted into the chromosome. An inversion does not involve a loss of genetic information but simply rearranges the linear gene sequence. An inversion requires two breaks along the length of the chromosome prior to the reinsertion of the inverted segment. Figure 12.13 illustrates how an inversion might arise. In this way, the newly created "sticky ends" are brought close together and rejoined.

The inverted segment may be short or quite long and may or may not include the centromere. If the centromere is not part of the rearranged chromosome segment, the inversion is said to be **paracentric.** If the centromere is a part of the inverted segment, the term **pericentric** describes the inversion (Figure 12.13).

Although the gene sequence has been reversed in the paracentric inversion, the ratio of arm lengths extending from the centromere is unchanged. In contrast, some pericentric inversions create chromosomes with arms of different lengths than those of the non-inverted chromosome. Thus, the **arm ratio** is often changed when a pericentric inversion is produced (Figure 12.14). The change in arm lengths may sometimes be detected during the metaphase stage of mitotic or meiotic divisions.

Although inversions do not produce new genes or alleles, their consequences are of interest to geneticists. Organisms heterozygous for inversions may produce aberrant gametes. Inversions may also result in

FIGURE 12.13
One possible origin of a pericentric inversion.

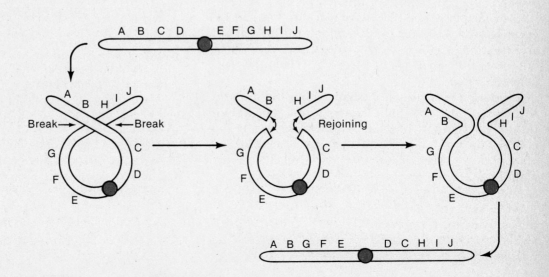

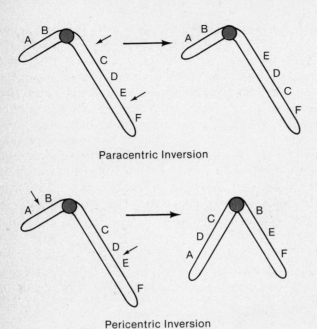

Paracentric Inversion

Pericentric Inversion

FIGURE 12.14
A comparison of the arm ratios before and after the occurrence of a paracentric and pericentric inversion.

position effects as well as play an important role in the evolutionary process.

Consequences of Inversions During Gamete Formation

If only one member of a homologous pair of chromosomes has an inverted segment, normal *linear synapsis* during meiosis is not possible. Organisms with one inverted chromosome and one noninverted homologue are called **inversion heterozygotes.** As shown in Figure 12.15, pairing between two such chromosomes in meiosis may be accomplished only if they form an **inversion loop.** In other cases, no loop can be formed and the homologues are seen to synapse everywhere but along the length of the inversion, where they appear separated.

If crossing over does not occur within the inverted segment of the inversion heterozygote, the homologues will segregate normally. When crossing over occurs within the inversion loop, abnormal chromatids are produced. The effects of single exchange events within paracentric and pericentric inversions are diagramed in Figure 12.15.

As in any meiotic tetrad, a single crossover produces two parental chromatids and two recombinant chromatids. In the case of a paracentric inversion, one recombinant chromatid is **dicentric** (two centro-

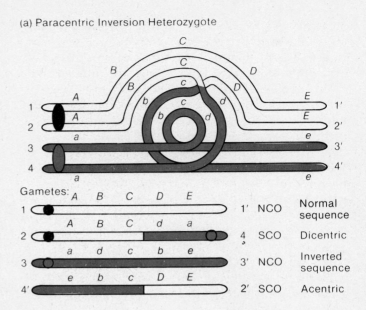

(a) Paracentric Inversion Heterozygote

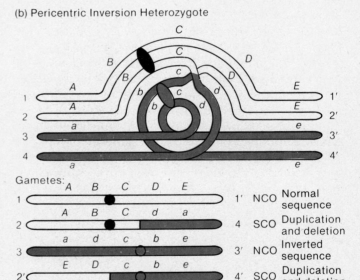

(b) Pericentric Inversion Heterozygote

FIGURE 12.15
The effects of a single-crossover event between nonsister chromatids at a point within a paracentric (a) and pericentric (b) inversion loop.

meres) and one recombinant chromatid is **acentric** (lacking a centromere). Both contain duplications and deletions of chromosome segments as well. During anaphase, an acentric chromatid moves randomly to one pole or the other or may be lost, while a dicentric chromatid is pulled in two directions. This polarized movement produces **dicentric bridges** that are cytologically recognizable. A dicentric chromatid will usually break at some point so that part of the chromatid goes into one gamete and part into another gamete during the reduction divisions. Therefore, gametes containing either recombinant chromatid are deficient in genetic material. When such a gamete participates in fertilization, the zygote most often develops abnormally.

A similar chromosomal imbalance is produced as a result of a crossover event between a chromatid bearing a pericentric inversion and its noninverted homologue. Following meiotic divisions, each tetrad yields two parental chromatids containing the complete chromosome complement of genes. The recombinant chromatids that are directly involved in the exchange have duplications and deletions. However, no acentric or dicentric chromatids are produced. Gametes receiving these chromatids also produce inviable embryos following their participation in fertilization.

Because fertilization events involving these aberrant chromosomes do not produce viable offspring, it appears as if the inversion suppresses crossing over since crossover gametes are not recovered in the offspring. Actually, in inversion heterozygotes, the inversion has the effect of **suppressing the recovery of crossover products when chromosome exchange occurs within the inverted region.** If crossing over always occurred within a paracentric or pericentric inversion, 50 percent of the gametes would be ineffective. The viability of the resulting zygotes is therefore greatly diminished. Furthermore, up to one-half of the viable gametes have the inverted chromosome, and the inversion will be perpetuated within the species. The cycle will be repeated continuously during meiosis in future generations.

Position Effects of Inversions

Another consequence of inversions involves the new positioning of genes relative to other genes and particularly to areas of heterochromatin. If the expression of the gene is altered as a result of its relocation, a change in phenotype may result. Such a change is another example of what is called a **position effect.**

In *Drosophila* females heterozygous for the sex-linked recessive mutation *white* eye (w^+w), the X-chromosome bearing the wild-type allele (w^+) may be inverted. In this particular inversion, the *white* locus moves to a point adjacent to centromeric heterochromatin. If the inversion is not present, a heterozygous female has wild-type red eyes, since the *white* allele is recessive. Females with the X chromosome inversion have eyes that are mottled or variegated (i.e., with red and white patches). Placement of the w^+ allele next to a heterochromatic area apparently causes the loss of

complete dominance over the *w* allele. Other genes, also located on the X chromosome, behave in a similar manner when they are relocated. Reversion to wild-type expression has sometimes been noted. When this has occurred, cytological examination has shown that the inversion has been reversed to give the normal gene sequence.

Evolutionary Consequences of Inversions

One major effect of an inversion is the maintenance of a set of specific alleles at a series of adjacent loci, provided they are contained within the inversion. Because the recovery of crossover products is suppressed in inversion heterozygotes, a particular gene sequence is preserved intact in the viable gametes. If this gene order provides a survival advantage to organisms maintaining it, the inversion is beneficial to the evolutionary survival of the species.

For example, if the set of alleles *ABcDef* is more adaptive than the sets *AbCdeF* or *abcdEF*, the favorable set will not be disrupted by crossing over if it is maintained within a heterozygous inversion. As we will see in Chapter 26, there are documented examples where inversions are adaptive in this way. Specifically, Theodosius Dobzhansky has shown that the maintenance of different inversions on chromosome 3 of *Drosophila pseudoobscura* through many generations has been highly adaptive to this species. Certain inversions seem to be characteristic of enhanced survival under specific environmental conditions.

TRANSLOCATIONS

Translocation, as the name implies, involves the movement of a segment of a chromosome to a new place in the genome. Translocation may occur within a single chromosome or between nonhomologous chromosomes. The exchange of segments between two nonhomologous chromosomes is a type of structural variation called a **reciprocal translocation.** The origin of a relatively simple reciprocal exchange is illustrated in Figure 12.16(a). The least complex way for this event to occur is for two nonhomologous chromosome arms to come close to each other so that an exchange is facilitated. For this type of translocation, only two breaks are required. If the exchange includes internal chromosome segments, four breaks are required, two on each chromosome.

The genetic consequences of reciprocal translocations are, in several instances, similar to those of inversions. For example, genetic information is not lost

FIGURE 12.16

Origin, synapsis, and gamete formation of a simple reciprocal translocation.

(a)
Possible origin of a reciprocal translocation (zygotene stage)

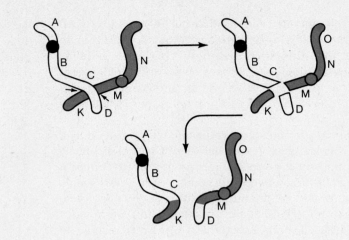

(b)
Heterozygous pairing and gamete formation (pachytene stage)

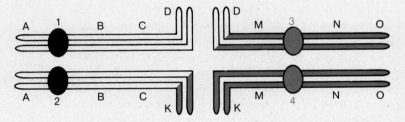

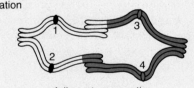

Gamete formation

(c) Two possible pairing arrangements leading to gamete formation

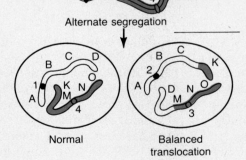

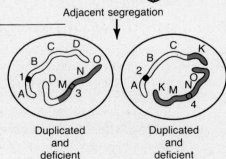

or gained. Rather, there is only a rearrangement of genetic material. The presence of a translocation does not, therefore, directly alter viability following fertilization. Like inversions, translocations may also produce position effects, because they may realign certain genes in relation to other genes or to heterochromatic areas. This exchange does, however, create new ge-

netic linkage relationships that can be detected experimentally.

In Figure 12.16(b), homologues heterozygous for a reciprocal translocation are shown to undergo unorthodox synapsis during meiosis, resulting in a cross-like configuration. As with inversions, genetically unbalanced gametes are also produced. In the case of translocations, however, aberrant gametes are not necessarily the result of crossing over. To see how unbalanced gametes are produced, focus on the homologous centromeres in Figure 12.16(b). According to the principle of independent assortment, the chromosome containing centromere 1 will migrate randomly toward one pole of the spindle during the first meiotic anaphase; it will travel with **either** the chromosome having centromere 3 or centromere 4. The chromosome with centromere 2 will move to the other pole along with **either** the chromosome containing centromere 3 **or** centromere 4. This results in four potential meiotic products. The 1,4 combination contains chromosomes uninvolved in the translocation. The 2,3 combination, however, contains translocated chromosomes. These contain a complete complement of genetic information and are balanced. The other two potential products, the 1,3 and 2,4 combinations, will contain chromosomes displaying duplicated and deleted segments. When incorporated into gametes, the resultant meiotic products are genetically unbalanced. If they participate in fertilization, lethality is the usual result. Therefore, only about 50 percent of the progeny of parents heterozygous for a reciprocal translocation survive. This condition is called **semisterility.**

During the transition from the diplotene stage to diakinesis in meiotic prophase 1, the chiasmata terminalize as the chromosomes begin to pull apart (see Chapter 2). When nonhomologous chromosome pairs contain a series of reciprocal translocations, chromosome rings may be produced. The pairing configuration in this situation becomes quite complex. Depending on the alignment of centromere pairs and the positions of chiasmata, various configurations may be seen prior to the first anaphase. In Figure 12.17, an alternating alignment has occurred. Many other variations have been detected cytologically in a variety of plants. Such a phenomenon has been particularly well studied in the evening primrose, *Oenothera*.

Translocations in Humans: Polydysspondylie and Familial Down Syndrome

Research performed since 1959 has revealed numerous translocations in members of the human population. One common type of translocation involves breaks at the extreme ends of the short arms of two nonhomologous acrocentric chromosomes. These small segments are lost, and the larger segments fuse at their centromeric region (Figure 12.18). This type of translocation produces a new, large submetacentric or metacentric chromosome and is called a **centric fu-**

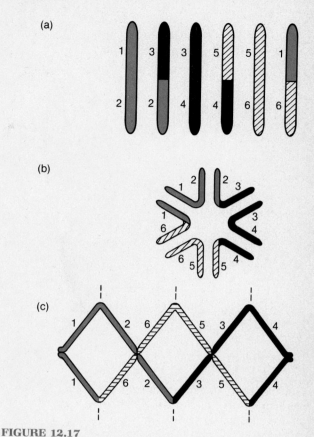

FIGURE 12.17

A set of reciprocal translocations involving three nonhomologous pairs of chromosomes. (a) Translocation heterozygote. (b) Synaptic arrangement in pachytene of meiosis. (c) Metaphase I (zigzag configuration).

sion translocation. Such an occurrence in other organisms is called **Robertsonian fusion.**

This aberration may produce only minor deficiencies and, in many cases, the affected individual may survive. This type of translocation in humans was first discovered in 1959 in an abnormal child who was mentally retarded and suffered from severe spinal malformations. The condition was named **polydysspondylie** by Raymond Turpin. Karyotypic analysis showed that the child's cell nuclei had only 45 chromosomes because of centric fusion of a small acrocentric chromosome of the G group and a large acrocentric chromosome of the D group. Deletions occurring during this fusion are apparently responsible for this phenotype.

A similar translocation accounts for cases in which Down syndrome is inherited or familial. Earlier in this chapter we pointed out that most instances of Down syndrome are due to trisomy 21. This chromosome

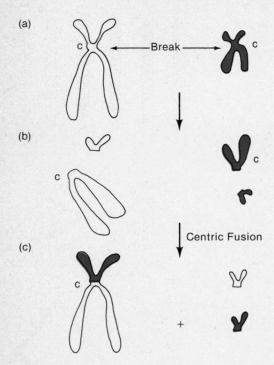

FIGURE 12.18
The possible origin of a Robertsonian translocation. (a) Two independent breaks occur at the centromeric region on two nonhomologous chromosomes. (b) Fragments produced. (c) Centric fusion of the long arms of the two acrocentric chromosomes.

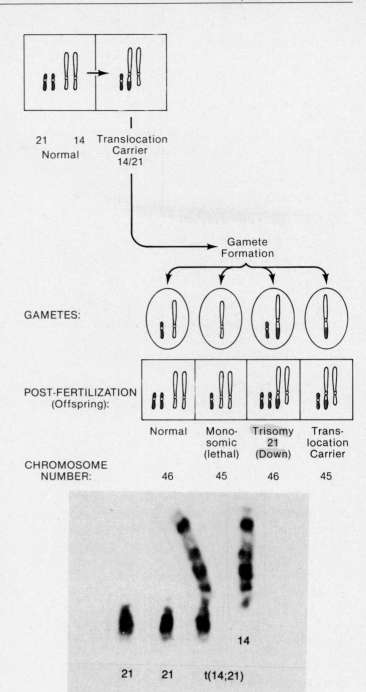

FIGURE 12.19
The origin of a D/G 14/21 translocation leading to familial Down syndrome. (Photo from Yunis and Chandler, 1979.)

composition results from nondisjunction during meiosis of one parent. Trisomy accounts for approximately 95 percent of all cases of Down syndrome. In such cases, the chance of the same parents producing a second afflicted child is very low. However, in about 5 percent of families with a Down child, the syndrome occurs in a much higher frequency over several generations.

Cytogenetic studies of the parents and their offspring from these unusual cases have explained the cause of **familial Down syndrome.** Analysis has revealed that one of the parents contains a **14/21 D/G translocation.** That is, one parent has the majority of the G-group chromosome 21 translocated to the D-group chromosome 14. This individual has only 45 chromosomes, but apparently has a nearly complete diploid content of genetic information. During meiosis (Figure 12.19), one-fourth of the individual's gametes will have two copies of chromosome 21: a normal chromosome and a second copy translocated to chromosome 14. When such a gamete is fertilized by a standard haploid gamete, the resulting zygote has 46 chromosomes but three copies of chromosome 21.

These individuals exhibit Down syndrome. Other potential surviving offspring contain either the standard diploid genome without a translocation or the balanced translocation like the parent. Both cases result in normal individuals, but the surviving offspring of the latter have a 1 in 3 chance of exhibiting Down syndrome. Knowledge of translocations has allowed geneticists to resolve the seeming paradox of an inherited trisomic phenotype in an individual with an apparent diploid number of chromosomes.

Translocations and Human Cancer: The Philadelphia Chromosome, Burkitt's Lymphoma, and Oncogenes

When one thinks about the potential causes of cancer in the human population, three possible etiologies seem most apparent: (1) viruses, (2) environmental substances, and (3) genetic factors. In fact, all three, either independently or in various combinations, have been linked to a variety of malignancies.

From a genetic standpoint, the discovery of **oncogenes** has provided one of the most direct associations between heritable factors and cancer. Every human cell contains a substantial number of such genes, each having the potential to cause cancer. There are twenty currently known, and the list is growing. These genes share DNA sequence homology with certain viral genes associated with transformation of host cells to malignant states. However, human oncogenes apparently carry out normal embryonic functions associated with cell growth and proliferation. It is only when some alteration of the gene occurs that malignant behavior is later induced. Such changes may include point mutations, gene amplification, and incorporation of the gene into a virus infecting the cell.

Most recently, it has become clear that another mechanism—a change in the oncogene's location relative to other genes—can activate its aberrant function. We will discuss two cases where translocations alter the location and affect the function of oncogenes.

The first case involves a karyotypic abnormality called the **Philadelphia chromosome (Ph).*** Discovered in the 1960s in Philadelphia (thus its name), the aberration was initially assessed as a deletion of the long arm of chromosome 22. Such a karyotype, 46,22q- is demonstrated by most bone marrow and white blood cells of the vast majority of patients suffering from **chronic myelocytic leukemia (CML),** and to a much lesser extent by patients with **acute leukemia (AL).**

More thorough analysis involving chromosome banding has revealed that the abbreviated nature of

chromosome 22 is actually the result of a reciprocal translocation between the long arm of chromosome 22 and the long arm of chromosome 8 or 9 (most often, chromosome 9 is involved). The break in chromosome 9 is at band q34, close to the telomere, while the break in chromosome 22 is at band q11, below the centromere. While the above description is of the standard Ph chromosome, which is illustrated in Figure 12.20, other break points have been identified.

There are several curious things about this reciprocal translocation. It appears to be balanced, in the sense that no major genetic information is missing. Why, then, does it lead to malignancy? And, why is it only present in certain somatic tissues? The answers were forthcoming as a result of the discovery and study of oncogenes. Two of these are involved in this translocation. The oncogenes *c-sis*, normally on the long arm of chromosome 22, and *c-abl*, normally on the long arm of chromosome 9, have their positions switched within the karyotype. Apparently, the repositioning of these genes results in the abnormal expression of one or both of them, leading to malignancy in cells bearing the translocation. The most recent support for this hypothesis has come from an analysis of the messenger RNA and protein products reflecting the *abl* gene in CML cells. Both are much larger than their counterparts in normal cells. The additional RNA and protein material is derived from a chromosome 22 gene, *bcr (breakpoint cluster region)*, which is adjacent to and fused with the *abl* gene following translocation. Apparently, the expression of the newly formed *bcr-abl* DNA sequence is regulated by a control unit of the original *bcr* gene. How the altered gene product is involved in the development of leukemia is not clear, however.

The second translocation that rearranges the location of an oncogene occurs in individuals with **Burkitt's lymphoma,** a malignancy of the immune system. Within the immune system are B cells, which make highly specific antibodies against foreign antigenic substances. When these cells become malignant, several different types of chromosomal translocations can be found. For example, the long arms of chromosomes 8 and 14 are most commonly involved in a translocation in such cells. Other rearrangements include reciprocal exchanges between chromosomes 8 and 22 and between chromosomes 2 and 8.

The following observations have helped to explain these B-cell malignancies. First, chromosomes 2, 14, and 22 all contain genes encoding portions of the antibody molecules made by B cells. Second, chromosome 8 contains an oncogene, *c-myc*. Detailed study has revealed that in the case of the 8/14 exchange, this oncogene has been translocated to chromosome 14

---

*The designations Ph¹, Ph', Ph1, and Ph₁ have in the past been used to indicate the Philadelphia chromosome. Most recently, the simpler Ph has been suggested as the standard designation.

FIGURE 12.20
The standard Philadelphia chromosome (Ph) resulting from a reciprocal translocation between one member of chromosome 9 and one member of chromosome 22. The bands shown schematically are derived from preparations of prophase chromosomes.

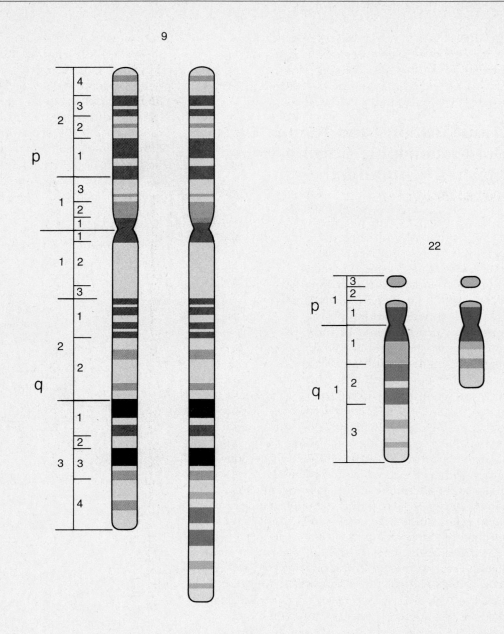

near a region that regulates the expression of the antibody-producing gene. This region, called an **enhancer,** is a DNA sequence that increases the activity of a gene, even though it is itself some distance away from that gene. Enhancers have been discovered in a number of locations in the genomes of a variety of organisms (see Chapter 21).

Apparently, the enhancer dramatically increases the activity of the oncogene, leading to malignant behavior in all B cells bearing this translocation. Since enhancer regions are known to be present on chromosomes 2 and 22, a similar mechanism of oncogene translocation has been proposed for the 8/22 and 2/8 translocations associated with malignant B cells. Given the number of known oncogenes, similar findings can be expected in other malignancies where chromosomal rearrangements have occurred.

FRAGILE SITES IN HUMANS

We conclude this chapter by discussing the results of an interesting discovery made around 1970 during observations of metaphase chromosomes prepared following human cell culture. In cells derived from certain individuals, a specific area along one of the chromosomes failed to stain, giving the appearance of a gap. In different individuals whose chromosomes displayed or expressed such morphology, the gap appeared in different positions within the set of chro-

FIGURE 12.21
A G-banded karyotype of the human genome. Fragile sites are marked by arrows. Solid arrows: Group I folate-sensitive sites at 2q13, 3p14, 6p23, 7p11, 8q22, 9p21, 9q32, 10q23, 11q13, 11q23, 12q13, 16p12, 17p12, 20p11, and Xq27; Open arrow: Group II site at 16q22; Large arrow: Group III site at 10q25.

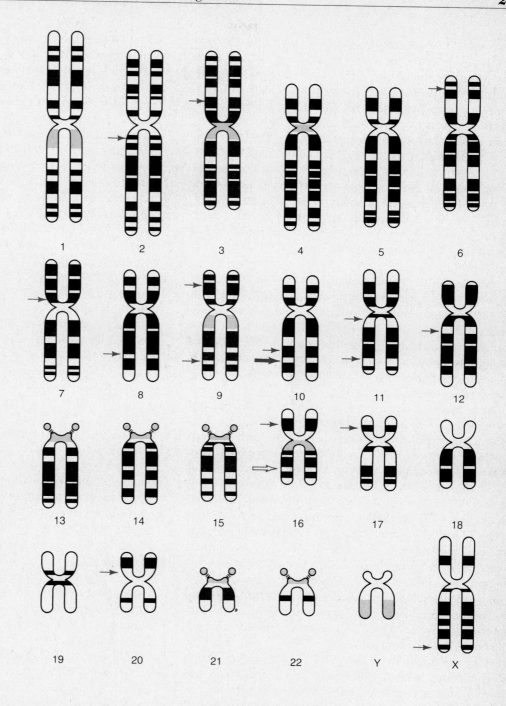

mosomes. Such areas eventually become known as **fragile sites,** since they were considered susceptible to chromosome breakage. Fragile sites were at first considered curiosities until a strong association was subsequently shown to exist between one of the sites and a form of mental retardation. In the following section, we will review this general topic.

Fragile sites are heritable morphological features of mitotic chromosomes that are expressed only under specific conditions of cell culture. Such sites are characterized by:

1 A nonstaining gap or break at a characteristic position along the chromosome. Such breaks usually involve both chromatids of a mitotic chromosome.

2 Inheritance in a codominant, Mendelian fashion, where individuals typically contain an affected and an unaffected chromosome, and transmit one or the other to each offspring.

3 The production of acentric chromosome fragments, deleted chromosomes, and other aberrations in subsequent mitotic divisions.

To date, 17 heritable fragile sites have been identified in the human genome (Figure 12.21). These sites have been grouped into three classes based on the chemical compounds used to induce their expression. The first group consists of 15 sites induced by the removal of folic acid or thymidine from the culture medium and includes positions at 2q13, 3p14, 6p23, 7p11, 8q22, 9p21, 9q32, 10q23, 11q13, 11q23, 12q13, 16p12, 17p12, 20p11, and Xq27. The second group consists of a single site at 16q22, whose expression is enhanced by the addition of distamycin A to the culture medium. The third group consists of a site at 10q25, which is induced by the addition of bromodeoxyuridine (BrdU) or bromodeoxycytidine (BrdC) to the culture medium.

The molecular nature of fragile sites is unknown. Because they appear to represent points along the chromosome susceptible to breakage, these sites may indicate regions that are not tightly coiled or compacted. It should be noted that almost all studies of fragile sites have been carried out in mitotically dividing cells, and it is unknown whether such sites are also expressed in meiotic cells.

Most fragile sites do not appear to be associated with any clinical syndrome except for the folate-sensitive site on the X chromosome (Xq27). Such a chromosome is now called the **fragile X chromosome.** This site is associated with an X-linked form of mental retardation known as **fragile X-linked mental retardation** or the **Martin-Bell syndrome (MBS).** Males afflicted with this condition exhibit long narrow faces with protruding chins, large ears, increased testicular volume, and varying degrees of mental retardation. About 3 to 5 percent of males institutionalized for mental retardation have a fragile X chromosome. Female carriers who have one fragile X and one normal X chromosome show no clearcut physical characteristics, but have a higher incidence of mental retardation than normal individuals.

It is uncertain whether the fragile site on the X chromosome is itself the cause of the physical and mental characteristics seen in the syndrome, or whether the site serves as a marker for a closely linked gene that causes retardation. Fragile X chromosomes have been found in normal males and no other fragile site is associated with a clinical syndrome. Therefore, it seems safer, for the present, to regard fragile sites as structural markers for chromosome with a possible effect on the rate of mitotic segregation, rather than as aberrations responsible for human genetic disease.

SUMMARY AND CONCLUSIONS

Variation in chromosome number and structure has been investigated from the time cytologists began detailed studies of chromosomes. Modern techniques of chromosome spreading and staining have resulted in renewed interest in the subject, particularly as applied to humans. In this chapter, we surveyed the various types of chromosome aberrations and reviewed their genetic consequences. Clearly, chromosome aberrations have been important to the evolutionary process as a source of genetic variation.

Aneuploidy is the gain or loss of one or more chromosomes from the diploid content. When only one chromosome is lost or gained, the conditions of monosomy and trisomy are created, respectively. In this chapter, we have discussed the human trisomies called the Down, Patau, and Edwards syndromes. By investigating these disorders, geneticists have established that the genetic information is present in a delicate equilibrium and cannot be altered substantially without deleterious consequences.

When complete sets of chromosomes are added to the diploid genome, polyploidy is created. The sets may have identical or diverse genetic origin, creating either autopolyploidy or allopolyploidy, respectively. Relatively rare in animals, polyploidy has been an important force in the evolution and domestication of plants.

Variation in chromosome arrangement includes deletions, duplications, inversions, and translocations of genes on larger segments of chromosomes. Both deletions and duplications have, in some cases, been mistaken for mutations when they produce altered phenotypes. If deletions are substantial, serious conditions such as the cri-du-chat syndrome result. Duplications have been particularly important during evolution as a source of genetic material which gives rise to redundant as well as unique genes.

Inversions involve rearrangements of gene order within a chromosome. Because no genetic information is gained or lost, an inversion is not deleterious to the individual. The same may be said for translocations, where segments of chromosomes are transferred to or exchanged with another chromosome. However, when heterozygous, both inversions and translocations may cause genetically abnormal gametes following their participation in meiosis. If crossing over occurs within an inversion, chromatids may be produced that are duplicated and deleted. They may also be acentric or dicentric. Such aberrant gametes cannot produce viable zygotes. Thus, the effect is to suppress the recovery of crossover products. Translocation heterozygotes also produce gametes containing duplicated or deficient chromatids. Thus, both inversions and translocations reduce the fecundity of organisms.

In humans, we have discussed two examples where translocations are responsible for the inheritance of abnormal phenotypes: polydysspondylie and familial

Down syndrome. The latter disorder is usually caused by nondisjunction resulting in trisomy 21. However, in 5 percent of the cases the affected individual is the offspring of a parent carrying a reciprocal translocation involving chromosome 21. Additionally, we have seen that translocations of oncogenes can activate them to induce malignancies. The cases of the Philadelphia chromosome (and chronic myelocytic leukemia) and Burkitt's lymphoma both involve oncogenes that are relocated following a reciprocal translocation.

We have concluded this chapter with a short discussion of fragile sites in human mitotic chromosomes. Thus far, it is not clear if these sites are of great significance in somatic cells. However, the association between one such site on the X chromosome and mental retardation has increased researchers' interest in this topic.

PROBLEMS AND DISCUSSION QUESTIONS

1 Define and distinguish between the following pairs of terms:
   aneuploidy/euploidy
   monosomy/trisomy
   Patau syndrome/Edwards syndrome
   autopolyploidy/allopolyploidy
   polyteny/endopolyploidy
   paracentric inversion/pericentric inversion.

2 Contrast the relative survival times of individuals with Down syndrome, Patau syndrome, and Edwards syndrome. Why do you think such differences exist?

3 What evidence suggests that Down syndrome is more often the result of nondisjunction during oogenesis rather than during spermatogenesis?

4 Why are humans with aneuploid karyotypes usually inviable?

5 What conclusions have been drawn about human aneuploidy as a result of karyotypic analyses of abortuses?

6 Compare the fertility of allopolyploids and autopolyploids.

7 Discuss the role of polyploidy in evolution. How is its role different in animal and plant species?

8 When two plants belonging to the same genus but different species were crossed together, the F_1 hybrid was more viable and had more ornate flowers. Unfortunately, this hybrid was sterile and could only be propagated by vegetative cuttings. Explain the sterility of the hybrid. How might a horticulturist attempt to reverse its sterility?

9 What experimental techniques may be used to convert diploid plants to tetraploids?

10 In Chapter 5, we discussed Calvin Bridges' work on sex determination in *Drosophila*. Based on the information on euploidy in this chapter, describe how Bridges used triploid females to obtain the flies in Figure 5.8?

11 Discuss the origin of cultivated American cotton.

12 For a species with a diploid number of 18, indicate how many chromosomes will be present in the somatic nuclei of individuals who are haploid, triploid, tetraploid, trisomic, and monosomic.

13 Discuss the possible mechanisms involved in the production of deletions, duplications, inversions, and translocations.

14 Contrast the synaptic configurations of homologous pairs of chromosomes in which one member is normal and the other member has a sustained deletion or duplication.

15 Contrast the synaptic configurations in an inversion heterozygote and a translocation heterozygote.

16 Inversions are said to "suppress crossing over." Is this terminology technically correct?

17 Contrast the genetic composition of gametes derived from tetrads of inversion heterozygotes where crossing over occurs with a paracentric and pericentric inversion.

18 In a cross in *Drosophila*, a female heterozygous for the autosomally linked genes *a*, *b*, *c*, *d*, and *e* (*abcde*/+ + + + +) was test-crossed to a male homozygous for all recessive alleles. Even though the distance between each of the above loci was at least 3 map units, only four phenotypes were recovered:

Phenotype	No. of Flies
+ + + + +	440
a b c d e	460
+ + + + e	48
a b c d +	52
Total	1000

Why are many expected crossover phenotypes missing? Can any of these loci be mapped from the data given here?

19 Contrast the *Notch* locus with the *Bar* locus in *Drosophila*. What phenotypic ratios would be produced in a cross between *Notch* females and *Bar* males?

20 Under what circumstances can gene duplication be essential to an organism's survival?

21 Discuss Ohno's hypothesis on the role of gene duplication in the process of evolution.

22 What roles have inversions and translocations played in the evolutionary process?

23 A human female with Turner syndrome also expresses the sex-linked trait hemophilia, as her father did. Which parent underwent nondisjunction during meiosis, giving rise to the gamete responsible for the syndrome?

24 In a plant species where 2*n* = 8, individuals were discovered that were a monosomic and a trisomic for the same chromosome. Determine the possible chromosome compositions of the offspring if these individuals were mated.

25 Contrast the phenomena of gene amplification and gene redundancy.

26 The primrose, *Primula kewensis*, has 36 chromosomes that are similar in appearance to the chromosomes in two related species, *Primula floribunda* (2*n* = 18) and *Primula verticillata* (2*n* = 18). How could *P. kewensis* arise from these species? How would you describe *P. kewensis* in genetic terms?

27 Varieties of chrysanthemums are known that contain 18, 36, 54, 72, and 90 chromosomes; all are multiples of a basic set of 9 chromosomes. How would you describe these varieties genetically? What feature is shared by the karyotypes of each variety? A variety with 27 chromosomes was discovered, but it was sterile. Why?

28 What is the effect of a rare double crossover within either a pericentric or paracentric inversion present heterozygously?

29 If the balanced reciprocal translocation shown below occurred in a gametic cell, would it still be possible for the translocated chromosomes to synapse with their normal homologues during meiosis? If so, draw the synaptic configuration.

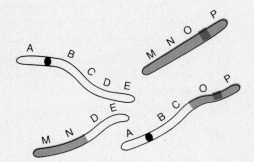

30 Many individuals who are heterozygous for a balanced reciprocal translocation are semisterile since 50 percent of their gametes contain duplications and deletions. The other 50 percent of their gametes are either normal or balanced like the parent (see Figure 12.16). Will this be the case in the gametes formed by the individual described in the previous problem (12.29)? Defend your answer.

SELECTED READINGS

BAKER, W. K. 1968. Position effect variegation. *Adv. in Genet.* 16: 133–69.

BEASLEY, J. O. 1942. Meiotic chromosome behavior in species, species hybrids, haploids and induced polyploids of *Gossypium. Genetics* 27: 25–54.

BLAKESLEE, A. F. 1934. New jimson weeds from old chromosomes. *J. Hered.* 25: 80–108.

BORGAONKER, D. S. 1984. *Chromosome variation in man: A catalogue of chromosomal variants and anomalies.* 4th ed. New York: Alan R. Liss.

BOUE, A. 1985. Cytogenetics of pregnancy wastage. *Adv. Hum. Genet.* 14: 1–58.

BURGIO, G. R., et al., eds. 1981. *Trisomy 21.* New York: Springer-Verlag.

CARR, D. H. 1971. Genetic basis of abortion. *Ann. Rev. Genet.* 5: 65–80.

CROCE, C. M., and KLEIN, G. 1985. Chromosome translocations and human cancer. *Scient. Amer.* (March) 252: 54–60.

DeARCE, M. A., and KEARNS, A. 1984. The fragile X syndrome: The patients and their chromosomes. *J. Med. Genet.* 21: 84–91.

de GOUCHY, J., and TURLEAU, C. 1977. *Clinical atlas of human chromosomes.* New York: Wiley.

DOBZHANSKY, T., AYALA, F., STEBBINS, G., and VALENTINE, J. *Evolution.* San Francisco: W. H. Freeman.

ERICKSON, J. D. 1979. Paternal age and Down syndrome. *Amer. J. Hum. Genet.* 31: 489–97.

FELDMAN, M., and SEARS, E. R. 1981. The wild gene resources of wheat. *Scient. Amer.* (Jan.) 244: 102–12.

GERMAN, J., ed. 1974. *Chromosomes and cancer.* New York: Wiley.

GUPTA, P. K., and PRIYADARSHAN, P. M. 1982. *Triticale:* Present status and future prospects. *Adv. in Genet.* 21: 256–346.

HAMERTON, J. L. 1971. *Human cytogenetics.* Vol. 2. New York: Academic Press.

HASSOLD, T., et al. 1980. Effect of maternal age on autosomal trisomies. *Ann. Hum. Genet. Lond.* 44: 29–36.

HSU, T. C. 1979. *Human and mammalian cytogenetics: A historical perspective.* New York: Springer-Verlag.

HULSE, J. H., and SPURGEON, D. 1974. Triticale. *Scient. Amer.* (Aug.) 231: 72–81.

KAISER, P. 1984. Pericentric inversions: Problems and significance for clinical genetics. *Hum. Genet.* 68: 1–47.

KHUSH, G. S. 1973. *Cytogenetics of aneuploids.* New York: Academic Press.

LEWIS, E. B. 1950. The phenomenon of position effect. *Adv. in Genet.* 3: 73–115.

LEWIS, W. H., ed. 1980. *Polyploidy: Biological relevance.* New York: Plenum Press.

LIMA-de-FARIA, A. 1983. *Molecular evolution and organization of the chromosome.* New York: Elsevier North-Holland.

MANNING, C. H., and GOODMAN, H. O. 1981. Parental origin of chromosomes in Down's syndrome. *Hum. Genet.* 59: 101–3.

OHNO, S. 1970. *Evolution by gene duplication.* New York: Springer-Verlag.

PRAKASH, O., and YUNIS, J. 1984. High resolution chromosomes of the t(9:22) positive leukemias. *Cancer Genet. Cytogenet.* 11: 361–67.

PRIEST, H. 1969. *Cytogenetics.* Philadelphia: Lea and Febiger.

ROWLEY, J. D. 1983. Human oncogene locations and chromosome aberrations. *Nature* 301: 290–91.

SANDBERG, A. A. 1980. *The chromosomes in human cancer and leukemia.* New York: Elsevier North-Holland.

SANDBERG, A. A., et al. 1985. Nomenclature: The Philadelphia chromosome or Ph without superscript. *Cancer Genet. Cytogenet.* 14: 1.

SCHIMKE, R. T., ed. 1982. *Gene amplification.* Cold Spring Harbor, N.Y.: Cold Spring Harbor Laboratory.

SERRA, J. A. 1968. *Modern genetics.* New York: Academic Press.

SHTIVELMAN, E., et al. 1985. Fused transcript of *abl* and *bcr* genes in chronic myelogenous leukemia. *Nature* 315: 550–54.

SIMMONDS, N. W., ed. 1976. *Evolution of crop plants.* London: Longman.

SLAMON, D. J., et al. 1984. Expression of cellular oncogenes in human malignancies. *Science* 224: 256–62.

SMITH, G. F., ed. 1984. *Molecular structure of the number 21 Chromosome and Down syndrome.* New York: New York Academy of Sciences.

STEBBINS, G. L. 1966. Chromosome variation and evolution. *Science* 152: 1463–69.

STENE, J., et al. 1981. Paternal age and Down's syndrome—Data from prenatal diagnoses. *Hum. Genet.* 59: 119–24.

STERN, C. 1977. *Principles of human genetics.* 3rd ed. San Francisco: W. H. Freeman.

SUTHERLAND, G. 1984. The fragile X chromosome. *Int. Rev. Cytol.* 81: 107–43.

————. 1985. The enigma of the fragile X chromosome. *Trends in Genetics* 1: 108–11.

SWANSON, C. P., MERZ, T., and YOUNG, W. J. 1981. *Cytogenetics: The chromosome in division, inheritance, and evolution.* 2nd ed. Englewood Cliffs, N. J.: Prentice-Hall.

TAYLOR, A. I. 1968. Autosomal trisomy syndromes: A detailed study of 27 cases of Edwards' syndrome and 27 cases of Patau's syndrome. *J. Med. Genet.* 5: 227–52.

THERMAN, E. 1980. *Human chromosomes.* New York: Springer-Verlag.

TURPIN, R., and LeJEUNE, J. 1969. *Human afflictions and chromosomal aberrations.* Oxford, England: Pergamon Press.

TURNER, G., and JACOBS, P. 1983. Marker(X) linked mental retardation. *Adv. Hum. Genet.* 13: 53–112.

WHARTON, K. A., et al. 1985. *opa:* A novel family of transcribed repeats shared by the *Notch* locus and other developmentally regulated loci in *D. melanogaster. Cell* 40: 55–62.

WILKINS, L. E., BROWN, J. A., and WOLF, B. 1980. Psychomotor development in 65 home-reared children with cri-du-chat syndrome. *J. Pediatr.* 97: 401–5

WINCHESTER, A. M., and MERTENS, T. R. 1983. *Human genetics.* 4th ed. Columbus, Ohio: Charles E. Merrill.

YUNIS, J. J., ed. 1977. *New chromosomal syndromes.* New York: Academic Press.

13

Mutation and Mutagenesis

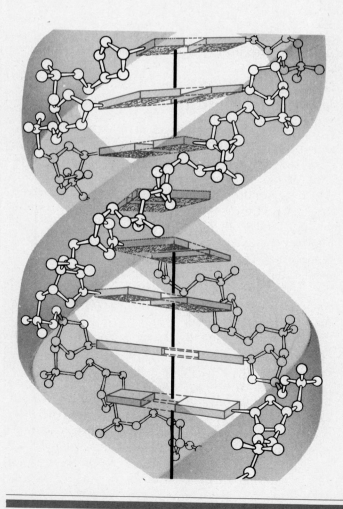

In Chapter 8, we defined the four characteristics or functions ascribed to the genetic information: **replication, storage, expression,** and **variation by mutation.** In a sense, mutation is a failure to store the genetic information faithfully. If a change occurs in the stored information, it will be reflected in the expression of that information and propagated following replication. Historically, the term **mutation** included the category of changes now referred to as **chromosomal aberrations** (see Chapter 12). Now it is often reserved for alteration of information within a single gene. The term **point mutation** is also used because the change usually involves a single nucleotide substitution.

The term *mutation* was coined in 1901 by Hugo DeVries to explain the variation he observed in crosses involving the evening primrose, *Oenothera lamarckiana*. Most of the variation was actually due to multiple translocations, but two cases were subsequently shown to be caused by point mutations. As the studies of mutations progressed, it soon became clear that mutations serve as the source of most alleles and thus are the origin of much of the existing genetic variability within populations. As new alleles arise, whether they are detrimental, neutral, or beneficial, they constitute "raw material" to be tested by the evolutionary process of natural selection.

Mutations are the working tool of the geneticist. The resulting phenotypic variability allows the geneticist to identify and study the genes that control the modified traits. Without the phenotypic variability that mutations provide, all genetic crosses would be meaningless. For example, if all pea plants displayed a uniform phenotype, Mendel would have had no basis for his experimentation. Because of the importance of mutations, great attention has been given to their origin, induction, and classification.

Certain organisms lend themselves to induction of mutations that can be easily detected and studied throughout reasonably short life cycles. Viruses, bacteria, fungi, fruit flies, certain plants, and mice fit these criteria to various degrees. Thus, these organisms have often been used in studying mutation and mutagenesis and through other studies have also contributed to more general aspects of genetic knowledge.

In this chapter we shall consider the classification, detection, and origin of mutations. Additionally, we will discuss the potential dangers posed to human populations by environmental factors that increase mutation rates. These same environmental sources are also believed to induce a wide variety of human cancers.

CLASSIFICATION OF MUTATIONS

There are various schemes by which mutations are classified. They are not mutually exclusive, but instead depend simply on which aspects of mutation are being investigated or discussed. In this section we will describe three sets of distinctions concerning mutations.

Spontaneous versus Induced Mutations

All mutations are described as either **spontaneous** or **induced.** While these two categories overlap to some degree, **spontaneous mutations** are considered as those that arise in nature. No specific agent—other than natural forces—is associated with their occurrence, and they are assumed to arise strictly by chance. They may, therefore, be considered as random changes in the nucleotide sequences of genes.

What causes spontaneous mutations? Although their origin is not fully understood, many such mutations can be linked to normal chemical processes or phenomena that result in rare errors. Such errors may cause alterations in the chemical structure of nitrogenous bases that are part of existing genes; more of-

ten, they may occur during the enzymatic process of DNA replication. We will explore examples of such errors later in this chapter. It is generally agreed that any natural phenomenon that heightens chemical reactivity in cells will lead to more errors. For example, background radiation from cosmic and mineral sources and ultraviolet light from the sun are energy sources to which most organisms are exposed. As such, they may be factors leading to spontaneous mutations. Once an error is present in the genetic code, it may be reflected in the amino acid composition of the specified protein. If the changed amino acid is present in a part of the molecule critical to its biochemical activity, a functional alteration may result.

In contrast to such spontaneous events, those that arise as a result of the influence of any artificial factor are considered to be **induced mutations.** The earliest demonstration of the induction of mutation occurred in 1927, when Herman J. Muller reported that X-rays could cause mutations in *Drosophila.* In 1928, Lewis J. Stadler reported the same finding in barley. In addition to various forms of radiation sources, a wide spectrum of chemical agents is also known to be mutagenic. We will examine specific aspects of mutagenic agents later in this chapter.

Gametic versus Somatic Mutations

When considering the effects of mutation in eukaryotic organisms, it is important to distinguish whether the change occurs in somatic cells or gametes. Mutations arising in somatic cells are not transmitted to future generations. In tissues of an adult organism, thousands upon thousands of cells may be performing a similar function. Thus, a mutation in a single cell may not impair the organism, even if the mutation is detrimental. First, a random mutation might occur in a gene that is not active or essential to the function of that cell. Second, even if a critical gene is affected, there are still thousands of unaffected cells to perform the function of the tissue in question.

Mutations in gametes are of greater concern because they have the potential of being expressed in all cells of an offspring. Such a mutation may also be transmitted to future generations, gradually increasing in frequency within the population. **Dominant autosomal mutations** will be expressed phenotypically in the first generation. **Sex-linked recessive mutations** arising in the gametes of a heterogametic female may be expressed in hemizygous male offspring. This will occur provided the male offspring receives the affected X chromosome. Because of heterozygosity, the occurrence of an **autosomal recessive mutation** in the gametes of either males or females, even one resulting in a lethal allele, may go unnoticed for many generations until it has become widespread in the population. The new allele will become evident only when a chance mating brings two copies of it together in the homozygous condition.

Categories of Mutation

Various types of mutations are classified on the basis of their effect on the organism. Even though some of these categories were introduced earlier, we will briefly review several of them again. A single mutation may well fall into more than one category.

The most obvious mutations are those affecting a **morphological trait.** For example, all of Mendel's pea characters and most genetic variations encountered in the study of *Drosophila* fit this designation. All such morphological variations deviate from the normal or wild-type phenotype.

A second broad category includes mutations that are **nutritional** or **biochemical variations** from the norm. In bacteria and fungi, the inability to synthesize an amino acid or vitamin is an example of a typical nutritional mutation. In humans, sickle-cell anemia and hemophilia are examples of biochemical mutations. While such mutations in these organisms are not visible and do not usually affect specific morphological characters, they can be detected through laboratory analysis.

A third category consists of mutations that affect behavior patterns of an organism. For example, mating behavior or circadian rhythms may be altered. The primary effect of **behavior mutations** is often difficult to discern. For example, the mating behavior of a fruit fly may be impaired if it cannot beat its wings. However, the defect may be in: (1) the flight muscles, (2) the nerves leading to them, or (3) the brain, where the nerve impulses that initiate wing movements originate. The study of behavior and the genetic factors influencing it has benefited immensely from investigations of behavior mutations.

Still another type of mutation may affect the regulation of genes. A regulatory gene may produce a product that controls the transcription of another gene. In other instances, a region of DNA either close to or far away from a gene may modulate its activity. In either case, **regulatory mutations** may disrupt this process and permanently activate or inactivate a gene. Our knowledge of genetic regulation has been dependent on the study of mutations that disrupt this process (see Chapter 21).

Another group consists of **lethal mutations.** Nutritional and biochemical mutations may also fall into

this category. A mutant bacterium that cannot synthesize a specific amino acid needed for its growth will die if plated on a medium lacking that amino acid. Various human biochemical disorders, when untreated, may also be lethal.

Finally, any of the above groups can contain **conditional mutations.** Even though a mutation is present in the genome of an organism, it may not be expressed under certain conditions. The best examples are the **temperature-sensitive mutations** found in a variety of organisms. At certain temperatures, a mutant gene product functions normally, only to lose its functional capability at a different temperature. The study of conditional mutations has been extremely important in experimental genetics.

DETECTION OF MUTATION

Before geneticists can study directly the mutational process or obtain mutant organisms for genetic investigations, they must be able to detect mutations. The ease and efficiency of detecting mutations in a particular organism has by and large determined the organism's usefulness in genetic studies. In this section, we will use several examples to illustrate how mutations are detected.

Detection in Bacteria and Fungi

Detection of mutations is most efficient in haploid microorganisms such as bacteria and fungi. Detection depends on a selection system where mutant cells are isolated easily from nonmutant cells. The general principles are similar in bacteria and fungi. To illustrate, we will describe how nutritional mutations in the fungus *Neurospora crassa* (see Figure 5.3) are detected.

Neurospora is a pink mold that normally grows on bread. It may also be cultured in the laboratory. This eukaryotic mold is haploid in the vegetative phase of its life cycle. Thus, mutations may be detected without the complications generated by heterozygosity in diploid organisms. Pioneering biochemical genetic studies with *Neurospora* were performed by George Beadle and Edward Tatum around 1940.

Visible mutants such as *albino* have been well studied, but the full potential of *Neurospora* genetics was attained during the investigation of nutritional mutants. Wild-type *Neurospora* grows on a **minimal culture medium** of glucose, a few inorganic acids and salts, a nitrogen source such as ammonium nitrate, and the vitamin biotin. Induced nutritional mutants will not grow on minimal medium, but will grow on a supplemented or **complete medium** that also contains numerous amino acids, vitamins, nucleic acid derivatives, and so forth. Microorganisms that are nutritional wild types (requiring only minimal medium) are called **prototrophs,** while those mutants that require a specific supplement to the minimal medium are called **auxotrophs.**

The procedural details for the detection of nutritional mutants are shown in Figure 13.1. Nutritional mutants may be detected and isolated by their failure to grow on minimal medium and their ability to grow on complete medium. The mutant cells can no longer synthesize some essential compound absent in minimal medium but present in complete medium. Once a nutritional mutant is detected and isolated, the missing compound is determined by attempts to grow the mutant strain in a series of tubes, each containing minimal medium supplemented with a single compound. Once the missing compound is found, crosses with the wild type are made, and individual asco-

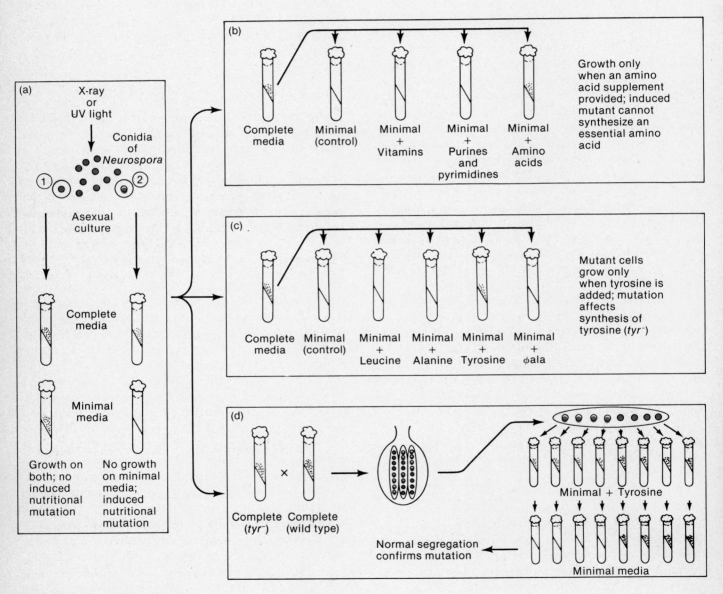

FIGURE 13.1
Induction, isolation, and characterization of a nutritional auxotrophic mutation in *Neurospora.* In (a), conidia 1 is not affected, but conidia 2 (shown in color) contains such a mutation. In (b) and (c), the precise nature of the mutation is determined to involve the biosynthesis of tyrosine. In (d), normal segregation in a cross between mutant and wild-type cells confirms that the *tyr⁻* mutation is nuclear in origin.

spores are dissected out for testing on supplemented and minimal media. Normal segregation yields a ratio of 1 wild type:1 nutritional mutant, confirming that the mutant is of nuclear origin.

In Chapter 14, where the topics of bacterial and viral genetics are introduced, we will return to the topic of mutation. There, we will see that it was not until 1943 that spontaneous mutations were believed to occur in bacteria. Techniques for the detection and study of such mutations will be discussed. Analysis of mutations in microorganisms has been particularly important to the study of molecular genetics (see Part 4).

Detection in *Drosophila*

Muller, in his studies demonstrating that X-rays are mutagenic, developed a number of detection systems in *Drosophila melanogaster*. These systems allow the estimation of the spontaneous and induced rates of sex-linked and autosomal recessive lethal mutations. We will consider two techniques: the **ClB** and the **attached-X-procedures,** which Muller devised. The *ClB* system (Figure 13.2) detects the rate of induction of sex-linked recessive lethal mutations. The *ClB* stock involves *C*, an inversion that suppresses the recovery of crossover products; *l*, a recessive lethal mutation; and *B*, the dominant gene duplication causing *Bar* eye. All of these traits are located on the X chromosome.

In this technique, wild-type P_1 males are treated with a mutagenic agent—in this case, X-rays—and mated to untreated, heterozygous *ClB* females. The F_1 offspring are of four types. One of them, the *ClB* males, dies because it contains the recessive lethal *l*, which is expressed in the hemizygous condition. The remaining class of F_1 males are wild type. They are mated with those F_1 females with *Bar* eyes. The F_1 wild-type females are discarded.

The F_1 *Bar*-eyed females receive the *ClB* chromosome from their mothers and the X-irradiated chromosome from their fathers. Following mating, the females are placed individually in culture bottles and allowed to lay eggs, which will yield an F_2 generation. If a lethal mutation has been produced in the P_1 generation sperm, the F_2 generation will not include any males. One-half of the males die because of the *l* allele on the *ClB* chromosome, and the other half die because of the induced lethal allele *l**. By inspecting the culture bottles for female populations only, the number of induced, sex-linked, recessive lethal mutations is found. If 500 such F_2 culture bottles were inspected, and 25 contained only females, the induced rate of such mutations would be 5 percent. This method also permits the detection of sex-linked, morphological mutations. If a mutation of this type has been induced, all surviving males will show the trait.

The second technique, using females with attached-X chromosomes, is even simpler to use in detecting

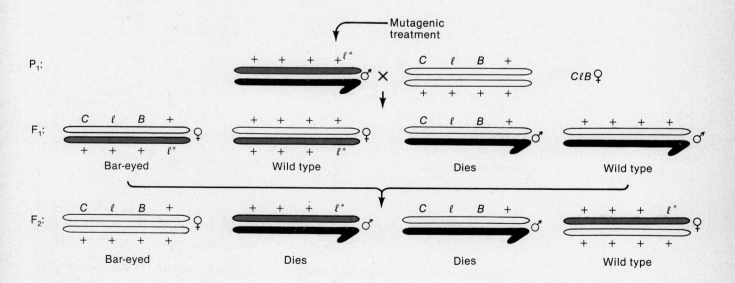

(l^* = induced sex-linked recessive mutation)

FIGURE 13.2

Muller's classical *ClB* technique for the detection of induced, sex-linked recessive lethal mutations in *Drosophila*.

recessive morphological mutations because it requires only one generation. These females have two X chromosomes attached to a single centromere and one Y chromosome, in addition to the normal diploid complement of autosomes. When attached-X females are mated to males with normal sex chromosomes (XY), four types of progeny result: triplo-X females that die; viable attached-X females; YY males that also die; and viable XY males. Figure 13.3 illustrates how P_1 males that have been treated with a mutagenic agent produce F_1 male offspring that express any induced sex-linked recessive mutation. In contrast to the *ClB* technique, which tests only a single X chromosome, the attached-X method tests numerous X chromosomes at one time in a single cross.

Detection techniques have also been devised for recessive autosomal lethals in *Drosophila*. In these techniques, dominant marker mutations are followed through a series of three generations. While more cumbersome to perform, these techniques are also fairly efficient.

Detection in Plants

Genetic variation in plants is extensive. Mendel's peas, for example, were the basis for the fundamental postulates of transmission genetics. Studies of plants have also enhanced our understanding of gene interaction, polygenic inheritance, linkage, sex determination, chromosome rearrangements, and polyploidy. Most variations are detected simply by visual observation. However, there are also techniques for the detection of biochemical mutations in plants. The first is the analysis of biochemical composition of plants. For exam-

ple, the isolation of proteins from maize endosperm, hydrolysis of the proteins, and determination of the amino acid composition have revealed that the **opaque-2 mutant strain** contains significantly more lysine than do other, nonmutant lines. As a result of this discovery, plant geneticists and other specialists have begun to analyze the amino acid compositions of various strains of several grain crops, including corn, rice, wheat, barley, and millet. When the results of these analyses are completed and catalogued, the information will be useful in combating malnutrition diseases resulting from inadequate protein or the lack of the essential amino acids in the diet.

The second detection technique involves tissue culture of plant cell lines in defined medium. The plant cells are handled as microorganisms, and biochemical requirements may be determined by adding or deleting nutrients in the culture medium. There are other advantages to this method. Techniques associated with conditional lethal mutants can be used on plant cells in tissue culture and then applied to the genetics of higher plants. Also, this method provides a detection system that is generally not useful with the intact plant. Temperature-sensitive mutations in plants are beginning to be explored, particularly in tobacco. These studies may add significantly to our understanding of plant growth, metabolism, and genetics.

Detection in Humans

Since designed matings are not possible or desirable in humans, many techniques that have been developed for the detection of mutations in organisms such as *Drosophila* are not available to human geneticists.

FIGURE 13.3
The attached-X method for detection of induced morphological mutations in *Drosophila.*

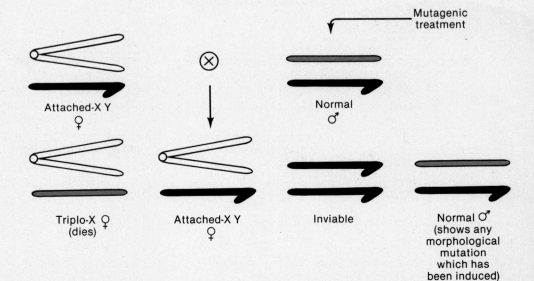

To determine the mutational basis for any human characteristic or disorder, geneticists must first analyze a pedigree that traces the family history as far back as possible. Once any trait has been shown to be inherited, it is possible to predict whether the mutant allele is behaving as a dominant or a recessive and whether it is sex-linked or autosomal.

Dominant mutations are the simplest to detect. If they are present on the X chromosome, affected fathers pass the phenotypic trait to all their daughters. If dominant mutations are autosomal, approximately 50 percent of the offspring of an affected heterozygous individual are expected to show the trait. Figure 13.4 shows a hypothetical pedigree illustrating the initial occurrence of an autosomal dominant allele for cataracts of the eye. The parents in generation I were unaffected, but one of three offspring (Generation II) developed cataracts. This female, the proposita, produced two children, of which the male child was affected. Of his six offspring (IV), four of six were affected (Generation III). These observations are consistent with, but do not prove, an autosomal dominant mode of inheritance. However, the high percentage of affected offspring in generation IV favors this conclusion. Also, the unaffected daughter in this generation argues against sex linkage because she received her X chromosome from her affected father. Provided that the mutant allele is completely penetrant, such a conclusion is soundly based.

Sex-linked recessive mutations may also be detected by pedigree analysis, as discussed in Chapter 5. The most famous case of a sex-linked mutation in humans is that of **hemophilia,** which was found in the descendants of Queen Victoria. The recessive mutation

for hemophilia has occurred many times in human populations, but the political consequences of the mutation that occurred in the royal family have been sweeping. Inspection of the pedigree in Figure 13.5 leaves little doubt that Victoria was heterozygous *(Hh)* for the trait. Her father was not affected, and there is no reason to believe that her mother was a carrier, as was Victoria. Robert Massie's *Nicholas and Alexandra* and Robert and Suzanne Massie's *Journey* (see the list of selected readings at the end of this chapter) provide fascinating reading on the topic of hemophilia.

In a similar manner, it is possible to detect autosomal recessive alleles. Because this type of mutation is "hidden" when heterozygous, it is not unusual for the trait to appear only intermittently through a number of generations. An affected individual and a homozygous normal individual will produce unaffected carrier children. Matings between two carriers will produce, on the average, one-fourth affected offspring.

In addition to pedigree analysis, human cells may now be routinely cultured *in vitro*. This procedure has allowed the detection of many more mutations than any other form of analysis. Analysis of enzyme activity, protein migration in electrophoretic fields, and direct sequencing of DNA and proteins are among the techniques that have demonstrated wide genetic variation between individuals in human populations.

SPONTANEOUS MUTATION RATE

The types of detection systems just described allow geneticists to estimate mutation rates. It is of considerable interest to determine the rate of spontaneous mutation. Not only does such information provide insights into evolution, but it also provides the baseline for measuring the rate of experimentally induced mutation. Induction of mutation can only be ascertained when the induced rate clearly exceeds the spontaneous rate for the organism under study.

Table 13.1 shows the spontaneous mutation rate determined in a variety of organisms. There are three particularly striking features evident in these data. First, the rate is exceedingly low for all organisms studied. Second, the rate is seen to vary considerably in different organisms. Third, even within the same species, the spontaneous mutation rate varies from gene to gene.

Viral and bacterial genes undergo spontaneous mutation on an average of about 1 in 100 million (10^{-8}) cell divisions. While *Neurospora* exhibits a similar rate, maize, *Drosophila*, and humans demonstrate a rate several orders of magnitude higher. The genes studied in these groups average between 1/1,000,000 to

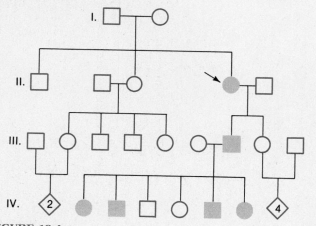

FIGURE 13.4
A hypothetical pedigree of inherited cataract of the eye in humans.

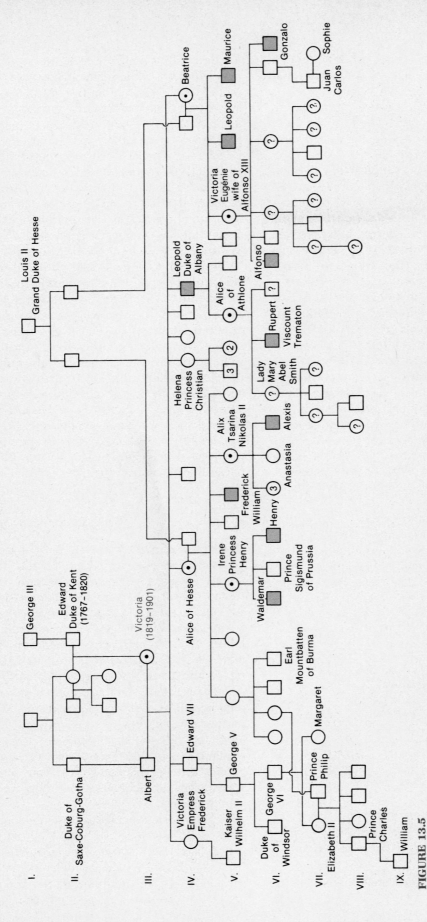

FIGURE 13.5

Pedigree of hemophilia in the royal family descended from Queen Victoria. The pedigree is typical of the transmission of sex-linked recessive traits. Circles with a dot in them (⊙) indicate presumed female carriers heterozygous for the trait. Circles with a question mark ⊙ indicate females whose status is uncertain. (From Victor A. McKusick, *Human Genetics*, 2nd ed., © 1969, p. 56. Reprinted by permission of Prentice-Hall, Inc., Englewood Cliffs, N.J.)

TABLE 13.1
Rates of spontaneous mutations at various loci in different organisms.

Organism	Character	Gene	Rate	Units
Bacteriophage T2	Lysis inhibition	$r \rightarrow r^+$	1×10^{-8}	Per gene
	Host range	$h^+ \rightarrow h$	3×10^{-9}	replication
E. coli	Lactose fermentation	$lac^- \rightarrow lac^+$	2×10^{-7}	
	Lactose fermentation	$lac^+ \rightarrow lac^-$	2×10^{-6}	
	Phage T1 resistance	$T1\text{-}s \rightarrow T1\text{-}r$	2×10^{-8}	
	Histidine requirement	$his^+ \rightarrow his^-$	2×10^{-6}	
	Histidine independence	$his^- \rightarrow his^+$	4×10^{-8}	
	Streptomycin dependence	$str\text{-}s \rightarrow str\text{-}d$	1×10^{-9}	Per cell division
	Streptomycin sensitivity	$str\text{-}d \rightarrow str\text{-}s$	1×10^{-8}	
	Radiation resistance	$rad\text{-}s \rightarrow rad\text{-}r$	1×10^{-5}	
	Leucine independence	$leu^- \rightarrow leu^+$	7×10^{-10}	
	Arginine independence	$arg^- \rightarrow arg^+$	4×10^{-9}	
	Tryptophan independence	$try^- \rightarrow try^+$	6×10^{-8}	
Salmonella typhimurium	Tryptophan independence	$try^- \rightarrow try^+$	5×10^{-8}	Per cell division
Diplococcus pneumoniae	Penicillin resistance	$pen^s \rightarrow pen^r$	1×10^{-7}	Per cell division
Chlamydomonas reinhardi	Streptomycin sensitivity	$str^r \rightarrow str^s$	1×10^{-6}	Per cell division
Neurospora crassa	Inositol requirement	$inos^- \rightarrow inos^+$	8×10^{-8}	Mutant frequency among asexual spores
	Adenine independence	$ade^- \rightarrow ade^+$	4×10^{-8}	
Zea mays	Shrunken seeds	$sh^+ \rightarrow sh^-$	1×10^{-6}	
	Purple	$pr^+ \rightarrow pr^-$	1×10^{-5}	Per gamete per generation
	Colorless	$c^+ \rightarrow c$	2×10^{-6}	
	Sugary	$su^+ \rightarrow su$	2×10^{-6}	
Drosophila melanogaster	Yellow body	$y^+ \rightarrow y$	1.2×10^{-6}	
	White eye	$w^+ \rightarrow w$	4×10^{-5}	
	Brown eye	$bw^+ \rightarrow bw$	3×10^{-5}	Per gamete per generation
	Ebony body	$e^+ \rightarrow e$	2×10^{-5}	
	Eyeless	$ey^+ \rightarrow ey$	6×10^{-5}	
Mus musculus	Piebald coat	$s^+ \rightarrow s$	3×10^{-5}	
	Dilute coat color	$d^+ \rightarrow d$	3×10^{-5}	Per gamete per generation
	Brown coat	$b^+ \rightarrow b$	8.5×10^{-4}	
	Pink eye	$p^+ \rightarrow p$	8.5×10^{-4}	
Homo sapiens	Hemophilia	$h^+ \rightarrow h$	2×10^{-5}	
	Huntington's disease	$Hu^+ \rightarrow Hu$	5×10^{-6}	
	Retinoblastoma	$R^+ \rightarrow R$	2×10^{-5}	Per gamete per generation
	Epiloia	$Ep^+ \rightarrow Ep$	1×10^{-5}	
	Aniridia	$An^+ \rightarrow An$	5×10^{-6}	
	Achondroplasia	$A^+ \rightarrow A$	5×10^{-5}	

$1/100,000$ (10^{-6} to 10^{-5}) mutations per gamete formed. Mouse genes are still another order of magnitude higher in their spontaneous mutation rate, $1/100,000$ to $1/10,000$ (10^{-5} to 10^{-4}). It is not clear why such a large variation occurs in mutation rate. The variation might reflect the relative efficiency of enzyme systems whose function is to repair errors created during replication. Repair systems will be discussed later in this chapter.

THE MOLECULAR BASIS OF MUTATION

For the purposes of the following discussion, we will use a *simplified* definition of a gene in the description of the molecular basis of mutation. In this context, it is easiest to consider a gene as a linear sequence of nucleotide pairs representing stored chemical information. Since the genetic code is a triplet, each sequence of three nucleotides specifies a single amino acid in the corresponding polypeptide. Any change that disrupts the coded information provides sufficient basis for a mutation. The simplest change would be the substitution of a single different nucleotide pair. In Figure 13.6, such a change is compared with our own written language, using three-letter words to be consistent with the genetic code. As you can see, a single change can alter the meaning of the sentence

"THE CAT SAW THE DOG," creating what is called missense. These are analogies to what are most appropriately referred to as **point mutations.** The mutation has turned information that makes sense into various forms of missense. Two terms are often used to describe nucleotide substitutions. If a pyrimidine replaces a pyrimidine or a purine replaces a purine, a **transition** has occurred. If a purine and a pyrimidine are interchanged, a **transversion** has occurred.

A second type of change that could occur is the insertion or deletion of a single nucleotide at any point along the gene. As also shown in Figure 13.6, wherever this occurs, the remainder of the information becomes nonsense. In our analogy, the sentence becomes garbled when one letter is inserted or deleted. These examples are called **frameshift mutations** because the frame of reading has become altered. In either the point or frameshift mutations, the end result is a change in the amino acid sequence of the protein encoded by the altered gene.

Tautomeric Shifts

In 1953, after they had proposed a molecular structure of DNA, Watson and Crick published a paper in which they discussed the genetic implications of this structure. They recognized that the purines and pyrimi-

FIGURE 13.6
The impact of the change, addition, or deletion of one letter in a sentence composed of three-letter words.

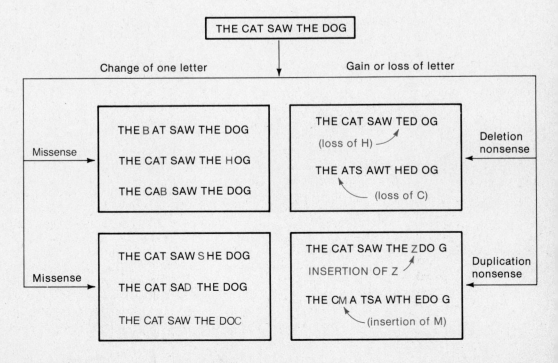

dines found in DNA could exist in **tautomeric forms;** that is, each can exist in several chemical forms, differing by only a single proton shift in the molecule. Watson and Crick suggested that **tautomeric shifts** could result in base pair changes or mutations. The biologically important tautomers involve keto-enol pairs for thymine and guanine, and amino-imino pairs for cytosine and adenine (Figure 13.7).

The infrequent tautomer is capable of hydrogen bonding with a normally noncomplementary base. However, the pairing is always between a pyrimidine and a purine. The end result is the transition where an A=T pair is replaced by a G≡C pair, or vice versa.

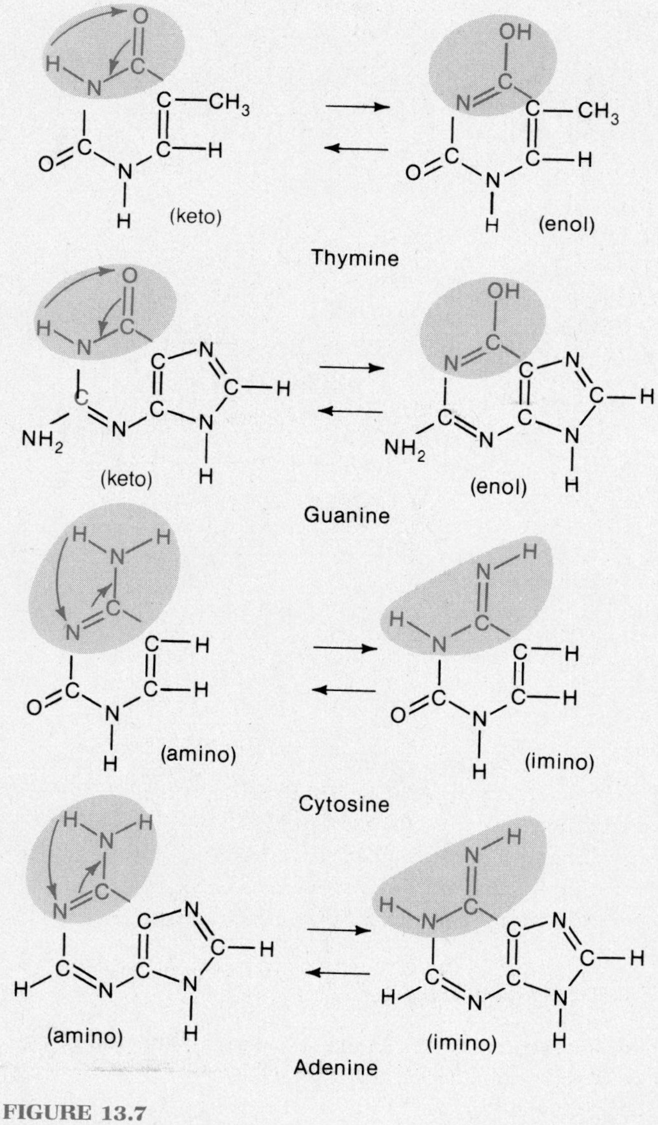

Figure 13.8 compares the normal base-pairing relationships with transition pairings. The effect of the rare tautomers comes at the time of DNA replication: A rare tautomer in the template strand matches with a noncomplementary base. In the next round of replication, the "mismatched" members of the base pair are separated, and each specifies its normal complementary base. As shown in Figure 13.9, the end result is a mutation.

Base Analogues

The **base analogues,** another group of mutagenic chemicals, are molecules that may substitute for purines or pyrimidines during nucleotide and DNA biosynthesis. They were recognized in 1956 to be efficiently incorporated into the DNA of the bacteriophage T4. The analogues were halogenated derivatives of uracil in the number-5 position of the pyrimidine ring: 5-bromouracil, 5-chlorouracil, and 5-iodouracil. Figure 13.10 compares the structure of a thymine analogue, **5-bromouracil** (5-BU),* with the structure of thymine. The presence of the bromine atom in place of the methyl group increases the probability that a tautomeric shift will occur. If 5-BU is incorporated into DNA in place of thymine and then a tautomeric shift occurs, the result is an A=T to G≡C transition. If the tautomeric shift to the enol form occurs before the analogue is incorporated into DNA and 5-BU is mistaken for C, the transition is from G≡C to A=T.

There are other base analogues that are mutagenic. One, **2-amino purine (2-AP),** can serve successfully as an analogue of adenine. In addition to its base-pairing affinity with thymine, 2-AP can also base pair with cytosine. As such, transitions from A=T to G≡C may result following replication.

Because of the specificity by which base analogues such as 2-AP induce transition mutuations, base analogues may also be used to induce reversion to the wild-type nucleotide sequence. This alteration is called **reverse mutation.** The process may also occur spontaneously, but at a much lower rate.

Nitrous Acid

Nitrous acid is one of a number of chemical compounds known to be mutagenic. It is capable of oxidatively deaminating purines and pyrimidines. In this reaction, an amino group is converted to a keto group

FIGURE 13.7
Rare tautomeric shifts that occur in the chemical structure of the four nitrogenous bases of DNA.

*If 5-BU is chemically linked to d-ribose, the analogue 5-bromouracildeoxy-ribonucleoside (BrdU), which has been introduced previously, is formed.

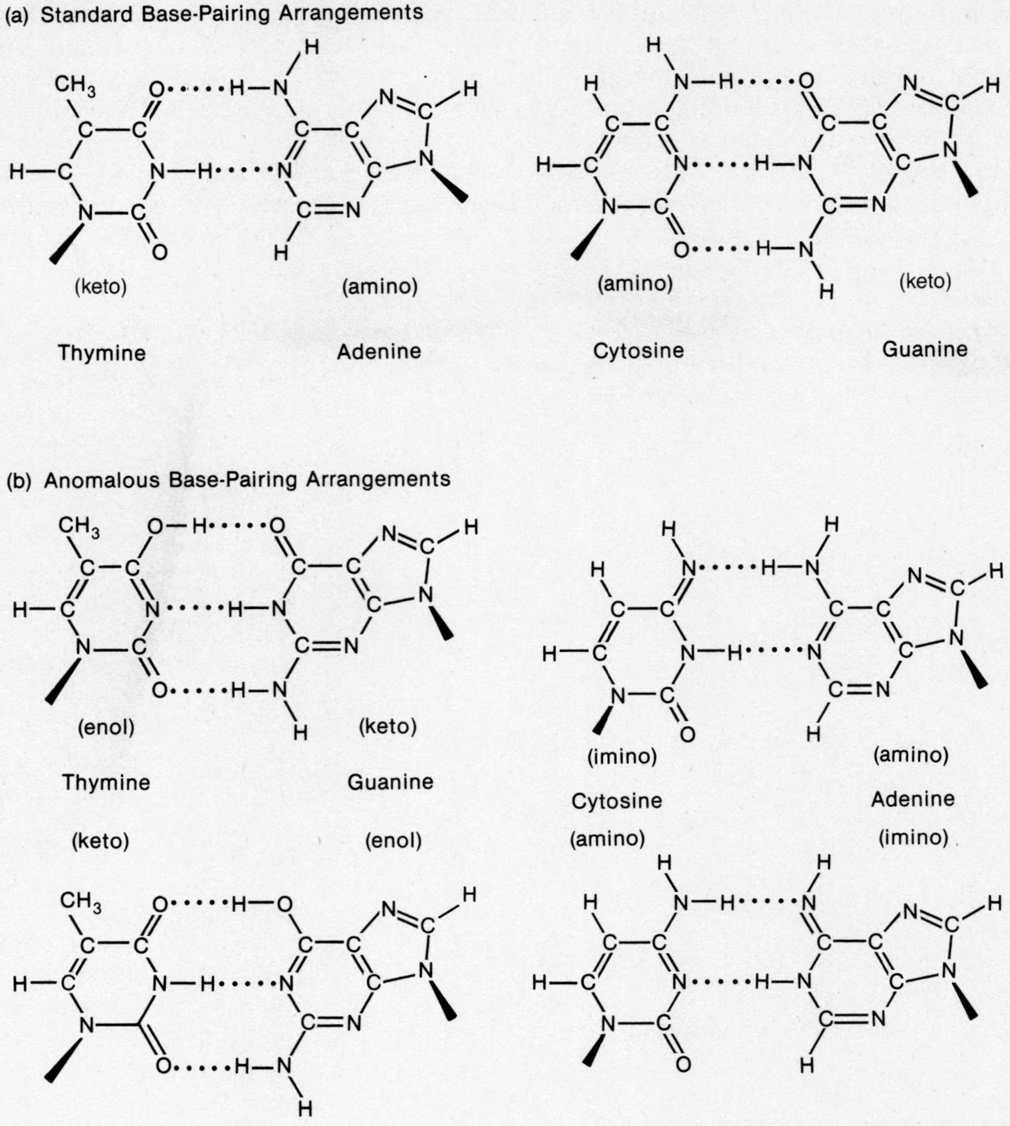

FIGURE 13.8
The standard base-pairing relationships compared with anomalous arrangements occurring as a result of tautomeric shifts. The dense arrow indicates the point of bonding to the pentose sugar.

(a) Standard Base-Pairing Arrangements

(keto)   (amino)

Thymine   Adenine

(amino)   (keto)

Cytosine   Guanine

(b) Anomalous Base-Pairing Arrangements

(enol)   (keto)

Thymine   Guanine

(keto)   (enol)

(imino)   (amino)

Cytosine   Adenine

(amino)   (imino)

in cytosine and adenine (Figure 13.11). In these two cases, cytosine is converted to uracil, and adenine is changed to hypoxanthine.

The major effect of these changes is to alter the base-pairing specificities of these two molecules during DNA replication. For example, cytosine normally pairs with guanine. Following conversion to uracil, which pairs with adenine, the original G≡C pair converted to an A=U pair and, following an additional replication, to an A=T pair. When adenine is deaminated, an original A=T pair is converted to a G≡C pair because hypoxanthine pairs naturally with cytosine.

Guanine is converted to the purine xanthine following deamination by nitrous acid (Figure 13.11). Xanthine does not have a pairing affinity with either of the pyrimidines, thymine or cytosine. Thus, replication at this point along the nucleotide chain cannot occur.

Hydroxylamine

Hydroxylamine (NH_2OH) reacts specifically with cytosine by adding a hydroxyl group to its existing amino group (Figure 13.12). The new product, hydroxylaminocytosine, may now undergo a tautomeric shift, allowing it to pair with adenine. Following two repli-

FIGURE 13.9
Formation of an A-T to a G-C base-pair transition as a result of a tautomeric shift in A during replication and subsequent anomalous base pairing.

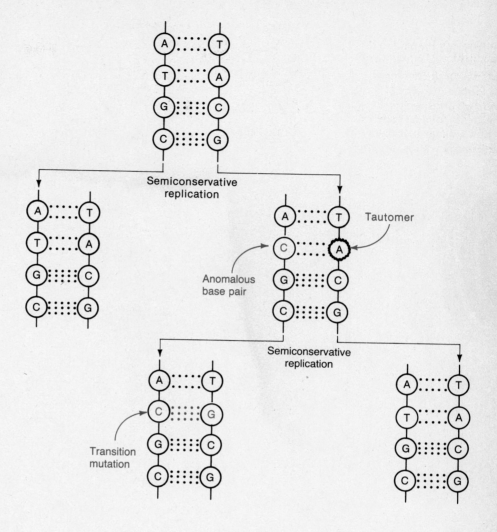

cations, a G≡C pair will be converted to an A=T pair. In this example, a specific, predictable mutation is induced.

Alkylating Agents

The sulfur-containing **mustard gases** were one of the first groups of chemical mutagens discovered. This discovery was made as a result of studies concerned with chemical warfare during World War II. Mustard gases are **alkylating agents;** that is, they donate an alkyl group such as CH_3— or CH_3—CH_2— to amino or keto groups in nucleotides. **Ethylmethane sulfonate (EMS),** for example, alkylates the keto groups in the number-6 position of guanine and in the number-4 position of thymine (Figure 13.13). As with nitrous acid and hydroxylamine, base-pairing affinities are altered and transition mutations result. In the case of 6-ethyl

guanine, this molecule acts like a base analogue of adenine, causing it to pair with thymine. Table 13.2 lists the chemical names and structures of several frequently used alkylating agents known to be mutagenic.

Acridine Dyes and Frameshift Mutations

Other chemical mutagens cause **frameshift mutations.** These result from the addition or removal of one or more base pairs in the polynucleotide sequence of the gene. Inductions of frameshift mutations have been studied in detail with a group of aromatic molecules known as **acridine dyes** or **acridines.** The structures of **proflavin,** the most widely studied acridine mutagen, and **acridine or-**

FIGURE 13.10
Similarity of 5-bromouracil (5-BU) structure to thymine structure. In the common keto form, 5-BU pairs normally with adenine, behaving as an analogue. In the rare enol form, it pairs anomalously with guanine.

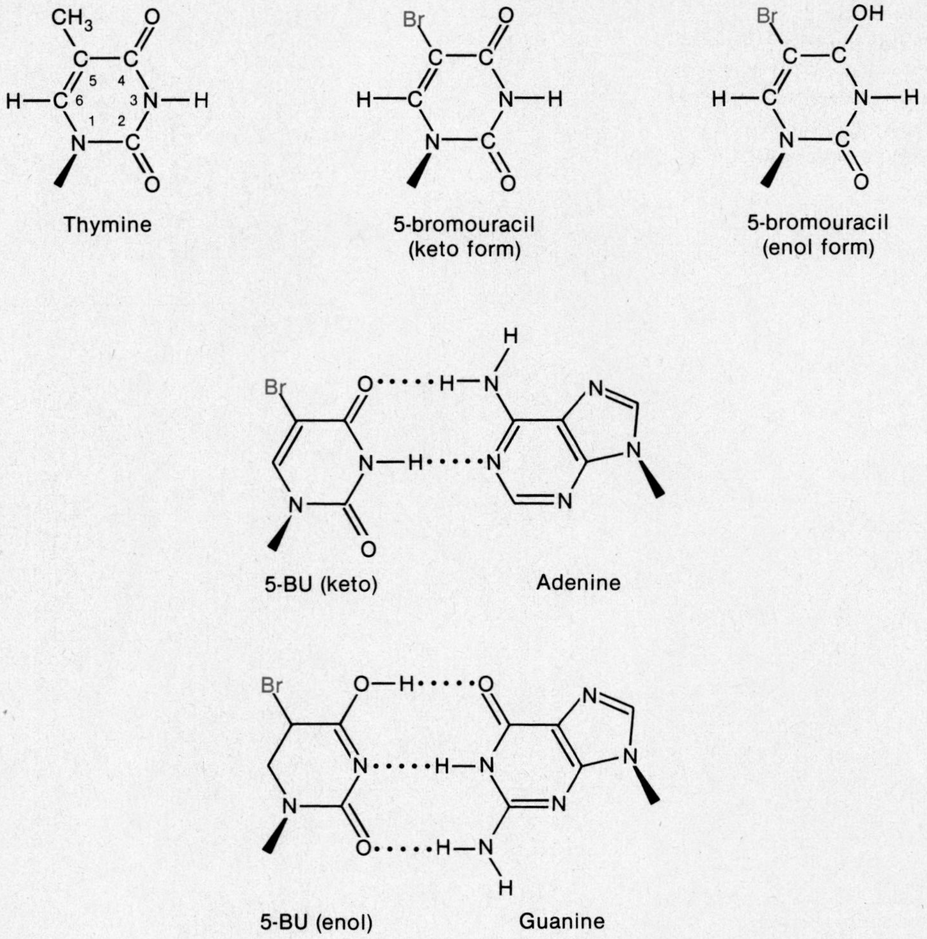

Thymine

5-bromouracil (keto form)

5-bromouracil (enol form)

5-BU (keto)   Adenine

5-BU (enol)   Guanine

FIGURE 13.11
Deaminations caused by nitrous acid (HNO₂) leading to new base-pairing arrangements and mutations. G≡C to A=T and A=T to G≡C transitions result.

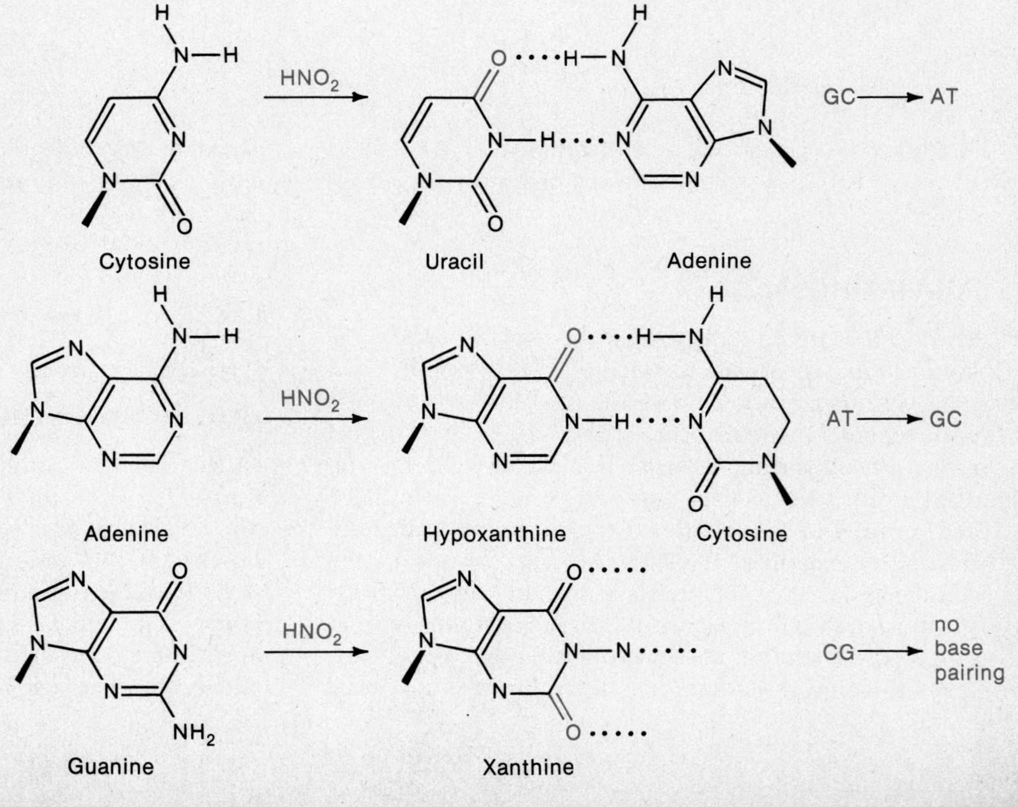

Cytosine   Uracil   Adenine

$GC \longrightarrow AT$

Adenine   Hypoxanthine   Cytosine

$AT \longrightarrow GC$

Guanine   Xanthine

$CG \longrightarrow$ no base pairing

FIGURE 13.12
Reaction of hydroxylamine with cytosine, converting it to a form that base pairs with adenine. The result is the G≡C to A=T transition mutation.

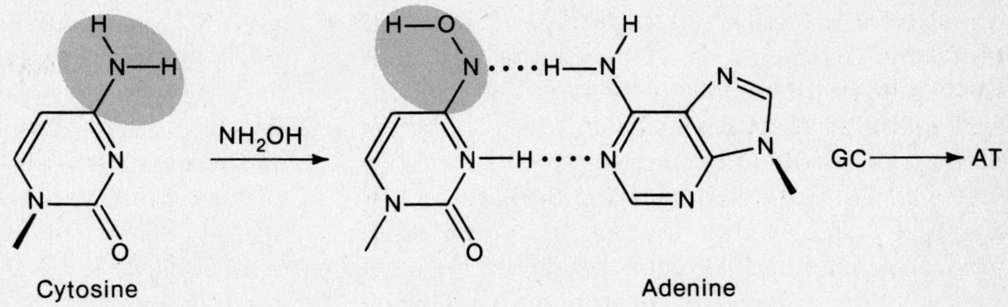

Cytosine   Adenine

FIGURE 13.13
Conversion of guanine to 6-ethylguanine by the alkylating agent ethylmethane sulfonate (EMS). 6-ethylguanine base pairs with thymine.

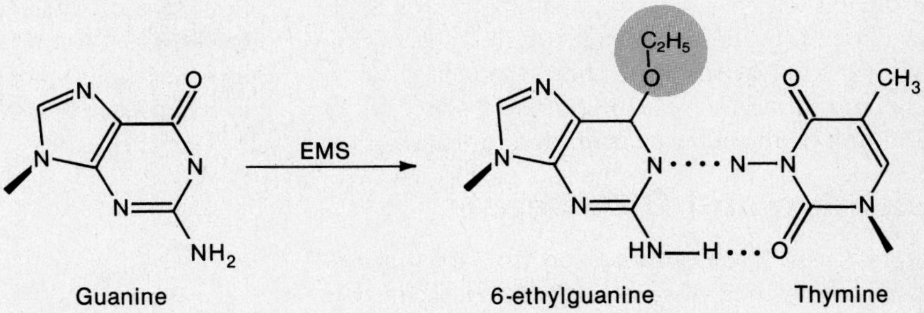

Guanine   6-ethylguanine   Thymine

TABLE 13.2
Alkylating agents.

Common Name or Symbol	Chemical Name	Chemical Structure
Mustard gas (sulfur)	Di-(2-chloroethyl)sulfide	Cl—CH₂—CH₂—S—CH₂—CH₂—Cl
EMS	Ethylmethane sulfonate	$CH_3-CH_2-O-\overset{O}{\underset{O}{\overset{\|\|}{\underset{\|}{S}}}}-CH_3$
EES	Ethylethane sulfonate	$CH_3-CH_2-O-\overset{O}{\underset{O}{\overset{\|\|}{\underset{\|}{S}}}}-CH_2-CH_3$
MMS	Methylmethane sulfonate	$CH_3-O-\overset{O}{\underset{O}{\overset{\|\|}{\underset{\|}{S}}}}-CH_3$
DES	Diethylsulfate	$CH_3-CH_2-O-\overset{O}{\underset{O}{\overset{\|\|}{\underset{\|}{S}}}}-O-CH_2-CH_3$
NG	N-methyl-N′-nitro-N-nitrosoguanidine	HN=C—NH—NO₂ ⎮ O=N—N—CH₃

ange are shown in Figure 13.14. Acridine dyes are of about the same dimension as a nitrogenous base pair and are known to intercalate or wedge between purines and pyrimidines of intact DNA. Intercalation of acridine dyes is considered to induce contortions in the DNA helix, causing deletions and additions.

One model suggests that the resultant frameshift mutations are generated at gaps produced in DNA during replication, repair, or recombination. During these events, there is the possibility of slippage and improper base pairing of one strand with the other. The model suggests that intercalation of the acridine into an improperly base-paired region may prolong the time during which these slippage structures exist. If so, the probability increases that the mispaired configuration will exist at the time that synthesis and rejoining occurs, thereby resulting in an addition or deletion of one or more bases from one of the strands.

UV Radiation and DNA Repair

In Chapters 8 and 9, we emphasized the fact that purines and pyrimidines absorb **ultraviolet light (UV)** most intensely at a wavelength around 260 nm. This property has been used extensively in the detection and analysis of nucleic acids. In 1934, as a result of studies involving *Drosophila* eggs, it was discovered that ultraviolet light is mutagenic. By 1960, several studies concerning the *in vitro* effect of UV light on the components of nucleic acids had been completed with the following conclusions. The major effect of UV light is on pyrimidines at a point where the addition of water to the ring structure is initiated. Subse-

quently, pyrimidine dimers are formed, particularly between two thymine residues (Figure 13.15). While cytosine-cytosine and thymine-cytosine dimers may also be formed, they are less prevalent. It is believed that the dimers distort the DNA conformation and inhibit normal replication. This inhibition seems to be responsible for the killing effects of UV light on microorganisms.

Other studies have provided more complete information on the relationship of UV light to genetic phenomena. These studies have demonstrated that an intricate set of repair processes function to counteract the lesions produced in DNA as a result of UV irradiation. It is during the repair process that most errors leading to UV-induced mutations occur. In the following discussion, we will examine DNA repair systems, as elucidated in the bacterium *E. coli*.

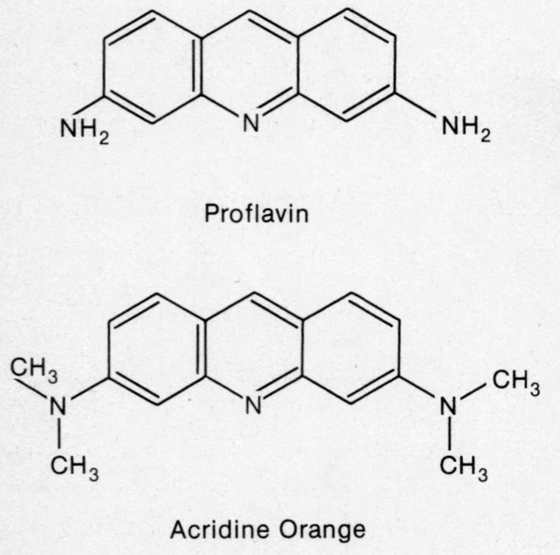

FIGURE 13.14
Chemical structures of proflavin and acridine orange, which intercalate with DNA and cause frameshift mutations.

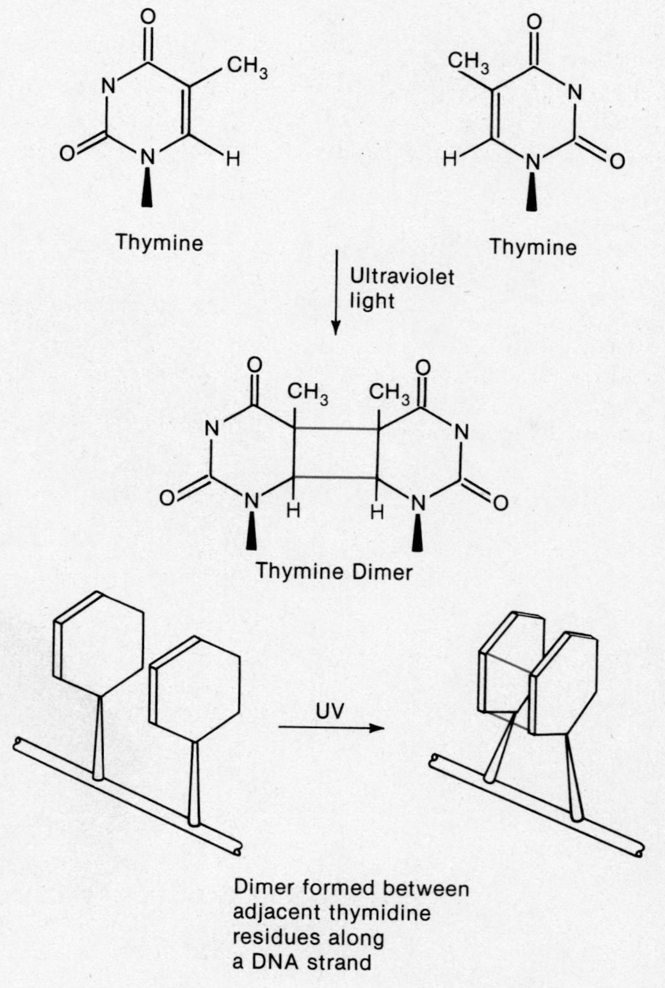

FIGURE 13.15
Formation of thymine dimers induced by ultraviolet light.

The first relevant discovery concerning UV repair in bacteria was made in 1949 when Albert Kelner observed the phenomenon of **photoreactivation.** He showed that the UV-induced damage to *E. coli* DNA could be partially reversed if, following irradiation, the cells were briefly exposed to light in the blue range of the visible spectrum. The photoreactivation process has subsequently been shown to be temperature-dependent, thereby suggesting that the light-induced mechanism involves an enzymatically controlled chemical reaction. Visible light appears to induce a process of repair in the DNA damaged by UV light.

Further studies of photoreactivation have revealed that the process is due to an enzyme, called the **photoreactivation enzyme (PRE).** This molecule may be isolated from extracts of *E. coli* cells, and its repair activity is destroyed by heat. The enzyme's mode of action is to cleave the bonds between thymine dimers, thus reversing the effect of UV light on DNA [Figure 13.16(a)]. While the enzyme will associate with a dimer in the dark, it must absorb a photon of light to cleave the dimer.

Investigations in the early 1960s suggested that a repair system or systems that do not require light also existed in *E. coli*. Paul Howard-Flanders and coworkers isolated several independent mutants demonstrat-

ing increased sensitivity to ultraviolet radiation. One group was designated *uvr* (ultraviolet repair) and included the *uvrA, uvrB,* and *uvrC* mutations. These genes were subsequently shown to be involved in a process called **excision repair** [Figure 13.16(b)]. During this process in wild-type *E. coli*, UV-induced dimers are cut out of the DNA strand along with some nucleotides on either side. The gap is then filled in by repair synthesis, which can be carried out by **DNA polymerase I.** The final step in excision repair is the sealing of the phosphodiester backbone by **DNA ligase,** the joining enzyme. Recall from Chapter 10 that DNA polymerase I is the enzyme discovered by Kornberg and long assumed to be the universal DNA-replication enzyme. The discovery of the *pol A1* mutation demonstrated that this was not the case. *E. coli* cells carrying the *pol A1* mutation lack functional polymerase I. However, replication of DNA takes place normally. Cells with this mutation are unusually sensitive to UV light. Apparently, such cells are unable to excise thymine dimers. This finding demonstrated the importance of excision repair mechanisms in counteracting the effects of UV light.

A second light-independent repair mechanism was discovered in 1968 and is called **post-replication repair.** This mechanism was discovered in an excision-

FIGURE 13.16
Contrasting diagrams of photoreactivation and excision repair of UV-induced thymine dimers.

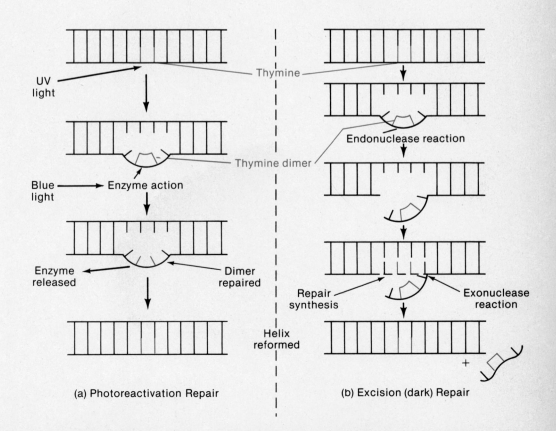

(a) Photoreactivation Repair

(b) Excision (dark) Repair

defective strain of *E. coli*. DNA strands in cells of this strain possess over 100 pyrimidine dimers per chromosome after a single dose of UV irradiation. Following a single round of replication, discontinuous gaps are found opposite the pyrimidine dimers, along both strands of the DNA helix. However, in the first hour following the initial replication, the degree of discontinuity of the strands diminishes as if the gaps are gradually disappearing. The cells that show this phenomenon may survive as viable progeny, but their re-covery is dependent on the product of a gene, *rec A*, which is also essential to recombinational phenomena in *E. coli*.

Figure 13.17 shows one model that attempts to explain how post-replication repair occurs to counteract the effect of just one dimer. Undoubtedly, some dimers escape repair even in cells with their PRE and excision repair systems intact. When unrepaired by either the photoreactivation or excision repair process, the dimer severely interferes with the ability of DNA

FIGURE 13.17
A model illustrating the proposed mechanisms involved in post-replication repair of one UV-induced thymine dimer.

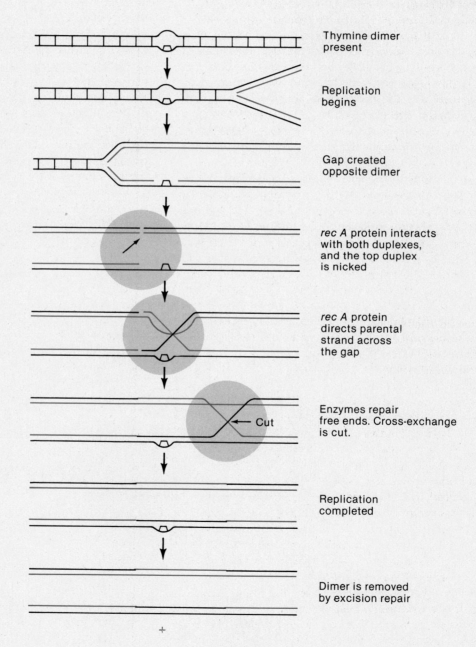

Thymine dimer present

Replication begins

Gap created opposite dimer

rec A protein interacts with both duplexes, and the top duplex is nicked

rec A protein directs parental strand across the gap

Enzymes repair free ends. Cross-exchange is cut.
← Cut

Replication completed

Dimer is removed by excision repair

polymerase III to synthesize the complementary strand. The enzyme skips over the dimer, creating a gap. In the absence of post-replication repair, the DNA remains broken and disrupts further rounds of replication. To overcome this effect, the *rec A* protein interacts with both duplexes, serving to align them. An enzyme nicks the duplex that is free of the dimer. Then, the parental member of the nicked duplex, which is complementary to the strand containing the dimer, is aligned across the gap. Through a recombinational process, still influenced by the *rec A* protein, the DNA strands are realigned and all loose ends are rejoined. Replication is now complete; however, the dimer is still present. It can now be repaired by the excision repair system.

The preceding discussion demonstrates that post-replication repair is really a rescue operation. It allows replication to occur when it would otherwise be unsuccessful. Philip Hanawalt and Paul Howard-Flanders, among others, have established by careful study that many different gene products—perhaps 50 or more—are involved. Of particular interest is the *lex A* product, which serves to repress the function of the *rec A* and *uvr* genes. However, when a *rec A* protein binds to DNA in the area of a post-replication gap, this binding somehow activates a second function of the *rec A* protein—the ability to cleave the *lex A* repressor molecule. The absence of a functional repressor molecule allows the activation of the *rec A* and *uvr* genes, leading to an increased production of the proteins for which they code. The activation of the repair system in this way has been called the **SOS response,** which counteracts most of the dimers through the process of excision and post-replication repair.

One disadvantage of the repair process associated with the SOS response is a diminution of an important function performed by DNA polymerase III. Normally, the enzyme **proofreads** as it proceeds along the DNA strand. This editing function is possible because of 3′-exonuclease function of the enzyme. Proofreading occurs at the 3′ terminus of the growing DNA chain; a nucleotide is excised immediately if it is not correctly base paired with its corresponding nucleotide present in the template strand. However, during the SOS response, the proofreading function is relaxed so that polymerization does not halt when it encounters a dimer that has distorted the shape of the double helix. While the mechanism by which the editing system is shut down is not yet understood, the effect is to prevent the correction of spontaneous errors at sites other than dimers. Thus, the fidelity of DNA replication is diminished. As a result, the SOS response to the repair of DNA is said to be an **error-prone system.**

Fortunately, the SOS repair system is not constantly engaged; instead it only becomes active when induced by UV-light damage to DNA.

UV Light and Human Skin Cancer: Xeroderma Pigmentosum

Xeroderma pigmentosum (XP), a rare autosomal recessive disorder in humans, predisposes individuals to epidermal pigment abnormalities. Exposure to sunlight results in malignant growth of the skin. The condition is very severe and leads to a reduced life span. Since sunlight contains UV radiation, scientists have suspected a causal relationship between thymine dimer production and XP. They have also been trying to determine which of the three forms of repair processes counteracting the effects of UV-induced damage to DNA, if any, are operating in humans. Furthermore, the question has been raised whether XP individuals might lack one or more repair systems, which may cause them to be susceptible to UV-induced skin damage.

The three modes of repair of UV-induced lesions—**photoreactivation, excision repair,** and **post-replication repair**—have been investigated in human fibroblast cultures derived from XP and normal individuals. Fibroblasts are undifferentiated connective tissue cells. The results, most of which have been obtained since 1968, are varied and suggest that the XP phenotype may be caused by more than one mutant gene, as the following summary shows:

1. In 1968, James Cleaver showed that cells from XP patients were deficient in the **unscheduled DNA synthesis** elicited in normal cells by UV light. This assay is thought to represent activity of the excision repair system, suggesting that XP cells are deficient in this form of repair.

2. There are two clinical forms of XP, the classical disease and the **DeSanctis-Cacchione syndrome** characterized by the additional involvement of neurological symptoms. Both types have been shown to be deficient in the excision repair process.

3. In 1974, the presence of a **photoreactivation enzyme** (PRE) was established in normal human cells. Betsy Sutherland identified the enzyme first in leukocytes and subsequently in fibroblast cells.

4. Sutherland further demonstrated that some XP cultures contained a lower PRE activity than control cultures. The activity in various XP cell strains ranged from zero to 50 percent in cultures established from different patients.

5   Cultured cells from any two XP variants may be induced to fuse together, forming a **heterokaryon** where the two nuclei share a common cytoplasm. This process is one aspect of **somatic cell genetics** (see Chapter 23). When fusion is performed with cells of different XP variants, excision repair, as assayed by unscheduled DNA synthesis, is sometimes reestablished. When this occurs, the two variants are said to undergo **complementation.** Alone, neither demonstrates excision repair, but together in a heterokaryon the process occurs. In genetic terms, this is taken as evidence that different genes are involved in the defect of each variant. Complementation occurs because the heterokaryon has at least one normal copy of each gene. In one study, fifteen variants were divided into five complementation groups, suggesting that as many as five different genes may be involved in causing XP.

6   One additional form of XP, with classical symptoms but apparently normal excision repair activity, has been described. In these variants, evidence suggests that while excision repair is normal, the process of post-replication repair is impaired. Impaired post-replication repair may be detected by the failure of cells to convert low molecular weight DNA strands to longer, high molecular weight strands.

These findings are of great interest for two reasons. First, they suggest that XP results if any of the three repair systems is interrupted as a result of mutation. Second, they establish that normal individuals are probably susceptible to UV-induced damage of DNA by exposure to sunlight. However, this damage activates the repair systems that counteract it. We can expect that future work will clarify the precise role and mechanism of repair of UV-induced damage in humans.

High-Energy Radiation

Within the electromagnetic spectrum, the inherent energy varies inversely with wavelength. Figure 13.18 compares the relative wavelengths of the various portions of the electromagnetic spectrum. **X-rays, gamma rays,** and **cosmic rays** have even shorter wavelengths and are therefore more energetic than ultraviolet light. As a result, they are strong enough to penetrate deeply into tissues, causing **ionization** of the molecules encountered along the way. It might be anticipated that these sources of **ionizing radiation** would be mutagenic. Indeed, Herman Muller detected such effects in 1927 while working with *Drosophila.* Since that time, the effects of ionizing radiation, particularly X-rays, have been studied intensely.

As X-rays penetrate cells, electrons are ejected from the atoms of molecules encountered by the radiation. Thus, stable molecules and atoms are transformed into free radicals and reactive ions. Along the path of a high-energy ray, a trail of ions is left that can initiate a variety of chemical reactions. These reactions can directly or indirectly affect the genetic material, altering the purines and pyrimidines in DNA and resulting in point mutations. Such ionizing radiation is also capable of breaking chromosomes, resulting in a variety of aberrations.

Figure 13.19 shows a plot of induced sex-linked recessive lethals versus the dose of X-rays administered. The graph shows a straight line which, if extrapolated, intersects near the zero axis. Thus, a linear relation-

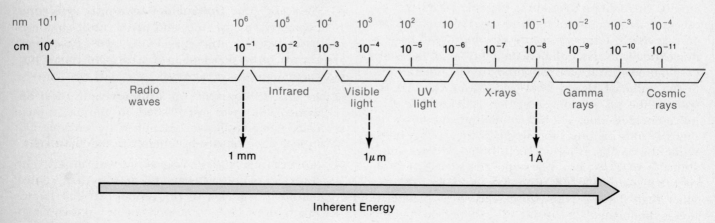

FIGURE 13.18
The spectrum of electromagnetic wavelengths shown in various metric units.

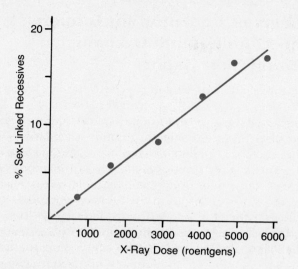

FIGURE 13.19

Plot of sex-linked recessive mutations induced by increasing doses of X-rays. If extrapolated, the graph intersects the zero axis.

ship exists between most X-ray doses and the induction of mutation. For each doubling of dose, twice as many mutations are induced. Because the line intersects near the zero axis, this graph suggests that there is no minimal dose of irradiation that is not mutagenic.

These observations may be interpreted in the form of the **target theory,** first proposed in 1924 by J. A. Crowther and F. Dessauer. The theory proposes that there are one or more sites, or targets, within cells and that a single event of irradiation at one site will bring about a damaging effect, or mutation. In a simple form, the target theory says that one "hit" of irradiation will cause one "event" or mutation, suggesting that the X-rays interact directly with the genetic material.

The target theory is no longer completely applicable because it has been shown that indirect effects of irradiation may also cause mutations. For example, if prior to inoculation of a bacterial culture the medium upon which the cell will grow is treated with X-rays, the bacteria subsequently demonstrate an induced mutation rate. Thus, reactive molecules in the medium are produced that can secondarily interact with DNA of bacterial cells. It has also been observed in several organisms, including *Drosophila*, that reduced oxygen concentration in tissue lowers the induced mutation rate. Oxygen is believed to be critical in the formation of highly reactive peroxides in tissues. Peroxides may be formed in the presence of oxygen following irradiation of water, which is split into its hydrogen and hydroxyl groups. These findings support the concept that the indirect effects of high-energy radiation may be extremely important in the observed mutagenic effect.

Two other observations concerning irradiation effects are of particular interest. In general, the **intensity of the dose** administered seems to make little difference in mutagenic effect for most organisms studied. That is, a 2000-roentgen exposure (a **roentgen** is a measure of dose equal to the production of two ionizations per cubic micron), whether occurring in a single acute dose or cumulatively in many smaller chronic doses, seems to produce the same mutagenic effect. *Drosophila*, for example, shows this response. Mice, however, are an important exception to this general rule. In this organism, a healing process seems to occur during the intervals between irradiation. Thus, several smaller doses are not so potent as a single large dose.

The second observation is that certain portions of the cell cycle are more susceptible to irradiation effects. As mentioned above, in addition to a mutagenic effect, X-rays will also break chromosomes, resulting in terminal or intercalary deletions, translocations, and general chromosome fragmentation. Damage occurs most readily when chromosomes are greatly condensed in mitosis. This property consitutes one of the reasons why radiation is used to treat human malignancy. Since tumorous cells are more often undergoing division than their nonmalignant counterparts, they are more susceptible to the immobilizing effect of radiation.

RADIATION SAFETY

Technological advances during the past 50 years have led to widespread applications of radiation in our society. Largely due to the pioneering efforts of Muller, the potential dangers of radiation to humans are known and radiation safety has received increasing attention. In addition to natural sources of radiation, therapeutic and diagnostic use, industrial application, and atomic and nuclear fallout represent the sources of major concern.

Human Radiation Safety Standards: National Academy of Sciences Report

It is now known that a dose of radiation equivalent to 50 to 150 roentgens (r) is lethal to 50 percent of human cells in culture. Whole-body irradiation of 400 r is usually fatal to individuals, and 100 r is sufficient to cause radiation sickness. Sublethal doses of radiation may also significantly shorten the human life span and increase the occurrence of various forms of cancer, particularly leukemia.

Since there seems to be no lower limit for the amount of radiation that will induce mutation and since the effects of ionizing radiation can be cumulative, precautionary measures are essential to the human population. In 1970, a National Academy of Sciences panel was commissioned to investigate radiation safety standards. At that time, the federal guidelines stated that the general population should not receive more than 170 millirems of artificial radiation per year, exclusive of medical sources. A **rem** is a "roentgen equivalent in man," and a millirem (mrem) is 1/1000 of a rem. The committee's report included the following statement:

> The present guides of 170 millirems per year grew out of an effort to balance societal needs against genetic risks. It appears that these needs can be met with far lower average exposures and lower genetic and somatic risk than permitted by the current radiation protection guide. To this extent, the current guide is unnecessarily high.

The committee strongly admonished the medical profession and urged it to improve the efficiency of X-ray equipment; increase shielding, particularly of the reproductive organs; and avoid unnecessary mass screening for many diseases. The panel estimated that an exposure of 170 millirems per year increases the death rate in the United States by between 3000 and 15,000 people each year. For each additional 100-millirem increase, 3500 added deaths from cancer were estimated per year. Table 13.3 summarizes the report's estimate of annual exposure from various sources of radiation.

TABLE 13.3

Estimates of annual whole-body exposure to radiation from various sources in the United States (1970).

Source	Average Dose Rate (mrem/yr)
Natural	102
Global fallout	4
Nuclear power	0.003
Diagnostic	72
Radiopharmaceuticals	1
Occupational	0.8
Miscellaneous	2
TOTAL	182

Effects of Radiation on Humans: The Atomic Bomb Casualty Commission Report

The information presented by the National Academy of Sciences is sufficient to cause concern about the future well-being of the human species. Many scientists have projected that the human population will continue to accumulate deleterious mutations as a result of exposure to mutagens such as radiation. The concept of **genetic load** refers to deleterious alleles that are already present in the human gene pool. It is estimated that the average individual carries two to ten recessive lethal alleles in the heterozygous condition. Before absolute conclusions concerning the dangers of radiation to the human population can be drawn, more direct evidence is needed. Tragically, the atomic bomb attacks on Japan in 1945 have made this possible.

Immediately following World War II, the **Atomic Bomb Casualty Commission (ABCC)** was established. Its large-scale study began in 1948 as a joint venture between the U.S. National Research Council and the Japanese National Institute of Health. The overall study considered various parameters of radiation effects on somatic and germinal tissue.

Children and adults who received direct somatic tissue exposure to radiation were studied along with individuals who were conceived prior to exposure but were developing *in utero* at the time of the attack. In comparison with the control groups, both groups demonstrated highly significant increases in cytogenetic abnormalities, including chromosome breakage, translocations, pericentric inversions, and deletions. Table 13.4 shows that persons over 30 years of age were most severely affected. Abnormalities were shown in the cells of 60 percent of the individuals, compared with a control figure of 16 percent. For those under 30 years of age, a cytogenetic effect was also noted in comparison with control groups. Those individuals developing *in utero* showed an increase in cytogenetic abnormalities similar to those under 30 if the mother received at least the equivalent of a 100-r exposure. It is clear that large doses of radiation greatly affect the genetic material in nuclei of somatic tissues. It is important to note that these data were gathered approximately 20 years after exposure and that the induced chromosomal abnormalities are representative of long-lasting damage in all age groups examined.

Other effects were also disclosed. A decrease in head circumference and an increase in mental retardation were noted in newborns who were exposed prior to their fifteenth week of gestation. The greatest

TABLE 13.4
Frequency of cytogenetic abnormalities of exposed and control populations among Japanese survivors of the atomic bombs.

Age at Drop	Dose (rad)	Exposed		Controls	
		Examined	Affected	Examined	Affected
30 years or less	200+	94	34%	94	1%
Over 30 years	200+	77	61%	80	16%
In utero	100+	38	39%	48	4%

SOURCE: Data from Miller, 1969, Table 1. Copyright 1969 by the American Association for the Advancement of Science.

effect was on those embryos and fetuses whose mothers were closest to the center of the atomic attack.

Leukemogenesis, or the development of **leukemia,** was suspected because of previous reports showing a correlation between this disease and radiation exposure. In those reports, physicians, particularly radiologists, who were occupationally exposed to X-rays died significantly more often from leukemia than the members of the general population. This correlation was upheld in the survivors of the atomic bomb attack. The increased incidence of acute leukemia peaked in 1951, six years after direct exposure, and continued to occur at higher than usual rates through 1966. A dose response (the greater the exposure, the higher the frequency of leukemia) was also confirmed. This study leaves little doubt that whole-body exposure to high doses of ionizing radiation can induce leukemia in humans.

Other somatic damage, both specific and of a more general nature, was evident. Although infrequent, corneal and lens abnormalities of the eye were observed. Mortality statistics were compiled between 1950 and 1960 for almost 100,000 survivors of all ages. Persons exposed within 1200 meters of the attack center showed a 15 percent increase in general mortality compared with control groups. Again, the evidence is strong that whole-body exposure to radiation leads to cellular damage and that the overall effect is very detrimental. In the words of a news report issued by the National Academy of Sciences, the findings have been "grim beyond doubt, yet not as grim as many had feared."

The most encouraging findings came from investigations to assess the impact of radiation exposure on future generations. In order to assess future generations, scientists studied children conceived after August, 1945, who had at least one parent exposed in Hiroshima or Nagasaki. From 1948 to 1953, over 70,000 such cases were located. All newborns were examined, and 40 percent of the children were reexamined when they were 8 to 10 months old. Each case was classified into one of five levels of radiation exposure.

The results of the study through 1953 revealed no statistically significant effect of increasing levels of radiation exposure. Increases in congenital malformations, stillbirths, altered birth weights, or abnormal physical development at age 8 to 10 months could not be detected. In an ongoing investigation, members of this group as well as other children born after 1953 to exposed parents are being examined. The focus of this study has been on mortality, chromosome abnormalities, and alterations in specific proteins. The findings parallel the results of the examination of newborns. No statistically significant increases have been observed as a result of radiation exposure. However, as noted by William Schull, Masanori Otake, and James Neel, major participants in this investigation, "for all indicators the observed effect is in the direction suggested by the hypothesis that genetic damage resulted from exposure."

Since previous experimental findings leave no doubt that radiation does have a genetic effect, we can assume that some mutations have been induced in exposed parents and transmitted to their offspring. Thus, even though no immediate genetic impairment has been observed in this group, the data have been used to estimate the effect of radiation on the induction of gametic mutations. A significant approach has been to calculate the **doubling dose of radiation for mutations.** Using three parameters, Schull, Otake, and Neel have estimated this to be about 150 rems. Their calculations also suggest that this figure would be considerably higher if the radiation exposure occurred in a series of smaller chronic doses rather than a single acute dose.

This estimated doubling dose is an encouraging one. It is approximately four times higher than that derived from extensive studies of radiation effects on mice. Thus, it is much higher than predicted, suggesting that human gametic tissue is more resistant to radiation damage than previously suspected. This and future improved estimates will undoubtedly have an impact on regulations involving permissible exposures of radiation in humans.

To summarize the findings of ABCC, the long-range effect of radiation on future generations may not be so severe as suspected from studies of other organisms. Nevertheless, the consequences of atomic and nuclear warfare are intolerable to the existing population. At Hiroshima alone, one plane carrying a single primitive atomic bomb resulted in 64,000 to 240,000 fatalities, depending on the accuracy of several estimates. Survivors within a 3-mile radius of this holocaust suffered immediate and long-range radiation-induced damage to their genetic material. Chromosome abnormalities, leukemia, mental retardation, and decreased life spans were observed. It is imperative that humankind intensify its efforts to prevent nuclear warfare and closely monitor and evaluate the relative benefits of technological advances that involve even minor amounts of radiation exposure to the general populace.

ENVIRONMENTAL MUTAGENS AND CARCINOGENS

Earlier in this chapter, we discussed chemical compounds that induce specific types of genetic change. These are called **mutagens.** Because these compounds can increase the spontaneous mutation rate, various environmental substances generated by our advanced technology have been closely scrutinized. Although certain observations might suggest that a specific compound is mutagenic in humans, precise information comes from tests on other experimental animals. Currently, many mutagenic compounds are also suspected of being **carcinogenic,** or **cancer-producing,** in humans. Tests have been devised to simultaneously predict both the mutagenic and carcinogenic effects of a given chemical compound.

Mutagenicity

Any unusual chemical that enters the human body by way of the digestive or respiratory system is considered a potential mutagen. For example, residual materials from air and water pollution, food preservatives, artificial sweeteners, pesticides, and pharmaceutical drugs routinely reach somatic cells, but the extent to which they affect germinal cells is not known. Although experimentation is obviously impossible in humans, increases in spontaneous abortions, stillbirths, and congenital malformations in populations exposed to noxious compounds can be assessed. However, a positive association provides only indirect or circumstantial evidence that a specific compound is harmful.

A more precise survey method uses biochemical analysis of the electrical charge and amino acid sequence of selected proteins of newborns. Sequence al-

terations provide evidence that point mutations, or changes in triplet codons, may have occurred. A number of proteins, including hemoglobin, haptoglobin, acid phosphatase, glucose-6-phosphate dehydrogenase, and lactic dehydrogenase, can be examined from a sample of blood. Amino acid sequences of specific proteins in newborns can also be compared with those in proteins of the parents as a direct test for mutation. This approach is theoretically useful when specific chemicals are suspected and large numbers of individuals who have been exposed are available for study.

It is much easier to expose experimental organisms or cultured cells to suspected mutagens and assay for genetic effects. Table 13.5 lists some organisms that have been used and the type of genetic change detected for each. Testing these organisms provides more specific information than the methods discussed for humans. In addition, these screening systems may be performed more rapidly with less expense and greater accuracy. Chemicals that are mutagenic in bacteria, *Drosophila*, or cultured mammalian cells, for example, are suspected to have similar effects in humans. Table 13.6 summarizes various practical parameters for the different screening systems.

The potential hazard of mutagenicity seems great. For example, more than 2500 substances are currently used as food additivies. Also, certain pesticides such as DDT are stable chemicals and are transferred intact through food chains to human consumers. Industrial pollution on the job site and in the atmosphere is well known. Are we unnaturally polluting our own bodies as well as accumulating deleterious mutations in somatic and gametic tissue? To some extent, we are, but we do not know yet whether to expect shorter life spans or general health problems on a large scale. We must also be concerned for future generations. A gradual deterioration in gene quality can only lead to greater numbers and kinds of heritable disorders. Prevention in the form of detection and subsequent control of mutagens is likely to be the best approach.

Carcinogenicity

Because cancer is a leading cause of death in humans, an enormous research effort has been mounted to define the causes of this dreaded disease. Hopefully, research will determine why cells undergo a transformation leading to uncontrolled growth and how the regulatory mechanisms controlling the cell cycle have been modified in these cells. In the previous chapter, we indicated that three factors have been linked to the etiology of cancer: (1) viral agents, (2) environmental factors, and (3) genetic influences. In the latter cate-

TABLE 13.5

Types of genetic damage detected by currently employed mutagen screening systems.

Screening System		Chromosome Aberrations				Mutations		Induced Recombination
Category	Organism	Dominant Lethality	Trans-locations	Deletions and Duplications	Nondis-junction	Forward or Reverse or Both	Multiple Specific Locus	
Bacterial	*Salmonella typhimurium*					+		
	Escherichia coli					+		
Fungal	*Neurospora crassa*				+	+	+	
	Aspergillus nidulans				+	+	+	+
	Yeasts	+		+	+	+	+	+
Plant	*Vicia faba*		+	+	+			
	Tradescantia paludosa		+	+	+	+		
Insect	*Drosophila melanogaster*	+				+	+	
	Habrobracon juglandis	+	+			+	+	+
	Bombyx mori	+	+	+	+	+	+	+
Mammalian cell culture	Chinese hamster		+	+	+	+		
	Mouse lymphoma		+	+	+	+		
Intact mammal	Mouse		+	+	+			
	Rat	+	+	+	+			
	Human	+	+	+	+		+	

SOURCE: From Drake et al., 1975, p. 506. Copyright 1975 by the American Association for the Advancement of Science.

gory, the subject of **oncogenes** was introduced. Because these potential cancer-causing genes must be altered before they induce malignancy, there is now an increased concern about the chemicals present in the environment and the work place that are potential or known mutagens.

It has long been known that some forms of cancer have a higher probability of occurring in people with certain occupations. This association is due to high concentrations of specific chemicals at work sites. Recognized human carcinogens include asbestos, some metals, aromatic amines, tar and pitch products, mustard gas, and polyvinylchloride. It seems probable that large numbers of other chemicals are also cancer-inducing. Thus, it is important to develop screening techniques to detect these compounds.

Chemicals that cause cancer in humans apparently first undergo metabolic conversion to more chemically reactive electrophilic products. It is this converted form of a molecule that interacts with cells, causing transformation to malignant forms. Since the genetic material may be the primary target in both mutagenicity and carcinogenicity, studies have sought to establish a correlation between the two processes. As a result, many known carcinogens have now been shown to be mutagenic also. These investigations have employed bacteria.

Bacteria, however, apparently lack the necessary metabolic systems to "activate" potential carcinogens to their most reactive form. Several ingenious techniques have been devised to overcome this problem. Michael Gabridge and Marvin Legator have developed the **host-mediated assay,** in which the indicator bacteria are injected into the peritoneal cavity of rodents and the suspected compound is administered independently. The rodent's metabolic systems activate the potential carcinogen, which subsequently is encountered by the bacteria. The bacteria are then recovered and screened for the development of mutations. Several chemicals that were inactive when tested directly were shown to be mutagenic in the host-mediated assay.

TABLE 13.6
Operational characteristics of mutagen screening systems.

Test System	Time to Run Test	Operating Costs*	Initial Investment Costs	Relative Ease of Detection†	
				Gene Mutations	Chromosome Aberrations
Microorganisms with metabolic activation:					
Salmonella typhimurium	2 to 3 days	Very low	Low	Excellent	
Escherichia coli	2 to 3 days	Very low	Low	Excellent	
Yeasts	3 to 5 days	Very low	Low	Good	Unknown
Neurospora crassa	1 to 3 weeks	Moderate	Moderate	Very good	Good
Cultured mammalian cells with metabolic activation	2 to 5 weeks	Moderate to high	Moderate	Excellent to fair	Unknown
Host-mediated assay with:					
Microorganisms	2 to 7 days	Low to moderate	Low to moderate	Good	
Mammalian cells	2 to 5 weeks	Moderate to high	Moderate	Unknown	Good
Body fluid analysis	Variable	Variable	Low to moderate	Variable	
Plants:					Relevance unclear
Vicia faba	3 to 8 days	Low	Low	Potentially excellent	
Tradescantia paludosa	2 to 5 weeks	Low to moderate	Moderate		
Insects:					
Drosophila melanogaster:	2 to 7 weeks	Moderate	Moderate	Good to excellent	Good to excellent
Mammals:					
Dominant lethal mutations	2 to 4 months	Moderate to high	Moderate		Unknown
Translocations	5 to 7 months	Moderate to high			Potentially very good
Blood or bone marrow cytogenetics	1 to 5 weeks	Moderate	Moderate		Potentially good
Specific locus mutations	2 to 3 months	High to very high	High to very high	Unknown	

*Operating costs vary widely depending upon the protocol specified and upon the number of substances tested simultaneously. Very approximately, very low is $1000; low is $1000 to $5000; moderate is $3000 to $10,000; high is $10,000 to $20,000; and very high is $25,000 upward.

†Since most of these test systems do not detect all classes of gene mutations or chromosome aberrations, these columns refer only to the detectable mutations.

SOURCE: From Drake et al., 1975, p. 507. Copyright 1975 by the American Association for the Advancement of Science.

Bruce Ames and his colleagues have developed another form of assay that is performed completely *in vitro*. In the **Ames assay**, the test compound is incubated in the presence of a mammalian liver extract containing enzymes capable of activating the test compound. This mixture and the suitable indicator microorganisms are placed together on agar culture plates and assayed. In an initial screening of over 150 known carcinogens, over 80 percent were shown also to behave as potent mutagens! This finding is significant because of the correlations between the carcinogenic and mutagenic properties of certain molecules. Since many other molecules have shown a similar correlation, the Ames assay now serves as a useful screening technique for potential carcinogenicity, even though it tests only mutagenicity. The assay has the advantage of being much more rapid and less expensive than formal laboratory testing with mammals. However, a positive response in the Ames assay does not prove a compound's carcinogenicity.

The Ames system utilizes four tester strains of the bacterium *Salmonella typhimurium* that were selected for sensitivity and specificity for mutagenesis. One strain is used to detect base-pair substitutions, and the other three detect various frameshift mutations. Each strain requires histidine for growth (his^-). The assay measures the frequency of reverse mutation, which yields wild-type (his^+) bacteria. Greater sensitivity to mutagens occurs because these strains bear mutations that eliminate the DNA excision repair system and the lipopolysaccharide barrier that coats the surface of the bacteria.

These techniques are also used to test the mutagenicity and carcinogenicity of chemical substances in urine samples of experimental animals that have been injected with test substances. The Ames group has shown that cigarette tars, most hair dyes, and many other compounds increase mutation rates, suggesting their potential carcinogenicity.

MUTATOR GENES

In certain strains of *Escherichia coli*, the spontaneous mutation rate is several orders of magnitude greater than in wild-type organisms. Investigation of these **mutator strains** has revealed that specific mutant genes, **mutator genes,** are responsible for the increased rate of mutation.

One such gene, *mut T1*, was discovered by Howard Treffers in 1954. Table 13.7 demonstrates the range of mutation frequencies at a number of specific *E. coli*

loci with and without the presence of the *mut T1* allele. Increases in the rate occur from 40 to 15,000 times.

The work on Charles Yanofsky and his coworkers has shown that the *mut T1*-induced mutations in the *trpA23* protein were specifically A≡T to C≡G transversions. Other alleles of the *mut T1* locus have been shown to induce this base-pair change. Other evidence suggests that the *mut T1* allele results in a mutant product that functions during DNA replication.

Other organisms have displayed mutator genes that are known to be related to DNA polymerase infidelity during replication. In phage T4, a temperature-sensitive mutation, *tsL88*, has been shown to increase the frequency of mutation in other genes. This mutator gene produces a thermolabile DNA polymerase that inserts incorrect nucleotides during replication at a higher frequency than the wild-type enzyme. In this case, both transversions and transitions have been observed.

A somewhat different but analogous situation exists with the *A1* locus on chromosome 3 in corn, discovered by Marcus Rhoades and studied by the famous corn geneticists Barbara McClintock and Alexander Brink. This locus controls anthocyanin pigment in the aleurone layer of kernels as well as in other parts of the plant. A recessive mutant allele, *a*, eliminates the presence of the dominant purple pigment if the triploid endosperm contains the genotype *aaa*. When such a genotype exists, another mutation, the *Dotted (Dt)* allele on chromosome 9, has the effect of causing both germinal and somatic mutations at the *A1* locus. There can be 0, 1, 2, or 3 *Dt* alleles in somatic triploid tissue. As shown in Table 13.8, the number of *Dt* alleles correlates with the rate of mutation of the *a* to *A* allele, and thus with the number of colored spots, or "dots," in the aleurone layer of each kernel. How this effect is imparted is unknown, but the *Dotted* locus does not seem to be involved in pigment production.

TABLE 13.7

Mutation frequencies in *E. coli* with and without the *mut T1* allele.

Mutation	Wild Type	Frequency of *mut T1*	Rate of Increase
$Sm^S \rightarrow Sm^R$	7×10^{-10}	1.4×10^{-7}	200
$Arg^- \rightarrow Arg^+$	1×10^{-9}	1.5×10^{-5}	15,000
$susP \rightarrow +$	5×10^{-8}	2.0×10^{-6}	40
$trpA23 \rightarrow Trp^+$	5×10^{-9}	2.1×10^{-5}	4,200

SOURCE: Cox, D. C. 1973. *Genetics*, 73: 67–80.

TABLE 13.8

The effect of *Dt* dose on the mutation rate of the *A1* locus.

A Locus Genotype	*Dotted* Locus Genotype	Mutations per Kernel
aaa	*dt dt dt*	0
aaa	*Dt dt dt*	7.2
aaa	*Dt Dt dt*	22.2
aaa	*Dt Dt Dt*	121.9

SUMMARY AND CONCLUSIONS

The phenomenon of mutation is extremely important in the study of genetics. Not only do mutations give rise to most of the inherent variation present in all organisms, but they constitute the working tools of the geneticist. Although mutations arise very rarely as spontaneous events, their rate of occurrence may be increased by a variety of agents. This induction process, combined with adequate detection systems for the recovery of mutant alleles, has allowed the in-depth investigation of the gene and genetic processes.

Mutations are classified in a variety of ways. Somatic mutations affect an individual but are not heritable. When mutations arise in gametic tissue, however, the new alleles may be passed to the offspring and enter the gene pool. In addition to the somatic/gametic classification, mutations may be differentiated according to their effect. The most obvious category includes the morphological mutations, which may be visibly detected. Four other categories have been discussed—biochemical, lethal, conditional, and regulatory mutations. They are not mutually exclusive.

Organisms in which newly arisen mutations can be detected easily are used most often in genetic studies. Bacteria, fungi, and *Drosophila* have relatively short life cycles, and detection systems are easily applied to them. Thus, a great deal more is known about the genetics of these organisms than is known for plants and other animals, including humans, where detection is much more difficult.

The elucidation of the molecular mechanisms involved in mutagenesis has been an active area of investigation. Spontaneous mutations may arise naturally as a result of rare chemical rearrangements of atoms, or tautomeric shifts, and by the infidelity of copying DNA during replication. However, much has also been learned through studying specific muta-

genic agents. Nitrous acid, hydroxylamine, alkylating agents, and base analogues all change the purine or pyrimidine structure in DNA, thus altering the base-pairing affinities. During DNA replication, then, copy errors are made which, after several rounds of replication, produce transitions or transversions. Acridine dyes, another group of chemical mutagens, cause copying errors in the form of frameshift mutations by intercalating with the DNA helix.

Ultraviolet and high-energy radiation are also potent mutagenic agents. UV light has a highly specific effect on DNA: It induces the formation of pyrimidine dimers. Although an intricate set of mechanisms is capable of repairing the dimer lesions, unrepaired lesions will result in transitions following replication. X-rays as well as gamma and cosmic rays are much more energetic than UV light. As a result, they penetrate tissue much more deeply and cause ionization of the molecules in their path. Such chemical reactivity has a nonspecific effect on DNA and results in mutations. Other affected molecules become highly reactive and secondarily interact with DNA, causing mutation.

The NAS report on and the AABC study of the mutagenic effect of high-energy radiation have caused concern about radiation safety in our society. The potential mutagenicity and carcinogenicity of chemical agents to which segments of the human population may be exposed are areas of investigation and concern. Rapid and inexpensive test systems using bacteria have been devised and have provided a correlation between known carcinogens and mutagenicity. It is possible that the mechanisms of environmentally induced mutagenicity and carcinogenicity are similar.

In addition to the effects of spontaneous and induced mutations, certain strains of an organism may carry mutator genes that affect the mutation rate of other genes.

PROBLEMS AND DISCUSSION QUESTIONS

1 What is the difference between a chromosomal aberration and a point mutation?

2 Discuss the importance of mutations to the successful study of genetics.

3 Describe the technique for detection of nutritional mutants in *Neurospora*.

4 Most mutations are thought to be deleterious. Why, then, is it reasonable to state that mutations are essential to the evolutionary process?

5 Why do you suppose that a random mutation is more likely to be deleterious than beneficial?

6 Most mutations in a diploid organism are recessive. Why?

7 What is meant by a conditional lethal mutation?

8 Contrast the concerns about mutation in somatic and gametic tissue.

9   In the *ClB* technique for detection of mutation in *Drosophila*:
    (a) What type of mutation may be detected?
    (b) What is the importance of the *l* gene?
    (c) Why is it necessary to have the crossover suppressor (*C*) in the genome?
    (d) What is *C*?

10 In an experiment with the *ClB* system, a student irradiated wild-type males and subsequently scored 500 F_2 cultures, finding all of them to have both males and females present. What conclusions can you draw about the induced mutation rate?

11 In *Drosophila*, induced recessive autosomal lethal mutations may be detected using a second chromosomal stock, *Curly, Lobe/Plum* (*Cy L/Pm*). These alleles are all dominant and lethal in the homozygous condition. Detection is performed by crossing *Cy L/Pm* females to wild-type males that have been subjected to a mutagen. Three generations are required. Demonstrate theoretically how this system of detection works.

12 Describe tautomerism and the way in which this chemical event may lead to mutation.

13 Contrast and compare the mutagenic effects of nitrous acid, hydroxylamine, alkylating agents, and base analogues.

14 Acridine dyes induce frameshift mutations. Is such a mutation likely to be more detrimental than a point mutation where a single pyrimidine or purine has been substituted?

15 Why is X-radiation a more potent mutagen than UV light?

16 Contrast the induction of mutations by UV light and X-radiation.

17 Contrast the three types of repair mechanisms known to counteract the effects of UV light. What is the role of visible light in photoreactivation?

18 Mammography is an accurate screening technique for the early detection of breast cancer in humans. Because this technique uses X-rays diagnostically, it has been highly controversial. Can you explain why?

19 Summarize the conclusions of the NAS and the ABCC reports concerning radiation safety. In the ABCC report, what important question remains unanswered?

20 Describe the Ames assay for screening potential environmental mutagens. Why is it thought that a compound that tests positively in the Ames assay may also be carcinogenic?

21 Why is it logical to suspect that the product of a mutator gene may be involved in replication?

22 Describe the disorder xeroderma pigmentosum in humans.

23 What is known about repair mechanisms of UV-induced lesions in humans? Relate this information to what has been learned about the genetic basis of xeroderma pigmentosum.

24 Presented below are theoretical findings from studies of heterokaryons formed from human xeroderma pigmentosum cell strains:

	XP1	XP2	XP3	XP4	XP5	XP6	XP7
XP1	0						
XP2	0	0					
XP3	0	0	0				
XP4	+	+	+	0			
XP5	+	+	+	+	0		
XP6	+	+	+	+	0	0	
XP7	+	+	+	+	0	0	0

NOTE: + = complementing; 0 = noncomplementing.

These data represent the occurrence of unscheduled DNA synthesis in the fused heterokaryon when neither of the strains alone showed synthesis. What does unscheduled DNA synthesis represent? Which strains fall into the same complementation groups? How many groups are revealed based on these limited data? How does one interpret the presence of these complementation groups?

25 If the human genome contains 100,000 genes and the mutation rate at each of these loci is 5×10^{-5} per gamete formed, what is the average number of new mutations that exist in each individual? If the current population is 4.3 billion people, how many newly arisen mutations exist in the current populace?

26 In a bacterial culture where all cells were unable to synthesize leucine (leu^-) a potent mutagen was added, and the cells were allowed to undergo one round of replication. At that point, samples were taken and a series of dilutions was made prior to plating the cells on either minimal medium or minimal medium to which leucine was added. The first culture condition (minimal medium) allows the detection of mutations from leu^- to leu^+, while the second culture condition allows the determination of total cells, since all bacteria can grow. From the following results, determine the frequency of mutant cells. What is the rate of mutation at the locus involved with leucine biosynthesis?

Culture Condition	Dilution	Colonies
minimal medium	10^{-1}	18
minimal + leucine	10^{-7}	6

SELECTED READINGS

AMES, B. N. 1979. Identifying environmental chemicals causing mutations and cancer. *Science* 204: 587–93.

AMES, B. N., McCANN, J., and YAMASAKI, E. 1975. Method for detecting carcinogens and mutagens with the *Salmonella*/mammalian microsome mutagenicity test. *Mut. Res:* 31: 347–64.

AUERBACH C. 1976. *Mutation research: Problems, results,* and *perspectives.* London: Chapman & Hall.

————. 1978. Forty years of mutation research: A pilgrim's progress. *Heredity* 40: 177–87.

AUERBACH, C., and KILBEY, B. J. 1971. Mutations in eukaryotes. *Ann. Rev. Genet.* 5: 163–218.

BALTIMORE, D. 1981. Somatic mutation gains its place among the generators of diversity. *Cell* 26: 295–96.

BEADLE, G. W., and TATUM, E. L. 1945. Neurospora II. Methods of producing and detecting mutations concerned with nutritional requirements. *Amer. J. Bot.* 32: 678–86.

BRENNER, S. L., BARNETT, F. H. C., and ORGEL, L. 1961. The theory of mutagenesis. *J. Mol. Biol.* 3: 121–24.

CARLSON, E. A. 1981. *Genes, radiation, and society.* The life and work of H. J. Muller. Ithaca, N.Y.: Cornell University Press.

CLEAVER, J. E. 1968. Defective repair replication of DNA in xeroderma pigmentosum. *Nature* 218: 652–56.

COX, D. C. 1973. Mutator gene studies in *Escherichia coli:* The *mut* T gene. *Genetics* 73 (supplement): 67–80.

CROW J. F., and DENNISTON, C. 1985. Mutation in human populations. *Adv. Hum. Genet.* 14: 59–216.

DEERING, R. A. 1962. Ultraviolet radiation and nucleic acid. *Scient. Amer.* (Dec.) 207: 135–44.

DENNISTON, C. 1982. Low level radiation and genetic risk estimation in man. *Ann. Rev. Genet.* 16: 329–56.

DEVORET, R. 1979. Bacterial tests for potential carcinogens. *Scient. Amer.* (Aug.) 241: 40–49.

DRAKE, J. W. 1969. Mutagenic mechanisms. *Ann. Rev. Genet.* 3: 247–68.

————. 1970. *Molecular basis of mutation.* San Francisco: Holden-Day.

————, ed. 1973. The genetic control of mutation. *Genetics* 73 (supplement): 1–205.

————. et al. 1975. Environmental mutagenic hazards. *Science* 187: 505–14.

DRAKE, J. W., GLICKMAN, B. W., and RIPLEY, L. S. 1983. Updating the theory of mutation. *Amer. Scient.* 71: 621–30.

FISHBEIN, L., FLAMM, W. G., and FALK, H. L. 1970. *Chemical mutagens.* New York: Academic Press.

FREESE, E. 1963. Molecular mechanism of mutations. In *Molecular genetics: Part 1,* ed. J. H. Taylor, pp. 207–69. New York: Academic Press.

FRIEDBERG, E.C. 1985. *DNA repair.* New York: W.H. Freeman.

HANAWALT, P. C., and HAYNES, R. H. 1967. The repair of DNA. *Scient. Amer.* (Feb.) 216: 36–43.

HANAWALT, P. C., et al. 1979. DNA repair in bacteria and mammalian cells. *Ann. Rev. Biochem.* 48: 783–836.

HARRIS, M. 1971. Mutagenicity of chemicals and drugs. *Science* 171: 51–52.

HASELTINE, W. A., 1983. Ultraviolet light repair and mutagenesis revisited. *Cell* 33: 13–17.

HEDDLE, J. A., ed. 1982. *Mutagenicity: New horizons in genetic toxicology.* New York: Academic Press.

HOWARD-FLANDERS, P. 1981. Inducible repair of DNA. *Scient. Amer.* (Nov.) 245: 72–80.

KELNER, A. 1949. Effect of visible light on the recovery of *Streptomyces griseus* conidia from ultraviolet irradiation injury. *Proc. Natl. Acad. Sci.* 35: 73–79.

————. 1951. Revival by light. *Scient. Amer.* (May) 184: 22–25.

KNUDSON, A. G. 1979. Our load of mutations and its burden of disease. *Am. J. Hum. Genet.* 31: 401–13.

KRAEMER, F. H., et al. 1975. Genetic heterogeneity in xeroderma pigmentosum: Complementation groups and their relationship to DNA repair rates. *Proc. Natl. Acad. Sci.* 72: 59–63.

LINDAHL, T. 1982. DNA repair enzymes. *Ann. Rev. Biochem.* 51: 61–88.

LITTLE, J. W., and MOUNT, D. W. 1982. The SOS regulatory system of *E. coli. Cell* 29: 11–22.

MASSIE, R. 1967. *Nicholas and Alexandra.* New York: Atheneum.

MASSIE, R., and MASSIE, S. 1975. *Journey.* New York: Knopf.

McCANN, J., CHOI, E., YAMASAKI, E., and AMES, B. 1975. Detection of carcinogens as mutagens in the *Salmonella*/microsome test: Assay of 300 chemicals. *Proc. Natl. Acad. Sci.* 72: 5135–39.

McKUSICK, V. A. 1965. The royal hemophilia. *Scient. Amer.* (Aug.) 213: 88–95.

MILLER, R. W. 1969. Delayed radiation effects in atomic-bomb survivors. *Science* 166: 569–74.

MULLER, H. J. 1927. Artificial transmutation of the gene. *Science* 66: 84–87.

MULLER, H. J. 1955. Radiation and human mutation. *Scient. Amer.* (Nov.) 193: 58–68.

NEEL, J. V., et al. 1980. Search for mutations affecting protein structure in children of atomic bomb survivors: Preliminary report. *Proc. Natl. Acad. Sci.* 77: 4221–25.

NEWCOMBE, H. B. 1971. The genetic effects of ionizing radiation. *Adv. in Genet.* 16: 239–303.

RHOADES, M. M. 1941. The genetic control of mutability in maize. *Cold Spr. Harb. Symp.* 9: 138–44.

SCHULL, W. J., OTAKE, M., and NEEL, J. V. 1981. Genetic effects of the atomic bombs: A reappraisal. *Science* 213: 1220–27.

SHORTLE, D., DiMARIO, D., and NATHANS, D. 1981. Directed mutagenesis. *Ann. Rev. Genet.* 15: 265–94.

SIGURBJORHSSON, B. 1971. Induced mutations in plants. *Scient. Amer.* (Jan.) 224: 86–95.

SILBERNER, J., 1981. Hiroshima and Nagasaki: Thirty-six years later, the struggle continues. *Science News* 120: 284–86.

SINGER, B., and KUSMIEREK, J. T. 1982. Chemical mutagenesis. *Ann. Rev. Biochem.* 51: 665–94.

STADLER, L. J. 1928. Mutations in barley induced by X-rays and radium. *Science* 66: 84–87.

STENT, G. S., and CALENDAR, R., 1978. *Molecular genetics—An introductory narrative.* 2nd ed. San Francisco: W. H. Freeman.

SUEOKA, N. 1967. Mechanisms of replication and repair of nucleic acid. In *Molecular genetics: Part 2*, ed. J. H. Taylor, pp. 1–46. New York: Academic Press.

SUGIMURA, T., KONDO, S., and TAKEBE, H., eds. 1982. *Environmental mutagens and carcinogens.* New York: Alan R. Liss.

SUTHERLAND, B. M. 1981. Photoreactivation. *Bioscience* 31: 439–44.

TOPAL, M. D., and FRESCO, J. R. 1976a. Complementary base pairing and the origin of substitution mutations. *Nature* 263: 285–89.

———. 1976b. Base pairing and fidelity in codon-anticodon interaction. *Nature* 263:289–93.

TREFFERS, H. P., SPINELLI, V., and BELSER, N. O. 1954. A factor influencing mutation rates in *E. coli. Proc. Natl. Acad. Sci.* 40: 1064–71.

VENITT, S., and PARRY, J.M., eds. 1984. Multagenicity testing: A practical approach. Oxford: IRL Press.

VOGEL, F. 1970. *Chemical mutagenesis in mammals and man.* New York: Springer-Verlag.

WALLACE, B., and DOBZHANSKY, T. 1959. *Radiation, genes and man.* New York: Holt, Rinehart and Winston.

WATSON, J. O., and CRICK, F. C. 1953. Genetic implications of the structure of deoxyribose nucleic acid. *Nature* 171: 964.

WILLS, C. 1970. Genetic load. *Scient. Amer.* (March) 222: 98–107.

WITKIN, E. M. 1966. Radiation-induced mutations and their repair. *Science* 152: 1345–53.

———. 1969. Ultraviolet mutation and DNA repair. *Ann. Rev. Genet.* 3: 525–52.

WOLFF, S. 1967. Radiation genetics. *Ann. Rev. Genet.* 1: 221–44.

YANOFSKY, C., COX, E. C., and HORN, V. 1966. The unusual specificity of an *E. coli* mutator gene. *Proc. Natl. Acad. Sci.* 55: 274–81.

14

Bacterial and Viral Genetics

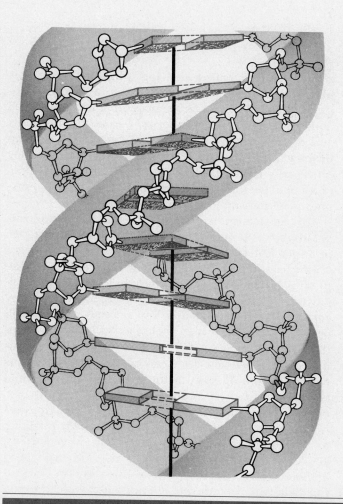

The use of bacteria and viruses has been essential to the accumulation of knowledge in many areas of genetic study. For example, much of what is known about molecular genetics, recombinational phenomena, and gene structure has been derived from experimental work with these organisms. Their successful use in genetic studies is due to numerous factors. Both bacteria and viruses have extremely short reproductive cycles. Literally hundreds of generations giving rise to millions of organisms will be produced in short periods of time. Importantly, they can be studied in **pure cultures.** That is, a single species or mutant strain of bacteria or one type of virus can be investigated independently of other similar organisms. Pure culture techniques for the study of microorganisms were developed in the latter part of the nineteenth century.

It was not until the 1940s and thereafter that genetic studies with microorganisms became productive. In 1943, Salvadore Luria and Max Delbruck established that bacteria, like eukaryotic organisms, undergo spontaneous mutation. These mutations were demonstrated to be the source of variation observed in bacteria, and this finding paved the way for genetic investigation. Mutant cells that arise in an otherwise pure culture can be isolated and established independently from the parent strain by using selection techniques. Given proper selection techniques, mutations for almost any desired characteristic can now be induced and isolated in pure culture. Since bacteria and viruses are haploid, all mutations are expressed directly in the descendants of mutant cells, adding to the ease with which these microorganisms are studied.

In this chapter, we will review a number of the historical developments leading to the extensive use of bacteria and viruses in genetics. Additionally, we will examine the processes by which bacteria and viruses exchange genetic information.

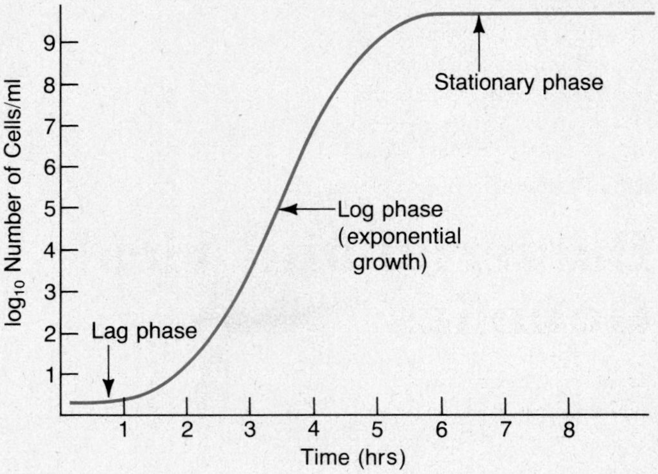

FIGURE 14.1
A typical bacterial population growth curve illustrating the initial lag phase, the log phase, where exponential growth occurs, and the stationary phase that results as nutrients are exhausted.

BACTERIAL GROWTH

Bacteria can be grown in either a liquid culture medium or in a Petri dish on a semisolid agar surface. If the nutrient components of the growth medium are very simple and consist only of an organic carbon source (such as a glucose or lactose) and a variety of ions, including Na^+, K^+, Mg^{++}, Ca^{++}, and NH_3^+ present as inorganic salts, it is called **minimal medium.** In order to grow on such a medium, a bacterium must be able to synthesize all essential organic compounds, e.g., amino acids, purines, pyrimidines, sugars, vitamins, fatty acids, etc. A bacterium that can accomplish this remarkable biosynthetic feat—one that we ourselves cannot duplicate—is termed a **prototroph.** It is said to be wild type for all growth requirements. On the other hand, if a bacterium loses the ability to synthesize one or more organic components, it is said to be an **auxotroph.** For example, if it loses the ability to make histidine, then this amino acid must be added as a supplement to the minimal medium in order for growth to occur. Such a loss of ability can occur as the result of a mutation, and the resulting bacterium would be designated as a *his⁻* auxotroph, as opposed to its prototrophic *his⁺* counterpart.

When an inoculum of bacteria is placed in liquid culture medium, a characteristic growth pattern is exhibited, as illustrated in Figure 14.1. Initially, during the **lag phase,** growth is slow. Then, a period of rapid

growth ensues called the **log phase.** During this phase, cells divide many times within a fixed time interval, resulting in logarithmic growth. When a cell density of about 10^9 cells per ml is reached, nutrients and oxygen become limiting and cells enter the **stationary phase.** Since the doubling time during the log phase may be as short as 20 minutes, an initial inoculum of a few thousand cells can easily achieve a maximum cell density in an overnight culture.

We can determine the approximate number of cells present in each milliliter by performing serial dilutions of an aliquot of the liquid culture and then plating a 0.1-ml sample of the dilutions on nutrient agar medium (Figure 14.2). Each bacterium so plated will, following incubation, give rise to a colony of progeny cells. If the dilution is too low, a **lawn of cells** will occur because the many colonies will fuse together during their growth. With greater dilution, the colonies do not fuse, and they can be counted. When this count and the dilution factor are known, the original cell density can be easily calculated. In Figure 14.2, one milliliter of medium is first removed from an overnight culture. It is serially diluted, ultimately attaining a final dilution of one in ten million. In other terms, a 10^7 dilution factor or a 10^{-7} dilution has been achieved. Two colonies grew at this dilution, indicating that two bacteria per 0.1 milliliter were present. Therefore, there were an estimated 20 bacteria per ml

FIGURE 14.2

Diagrammatic illustration of the serial dilution technique and subsequent culture of bacteria from several of the dilutions. The dilution value for each tube indicates how diluted the initial 1-ml sample has become after each transfer. For example, a dilution of 10^{-3} means that the original 1 ml is now equivalent to 1 ml in 1000 ml. The dilution factor indicates the degree of dilution, e.g., 10^3 equals a $1000\times$ dilution. When 0.1 ml of the 10^{-3} dilution is plated, so many bacteria are still present that a lawn of cells is formed as the individual colonies fuse. Plating of 10^{-5} and 10^{-7} dilutions yields 203 and 2 colonies, respectively. The original cell density may be calculated by multiplying the number of colonies per ml times the dilution factor: 2030×10^5 (20.3×10^7) or 20×10^7, respectively. Each colony represents a clone of genetically identical bacteria arising from a single bacterium.

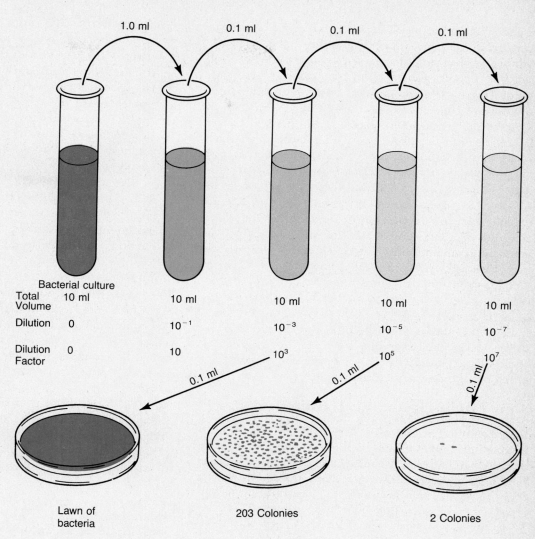

1.0 ml	0.1 ml	0.1 ml	0.1 ml	

Bacterial culture

Total Volume	10 ml	10 ml	10 ml	10 ml	10 ml
Dilution	0	10^{-1}	10^{-3}	10^{-5}	10^{-7}
Dilution Factor	0	10	10^3	10^5	10^7

0.1 ml

0.1 ml

0.1 ml

Lawn of bacteria

203 Colonies

2 Colonies

at that dilution. The original density is then calculated as

$$(\text{colony number/ml}) \times (\text{dilution factor})$$
$$\text{or}$$
$$20 \times 10^7 \text{ bacteria/ml}$$

If we examine the plate derived from the 10^{-5} dilution, we expect one hundred times the number of colonies present in the 10^{-7} dilution. The observation of 203 colonies fulfills this expectation. From this plate, the original density is calculated as:

$$(2030)\,(10^5) = 20.3 \times 10^7 \text{ cells/ml}$$

This value is very close to that obtained using the 10^{-7} dilution.

Calculations such as these make possible the quantitative analysis of a variety of parameters of a given bacterial population. As we will see, such analysis is particularly useful in mutational studies designed to determine mutation rates. Additionally, serial dilutions allow for the isolation of individual colonies. Since each colony contains only genetic clones of the original cell, true-breeding or pure strains can be isolated, propagated, and studied.

MUTATION IN BACTERIA

While it was known well before 1943 that pure cultures of bacteria could give rise to small numbers of cells exhibiting heritable variation, particularly with respect to survival under different environmental conditions, the source of the variation was hotly debated. The majority of bacteriologists believed that environmental factors induced changes in certain bacteria that led to their survival or adaptation to the new conditions. For example, strains of *E. coli* are known to be *sensitive* to infection by the bacteriophage T1. Infection by the bacteriophage leads to the reproduction of the virus at the expense of the bacterial cell, which is lysed or destroyed (see Figure 8.5). If a plate of *E. coli* is homogeneously sprayed with T1, almost all cells are lysed. Rare *E. coli* cells, however, survive infection and are not lysed. If these cells are isolated and established in pure culture, all descendants are *resistant* to T1 infection. The **adaptation hypothesis,** put forth to explain this type of observation, implied that the interaction of the phage and bacterium was essential to the acquisition of immunity. In other words, the phage had "induced" resistance in the bacteria.

The occurrence of **spontaneous mutations** provided an alternative "preadaptation" hypothesis to explain the origin of T1 resistance in *E. coli*. In 1943, Luria and Delbruck presented the first direct evidence that bacteria, like eukaryotic organisms, are capable of spontaneous mutation. This experiment marked the initiation of modern bacterial genetic study.

The Luria-Delbruck Experiment

In a beautiful example of analytical and theoretical work, Luria and Delbruck carried out experiments to differentiate between the adaptation and spontaneous mutation hypotheses. In their work, they used the *E. coli*/T1 systems just described. Many individual liquid cultures of phage-sensitive *E. coli* were grown up. Then, numerous aliquots from each culture were added to Petri dishes containing agar medium to which T1 bacteriophages had been previously added. Following incubation, each plate was scored for the number of phage-resistant bacterial colonies. The precise number of total bacteria added to each plate prior to incubation was also determined so that quantitative data could be obtained.

The experimental rationale for distinguishing between the two hypotheses was as follows:

1 **Adaptation:** Every bacterium has a small but constant probability of acquiring resistance as a result of contact with the phages. Therefore, the number of resistant cells will depend only on the number of bacteria and phages added to each plate. The final results should be independent of all other experimental conditions.

The adaptation hypothesis predicts, therefore, that if a constant number of bacteria and phages is used for each culture and if incubation time is constant, then little fluctuation in the number of resistant cells should be noted from experiment to experiment.

2 **Spontaneous Mutation:** On the other hand, if resistance is acquired as a result of mutation, resistance will occur at a low but constant rate during the incubation in liquid medium prior to plating. If mutations occur early during incubation, which they may do randomly, subsequent reproduction of the mutant bacteria will produce large numbers of resistant cells. If the mutations occur relatively late during incubation, far fewer resistant cells will be present.

The mutation hypothesis predicts, therefore, that significant fluctuation in the number of resistant cells will be observed from experiment to experiment, depending on the time when most spontaneous mutations occurred while in liquid culture.

Table 14.1 shows a representative set of data from the Luria-Delbruck experiments. The middle column shows the number of mutants recovered from a series of aliquots derived from one individual liquid culture. These data serve as a control of experimental error because the number in each aliquot should be nearly identical. Little fluctuation is observed. The right-hand column, however, shows the number of resistant mutants recovered from a single aliquot from each of ten different liquid cultures. The amount of fluctuation in the data will support only one of the two alternate hypotheses. For this reason, the experiment has been designated the **fluctuation test.** A great fluctuation *is*

TABLE 14.1

The Luria-Delbruck experiment demonstrating that spontaneous mutations are the source of phage-resistant bacteria.

Sample No.	Number of T1-Resistant Bacteria	
	Same Culture (Control)	Different Cultures
1	14	6
2	15	5
3	13	10
4	21	8
5	15	24
6	14	13
7	26	165
8	16	15
9	20	6
10	13	10
Mean	16.7	26.2
Variance	15.0	2178.0

SOURCE: After Luria and Delbruck, 1943.

observed between cultures, thus supporting the spontaneous mutation theory. Fluctuation is measured by the amount of variance, a statistical calculation.

The conclusion reached in the Luria-Delbruck experiment received further support from other experimentation. Nevertheless, until the 1950s, staunch supporters of the adaptation theory were not immediately convinced of the spontaneous mutation origin of variation. At that time, Joshua and Esther Lederberg and their coworkers developed the **replica plating technique.**

Replica Plating: Selection of Bacterial Mutants

In the replica plating technique, a dilute suspension of bacterial cells is plated on a Petri dish containing agar and then incubated until discernible colonies appear. An absorbent material, usually velvet, in the circular shape of the Petri dish is then pressed onto the surface of the growing bacterial culture, as shown in Figure 14.3. A few cells from each colony are picked up on the absorbent surface, which is then pressed lightly onto a second fresh agar plate. The transferred cells produce colonies on the replica plate in a pattern of growth similar to that of the original Petri dish. Consecutive replicas may be prepared in the same way.

In their original publication in 1956, the Lederbergs described work using the replica plating technique that provided strong support for Luria and Delbruck's conclusion. In one experiment, bacteria were added to agar medium and allowed to reproduce until a dense lawn of cells covered the plate surface. Several replicas were then made on agar medium to which T1 bacteriophages had been added. As in the Luria-Delbruck

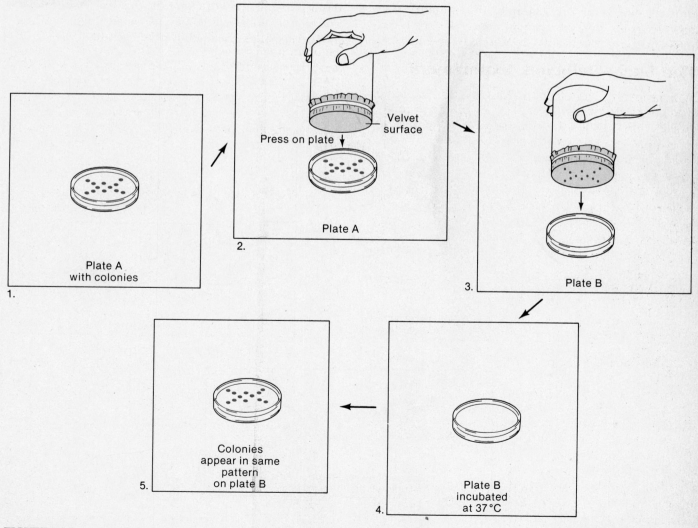

FIGURE 14.3
Replica plating technique.

experiment, only phage-resistant mutant cells could grow on these plates. As shown in Figure 14.4, many phage-resistant cells appeared at the same places on all replica plates. This observation suggests that all resistant cells were present at congruent spots on the original lawn of bacteria, prior to any exposure to the bacteriophage. This conclusion was directly verified by returning to the original plate and "picking" individual colonies from the locations revealed to contain resistant cells on the replica plates. Such cells were expected to be phage-resistant mutants, and they were.

The replica plating technique is now routinely used in microbial genetic studies. It has been particularly useful in isolating spontaneous and induced mutations. For example, the Lederbergs showed that if streptomycin-sensitive bacteria were used and if streptomycin was added to the replica plates, cells resistant to this antibiotic could be isolated. The technique has also been used to isolate **auxotrophic** mutations, which interfere with the biosynthesis of an essential nutrient molecule. For example, the mutant strain may be unable to synthesize an amino acid or a vitamin. Thus, the mutation is conditional, and the bacteria harboring it will grow only if minimal medium is supplemented with the missing component. Prototrophs on the other hand, are wild-type strains of bacteria that will grow on minimal medium.

Enrichment Techniques

Other procedures used in conjunction with replica plating facilitate the recovery of greater numbers of mutant cells. One technique uses penicillin, which inhibits cell synthesis in dividing gram-positive bacteria. For example, if *E. coli* is plated on minimal medium containing penicillin, all prototrophs will enter the division stage and die. Auxotrophs cannot divide and thus are not killed. After penicillin treatment, cells are transferred to a penicillin-free medium supplemented with a variety of growth factors, where the surviving mutant cells grow and form colonies. Replica plating is then performed, transferring these colonies to a number of agar plates, each containing minimal medium plus one growth factor. In this way, the specific nutritional requirement may be determined, and the mutant strain established in pure culture.

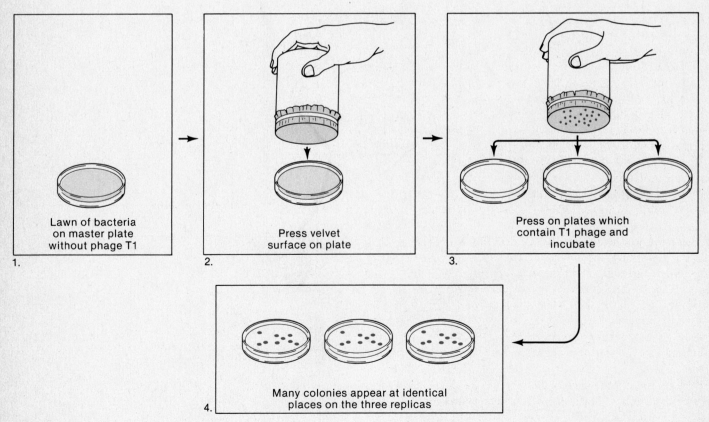

1. Lawn of bacteria on master plate without phage T1

2. Press velvet surface on plate

3. Press on plates which contain T1 phage and incubate

4. Many colonies appear at identical places on the three replicas

FIGURE 14.4
Demonstration of phage T1-resistant bacteria using the replica plating technique.

Another procedure employs the phenomenon of **tritium suicide.** In this technique, bacterial cells are incubated on a minimal medium containing highly radioactive ^{3}H-labeled DNA precursors. These radioactive molecules are actively incorporated into the DNA of growing cells. Nutritional mutants that have arisen spontaneously or that have been induced do not divide on the minimal medium and do not incorporate the tritiated precursors. The total cell population is then stored in the cold for several weeks. During this time, little growth occurs, but the radioactive decay of the tritium-labeled DNA disrupts the genetic material of the nonmutant cells. However, the mutant cells survive and may be recovered when returned to growth medium containing the proper supplements. The radioisotope phosphorus-32 (^{32}P), which has a much shorter half-life and decays much more rapidly than tritium, may be used in a similar manner.

Phenotypic and Segregation Lag

Two phenomena, **phenotypic lag** and **segregation lag,** operate to delay the immediate expression and detection of newly arisen mutations. In phenotypic lag, the mutant trait may only be expressed several generations after the mutation actually occurred. The reason for this can be illustrated by considering phage resistance.

Phage-sensitive bacteria contain specific receptors in the cell wall where phage particles are adsorbed prior to infection. Phage-resistant mutations eliminate the ability to establish any new functional receptor sites. However, those sites already present still allow phage attachment, and as the mutant bacterium divides, these existing sites are gradually diluted in the descendant cells. Thus, phenotypic expression of resistance occurs only after descendant cells are produced containing too few receptors to permit significant phage adsorption.

Segregation lag, a related phenomenon, is explained in the following way. Bacteria in the log phase of exponential growth divide more slowly than chromosome replication occurs. As a result, a single cell may contain two, four, or even eight chromosomes. If a spontaneous auxotrophic mutation arises, for example, in one of the two, four, or eight chromosomes present, the mutation will not be expressed until the nonmutant chromosomes are subsequently distributed to daughter cells (Figure 14.5). As long as one chromosome is still present containing the nonmutant gene in question, prototrophic growth will occur.

Observations such as these, indicative of either phenotypic or segregation lag, must be taken into ac-

count when quantitative studies of mutation are performed. For example, studies to determine the spontaneous mutation rate to phage resistance are likely to be underestimates because of phenotypic or segregation lag.

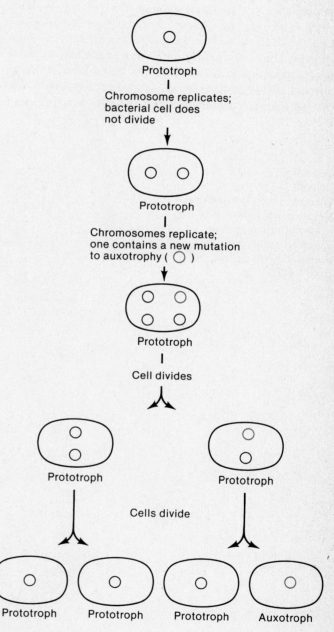

FIGURE 14.5

Explanation of segregation lag, where the chromosome replicates more quickly than the cell divides. If an auxotrophic mutation arises, it may not be expressed immediately if a second chromosome containing the wild-type gene is present. Eventually, the cell division catches up with replication, the auxotroph-bearing chromosome is present alone in the cytoplasm, and the mutation is expressed.

GENETIC EXCHANGE AND RECOMBINATION IN BACTERIA

The development of the isolation techniques just described and the information derived from the subsequent study of bacterial mutagenesis led to detailed investigations of the arrangement of genes on the bacterial chromosome. These studies were initiated in 1946 by Joshua Lederberg and Edward Tatum. They showed that bacteria undergo **conjugation,** a parasexual process in which the genetic information of one bacterium is transferred to and recombined with that of another bacterium. Like meiotic crossing over in eukaryotes, genetic recombination in bacteria led to methodology for chromosome mapping.

Two other phenomena, **transformation** and **transduction,** also result in the transfer of genetic information from one bacterium to another. These processes also have served as a basis for determining the arrangement of genes on the bacterial chromosome. Transformation (see Chapter 8) involves the entry and integration of a piece of DNA from one bacterium into the chromosome of another intact organism. Transduction, on the other hand, is a phage-mediated transfer of small pieces of DNA from one bacterial cell to another.

FIGURE 14.6
Recombination between two auxotrophic strains resulting in prototrophs. Neither auxotroph will grow on minimal medium, but prototrophs will.

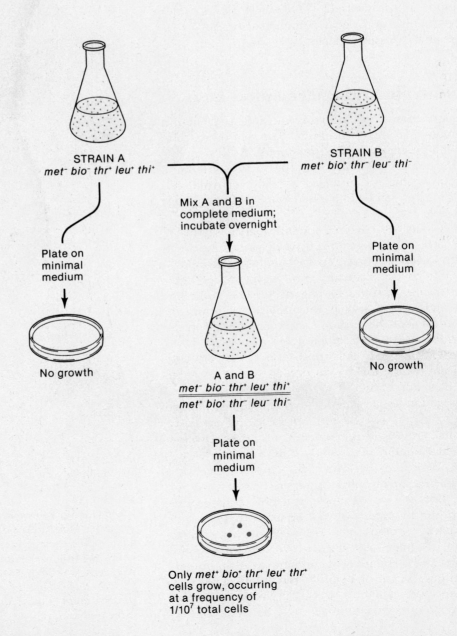

STRAIN A
met⁻ bio⁻ thr⁺ leu⁺ thi⁺

STRAIN B
met⁺ bio⁺ thr⁻ leu⁻ thi⁻

Mix A and B in complete medium; incubate overnight

Plate on minimal medium

Plate on minimal medium

No growth

No growth

A and B
met⁻ bio⁻ thr⁺ leu⁺ thi⁺
met⁺ bio⁺ thr⁻ leu⁻ thi⁻

Plate on minimal medium

Only *met⁺ bio⁺ thr⁺ leu⁺ thr⁺*
cells grow, occurring
at a frequency of
1/10⁷ total cells

BACTERIAL CONJUGATION

Lederberg and Tatum's initial experiments were performed with two multiple auxotrophic strains of *E. coli* K12. Strain A required methionine and biotin in order to grow, while strain B required threonine, leucine, and thiamine (Figure 14.6). Neither strain would grow on minimal medium. The two strains were first grown separately in supplemented media, and then cells from both were mixed and grown together for several more generations. They were then plated on minimal medium. Any bacterial cells that grew on minimal medium would be prototrophs. It was highly improbable that any of the cells that contained two or three mutant genes would undergo spontaneous mutation simultaneously at two or three locations. Therefore, any prototrophs recovered must have arisen as a result of some form of genetic exchange and recombination.

In this experiment, prototrophs were recovered at a rate of $1/10^7$ (10^{-7}) cells plated. The controls for this experiment involved separate plating of cells from the strains A and B on minimal medium. No prototrophs were recovered. Lederberg and Tatum therefore concluded that genetic exchange had occurred!

In the following decade, many experiments designed to elucidate the mechanism of bacterial recombination were performed. Because the description of these experiments is lengthy and the results and conclusions complex, they will be discussed in a stepwise fashion.

F^+ and F^- Bacteria: Observations and Conclusions

1   Lederberg and Tatum showed that filtrates of lysed cultures of either strain A or B would not lead to the production of prototrophs when added to intact cells of the other strain. Therefore, it appeared that the cells must interact directly for genetic recombination to occur. This idea received strong support from an experiment by Bernard Davis using a U-tube (Figure 14.7). At the base of the tube was a sintered glass filter with a pore size that allowed the passage of the liquid medium, but was too small to allow the passage of bacteria. Strain A was placed on one side and strain B on the other side of the filter. The medium was pulled back and forth by suction during bacterial incubation. Samples from both sides of the tube were then plated on minimal medium, but no prototrophs were found. Thus, it was concluded that **physical contact is essential to genetic recombination.** This physical interaction is the initial stage of the process of conjugation and is mediated through structures called **pili.** Bacteria often have many pili, which are microscopic extensions of the cell. After contact has been initiated, a **conjugation tube** is formed between mating pairs (Figure 14.8), and transfer of DNA begins. While *pilus* and *conjugation tube* are not synonymous terms, some workers believe that a single sex pilus may give rise to a conjugation tube. Others disagree, however.

FIGURE 14.7
When strain A and B auxotrophs are grown in a common medium but are separated by a filter, no recombination occurs and no prototrophs are produced. This arrangement is called a Davis U-tube.

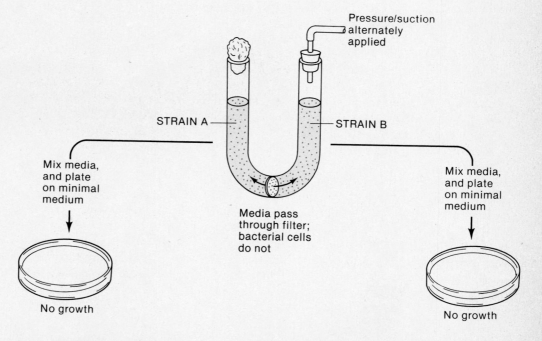

Pressure/suction alternately applied

STRAIN A — STRAIN B

Mix media, and plate on minimal medium

Media pass through filter; bacterial cells do not

Mix media, and plate on minimal medium

No growth

No growth

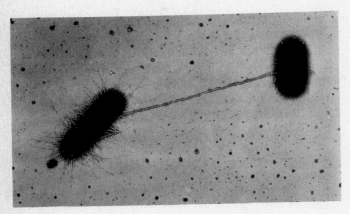

FIGURE 14.8

Conjugation in *E. coli*. The bacterial cell covered with numerous appendages is a genetic donor (F⁺) connected to a recipient cell (F⁻), which lacks the appendages. The appendages are called pili, and the connection between the cells is the F pilus, sometimes called the conjugation tube. Its formation by the F⁺ cell is under the control of genetic information contained in the cytoplasmic F plasmid. The F pilus is labeled along its length by specific virus particles that infect donor cells through this pilus. (Reproduced with the permission of Charles C. Brinton, Jr., and Judith Carnahan.)

2   The initial interpretation of sexuality in bacteria was based on what was already known about single-celled algae and fungi: cell contact, fusion of nuclei forming a diploid zygote, and a meiotic division reestablishing haploidy. In 1952, William Hayes performed experiments with auxotrophs similar to those used by Lederberg and Tatum. His findings established that **transfer of the genetic material is unidirectional.** Hayes found that the number of prototrophs was not diminished greatly if strain A cells were inactivated, or sterilized, by exposing them to high concentrations of streptomycin prior to the cross. Streptomycin is a potent antibiotic that inhibits protein synthesis and prevents subsequent cell divisions. However, no prototrophs were recovered when strain B was similarly treated with streptomycin.

The conclusion drawn was that conjugation involves a **donor** and a **recipient** cell. In Hayes' experiment, cells of strain A were serving as donors, and cells of strain B were the recipients. The reasoning used in drawing this conclusion was that streptomycin, while inhibiting bacterial reproduction of the donor cells, will not prevent them from transferring the genetic material to the recipient. On the other hand, by inhibiting the growth and division of cells serving as recipients, streptomycin effectively terminates the process leading to the recovery of prototrophs. Therefore, these observations support the concept of the occurrence of a nonreciprocal transfer of the bacterial chromosome where cells of strain A serve as "male" donors and those of strain B serve as "female" recipients.

3   In subsequent independent experiments performed by Hayes, the Lederbergs, and Luca Cavalli-Sforza, it was shown that certain conditions could eliminate donor ability in otherwise fertile cells. However, if these cells were then grown with fertile donor cells, fertility was reestablished. The conclusion drawn was that a **fertility factor,** or **F factor,** controls donor ability. It can be lost, converting the donor cell to a recipient, and can also be regained during conjugation. Cells containing the F factor are designated **F⁺,** and those lacking it are designated **F⁻.** On this basis, the initial crosses of Lederberg and Tatum can be clarified:

$$\begin{array}{ccc} \text{Strain A} & & \text{Strain B} \\ F^+ & \times & F^- \\ \text{Donor} & & \text{Recipient} \end{array}$$

4   **The F factor has been subsequently shown to consist of a circular, double-stranded DNA molecule.** It exists independently from the bacterial chromosome and contains 6×10^4 nucleotide base pairs. This amount of DNA, equivalent to about 2 percent of that making up the bacterial chromosome, is sufficient to code for about 11 genes. Eight of these are involved in the formation of the **sex pilus** that initiates contact between an F⁺ and an F⁻ cell.

5   Following conjugation, **virtually every recipient cell becomes F⁺,** suggesting that the F factor is passed through the conjugation tube during each mating, as diagramed in Figure 14.9. In only about $1/10^7$ cells, however, is any part of the donor's chromosome passed to the recipient cell. Apparently, when this occurs, the F factor carries a small piece of the donor chromosome with it during transfer. The part of the chromosome transferred appears to be a random event. All genes thus have a very low, random probability of entering the recipient cell. Following conjugation, the recipient bacterium is partially diploid for the genes transmitted, since two copies are now present. This creates a temporary condition that may lead to genetic recombination. For recombination to occur, the donor genes must be exchanged with the ho-

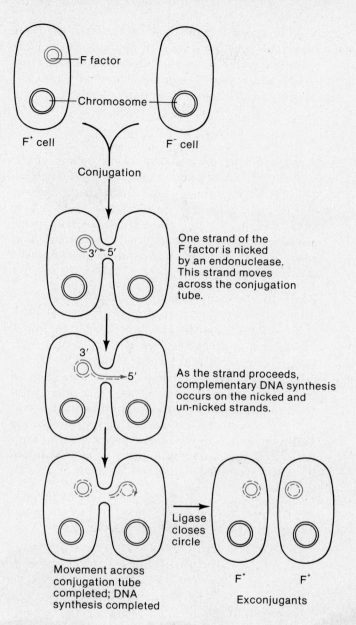

FIGURE 14.9
An $F^+ \times F^-$ mating demonstrating how the recipient F^- cell is converted to F^+. The F factor is transferred across the conjugation tube during its replication.

Labels within Figure 14.9:
- F factor
- Chromosome
- F^+ cell
- F^- cell
- Conjugation
- One strand of the F factor is nicked by an endonuclease. This strand moves across the conjugation tube.
- As the strand proceeds, complementary DNA synthesis occurs on the nicked and un-nicked strands.
- Movement across conjugation tube completed; DNA synthesis completed
- Ligase closes circle
- F^+ F^+
- Exconjugants

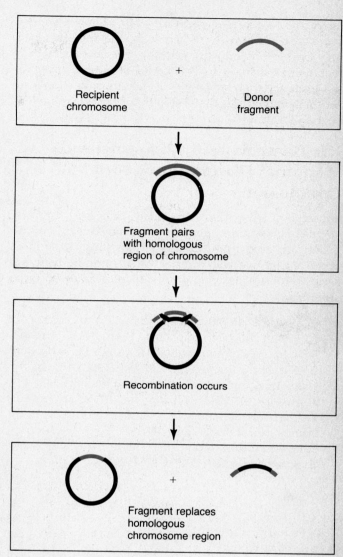

Labels within Figure 14.10:
- Recipient chromosome
- Donor fragment
- Fragment pairs with homologous region of chromosome
- Recombination occurs
- Fragment replaces homologous chromosome region

FIGURE 14.10
One possible mechanism by which a piece of a donor chromosome can recombine with the recipient chromosome in an $F^+ \times F^-$ mating.

mologous region of the recipient chromosome. This exchange is most simply visualized as an event involving two crossover sites (Figure 14.10). As a result, the donor genes are integrated into the recipient chromosome, and a fragment of this chromosome is produced. In most cases, the fragment is degraded in the cell.

F^+ and F^-: Summary

In *E. coli*, all cells are either F^+ or F^-, depending on the presence or absence of a small, circular DNA fragment called the F factor. When this factor is present, the cell is able to form a conjugation pilus and serve as a donor of genetic information. During conjugation, a copy of the F factor is always transferred to the F^- recipient, converting it to the F^+ state. Only infrequently is a random portion of the donor chromosome also transferred. When it is, the recipient cell may integrate it into its own chromosome through a recombinational event. This series of events serves as the basis for the conversion of auxotrophic strains to prototrophic strains, as originally observed by Lederberg and Tatum.

Hfr Bacteria and Chromosome Mapping: Observations and Conclusions

Subsequent discoveries revealed that the circumstances surrounding genetic exchange during conjugation are even more complex than those described above. Again, we will approach this topic by listing a series of observations and conclusions, followed by a summary.

1 In 1950, Cavalli-Sforza treated an F^+ strain of *E. coli* K12 with nitrogen mustard. From these treated cells he recovered a culture of donor bacteria that underwent recombination at a rate of $1/10^4$ (10^{-4}), 1000 times more frequently than the original F^+ strains. In 1953, Hayes isolated a similar strain that yielded recombinants at a similar frequency. Both strains were designated **Hfr,** or **high-frequency recombination.** Since Hfr cells behave as donors, they are a special class of F^+ cells.

2 Another important difference was noted between Hfr strains and the original F^+ strains. The recipient cells, while sometimes displaying genetic recombination, rarely became Hfr; they remained F^-. In comparison, then,

$$F^+ \times F^- \rightarrow F^+$$
$$Hfr \times F^- \rightarrow F^-$$

3 Perhaps the most significant characteristic of Hfr strains is the nature of recombination. While an F^+ cell donates a random segment of its chromosome, genetic recombination involving an Hfr cell is strain-specific. In any given strain, certain genes are more frequently recombined than others, and some not at all. This **nonrandom pattern** was shown to vary from Hfr strain to Hfr strain.

4 While these results were puzzling, Hayes interpreted them to mean that some physiological alteration of the F factor had occurred, resulting in the production of Hfr strains of *E. coli*. In the mid-1950s, experimentation by Ellie Wollman and Francois Jacob clarified the basis of Hfr and showed how these strains allowed genetic mapping of the *E. coli* chromosome. In their experiments, Hfr and F^- strains with suitable marker genes were mixed and recombination of specific genes assayed with time. To accomplish this, cultures of an Hfr and an F^- strain were first incubated together. At various

FIGURE 14.11
The progressive transfer during conjugation of various genes from a specific Hfr strain of *E. coli* to an F^- strain. Certain genes (thr^+ and leu^+) are quickly transferred and recombine with high frequency. Others (lac^+ and gal^+) take longer to be transferred and recombine with a lower frequency.

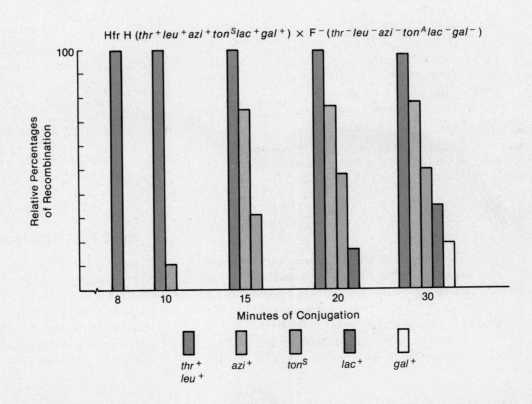

Hfr H ($thr^+ leu^+ azi^+ ton^S lac^+ gal^+$) × F$^-$ ($thr^- leu^- azi^- ton^A lac^- gal^-$)

Relative Percentages of Recombination

Minutes of Conjugation

thr^+ leu^+ azi^+ ton^S lac^+ gal^+

intervals, samples were removed and placed in a blender. The shear forces created in the blender separated conjugating bacteria so that the transfer of the chromosome was effectively terminated. The cells were then assayed for genetic recombination. This process, called the **interrupted mating technique,** demonstrated that specific genes of a given Hfr strain were transferred and recombined sooner than others. Figure 14.11 illustrates this point. During the first 8 minutes after the two strains were initially mixed, no genetic recombination could be detected. Between 8 and 10 minutes, recombination in high frequency occurred for the thr^+ and leu^+ genes. At 10 minutes, only 10 percent of the thr^+, leu^+ recombinants also showed recombination of the azi^+ gene, and no transfer of the ton^s, lac^+, or gal^+ genes was noted. By 15 minutes, 70 percent of the recombinants were also azi^+; 30 percent were also ton^s; but none was lac^+ or gal^+. By 20 minutes, the lac^+ gene was found among the recombinants; and by 30 minutes, gal^+ was also being transferred. Therefore, Wollman and Jacob had demonstrated an **oriented transfer** of genes that was correlated with the length of time conjugation was allowed to proceed.

It appeared that the chromosome of the Hfr bacterium was transferred linearly and that the sequence and distance between the genes, as measured in minutes, could be predicted from such experiments (Figure 14.12). This information served as the basis for the first genetic map of the E. coli chromosome. These results also explained why only certain genes were transferred in Hfr $\times$ F$^-$ matings. Initially, those genes close to a point of origin of transfer (O) are passed across the conjugation tube; apparently, conjugation does not usually last long enough to allow the entire chromosome to pass across.

Wollman and Jacob then repeated the same type of experimentation with other Hfr strains, obtaining similar results with one important difference. While genes were always transferred linearly with time, as in their original experiment, which genes entered first and which followed later seemed to vary from Hfr strain to Hfr strain [Figure 14.13(a)]. When they reexamined the rate of entry of genes, and thus the different genetic maps for each strain, a definite pattern emerged. The major differences between all strains were simply the point of the origin (O) and the direction in which entry proceeded from that point [Figure 14.13(b)].

In order to explain these results, Wollman and Jacob postulated that **the E. coli chromosome is circular.** If the point of origin (O) varied from strain to strain, a different sequence of genes would be transferred in each case. But what determines O? They proposed that in various Hfr strains, **the F factor integrates into the chromosome at different points.** Its position determines the O site. During conjugation (Figure 14.14), the circular chromosome breaks open at that point. Those genes adjacent to O are transferred first, leading the mobilized chromosome into the F$^-$ cell. The F factor becomes the end of the broken chromosome and is the last part to be transferred. Apparently, conjugation rarely, if ever, lasts long enough to allow the entire chromosome to pass across the conjugation tube. This proposal explains why recipient cells, when mated with Hfr cells, remain F$^-$. The current thinking is that only one of the two DNA strands of the bacterial chromosome enters the recipient cell. This and the strand left behind undergo semiconservative replication as transfer occurs, creating two identical double helices, one entering the F$^-$ cell and one remaining in the Hfr cell.

The use of the interrupted mating technique with different Hfr strains has provided the basis for mapping the entire E. coli chromosome. Mapped in time units, strain K12 (or E. coli K12) is 90 minutes long. Over 400 genes have now been placed on the map. This technique is accurate for genes that are 3 or more minutes apart on the chromosome. For genes that are even closer together, the approximate location but not the precise sequence or time distance can be determined. In such cases, other forms of recombination in E. coli allow closely linked genes to be mapped more precisely.

FIGURE 14.12
A time map of the genes studied in the experiment depicted in Figure 14.10.

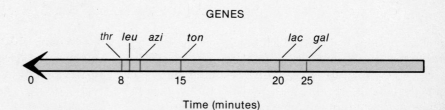

GENES

thr leu azi   ton   lac gal

0   8   15   20 25

Time (minutes)

(a)

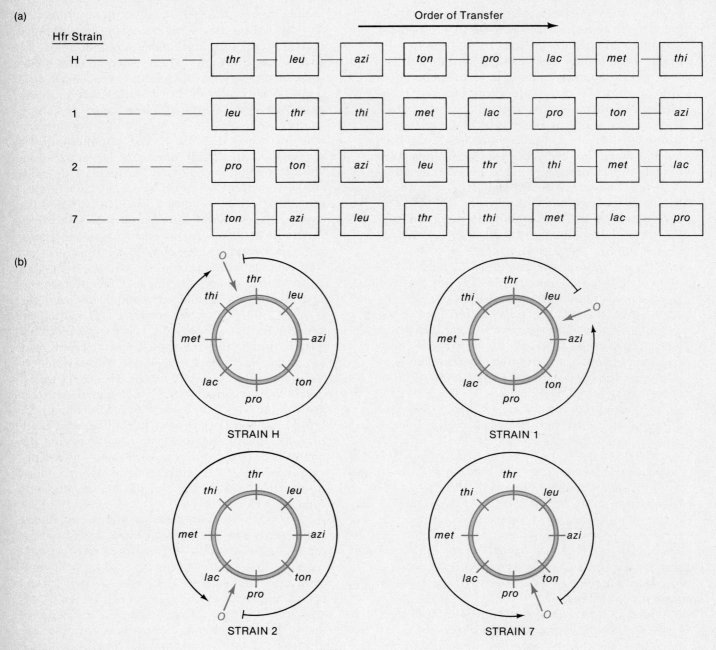

(b)

FIGURE 14.13
The order of gene transfer in various Hfr strains suggests that the *E. coli* chromosome is circular. The position designated *O* in each strain represents the origin of transfer. Note that transfer can proceed in either direction, depending on the strain. The origin is determined by the point of integration into the chromosome of the F factor, and the direction of transfer is determined by the orientation of the F factor as it integrates.

6   The above model has helped geneticists to understand better how genetic recombination occurs in the $F^+ \times F^-$ matings. Recall that recombination occurs less frequently in them than in Hfr $\times$ F^- matings, and that random gene transfer is involved.

The current belief is that when F^+ and F^- cells are mixed, conjugation occurs readily and the F^- cell receives a copy of the F factor, but that no genetic recombination occurs. However, in a very low frequency in a population of F^+ cells, the F factor

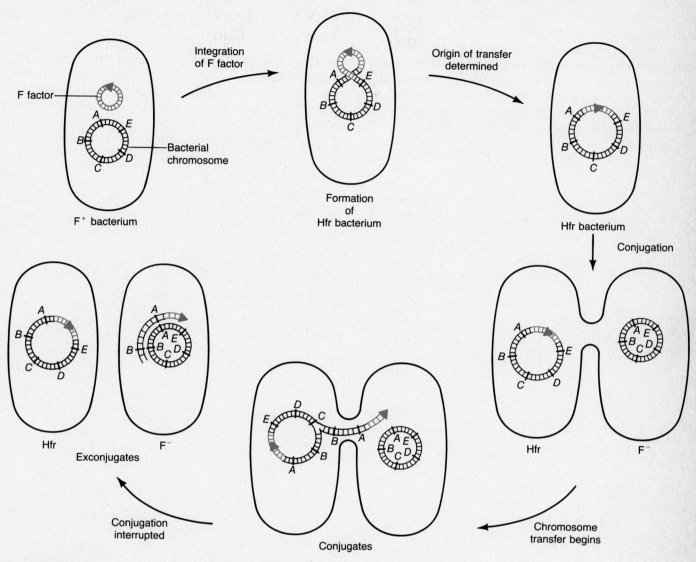

FIGURE 14.14

Conversion of F^+ to an Hfr state occurs by the integration of the F factor into the bacterial chromosome. During conjugation, the F factor is nicked by an enzyme, creating the origin of transfer of the chromosome. The major portion of the F factor is now on the end of the chromosome opposite the origin and is the last part to be transferred. DNA replication occurs simultaneously as transfer proceeds. Conjugation is usually interrupted prior to complete transfer. Above, only the A and B genes were transferred.

spontaneously integrates at a random point into the bacterial chromosome. This integration converts the F^+ cell to the Hfr state. Therefore, in F^+ × F^- crosses, the very low frequency of genetic recombination (10^{-7}) is attributed to the rare, newly formed Hfr cells. Since the point of integration is random, a nonspecific gene transfer is noted.

Hfr Bacteria and Chromosome Mapping: Summary

Hfr bacteria are produced by the spontaneous and random integration of the cytoplasmic F factor into the bacterial chromosome. In the Hfr condition, conjugation is initiated with F^- bacteria. Depending on the site of integration of the F factor, the chromosome is passed across the conjugation tube in a linear fashion, and genetic recombination occurs at a rate of about 10^{-4}. The sequence of gene entry, which can be determined experimentally, is used to construct maps of bacterial chromosomes. Mapping of various Hfr strains has shown the *E. coli* chromosome to be circular. The spontaneous conversion of F^+ cells to Hfr cells serves as the basis for genetic exchange in $F^+ \times F^-$ crosses.

The F′ State, Merozygotes, and Sexduction

In 1959 during experiments with Hfr strains of *E. coli*, Edward Adelberg discovered that the F factor could lose its integrated status, causing reversion to the F^+ state (Figure 14.15). When this occurs, the F factor frequently carries several adjacent bacterial genes along with it. He labeled this condition **F′** to distinguish it

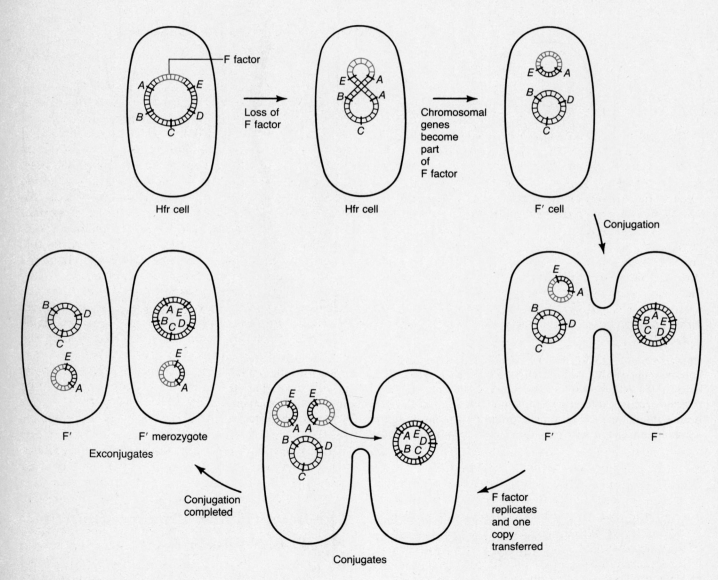

FIGURE 14.15
Conversion of an Hfr bacterium to F′ and its subsequent sexduction of an F^- cell. The conversion occurs when the F factor loses its integrated status and carries with it one or more chromosomal genes (A and E). Following conjugation with such an F′ cell, an F^- recipient becomes partially diploid and is called a merozygote. It also becomes F′.

from F^+ and Hfr. F' is thus a special case of F^+.

The presence of bacterial genes within a cytoplasmic F factor creates an interesting situation. An F' bacterium behaves as an F^+ cell, initiating conjugation with F^- cells. When this occurs, the F factor, along with the chromosomal genes, is always transferred to the F^- cell. This process is called **sexduction.** As a result, whatever chromosomal genes are part of the F factor are now present in duplicate in the recipient cell because they are also present in the recipient cell's chromosome. This creates a partially diploid cell called a **merozygote.** Pure cultures of F' merozygotes may be established. They have been extremely useful in the study of bacterial genetics, particularly in the area of genetic regulation (see Chapter 21).

PLASMIDS, INSERTION SEQUENCES, AND TRANSPOSONS

In the above sections, we introduced and discussed extensively an extrachromosomal hereditary unit in bacteria—the **F factor.** When existing autonomously in the cytoplasm, the F factor takes the form of a double-stranded closed circle of DNA and is self-replicating. As such, it meets the criteria that define the more general category of molecules called **plasmids.** Plasmids, such as the F factor, which can also integrate into the bacterial chromosome are called **episomes.** Other plasmids maintain an independent existence in the bacterial cytoplasm, and multiple copies may be present in a single cell. Each may contain as few as three or four genes or as many as several hundred genes.

Plasmids are generally classified according to the genetic information specified by their DNA. The F factor, as you will recall, is a fertility factor and contains genes essential for sex pilus formation as well as for DNA replication. Other plasmids contain information conferring resistance to antibiotics (the **R plasmids**), heavy metals, and ultraviolet radiation, and leading to the production of toxins (the **Col plasmids**).

While plasmids as extrachromosomal units are valid topics for genetic study, two recent discoveries have dramatically increased the interest in their investigation. The first involves segments of DNA called **insertion sequences (IS).** They consist of specific nucleotide sequences and are capable of moving in and out of DNA molecules. Insertion sequences can combine with other genes to form a vehicle, the **transposon (Tn),** which allows for transfer of genetic information from one DNA source to another. For example, transposons can move between plasmids and bacterial and viral chromosomes. As a result, they have sometimes been called "jumping genes."

Interest in plasmids has also been enhanced by the phenomenon of **recombinant DNA**—one aspect of the more general topic of **genetic engineering.** In genetic engineering, specific genes from a variety of organisms may be inserted into plasmids by experimental manipulation. As such, these genes may be **cloned** as the plasmid replicates and copies of it are distributed to progeny cells. As we will discuss in Chapter 20, recombinant DNA technology is undoubtedly the most important development in genetics to occur in the 1970s.

Insertion Sequences and Transposons in Bacteria

In the early 1970s, a number of independent workers, including Peter Starlinger and James Shapiro, discovered a unique class of mutations in *E. coli*. These mutations affected several genes in different strains of bacteria. For example, transcription of a cluster of genes (or **operon**) controlling galactose metabolism in *E. coli* was repressed or shut off as a result of one of these mutations. This phenotypic effect was heritable, but was found not to be caused by a nucleotide change characteristic of conventional mutations. Instead, it was shown that a short, specific DNA segment had been inserted into the bacterial chromosome at the beginning of the galactose gene cluster. When this segment was spontaneously excised from the bacterial chromosome, wild-type function was restored.

It was subsequently revealed that several other distinct DNA segments could behave in a similar fashion, inserting into the chromosome and affecting gene function. These DNA segments are relatively short, not exceeding 2000 base pairs (2 kb, or 2 kilobases), and are now called **insertion sequences (IS).** The genetic information contained within an IS unit appears to be involved only in the insertion process itself.

Analysis of the DNA sequences of IS units has revealed a most interesting feature. At each end of any given double-stranded unit, the nucleotide sequence consists of a **perfect inverted repeat** of the other end. While Figure 14.16 shows this terminal repeating unit to consist of only a few nucleotides, many more are actually involved. For example, in *E. coli*, the IS1 termini contain about 30 nucleotide pairs, IS2 about 40 pairs, and IS4 about 18 pairs. While it is not known for certain, it seems likely that these terminal sequences are involved closely in the mechanism of insertion of IS units into DNA. This contention is supported by the fact that insertion of IS units is not random but instead is more likely to occur at certain DNA regions. This suggests that the termini may be

FIGURE 14.16
Diagrammatic representation
of an insertion sequence (IS).

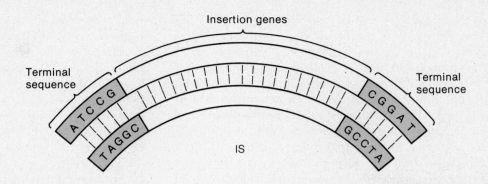

involved in recognition sites during the insertion process.

Careful investigation has revealed that IS units are
present in the wild-type *E. coli* chromosome as well
as in various plasmids. Thus, their presence does not
always result in mutation. In the *E. coli* chromosome,
multiple copies of IS1 and IS2 are known to be present, but the number of IS3 and IS4 is yet to be determined. The position of IS units is not fixed; rather,
they move from point to point, inserting at preferential
positions. The F factor also contains IS units, as shown
in Figure 14.17. Two copies of IS3, one IS2, and one
designated γδ are present. It appears that these IS
sites are involved in the integration of the F factor into
the *E. coli* chromosome during Hfr formation. A possible mode of integration and excision of the F factor

into and out of the bacterial chromosome is illustrated
in Figure 14.18

In addition to its mutational effects, the insertion of
IS units has a significant role in the formation and
movement of the larger **transposon (Tn) elements.**
Transposons consist of IS units that are covalently

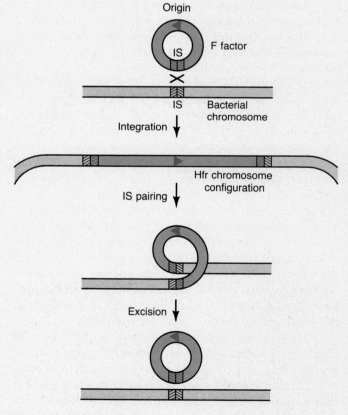

FIGURE 14.18
A model depicting the role of insertion sequences (IS) in
the integration and excision of the F factor into and out of
the bacterial chromosome. Insertion leads to the conversion
of an F$^+$ bacterium to an Hfr bacterium.

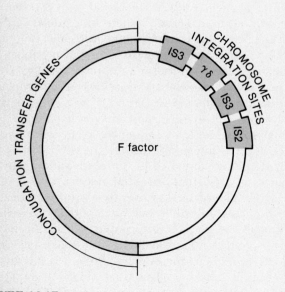

FIGURE 14.17
Diagram of the F factor of *E. coli* showing four insertion
sequences that serve as integration sites into the bacterial
chromosome.

joined to other genes whose functions are unrelated to the insertion process. Like IS units, Tn elements can insert into bacterial and viral chromosomes as well as plasmids. This property is undoubtedly related to the presence of IS units in each transposon. Thus, Tn elements provide a mechanism for movement of genetic information from place to place both within and between organisms.

Transposons were first discovered to move between DNA molecules as a result of observations of anti-biotic-resistant bacteria. In the mid-1960s, Susumu Mitsuhashi suggested that an R plasmid containing a gene responsible for resistance to the antibiotic chlor-amphenicol had been transferred to the bacterial chromosome. Since then, it has been demonstrated that antibiotic-resistant genes can move from plasmid to plasmid as well as between plasmids and chromosomes. Electron microscopic studies have confirmed that the two terminal inverted repeat sequences are

indeed present within plasmids harboring transposons. As shown in Figure 14.19, when the strands of DNA from such a plasmid are separated and allowed to reanneal separately, the inverted repeat units are complementary, and a heteroduplex is formed as expected. All areas other than the repeat units remain single stranded and form loops on either end of the double-stranded stem.

One of the most significant aspects of these findings is that transposition does not appear to require homologous pairing in order for insertion to occur. Recombination of a more conventional nature—conjugation and transformation, for example—is thought to rely on homologous pairing prior to the recombination event. In *E. coli*, such events are dependent upon the product of the *rec A* locus. In *rec A* mutants, such recombination is abolished. However, transposon movement can occur in such mutants. Therefore, transposition as a form of recombination would seem

FIGURE 14.19
Heteroduplex formation as a result of inverted repeat sequences within a transposon inserted into a bacterial plasmid. (From "Transposable Genetic Elements" by S. N. Cohen and J. A. Shapiro. Copyright © 1980 by Scientific American, Inc. All rights reserved.)

IR = Inverted Repeat Sequence

IG = Internal Genes

Tn = Transposon

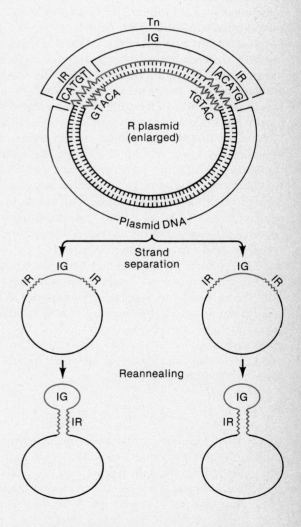

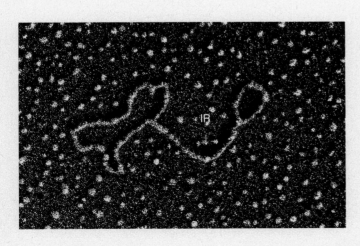

not to involve or be dependent on the homologous pairing-breakage-rejoining mechanism. Because of this characteristic, transposition is described as a form of **illegitimate recombination.** Nevertheless, the precise mechanism of insertion remains unknown.

Given this information, it can be predicted that DNA segments that are evolutionarily unrelated to one another may potentially recombine. In fact, it is believed that the wide variety of **R plasmids** that confer antibiotic resistance in bacteria are formed as a result of transposon insertion. Because of the severe medical problems created by multiply antibiotic-resistant bacteria, the formation of R plasmids has been extensively investigated.

R plasmids consist of two components: the **RTF (resistance transfer factor)** and one or more **r-determinants.** The RTF contains information essential to transfer of the plasmid between bacteria, and r-determinants are genes conferring antibiotic resistance. While the DNA of RTFs is quite similar in a variety of plasmids from different bacteria species, there is a wide variation in r-determinants. Each is specific for resistance to one antibiotic. If a bacterial cell contains r-determinant plasmids but no RTF is present, the cell is resistant but not infectious; that is, resistance cannot be transferred. However, the most commonly studied plasmids consist of the RTF as well as one or more r-determinants. Resistances to tetracycline, streptomycin, ampicillin, sulfonamide, kanamycin, and chloramphenicol are most frequently encountered.

Another unique observation is that there seems to be a modular construction of complex R plasmids that depends upon IS unit integration and excision events. As diagramed in Figure 14.20, r-determinant genes are part of transposons that contain inverted repeat termini of the IS unit. As a result, r-determinants move readily between plasmids as well as integrate into bacterial or viral chromosomes. Figure 14.21 shows one such complex R plasmid thought to arise in this way.

The implications of these findings are great. Not only has a new and unique recombinational mechanism been revealed, but the way in which prokaryotic genomes have evolved may possibly have been discovered. Furthermore, it appears that similar events may occur in eukaryotic systems. We will return to this topic in Chapter 19.

BACTERIAL TRANSFORMATION

In Chapter 8 we described how the phenomenon of transformation provided direct evidence that DNA is the genetic material. Transformation, like conjugation, provides a mechanism for the recombination of genetic information in some bacteria. Transformation

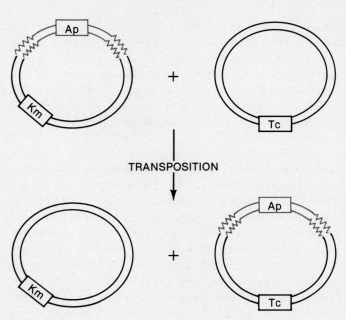

FIGURE 14.20

Transposition event creating an ampicillin (Ap)-tetracycline (Tc)-resistant R plasmid.

can also be used to map bacterial genes, although in a more limited way.

Transformation consists of numerous steps that can be divided into two main categories: (1) entry of exogenous DNA into a recipient cell, and (2) recombination of the donor DNA with its homologous region in the recipient chromosome. In a population of cells, only those in a particular physiological state, referred to as **competence,** take up DNA. Entry apparently occurs at a limited number of receptor sites on the surface of

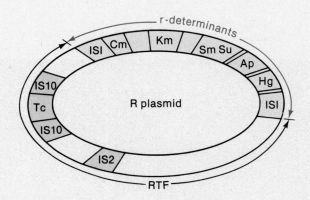

FIGURE 14.21

R plasmid containing resistance transfer factors (RTF) and multiple r-determinants (Tc—tetracycline; Km—kanamycin; Sm—streptomycin; Su—sulfonamide; Ap—ampicillin; Hg—mercury). (After Cohen and Shapiro, 1980.)

the bacterial cell. The efficient length of transforming DNA is about 10,000 to 20,000 base pairs, an amount equal to about 1/200 of the *E. coli* chromosome. Passage across the cell wall and membrane is an active process requiring energy and specific transport molecules. This concept is supported by the fact that substances that inhibit energy production or protein synthesis in the recipient cell also inhibit the transformation process.

During the process of entry, one of the two strands of the double helix is digested by nucleases, leaving only a single strand to participate in transformation. This DNA segment aligns with its complementary region of the bacterial chromosome. In a process in-

volving several enzymes, the segment replaces its counterpart in the chromosome, which is excised and degraded (Figure 14.22).

In order for recombination to be detected, the transforming DNA must be derived from a genetically different strain of bacteria. Once it is integrated into the chromosome, that region contains one strand originally present and one mutant strand. Since these strands are not genetically identical, this helical region is referred to as a **heteroduplex.** Following one round of semiconservative replication, one chromosome is identical to that of the original recipient, and the other contains the mutant gene. Following cell division, one nonmutant and one mutant cell are produced.

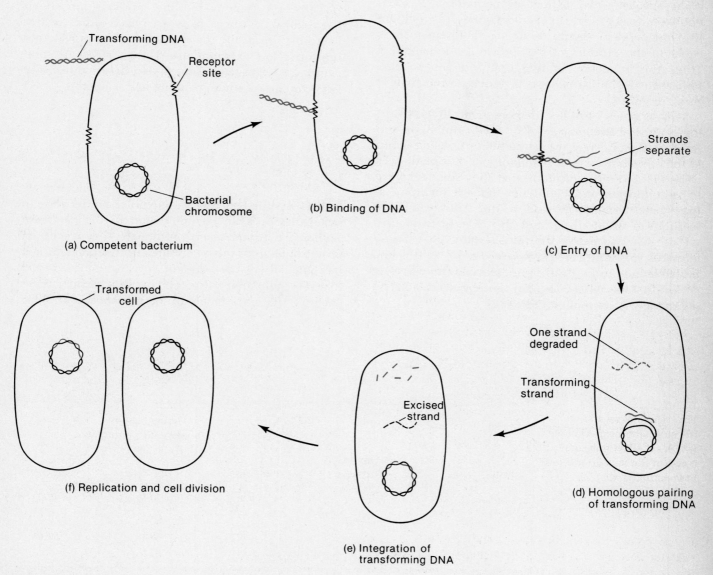

FIGURE 14.22

Proposed steps leading to transformation of a bacterial cell by exogenous DNA.

Transformation and Linked Genes

Because transforming DNA is usually about 10,000 to 20,000 base pairs long, it contains a sufficient number of nucleotides to encode several genes. Genes that are adjacent or very close to one another on the bacterial chromosome may be carried on a single piece of DNA of this size. Because of this fact, a single event may result in the **cotransformation** of several genes simultaneously. Genes that are close enough to each other to be cotransformed are said to be **linked.** Note that here *linkage* refers to the proximity of genes, in contrast to the use of this term in eukaryotes to indicate all genes on a single chromosome.

If two genes are not linked, simultaneous transformation can occur only as a result of two independent events involving two distinct segments of DNA. As in double crossing over in eukaryotes, the probability of two independent events occurring simultaneously is equal to the product of the individual probabilities. Thus, the frequency of two unlinked genes being transformed simultaneously is much lower than if they are linked.

Linked genes were first demonstrated in 1954 during studies of *Pneumococcus* by Rollin Hotchkiss and James Marmur. They were examining transformation at the *streptomycin* and *mannitol* loci. Recipient cells were str^s and mtl^-, meaning that they were sensitive to streptomycin and could not ferment mannitol. If these genes are unlinked and $str^r \, mtl^+$ cells, which are resistant to streptomycin and able to ferment mannitol, are used to derive the transforming DNA, transformants for both genes would occur in such a low probability as to be nearly undetectable. However, as shown in Table 14.2, a low but detectable rate (0.17%) of double transformation did occur.

In order to confirm that the genes were indeed linked, a second experiment was performed. Instead of using $str^r \, mtl^+$ DNA, a mixture of $str^r \, mtl^-$ and $str^s \, mtl^+$ DNA simultaneously served as the donor DNA. Thus, both the str^r and mtl^+ genes were available to recipient cells, but on separate segments of DNA. If the observed rate of double transformants in the previous experiment had been due to two independent events and the genes unlinked, then a similar rate would also be observed under these conditions. Instead, the number of double transformants was 25 times fewer, thus confirming linkage (Table 14.2).

Subsequent studies of transformation have shown that a variety of bacteria undergo this process of recombination, including *Bacillus subtilis*, *Hemophilus influenzae*, *Shigella paradysenteriae*, and *Escherichia coli*. Under certain conditions, studies have shown that relative distances between linked genes can be determined from the recombination data provided by transformation experiments. While analysis is more complex, such data are interpreted in a manner analogous to chromosome mapping in eukaryotes.

TRANSDUCTION: VIRUS-MEDIATED BACTERIAL RECOMBINATION

In 1952, Joshua Lederberg and Norton Zinder were investigating possible recombination in the bacterium *Salmonella typhimurium*. Although they recovered prototrophs from mixed cultures of two different auxotrophic strains, subsequent investigations revealed that recombination was occurring in a manner different from that attributable to the presence of an F fac-

---

TABLE 14.2
The results of several transformation experiments, which establish linkage between the *str* and *mtl* loci in *Pneumococcus*.

Donor DNA	Recipient Cell Genotype	Percentage of Transformed Genotypes		
		$str^r \, mtl^-$	$str^s \, mtl^+$	$str^r \, mtl^+$
$str^r \, mtl^+$	$str^s \, mtl^-$	4.3	0.40	0.17
$str^r \, mtl^-$ and $str^s \, mtl^+$		2.8	0.85	0.0066

SOURCE: Data from Hotchkiss and Marmur, 1954, p. 55.

tor, as in *E. coli*. What they were to discover was still a third mode of bacterial recombination, one mediated by bacteriophages and now called **transduction.**

Bacteriophages and Lysogeny

Before discussing these experiments, we should review the life cycle of bacteriophages (see Figure 8.5) as well as observations of the association of these viruses with their bacterial hosts. Under normal conditions, the relationship is parasitic, with the phage reproducing at the expense of the host cell. The phage attaches to the bacterial cell and injects its genetic material, usually DNA, into the host. The genetic information of the phage uses the cellular machinery and substrate molecules of the bacterium to replicate many copies of its DNA and produce all necessary viral proteins. During packaging, a self-assembly process involving the DNA and coat proteins occurs, resulting in hundreds of mature virus particles that are released following lysis of the bacterial cell.

If phage are added to a confluent bacterial plate, each new virus particle will repeat this cycle, attacking an adjacent bacterial cell. In an overnight culture this process is repeated many times, producing areas of lysed bacterial cells called **plaques.** Each plaque represents the point where a single phage began reproduction. Under these circumstances, the phage has pursued the **lytic pathway.**

The relationship between virus and bacterium does not always result in viral reproduction and lysis. As early as the 1920s it was known that a virus could enter a bacterial cell and establish a symbiotic relationship with it. The precise molecular basis of this symbiosis is now well understood. Upon entry, the viral DNA, instead of replicating in the bacterial cytoplasm, is integrated into the bacterial chromosome. This is the first step in the **lysogenic pathway.** Each time the bacterial chromosome is replicated, the viral DNA is also replicated and passed to daughter bacterial cells following division. Since no new viruses are produced, no lysis of the bacterial cell occurs. However, under certain stimuli, such as chemical or ultraviolet light treatment, the viral DNA may lose its integrated status and initiate replication, phage reproduction, and lysis of the bacterium.

Several terms are used to describe this relationship. The viral DNA, integrated in the bacterial chromosome, is called a **prophage.** Viruses that can either lyse the cell or behave as a prophage are called **temperate.** Those that can only lyse the cell are referred to as **virulent.** A bacterium harboring a prophage is said to be **lysogenic;** that is, it is capable of being lysed as a result of induced viral reproduction. When this occurs, the virus is said to enter the **vegetative** state. The viral DNA, which can replicate either in the bacterial cytoplasm or as part of the bacterial chromosome, may be classified as an **episome.**

The relationship of a prophage and its host cell is symbiotic because lysogenic bacteria are immune to viral attacks by the same phage whose genetic information it harbors. As we shall see in Chapter 21, this immunity is due to the synthesis of a prophage repressor molecule that regulates the expression of the viral DNA.

The Lederberg-Zinder Experiment

Lederberg and Zinder mixed the *Salmonella* auxotrophic strains LA-22 and LA-2 together and recovered prototroph cells when the mixture was plated on minimal medium. LA-22 was unable to synthesize the amino acids phenylalanine and tryptophan (*phe⁻ trp⁻*), and LA-2 could not synthesize the amino acids methionine and histidine (*met⁻ his⁻*). Prototrophs (*phe⁺ trp⁺ met⁺ his⁺*) were recovered at a rate of about $1/10^5$ (10^{-5}) cells.

Although these observations at first appeared to suggest that the type of recombination involved was the kind observed earlier in *E. coli*, experiments using the Davis U-tube soon showed otherwise (Figure 14.23). When the two auxotrophic strains were separated by a glass-sintered filter, thus preventing cell contact but allowing growth to occur in a common medium, a startling observation was made. When samples were removed from both sides of the filter and plated independently on minimal medium, prototrophs were recovered, but only from one side of the tube. Prototrophs were recovered only when cells from the side of the tube containing LA-22 bacteria were plated. Obviously, the presence of LA-2 cells on the other side of the tube was essential for recombination, since LA-2 cells were the source of the new genetic information. Because the genetic information responsible for recombination had somehow to pass across the filter, but in an unknown form, it was initially designated simply as a **filterable agent** (FA).

Three subsequent observations made it clear that this recombination was quite distinct from the F factor-mediated gene transfer observed earlier in *E. coli*:

1 The FA would not pass across filters with a pore diameter of less than 100 nm, a size that normally allows passage of small DNA molecules.

2 Testing the FA in the presence of DNase, which will enzymatically digest DNA, showed that the FA was not destroyed. These first two observations demonstrate that FA is not naked DNA.

FIGURE 14.23
The Lederberg–Zinder experiment using *Salmonella*. After mixing two auxotrophic strains in a Davis U-tube, Lederberg and Zinder recovered prototrophs from the side containing LA-22 but not from the side containing LA-2. These initial observations led to the discovery of the phenomenon called transduction.

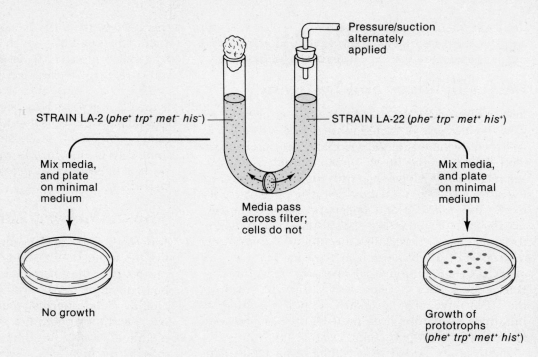

STRAIN LA-2 (*phe⁺ trp⁺ met⁻ his⁻*)

STRAIN LA-22 (*phe⁻ trp⁻ met⁺ his⁺*)

Pressure/suction alternately applied

Mix media, and plate on minimal medium

Media pass across filter; cells do not

Mix media, and plate on minimal medium

No growth

Growth of prototrophs (*phe⁺ trp⁺ met⁺ his⁺*)

3 The third observation was particularly important. It was observed that FA was produced by the LA-2 cells only when they were grown in association with LA-22 cells. If LA-2 cells were grown independently and that culture medium was then added to LA-22 cells, recombination was not observed. Therefore, LA-22 cells play some role in the production of FA by LA-2 cells and do so only when sharing common growth medium.

These observations were explained by the presence of a prophage (P22) present in the LA-22 *Salmonella* cells. Rarely, P22 prophages entered the vegetative or lytic phase, reproduced, and lysed some of the LA-22 cells. This phage, being much smaller than a bacterium, was then able to cross the filter and lyse some of the LA-2 cells, because this strain was not immune to attack. Apparently, in the process, the P22 phages produced in LA-2 often carried with them a region of the LA-2 chromosome. If this region contained the *phe⁺*, and *trp⁺* genes present in LA-2, and if the phages subsequently passed back across the filter and reinfected LA-22 cells, prototrophs were produced. Just how this occurred was not immediately clear.

The Nature of Transduction

Further studies revealed the existence of transducing phages in other species of bacteria. For example, *E. coli*, *B. subtilis*, and *Pseudomonas aeruginosa* can be transduced by the phages P1, SP10, and F116, respec-

tively. The precise mode of transfer of DNA during transduction has also been established. The process begins when a prophage enters the lytic cycle and progeny viruses subsequently infect and lyse other bacteria. During infection, the bacterial DNA is degraded into small fragments and the viral DNA is replicated. As packaging of the viral chromosomes in the protein head of the phage occurs, errors are sometimes made. Rarely, instead of viral DNA, segments of bacterial DNA are packaged.

Regions as large as 1 percent of the bacterial chromosome may become enclosed randomly in the viral head. Following lysis, these aberrant phages, lacking their own genetic material, are released in the culture medium. Because the ability to infect is a property of the protein coat, they can initiate infection of other unlysed bacteria. When this occurs, bacterial rather than viral DNA is injected into the bacterium and can either remain in the cytoplasm or recombine with its homologous region of the bacterial chromosome. If the bacterial DNA remains in the cytoplasm, it does not replicate but may remain in one of the progeny cells following each division. When this happens, only a single cell, partially diploid for the transduced genes, is produced—a phenomenon called **abortive transduction.** If the bacterial DNA recombines with its homologous region of the bacterial chromosome, the transduced genes are replicated as part of the chromosome and passed to all daughter cells. This process is called **complete transduction.** Both abortive and complete transduction are subclasses of the broader category of **generalized transduction.** As described above, transduction is characterized by the random nature of DNA fragments and genes transduced. Each fragment has a finite but small chance of being packaged in the phage head. Most cases of generalized transduction are of the abortive type; some data suggest that complete transduction occurs 10 to 20 times less frequently. This finding may be related to the fact that double-stranded DNA is involved. In comparison with single-stranded DNA, which is integrated during transformation, it may be much more difficult for double-stranded DNA to become integrated.

Mapping and Specialized Transduction

As with transformation, transduction has been used in linkage studies and mapping of the bacterial chromosome. Cotransduced genes must be aligned closely to one another along the chromosome. By concentrating on two or three linked genes, transduction studies can also determine the precise order of genes. Such an analysis is based on the number of recombinational exchanges necessary to integrate the various combinations of linked genes. For example, depending on the linear order of three linked genes, single and double recombination events are required to transduce one or the other genes. By calculating the probabilities of these events, as in single and double crossover events in eukaryotic mapping experiments, the arrangement of genes can be determined by careful analysis.

Another aspect of virally mediated bacterial recombination involves **specialized transduction.** Compared with generalized transduction, where all genes have an equal probability of being transduced, specialized transduction is restricted to certain genes. One of the best examples involves transduction of *E. coli* by the prophage λ. In this case, only the *gal* (galactose) or *bio* (biotin) genes are transduced.

The reason why specialized transduction occurs became clear when it was learned how the λ DNA integrates into the *E. coli* chromosome during lysogeny. A region of λ DNA, designated *att*, is some 15 nucleotides long and is responsible for integration. A precisely homologous region exists on the *E. coli* chromosome, which is flanked by the *gal* and *bio* loci on either end. Therefore, λ DNA always integrates at a location in the chromosome between these genes (Figure 14.24).

Phage λ, like other phages, can be induced to detach from the chromosome and become lytic. Sometimes the detachment process occurs incorrectly and

carries either the *gal* or *bio E. coli* genes in place of part of the viral DNA. This happens when the recombinational event leading to detachment occurs incorrectly, outside the *att* region of the λ chromosome. The resulting chromosome is defective because it has lost some of its genetic information, but it is nevertheless replicated and packaged during the formation of mature phage particles. Once the previously lysogenized cell is lysed, the virus can inject the defective chromosome into another bacterial cell.

In a process involving lysogeny by a nondefective chromosome, the defective viral chromosome may also be integrated into the bacterial chromosome and is replicated along with it. Such cells are diploid for the *gal* or *bio* genes. If the recipient cells are *gal⁻* and are unable to use galactose as a carbon source, the presence of the transducing *gal⁺* DNA will cause them to revert to a *gal⁺* phenotype, where they can use this carbohydrate. In a similar way, *bio⁻* cells can be transduced to a *bio⁺* phenotype.

As is evident from this discussion, specialized transduction occurs in quite a different way from generalized transduction. Because it is limited to certain genes, it is not useful in linkage or mapping studies.

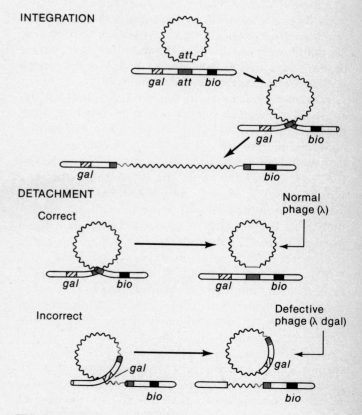

FIGURE 14.24

Production of a defective phage λ, leading to specialized transduction. (Redrawn from Fristrom and Spieth, 1980.)

VIRAL GENETICS

So far, we have considered recombination involving bacterial genes. Recombination has also been observed in bacterial viruses, but only when genetically distinguishable viral strains have simultaneously infected the same host cell. In this section, we will first review briefly the structure and life cycle of a typical T-even bacteriophage. Then we will consider various strategies which have been evolved by viruses to replicate their genetic material. With this information as background, we can then consider mutations and genetic recombination in viruses. Such recombinational studies have led to chromosome mapping and analysis of individual genes.

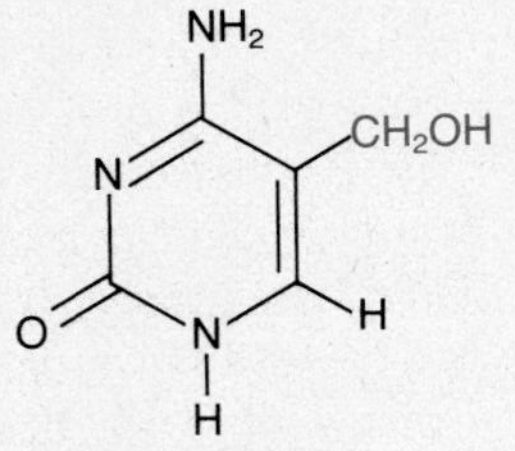

FIGURE 14.25
The chemical structure of 5-hydroxymethylcytosine, which is present in some bacteriophage DNA. This modified nitrogenous base protects the viral DNA from the enzymatic degradation that affects the bacterial host DNA during infection.

The T4 Phage Life Cycle

In Chapter 8, we discussed the Hershey-Chase experiment, which utilized bacteriophage T2 in providing support for the concept that DNA is the genetic material. Phage T4 is very similar and typical of the family of T-even phages. It contains double-stranded DNA in a quantity sufficient to encode more than 150 average-sized genes. The genetic material is enclosed by an icosahedral protein coat making up the head of the virus. This is connected to a tail that contains a collar and a contractile sheath that surrounds a central core. Six tail fibers protrude from the tail; their tips contain binding sites that specifically recognize unique areas of the external surface of *E. coli* cell wall.

Binding of tail fibers to these unique areas of the cell wall is the first step of infection. Then, an ATP-driven contraction of the tail sheath causes the central core to penetrate the cell wall. The DNA in the head is extruded through the core where it then moves across the cell membrane into the bacterial cytoplasm. Within minutes, all bacterial DNA, RNA, and protein synthesis is inhibited, and synthesis of viral macromolecules begins. This initiates the **latent** or **eclipse period,** in which the production of viral components occurs prior to the assembly of virus particles.

RNA synthesis is initiated by the transcription of viral genes by the host cell RNA polymerase. The viral genes are divided into three groups: (1) **immediate early genes,** (2) **delayed early genes,** and (3) **late genes.** The products of the first two groups are made prior to the initiation of T4 DNA synthesis. One of the earliest gene products is a DNase enzyme that degrades the *E. coli* chromosome. The T4 DNA molecule is protected from digestion by this enzyme because its cytosine residues are modified to form **5-hydroxymethylcytosine** (Figure 14.25); as a result, this DNase is unable to use T4 DNA as a substrate.

A plethora of early and late gene products are subsequently synthesized using the host cell ribosomes. For example, about 20 different proteins are part of the protein capsid of the head. Many others are structural components of the tail and its fibers. Additionally, numerous enzymes participate in assembly of mature bacteriophages, but are not themselves structural components. One late gene product, **lysozyme,** digests the bacterial cell wall, leading to its rupture and release of mature phages.

While the viral products are being made, semiconservative DNA replication begins, leading to a pool of viral DNA molecules that are available to be packaged into phage heads. It is at this time that recombination may occur between DNA molecules.

Assembly of mature viruses is a complex process that has been well studied by William Wood, Robert Edgar, and others. Three major independent pathways occur, leading to: (1) DNA packaged into heads; (2) construction of tails; and (3) synthesis and assembly of tail fibers. A mutation that interrupts any one of the three pathways does not inhibit the other two. As shown in Figure 14.26, the head is assembled and DNA is packaged into it. This complex then combines with the tail, and only then are the tail fibers added. Total construction is a combination of self-assembly and enzyme-directed processes. When approximately 200 viruses are assembled, the bacterial cell is ruptured by the action of lysozyme and the mature phages are released. If this occurs on a lawn of bacteria, the 200 phages will infect other bacterial cells and the process will be repeated over and over again. On the lawn, this leads to a clear area where all bacteria have been lysed. This area is called a **plaque** and represents multiple clones of the single infecting T4 bacteriophage (see Figure 14.27).

FIGURE 14.26
The assembly of
bacteriophage T4 following
infection of *E. coli*. Three
separate pathways lead to
the formation of an
icosahedral head filled with
DNA, a tail containing a
central core, and tail fibers.
The tail is added to the head
and then tail fibers are
added. Each step of each
pathway is influenced by
one or more phage genes.

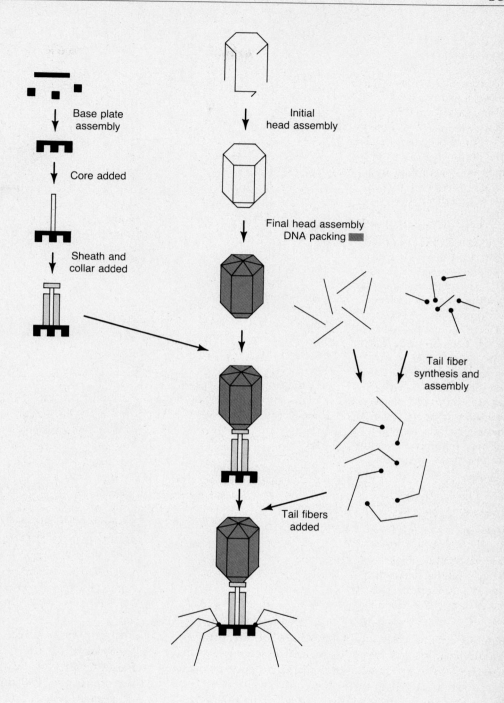

The Viral Plaque Assay

The experimental study of bacteriophages and other
viruses has played a critical role in our understanding
of molecular genetics. During infection of bacteria,
enormous quantities of bacteriophages may be obtained for investigation. Often, over 10^{10} viruses per
milliliter of culture medium are produced. Many genetic studies have relied on the ability to quantitate
the number of phage produced following infection un-

der specific culture conditions. The technique that
works nicely is called the **plaque assay.**

This assay is illustrated in Figure 14.27. A serial dilution of the original virally infected bacterial culture
is first performed. Then, a 0.1-ml aliquot from one or
more dilutions is added to a small volume of melted
nutrient agar (about 3 ml) to which a few drops of a
healthy bacterial culture have been mixed. The solution is then poured evenly over a base of solid nutrient

FIGURE 14.27
Diagrammatic illustraton of the plaque assay for bacteriophage. Serial dilutions of a bacterial culture infected with bacteriophage are first made. Then, three of the dilutions (10^{-3}, 10^{-5}, and 10^{-7}) are analyzed using the plaque assay technique. In each case 0.1 ml of the bacterial-viral culture is mixed with a few drops of a healthy bacterial culture in nutrient agar. This mixture is spread over and allowed to solidify on a base of agar. Each clear area that arises is a plaque and represents an initial infection of one bacterial cell by one bacteriophage. Each generation of viral reproduction involves more bacteria, creating a visible plaque. In the 10^{-3} dilution, so many phage are present that all bacteria are lysed. In the 10^{-5} dilution, 23 plaques are produced. In the 10^{-7} dilution, the dilution factor is so great that no phage are present in the 0.1-ml sample, and thus no plaques form. From the 0.1-ml aliquot of the 10^{-5} dilution, the original bacteriophage density can be calculated as 230×10^5 phage/ml. This calculation is described in the text.

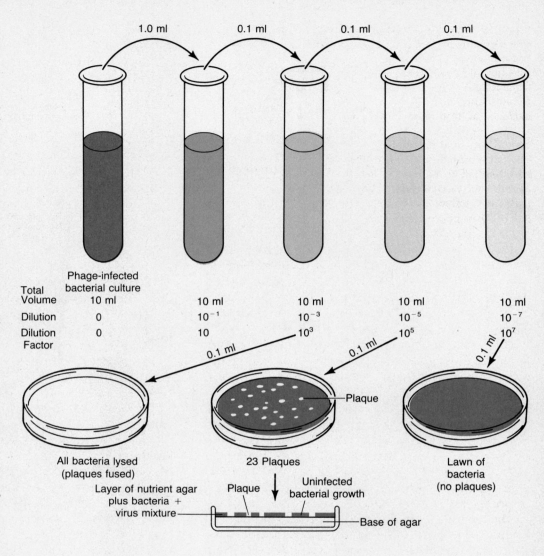

agar in a Petri dish and allowed to solidify prior to incubation. As described in the above section and illustrated in Figure 14.27, viral plaques occur at each place where a single virus has initially infected one bacterium in the lawn that has grown up during incubation. If the dilution factor is too low, the plaques are plentiful and will fuse, lysing the entire lawn. This has occurred in the 10^{-3} dilution in Figure 14.27. On the other hand, if the dilution factor is increased, plaques can be counted and a density of viruses in the initial culture can be estimated. The calculation is the same type as that used for determining bacterial density by counting colonies following serial dilution of an initial culture:

$$(\text{plaque number/ml}) \times (\text{dilution factor})$$

Using the results shown in Figure 14.27, it is observed that there are 23 phage plaques derived from the 0.1-ml aliquot of the 10^{-5} dilution. Therefore, we can estimate that there are 230 phage per ml at this dilution. Thus, the initial viral density in the undiluted sample is calculated as:

$$(230/\text{ml}) \times (10^5) = 230 \times 10^5/\text{ml}$$

Since this figure is derived from the 10^{-5} dilution, we can estimate that there will be only 0.23 phage per 0.1 ml in the 10^{-7} dilution. As a result, when 0.1 ml from this tube is assayed, it is likely that no phage particles will be present. This prediction is borne out in Figure 14.27, where an intact lawn of bacteria exists. The dilution factor is simply too great.

The use of the plaque assay has been invaluable in mutational and recombinational studies of bacteriophages. We will apply the above technique more directly in Chapter 19 when Seymour Benzer's elegant genetic analysis of a single gene in phage T4 is discussed.

Strategies for Viral Reproduction

Before turning to the study of viral mutations and recombination, we will pursue a short discussion of the various forms in which the genetic material is contained within different types of viruses. In order to reproduce, different types of viruses have evolved various strategies leading to the replication of their genetic material.

The T-even phages have been well studied. They contain double-stranded DNA that replicates semiconservatively. The process occurs geometrically, where one chromosome yields two daughter chromosomes, which replicate yielding four daughters, etc. New DNA molecules are produced using the host polymerase.

This mode is in contrast to that occurring during the infection of *E. coli* by the smaller **bacteriophage ϕX174.** This virus, as you will recall from the discussion of replication in Chapter 10, contains only a circular single strand of DNA, called the plus (+) strand. It infects the host and immediately serves as the template for the synthesis of the complementary negative (−) strand. This results in a circular double-stranded molecule called the **replicative form** (RF). The RF is itself replicated semiconservatively, producing about 50 progeny molecules. These then serve as templates for the **rolling circle mode** of production of only (+) strands (see Figure 10.11). As hundreds of these DNA strands are produced, they are packaged into protein capsids synthesized under the direction of the viral genetic information.

While DNA viruses can pirate the use of the host DNA polymerase in order to replicate their nucleic acid, no comparable enzyme exists when RNA-containing viruses invade their host cells. Some RNA viruses require an RNA-dependent RNA polymerase, which is sometimes more simply called **RNA replicase.** For example, when the small **bacteriophage Qβ** infects *E. coli* as its host cell, the infective single-stranded RNA molecule first serves as a messenger RNA (mRNA), encoding three proteins. Two are components of the coat protein, while one is a subunit of RNA replicase. This subunit combines with three other subunits that are bacterial in origin. The resultant enzyme is specific for the replication of the phage RNA (as compared with the bacterial RNAs that are present). First, complementary RNA virus (−) strands are created. These serve as the template for the synthesis of many (+) strands. Ultimately, the (+) strands are encapsulated by coat proteins as mature viruses are

formed. A similar mode of replication occurs in the eukaryotic, RNA-containing **poliovirus.**

The final type of viral reproduction to be discussed was the basis of an extraordinary finding made in the early 1960s. A long series of investigations by Howard Temin led to the discovery of **reverse transcriptase,** an enzyme capable of synthesizing DNA on an RNA template. This enzyme is also called **RNA-directed DNA polymerase.** Temin proposed that the RNA that is the genetic material of certain RNA-containing animal tumor viruses is transcribed reversibly into DNA, which then serves as the template for replication and transcription of RNA during viral infection. These viruses are called **oncogenic** because they induce malignant growths.

Temin's proposed reverse transcriptase was largely ignored because it violated the **central dogma,** which says DNA makes RNA which makes protein. The central dogma did not recognize any reversal of the process. In addition, other RNA-containing viruses (polio virus and double- and single-stranded RNA bacteriophages), whose RNA is replicated by the enzyme RNA replicase, did not demonstrate reverse transcriptase activity. Thus, molecular biologists questioned why the tumor viruses should require this enzyme.

Working with the RNA-containing oncogenic **Rous sarcoma virus (RSV),** which produces cancer in chickens, Temin added the antibiotic actinomycin D to cultures of cells that had been infected. He found that the antibiotic inhibited viral multiplication. Actinomycin D is known to specifically inhibit DNA-directed RNA synthesis and not to affect RNA-directed RNA synthesis. Temin also found that inhibitors of DNA synthesis such as 5-fluorodeoxyuridine also blocked infection. Thus, he reasoned that DNA must therefore serve as an essential intermediate molecule during infection.

Temin subsequently demonstrated directly that DNA is synthesized by RSV during infection. He announced these results in 1970 and found that other workers, using a mouse leukemia virus, had arrived at the same conclusions. These reports stimulated a tremendous amount of work on the enzyme and its role in tumor production. Reverse transcriptase was subsequently isolated independently by Temin and David Baltimore. It has since been found in all RNA oncogenic viruses. Because they "reverse" the flow of genetic information they are called **retroviruses.**

The enzyme uses the infecting (+) RNA strand as a template, synthesizing DNA in the 5'-to-3' direction. An RNA primer is first synthesized. Then, the enzyme directs synthesis of the complementary DNA strand, creating an RNA/DNA hybrid molecule. Reverse transcriptase possesses RNase activity, and as a result, it degrades the RNA portion of the hybrid molecule. The enzyme then uses the remaining DNA strand as a template and synthesizes the complementary strand, creating a double-stranded DNA molecule. This is capable of integrating into the genome of the host cell, forming what is called a **provirus.** When the host genome is transcribed into RNA, so is the proviral DNA. This produces (+) RNA strands, some of which are translated into viral proteins. Packaging of other newly formed RNA molecules into the viral capsid occurs to form mature viruses. Such a host cell is said to be **transformed.** A similar life cycle occurs in all retroviruses.

Mutations in Viruses

Much of what is known concerning viral genetics has been derived from studies of bacteriophages. Phage mutations often affect the morphology of the plaques formed following lysis of the bacterial cells. For example, in 1946 Alfred Hershey observed unusual T2 plaques on plates of *E. coli* strain B. While the normal T2 plaques are small and have a clear center surrounded by a turbid, diffuse halo, the unusual plaques were larger and possessed a sharp outer perimeter (Figure 14.28). When the viruses were isolated from these plaques and replated on *E. coli* B cells, an identical plaque appearance was noted. Thus, the plaque phenotype was an inherited trait resulting from the reproduction of mutant phages. Hershey named the mutant *rapid lysis* (*r*) because the plaques were larger, apparently resulting from a more rapid or more efficient life cycle of the phage. It is now known that wild-type phages undergo an inhibition of reproduction once a particular-sized plaque has been formed. The *r* mutant T2 phages are able to overcome this inhibition, producing larger plaques with a sharp perimeter.

Another bacteriophage mutation, *host range* (*h*), was

discovered by Luria. This mutation extends the range of bacterial hosts that the phage can infect. Although wild-type T2 phages can infect *E. coli* B, they cannot normally attach or be absorbed to the surface of *E. coli* B-2. The *h* mutation, however, provides the basis for adsorption and subsequent infection.

Table 14.3 lists other representative types of mutations that have been isolated and studied in the T-even series of bacteriophages (T2, T4, T6, etc.). These mutations are important to the study of genetic phenomena in bacteriophages. Conditional, temperature-sensitive mutations have been particularly valuable in the study of essential genes where mutations eliminate the ability of the phage to reproduce.

Recombination in Viruses

Around 1947, several research teams demonstrated that recombination occurs between viruses. This discovery was made during experiments in which two mutant strains of bacteriophage were allowed to simultaneously infect the same bacterial culture. These **mixed infection experiments** were designed so that the number of viral particles sufficiently exceeded the number of bacterial cells so as to ensure simultaneous infection of most cells by both viral strains.

For example, in one study using the T2/*E. coli* system, the viruses were of either the h^+r or $h\ r^+$ genotype. If no recombination occurred, these two parental genotypes (wild-type host range restriction, rapid lysis; and extended host range, normal lysis) would be the only expected phage progeny. However, the recombinant h^+r^+ and $h\ r$ were detected in addition to the parental genotypes (Figure 14.29).

Similar recombinational studies have been performed with large numbers of mutant genes in a variety of bacteriophages. Data are analyzed in much the same way as they are in eukaryotic mapping experiments. Two- and three-point mapping crosses are possible, and the percentage of recombinants in the total number of phage progeny is calculated. This value is proportional to the relative distance between two genes along the chromosome. Analysis has shown that most phage chromosomes are circular. This is the case, for example, in the phages ϕX174 and λ and in the T-even series.

An interesting observation in phage crosses is **negative interference.** Recall that in eukaryotic mapping crosses, positive interference is the rule. Fewer than expected double-crossover events are observed. In phage crosses, often just the reverse occurs. In three-point analysis, a greater than expected frequency of double exchanges is observed. Negative interference is explained on the basis of the dynamics of the condi-

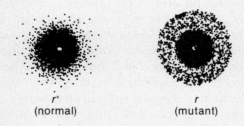

FIGURE 14.28
Drawing of normal (*r⁺*) and mutant rapid lysis (*r*) T2 plaques.

TABLE 14.3

Some mutant types of T-even phages.

Name	Description
Minute	Small plaques
Turbid	Turbid plaques on *E. coli* B
Star	Irregular plaques
UV-sensitive	Alters UV sensitivity
Acriflavin-resistant	Forms plaques on acriflavin agar
Osmotic shock	Withstands rapid dilution into distilled water
Lysozyme	Does not produce lysozyme
Amber	Grows in *E. coli* K12, but not B
Temperature-sensitive	Grows at 25°C, but not at 42°C

tions leading to recombination within the bacterial cell. The available evidence supports the concept that recombination between phage chromosomes involves a breakage and reunion process similar to that of eukaryotic crossing over. The process is facilitated by nucleases that nick and reseal DNA strands. A fairly clear picture of the dynamics of viral recombination has emerged.

Following the early phase of infection, the chromosomes of each phage begin replication. As this stage progresses, a pool of chromosomes accumulates in the bacterial cytoplasm. If double infection by phages of two genotypes has occurred, then the pool of chromosomes initially consists of the two parental types. Genetic exchange between these two types will occur, producing recombinant chromosomes.

In the case of the h^+r and $h\ r^+$ example just dis-

cussed, h^+r^+ and $h\ r$ chromosomes are produced. Each of these recombinant chromosomes may undergo replication and is also free to undergo new exchange events with the other and with the parental types. Furthermore, recombination is not restricted to exchange between two chromosomes—three or more may be involved. As phage development progresses, chromosomes are randomly removed from the pool and packed into the phage head, forming mature phage particles. Thus, parental and recombinant genotypes are produced (Figure 14.30).

Recombination events in viruses are not restricted to intergenic exchanges. With powerful selection systems, it is possible to detect intragenic recombination, where exchanges occur within a single gene. Such studies have led to the fine-structure analysis of the gene (see Chapter 19).

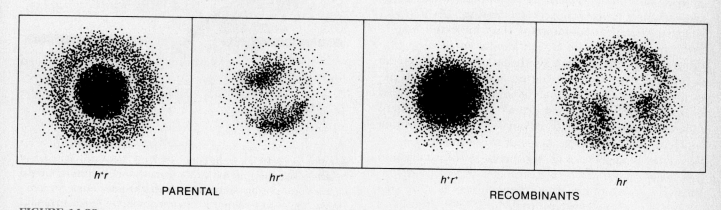

h^+r hr^+

PARENTAL

h^+r^+ hr

RECOMBINANTS

FIGURE 14.29

Plaque phenotypes observed from simultaneous infection of *E. coli* by two T2 strains, h^+r and $h\ r^+$. In addition to these parental phenotypes, recombinant h^+r^+ and $h\ r$ plaques are recovered.

FIGURE 14.30

Results of a cross involving the h and r genes in phage T2. (Data derived from Hershey and Rotman, 1949.)

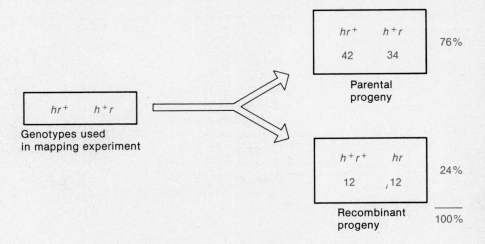

Breakage-Rejoining vs. Copy Choice

While we might assume that recombination in viruses occurs as a result of breakage and rejoining, as is thought to occur during crossing over in eukaryotes, some evidence argues against this mechanism. Foremost was the original observation by Alfred Hershey that, while equal proportions of reciprocal recombinant products occur in experiments where large sample sizes are obtained (Figure 14.30), lysates from single bacteria often fail to demonstrate the expected equal proportions. Breakage and reunion also provides the expectation of positive interference, particularly where genes are close to one another (see Chapter 6). As we have just discussed above, negative interference is usually the rule during viral recombination.

As an alternative to breakage-rejoining, the model of **copy-choice** was proposed by Hershey. This model suggests that the recombinant DNA molecule results from a hybrid copy of both parental molecules. The event occurs during DNA synthesis where the enzyme is presumed to jump during replication from one parental template to the other, as shown in Figure 14.31.

Because evidence exists that some recombinant molecules actually consist of portions of both parental molecules (and are not copies of these molecules), favoring breakage-rejoining, and because recombination has been shown to occur in the absence of DNA synthesis, it is doubtful that copy-choice is the sole source of viral recombinant molecules. While modified proposals involving both modes have been put forth, the question remains largely unanswered.

Packaging and Concatemer Formation

Once a sufficient pool of phage chromosomes is produced as a result of replication, they are individually packaged into the head of the virus particle. Most often, this packaging involves a complex of phage chromosomes called a **concatemer** (Figure 14.32). This structure represents a linear series of chromosomes covalently bonded to one another. A concatemer re-

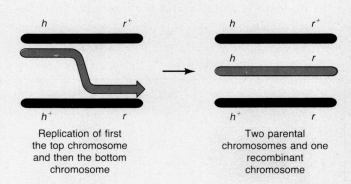

FIGURE 14.31

The proposed mechanism of the copy-choice model of genetic recombination in bacteriophage. DNA polymerase begins replication of one viral chromosome and then jumps to a second chromosome. If the two chromosomes contain unlike alleles at different loci, a recombinant chromosome may result. The arrow indicates the direction of replication by the DNA polymerase.

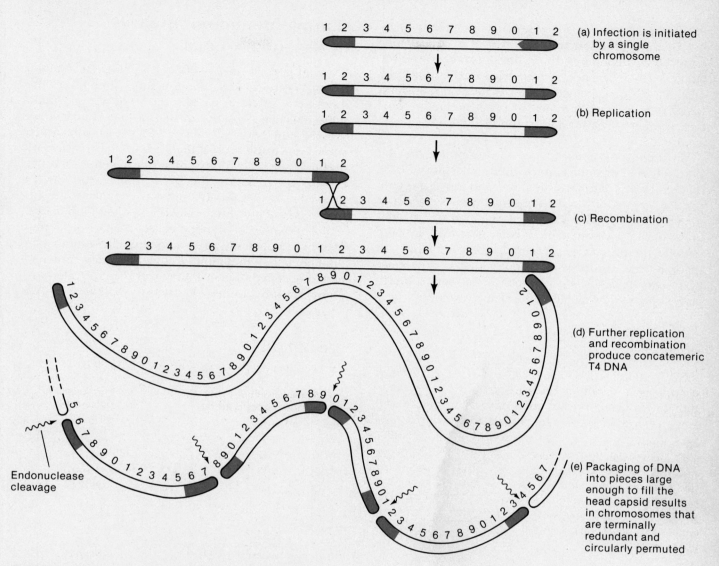

FIGURE 14.32
Concatemer formation and the production of circularly permuted chromosomes prior to packaging during bacteriophage maturation. (Reprinted by permission from Ayala, F. J., and Kiger, J. A., *Modern Genetics*, Menlo Park, California, The Benjamin/Cummings Publishing Company, 1980, pp. 216–17, fig. 6.21.)

sults from a series of recombinational events between the terminally redundant portions of individual chromosomes. This process may be illustrated by considering the events that occur in the T-even series of phages.

Following several rounds of replication of the phage chromosome in the bacterial cytoplasm, numerous individual chromosomes are produced. Each one contains redundant terminal regions that are homologous

to one another. Recombination between these ends results in concatemer formation. In this group of phages, the concatemer is further replicated, and it is in this form that intergenic and intragenic recombination occurs.

When replication is completed, packaging is initiated. It is believed that the phage head is just large enough to hold a length of DNA equivalent to one chromosome, including its terminally redundant ends.

Packaging is accomplished by what is referred to as the **headful mechanism.** The end of a concatemer enters and is condensed into the phage head until it is filled. Then, an endonuclease cleaves the molecule, producing a concatemer shorter by one chromosome length. The process is repeated until the entire complex has been cleaved and the individual units packaged.

An interesting consequence of this process is the production of **circularly permuted** molecules of DNA (Figure 14.32). Assuming that the endonuclease cuts precise lengths of the chromosome, each subsequent molecule will contain a different region at its terminal ends. Until the mechanism of packaging was clarified, circular permutation presented great difficulty in mapping the chromosome. However, it is now clear that each chromosome contains a full set of phage genes plus the terminally redundant sequences.

Complementation Analysis of Essential Genes

If a viral gene is responsible for an essential function during reproduction, a mutation eliminating that function is lethal. In order to be studied, such mutations must be conditional; that is, they must be expressed only under certain conditions. **Temperature-sensitive conditional lethals** have been particularly useful in the study of essential phage genes. At a permissive temperature, the gene under study is expressed normally, but at a restrictive temperature, mutant expression occurs and wild-type function is lost.

It is possible to simultaneously infect a bacterium with two viral strains, each containing a separate conditional lethal mutation. If the mutations are contained in different genes, they will demonstrate **complementation** under restrictive conditions. The

FIGURE 14.33
Comparison of complementation and noncomplementation during the simultaneous infection of a bacterial cell by two phage strains. In (a) neither strain contains a mutation. In (b) and (c), each strain contains a mutation interfering with phage reproduction.

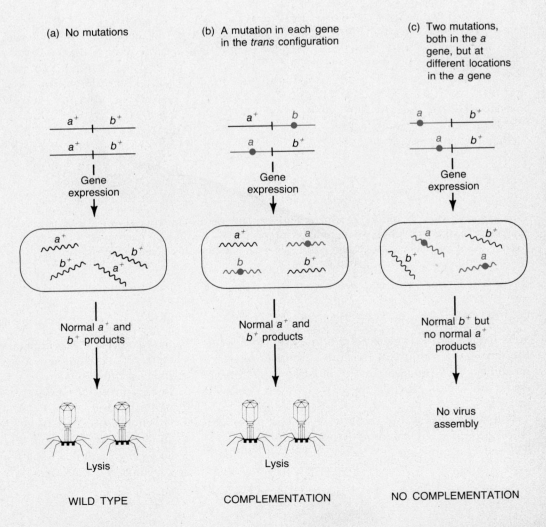

chromosome of each strain contains the normal allele of the gene that is mutant in the other strain. As a result, a pool of all normal gene products will be produced in the bacterial cytoplasm during simultaneous infection, normal phage assembly will occur, and the bacterium will be lysed. The concept of complementation is illustrated in part (b) of Figure 14.33.

On the other hand, if two mutations present in the same gene are studied simultaneously, complementation will not occur under restrictive conditions. Each type of chromosome will produce the same defective gene product. As a result, reproduction cannot be completed. This case is illustrated in part (c) of Figure 14.33.

If a large number of conditional lethal mutations is isolated, it is possible to saturate the chromosome so that all regions contain at least one mutation. When mutations are studied in pairs, complementation shows that the two mutations are in separate genes. Noncomplementation, on the other hand, shows that they are in the same gene.

About 40 conditional lethal mutations in the phage φX174 have been investigated in this way. They were shown to fall into eight complementation groups. This finding suggests that there are eight essential genes in the small genome of this phage. In the larger genome of phage λ, 25 complementation groups or genes are recognized. These include genes required for head and tail production, DNA replication, cell lysis, and regulation of gene expression. Complementation analysis thus allows a determination of the number of essential genes in certain bacteriophages.

SUMMARY AND CONCLUSIONS

The study of the genetics of bacteria and viruses is a relatively recent one, having commenced in the 1940s. It is not a coincidence that the great strides and accomplishments made in molecular biology paralleled the expansion of knowledge in bacterial and viral genetics. The ease of obtaining large quantities of pure cultures of bacteria and mutant strains of viruses as experimental material has been important to their successful study. This advantage, in addition to the efficient selection and isolation of naturally occurring and induced mutations, has made these the organisms of choice in numerous types of genetic study.

In this chapter, we have focused on genetic exchange or recombination of genetic information between bacteria and between viruses. In bacteria, three different modes of recombination are known: conjugation, transformation, and transduction.

Conjugation, initiated by a bacterium harboring an F factor, requires the physical contact established by the conjugation tube. Genetic information flows unidirectionally, and upon its entry into the recipient cell, recombination with the recipient chromosome must occur. Efficient transfer of the donor chromosome occurs when the F factor has integrated into it. The point and polarity of integration determines the orientation of transfer, which serves as the basis for time mapping of the donor chromosome.

The F factor is just one example of a bacterial plasmid. We have also discussed the R plasmids that confer resistance to antibiotics and the Col plasmids that carry genetic information that leads to the production of bacterial toxins. Both plasmids and the bacterial chromosome have been shown to contain unique DNA sequences facilitating movement of genetic information such as the insertion and excision of the F factor into and out of the chromosome. These are called insertion sequences; they may combine with genes, producing transposons, which also move from one genetic molecule or position to another. The discovery of these elements has provided great insights into the organization and evolution of the genetic information in bacteria.

Other types of bacterial recombination do not require cell contact. Transformation, for example, involves exogenous DNA, which enters a bacterial cell in single-stranded form, recombining with the host chromosome. Linkage mapping of closely aligned genes may be performed using transformation.

Transduction is a third mode of bacterial recombination and is mediated by bacterial viruses. An understanding of transduction required the knowledge that viruses may establish a symbiotic relationship with bacteria. The viral genetic material may become integrated into the bacterial chromosome as a prophage and lysogenize the host cell. In the process of generalized transduction, prophages lose their integrated status, and the resultant viruses mediate the transfer of a random part of the bacterial chromosome between organisms. In specialized transduction, as illustrated in phage λ, only specific genes adjacent to the point of insertion of the prophage into the bacterial chromosome are transferred. Bacterial linkage and mapping studies may also be performed using transduction.

Before discussing recombination in viruses, we have focused on the general life cycle of a bacteriophage and how phage are studied using the plaque assay. Different strategies of viral reproduction were also examined. The discovery of visible mutations in plaques and the host range of infectivity have allowed the analysis of recombination between viruses. Discovery of such genetic exchange occurred during simultaneous or mixed infection of bacterial cells by two distinct mutant viral strains. While the viral nucleic acid molecules are present in the bacterial cytoplasm, recombination may occur. As with bacteria, observations of this recombination may be used to map the chromosome. Many viral chromosomes, like bacterial structures, have been shown to be circular.

Replication, recombination, and packaging of phage chromosomes most often involves the formation of concatemers. These linear complexes are usually formed as the result of recombination between terminally redundant regions of the phage chromosomes that comprise them. The headful mechanism enzymatically cleaves the concatemers during packaging into the phage heads.

We have concluded the discussion of viral genetics with a consideration of complementation analysis of essential conditional lethal mutations. This type of study can elucidate the number of distinct genes essential to the reproduction of mature, viable phages.

PROBLEMS AND DISCUSSION QUESTIONS

1 Contrast the two major hypotheses explaining the origin of heritable variation in bacteria.

2 Discuss the experiments performed by Luria and Delbruck that clarified the origin of heritable variation in bacteria. What was the rationale for their conclusions? Why were these experiments called the "fluctuation tests"?

3 Describe the replica plating technique. Why is this technique so valuable experimentally?

4 How did the use of replica plating strengthen the conclusions drawn in the Luria-Delbruck experiment?

5 Describe and indicate the significance of enrichment techniques for isolating bacterial mutants.

6 Differentiate between phenotypic lag and segregation lag. Do these phenomena lead to overestimates or underestimates of mutation rates?

7 Distinguish between the three modes of recombination in bacteria.

8 With respect to F^+ and F^- bacterial matings, answer the following questions:
   (a) How was it established that physical contact was necessary?
   (b) How was it established that chromosome transfer was unidirectional?
   (c) What is the physical-chemical basis of a bacterium being F^+?

9 List all major differences between (a) the $F^+ \times F^-$ and the Hfr $\times F^-$ bacterial crosses and (b) Hfr and F^+ bacteria.

10 Describe the basis for chromosome mapping in the Hfr $\times F^-$ crosses.

11 When the interrupted mating technique was used with five different strains of Hfr bacteria, the following order of gene entry and recombination was observed. On the basis of these data, draw a map of the bacterial chromosome. Do the data support the concept of circularity?

Hfr Strain	Order
1	T C H R O
2	H R O M B
3	M O R H C
4	M B A K T
5	C T K A B

12 Why are the recombinants produced from an Hfr $\times F^-$ cross almost never F^+?

13 Define *F′ bacteria* and *merozygotes*.

14 Describe what is known about the mechanisms of the transformation process.

15 In a transformation experiment involving a recipient bacterial strain of genotype *a b*, the following results were obtained:

Transforming DNA	% Transformants		
	$a^+ b$	$a b^+$	$a^+ b^+$
$a^+ b^+$	3.1	1.2	0.04
$a^+ b$ and $a b^+$	2.4	1.4	0.03

What can you conclude about the location of the *a* and *b* genes relative to each other?

16 In a transformation experiment, donor DNA was obtained from a prototroph bacterial strain ($a^+ b^+ c^+$), and the recipient was a triple auxotroph (*a b c*). The following transformant classes were recovered:

$$
\begin{array}{ll}
a^+ \; b \;\; c & 180 \\
a \;\; b^+ \; c & 150 \\
a^+ \; b^+ \; c & 210 \\
a \;\; b \;\; c^+ & 179 \\
a^+ \; b \;\; c^+ & 2 \\
a \;\; b^+ \; c^+ & 1 \\
a^+ \; b^+ \; c^+ & 3
\end{array}
$$

What general conclusions can you draw about the linkage relationships among the three genes?

17 Explain the observations that led Zinder and Lederberg to conclude that the prototrophs recovered in their transduction experiments were not the result of F-mediated conjugation.

18 Define *plaque, lysogeny,* and *prophage.*

19 Differentiate between generalized and restricted transduction. Which can be used in mapping and why?

20 Two theoretical genetic strains of a virus (*a b c* and $a^+ b^+ c^+$) are used to simultaneously infect a culture of host bacteria. Of 10,000 plaques scored, the following genotypes were observed:

$$
\begin{array}{llll}
a^+ \; b^+ \; c^+ & 4100 & \quad a \;\; b^+ \; c & 160 \\
a \;\; b \;\; c & 3990 & \quad a^+ \; b \;\; c^+ & 140 \\
a^+ \; b \;\; c & 740 & \quad a \;\; b \;\; c^+ & 90 \\
a \;\; b^+ \; c^+ & 670 & \quad a^+ \; b^+ \; c & 110
\end{array}
$$

Determine the genetic map of these three genes on the viral chromosome. Determine whether interference was positive or negative.

21 Describe the conditions under which viral recombination may occur.

22 Describe concatemer formation and its role in the headful packaging mechanism.

23 In complementation studies in phage T4, six mutations were tested in three separate experiments. In the first two experiments, the results were misplaced for one of the crosses. For these experiments, predict the results of the missing cross. From an analysis of all three experiments, determine which mutations are in the same gene. (Note: + = complementation; − = no complementation.)

Expt. 1	Expt. 2	Expt. 3
$d \times e \rightarrow +$	$g \times h \rightarrow -$	$d \times g \rightarrow +$
$d \times f \rightarrow -$	$g \times i \rightarrow -$	$e \times i \rightarrow -$
$e \times f \rightarrow ?$	$h \times i \rightarrow ?$	

24 A culture of an auxotrophic *leu⁻* strain of bacteria was irradiated and incubated until it reached the stationary phase. A control culture, which was not irradiated, was also studied. These cultures were then serially diluted, and 0.1 ml of various dilutions was plated on minimal medium plus leucine and on minimal medium. The results, shown below, were used to determine the spontaneous and X-ray induced mutation rate of *leu*⁻ to *leu*⁺.

Culture Condition	Culture Medium	Dilution	Number of Colonies
Irradiated	(1) Minimal medium plus leucine	10^{-9}	24
	(2) Minimal medium	10^{-2}	12
Control	(3) Minimal medium plus leucine	10^{-9}	12
	(4) Minimal medium	10^{-1}	3

(a) Describe what is represented by each value obtained. Which values should be approximately equal? Are they?

(b) Determine the induced and spontaneous mutation rate leading to prototrophic growth (*leu*⁻ to *leu*⁺).

25 Distinguish between insertion sequences and transposons in bacteria.

SELECTED READINGS

ADELBERG, E. A. 1960. *Papers on bacterial genetics.* Boston: Little, Brown.

ADELBERG, E. A., and PITTARD, J. 1965. Chromosome transfer in bacterial conjugation. *Bacteriol. Rev.* 29: 161–72.

BIRGE, E. A. 1981. *Bacterial and bacteriophage genetics: An introduction.* New York: Springer-Verlag.

BRODA, P. 1979. *Plasmids.* San Francisco: W. H. Freeman.

BUKHARI, A. I., SHAPIRO, J. A., and ADHYA, S. L. eds. 1977. *DNA insertion elements, plasmids, and episomes.* Cold Spring Harbor, N.Y.: Cold Spring Harbor Laboratory.

CAIRNS, J., STENT, G. S., and WATSON, J. D. *Phage and the origins of molecular biology.* Cold Spring Harbor, N.Y. Cold Spring Harbor Laboratory.

CAMPBELL, A.M. 1976. How viruses insert their DNA into the DNA of the host cell. *Scient. Amer.* (Dec.) 235:102–13.

CAVALLI-SFORZA, L. L., and LEDERBERG, J. 1956. Isolation of pre-adaptive mutants in bacteria by sib selection. *Genetics* 41: 367–81.

CLARK, A. J., and WARREN, G. J. 1979. Conjugal transfer of plasmids. *Ann. Rev. Genet.* 13: 99–125.

COHEN, S. 1976. Transposable genetic elements and plasmid evolution. *Nature* 263: 731–38.

COHEN, S., et al. 1973. Construction of biologically functional bacterial plasmids *in vitro. Proc. Natl. Acad. Sci.* 70: 3240–44.

COHEN, S. N., and SHAPIRO, J. A. 1980. Transposable genetic elements. *Scient. Amer.* (Feb.) 242: 40–49.

COLD SPRING HARBOR LABORATORY. 1981. Movable genetic elements. *Cold Spr. Harb. Symp.,* vol. 45, parts 1 and 2.

EARNSHAW, W. C., and CASJENS, S. R. 1980. DNA packaging by the double-stranded DNA bacteriophages. *Cell* 21: 319–31.

EDGAR, R. 1982. Max Delbruck. *Ann. Rev. Genet.* 16: 501–6.

EISENSTARK, A. 1977. Genetic recombination in bacteria. *Ann. Rev. Genet.* 11: 369–96.

FOX, M. S. 1966. On the mechanism of integration of transforming deoxyribonucleate. *J. Gen. Physiol.* 49: 183–96.

FRISTROM, J. W., and SPIETH, P. T. 1980. *Principles of genetics.* New York: Chiron Press.

GLOVER, S. W., and HOPWOOD, D. A., eds. 1981. *Genetics as a tool in microbiology.* New York: Cambridge University Press.

HARDY, K. 1981. *Bacterial plasmids.* Washington, D.C.: American Society of Microbiology.

HAYES, W. 1953. The mechanisms of genetic recombination in *Escherichia coli. Cold Spr. Harb. Symp.* 18: 75–93.

————. 1968. *The genetics of bacteria and their viruses.* 2nd ed. New York: Wiley.

HERSHEY, A. D. 1946. Spontaneous mutations in bacterial virus. *Cold Spr. Harb. Symp.* 11: 67–76.

HERSHEY, A. D., and CHASE, M. 1951. Genetic recombination and heterozygosis in bacteriophage. *Cold Spr. Harb. Symp.* 16: 471–79. (Reprinted in H. J. Taylor, 1965, *Selected papers on molecular genetics.*)

HERSHEY, A. D., and ROTMAN, R. 1949. Genetic recombination between host range and plaque-type mutants of bacteriophage in single cells. *Genetics* 34: 44–71.

HOTCHKISS, R. D., and MARMUR, J. 1954. Double marker transformations as evidence of linked factors in deoxyribonucleate transforming agents. *Proc. Natl. Acad. Sci.* 40: 55–60.

JACOB, F., and WOLLMAN, E. L. 1961a. *Sexuality and the genetics of bacteria.* New York: Academic Press.

————. 1961b. Viruses and genes. *Scient. Amer.* (June) 204: 92–106.

KAZAZIAN, H. H. 1985. The nature of mutation. *Hosp. Pract.* 20: 55–69.
K12. *Genetics* 41: 141–56.

LANDY, A., and ROSS, W. 1977. Viral integration and excision: Structure of the lambda *att* sites. *Science* 197: 1147–60.

LEDERBERG, J., and LEDERBERG, E. M. 1952. Replica plating and indirect selection of bacterial mutants. *J. Bacteriol.* 63: 399–406.

LEWIN, B. 1977a. *Gene expression,* vol. 1. *Bacterial genomes.* New York: Wiley.

————. 1977b. *Gene expression,* vol. 3. *Plasmids and phages.* New York: Wiley.

LOW, B., and PORTER, R. 1978. Modes of genetic transfer and recombination in bacteria. *Ann. Rev. Genet.* 12: 249–87.

LURIA, S. E., and DELBRUCK, M. 1943. Mutations of bacteria from virus sensitivity to virus resistance. *Genetics* 28: 491–511.

LWOFF, A. 1953. Lysogeny. *Bacteriol. Rev.* 17: 269–337.

MESELSON, M., and WEIGLE, J. J. 1961. Chromosome breakage accompanying genetic recombination in bacteriophage. *Proc. Natl. Acad. Sci.* 47: 857–68.

MORSE, M. L., LEDERBERG, E. M., and LEDERBERG, J. 1956. Transduction in *Escherichia coli* K12. *Genetics* 41: 141–56.

NOVICK, R. P. 1980. Plasmids. *Scient. Amer.* (Dec.) 243: 102–27.
Genetics 38: 5–33.

OZEKI, H., and IKEDA, H. 1968. Transduction mechanisms. *Ann. Rev. Genet.* 2: 245–78.

PETERS, J. A. 1969. *Classic papers in genetics.* Englewood Cliffs, N.J.: Prentice-Hall.

SMITH, H. O., DANNER, D. B., and DEICH, R. A. 1981. Genetic transformation. *Ann. Rev. Biochem.* 50: 41–68.

STAHL, F. W. 1969. *The mechanics of inheritance.* 2nd ed. Englewood Cliffs, N.J.: Prentice-Hall.

————. 1979. *Genetic recombination: Thinking about it in phage and fungi.* San Francisco: W. H. Freeman.

STARLINGER, P. 1980. IS elements and transposons. *Plasmid* 3: 241–59.

STENT, G. S. 1963. *Molecular biology of bacterial viruses.* San Francisco: W. H. Freeman.

————. 1966. *Papers on bacterial viruses.* 2nd ed. Boston: Little, Brown.

STENT, G. S., and CALENDAR, R. 1978. *Molecular genetics: An introductory narrative.* 2nd ed. San Francisco: W. H. Freeman.

TAYLOR, J. H., ed. 1965. *Selected papers on molecular genetics.* New York: Academic Press.

TEMIN, H. 1972. RNA-directed DNA synthesis. *Scient. Amer.* (Jan.) 24–33.

TEMIN, H. M., and MIZUTANI, S. 1970. RNA-dependent DNA polymerase in virions of Rous sarcoma virus. *Nature* 226: 1211–13.

VISCONTI, N., and DELBRUCK, M. 1953. The mechanism of genetic recombination in phage. *Genetics* 38: 5–33.

WOLLMAN, E. L., JACOB, F., and HAYES, W. 1956. Conjugation and genetic recombination in *Escherichia coli* K-12. *Cold Spr. Harb. Symp.* 21: 141–62.

ZINDER, N. D. 1953. Infective heredity in bacteria. *Cold. Spr. Harb. Symp.* 18: 261–69.

_____. 1958. Transduction in bacteria. *Scient. Amer.* (Nov.) 199: 38.

ZINDER, N. D., and LEDERBERG, J. 1952. Genetic exchange in *Salmonella*. *J. Bacteriol.* 64: 679–99.

15

Extrachromosomal Inheritance

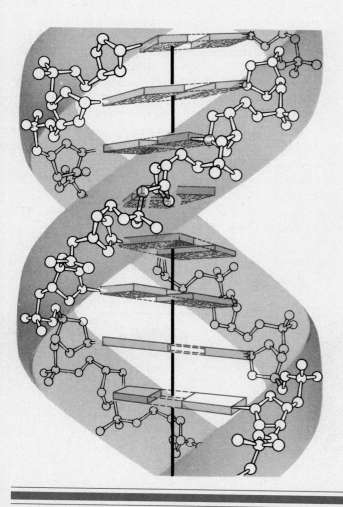

Throughout the history of genetics, occasional reports have challenged the basic tenets of transmission genetics—the production of the phenotype through the transmission of genes located on chromosomes carried in the nucleus. These reports have indicated an apparent extranuclear or extrachromosomal influence on the phenotype. Such reports have often been regarded with skepticism. However, with the increasing knowledge of molecular genetics and the discovery of DNA in mitochondria and chloroplasts, extrachromosomal inheritance is now recognized as an important aspect of genetics.

There are many diverse examples of these unusual modes of inheritance. The expression of such information can influence the phenotype in many ways. In this chapter we will focus on three general types of extrachromosomal genetic phenomena: (1) maternal influence resulting from the effect of stored products of nuclear genes of the female parent during early development; (2) organelle heredity resulting from the expression of DNA contained in mitochondria and chloroplasts; and (3) infectious heredity resulting from the symbiotic association of microorganisms with eukaryotic cells. Each has the effect of producing inheritance patterns that vary from those predicted by the concepts of Mendelian and neo-Mendelian genetics.

MATERNAL INFLUENCE

Maternal influence implies that an offspring's phenotype for a particular trait is strongly influenced by the genotype of the maternal parent. This is in contrast to most cases, where inheritance of traits is biparental. Crosses involving traits inherited maternally produce results that do not adhere to Mendelian or neo-Mendelian patterns. In such cases, the genetic information of the female gamete is transcribed and these genetic products are present in the egg cytoplasm. Following fertilization, these products influence patterns or traits established during early development. Three examples will illustrate the influences of the maternal genome on particular traits. The developmental consequences of such maternal effects will be discussed in Chapter 22.

Ephestia Pigmentation

In the flour moth, *Ephestia kuehniella*, the wild-type larva has a pigmented skin and brown eyes as a result of the dominent gene *A*. The pigment is derived from a precursor molecule, kynurenine, which is in turn a derivative of the amino acid tryptophan. A mutation, *a*, results in red eyes and little pigmentation in larvae when homozygous. As illustrated in Figure 15.1, different results are obtained in the cross *Aa* × *aa*, depending on which parent carries the dominant gene. When the male is the heterozygous parent, a 1:1 brown/red-eyed ratio is observed in larvae as predicted by Mendelian segregation. However, when the female is heterozygous for the *A* gene, all larvae are pigmented and have brown eyes. As these larvae develop into adults, one-half of them gradually develop red eyes.

One explanation of these results is that the *Aa* oocytes synthesize kynurenine or an enzyme necessary for its synthesis and accumulate it in the ooplasm prior to the completion of meiosis. Even in *aa* progeny (whose mothers were *Aa*), this pigment is distributed in the cytoplasm of the cells of the developing larvae—thus, they develop pigmentation and brown eyes. Eventually, the pigment is diluted among many cells and used up, resulting in the conversion to red eyes as adults. The *Ephestia* example demonstrates the maternal effect in which a cytoplasmically stored nuclear gene product influences the larval phenotype and at least temporarily overrides the genotype of the progeny.

Limnaea Coiling

Shell coiling in the snail *Limnaea peregra* represents a permanent rather than a transitory maternal effect. Some strains of this snail have left-handed or sinistrally coiled shells (*dd*), while others have right-handed or dextrally coiled shells (*DD* or *Dd*). These snails are hermaphroditic and may undergo either cross- or self-fertilization, thus providing a variety of types of matings.

As shown in Figure 15.2, the coiling pattern of the progeny snails is determined by **the genotype of the parent producing the female gamete, regardless of the parental phenotype.** Investigation of the developmental events in these snails reveals that the orientation of the spindle in the first cleavage division after fertilization apparently determines the direction of coiling. Spindle orientation is thought to be controlled by maternal genes acting on the developing eggs in the ovary. The orientation of the spindle, in turn, influences cell divisions following fertilization and quickly establishes the permanent adult coiling pattern. Therefore, the progeny phenotypes are determined by the maternal genotype.

The dextral allele (*D*) produces an active gene product that causes right-handed coiling. If ooplasm from dextral eggs is injected into uncleaved sinistral eggs, they cleave in a dextral pattern. However, in the converse experiment, sinistral ooplasm has no effect when injected into dextral eggs. Apparently, the sinistral allele is the result of a classical recessive mutation inactivating the gene product.

FIGURE 15.1

Illustration of maternal influence in the inheritance of eye pigment in the flour moth *Ephestia kuehniella.*

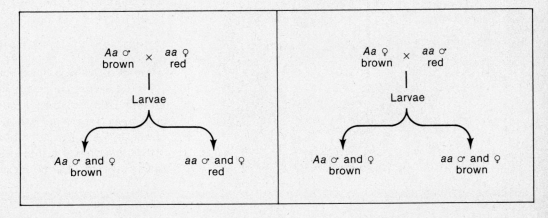

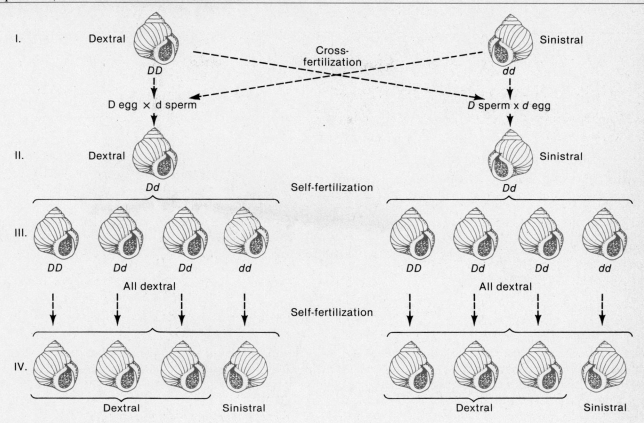

FIGURE 15.2

Inheritance of coiling in the snail *Limnaea peregra*. Coiling is either dextral (right-handed) or sinistral (left-handed). A maternal effect is evident in generations II and III, where the genotype of the maternal parent controls the phenotype of the offspring rather than the offspring's own genotype. (Reprinted with permission of Macmillan Publishing Co., Inc., from *Genetics*, 2nd Edition, by M. W. Strickberger. Copyright © 1976 by Monroe W. Strickberger.)

Drosophila Embryogenesis

A third example illustrates that the absence of certain maternal gene products may have a lethal effect following fertilization. In *Drosophila*, the sex-linked recessive gene *fused* (*fu*) causes partial sterility as well as the fusion of two longitudinal wing veins. As illustrated in Figure 15.3, at least one wild-type allele of the *fused* locus must be present in either the maternal parent or in the genome of the progeny in order for embryogenesis to proceed normally. As a result of this effect, progeny with identical genotypes (*fu/fu* or *fu/↑*) may either die or develop to maturity, depending on the genotype of their mother. Thus, during oogenesis in a +/+ or *fu*/+ female, an essential gene product appears to be made and stored in the egg, assuring complete development. This may occur even though the genotype of the progeny lacks the wild-type allele of *fused*.

ORGANELLE HEREDITY

In this section we will examine examples of inheritance patterns of phenotypes related to chloroplast and mitochondrial function. Prior to the discovery of DNA in these organelles and extensive characterization of genetic processes occurring within them, these patterns were grouped under the category of **cytoplasmic inheritance.** That is, certain mutant phenotypes seemed to be inherited through the cytoplasm rather than through the genetic information of chromosomes. Often transmission was from the maternal parent through the ooplasm, and thus the results of reciprocal crosses (as in Figure 15.1) varied. While such results are similar to the examples of maternal inheritance presented in the preceding section, a major difference exists. Effects of maternal influence (due to the presence of nuclear gene products) are transitory in the sense that they are not heritable. With cy-

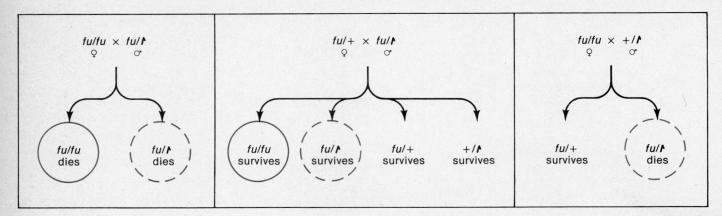

FIGURE 15.3
Results of various crosses in *Drosophila* involving the *fused* allele. The solid and dashed colored circles contrast identical genotypes with different outcomes of survival.

toplasmic inheritance, however, the phenotype is stable and is continually passed to future generations.

Analysis of the hereditary transmission of mutant alleles of chloroplast and mitochondrial DNA has been difficult owing to a number of factors. First, the function of these organelles is dependent upon gene products of both nuclear and organelle DNA. Second, the transmission of mitochondrial or chloroplast DNA to progeny is necessarily through the transmission of the organelle itself. Sometimes, it is difficult to determine the origin (maternal vs. paternal) of the organelle contributed to the zygote. This pattern often varies from organism to organism. Third, the usual number of organelles contributed to each progeny exceeds one. If many chloroplasts and/or mitochondria are contributed and only one or a few of them contain a mutant gene, the corresponding mutant phenotype may not be revealed. Taken together, these complexities have made analysis much more involved than for Mendelian characters.

In this section, for both chloroplasts and mitochondria, we shall discuss examples of inheritance patterns

seemingly related to the respective organelle. Then we shall explore the molecular information known concerning each organelle.

Chloroplasts: Variegation in Four O'Clock Plants

In 1908, Carl Correns (one of the rediscoverers of Mendel's work) provided the earliest example of inheritance linked to chloroplast transmission. Correns discovered a variety of the four o'clock plant, *Mirabilis jalapa*, which had branches with either white, green, or variegated leaves. As shown in Table 15.1, inheritance in all possible combinations of crosses is strictly determined by the phenotype of the ovule source. For example, if the seeds (representing the progeny) were derived from ovules on branches with green leaves, all progeny plants bore only green leaves regardless of the phenotype of the source of pollen.

Correns concluded that inheritance was through the cytoplasm of the maternal parent because the pollen, which contributes little or no cytoplasm to the zy-

TABLE 15.1

Crosses between flowers from various branches of variegated four o'clock plants. The progeny phenotypes are shown in the shaded boxes.

Source of Pollen	Source of Ovule		
	White Branch	Green Branch	Variegated Branch
White branch	White	Green	White, green, or variegated
Green branch	White	Green	White, green, or variegated
Variegated branch	White	Green	White, green, or variegated

gote, had no influence on the progeny phenotypes. Since the leaf coloration involves the chloroplast, either genetic information in that organelle or in the cytoplasm and influencing the chloroplast could be responsible for this inheritance pattern.

Iojap In Maize

A phenotype similar to *Mirabilis* but with a different pattern of inheritance has been analyzed in maize by Marcus M. Rhoades. In this case, green, colorless, or green-and-colorless striped leaves are under the influence not only of the cytoplasm but, in addition, of a nuclear gene located on the seventh linkage group. This locus is called *iojap*, after the recessive mutation *ij* located there. The wild-type allele is designated *Ij*. Plants homozygous for the mutation *ij/ij* may have green-and-white striped leaves. However, when reciprocal crosses are made between plants with striped leaves and green leaves, the results are seen to vary, depending upon which parent is mutant (Figure 15.4). If the female is striped *(ij/ij)* and the male is green *(Ij/Ij)*, plants with colorless, striped, and green leaves are observed as progeny. If the male parent is striped and the female green, only green plants are produced! In both types of cross, all offspring have identical genotypes *(Ij/ij)*. We may conclude that although a nuclear gene is somehow involved, the inheritance pattern is influenced maternally.

We can understand this pattern better by examining the offspring resulting from self-fertilization of these heterozygous plants (Figure 15.4). The striped plant gives rise to progeny with colorless, striped, and green leaves, regardless of their genotype. The green plants give rise to only green offspring. The results of these self-fertilizations substantiate that the mutant phenotype is controlled solely through the female cytoplasm, regardless of the plant's genotype.

The role of the nucleus in the origin of the mutant phenotype is unclear. However, the colorless areas of the leaf are due to the lack of the green chlorophyll pigment in chloroplasts. Once acquired, colorless chloroplasts are transmitted through the egg cytoplasm, establishing the phenotypes of leaves of progeny plants.

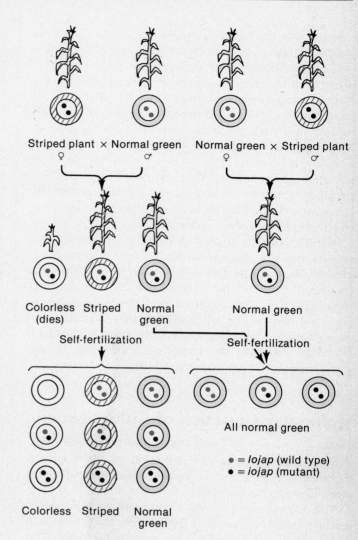

FIGURE 15.4
Inheritance of striping in maize. Regardless of the genotype of the *iojap* genes, offspring reflect the maternal phenotype in their appearance. If the maternal parent has green leaves, so do all progeny plants. If the maternal parent is striped (containing green *and* colorless areas), progeny are either colorless, green, or striped. Since color is due to chloroplasts in the leaves, inheritance is controlled by these organelles as they are passed through the maternal cytoplasm. (Reprinted with permission of Macmillan Publishing Co., Inc., from *Genetics*, 2nd Edition, by M. W. Strickberger. Copyright © 1976 by Monroe W. Strickberger.)

Chlamydomonas Mutations

The unicellular green alga *Chlamydomonas reinhardi* has provided an excellent system for the investigation of cytoplasmic inheritance. The organism is eukaryotic, contains a single large chloroplast as well as numerous mitochondria, undergoes matings which are followed by meiosis, yet can easily be cultured in the laboratory. The first cytoplasmic mutant, *streptomycin resistance (sr)*, was reported in 1954 by Ruth Sager. Although *Chlamydomonas'* two mating types (see Chapter 5)—mt^+ and mt^-—appear to make equal cytoplasmic contributions to the zygote, Sager determined that the *sr* phenotype was transmitted only through the mt^+ parent.

Since this discovery, a number of other *Chlamydo-monas* mutations (including an acetate requirement as well as resistance to or dependence on a variety of bacterial antibiotics) have been discovered that show a similar cytoplasmic inheritance pattern. These mutations have been linked to the transmission of the chloroplast, and their study has extended our knowledge of chloroplast inheritance.

Following fertilization, the single chloroplasts of the two mating types fuse. After the resulting zygote has undergone meiosis, it is apparent that the genetic information of the chloroplasts of progeny cells is derived only from the mt^+ parent.

Further, studies suggest that these chloroplast mutations, representing numerous gene sites, form a single circular linkage group—the first such extrachromosomal unit to be established in a eukaryotic organism. It is also apparent that the linkage groups from the two mating types are capable of undergoing recombination following zygote formation. All available evidence supports the hypothesis that this linkage group consists of DNA residing in the chloroplast.

Molecular Genetics of Chloroplasts

There is now a substantial amount of molecular information about the genetic function of chloroplasts. However, this is a current area of research and much more is yet to be learned.

The chloroplast contains a complete genetic system distinct from that found in the nucleus and cytoplasm of eukaryotic plant cells. This system includes DNA (**cpDNA**) as a source of genetic information and a complete protein-synthesizing apparatus. However, the molecular components of the translation apparatus are jointly derived from both nuclear and chloroplast genetic information. As we shall see, the chloroplast genetic system is very similar to that found in prokaryotic cells.

DNA isolated from chloroplasts (see Figure 11.16) is found to be circular, double-stranded, replicated semiconservatively, and free of the associated proteins characteristic of eukaryotic DNA. Compared with nuclear DNA of the same organism, it invariably shows a different buoyant density, base composition, and nearest-neighbor frequency. Furthermore, cpDNA lacks 5-methylcytosine, a modified base present in nuclear DNA of many plants.

In *Chlamydomonas*, there are about 75 copies of the chloroplast DNA molecule per organelle. Each copy consists of a length of DNA measuring 62 μm that contains 195,000 bases (195 kb), almost twice the size of a T-even bacteriophage genome. In higher plants such as the sweet pea, multiple copies of the DNA molecule are present in each organelle, but the molecule is considerably smaller than that in *Chlamydomonas*, consisting of 134 kb.

Molecular hybridization studies have demonstrated the relationship of some chloroplast DNA products to the protein-synthesis apparatus. In a variety of higher plants (beans, lettuce, spinach, maize, and oats), such studies reveal two sets of the genes coding for the ribosomal RNAs—5S, 16S, and 23S rRNA. These are present in two large DNA segments (greater than 20 kb), which are of opposite polarity and separated by a large spacer region (greater than 10 kb). Additionally, certain genes in chloroplast DNA code for at least 25 tRNA species and a number of ribosomal proteins specific to the chloroplast ribosomes. These ribosomes demonstrate a sedimentation coefficient slightly less than 70S, similar to that of bacteria.

Still other chloroplast genes have been identified that are specific to photosynthetic function. One of the major chloroplast gene products is the large subunit of the photosynthetic enzyme **ribulose-1-5-biphosphate carboxylase (RuBP).** * Interestingly, the small subunit of this enzyme is encoded by a nuclear gene. The gene for the small subunit is thus transcribed in the nucleus, the corresponding mRNA translated in the cytoplasm, and the resultant small subunit protein transported into the chloroplast before it can function in photosynthesis. The large subunit, like all other chloroplast-derived gene products, is synthesized and remains in the organelle.

Chloroplast ribosomes are sensitive to the same protein-synthesis-inhibiting antibiotics as bacterial ribosomes: chloramphenicol, erythromycin, streptomycin, and spectinomycin. Even though ribosomal proteins are derived from nuclear and chloroplast DNA, most, if not all, are distinct from the equivalent proteins of cytoplasmic ribosomes.

Given this information, it is not surprising that experiments have revealed mutations relating to chloroplast function. Some of the products essential to photosynthesis are coded by chloroplast DNA and synthesized in the chloroplast. Therefore, critical mutations in the chloroplast DNA can lead to inactivation of photosynthesis. If such chloroplasts are transmitted to progeny through the cytoplasm, offspring may similarly lack photosynthetic activity.

---

*The enzyme is also named ribulose-1-5-diphosphate carboxylase.

Mitochondria: *poky* in *Neurospora*

Mitochondria, like chloroplasts, play a critical role in cellular bioenergetics and contain a distinctive genetic system. Mutants affecting mitochondrial function have been discovered and studied. As with chloroplast mutants, these are transmitted through the cytoplasm and result in non-Mendelian inheritance patterns. Additionally, the mitochondrial genetic system has now been extensively characterized.

In 1952, Mary B. and Hershel K. Mitchell discovered a slow-growing mutant strain of the mold *Neurospora crassa* and called it *poky*. (It is also designated *mi-1* for maternal inheritance.) Studies have shown slow growth to be associated with impaired mitochondrial function specifically related to certain cytochromes essential to electron transport. The results of genetic crosses between wild-type and *poky* strains suggest that the trait is maternally inherited. If the female parent is *poky* and the male parent is wild type, all progeny colonies are *poky*. The reciprocal cross produces normal colonies.

Studies with *poky* mutants illustrate the use of **heterokaryon** formation during the investigation of maternal inheritance in fungi. Occasionally, hyphae from separate mycelia fuse with one another, giving rise to structures containing two or more nuclei in a common cytoplasm. If the hyphae contain nuclei of different genotypes, the structure is called a heterokaryon. The cytoplasm will contain mitochondria derived from both initial mycelia. A heterokaryon may give rise to haploid spores, or **conidia,** that produce new mycelia. The phenotypes of these structures may be determined.

Heterokaryons produced by the fusion of *poky* and wild-type hyphae initially show normal rates of growth and respiration. However, mycelia produced through conidia formation become progressively more abnormal until they show the *poky* phenotype. This occurs in spite of the presumed presence of both wild-type and *poky* mitochondria in the cytoplasm of the hyphae.

To explain the initial growth and respiration pattern, it is assumed that the wild-type mitochondria support the respiratory needs of the hyphae. The subsequent expression of the *poky* phenotype suggests that the presence of the *poky* mitochondria may somehow prevent or depress the function of these wild-type mitochondria. Perhaps the *poky* mitochondria replicate more rapidly and "wash out" or dilute wild-type mitochondria numerically. Another possibility is that *poky* mitochondria produce a substance that inactivates the wild-type organelle or interferes with the replication of its DNA (mtDNA). As a result of this type of interaction, mutations such as *poky* are labeled **suppressive.** This general phenomenon is characteristic of many other suspected mitochondrial mutations of *Neurospora* and yeast.

Petite in *Saccharomyces*

Another extensive study of mitochondrial mutations has been performed with the yeast *Saccharomyces cerevisiae*. The first such mutation, *petite*, was described by Boris Ephrussi and his coworkers in 1956. The mutant was so named because of the small size of the yeast colonies. Many independent *petite* mutations have since been discovered and studied. They all have a common characteristic: deficiency in cellular respiration involving abnormal electron transport. Fortunately, this organism is a facultative anaerobe and can grow by fermenting glucose through glycolysis. Thus, although colonies are small, the organism may survive the loss of mitochondrial function by generating energy anaerobically.

The complex genetics of *petite* mutations is diagramed in Figure 15.5. A small proportion of these mutants exhibit Mendelian inheritance and are called **segregational petites,** indicating that they are the result of nuclear mutations. The remainder demonstrate cytoplasmic transmission, producing one of two effects in matings. The **neutral petites,** when crossed to wild type, produce ascospores that give rise only to wild-type or normal colonies. The same pattern continues if progeny of this cross are back-crossed to *neutral petites*. These results are due to the fact that the majority of neutrals lack mtDNA completely or have lost a substantial portion of it. Thus, the wild-type gamete is the effective source of normal mitochondria capable of reproduction. Neutral petites have now been shown to lack most, if not all, mitochondrial DNA (mtDNA).

A third type, the **suppressive petites,** behaves similarly to *poky* in *Neurospora*. Crosses between mutant and wild type give rise to mutant diploid zygotes which, upon undergoing meiosis, immediately yield all mutant cells. Under these conditions, the *petite* mutation behaves "dominantly" and seems to suppress the function of the wild-type mitochondria. *Suppressive petites* also represent deletions of mtDNA, but not nearly to the extent of the *neutral petites*. Buoyant density studies show a lower G-C content in suppressive mtDNA compared with normal mtDNA.

Suppressiveness remains unexplained. Two major hypotheses have been advanced. One explanation suggests that the mutant (or deleted) mtDNA replicates more rapidly, and thus mutant mitochondria "take

FIGURE 15.5
The three types of *petite* mutations affecting mitochondrial function in the yeast *Saccharomyces cerevisiae*.

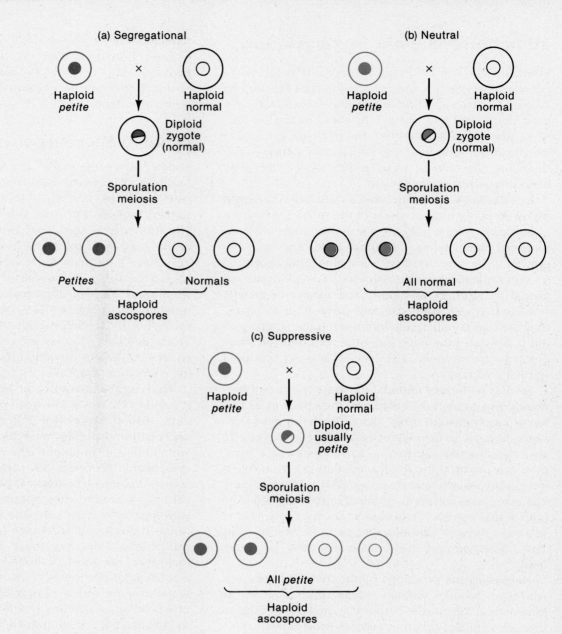

over" or dominate the phenotype by numbers alone. The second explanation suggests that recombination occurs between the mutant and wild-type mtDNA, introducing errors into or disrupting the normal mtDNA. It is not yet clear which, if either, of these explanations is correct.

Molecular Genetics of Mitochondria

Extensive information is now available about the molecular aspects of mitochondrial gene function. This information generally parallels what is known about chloroplasts, including the similarities to prokaryotes.

It is known that mtDNA (see Figure 11.16) is a circular duplex that replicates semiconservatively and is free of proteins in most all organisms studied. In size, mtDNA differs between plants and animals but is invariably much smaller than the DNA present in chloroplasts. This can be seen by examining the information presented in Table 15.2. In a variety of animals, mtDNA consists of about 16 to 18 kb. It seems that several copies are present in each organelle. There is a considerably greater amount of DNA present in plant mitochondria, where 100 kb is considered a minimal size in higher plants.

TABLE 15.2
The size of mtDNA in different organisms.

Organism	Size in Kilobases
Human	16.6
Mouse	16.2
Xenopus (frog)	18.4
Drosophila (fruit fly)	18.4
Saccharomyces (yeast)	84.0
Pisum sativum (pea)	110.0

Several general statements can now be made concerning mtDNA. There appear to be few or no gene repetitions, and replication is dependent upon nuclearly derived enzymes. Genes have been identified that code for the ribosomal RNAs, over 20 tRNAs, ribosomal proteins, and numerous products essential to the cellular respiratory functions of the organelles.

As with chloroplasts, the protein-synthesizing apparatus as well as the molecular components for cellular respiration are jointly derived from nuclear and mitochondrial DNA. Ribosomes found in the organelle are different from those present in the cytoplasm. In contrast to chloroplasts, the sedimentation coefficient of these particles is highly variable in different organisms. The data in Table 15.3 show that mitochondrial ribosomes can be either lower in their coefficients than in prokaryotes ($55S$ to $60S$ in vertebrates), the same general coefficient (about $70S$ in some algae and fungi), or as high as eukaryotic cytoplasmic ribosomes ($80S$ in certain protozoans and fungi).

TABLE 15.3
Sedimentation coefficients of mitochondrial ribosomes.

Kingdom	Examples	Sedimentation Coefficient (S)
Animals	Vertebrates	50–60
	Insects	60–71
Protists	*Euglena*	71
	Tetrahymena	80
Fungi	*Neurospora*	73–80
	Saccharomyces	72–80
Plants	Maize	77

Many nuclear-coded gene products are essential to biological activity within mitochondria: DNA and RNA polymerases, initiation and elongation factors essential for translation, ribosomal proteins, aminoacyl tRNA synthetases, and some tRNA species. As in chloroplasts, these imported components are generally regarded as distinct from their cytoplasmic counterparts, even though both sets are coded by nuclear genes. For example, the synthetases essential to charging tRNA molecules (a process essential to translation) show a distinct affinity for the mitochondrial tRNA species as compared with the cytoplasmic tRNAs. Similar affinity has been shown for the initiation and elongation factors. Furthermore, while bacterial and nuclear RNA polymerases are known to be composed of numerous subunits, the mitochondrial variety consists of only one polypeptide chain. This polymerase is generally susceptible to bacterial antibiotics specific to RNA synthesis but not to eukaryotic inhibitors. The relative contributions of nuclear and mitochondrial gene products are illustrated in Figure 15.6.

There is a surprising diversity between mtDNA derived from mammalian (including human) and yeast mitochondria. The mammalian DNA is compact, is replicated in a single orderly process, and is free of intergenic (noncoding) regions. Yeast mtDNA, on the other hand, demonstrates multiple origins of replication and extensive intergenic DNA sequences. Human mtDNA is free of **introns** (see Chapter 19), while yeast mtDNA contains such intragenic sequences. Introns are nucleotide sequences within genes that are not reflected in the protein following translation. To some extent, the presence of introns accounts for the greater amount of DNA in mitochondria of plants compared with animals.

That genes have been exchanged between mitochondrial and nuclear DNA during evolution is established by the case of the ATPase subunit 9 gene. In yeast this gene is found in the nucleus, and in *Neurospora* it is present in the mitochondrion. It has been observed that the genetic code used in mitochondria contains several minor variations compared with that used in the cytoplasmic translation of nuclearly derived mRNA (see Chapter 17).

The Endosymbiont Theory

Because the genetic systems of mitochondria and chloroplasts are distinctive compared with the nuclear-derived eukaryotic system, there has been much interest in and speculation about the evolutionary origin of these organelles. The **endosymbiont theory** has been proposed, based on the many genetic similarities between these organelles and prokaryotic organisms. Championed by Lynn Margulis, the theory

FIGURE 15.6
Nuclear gene products essential to mitochondria. Mitochondrial DNA, gene products encoded by it, and genetic processes occurring in the mitochondria are shown in color.

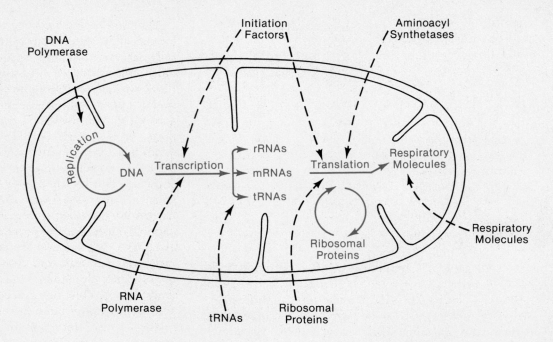

suggests that chloroplasts have descended from free-living blue-green algae, and that mitochondria were once aerobic bacteria. Thus, the origin of both organelles is ancient prokaryotic cells.

The similarities shared by the DNA and genetic machinery of both organelles and prokaryotes are rather striking:

1 DNA is circular and free of associated histone proteins.

2 No nuclear envelope encloses the DNA.

3 Antibiotics such as erythromycin, rifampicin, and chloramphenicol, which inhibit protein synthesis in prokaryotes, are also inhibitory in chloroplasts and mitochondria.

4 Antibiotics such as cycloheximide and α-amanitin, which inhibit eukaryotic protein synthesis, do not affect this process in these organelles.

5 The sedimentation coefficients of the ribosomal subunits and the rRNAs of these organelles are very similar to those of prokaryotes.

6 Substantial nucleotide sequence homology exists between rRNA of these organelles and that of *E. coli*. For example the 16*S* RNA component derived from maize chloroplast ribosomes shares 74 percent of its sequences with the same molecule derived from *E. coli*. The 16*S* RNA of wheat chloroplasts displays 88 percent homology with *E. coli*. Homology also exists with other prokaryotes such as blue-green algae.

Thus, with the above support for its general nature, the theory more specifically suggests that primitive eukaryotic cells originally were anaerobic. Some cells ingested or were invaded by an aerobic bacterium, which conferred on them the ability to utilize the more efficient aerobic cellular respiration process. A symbiotic relationship was thus established, and these modified primitive cells were the progenitors of the modern eukaryotic animal cells. Other primitive cells were further invaded by ancient blue-green algal cells, thus acquiring the ability to fix carbon through photosynthesis. These cells eventually gave rise to the modern plant cells. Obviously, if the theory is correct, evolution over a long period of time has altered and rearranged the genetic information contained initially in the invading prokaryotes. Many of the genetic functions of the organelles are now dependent on nuclear genes. Transfer of genetic information from the organelles to the nucleus is certainly not far-fetched given our knowledge of transposons (Chapter 14). While some biologists oppose the theory, it is intriguing and has stimulated a great deal of investigation.

INFECTIOUS HEREDITY

There are numerous examples of cytoplasmically transmitted phenotypes in eukaryotes that are due to an invading microorganism or particle. The foreign invader coexists in a symbiotic relationship, is usually passed through the maternal ooplasm to progeny cells or organisms, and confers a specific phenotype that may be studied. We shall consider several examples illustrating this phenomenon.

Kappa in *Paramecium*

First described by Tracy Sonneborn, certain strains of *Paramecium aurelia* are called **Killers** because they release a cytoplasmic substance called **paramecin** that is toxic and sometimes lethal to sensitive strains. This substance is produced by particles called **kappa** that replicate in the Killer cytoplasm, contain DNA and protein, and depend for their maintenance on a dominant nuclear gene, *K*. One cell may contain 100 to 200 such particles.

Paramecia are diploid protozoans that can undergo sexual exchange of genetic information through the process of **conjugation.** In some instances, cytoplasmic exchange also occurs. Thus, there is a variety of ways in which the *K* gene and kappa can be transmitted.

The life cycle of *Paramecium aurelia* is shown in Figure 15.7. Paramecia contain two diploid micronuclei, which are involved in cross-fertilization. Prior to conjugation, both micronuclei in each mating pair undergo meiosis, resulting in eight haploid nuclei. However, seven of these degenerate, and the remaining one undergoes a single mitotic division. During conjugation, each cell donates one of the two haploid nuclei to the other, recreating the diploid condition in both cells. As a result, exconjugates are of identical genotypes.

In a similar process involving only a single cell, **autogamy** occurs. Following meiosis of both micronuclei, seven products degenerate and one survives. This nucleus divides, and the resulting nuclei fuse to recreate the diploid condition. If the original cell was heterozygous, autogamy results in homozygosity because the newly formed diploid nucleus was derived solely from a single haploid meiotic product. In a pop-

FIGURE 15.7
Conjugation during the life cycle of *Paramecium aurelia*.

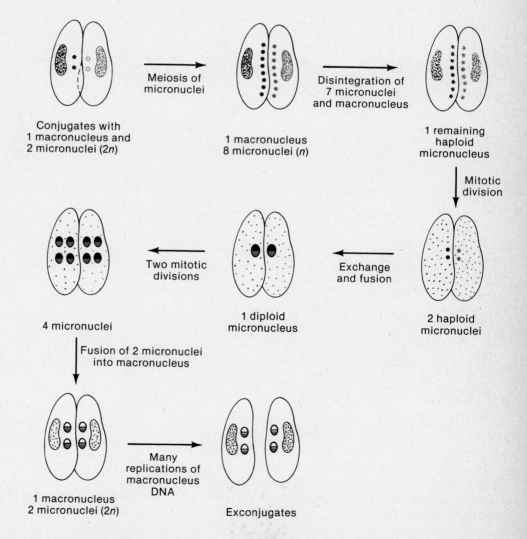

Conjugates with
1 macronucleus and
2 micronuclei (2*n*)

Meiosis of
micronuclei

1 macronucleus
8 micronuclei (*n*)

Disintegration of
7 micronuclei
and macronucleus

1 remaining
haploid
micronucleus

Mitotic
division

2 haploid
micronuclei

Exchange
and fusion

1 diploid
micronucleus

Two mitotic
divisions

4 micronuclei

Fusion of 2 micronuclei
into macronucleus

1 macronucleus
2 micronuclei (2*n*)

Many
replications of
macronucleus
DNA

Exconjugates

ulation of cells that were originally heterozygous, half of the new cells express one allele and half express the other allele.

Figure 15.8 illustrates the results of crosses between *KK* and *kk* cells, without and with cytoplasmic exchange. When no cytoplasmic exchange occurs, even though the resultant cells may be *Kk* (or *KK* following autogamy), they remain sensitive if no kappa particles are transmitted. When exchange occurs, the cells become Killers provided the kappa particles are supported by at least one dominant *K* allele.

Kappa particles are bacterialike and may contain temperate bacteriophages. One theory holds that these viruses of kappa may become vegetative; during this multiplication, they produce the toxic products that are released and kill sensitive strains.

Infective Particles in *Drosophila*

Two examples of similar phenomena are known in *Drosophila*: **CO₂ sensitivity** and **sex-ratio.** In the former, flies that would normally recover from carbon dioxide anesthetization instead become permanently paralyzed and are killed by CO_2. Sensitive mothers pass this trait to all offspring. Furthermore, extracts of sensitive flies induce the trait when injected into resistant flies. Phillip L'Heritier has postulated that sensitivity is due to the presence of a virus, **sigma.** The particle has been visualized and is smaller than kappa. Attempts to transfer the virus to other insects have been unsuccessful, demonstrating that specific nuclear genes support the presence of sigma in *Drosophila*.

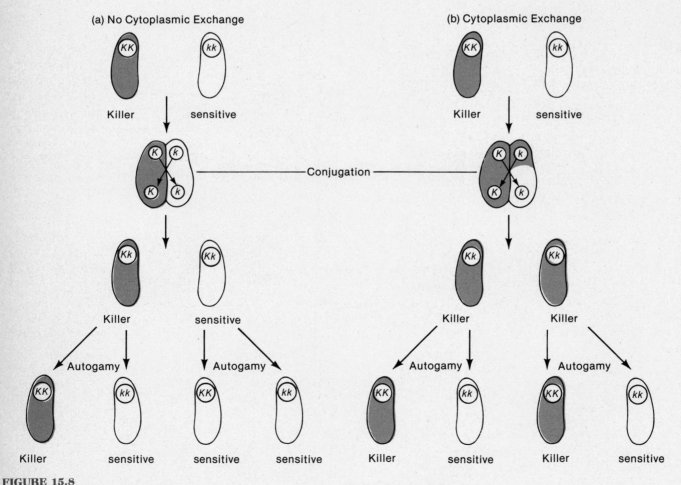

FIGURE 15.8
Results of crosses between Killer *(KK)* and sensitive *(kk)* strains of *Paramecium* with and without cytoplasmic exchange during conjugation. (Reprinted with permission from Macmillan Publishing Co., Inc., from *Genetics*, 2nd Edition, by M. W. Strickberger. Copyright © 1976 by Monroe W. Strickberger.)

A second example of infective particles comes from the study of *Drosophila bifasciata*. A small number of these flies were found to produce predominantly female offspring if reared at 21°C or lower. This condition, designated **sex-ratio,** was shown to be transmitted to daughters but not to the low percentage of males produced. This phenomenon was subsequently investigated in *Drosophila willistoni*. In these flies, the injection of ooplasm from sex-ratio females into normal females induced the condition. This observation suggested that an extrachromosomal element is responsible for the sex-ratio phenotype. The agent has now been isolated and shown to be a protozoan. While the protozoan has been found in both males and females, it is lethal primarily to developing male larvae. There is now some evidence that a virus harbored by the protozoan may be responsible for producing a male-lethal toxin.

SUMMARY AND CONCLUSIONS

In many instances, inheritance of phenotypes does not follow the pattern established for nuclear genes. In such cases, extrachromosomal inheritance is said to occur. Many higher organisms, for example, appear to exhibit genetic control originating from the cytoplasm of the maternal gamete. Investigation has indeed shown this to be true, and the basis for such inheritance is now better understood.

Maternal influence appears to stem from the presence of molecules critical to early development, which have been transcribed during oogenesis. Pigmentation in *Ephestia*, coiling in *Limnaea*, and embryogenesis in *Drosophila* illustrate such inheritance.

Other extrachromosomally inherited phenotypes have been related to mitochondrial and chloroplast function. Again, inheritance is through the cytoplasm, coinciding with the source of these organelles in the zygote. Study of mitochondria and chloroplasts has revealed a partially autonomous DNA-based genetic system in each organelle. These DNAs code for some of the components essential to translation and transcription within each organelle, including ribosomal and transfer RNA species and ribosomal proteins. The remainder of the molecular components of the genetic apparatus in each organelle are of nuclear origin. However, they are distinct from their cytoplasmic counterparts. In addition, the DNA of each organelle codes for certain products essential to either photosynthetic or respiratory function. The genetic systems of these organelles are more closely related to that of prokaryotes than to the nuclear system of the cells harboring the organelles. This observation has served as the basis of the endosymbiont theory which attempts to explain the evolutionary origin of chloroplasts and mitochondria.

Another type of extrachromosomal inheritance is due to the presence of infectious microorganisms existing in a symbiotic relationship with the host cell. Kappa particles in *Paramecium* and CO_2-sensitive and sex-ratio determinants in *Drosophila* are examples of such infective particles. The responsible microorganisms contain their own DNA, confer certain phenotypes, and are transmitted cytoplasmically. In all three examples, the presence of the symbionts is partially dependent on the host's nuclear genotype.

PROBLEMS AND DISCUSSION QUESTIONS

1 What genetic criteria distinguish a case of extrachromosomal inheritance from a case of Mendelian autosomal inheritance? from a case of sex-linked inheritance?

2 In *Limnaea*, what results would be expected in a cross between a *Dd* dextrally coiled and a *Dd* sinistrally coiled snail, assuming cross-fertilization occurs as shown in Figure 15.2? What results would occur if the *Dd* dextral produced only eggs and the *Dd* sinistral produced only sperm?

3 Streptomycin resistance in *Chlamydomonas* may result from a mutation in a chloroplast gene or in a nuclear gene. What phenotypic results would occur in a cross between a member of an mt^+ strain resistant in both genes and a member of an mt^- strain sensitive to the antibiotic? What results would occur in the reciprocal cross?

4 A plant may have green, white, or green and white (variegated) leaves on its branches owing to a mutation in the chloroplast (which produces the white leaves). Predict the results of the following crosses.

	Ovule Source		Pollen Source
(a)	Green branch	×	White branch
(b)	White branch	×	Green branch
(c)	Variegated branch	×	Green branch
(d)	Green branch	×	Variegated branch

5 In diploid yeast strains, sporulation and subsequent meiosis can produce haploid ascospores. These may fuse to reestablish diploid cells. When ascospores from a *segregational petite* strain fuse with those of a normal wild-type strain, the diploid zygotes are all normal. However, following meiosis, ascospores are ½ petite and ½ normal. Is the *segregational petite* phenotype inherited as a dominant or recessive gene?

6 Predict the results of a cross between ascospores from a *segregational petite* strain and a *neutral petite* strain. Indicate the phenotype of the zygote and the ascospores it may subsequently produce.

7 Contrast the molecular components constituting the genetic apparatus in mitochondria and chloroplasts.

8 The genetic apparatus of mitochondria and chloroplasts is similar to that of prokaryotic cells. This information has been the basis of the theory of the endosymbiotic origin of these organelles. That is, these organelles are thought to have originated as invading prokaryotic cells into primitive eukaryotic cells. What similarities and observations support this theory? Based on information presented in this chapter, what evidence or observations might dispute the theory?

9 Described below are the results of three crosses between strains of *Paramecium*. Determine the genotypes of the parental strains.
(a) Killer × sensitive → ½ Killer: ½ sensitive
(b) Killer × sensitive → all Killer
(c) Killer × sensitive → ¾ Killer: ¼ sensitive

10 *Chlamydomonas*, a eukaryotic green alga, is sensitive to the antibiotic erythromycin that inhibits protein synthesis in prokaryotes.
(a) Explain why.
(b) There are two mating types in this alga, mt^+ and mt^-. If an mt^+ cell sensitive to the antibiotic is crossed with an mt^- cell that is resistant, all progeny cells are sensitive. The reciprocal cross (mt^+ resistant and mt^- sensitive) yields all resistant progeny cells. Assuming that the mutation for resistance is in cpDNA, what can be concluded?

11 In *Limnaea*, a cross where the snail contributing the eggs was dextral but of unknown genotype mated with another snail of unknown genotype and phenotype. All F_1 offspring exhibited dextral coiling. Ten of the F_1 snails were allowed to undergo self-fertilization. One-half produced only dextrally coiled offspring, while the other half produced only sinistrally coiled offspring. What were the genotypes of the original parents?

12 In *Drosophila subobscura*, the presence of a recessive gene called *grandchildless* (*gs*) causes the offspring of homozygous females, but not homozygous males, to be sterile. Can you offer an explanation as to why females but not males are affected by the mutant gene?

SELECTED READINGS

BOGORAD, L. 1977. Genes for chloroplast ribosomal RNAs and ribosomal proteins. In *International cell biology*, eds. B. R. Brinkley and K. R. Porter, pp. 175–82. New York: Rockefeller University Press.

————. 1981. Chloroplasts. *J. Cell Biol.* 91: 256s–70s.

COHEN, S. 1973. Mitochondria and chloroplasts revisited. *Amer. Sci.* 61: 437–45.

FREDERICK, J. F., ed. 1981. Origins and evolution of eukaryotic intracellular organelles. New York: New York Academy of Science.

FREEMAN, G., and LUNDELIUS, J. W. 1982. The developmental genetics of dextrality and sinistrality in the gastropod *Lymnaea peregra*. *Wilhelm Roux Arch.* 191: 69–83.

GILLHAM, N. W. 1974. Genetic analysis of chloroplast and mitochondrial genomes. *Ann. Rev. Genet.* 8: 347–92.

———. 1978. *Organelle hereditary.* New York: Raven Press.

GOODENOUGH, U., and LEVINE, R. P. 1970. The genetic activity of mitochondria and chloroplasts. *Scient. Amer.* (Nov.) 223: 22–29.

GRIVELL, L. A. 1983. Mitochondrial DNA. *Scient. Amer.* (March) 248: 78–89.

LEVINE, R. P., and GOODENOUGH, U. 1970. The genetics of photosynthesis and of the chloroplast in *Chlamydomonas reinhardi. Ann. Rev. Genet.* 4: 397–408.

MARGULIS, L. 1970. *Origin of eukaryotic cells.* New Haven, Conn.: Yale University Press.

———. 1981. *Symbiosis in cell evolution.* San Francisco: W. H. Freeman.

MITCHELL, M. B., and MITCHELL, H. K. 1952. A case of maternal inheritance in *Neurospora crassa. Proc. Natl. Acad. Sci.* 38: 442–49.

NASS, M. M. K. 1976. Mitochondrial DNA. In *Handbook of genetics,* ed. R. C. King, vol. 5, pp. 477–533. New York: Plenum Press.

O'BRIEN, T. W. 1977. Transcription and translation in mitochondria. In *International cell biology,* eds. B. R. Brinkley and K. R. Porter, pp. 245–55. New York: Rockefeller University Press.

PREER, J. R. 1971. Extrachromosomal inheritance: Hereditary symbionts, mitochondria, chloroplasts. *Ann. Rev. Genet.* 5: 361–406.

SAGER, R. 1965. Genes outside the chromosomes. *Scient. Amer.* (Jan.) 212: 70–79.

———. 1972. *Cytoplasmic genes and organelles.* New York: Academic Press.

———. 1985. Chloroplast genetics. *BioEssays* 3:180–84.

SAGER, R., and SCHLANGER, G. 1976. Chloroplast DNA: Physical and genetic studies. In *Handbook of genetics,* ed. R. C. King, vol. 5, pp. 371–423. New York: Plenum Press.

SCHWARTZ, R. M., and DAYHOFF, M. O. 1978. Origins of prokaryotes, eukaryotes, mitochondria and chloroplasts. *Science* 199: 395–403.

SLONIMSKI, P. 1982. *Mitochondrial genes.* Cold Spring Harbor, N. Y.: Cold Spring Harbor Laboratory.

SONNEBORN, T. M. 1959. Kappa and related particles in *Paramecium. Adv. in Virus Res.* 6: 229–356.

SPENCER, D., SCHNARE, M., and GRAY, M. 1984. Pronounced structural similarities between the small subunit ribosomal RNA genes of wheat mitochondria and *Escherichia coli. Proc. Nat. Acad. Sci.* 81: 493–99.

STRATHERN, J. N., et al., eds. 1982. *The molecular biology of the yeast* Saccharomyces: *Life cycle and inheritance.* Cold Spring Harbor, N.Y.: Cold Spring Harbor Laboratory.

STURTEVANT, A. H. 1923. Inheritance of the direction of coiling in *Limnaea. Science* 58: 269–70.

TZAGOLOFF, A. 1982. *Mitochondria.* New York: Plenum Press.

TZAGOLOFF, A., MACINO, G., and SEBALD, W. 1979. Mitochondrial genes and translation products. *Ann. Rev. Biochem.* 48: 419–41.

WEEDEN, N. F. 1981. Genetic and biochemical implications of the endosymbiotic origin of the chloroplast. *J. Mol. Evol.* 17: 133–39.

PART FOUR GENE STRUCTURE, FUNCTION, AND REGULATION

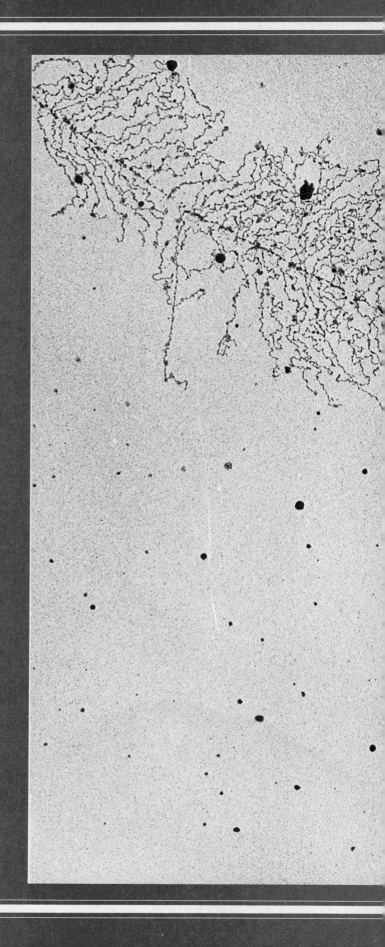

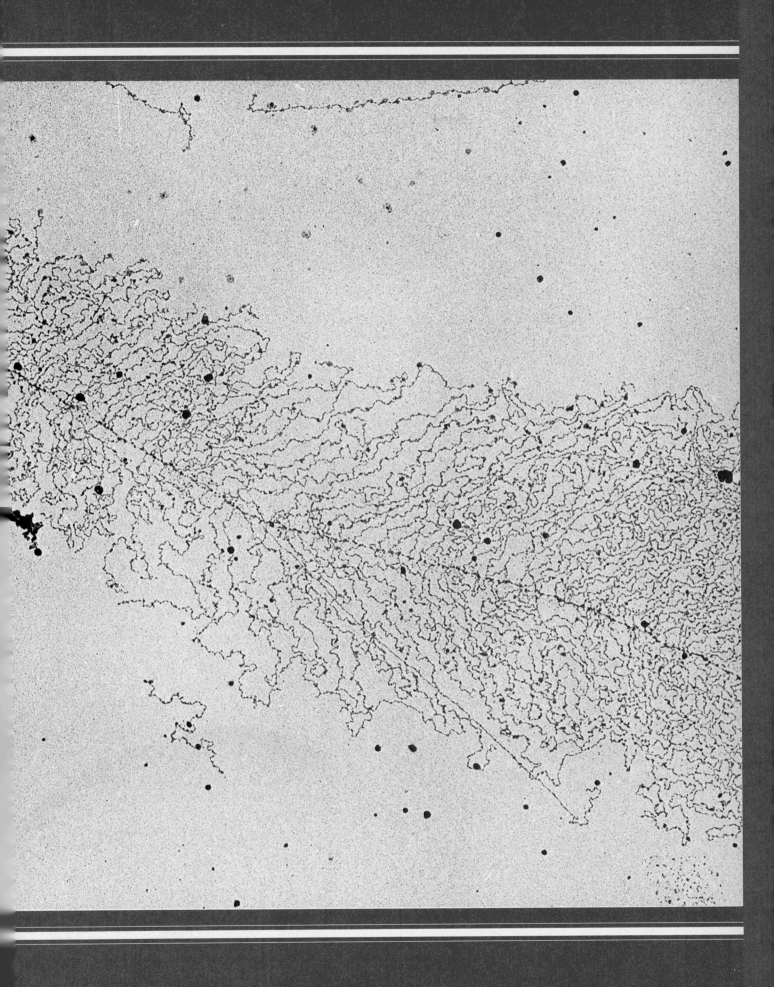

16

Genes and Proteins

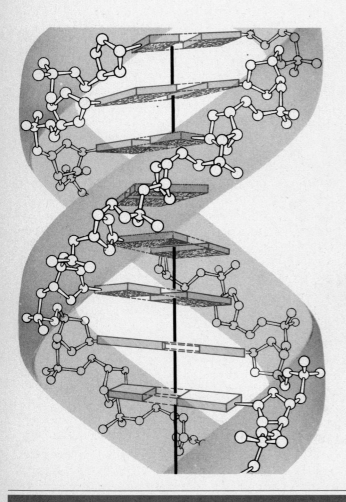

The central dogma of molecular genetics is that "DNA makes RNA which in turn makes protein." What this means, of course, is that DNA stores genetic information in a chemical language, which may be transcribed into complementary RNA molecules. In turn, RNA may be translated into protein molecules. While we have seen exceptions to this dogma—where RNA is the end product and where RNA serves as the template for DNA synthesis—in most instances proteins are the final products of information stored in the genes. This dogma implies that most mutations will be expressed at the level of protein function. However, while gene mutations resulting in distinct phenotypic effects had been studied early in the twentieth century, it was not known how the gene functions nor how a mutation might cause a phenotypic change. When it was discovered that proteins are the end products of gene activity, scientists began to understand mutational effects.

The first insight into the role of proteins in genetic processes was provided by observations made by Sir Archibald Garrod and William Bateson early in this century. Garrod established a correlation between certain inherited human disorders and abnormal metabolism of amino acids, the building blocks of proteins. Bateson was able to relate his own studies to Garrod's findings and link enzymes with genetic processes. However, it was not until around 1940 that solid evidence about the genetic role of proteins appeared. In a series of experiments, George Beadle and his co-workers were able to correlate genes and enzymes with the control of phenotypes and thus developed the **one-gene: one-enzyme theory.** Subsequently, Vernon Ingram and Linus Pauling showed that mutations alter the amino acid sequences of human hemoglobin molecules. Their work was particularly significant because it showed that nonenzymatic proteins are also produced as a result of transcription and translation, thus extending the one-gene: one-enzyme theory.

These findings are the foundation of what has been called the field of **biochemical genetics.** Today, research findings on the biochemistry of genes and the formation of their products comprise the field called **molecular genetics.** This general topic served as the basis for Part 2 of this text. In this section, we will continue this discussion.

In this chapter, we will focus on two major areas. First, we will present numerous examples to illustrate that proteins are the final products of most stored genetic information. Then, we will discuss the chemical structure of proteins. As we will see, the specific function of any protein molecule is related directly to its structure.

GARROD AND BATESON: INBORN ERRORS OF METABOLISM AND ENZYMES

As a practicing physician in England, Garrod became interested in several human disorders that seemed to be inherited. Although he also studied albinism and cystinuria, we will describe his investigation of the disorder **alkaptonuria.** Individuals afflicted with this disorder cannot metabolize the alkapton 2,5-dihydroxyphenylacetic acid, also known as **homogentisic acid.** As a result, an important metabolic pathway (Figure 16.1) is blocked. Homogentisic acid accumulates in cells and tissues and is excreted in the urine. The molecule's oxidation products are black and thus are easily detectable in the diapers of newborns and the urine of older individuals. The products tend to accumulate in cartilaginous areas, causing a darkening of the ears and nose. In joints, this deposition leads to a benign arthritic condition. This rare disease is not a serious one, but it persists throughout an individual's life.

Garrod studied alkaptonuria by increasing dietary protein or adding the amino acids phenylalanine or tyrosine, which are chemically related to homogentisic acid, to the diet. Under such regimes, homogentisic acid levels increase in the urine of alkaptonurics but not in unaffected individuals. He concluded that normal individuals break down, or catabolize, this alkapton but that afflicted individuals cannot. By studying the patterns of inheritance of the disorder, Garrod further concluded that alkaptonuria was inherited as a simple recessive trait.

On the basis of these conclusions, Garrod hypothesized that the hereditary information controls chemical reactions in the body and that the inherited dis-

orders he studied are the result of alternate modes of metabolism. While the terms *genes* and *enzymes* were not familiar during Garrod's work, the corresponding concepts of unit factors and ferments were. Ferments, now known to be enzymes, were recognized by 1900, but their chemical nature was not determined until 1926. Garrod published his initial observations in 1902 and his classic work *Inborn Errors of Metabolism* in 1909.

Only a few geneticists, including Bateson, were familiar with or referred to Garrod's work. His ideas fit nicely with Bateson's belief that inherited conditions were caused by the lack of some critical substance. In 1909, Bateson published *Mendel's Principles of Heredity,* in which he linked ferments with heredity. However, most geneticists failed to see the relationship between genes and enzymes for almost thirty years. Garrod and Bateson, like Mendel, were ahead of their time.

Phenylketonuria

Described first in 1934, **phenylketonuria (PKU)** results in mental retardation and is inherited as an autosomal recessive disease. Afflicted individuals are unable to convert the amino acid phenylalanine to the amino acid tyrosine (Figure 16.1). If we examine the chemical structures of these molecules, we see that they differ by only a single hydroxyl group (—OH) present in tyrosine. The reaction is catalyzed by the enzyme **phenylalanine hydroxylase,** which is inactive in affected individuals and active at about a 30 percent level in heterozygotes. The enzyme functions in the liver. While the normal blood level of phenylalanine is about 1 mg/100 ml, phenylketonurics show a level as high as 50 mg/100 ml.

As phenylalanine accumulates, it may be converted in phenylpyruvic acid and other derivatives. These are less efficiently resorbed by the kidney and tend to spill into the urine more quickly than phenylalanine. Both enter the cerebrospinal fluid, resulting in elevated levels in the brain. The presence of these substances during early development is thought to be responsible for retardation.

Retardation can be prevented by PKU screening of newborns. When the condition is detected in the analysis of an infant's blood, a strict dietary regimen is instituted. A low-phenylalanine diet can reduce such by-products as phenylpyruvic acid, and abnormalities characterizing the disease can be diminished. Screening of newborns occurs routinely in almost every state in this country. Phenylketonuria occurs in approximately 1 in 11,000 births.

Knowledge of inherited metabolic disorders such as alkaptonuria, phenylketonuria, and albinism has

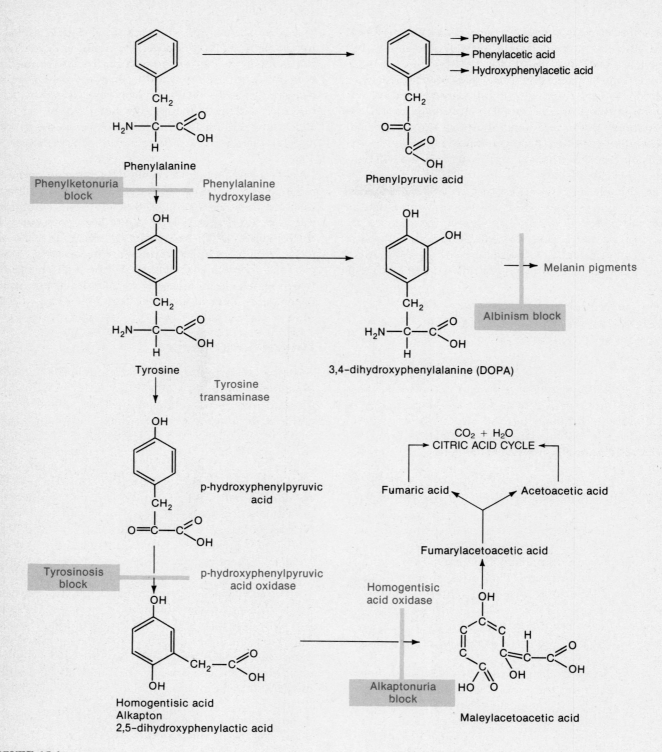

FIGURE 16.1
Metabolic pathway involving phenylalanine and tyrosine. Various metabolic blocks resulting from mutations lead to the disorders phenylketonuria, tyrosinosis, alkaptonuria, and albinism.

caused a revolution in medical thinking and practice. No longer is human disease attributed solely to the action of invading microorganisms, viruses, or parasites. We know now that literally hundreds of abnormal physiological conditions are caused by errors in metabolism that are the result of mutant genes. These human biochemical disorders are far-ranging and include all classes of organic biomolecules.

THE ONE-GENE: ONE-ENZYME HYPOTHESIS

In two separate investigations beginning in 1933, George Beadle was to provide the first convincing experimental evidence that genes are responsible for the synthesis of enzymes. The first investigation, conducted in collaboration with Boris Ephrussi, involved *Drosophila* eye pigments. Encouraged by these findings, Beadle then joined with Edward Tatum to investigate nutritional mutations in the pink bread mold *Neurospora crassa*. The latter investigation led to the one-gene: one-enzyme hypothesis.

Beadle and Ephrussi: *Drosophila* Eye Pigments

The studies of Beadle and Ephrussi involved **imaginal disks** in *Drosophila*. These embryonic tissues found in the larvae of insects, upon metamorphosis, differentiate into a variety of adult structures. Among others, adult eyes, legs, antennae, and genital structures are derived from these groups of cells. Beadle and Ephrussi found that if an imaginal disk was removed from one larva and transplanted into the abdomen of another, that disk would differentiate during metamorphosis and its corresponding adult structure could be recovered from the abdomen of the adult fly. For example, if an eye disk were transplanted, it would develop into an eyelike structure that could be recovered and analyzed.

These investigators wondered whether a mutant eye disk, implanted into a wild-type larva, would be altered by its new environment. Would it appear normally pigmented, or would it develop its characteristic mutant color? The first mutant used was *vermilion*, a sex-linked, bright red mutation. Nonautonomous development occurred. The transplanted *vermilion* eye disk was altered to produce a normally pigmented wild-type eye. The normal wild-type color is brick red, produced as a combination of brown and bright red pigments. The *vermilion* mutant lacks the brown pigment, indicating that its synthesis is inhibited.

Beadle and Ephrussi then proceeded to perform similar experimentation with 25 other mutants. Only one, *cinnabar*, another bright red eye mutation on chromosome 2, behaved nonautonomously. The remaining 24 behaved autonomously and exhibited their respective mutant color when differentiated.

They concluded that the *vermilion* and *cinnabar* phenotypes are due to the absence of some *diffusible substance* in the mutants. Because this substance was present in wild type, it enabled full pigment production in the transplant. However, in the case of the 24 mutations that developed autonomously, the substance that might have permitted normal pigment production was not diffusible but confined to the host cells forming the eye. Thus, no "correction" was noted.

Beadle and Ephrussi then asked whether the same substance was lacking in both the *vermilion* and *cinnabar* mutants. To answer this question, they transplanted *vermilion* eye disks into *cinnabar* larvae, and vice versa. What they found was intriguing. *Vermilion* disks were "cured" or converted to wild-type color, but *cinnabar* disks developed autonomously in *vermilion* hosts and retained their mutant phenotype. These experiments are diagramed in Figure 16.2.

To explain such results, they proposed that two tandem biochemical reactions are involved in normal synthesis of the brown pigment. One is inactive in flies with the *vermilion* mutation, and the other is inactive in flies with the *cinnabar* mutation. Each reaction produces a product that is diffusible in the tissues of wild type. As shown in Figure 16.2, the substance produced under the direction of the wild-type allele of the *vermilion* locus (substance Y) occurs first in the pathway. It is then converted to a second substance in the pathway. This conversion is controlled by the wild-type allele of the *cinnabar* locus (substance Z).

The order of this pathway explains the results of the reciprocal transplantations. Consider first the *vermilion (v)* disk in the *cinnabar (cn)* host. The *vermilion* disk lacks the *v* enzyme but has the *cn* enzyme. As the host, the *cinnabar* tissue can make substance Y, which is diffusible. Once substance Y enters the *vermilion* disk, the *cn* enzyme converts it to substance Z. Substance Z is then converted to the brown pigment, restoring the wild-type eye color.

Now consider the *cinnabar* disk in the *vermilion* host. The host tissue, lacking the *v* enzyme, cannot make substance Y. Even though the *cn* enzyme is present, because no Y is available to be converted to Z, no Z is produced either. Thus, the transplanted *cn* disk is still blocked. It can make substance Y but cannot complete the transition through Z to the brown pigment. Thus, it remains bright red!

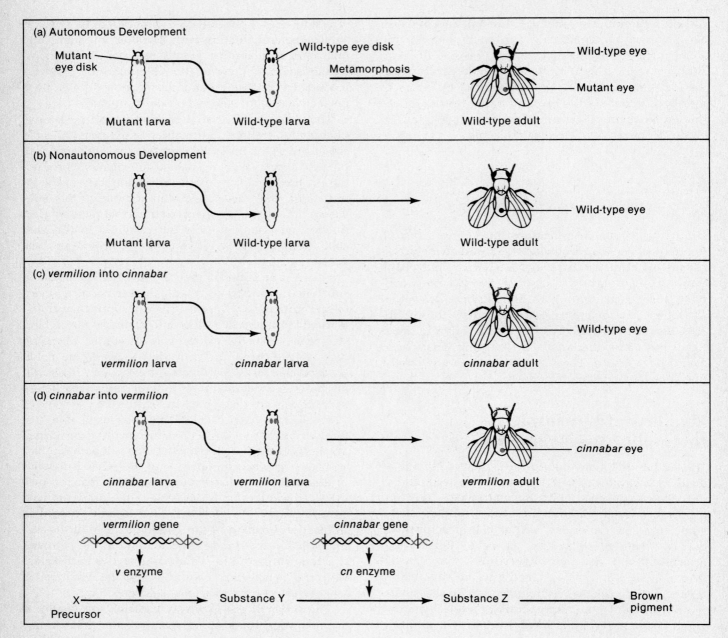

FIGURE 16.2
Beadle and Ephrussi's transplantation experiments involving imaginal disks that develop into eyes. When a mutant *vermilion* disk is transplanted into a *cinnabar* mutant larva, it behaves nonautonomously and develops normal wild-type pigment. The converse experiment results in autonomous development. On this basis, they concluded that the biosynthetic step controlled by the *vermilion* gene product occurs before that controlled by the *cinnabar* gene product.

Within a few years, the other workers had investigated the biochemical pathway leading to the brown xanthommatin pigment in *Drosophila*. As shown in Figure 16.3, the brown pigment is a derivative of the amino acid tryptophan. The reactions controlled by the specific enzymes altered by the *vermilion* and *cinnabar* mutations have been identified. Other autosomal recessive mutations, including *scarlet* and *cardinal*, have also been studied and shown to affect enzymes in this pathway.

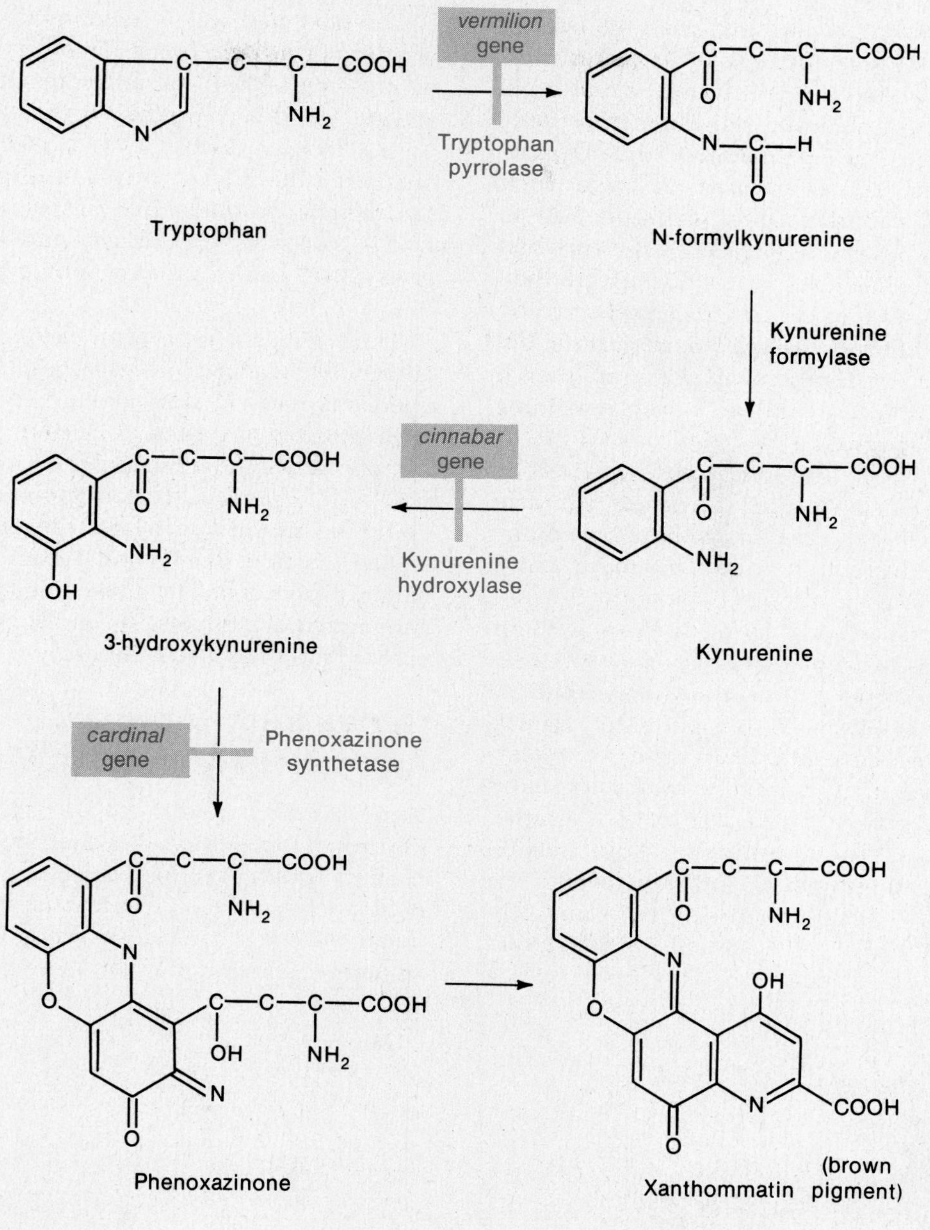

FIGURE 16.3

The biosynthetic pathway leading to the conversion of tryptophan to the brown eye pigment xanthommatin.

Beadle and Tatum: *Neurospora* Mutants

In the early 1940s, Beadle joined Tatum to seek more substantial evidence about the nature of gene products. They chose to work with the organism *Neurospora crassa* because much was known about its biochemistry, and mutations could be induced and isolated with relative ease. Through mutagenesis, they induced genetic blocks in reactions essential to the growth of the organism.

Beadle and Tatum knew that *Neurospora* could manufacture nearly everything necessary for normal development and function. For example, using rudimentary carbon and nitrogen sources, this organism can synthesize 9 water-soluble vitamins, 20 amino acids, numerous carotenoid pigments, and purines and pyrimidines. Beadle and Tatum irradiated conidia (asexually formed spores) with X-rays and used the treated spores to produce mycelia and sexual structures used in fertilization. After the completion of a sexual cycle (see Figure 5.3), they picked out individual ascospores and grew then on "complete" medium containing vitamins, amino acids, etc. (see Chapter 15). Under such growth conditions, a mutant strain would be able to grow by virtue of supplements present in the enriched complete medium. All single-spore cultures were then transferred to minimal medium. If growth occurred on minimal medium, then the single-spore culture did not contain a mutation. If no growth occurred, then it contained a mutation, and the only task remaining was to determine its type.

Literally thousands of individual spores derived by this procedure were isolated and grown on complete medium. In subsequent tests on minimal medium, several cultures failed to grow, indicating that a nutritional mutation had been induced. To identify the mutant type, the mutant strains were tested on a series of tubes containing minimal medium plus one of a series of vitamins, amino acids, purines, and pyrimidines until some single supplement that permitted growth was found. (The experimental procedures for determining the type of mutant and the specific compound required were discussed in Chapter 13 and summarized in Figure 13.1.) The first mutant strain isolated required vitamin B-6 (pyridoxin), and the second one required vitamin B-1 (thiamine). Using the same procedure, Beadle and Tatum eventually isolated and studied hundreds of mutants.

The next step was to demonstrate that each mutation was of nuclear origin. This was easily ascertained by crossing a given mutant with a wild type and observing Mendelian segregation in the ascospores following meiosis (Figure 16.4). A positive test for demonstrating the nuclear origin occurs when one-half of the meiotic products give rise to wild-type cultures, which can grow on minimal medium, and one-half gives rise to cultures that cannot grow on minimal medium.

The findings derived from testing over 80,000 spores convinced Beadle and Tatum that genetics and biochemistry had much in common. It seemed likely that each nutritional mutation caused the loss of the enzymatic activity that facilitates an essential reaction in wild-type organisms. It also appeared that a mutation could be found for nearly any enzymatically controlled reaction. Beadle and Tatum had thus provided sound experimental evidence that **one gene specifies one enzyme,** a hypothesis alluded to over thirty years earlier by Garrod and Bateson.

Genes and Enzymes: Analysis of Metabolic Pathways

Reporting their work in a 1941 paper entitled "Genetic Control of Biochemical Reactions in *Neurospora*," Beadle and Tatum described three mutant strains. The genetic analysis of one, *pyridoxineless*, was described in detail. This report and subsequent work in Beadle's laboratory stimulated much research and firmly estab-

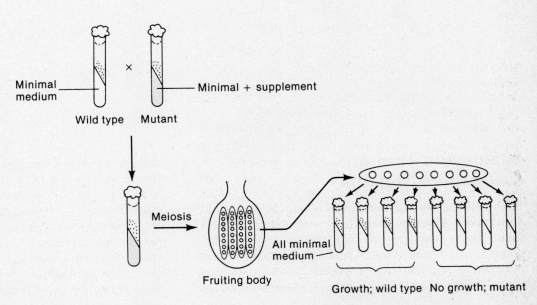

FIGURE 16.4
Method used to confirm that a nutritional mutation in *Neurospora* is of nuclear origin. The mutant is crossed with wild type. Following meiosis and ascus formation, the individual ascospores from one fruiting body are tested on minimal medium. Segregation of a nuclear mutation results in four ascospores that will grow and four that will not grow. (After Beadle, 1946.)

lished the link between genetics and metabolism. Although the initial report did not use the words "one-gene one-enzyme," this phrase gained immediate popularity. With modifications, this concept emerged as a major principle of biology.

The one-gene: one-enzyme concept and its attendant methods have been used over the years to work out many details of metabolism in *Neurospora*, *Escherichia coli*, and a number of other microorganisms. One of the first metabolic pathways to be investigated in detail was that leading to the synthesis of the amino acid arginine in *Neurospora*. By studying seven mutant strains, each requiring arginine for growth, Adrian Srb and Normam Horowitz were able to ascertain a partial biochemical pathway leading to the synthesis of this molecule. The rationale followed in their work illustrates how genetic analysis can be used to establish biochemical information.

Srb and Horowitz tested each strain's ability to re-establish growth if either **citrulline** or **ornithine**, two compounds with close chemical similarity to arginine, was used as a supplement to minimal medium. If either was able to substitute for arginine, they reasoned that it must be involved in the biosynthetic pathway of arginine. In fact, both molecules could be substituted in one or more strains.

Of the seven mutant strains, four of them (*arg 1–4*) grew if supplied with either citrulline, ornithine, or arginine. Two of them (*arg 5* and *6*) grew if supplied with citrulline or arginine. One strain (*arg 7*) would grow only if arginine was supplied. Neither citrulline nor ornithine could substitute for it. From these experimental observations, the following pathway and metabolic blocks for each mutation were deduced:

arg 1–4		*arg 5–6*		*arg 7*	
Precursor	→	Ornithine	→	Citrulline	→ Arginine
Enzyme A		Enzyme B		Enzyme C	

The reasoning supporting these conclusions was based on the following logic. If mutants *arg 1* through *4* can grow regardless of which of the three molecules is supplied as a supplement to minimal medium, the mutations preventing growth must cause a metabolic block that occurs prior to the involvement of ornithine, citrulline, or arginine in the pathway. When any of these three molecules is added, its presence bypasses the block. As a result, it can be concluded that both citrulline and ornithine are involved in the biosynthesis of arginine. However, the sequence of their participation in the pathway cannot be determined on the basis of these data.

On the other hand, the *arg 5* and *6* mutations grow if supplied citrulline but not ornithine. Therefore, ornithine must occur in the pathway prior to the block.

Its presence will not overcome the block. Citrulline, however, does overcome the block, so it must be involved beyond the point of blockage. Therefore, the conversion of ornithine to citrulline represents the correct sequence in the pathway.

Finally, it can be concluded that *arg 7* represents a mutation preventing the conversion of citrulline to arginine. Neither ornithine nor citrulline can overcome the metabolic block because both participate earlier in the pathway.

Taken together, these reasons support the sequence of biosynthesis outlined here. Since Srb and Horowitz's work in 1944, the detailed pathway has been worked out and the enzymes controlling each step characterized. The chemical structures and specific enzymes are shown in Figure 16.5.

The concept of one-gene: one-enzyme developed in the early 1940s was not immediately accepted by all geneticists. This is not surprising, because it was not yet clear how mutant enzymes could cause variation in many phenotypic traits. For example, *Drosophila* mutants demonstrated altered eye size, wing shape, wing vein pattern, and so on. Plants exhibited mutant varieties of seed texture, height, and fruit size. How an inactive, mutant enzyme could result in such phenotypes was puzzling to many geneticists. Another reason for their reluctance to accept this concept was the paucity of information then available in molecular genetics. It was not until 1944 that Avery, McCarty, and MacLeod showed DNA to be the transforming factor and not until the early 1950s that most geneticists believed that DNA serves as the genetic material. However, by this time, the evidence in support of the concept of enzymes as gene products was overwhelming, and the concept was accepted as valid. Nevertheless, the question of how DNA specifies the structure of enzymes remained unanswered.

ONE-GENE: ONE-PROTEIN/ ONE-GENE: ONE-POLYPEPTIDE

Two factors soon modified the one-gene: one-enzyme hypothesis. First, while all enzymes are proteins, all proteins are not enzymes. As the study of biochemical genetics proceeded, it became clear that all proteins are specified by the information stored in genes, leading to the more accurate phraseology **one-gene: one-protein.** Second, proteins were often shown to have a subunit structure consisting of two or more **polypeptide chains.** This is the basis of the **quaternary structure** of proteins, which we will discuss later in this chapter. Because each distinct polypeptide chain is encoded by a separate gene, a more modern state-

FIGURE 16.5
Arginine biosynthesis in
Neurospora.

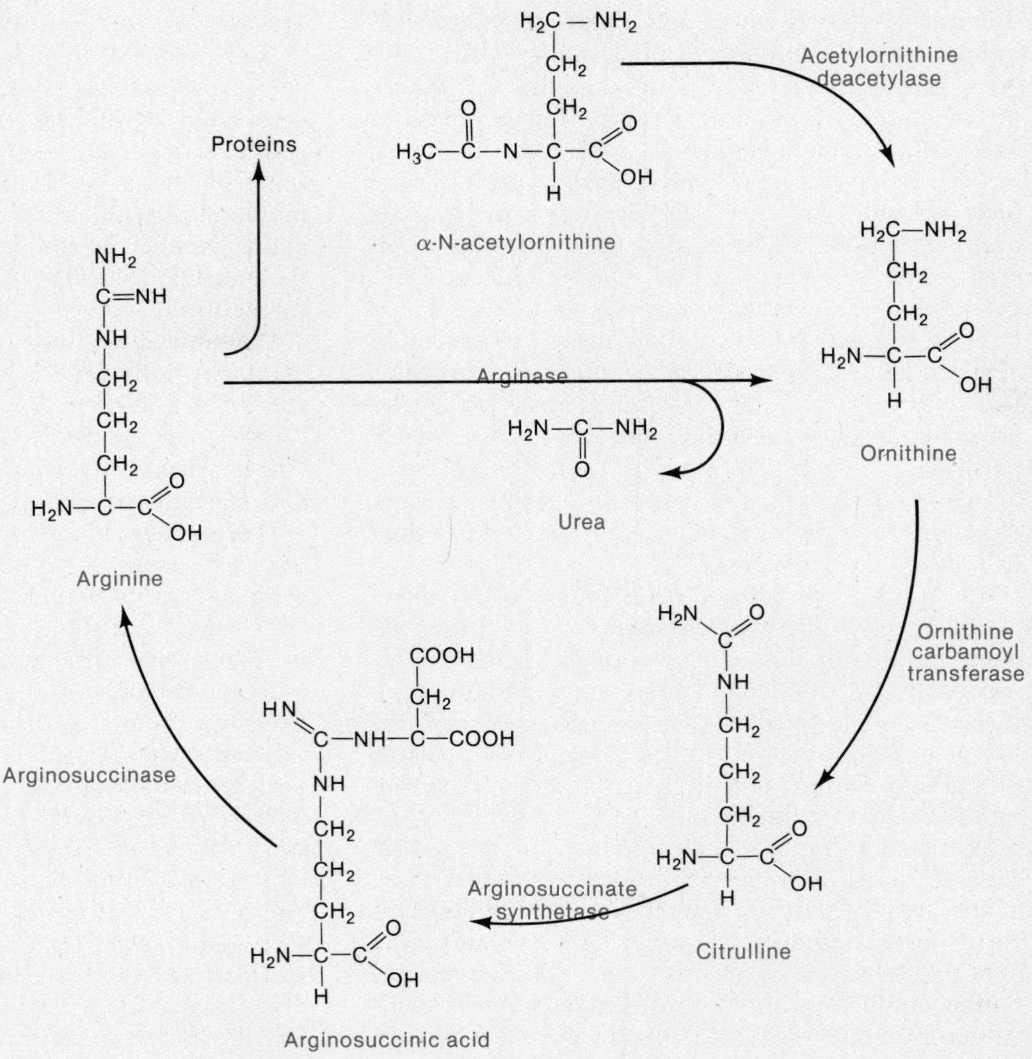

ment of Beadle and Tatum's basic principle is **one-gene: one-polypeptide chain.** These modifications of the original hypothesis became apparent during the analysis of hemoglobin structure in individuals afflicted with sickle-cell anemia.

Sickle-Cell Anemia

The first direct evidence that genes specify proteins other than enzymes came from the work on mutant hemoglobin molecules derived from humans afflicted with the disorder **sickle-cell anemia.** Affected individuals contain erythrocytes which, under low oxygen tension, become elongated and curved because of the polymerization of hemoglobin. The "sickle" shape of these erythrocytes is in contrast to the biconcave disc shape characteristic of normal individuals (Figure

16.6). Individuals with the disease suffer attacks when red blood cells aggregate in the venous side of capillary systems, where oxygen tension is very low. As a result, a variety of tissues may be deprived of oxygen and suffer severe damage. When this occurs, an individual is said to experience a sickle-cell crisis. If untreated, a crisis is often fatal. The kidneys, muscles, joints, brain, gastrointestinal tract, and lungs may be affected.

In addition to suffering crises, these individuals are anemic because their erythrocytes are destroyed more rapidly than normal red blood cells. Compensatory physiological mechanisms include increased red cell production by bone marrow and accentuated heart action. These mechanisms lead to abnormal bone size and shape as well as dilation of the heart.

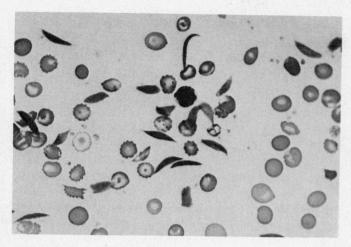

FIGURE 16.6
Red blood cells from an individual with sickle-cell anemia. Both normal unsickled and abnormal sickled cells are apparent. (Courtesy of Carolina Biological Supply Company.)

In 1949, James Neel and E. A. Beet demonstrated that the disease is inherited as a Mendelian trait. Pedigree analysis revealed three genotypes and phenotypes controlled by a single pair of alleles, *A* and *S*. Normal and affected individuals result from the homozygous genotypes *AA* and *SS*, respectively. The red blood cells of the *AS* heterozygote, which exhibits the **sickle-cell trait** but not the disease, are about one-half sickled and one-half normal. Although largely unaffected, such persons are carriers of the disorder.

In the same year, Linus Pauling and his coworkers provided the first insight into the molecular basis of the disease. They showed that hemoglobins isolated from diseased and normal individuals differed in their rates of electrophoretic migration. In this technique (see Appendix A), charged molecules migrate in an electric field. If the net charge of two molecules is different, the rate of migration will vary. On this basis, Pauling and his colleagues concluded that a chemical difference existed between the two types of hemoglobin. The two molecules are now designated **HbA** and **HbS.**

Figure 16.7 illustrates the migration pattern of hemoglobin derived from individuals of all three possible genotypes when subjected to **starch gel electrophoresis.** The gel provides the supporting medium for the molecules during migration. In this experiment, samples are placed at a point of origin between the cathode (−) and the anode (+), and an electric field is applied. The migration pattern reveals that all molecules move toward the anode, indicating a net negative charge. However, HbA migrates farther than HbS, suggesting that its negative charge is greater. The electrophoretic pattern of hemoglobin derived from carriers reveals the presence of both HbA and HbS, confirming their heterozygous genotype.

Pauling's findings suggested two possibilities. It was known that hemoglobin consists of four nonproteinaceous, iron-containing **heme groups** and a **globin portion** that contains four polypeptide chains. The alteration in net charge in HbS could be due, theoretically, to a chemical change in either component.

Work carried out between 1954 and 1957 by Vernon Ingram resolved this question. He demonstrated that the chemical change occurs in the primary structure of the globin portion of the hemoglobin molecule. Using the **fingerprinting technique,** he showed that HbS contains a single amino acid change in the sixth position (from the N-terminus) of the beta (β) chain (Figure 16.7). Human adult hemoglobin contains two identical alpha (α) chains of 141 amino acids and two identical beta chains of 146 amino acids in its quaternary structure.

The fingerprinting technique involves enzymatic digestion of the protein into peptide fragments. The mixture is then placed on absorbent paper and exposed to an electric field, where migration occurs according to net charge. The paper is then turned at a right angle and placed in a solvent, where chromatographic action causes migration of the peptides in the second direction. The end result is a two-dimensional separation of the peptide fragments into a distinctive pattern of spots or "fingerprints." Ingram's work revealed that HbS and HbA differed by only a single fragment (Figure 16.7). Further analysis then revealed a single amino acid change: Valine was substituted for glutamic acid at the sixth position of the β chain.

The significance of this discovery has been multifaceted. It clearly establishes that a single gene provides the genetic information for a single polypeptide chain. Studies of HbS also demonstrate that a mutation can affect the phenotype by directing a single amino acid substitution. Also, by providing the explanation for sickle-cell anemia, the concept of **molecular disease** was firmly established. Finally, this work led to a thorough study of human hemoglobins, which has provided valuable genetic insights.

Sickle-cell anemia is found almost exclusively in the black population. Extensive research is now underway to determine modes of treatment for those with the disease. It affects about one in every 625 black infants born in the United States. Currently, about 50,000 individuals are afflicted. About one of every 145 black married couples are heterozygous carriers, with the

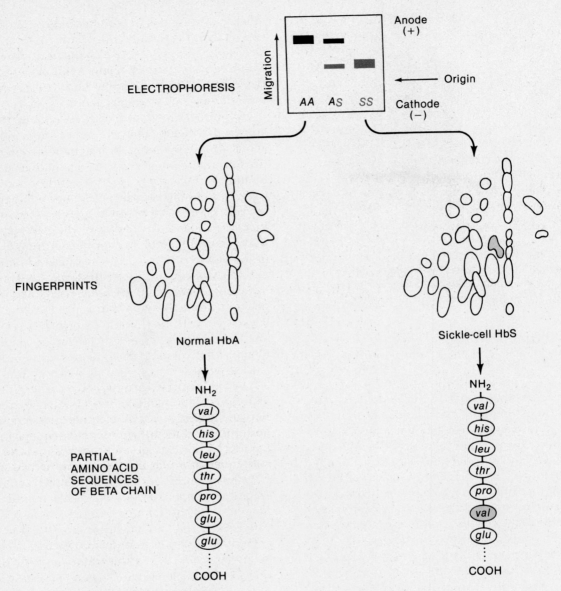

FIGURE 16.7
Investigation of hemoglobin derived from *AA*, *AS*, and *SS* individuals using electrophoresis, fingerprinting, and amino acids analysis. Hemoglobin from individuals with sickle-cell anemia (*SS*) migrates differently in an electrophoretic field, contains an altered peptide in fingerprinting, and contains an altered amino acid, valine, at the sixth position in the β chain.

result that each of their children has a 25 percent chance of having the disease.

Human Hemoglobins

Molecular analysis has revealed that a variety of hemoglobin molecules are produced in humans. All are tetramers consisting of numerous combinations of seven distinct polypeptide chains, each encoded by a separate gene. In our discussion of sickle-cell anemia, we learned that **HbA** contains two **alpha** (α) and two

beta (β) chains. HbA represents about 98 percent of all hemoglobin found in an individual's erythrocytes after the age of six months. The remaining 2 percent consists of **HbA$_2$**, a minor adult component. This molecule contains two alpha and two **delta** (δ) chains. The latter chain is very similar to the beta chain, consisting of 146 amino acids.

During embryonic and fetal development, quite a different set of hemoglobins is found. The earliest set to develop is called **Gower 1** and contains two **zeta** (ζ) chains, which are alphalike, and two **epsilon (ε)** chains, which are betalike. By eight weeks of gestation, this embryonic form is gradually replaced by still another hemoglobin molecule with different chains. This molecule is called **HbF,** or fetal hemoglobin, and consists of two alpha chains and two **gamma (γ)** chains. These gamma chains are of nearly identical types and are designated $^G\gamma$ and $^A\gamma$. Both are betalike and differ from each other by only a single amino acid.

The nomenclature and sequence of appearance of the five tetramers described so far are summarized in Figure 16.8. The genes coding for each of the seven chains have been mapped. Those coding for α and ζ are located on chromosome 16, while those coding for ε, $^G\gamma$ $^A\gamma$, δ. and β are located on chromosome 11. In each case, they are clustered together, constituting **gene families.**

Many mutant varieties of human hemoglobin have been discovered. Most demonstrate a single amino acid change in either the α or β chains. One of the first mutant hemoglobins found was HbC. As in HbS, an amino acid change at the sixth position of the β chain has occurred. However, instead of a valine substitution for glutamic acid, lysine is found in HbC. Fig-

ure 16.9 illustrates some of more than 100 variants now known. Most have been discovered in the heterozygous condition and thus produce no noticeable effect or only a minor anemia. Each of the changes reflects a nucleotide alteration in the respective gene.

Hemoglobin is one of the most carefully and extensively studied proteins, both in humans and in other organisms. Related findings from these studies will be described in subsequent chapters. For example, human hemoglobin genes have been cloned through recombinant DNA technology. Their analysis has added to our knowledge of gene structure (Chapter 19). The temporal sequence of gene activation and repression has been of great interest in studies of genetic regulation of development in eukaryotes (Chapter 22). The homologies of amino acid sequences between Hb polypeptide chains of different organisms have served as one approach in molecular evolutionary investigations (Chapter 26).

COLINEARITY

Once it was established that genes specify the synthesis of polypeptide chains, the next logical question was: How can this occur? How can genetic information contained in the nucleotide sequence of a gene be transferred to the amino acid sequence of a polypeptide chain? It seemed most likely that a **colinear relationship** would exist between the two molecules. That is, the order of nucleotides in the DNA of a gene would correlate directly with the order of amino acids in the corresponding polypeptide.

In his studies of the A subunit of the enzyme **tryptophan synthetase** in *E. coli*, Charles Yanofsky sought to demonstrate colinearity. He isolated many mutants that had lost activity of the enzyme. He was able to map these mutations and establish their location with respect to one another within the gene. Then, he determined where the amino acid substitution had occurred in each mutant protein. When the two sets of data were compared, the colinear relationship was apparent. The location of each mutation in the *trp* A gene correlated with the position of the altered amino acid in the A polypeptide of tryptophan synthetase. This comparison is illustrated in Figure 16.10. We will discuss the details of how information is transferred from DNA to protein (transcription and translation) in Chapter 18.

PROTEIN STRUCTURE AND FUNCTION

Having established that the genetic information is stored in DNA, yet influences cellular activities through the proteins it encodes, we turn now to a dis-

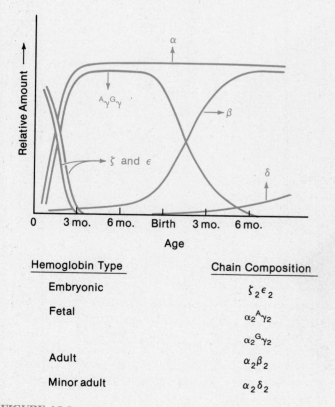

Hemoglobin Type	Chain Composition
Embryonic	$\zeta_2\epsilon_2$
Fetal	$\alpha_2{}^A\gamma_2$
	$\alpha_2{}^G\gamma_2$
Adult	$\alpha_2\beta_2$
Minor adult	$\alpha_2\delta_2$

FIGURE 16.8
Sequence of appearance and chain compositions of human hemoglobins.

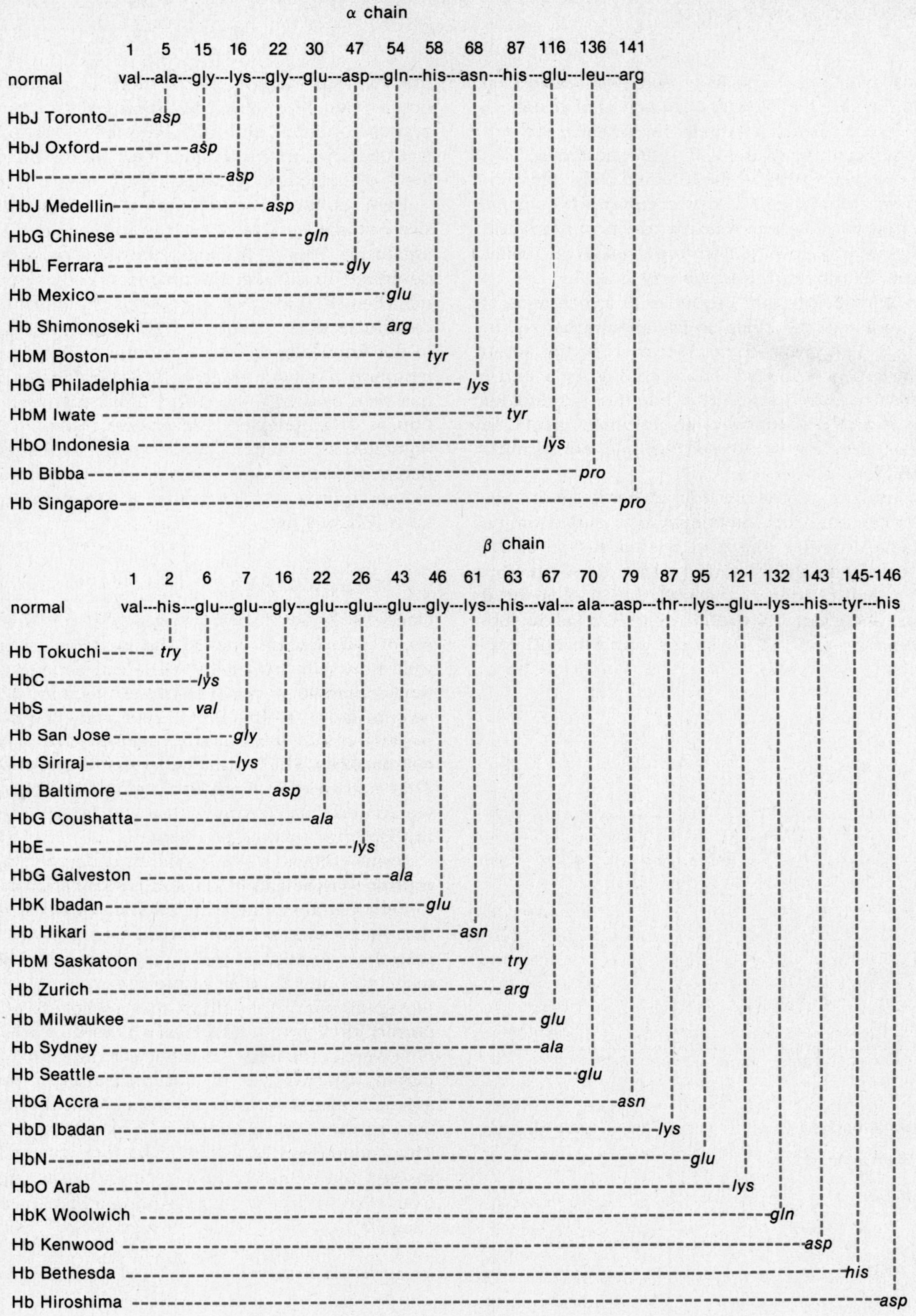

α chain

	1	5	15	16	22	30	47	54	58	68	87	116	136	141
normal	val	ala	gly	lys	gly	glu	asp	gln	his	asn	his	glu	leu	arg
HbJ Toronto		asp												
HbJ Oxford			asp											
HbI				asp										
HbJ Medellin					asp									
HbG Chinese						gln								
HbL Ferrara							gly							
Hb Mexico								glu						
Hb Shimonoseki								arg						
HbM Boston									tyr					
HbG Philadelphia										lys				
HbM Iwate											tyr			
HbO Indonesia												lys		
Hb Bibba													pro	
Hb Singapore														pro

β chain

	1	2	6	7	16	22	26	43	46	61	63	67	70	79	87	95	121	132	143	145-146
normal	val	his	glu	glu	gly	glu	glu	glu	gly	lys	his	val	ala	asp	thr	lys	glu	lys	his	tyr---his
Hb Tokuchi		try																		
HbC			lys																	
HbS			val																	
Hb San Jose				gly																
Hb Siriraj				lys																
Hb Baltimore					asp															
HbG Coushatta						ala														
HbE							lys													
HbG Galveston								ala												
HbK Ibadan									glu											
Hb Hikari										asn										
HbM Saskatoon											try									
Hb Zurich											arg									
Hb Milwaukee												glu								
Hb Sydney												ala								
Hb Seattle													glu							
HbG Accra														asn						
HbD Ibadan															lys					
HbN																glu				
HbO Arab																	lys			
HbK Woolwich																		gln		
Hb Kenwood																			asp	
Hb Bethesda																				his
Hb Hiroshima																				asp

FIGURE 16.9

Hemoglobin mutations in humans. The normal amino acid at each affected position is shown across the top for both the α and β chains. The number indicates the position of the amino acid in the chain from the N-terminus. The amino acid substitution in each mutation is determined by following down each vertical dashed line.

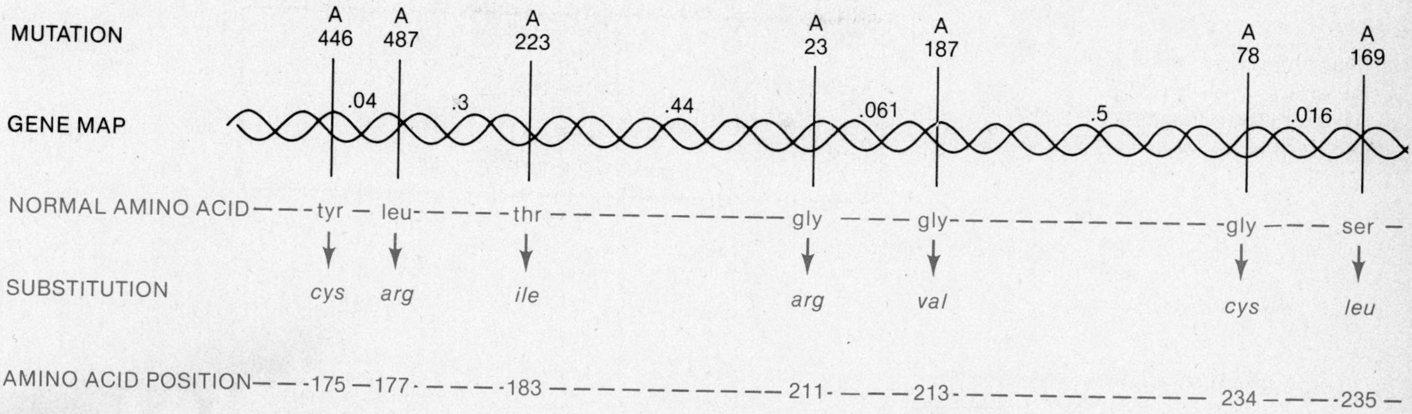

MUTATION	A 446	A 487		A 223				A 23		A 187				A 78		A 169

GENE MAP .04 .3 .44 .061 .5 .016

NORMAL AMINO ACID — — tyr — leu — — — — thr — — — — — — — — — gly — — — gly — — — — — — — — — — — gly — — — ser —

SUBSTITUTION cys arg ile arg val cys leu

AMINO ACID POSITION — — -175 —177· — — —183 — — — — — — — — — 211· — — —213· — — — — — — — — — 234 — — —235 —

FIGURE 16.10

Demonstration of colinearity between the genetic map of various *trp* A mutations in *E. coli* and the affected amino acids in the protein product. The values shown between mutations represent linkage distances.

cussion of protein structure and function. How is it that these molecules play such a critical role in determining the complexity of cellular activities? As we will see, the structure of proteins is intimately related to the functional diversity of these molecules.

Protein Structure

First, we should differentiate between the terms **polypeptides** and **proteins.** Both describe a molecule composed of amino acids. The molecules differ, however, in their state of assembly and functional capacity.

Polypeptides are the precursors of proteins. As assembled on the ribosome during translation, the molecule is called a polypeptide. When released from the ribosome following translation, a polypeptide folds up and assumes a higher order of structure. When this occurs, a three-dimensional conformation in space is produced. In many cases, several polypeptides interact to produce this conformation. Whether or not several polypeptides interact, the three-dimensional conformation is essential to the function of the molecule. When the functional state is achieved, the molecule is appropriately called a protein.

The polypeptide chains of proteins, like nucleic acids, are linear nonbranched polymers. There are approximately 20 amino acids found in biological material, and they serve as the building blocks or subunits of proteins. Each amino acid has a **carboxyl group,** an **amino group,** and an **R (radical) group** (or side chain) bound covalently to a central carbon atom (Figure 16.11). The R group gives each amino acid its chemical identity. Figure 16.11 illustrates the 20 different R groups, which show a variety of configurations

and may be divided into four main classes: (1) **nonpolar** or **hydrophobic;** (2) **polar** or **hydrophilic;** (3) **negatively charged;** and (4) **positively charged.** Because polypeptides are long polymers, and because each position may be occupied by any one of 20 amino acids with unique chemical properties, an enormous variation in chemical activity is possible. For example, if an average polypeptide is composed of 200 amino acids (molecular weight of about 20,000 daltons), 20^{200} different molecules, each with a unique sequence, can be created using 20 different building blocks.

Around 1900, the German chemist Emil Fischer determined the manner in which the amino acids are bonded together. He showed that the amino group of one amino acid can react with the carboxyl group of another amino acid in a dehydration reaction, releasing a molecule of H_2O. The resulting covalent bond is known as a **peptide bond** (Figure 16.12). Two amino acids linked together constitute a **dipeptide,** three a **tripeptide,** and so on. When more than ten amino acids are linked by peptide bonds, the chain is referred to as a **polypeptide.** Generally, no matter how long a polypeptide is, it will contain a free amino group at one end (the **N-terminus**) and a free carboxyl group at the other end (the **C-terminus**).

Four levels of protein structure are recognized: **primary (I°); secondary (II°); tertiary (III°);** and **quaternary (IV°).** The sequence of amino acids in the linear backbone of the polypeptide constitutes its primary structure. This sequence is specified by the sequence of deoxyribonucleotides in DNA. The primary structure of a polypeptide determines the specific characteristics of the higher orders of structure as a protein is formed.

FIGURE 16.11

Structural formulae of amino
acids. Amino acids are
divided into four categories
of chemical behavior.

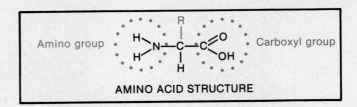

AMINO ACID STRUCTURE

1. Nonpolar-hydrophobic

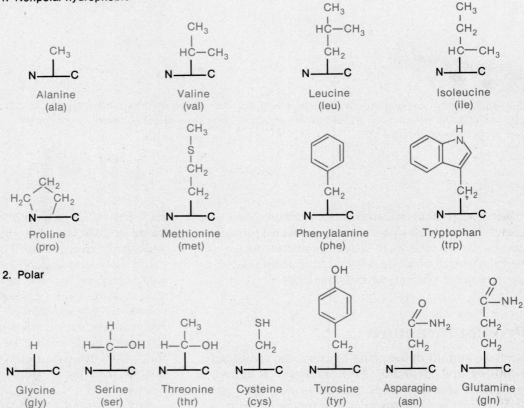

Alanine
(ala)

Valine
(val)

Leucine
(leu)

Isoleucine
(ile)

Proline
(pro)

Methionine
(met)

Phenylalanine
(phe)

Tryptophan
(trp)

2. Polar

Glycine
(gly)

Serine
(ser)

Threonine
(thr)

Cysteine
(cys)

Tyrosine
(tyr)

Asparagine
(asn)

Glutamine
(gln)

3. Negatively charged
   (acidic)

Aspartic acid
(asp)

Glutamic acid
(glu)

4. Positively charged
   (basic)

Lysine
(lys)

Arginine
(arg)

Histidine
(his)

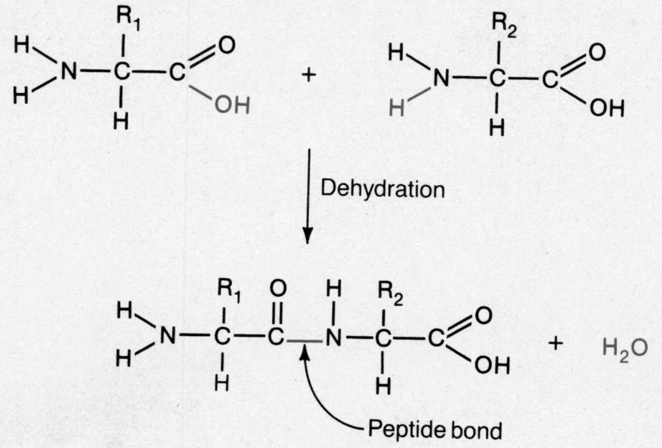

FIGURE 16.12
Peptide bond formation between two amino acids.

The secondary structure refers to a regular or repeating configuration in space assumed by amino acids closely aligned to one another in the polypeptide chain. In 1951, Linus Pauling and Robert Corey predicted, on theoretical grounds, an **α helix** as one type of secondary structure. The α-helix model (Figure 16.13) has since been confirmed by X-ray crystallographic studies. It is rodlike and has the greatest possible theoretical stability. The helix is composed of a spiral chain of amino acids stabilized by hydrogen bonds.

The side chains of amino acids extend outward from the helix, and each amino acid residue occupies a vertical distance of 1.5 Å in the helix. There are 3.6 residues per turn. While left-handed helices are theoretically possible, all proteins demonstrating an α helix are right-handed.

Also in 1951, Pauling and Corey proposed a second structure, the β-pleated-sheet configuration, where individual chains are in a more extended, zigzag conformation [Figure 16, 13(b)]. In this model, polypeptide chains run parallel or antiparallel to one another and are stabilized by hydrogen bonding between chains. In the β-pleated-sheet configuration, amino acid residues in a single chain are 3.5 Å apart. As a general rule, most proteins demonstrate a mixture of α and β secondary structure, although the α helix is the predominant form.

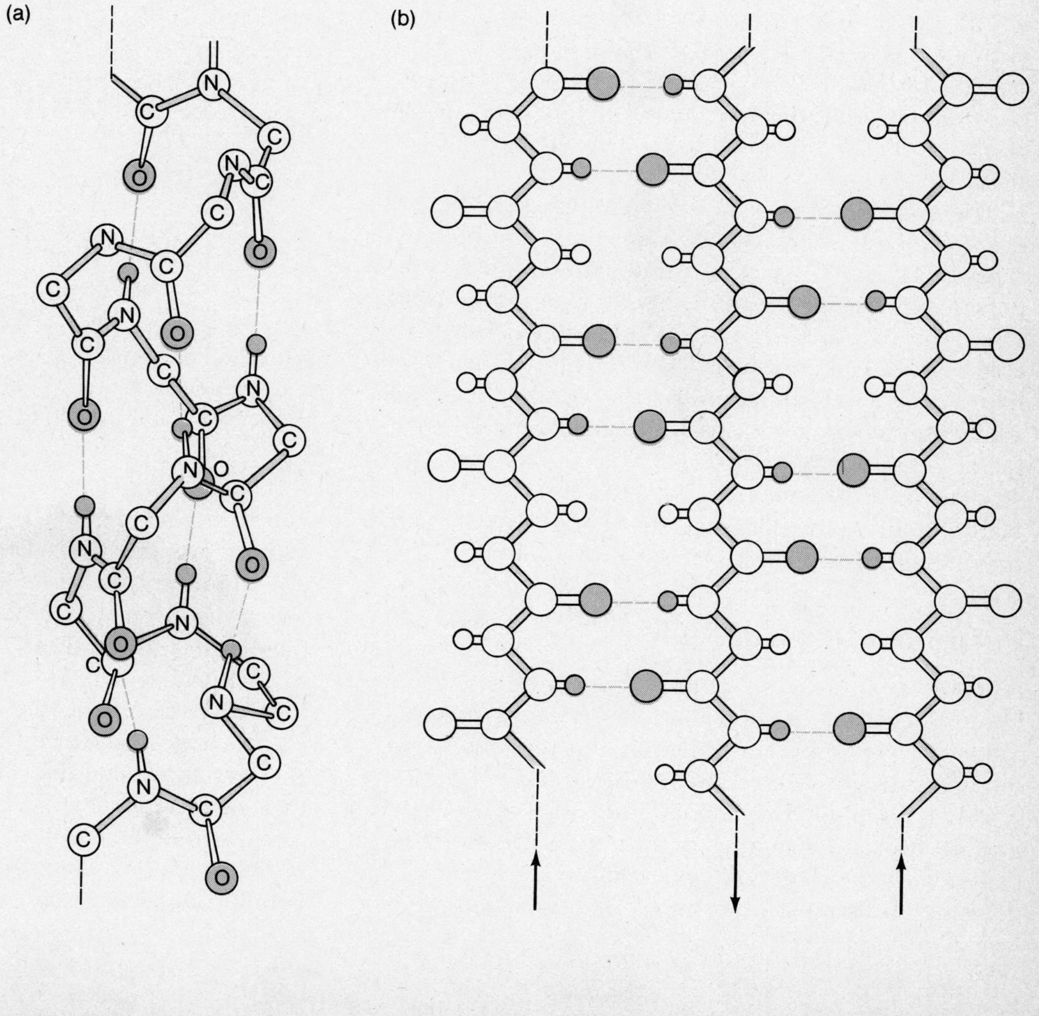

FIGURE 16.13
(a) The right-handed α helix, which represents the secondary structure of a protein chain. The atoms involved in hydrogen bonding are shown in color. The bonds essential to the linear continuity are shaded. In this simplified diagram, the R groups and hydrogen atoms not involved in hydrogen bonding are not shown. (b) The β-pleated-sheet configuration proposed for the secondary structure of polypeptide strands running antiparallel to one another. As in part (a), some atoms are not shown for the sake of clarity. (Adapted from Linus Pauling: *The Nature of the Chemical Bond,* 3rd ed. Copyright © 1960 by Cornell University. Used by permission of the publisher, Cornell University Press.)

Tertiary protein structure defines the three-dimensional conformation of the molecule in space. Beyond the secondary structure, polypeptides fold back on themselves to form a compact structure. Proteins that demonstrate III° structure are called **globular.** Three aspects of III° structure are most important in stabilization of the molecule:

1 Covalent disulfide bonds form between adjacent cysteine residues to form the amino acid cystine.

2 Nearly all of the polar, hydrophilic R groups are located on the surface, where they interact with water.

3 Nearly all of the nonpolar, hydrophobic R groups are located on the inside of the molecule, where they interact with one another.

It is important to emphasize that the three-dimensional conformation achieved by any protein is the direct result of the primary (I°) structure of the polypeptide. It should be obvious that the three stabilizing factors listed above depend on the location of each amino acid relative to all others in the chain. As folding occurs, the most thermodynamically stable conformation results.

The three-dimensional structures of two globular proteins, **myoglobin** and **ribonuclease,** are diagramed in Figure 16.14. This III° level of organization is extremely important because the specific function of any protein is the direct result of its three-dimensional conformation.

The quaternary level of organization is characteristic of proteins composed of more than one polypeptide chain. The IV° structure indicates the conformation of the various chains in relation to one another. This type of protein is called **oligomeric,** and each chain is called a **protomer.** The individual protomers have native conformations that fit together in a specific complementary fashion. Hemoglobin, an oligomeric protein of four polypeptide chains, has been studied in great detail. Its IV° structure is shown in Figure 16.15. Many enzymes, including DNA and RNA polymerase, demonstrate IV° structure.

Protein Function

Proteins are the most abundant organic molecules found in cells. As the end products of genes, they play many diverse roles. For example, the respiratory pigments **hemoglobin** and **myoglobin** transport oxygen, which is essential for cellular metabolism. **Collagen** and **keratin** are examples of structural proteins associated with the skin, connective tissue, and hair of organisms. **Actin** and **myosin** are contractile proteins,

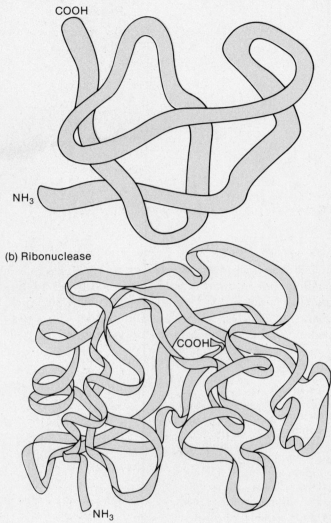

FIGURE 16.14
Schematic drawings of the tertiary structure of myoglobin and ribonuclease.

found in abundance in muscle tissue. Still other examples are the **immunoglobins,** which function in the immune system of vertebrates; **transport proteins,** involved in movement of molecules across membranes; **hormones,** which regulate various types of chemical activity; and **histones,** which bind to DNA in eukaryotic organisms.

The largest group of proteins with a related function are the **enzymes.** These molecules specialize in catalyzing biological reactions. Enzymes increase the rate at which a chemical reaction reaches equilibrium, but they do not alter the end point of the chemical equilibrium. Their remarkable, highly specific catalytic

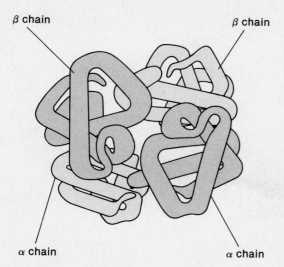

β chain   β chain

α chain   α chain

FIGURE 16.15
Schematic drawing of the quaternary structure of
hemoglobin. Four chains (2 alpha and 2 beta) interact with
four heme groups (not shown) to form the functional
molecule. (From Dickerson, 1964, Fig. 21.)

The catalytic properties and specificity of an enzyme are determined by the chemical configuration of
the molecule's **active site.** This site is associated with
a crevice, a cleft, or a pit on the surface of the enzyme
which binds the reactants, or substrates, enhancing
their interaction. Enzymatically catalyzed reactions
control metabolic activities in the cell. Each reaction is
either **catabolic** or **anabolic.** Catabolism is the degradation of large molecules into smaller, simpler ones
with the release of chemical energy. Anabolism is the
synthetic phase of metabolism, yielding nucleic acids,
proteins, lipids, and carbohydrates. Metabolic pathways that serve the dual function of anabolism and
catabolism are **amphibolic.**

To complete the discussion of the central dogma of
molecular biology, we must describe how the information specifying proteins in DNA is encoded and
how the information is expressed. These topics serve
as the subjects of Chapters 17 and 18.

Protein Structure and Function: The Collagen Fiber

In order to provide a more in-depth example of the
relationship between protein structure and function,
we conclude this chapter with a discussion of one of
the most interesting and abundant proteins found in
vertebrates: **collagen.** It has been extensively studied
and clearly illustrates the relationship between protein structure and function. Collagen is found in many
places in the body including tendons, ligaments, bone,
connective tissue, skin, blood vessels, teeth, and the
lens and cornea of the eye. In each case, the presence
of collagen increases the tensile strength of and provides support to the tissue. An inspection of the above
list makes it evident that collagen must be a versatile

properties largely determine the biomolecular nature
of any cell type. The specific functions of many enzymes involved in the genetic and cellular processes
of cells have been described throughout the text.

Catalysis is a process whereby the **energy of activation** for a given reaction is lowered (Figure 16.16).
The energy of activation is the increased kinetic energy state that molecules must usually reach before
they react with one another. It can be attained as a
result of elevated temperatures. Enzymes allow biological reactions to occur at lower physiological temperatures. In this way, enzymes make possible life as we
know it on this planet.

FIGURE 16.16
An uncatalyzed versus an
enzymatically catalyzed
chemical reaction. The
energy of activation
necessary to initiate the
reaction is substantially
lower as a result of catalysis.

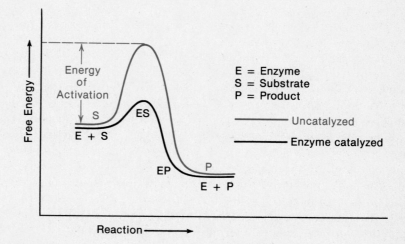

Energy
of
Activation

E = Enzyme
S = Substrate
P = Product

——— Uncatalyzed
——— Enzyme catalyzed

protein, providing a variable degree of flexibility and support depending on which tissue it comprises.

While there are several types of collagen produced in vertebrates, we will restrict our discussion to just the major one, called type I. Its main subunit or building block is called **tropocollagen** [Figure 16.17(a)]. This consists of three polypeptide chains wrapped around one another in a triple helix. This unit is extremely large, being 15 Å in diameter and 3000 Å long. Each polypeptide contains about 1000 amino acids, and the three interacting chains of the helix are stabilized by hydrogen bonds between them. Of the three polypeptides, two are identical and labeled **α-1(I).** The other chain is distinct and labeled **α-2.**

The amino acid composition of the chains is rather unusual. Almost one-third of the amino acid residues are glycine and a high percentage of them are proline. Additionally, two modified amino acids, **hydroxyproline** and **hydroxylysine** are abundant. The tripeptide sequence glycine-proline-hydroxyproline occurs frequently.

The maturation of the α-1(I) and α-2 polypeptides into tropocollagen and the subsequent condensation of these helical structures into the more densely coiled collagen fibers is a fascinating tale (Figure 16.17). The α-1(I) and α-2 polypeptides are first synthesized by fibroblasts as even longer units called **pro-α-1(I)** and **pro-α-2 collagen.** These are secreted into extracellular spaces, where they are cleaved and shortened at both the N-terminus and C-terminus by specific enzymes called **procollagen peptidases.** A genetic defect in this conversion leads to generalized human disorders of connective tissue such as the **Ehlers-Danlos syndrome.** Individuals exhibiting this disorder have fragile, stretchable skin and are loose-jointed, providing hypermobility. Some forms of the syndrome render individuals susceptible to arterial and colon ruptures as well as periodontal problems. They exhibit elevated levels of procollagen and decreased activity of the peptidase enzyme.

When the procollagen polypeptides are first synthesized, no hydroxyproline or hydroxylysine residues are part of them. Rather, the hydroxyl groups are added to the amino acids proline and lysine after they have become part of the polypeptide chains. This is an example of the phenomenon known as **post-translational modification** [Figure 16.17(a)]. This occurs as a result of two enzymes, **prolyl hydroxylase** and **lysl hydroxylase.** Each enzyme acts on its specific substrate amino acid only if that amino acid is on the amino side of an adjacent glycine residue. The enzymatic reactions occur prior to the formation of the triple helix. There are many fewer hydroxylysine residues than hydroxyproline residues formed. Some of the former amino acids are further modified by the addition of a disaccharide containing one molecule each of glucose and galactose. The number of carbohydrate units per tropocollagen molecule varies in different tissues. For example, the lens of the eye contains many more such units than tendons or ligaments.

The hydroxylation of proline is very important to normal collagen function. The enzyme that hydroxylates this amino acid contains an iron atom, which must be in the reduced ferrous state, near its active site. It is interesting to note that individuals with the vitamin-deficiency disease **scurvy,** where inadequate dietary amounts of ascorbic acid (vitamin C) are ingested, show symptoms related to defective collagen production. These include loss of strength, skin lesions, fragile blood vessels, and degeneration of gum tissue. It is now clear that in individuals with normal diets, ascorbic acid serves as a reducing agent for the prolyl hydroxylase enzyme, maintaining it in its active state. Collagen synthesized in the absence of ascorbic acid is under-hydroxylated and less stable, causing the various maladies associated with scurvy.

Assuming that *normal* tropocollagen has been formed, these triple-helical rods spontaneously associate to form dense collagen fibers. The association follows an orderly pattern where rows of end-to-end molecules line up in a staggered fashion next to one another [Figure 16.17(b)]. In each row a gap of approximately 400 Å exists between each tropocollagen unit. This complex structure is stabilized by covalent bonds formed within and between the tropocollagen units. First, the free terminal amino groups (called the ϵ-amino termini) of two lysine residues within the α helix of a tropocollagen molecule may react to form a covalent linkage. This structure, called an **aldol crosslink** can react with a histidine residue from another tropocollagen chain, forming another covalent bond. This modified structure can further interact with other amino acids forming still other cross-links. Many such linkages can form, creating great stability in the collagen fibers.

Taken together, this complex structural arrangement creates protein fibers that strengthen and support a variety of tissues. While it is not yet completely clear how certain tissues are strengthened or rendered flexible to a greater or lesser extent by collagen, the degree of cross-linkage between tropocollagen units is one determining factor. The frequency of carbohydrate side chains may be another. The fibers are usually embedded in an extracellular matrix specific to a given tissue. This undoubtedly adds to each unique environment in which collagen finds itself. Finally, while we have concentrated our discussion on

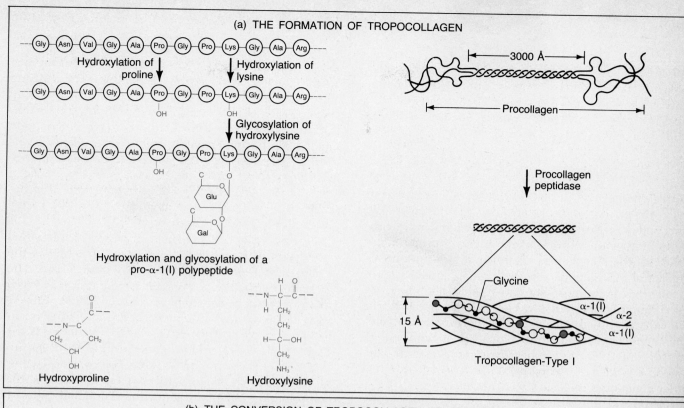

(a) THE FORMATION OF TROPOCOLLAGEN

Hydroxylation and glycosylation of a
pro-α-1(I) polypeptide

Hydroxyproline

Hydroxylysine

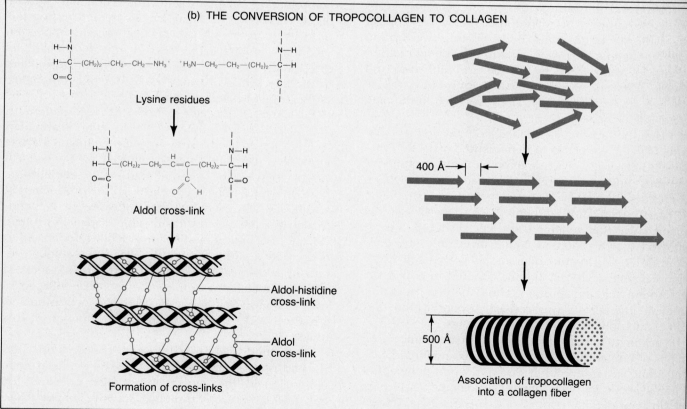

(b) THE CONVERSION OF TROPOCOLLAGEN TO COLLAGEN

Lysine residues

Aldol cross-link

Formation of cross-links

Association of tropocollagen
into a collagen fiber

FIGURE 16.17

Various steps and intermediate structures involved in the formation of collagen. In part (*a*),
the process of post-translational modification of polypeptides is illustrated during the
formation of tropocollagen, the precursor of collagen. In part (*b*), the formation of cross-links
leading to the assembly of the collagen fiber is illustrated.

type I collagen, there are four other minor types of collagen (II–V), which contain polypeptide chains other than α-1(I) and α-2. These chains, encoded by separate genes, probably impart slightly different properties to these forms of collagen.

SUMMARY AND CONCLUSIONS

In this chapter, we discussed the evidence leading to the concept that genes control the synthesis of specific polypeptide chains. The first insights were provided by Garrod's studies of the inborn human metabolic disorders cystinuria, albinism, and alkaptonuria. The latter two genetic diseases and phenylketonuria result from mutations that cause metabolic blocks in the phenylalanine-tyrosine biosynthetic pathway. In individuals with these disorders, the function of specific enzymes is impaired, biochemical conversions are prevented, and metabolic byproducts accumulate. Garrod's work was recognized by Bateson, but few other geneticists realized its significance.

Garrod did not have access to the necessary technology to support his hypothesis that genes control the synthesis of enzymes. However, the investigation of eye pigments in *Drosophila* and nutritional requirements in *Neurospora* by Beadle and Ephrussi and Beadle and Tatum made it clear that mutations cause the loss of enzyme activity. Their work led to the one-gene: one-enzyme concept.

Using the techniques established by Beadle and Tatum, workers were able to demonstrate many complex biosynthetic pathways in microorganisms. By studying mutants and determining the reaction affected by each, they have ordered the steps in many pathways.

Later investigations of hemoglobin derived from humans with the genetic disorder sickle-cell anemia showed that a mutation could affect only one polypeptide chain. Ingram demonstrated that the mutation causing sickle-cell anemia resulted in only a single amino acid substitution. Thus, the one-gene: one-enzyme hypothesis was revised, becoming the one-gene: one-polypeptide chain hypothesis.

Thorough investigations have revealed four major types of human hemoglobin molecules: embryonic, fetal, adult, and minor adult. Seven different polypeptide chains occur in these different molecules, and each chain is under the control of a specific gene. Careful scrutiny of human adult hemoglobin from different segments of the population has demonstrated many other mutations causing amino acid substitutions. Most result in less severe effects than that causing sickle-cell anemia.

Because available evidence supported the hypothesis that genes control the synthesis of the polypeptide chains composing enzymes and other proteins, it was further proposed that the gene's nucleotide sequence specifies in a colinear way the amino acid sequence of the polypeptide chain. Evidence confirming this hypothesis in *E. coli* was obtained by Yanofsky in his studies of tryptophan synthetase.

Proteins, the end products of genes, demonstrate four levels of structure. The primary structure is the amino acid sequence of the polypeptide chain. The secondary structure is a regular configuration in space that imparts stabilization to the molecule. Closely aligned amino acids interact to produce either an α-helix or a β-pleated sheet. The tertiary structure is the three-dimensional conformation produced by the interaction of parts of one or more polypeptide chains. As the chain or chains fold back on themselves, the conformation attained defines the functional capacity of the molecule. Those molecules demonstrating tertiary structure are called globular proteins. Both the secondary and tertiary structure are determined by the primary structure. The quaternary structure refers to the interaction of two or more polypeptide chains, as illustrated by hemoglobin.

Even though proteins perform varied functions in cells, the most influential, related function is assumed by enzymes. These highly specific cellular catalysts lower the activation energy required in biochemical reactions, speeding their attainment of equilibrium. Thus, enzymes play a central role in the production of all classes of molecules in living systems. We concluded this chapter by discussing a specific protein, collagen. Structural modifications following translation of the polypeptides making it up lead to a unique precursor, tropocollagen, which spontaneously associates with like molecules to form the cablelike collagen fiber. Our discussion emphasized the acquisition of structure and its relationship to the function of this molecule.

By the mid 1950s, two important concepts of genetics had been established: DNA serves as the genetic material, and the genetic information specifies the chemical structure of proteins. How the genetic material stores information and how it is expressed are the topics of Chapters 17 and 18, respectively.

PROBLEMS AND DISCUSSION QUESTIONS

1 Discuss the potential difficulties involved in designing a diet to alleviate the symptoms of phenylketonuria.

2 Phenylalanine is present in excessive amounts in phenylketonurics. Why don't these individuals exhibit a deficiency of tyrosine?

3 Phenylketonurics are often more lightly pigmented than normal individuals. Can you suggest a reason why this is so?

4 Consider the following hypothetical circumstances involving reciprocal transplantations of three mutant imaginal eye disks. The three mutations, *a*, *b*, and *c* are responsible for the enzymatic pathway shown below. However, it is not known which mutant gene controls which step in the reactions.

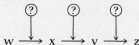

The results of the transplantations are shown below. Assuming that the *w*, *x*, *y*, and *z* substances are diffusible, determine which reactions are controlled by which genes.

Donor	Host	Transplant Eye Pigment	Development
a	*c*	Mutant	Autonomous
a	*b*	Mutant	Autonomous
c	*b*	Mutant	Autonomous
c	*a*	Wild type	Nonautonomous
b	*c*	Wild type	Nonautonomous
b	*a*	Wild type	Nonautonomous

5 What F_1 and F_2 ratios will occur in a cross between *vermilion* females and *cinnabar* males? Recall that *v* is sex-linked and *cn* is autosomal, and that these mutants produce identical bright red phenotypes. What ratios will occur in the reciprocal cross (*v* males and *cn* females)?

6 The synthesis of flower pigments is known to be dependent upon enzymatically controlled biosynthetic pathways. In the crosses shown below, postulate the role of mutant genes and their products in producing the observed phenotypes.
  (a) P_1: white strain A × white strain B
  F_1: all purple
  F_2: 9/16 purple: 7/16 white
  (b) P_1: white × pink
  F_1: all purple
  F_2: 9/16 purple: 3/16 pink: 4/16 white

7 A series of mutations in the bacterium *Salmonella typhimurium* result in the requirement of either tryptophan or some related molecule in order for growth to occur. From the data shown below, suggest a biosynthetic pathway for tryptophan.

	Growth Response Supplement				
Mutation	Minimal Medium	Anthranilic Acid	Indole Glycerol Phosphate	Indole	Tryptophan
trp-8	—	+	+	+	+
trp-2	—	−	+	+	+
trp-3	—	−	−	+	+
trp-1	—	−	−	−	+

8 The study of biochemical mutants in organisms such as *Neurospora* has demonstrated that some pathways are branched. The data shown below illustrate the branched nature of the pathway resulting in the synthesis of thiamine. Why don't the data support a linear pathway? Can you postulate a pathway for the synthesis of thiamine in *Neurospora?*

	Growth Response Supplement			
Mutation	Minimal Medium	Pyrimidine	Thiazole	Thiamine
thi-1	—	—	+	+
thi-2	—	+	−	+
thi-3	—	—	−	+

9 Explain why the one-gene: one-enzyme concept is not considered accurate today.

10 Why is an alteration of electrophoretic mobility interpreted as a change in the primary structure of the protein under study?

11 Contrast the polypeptide chain components of each of the hemoglobin molecules found in humans.

12 Using sickle-cell anemia as a basis, describe what is meant by a molecular or genetic disease. What are the similarities and dissimilarities between this type of a disorder and a disease caused by an invading microorganism?

13 Contrast the contributions of Pauling and Ingram to our understanding of the genetic basis for sickle-cell anemia.

14 Hemoglobins from two individuals are compared by electrophoresis and by fingerprinting. Electrophoresis reveals no difference in migration, but fingerprinting shows an amino acid difference. How is this possible?

15 When the amino acid sequences of the polypeptide Hb chains are compared, many homologous sequences are detected, but the degree of homology varies between any two chains. For example, a greater degree of homology exists between the β and γ chains than between the α and β chains. Relate this information to Ohno's theory of gene duplication (discussed in Chapter 12).

16 Describe what *colinearity* means.

17 Certain mutations called *amber* in bacteria and viruses result in premature termination of polypeptide chains during translation. Many *amber* mutations found at different points along the gene coding for a head protein in phage T4 have been detected. How might this system be further investigated to demonstrate and support the concept of colinearity?

18 Define and compare the four levels of protein organization.

19 List as many different protein functions as you can, with an example of each.

20 How does an enzyme function? Why are enzymes essential for living organisms on earth?

SELECTED READINGS

ALBERTS, B., et al. 1983. *Molecular biology of the cell.* New York: Garland.

BARTHOLOME, K. 1979. Genetics and biochemistry of phenylketonuria—Present state. *Hum. Genet.* 51: 241–45.

BATESON, W. 1909. *Mendel's principles of heredity.* Cambridge, England: Cambridge University Press.

BEADLE, G. W. 1945. Genetics and metabolism in *Neurospora. Physiol. Rev.* 25: 643.

———. 1946. Genes and the chemistry of the organism. *Amer. Scient.* 34: 31–53.

BEADLE, G. W., and EPHRUSSI, B. 1937. Development of eye colors in *Drosophila*: Diffusible substances and their interrelations. *Genetics* 22: 76–86.

BEADLE, G. W., and TATUM, E. L. 1941. Genetic control of biochemical reactions in *Neurospora*. *Proc. Natl. Acad. Sci.* 27: 499–506.

BEET, E. A. 1949. The genetics of the sickle-cell trait in a Bantu tribe. *Ann. Eugenics* 14: 279–84.

BEIGHTON, P. H., et al. 1969. Variants of the Ehlers-Danlos syndrome. Clinical, biochemical, haematological, and chromosome features of 100 patients. *Ann. Rheum. Dis.* 28: 228–45.

BORNSTEIN, R., and SAGE, H. 1980. Structurally distinct collagen types. *Ann. Rev. Biochem.* 49: 957–1003.

BOYER, P. D., ed. 1974. *The enzymes.* 3rd ed. Vol. 10. New York: Academic Press.

BRENNER, S. 1955. Tryptophan biosynthesis in *Salmonella typhimurium*. *Proc. Natl. Acad. Sci.* 41: 862–63.

BURLEY, S. K., and PETSKO, G. A. 1985. Aromatic–aromatic interaction: A mechanism of protein structure stabilization. *Science* 229: 23–28.

DICKERSON, R. E. 1964. X-ray analysis and protein structure. In *The proteins*, 2nd ed., ed. H. Neurath, vol. 2. New York: Academic Press.

DICKERSON, R. E., and GEIS, I. 1983. *Hemoglobin: Structure, function, evolution, and pathology.* Menlo Park, Calif.: Benjamin/Cummings.

DOOLITTLE, R. F. 1985. Proteins. *Scient. Amer.* (Oct.) 253: 88–99.

EPHRUSSI, B. 1942. Chemistry of eye color hormones of *Drosophila*. *Quart. Rev. Biol.* 17: 327–38.

EYRE, D. R. 1980. Collagen: Molecular diversity in the body's protein scaffold. *Science* 207: 1315–22.

FESSLER, J. H., and FESSLER, L. I. 1978. Biosynthesis of collagen. *Ann. Rev. Biochem.* 47: 129–62.

GARROD, A. E. 1902. The incidence of alkaptonuria: A study in chemical individuality. *Lancet* 2: 1616–20.

————. 1909. *Inborn errors of metabolism.* London: Oxford University Press. (Reprinted 1963, Oxford University Press, London.)

HARRIS, H. 1975. *Principles of human biochemical genetics.* 2nd ed. New York: Elsevier.

HOLLISTER, D. W., BYERS, P. H., and HOLBROOK, K. A. 1982. Genetic disorders of collagen metabolism. *Adv. Human Genet.* 12: 1–88.

INGRAM, V. M. 1957. Gene mutations in human hemoglobin: The chemical difference between normal and sickle cell hemoglobin. *Nature* 180: 326–28.

————. 1963. *The hemoglobins in genetics and evolution.* New York: Columbia University Press.

KOSHLAND, D. E. 1973. Protein shape and control. *Scient. Amer.* (Oct.) 229: 52–64.

LaDU, B. N., ZANNONI, V. G., LASTER, L., and SEEGMILLER, J. E. 1958. The nature of the defect in tyrosine metabolism in alkaptonuria. *J. Biol. Chem.* 230: 251.

LEHMANN, H., and HUNTSMAN, R. G. 1966. *Man's hemoglobins.* Amsterdam: North-Holland.

LESK, A. M., and HARDMAN, K. D. 1982. Computer-generated schematic diagrams of protein structure. *Science* 216: 539–40.

LIGHT, A. 1974. *Proteins: Structure and function.* Englewood Cliffs, N. J.: Prentice-Hall.

MANIATIS, T., et al. 1980. The molecular genetics of human hemoglobins. *Ann. Rev. Genet.* 14: 145–78.

MECHANIC, G. 1972. Cross-linking of collagen in a heritable disorder of connective tissue: Ehlers-Danlos syndrome. *Biochem. Biophys. Res. Commun.* 47: 267–72.

MURAYAMA, M. 1966. Molecular mechanism of red cell sickling. *Science* 153: 145–49.

NEEL, J. V. 1949. The inheritance of sickle-cell anemia. *Science* 110: 64–66.

PAULING, L., ITANO, H. A., SINGER, S. J., and WELLS, I. C. 1949. Sickle cell anemia, a molecular disease. *Science* 110: 543–48.

ROSSMAN, M. G., and ARGOS, P. 1981. Protein folding. *Ann. Rev. Biochem.* 50: 497–532.

SARABHAI, A., STRETTON, A., BRENNER, S., and BOLLE, A. 1964. Colinearity of the gene with the polypeptide chain. *Nature* 201: 13–17.

SCOTT-MONCRIEFF, R. 1936. A biochemical survey of some Mendelian factors for flower colour. *J. Genet.* 32: 117–70.

SCRIVER, C. R., and CLOW, C. L. 1980. Phenylketonuria and other phenylalanine hydroxylation mutants in man. *Ann. Rev. Genet.* 14: 179–202.

SIGMAN, D. S., and BRAZIER, M. A. B., eds. 1980. *The evolution of protein structure and function.* New York: Academic Press.

SRB, A. M., and HOROWITZ, N. H. 1944. The ornithine cycle in *Neurospora* and its genetic control. *J. Biol. Chem.* 154: 129–39.

STRYER, L. 1981. *Biochemistry.* 2nd ed. San Francisco: W. H. Freeman.

SYKES, B. 1985. The molecular genetics of collagen. *BioEssays* 3:112–17.

WAGNER, R. P., ed. 1975. *Genes and proteins.* New York: Halsted Press.

WAGNER, R. P., and MITCHELL, H. K. 1964. *Genetics and metabolism.* 2nd ed. New York: Wiley.

WELNER, D., and MEISTER, A. 1981. A survey of inborn errors of amino acid metabolism and transport in man. *Ann. Rev. Biochem.* 50: 911–68.

WOODS, R. A. 1980. *Biochemical genetics.* London: Chapman and Hall.

YANOFSKY, C., DRAPEAU, G., GUEST, J., and CARLTON, B. 1967. The complete amino acid sequence of the tryptophan synthetase A protein and its colinear relationship with the genetic map of the *A* gene. *Proc. Natl. Acad. Sci.* 57: 296–98.

ZIEGLER, I. 1961. Genetic aspects of ommochrome and pterin pigments. *Adv. in Genet.* 10: 349–403.

ZUBAY, G. L., and MARMUR, J., eds. 1973. *Papers in biochemical genetics.* 2nd ed. New York: Holt, Rinehart and Winston.

17

The Genetic Code

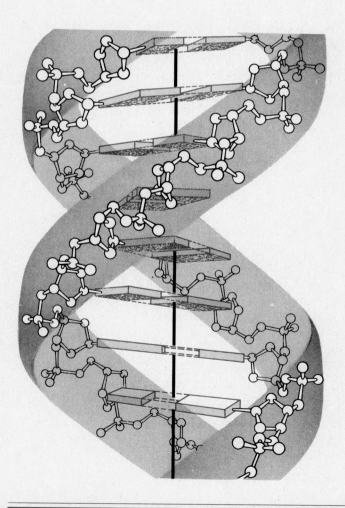

Having previously established that DNA serves as the genetic material and that each gene contains information essential to the synthesis of a polypeptide chain, we now consider **information storage.** How do a gene's deoxyribonucleotides specify the linear sequence of amino acids in a polypeptide chain? In Chapter 16 we learned that there is a colinear relationship between the molecular information in the gene and that of the polypeptide chain.

We also know, but have yet to discuss in detail, that the information in DNA is first transferred to a complementary messenger RNA (mRNA) through transcription. Once transferred to mRNA, the complementary ribonucleotides of this molecule must direct the insertion of amino acids into a polypeptide chain during translation on the ribosome.

The fundamental question, then, is how an RNA molecule consisting of only four different types of nucleotides (A, U, C, and G) can specify 20 different amino acids. This question poses an intriguing theoretical problem. While the earliest proposals were imaginative, it was not until ingenious analytical research was applied that a hypothesis was shown to be correct. It was shown that information is stored in a **genetic code** which is triplet in nature. Code words, or **codons,** consisting of three ribonucleotides direct the insertion of a single amino acid into a polypeptide chain during its synthesis.

In this chapter, we will describe the way in which the code was deciphered, the coding assignments, and the general properties exhibited by the code. The work leading to these discoveries occurred most intensively in the late 1950s and early 1960s, one of the most exciting periods in the study of molecular genetics. This research revealed the intricacies of the specific chemical language that serves as the basis of all life on earth.

AN OVERVIEW OF THE GENETIC CODE

Before considering the various analytical approaches used in arriving at our current understanding of the genetic code, we shall provide a summary of the general features that characterize it:

1 The code is written in linear form using the ribonucleotide bases that compose mRNA molecules as the letters. The ribonucleotide sequence is, of course, derived from its complement in DNA.

2 Each word within the mRNA contains three letters. Thus, the code is a **triplet,** and each group of *three* ribonucleotides specifies *one* amino acid.

3 The code is **unambiguous,** meaning that each triplet specifies only a single amino acid.

4 The code is **degenerate,** meaning that more than one triplet specifies a given amino acid. This is the case for 18 of the 20 amino acids.

5 The code contains "start" and "stop" punctuation signals. Certain triplets are necessary to initiate and to terminate translation.

6 No commas (or internal punctuation) are used in the code. Thus, the code is said to be **commaless.** Once translation of mRNA begins, each three ribonucleotides are read in turn, one after the other.

7 The code is **nonoverlapping.** Any single ribonucleotide at a specific location within the mRNA is part of only one triplet.

8 The code is almost **universal.** With only minor exceptions, a single coding dictionary is used by almost all prokaryotes and eukaryotes.

EARLY PROPOSALS: GAMOW'S CODE

The earliest proposals for the genetic code occurred before scientists knew that mRNA is an intermediate in information transfer. As a result, these proposals assumed that DNA served directly as the template for protein synthesis. While anticipated by P. C. Caldwell and Sir Cyril N. Hinshelwood in 1950 and by A. L. Dounce in 1952, the **diamond code** proposed by

George Gamow in 1954 generated the most interest. Gamow, a physicist and cosmologist, addressed directly the question of how DNA, written in four letters, could specify the 20 amino acids constituting proteins.

As illustrated in Figure 17.1, Gamow proposed that groups of nucleotides of both strands of the helix are arranged to form a series of "diamonds." These diamonds form geometric pockets, each specifying an amino acid. Notice that this model leads to an **overlapping** code, since any given nucleotide position is shared by two code words. Each diamond consists of a nucleotide on one DNA strand, the adjacent nucleotide pair, and the adjoining nucleotide on the opposite strand. Any of the four possible nucleotide pairs may occupy the middle of any diamond (A-T, T-A, C-G, and G-C). Each of these may be combined with any of the four nucleotides at either adjacent position, creating 64 possible codes. However, when geometric equivalents are taken into account, the number of codes is reduced to 20—precisely the number of amino acids that must be accounted for! Therefore, Gamow speculated that amino acids from the surrounding medium would have a chemical affinity for the pockets formed along the DNA duplex. Once sequestered in the correct sequence, they would unite to form a linear polypeptide chain.

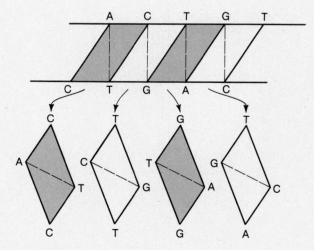

FIGURE 17.1
Gamow's diamond code. He proposed that each group of four adjacent and adjoining nucleotides directly specifies one amino acid. This code is overlapping.

Objections to the Diamond Code

Central to Gamow's proposal were the overlapping nature of the code and the use of a direct DNA template. Theoretical objections were quickly raised on both points. First, we will address the arguments against an overlapping code.

The assumption that the code was a triplet, with each amino acid specified by three nucleotides, was based on considerations set forth by Sidney Brenner. He argued that a triplet code represents the minimal use of four nucleotides to specify 20 amino acids. For example, four nucleotides, taken two at a time in all possible permutations, provide only 16 unique code words (4^2). While a triplet code provides 64 possibilities (4^3)—clearly more than the 20 needed—it seemed much more likely to Brenner than a four-letter code, where 256 code words (4^4) would exist.

Assuming a triplet code, Brenner considered restrictions that might be placed on it if it were overlapping. For example, he considered sequences within a protein consisting of three consecutive amino acids. In the nucleotide sequence GTACA, parts of the central triplet, TAC, are shared by the outer triplets, GTA and ACA. He reasoned that if this were the case, then only certain amino acids should be found adjacent to the one encoded by the central triplet, as shown below:

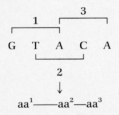

In fact, for any given central amino acid, only 16 combinations (2^4) of adjacent amino acids are theoretically possible.

Therefore, Brenner concluded that if the code were overlapping, tripeptide sequences of proteins should be somewhat limited. Looking at the available amino acid sequences of proteins that had been studied, he found no such restriction in tripeptide sequences. For any central amino acid, many more than 16 different tripeptides were found.

A second major argument against an overlapping code involved the effect of a single nucleotide change characteristic of a point mutation. With an overlapping code, two adjacent amino acids would be affected. But, mutations in the genes coding for the protein coat of tobacco mosaic virus (TMV), human hemoglobin, and the bacterial enzyme tryptophan synthetase invariably revealed only single amino acid changes.

The third argument against an overlapping code was presented by Francis Crick in 1957. He also argued against DNA serving as a direct template for the formation of proteins. Crick reasoned that any affinity between nucleotides and an amino acid would require hydrogen bonding. Chemically, however, such affinities seemed unlikely. Instead, he proposed that there must be an **adaptor molecule** that could covalently bind to the amino acid, yet be capable of hydrogen bonding to a nucleotide sequence. As we will see in Chapter 18, Crick's prediction was correct; transfer RNA (tRNA) serves as the adaptor in protein synthesis. Because various adaptors would somehow have to overlap one another at nucleotide sites, Crick reasoned that such a chemical interaction would be highly unlikely. Therefore, it seemed doubtful that the code was overlapping.

Crick's and Brenner's arguments, taken together, strongly suggested that the genetic code is **nonoverlapping.** Without exception, this concept has been upheld.

THE CODE: FURTHER DEVELOPMENTS

Between 1958 and 1960, information related to the genetic code continued to accumulate. In addition to his adaptor proposal, Crick hypothesized, on the basis of genetic evidence, that the code is **comma-free;** that is, he believed no punctuation occurs along the reading frame. He also speculated that only 20 of the 64 possible triplets specify an amino acid and that the remaining 44 carry no coding assignment.

At the time, however, there was as yet no experimental evidence that the code was indeed a triplet; nor had the concept of messenger RNA, as an intermediate between DNA and protein, been established. Thus, since ribosomes had already been identified, the current thinking was that information in DNA is transferred in the nucleus to the RNA of the ribosome, which serves as the template for protein synthesis in the cytoplasm.

This concept soon became untenable as accumulating evidence demonstrated that the template intermediate was unstable. The RNA of ribosomes, on the other hand, was found to be extremely stable. As a result, in 1961 François Jacob and Jacques Monod postulated the existence of **messenger RNA (mRNA).** The scene was thus set to demonstrate the triplet nature of the code, as contained in the intermediate mRNA, and to decipher the specific codon assignments.

Many questions beyond the triplet assignments also remained. Are there start-stop **punctuation signals?** Is the code **ambiguous,** with one triplet specifying more than one amino acid? Was Crick wrong with respect to the 44 "blank" codes? That is, is the code **degenerate,** with more than one triplet assignment for each amino acid? Is the code **universal?** As we shall see, these and other questions were answered in the next decade.

The Study of Frameshift Mutations

Before discussing the experimentation in which specific codon assignments were deciphered, we will consider the ingenious experimental work of Crick, Leslie Barnett, Brenner, and R. J. Watts-Tobin. Their work represented the first solid evidence for the triplet nature of the code.

In their work, these researchers induced insertion and deletion mutations in the B cistron of the rII locus of phage T4. The B cistron is one of two functional sites in a locus which, in mutant form, causes rapid lysis and distinctive plaques. Mutants in the rII locus will successfully infect *E. coli* B but cannot reproduce on *E. coli* K12 (λ). Crick and his colleagues used the acridine dye **proflavin** to induce mutations (see Figure 13.14). This mutagenic agent intercalates within the double helix of DNA, often causing the insertion of an extra nucleotide or the deletion of one nucleotide during replication. As shown in Figure 17.2, the frame of reading will therefore shift, changing all subsequent triplets to the right of the insertion or deletion. Upon translation, the protein subsequently synthesized will be garbled, so to speak, from the point of change. These mutations are called **frameshifts.** When they are present at the rII locus, T4 will not reproduce on *E. coli* K12.

They reasoned that if phages with these induced mutations were treated again with proflavin, still other insertions or deletions would occur. This second change might result in a revertant phage, which would behave like wild type and successfully infect *E. coli* K12. For example, if the original mutant contained an insertion (+), a second event causing a deletion (−) close to the insertion would restore the original reading frame (Figure 17.2). In the same way, an event resulting in an insertion (+) might correct an original deletion (−).

In studying many mutations of this type, these researchers were able to compare various mutant combinations together on the same DNA molecule, as shown in Figure 17.3. They found that various combinations of one (+) and one (−) indeed caused reversion to wild-type behavior. Although this observation shed no light on the number of nucleotides constituting the genetic code, analysis of other combinations of mutations did. When two (+)'s were together or when two (−)'s were together, the correct reading frame was not reestablished. This argued against a doublet (two-

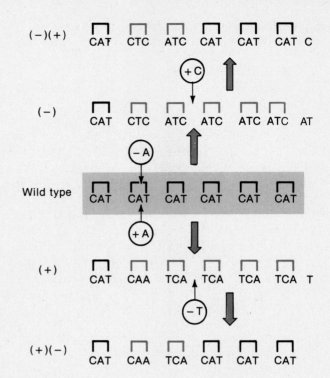

FIGURE 17.2

Schematic diagram of frameshift mutations caused by the addition (+) or deletion (−) of a single nucleotide in the middle of the reading frame. From that point on, all triplet code words are changed. In cases of an addition and a deletion [(+), (−)], the correct frame of reading is restored.

Restoration—Wild Type	No Restoration—Mutant
(+)(−)	(+)
(−)(+)	(−)
(+)(+)(+)	(+)(+)
(−)(−)(−)	(−)(−)

FIGURE 17.3

Effects of multiple additions (+) and deletions (−) on the restoration of the original reading frame in phage T4.

letter) code. However, when three (+)'s or three (−)'s were present together, the original frame was reestablished. These observations strongly supported the triplet nature of the code.

These data further suggested that the code is degenerate, which was contrary to Crick's earlier proposal. A degenerate code is one in which more than one codon specifies the same amino acid. The reasoning leading to this conclusion is as follows. In the cases where wild-type function is restored—(+) and (−), (+ + +) and (− − −)—the original frame of reading is also restored. However, there may be numerous triplets between the various additions and deletions that would still be out of frame. If 44 of the 64 possible triplets were blank and did not specify an amino acid, one of these so-called **nonsense triplets** would very likely occur in the length of nucleotides still out of frame. If a nonsense code were encountered during protein synthesis, it was reasoned that the process would stop or be terminated at that point. If so, the product of the rII-B locus would not be made, and restoration would not occur. Since the various mutant combinations were able to reproduce on *E. coli* K12, Crick and his colleagues concluded that, in all likelihood, most if not all of the remaining 44 codes were not blank. It follows that the genetic code is **degenerate.** As we shall see, this reasoning proved correct and was soon established on the basis of more direct experimentation.

DECIPHERING THE CODE: INITIAL STUDIES

In 1961 Marshall Nirenberg and J. Heinrich Matthaei published results characterizing the first specific coding sequences. These results served as a cornerstone for the complete analysis of the code. Their success was dependent on the use of two experimental tools: **a cell-free protein-synthesizing system,** and the use of an enzyme, **polynucleotide phosphorylase,** which allowed the production of synthetic mRNAs. These mRNAs served as templates for polypeptide synthesis in the cell-free system.

In the cell-free (*in vitro*) system, amino acids can be incorporated into polypeptide chains. This *in vitro* mixture, as might be expected, must contain the essential factors for protein synthesis in the cell: ribosomes, tRNAs, amino acids, and enzymes necessary for the translation process. Additionally, ATP, GTP, Mg^{++} and K^+ ions, and a sulfhydryl compound to stabilize the system are needed. These components and their function in cell-free protein synthesis are summarized in Table 17.1.

Many of the enzymes are found in the supernatant

TABLE 17.1

Components of cell-free protein synthesis.

Component	Function
Ribosomes	Workbench for translation
tRNA	Adaptor molecule
Amino acids	Building blocks of protein
mRNA	Template
Aminoacyl transferases	Charging tRNA
Initiation factors	Initiation of polypeptide synthesis
Peptidyl transferase	Elongation and translocation of growing polypeptide chain
Mg^{++}, K^+	Stabilizes conformation of ribosomes
ATP	Energy source; essential for charging tRNA
GTP	Energy source for translocation

following isolation of ribosomes from the cell homogenate by high-speed centrifugation. In order to follow protein synthesis, one or more of the amino acids must be radioactive. Finally, an mRNA pool is needed. While mRNA may be included in, or endogenous to, the system during cell homogenization, it is soon conveniently degraded by nucleases and must be added exogenously. This step is essential if a known template is to be translated.

In 1961 mRNA had yet to be isolated. However, the use of the enzyme **polynucleotide phosphorylase** allowed artificial synthesis of RNA templates, which could be added to the cell-free system. This enzyme, isolated from bacteria, catalyzes the reaction shown in Figure 17.4. Discovered in 1955 by Marianne Grunberg-Manago and Severo Ochoa, the enzyme functions metabolically in bacterial cells to degrade RNA. However, *in vitro*, with high concentrations of ribonucleoside diphosphates, the reaction can be "forced" in the opposite direction to synthesize RNA.

In contrast to RNA polymerase, polynucleotide phosphorylase requires no DNA template. As a result, ribonucleotides are assembled at random, according to the relative concentration of the four ribonucleoside diphosphates added to the reaction mixture. **This point is absolutely critical to understanding the work of Nirenberg and others in the ensuing discussion.**

Taken together, the cell-free system for protein synthesis and the availability of synthetic mRNAs provided a means of deciphering the composition of various triplet codes.

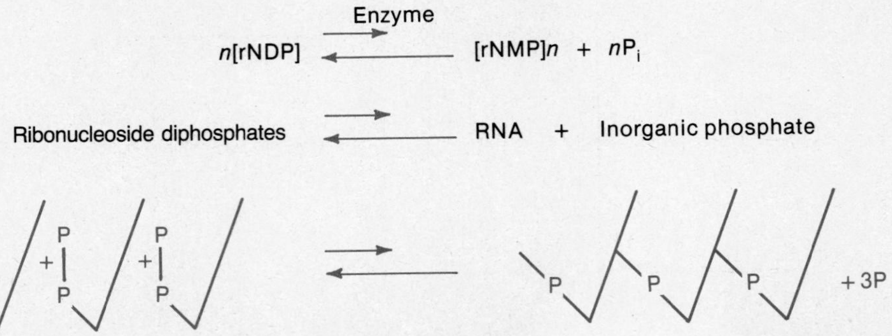

FIGURE 17.4
The reaction catalyzed by the enzyme polynucleotide phosphorylase.

Nirenberg and Matthaei's Homopolymer Codes

In their initial experiments, Nirenberg and Matthaei synthesized **RNA homopolymers,** each consisting of only one ribonucleotide. Therefore, the mRNA added to the *in vitro* system was either UUUUUU. . . , AAAAAA. . . , CCCCCC. . . , or GGGGGG. . . . In testing each mRNA, they were able to determine which, if any, amino acids were incorporated into newly synthesized proteins. They determined this by labeling one of the 20 amino acids added to the *in vitro* system and conducting a series of experiments, each with a different amino acid made radioactive.

For example, consider one of the initial experiments, where ^{14}C-phenylalanine was used (Table 17.2). From these and related experiments, Nirenberg and Matthaei concluded that the message poly U (polyuridylic acid) directs the incorporation of only phenylalanine into the homopolymer polyphenylalanine. As-

suming a triplet code, they had determined the first specific codon assignment! UUU codes for phenylalanine.

In the same way, they quickly found that AAA codes for lysine and CCC codes for proline. Poly G did not serve as an adequate template, probably because the molecule folds back on itself. Thus, the assignment for GGG had to await other approaches. Note that these specific codon assigments were possible only because of the use of homopolymers. This method yields only the composition of triplets, not the sequence. However, three U's, C's, or A's can have only one possible sequence (i.e., UUU, CCC, and AAA).

The Use of Heteropolymers

With these techniques in hand, Nirenberg and Matthaei, as well as Ochoa and coworkers, turned to the use of **RNA heteropolymers.** In this technique, two or more different ribonucleoside diphosphates are added in combination to form the message. These researchers reasoned that if the relative proportion of each type of ribonucleoside diphosphate is known, the frequency of any particular codon occurring in the synthetic mRNA can be predicted. If the mRNA is then added to the cell-free system and the percentage of any particular amino acid present in the new protein is ascertained, composition assignments may be predicted.

This concept is illustrated in Figure 17.5. Suppose that A and C are added in a ratio of 1A:5C. Now, the insertion of a ribonucleotide at any position along the RNA molecule during its synthesis is determined by the ratio of A:C. Therefore, there is a 1/6 possibility for an A and a 5/6 chance for a C to occupy each position. On this basis, we can calculate the frequency of any given triplet appearing in the message.

For AAA, the frequency is $(1/6)^3$, or about 0.4 percent. For AAC, ACA, and CAA, the frequencies are identical—that is, $(1/6)^2(5/6)$, or about 2.3 percent. Together,

TABLE 17.2
Incorporation of ^{14}C-phenylalanine into protein.

Artificial mRNA	Radioactivity (counts/min)
None	44
Poly U	39,800
Poly A	50
Poly C	38
Poly inosinic acid	57
Poly AU	53
Poly U + Poly A	60

SOURCE: After Nirenberg and Matthaei, 1961, p. 1595.

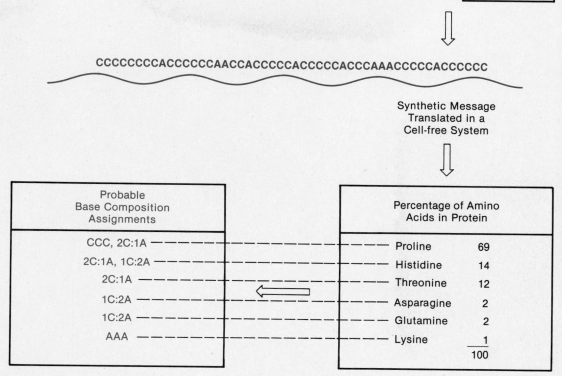

Experimental Conditions: 1/6A:5/6C

Possible Compositions	Probability of Occurrence of Any Triplet	Possible Triplets	Final %
3A	$(1/6)^3 = 1/216 = 0.4\%$	AAA	0.4
2A:1C	$(1/6)^2(5/6) = 5/216 = 2.3\%$	AAC ACA CAA	6.9
1A:2C	$(1/6)(5/6)^2 = 25/216 = 11.6\%$	ACC CAC CCA	34.8
3C	$(5/6)^3 = 128/216 = 57.9\%$	CCC	57.9
			100.0%

CCCCCCCCACCCCCCAACCACCCCCACCCCCACCCAAACCCCCACCCCCC

Synthetic Message
Translated in a
Cell-free System

Probable Base Composition Assignments		Percentage of Amino Acids in Protein	
CCC, 2C:1A		Proline	69
2C:1A, 1C:2A		Histidine	14
2C:1A		Threonine	12
1C:2A		Asparagine	2
1C:2A		Glutamine	2
AAA		Lysine	1
			100

FIGURE 17.5

Results and interpretation of a mixed copolymer experiment.

all three 2A:1C triplets account for 6.9 percent of the total three-letter sequences. In the same way, each of three 1A:2C triplets accounts for $(1/6)(5/6)^2$, or 11.6 percent (or a total of 34.8 percent). CCC is represented by $(5/6)^3$, or 57.9 percent of the triplets.

By examining the percentages of any given amino acid incorporated into the protein synthesized under the direction of this message, we may make tentative composition assignments. Since proline appears 69 percent of the time (Figure 17.5), we can deduce that it is coded by CCC 58 percent of the time and by one triplet of the 2C:1A variety 11 percent of the time. Histidine, at 14 percent, is probably coded by one 2C:1A

(11 percent) and one 1C:2A (2 percent). Threonine, at 12 percent, is likely coded by only one 2C:1A. Asparagine and glutamine appear each to be coded by one of the 1A:2C triplets, and lysine appears to be coded by AAA.

In similar experiments, some using as many as all four ribonucleotides to construct the mRNA, many possible combinations were studied. Thus, composition assignments were determined for all amino acids. Although this represented a very significant breakthrough, specific sequences of triplets were still unknown. Their determination awaited still other approaches.

The Triplet Binding Technique

It was not long before more advanced techniques were developed. In 1964 Nirenberg and Phil Leder developed the **triplet binding assay,** which led to specific assignments of triplets. The technique took advantage of the observation that ribosomes, when presented with an RNA sequence as short as three ribonucleotides, will bind to it and attract the correct charged tRNA. For example, if ribosomes are presented with an RNA triplet UUU and tRNA[phe], a complex will form (Figure 17.6). While it was not yet feasible to chemically synthesize long stretches of RNA, triplets of known sequence could be constructed in the laboratory.

All that was needed now was a method to determine which tRNA-amino acid bound to the mRNA-ribosome complex. The test system devised was quite simple. The amino acid to be tested is made radioactive, and a charged tRNA (where the amino acid is bonded to tRNA) is produced. Since code compositions were known, it was possible to narrow the decision as to which amino acids should be tested for each specific triplet.

The radioactive charged tRNA, the RNA triplet, and ribosomes are incubated together with a nitrocellulose filter, which will retain the ribosomes but not the other components, such as charged tRNA. If radioactivity is not retained on the filter, an incorrect amino acid has been tested. If radioactivity remains on the filter, it is retained because the charged tRNA has bound to the triplet associated with the ribosome. In such a case, a specific codon assignment may be made.

Work proceeded in several laboratories, and in many cases clearcut, unambiguous results were obtained. Table 17.3, for example, shows 25 triplets assigned to 10 amino acids. However, in some cases the degree of binding was insufficient, and unambiguous assignments were not possible. Eventually, about 50 of the 64 triplets were assigned. The binding technique was a major innovation in deciphering the code. Based on these specific assignments of triplets to amino acids, two major conclusions were drawn. The genetic code is **degenerate;** that is, one amino acid may be specified by more than one triplet. The code is also **unambiguous;** that is, a single triplet specifies only one amino acid. As we shall see later in this chapter, these conclusions have been upheld with only minor exceptions.

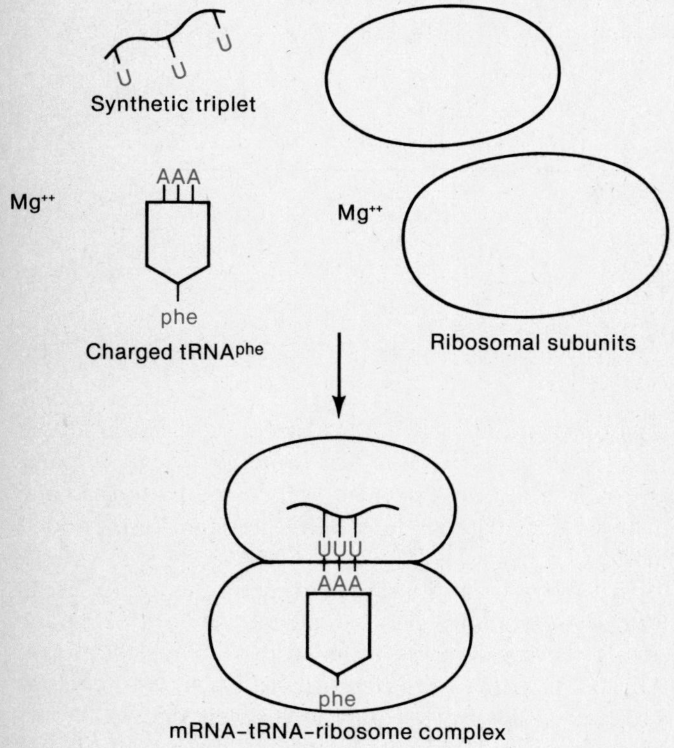

FIGURE 17.6
Molecular components used in the triplet binding assay.

TABLE 17.3
Amino acid assignments to specific trinucleotides derived from the triplet binding assay.

Trinucleotide	Amino Acid
UGU UGC	Cysteine
GAA GAG	Glutamic acid
AAU AUC AUA	Isoleucine
UUA UUG CUU	Leucine
CUC CUA CUG	Leucine
AAA AAG	Lysine
AUG	Methionine
UUU UUC	Phenylalanine
CCU CCC	Proline
CCG CCA	Proline
UCU UCC	Serine
UCA UCG	Serine
UCU UCC	Valine
UCA UCG	Valine

The Use of Regular Copolymers

Still another innovative technique used to decipher the genetic code was developed by Gobind Khorana. He was able to chemically synthesize long repeating RNA sequences, which could be used in the cell-free protein synthesizing system. First, he created shorter sequences (e.g., di-, tri-, and short tetranucleotides), which are replicated many times and finally joined enzymatically to form the long polynucleotides.

As illustrated in Figure 17.7 a dinucleotide made in this way is converted to a message with two repeating triplets. A trinucleotide is converted to one with three potential triplets, depending on the point at which initiation occurs. Similarly, a tetranucleotide creates four repeating triplets.

When these synthetic mRNAs were added to a cell-free system, the predicted number of different amino acids incorporated was upheld. Several examples are shown in Table 17.4. When such data were combined with conclusions drawn from other approaches (composition assignment, triplet bindings), specific assignments were possible.

One example of a specific assignment made from such data will illustrate the value of Khorana's approach. The trinucleotide sequence UUCUUCUUC . . . produces three possible triplets—UUC, UCU, and CUU, depending on the first nucleotide to initiate reading. When placed in a cell-free translation system, the polypeptides containing phenylalanine (phe), serine (ser), and leucine (leu) are produced. On the other hand, the dinucleotide sequence UCUCUCUC . . . produces the triplets UCU and CUC with the incorporation of leucine and serine. Therefore, the triplets UCU and CUC specify leucine and serine, but it cannot be determined which is which. By a process of elimination, the remaining trinucleotide UUC is assigned to phenylalanine.

Of UCU and CUU, which codes for serine and which for leucine? To answer this question, we must examine the results of using the repeating tetranucleotide sequence UUAC, which produces the triplets UUA, UAC, ACU, and CUU. The CUU triplet is one of the two in which we are interested. Three amino acids are incorporated: leucine, threonine, and tyrosine. Because CUU must specify only serine or leucine, and because, of these two, only leucine appears, we may conclude that CUU specifies leucine. By the process of elimination, UCU would seem to code for serine.

From such interpretations, Khorana reaffirmed triplets already deciphered and filled in gaps left from other approaches. For example, the use of two tetranucleotide sequences, GAUA and GUAA, suggested that at least two triplets were termination signals. This conclusion was drawn because neither of these sequences directed the incorporation of any amino acids into a polypeptide (Table 17.4). Since there are no triplets common to both messages, it was predicted

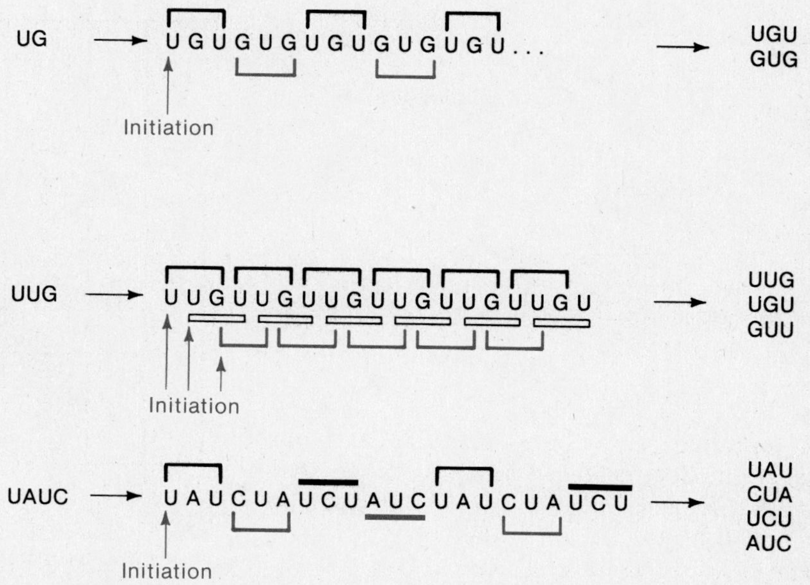

FIGURE 17.7
The conversion of di-, tri-, and tetranucleotides into repeating copolymers.

TABLE 17.4
Amino acids incorporated using synthetic copolymers of RNA.

Copolymer	Codons Produced	Amino Acids in Polypeptide
UG	UGU	Cysteine
	GUG	Valine
UC	UCU	Serine
	CUC	Leucine
AC	ACA	Threonine
	CAC	Histidine
AG	AGA	Arginine
	GAG	Glutamine
UUC	UUC	Phenylalanine
	UCU	Serine
	CUU	Leucine
UUG	UUG	Leucine
	UGU	Cysteine
	GUU	Valine
AUC	AUC	Isoleucine
	UCA	Serine
	CAU	Histidine
UAC	UAC	Tyrosine
	ACU	Threonine
	CUA	Leucine
UUAC	UUA	Leucine
	CUU	Threonine
	ACU	Tyrosine
	UAC	
UAUC	UAU	Tyrosine
	CAU	Leucine
	UCU	Serine
	AUC	Isoleucine
GAUA	GAU	
	AGA	
	UAG	None
	AUA	
GUAA	GUA	
	AGU	
	AAG	None
	UAA	

THE CODING DICTIONARY

Taken together, the various techniques applied to decipher the genetic code have yielded a dictionary of 61 triplet codon–amino acid assignments. The remaining three triplets are termination signals, not specifying any amino acid. Figure 17.8 designates the assignments in a particularly illustrative form first suggested by Crick.

Degeneracy and Wobble

The degenerate nature of the code is immediately apparent when we inspect this presentation of the genetic code. While the amino acids tryptophan and methionine are encoded by only single triplets, most amino acids are specified by two, three, or four triplets. Three amino acids (serine, arginine, and leucine) are coded by as many as six triplets each.

What also becomes evident is the pattern of degeneracy. Most often in a set of codes specifying the same amino acid, the first two letters are the same, with only the third differing. This interesting pattern prompted Crick to postulate the **wobble hypothesis.** He proposed that during translation, the first two positions of each codon must be very precise. The third position, however, is less specific. The wobble hypothesis refers to the interaction between the codons of mRNA and the **anticodons** of tRNA. The anticodon is the portion of each tRNA that is complementary to the

		Second Position				Third Position (3′ end)
		U	C	A	G	
First Position (5′ end)	U	phe	ser	tyr	cys	U
		phe	ser	tyr	cys	C
		leu	ser	term	term	A
		leu	ser	term	trp	G
	C	leu	pro	his	arg	U
		leu	pro	his	arg	C
		leu	pro	gln	arg	A
		leu	pro	gln	arg	G
	A	ileu	thr	asn	ser	U
		ileu	thr	asn	ser	C
		ileu	thr	lys	arg	A
		met	thr	lys	arg	G
	G	val	ala	asp	gly	U
		val	ala	asp	gly	C
		val	ala	glu	gly	A
		val	ala	glu	gly	G

FIGURE 17.8
The coding dictionary.

that each repeating sequence contains at least one triplet that terminates protein synthesis. As we shall see, there are three such triplets, two of which are included in poly-(GAUA) and poly-(GUAA).

codon and is vital to the participation of the correct tRNA at each step during translation. (This concept will be discussed in greater detail in Chapter 18.) Crick proposed that the third position in the codon-anticodon interaction is less restricted in adhering to the base-pairing rules establishing complementarity. This "wobble" represents greater flexibility at this position.

Such a wobble would allow the anticodon of a single tRNA species to pair with more than one triplet in mRNA. The code's degeneracy would often allow this to occur without changing the amino acid. Crick proposed that U at the third position of the anticodon of tRNA may pair with A or G at the third position of the triplet in mRNA, and that G may likewise pair with U or C. Inosine, one of the "odd" bases found in tRNA, may pair with C, U, or A. To date, Crick's prediction has been upheld. A tRNA molecule of specific anticodon sequence can indeed bind to more than one codon. The significance of this finding will be clarified with further study.

Initiation, Termination, and Suppression

Initiation of protein synthesis is a highly specific process. In bacteria, the initial amino acid inserted into all polypeptide chains is a modified form of methionine—**N-formylmethionine (fmet).** Only one codon, AUG, codes for methionine, and it is sometimes called the **initiator codon.** However, when AUG appears internally in mRNA, unformylated methionine is inserted into the polypeptide chain.

In bacteria, either the formyl group is removed from the initial methionine upon completion of synthesis of a protein, or the entire formylmethionine residue is removed. In eukaryotes, methionine is also the initial amino acid during polypeptide synthesis. However, it is not formylated.

As mentioned in the preceding section, three other triplets (UAA, UAG, and UGA) serve as punctuation signals and do not code for any amino acid. They are not recognized by a tRNA molecule, and termination of translation occurs when they are encountered. Mutations occurring internally in a gene that produce any of the three triplets also result in termination. As a result, only a partial polypeptide is synthesized, since it is prematurely released from the ribosome. When such a change occurs, it is called a **nonsense mutation.** The terms **amber** (UAG), **ochre** (UAA), and **opal** (UGA) have been used to distinguish the three possibilities.

Interestingly, a distinct mutation in a second gene may cause **suppression** of premature termination.

These mutations cause the chain-termination signal to be read as a sense codon. The "correction" usually inserts an amino acid other than that found in the wild-type protein. However, if the protein's structure is not altered drastically, it may function almost normally. Therefore, this second mutation has "suppressed" the mutant character resulting from the initial change to the termination codon.

Investigation has shown that some suppressor mutations occur in genes specifying tRNAs. If the mutation results in a change in the anticodon such that it becomes complementary to a termination code, there is the potential for insertion of an amino acid and suppression.

Other types of suppression involve mutations in genes coding for aminoacyl synthetases, which are responsible for attaching the amino acid to tRNA (the process called *charging*), and in genes coding for ribosomal proteins. In both cases, suppression occurs as a result of misreading either a nonsense or missense mutation. That a ribosomal protein can cause ambiguity of translation demonstrates the intimate relationship between the ribosome, mRNA, and tRNA during translation.

CONFIRMATION OF CODE STUDIES: PHAGE MS2

Taken together, all aspects of the genetic code discussed so far yield a fairly complete picture. The code is triplet in nature, degenerate, unambiguous, and comma-free, but contains punctuation with respect to start and stop signals. These individual principles have been confirmed by the detailed analysis of the RNA-containing bacteriophage MS2 by Walter Fiers and his coworkers.

MS2 is a bacteriophage that infects *E. coli*. Its nucleic acid (RNA) contains only about 3500 ribonucleotides, making up only three genes. These genes specify a coat protein, an RNA-directed replicase, and a maturation protein (the A protein). This simple system of a small genome and few gene products allowed Fiers and his colleagues to sequence the genes and their products. The amino acid sequence of the coat protein was completed in 1970, and the nucleotide sequence of the gene and a number of nucleotides on each end of it were reported in 1972. By 1976, this same research effort led to the complete sequencing of all three genes.

The coat protein contains 129 amino acids, and the gene contains 387 nucleotides, as expected for a triplet code. Each amino acid and triplet corresponds in linear sequence to the correct codon in the RNA code word dictionary, providing direct proof of **colinearity.** The codon for the first amino acid is preceded by AUG, the common initiator codon; and the codon for the last amino acid is succeeded by two consecutive termination codons, UAA and UAG.

The analysis clearly shows that the genetic code as established in bacterial systems is identical in this virus. We shall now briefly consider other evidence suggesting that the code is also identical in eukaryotes.

UNIVERSALITY OF THE CODE?

Between 1960 and 1978, it was generally assumed that the genetic code would be found to be universal, applying equally to viruses, bacteria, and eukaryotes. Certainly, the nature of mRNA and the translation machinery seemed to be very similar in these organisms. For example, cell-free systems derived from bacteria could translate eukaryotic mRNAs. Poly U was shown to stimulate translation of polyphenylalanine in cell-free systems when the components were derived from eukaryotes. Many recent studies involving recombinant DNA technology (see Chapter 20) have revealed that eukaryotic genes can be inserted into bacterial cells and transcribed and translated. Within eukaryotes, mRNAs from mice and rabbits have been injected into amphibian eggs and efficiently translated. For those eukaryotic genes that have been sequenced, notably those for hemoglobin molecules, the amino acid sequence of the encoded proteins adheres to the coding dictionary established from bacterial studies.

However, several 1979 reports on the coding properties of DNA derived from yeast and human mitochondria (mtDNA) altered the principle of universality of the genetic language. Since then, mtDNA has been examined in other organisms, including the fungus *Neurospora*.

Recall that mitochondria contain DNA (see Chapter 15), and that transcription and translation occur within these organelles. Cloned mtDNA fragments were sequenced and compared with the amino acid sequences of various mitochondrial proteins. Several exceptions to the coding dictionary were revealed. Most surprising was that the codon UGA, normally causing termination, specifies the insertion of tryptophan during translation in yeast and human mitochondria. In human mitochondria, AUA, which normally specifies isoleucine, directs the insertion of methionine. In yeast mitochondria, threonine is inserted instead of leucine when CUA is encountered in mRNA.

More recently, in 1985, several other code alterations have been discovered. These and prior aberrant codes are summarized in Table 17.5. These changes have been observed in the bacterium, *Mycoplasma capricolum*, and in the protozoan ciliates *Paramecium*, *Tetrahymena*, and *Stylonychia*. As shown, each change converts one of the termination codons to glutamine (gln) or tryptophan (trp). These changes are felt to be more significant than those in mitochondria because both a prokaryote and several eukaryotes are involved, representing distinct species that have evolved over a long period of time.

Note the apparent pattern in several of the altered codon assignments. The change in coding capacity involves only a shift in recognition of the third, or wobble, position. For example, AUA specifies isoleucine during translation in the cytoplasm and methionine in the mitochondrion. In cytoplasmic translation, methionine is specified by AUG. In a similar way, UGA calls for termination in the cytoplasmic system but tryptophan in the mitochondrion. In the cytoplasm, tryptophan is specified by UGG. Although it has been suggested that such changes in codon recognition may represent an evolutionary trend toward reducing the number of tRNAs needed in mitochondria, the significance of these findings is not yet clear. Until still other examples are revealed, the differences must be considered as exceptions to the previously established general coding rules.

TABLE 17.5
Exceptions to the universal code.

Triplet	Normal Code Word	Aberrant Code Word	Source
UGA	Termination	trp	Human mitochondria Yeast mitochondria *Mycoplasma*
UAG	Termination	gln	*Paramecium*
UAA	Termination	gln	*Paramecium* *Tetrahymena* *Stylonychia*
CUA	leu	thr	Yeast mitochondria
AUA	ileu	met	Human mitochondria

FIGURE 17.9
Illustration of the concept of overlapping genes. (a) An mRNA sequence initiated at two different AUG positions out of frame with one another will give rise to two distinct amino acid sequences. (b) The relative positions of the sequences encoding seven polypeptides of the phage φX174. Those unshaded (B, K, and E) are out of frame with those that are shaded.

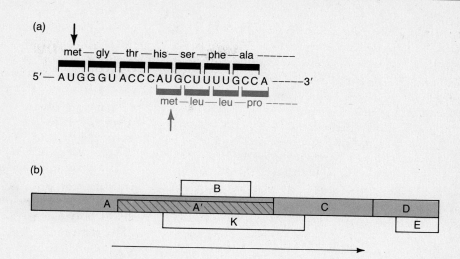

READING THE CODE: THE CASE OF OVERLAPPING GENES

In this chapter we established that the genetic code is nonoverlapping. This means that each ribonucleotide of an mRNA which specifies a polypeptide chain is part of only one triplet. However, this characteristic of the code does not rule out the possibility that a single mRNA may have multiple initiation points for translation. If so, these points could theoretically create several different frames of reading within the same mRNA, thus specifying more than one polypeptide. This concept, which would create **overlapping genes,** is illustrated in Figure 17.9(a).

That this might actually occur in some viruses was suspected when phage φX174 was carefully investigated. As you may recall from Chapter 14, the circular DNA chromosome consists of 5386 nucleotides, which should encode a maximum of 1795 amino acids, sufficient for five or six proteins. However, it was realized that this small virus in fact synthesizes 11 proteins consisting of more than 2300 amino acids! Comparison of the nucleotide sequence of the DNA and the amino acid sequences of the polypeptides synthesized has clarified this paradox. At least four cases of multiple initiation have been discovered, creating overlapping genes [Figure 17.9(b)].

The sequences specifying the K and B polypeptides are initiated with separate reading frames within the sequence specifying the A polypeptide. The K sequence overlaps into the adjacent sequence specifying the C polypeptide. The E sequence is out of frame with, but initiated in, that of the D polypeptide. Finally, the A' sequence, while in frame, begins in the middle of the A sequence. They both terminate at the identical point. In all, seven different polypeptides are created from a DNA sequence that might otherwise have specified only three (A, C, and D).

A similar situation has been observed in other viruses, including phage G4 and the animal virus SV40. The use of overlapping reading frames optimizes the use of a limited amount of DNA present in these small viruses. However, such an approach to storing information has the distinct disadvantage that a single mutation may affect more than one protein and thus increase the chances that the change will be deleterious or lethal. In the case discussed above, a single mutation at the junction of genes A and C will affect three proteins (the A, C, and K proteins). It may be for this reason that such an approach has not been utilized by other organisms.

SUMMARY AND CONCLUSIONS

In this chapter, we have recounted the historical developments leading to our current understanding of the genetic code. The earliest proposals suggested that DNA serves as the template directing the insertion of amino acids into a polypeptide chain. Gamow's overlapping triplet diamond code generated the greatest interest. Although the triplet nature of the code has proved to be correct, its overlapping nature was quickly shown to be wrong. Crick also argued that DNA could not serve as the direct template for translation. Instead, he proposed that an adaptor molecule was needed between DNA and protein. This is now the role ascribed to tRNA.

As evidence accumulated linking mRNA and tRNA to the transfer of information from DNA to protein, attention turned to more specific details of the code. In an ingenious set of experiments, Crick, Barnett, Brenner, and Watts-Tobin demonstrated that the code consists of a triplet, is comma-free, and is degenerate. Their work involved the induction of insertion and deletion mutants in the rII locus of phage T4.

Beginning in 1961, published experimental data assigned specific triplets to individual amino acids. This work, performed by Nirenberg, Matthaei, Ochoa, Khorana, and others, relied on a cell-free protein-synthesizing system. When added to this system, synthetic mRNAs directed the synthesis of polypeptide chains. At first, only triplet nucleotide compositions were determined. Later, specific triplet codes were deciphered. Work proceeded, first using polynucleotide phosphorylase to assemble mRNAs and then using specific triplets and regular copolymers as mRNAs.

From these experiments, a complete coding dictionary was produced. Of the 64 possible triplet codes, 61 received amino acid assignments. The remaining three—UAA, UAG, and UGA—specify no amino acid and serve in the termination of translation. One codon, AUG, serves as the initiator *in vivo* by specifying N-formylmethionine in bacteria and methionine in eukaryotes. The code was further shown to be basically unambiguous and degenerate. The pattern of degeneracy frequently involves only the third letter of a series of triplets specifying the same amino acid. This observation led Crick to propose the wobble hypothesis, which suggests that the base pairing between the third letter of the triplet in mRNA and its complement

in tRNA is less specific than in the first two nucleotides of the code.

When a mutation occurs internally in a gene and produces one of the termination codons, the polypeptide chain is prematurely released from the ribosome during translation. Interestingly, a second mutation may suppress this premature termination. Studies have shown that the second mutation often involves an altered tRNA that can read the nonsense codon as sense, thus inserting an amino acid. Other modes of suppression also occur.

When complete nucleotide sequencing of one of the genes of phage MS2 was accomplished and compared with the amino acid sequence of the corresponding proteins, the features of the genetic code suggested by these studies were confirmed. Each triplet amino acid combination was shown to adhere to the established coding dictionary, including those for initiation and termination. Thus, not only have the *in vitro* studies been verified, but the code, which was deciphered with bacterial components and synthetic mRNAs, is identical in a virus. Still other findings have led us to believe that the code is universal in all organisms, with only a few exceptions.

In some cases, multiple initiation points for reading the genetic code have been observed in the RNAs of some viruses. Since RNAs are simply the reflection of DNA, these cases represent genes within other genes, or overlapping genes. Such an approach to information storage is interesting, but appears to be confined to certain viruses.

PROBLEMS AND DISCUSSION QUESTIONS

1 In considering Gamow's diamond code, Brenner argued that if it were overlapping, only 16 different tripeptides involving any central amino acid could exist in proteins. Assuming 20 different amino acids, how many tripeptides containing a central amino acid can occur if the code is nonoverlapping?

2 Crick, Barnett, Brenner, and Watts-Tobin, in their studies of frameshift mutations, found that either 3 (+)'s or 3 (−)'s restored the correct reading frame. If the code were a sextuplet (consisting of six nucleotides), would the reading frame be restored by either of the above combinations?

3 If, in Problem 2, 6 (+)'s or 6 (−)'s restored the original reading frame, could you decide on this basis if the genetic code was a triplet or a sextuplet? Why?

4 In a mixed copolymer experiment using polynucleotide phosphorylase, 3/4 G:1/4 C was added to form the synthetic message. The resulting amino acid composition of the ensuing protein was determined:

Glycine	36/64 (56%)
Alanine	12/64 (19%)
Arginine	12/64 (19%)
Proline	4/64 (7%)

From this information,

(a) Indicate the percentage (or fraction) of the time each possible triplet will occur in the message.

(b) Determine one consistent base composition assignment for the amino acids present.

(c) Considering the wobble hypothesis, predict as many specific triplet assignments as possible.

5   In a mixed copolymer experiment, messengers were created with either 4/5 C:1/5 A or 4/5 A:1/5 C. These messages yielded proteins with the amino acid compositions shown below. Using these data, predict the most specific coding composition for each amino acid.

4/5 C:1/5 A			4/5 A:1/5 C	
Proline	63.0%		Proline	3.5%
Histidine	13.0%		Histidine	3.0%
Threonine	16.0%		Threonine	16.0%
Glutamine	3.0%		Glutamine	13.0%
Asparagine	3.0%		Asparagine	13.0%
Lysine	0.5%		Lysine	50.0%
	98.5%			98.5%

6   When the amino acid sequences of insulin isolated from different organisms were determined, some differences were noted. For example, alanine was substituted for threonine, serine was substituted for glycine, and valine was substituted for isoleucine at corresponding positions in the protein. List the single base changes that could occur in triplets to produce these amino acid changes.

7   When copolymers are used to form synthetic mRNAs, dinucleotides produce a single type of polypeptide that contains only two different amino acids. On the other hand, using a trinucleotide sequence produces three different polypeptides, each consisting of only a single amino acid. Why? What will be produced when a repeating tetranucleotide is used?

8   Refer to Table 17.4. The repeating mRNA formed from the tetranucleotide UUAC incorporated only three amino acids, but the use of UAUC incorporated four amino acids. Why?

9   In studies using repeating copolymers, ACAC. . . incorporates threonine and histidine, and CAACAA. . . incorporates glutamine, asparagine, and threonine. What triplet code can definitely be assigned to threonine?

10   In a coding experiment using copolymers (as shown in Table 17.4), the following data were obtained:

Copolymer	Codons Produced	Amino Acids in Polypeptide
AG	AGA, GAG	Arg, Glu
AAG	AGA, AAG, GAA	Lys, Arg, Glu

AGG is known to code for arginine. Taking into account the wobble hypothesis, assign each of the four remaining different triplet codes to its correct amino acid.

11   Why doesn't polynucleotide phosphorylase synthesize RNA *in vivo*?

12   In the triplet binding technique (Figure 17.6), radioactivity remains on the filter when the amino acid corresponding to the triplet is labeled. Explain the basis of this technique.

13   Differentiate between the cause of suppression of a frameshift mutation and suppression of an amber mutation.

14   Why is Fiers' work with phage MS2 a more direct evidence in support of colinearity than Yanofsky's work with the *trp* A locus in *E. coli* (described in Chapter 16)?

15   Of the changes noted in the coding dictionary for mitochondrial mRNAs, which represents the most surprising alteration?

16   Using the amino acid substitutions shown in Figure 16.9 for the α and β chains of human hemoglobin and the code table, determine how many of them can occur as a result of a single nucleotide change.

17 In studies of the amino acid sequence of wild-type and mutant forms of tryptophan synthetase in *E. coli*, the following changes have been observed:

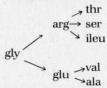

Determine a set of triplet codes in which a single nucleotide change produces each amino acid change.

18 Refer to Table 17.2. Can you hypothesize why Poly U + Poly A does not stimulate incorporation of ^{14}C-phenylalanine into protein?

19 What does this hypothesis (Problem 18) indicate about the RNA directing protein synthesis?

20 Predict the amino acid sequence produced during translation by the following short theoretical mRNA sequences. Note that the second sequence was formed from the first by a deletion of only one nucleotide.

Sequence 1: AUGCCGGAUUAUAGUUGA
Sequence 2: AUGCCGGAUUAAGUUGA

What type of mutation gave rise to Sequence 2?

21 A glycine residue exists at position 210 of the tryptophan synthetase enzyme of wild-type *E. coli*. If the codon specifying glycine is GGA, how many single base substitutions will result in an amino acid substitution at position 210? What are they? How many will result if the wild-type codon is GGU?

22 (a) Shown below is a theoretical viral mRNA sequence. Assuming that it could arise from overlapping genes, how many different polypeptide sequences can be produced? Using Figure 17.8, what are the sequences?
       5'-AUGCAUACCUAUGAGACCCUUGGGA-3'
    (b) A mutation at one position in the DNA giving rise to the above sequence occurred, eliminating the synthesis of all but one polypeptide. The base substitution of the mutant mRNA is shown below. Using Figure 17.8, determine why.
       5'-AUGCAUACCUAUGUGACCCUUGGGA-3'

SELECTED READINGS

BARRELL, B. G., et al. 1980. Different pattern of codon recognition by mammalian mitochondrial tRNAs. *Proc. Natl. Acad. Sci.* 77: 3164–66.

BARRELL, B. G., AIR, G., and HUTCHINSON, C. 1976. Overlapping genes in bacteriophage φX174. *Nature* 264: 34–40.

BARRELL, B. G., BANKIER, A. T., and DROUIN, J. 1979. A different genetic code in human mitochondria. *Nature* 282: 189–94.

BONITZ, S. G., et al. 1980. Codon recognition rules in yeast mitochondria. *Proc. Natl. Acad. Sci.* 77: 3167–70.

BRENNER, S. 1957. On the impossibility of all overlapping triplet codes in information transfer from nucleic acids to proteins. *Proc. Natl. Acad. Sci.* 43: 687–94.

BRENNER, S., STRETTON, A. O. W., and KAPLAN, D. 1965. Genetic code: The nonsense triplets for chain termination and their suppression. *Nature* 206: 994–98.

CALDWELL, P. C., and HINSHELWOOD, C. N. 1950. Some considerations on autosynthesis in bacteria. *J. Chem. Soc.*, pp. 3156–59.

CARON, F., and MEYER, E. 1985. Does *Paramecium primaurelia* use a different genetic code in its macronucleus? *Nature* 314: 185–88.

COLD SPRING HARBOR LABORATORY. 1966. The genetic code. *Cold Spr. Harb. Symp.*, Vol. 31.

CRICK, F. H. C. 1962. The genetic code. *Scient. Amer.* (Oct.) 207: 66–77.

————. 1966. Codon-anticodon pairing: The wobble hypothesis. *J. Mol. Biol.* 19: 548–55.

————. 1966. The genetic code: III. *Scient. Amer.* (Oct.) 215: 55–63.

————. 1967. The Croonian lecture: The genetic code. *Proc. Roy. Soc. Biol.* 167: 331.

CRICK, F. H. C., BARNETT, L., BRENNER, S., and WATTS-TOBIN, R. J. 1961. General nature of the genetic code for proteins. *Nature* 192: 1227–32.

DICKERSON, R.E. 1983. The DNA helix and how it is read. *Scient. Amer.* (Dec.) 249: 94–111.

FIERS, W., et al. 1976. Complete nucleotide sequence of bacteriophage MS2 RNA: Primary and secondary structure of the replicase gene. *Nature* 260: 500–507.

GAMOW, G. 1954. Possible relation between DNA and protein structures. *Nature* 173: 318.

GAREN, A. 1968. Sense and nonsense in the genetic code. *Science* 160: 149–59.

JUKES, T. H. 1963. The genetic code. *Amer. Scient.* 51: 227–45.

KHORANA, H. G. 1967. Polynucleotide synthesis and the genetic code. *Harvey Lectures* 62: 79–105.

LAGERKUIST, U. 1980. Codon misreading: A restriction operative in the evolution of the genetic code. *Amer. Scient.* 68: 192–98.

MIN JOU, W., HAGEMAN, G., YSEBART, M., and FIERS, W. 1972. Nucleotide sequence of the gene coding for bacteriophage MS2 coat protein. *Nature* 237: 82–88.

NIRENBERG, M. W. 1963. The genetic code: II. *Scient. Amer.* (March) 190: 80–94.

NIRENBERG, M. W., and LEDER, P. 1964. RNA codewords and protein synthesis. *Science* 145: 1399–1407.

NIRENBERG, M. W., and MATTHAEI, H. 1961. The dependence of cell-free protein synthesis in *E. coli* upon naturally occurring or synthetic polyribosomes. *Proc. Natl. Acad. Sci.* 47: 1588–1602.

PREER, J.R., et al. 1985. Deviation from the universal code shown by surface protein 51A in *Paramecium. Nature* 314: 188–90.

SPEYER, J. F. 1967. The genetic code. In *Molecular genetics, Part II*, ed. J. H. Taylor, pp. 137–91, New York: Academic Press.

WATSON, J. D. 1976. *Molecular biology of the gene.* 3rd ed. Menlo Park, Calif.: W. A. Benjamin.

YAMAO, F., et al. 1985. UGA is read as tryptophan in *Mycoplasma capricolum. Proc. Natl. Acad. Sci.* 82: 2306–09.

YANOFSKY, C. 1967. Gene structure and protein structure. *Scient. Amer.* (May) 216: 80–95.

18

Synthesis of RNA and Protein: Transcription and Translation

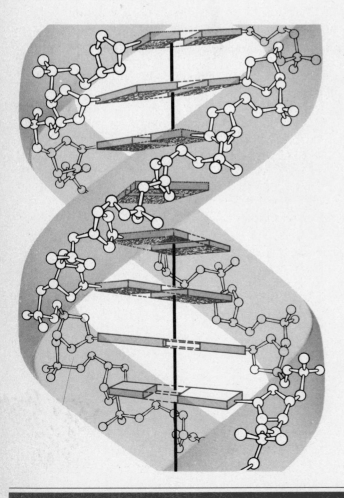

At the same time the genetic code was being studied, extensive research was being conducted to elucidate the nature of genetic expression. The central question was, How is DNA able to specify proteins? By the time the genetic code was deciphered, the basic aspects of how **information transfer** occurs between DNA and protein were revealed.

Genetic information, stored in DNA, was shown to be transferred to RNA during the initial stage of gene expression. The process by which RNA molecules are synthesized on a DNA template is called **transcription.** The ribonucleotide sequence of RNA, written in a genetic code, is then capable of directing the process of **translation.** During translation, polypeptide chains—the precursors of proteins—are synthesized. Protein synthesis is dependent on a series of **transfer RNA (tRNA)** molecules, which serve as adaptors between the codons of mRNA and the amino acids specified by them. In addition, the process occurs only in conjunction with an intricate cellular organelle, the **ribosome.**

The processes of transcription and translation are complex molecular events. Like the replication of DNA, both rely heavily on base-pairing affinities between complementary nucleotides. The initial transfer from DNA to mRNA produces a molecule complementary to the gene sequence of one of the two strands of the double helix. Then, each triplet codon is complementary to the anticodon region of tRNA as the corresponding amino acid is corrrectly inserted into the polypeptide chain during translation. In this chapter, we will describe in detail how these processes were discovered and how they are executed.

TRANSCRIPTION: RNA SYNTHESIS

The concept of an intermediate RNA molecule participating in protein synthesis is suggested by several observations:

1 DNA in eukaryotes is largely restricted to the nucleus, and proteins are synthesized in association with the RNA-containing ribosomes in the cytoplasm. Therefore, DNA does not participate directly in the synthesis of proteins.

2 The amount of protein in a cell is generally proportional to the amount of RNA, but not to the amount of DNA.

3 Cellular RNA in eukaryotes is synthesized in the nucleus, where DNA is found.

Collectively, these observations suggest that genetic information in DNA is transferred to RNA, which in turn directs the synthesis of proteins at the ribosomal level.

Experimental Evidence for the Existence of mRNA

In two papers published in 1956 and 1958, Elliot Volkin and his colleagues reported their analysis of RNA produced immediately after bacteriophage infection of *E. coli*. Using the isotope ^{32}P to follow newly synthesized RNA, they found that its base composition closely resembled that of the phage DNA but was different from that of bacterial RNA (Table 18.1). Although this newly synthesized RNA was unstable, or short-lived, its production was shown to precede the synthesis of new phage proteins. Thus, they considered the possibility that synthesis of RNA is a preliminary step in the process of protein synthesis.

Although ribosomes were known to participate in protein synthesis, their role in this process was not clear. One possible role was that each ribosome is specific for the protein synthesized in association with it. That is, perhaps genetic information in DNA is transferred to the RNA of a ribosome during its synthesis so that each class of ribosome specifies a particular protein. The alternative hypothesis was that ribosomes are nonspecific "workbenches" for protein synthesis and that specific genetic information rests with a "messenger" RNA.

In an elegant experiment using the *E. coli*–phage system, the results of which were reported in 1961, Sidney Brenner, François Jacob, and Matthew Meselson clarified this question. They labeled uninfected *E. coli* ribosomes with "heavy" isotopes and then allowed phage infection to occur in the presence of radioactive RNA precursors. They demonstrated that phage proteins were synthesized on bacterial ribosomes that were present prior to infection. Therefore, the ribosomes appeared to be nonspecific, strengthening the case that another type of RNA serves as an intermediary in the process of protein synthesis.

That same year, Sol Spiegelman and his colleagues isolated ^{32}P-labeled phage RNA following infection of bacteria and used it in molecular hybridization studies. They tried hybridizing this RNA to the DNA of both phages and bacteria in separate experiments. The RNA hybridized only with the phage DNA, showing that it was complementary in base sequence to the viral genetic information.

The results of all these experiments agree with the concept of a **messenger RNA (mRNA)** being made on a DNA template and then directing the synthesis of specific proteins in association with ribosomes. This concept was formally proposed by François Jacob and

TABLE 18.1

Base compositions (in mole percents) of RNA produced immediately following infection of *E. coli* by the bacteriophages T2 and T7 in contrast to the composition of RNA of uninfected *E. coli*.

	Adenine	Thymine	Uracil	Cytosine	Guanine
Postinfection RNA in T2-infected cells	33	—	32	18	18
T2 DNA	32	32	—	17*	18
Postinfection RNA in T7-infected cells	27	—	28	24	22
T7 DNA	26	26	—	24	22
E. coli RNA	23	—	22	18	17

*5-hydroxymethyl cytosine.

SOURCE: From Volkin and Astrachan, 1956; and Volkin, Astrachan, and Countryman, 1958.

Jacques Monod in 1961 as part of a model for gene regulation in bacteria. Since then, mRNA has been isolated and thoroughly studied. There is no longer any question about its role in genetic processes.

RNA Polymerase

In order to prove that mRNA, as well as rRNA and tRNA, may be synthesized on a DNA template, it was necessary to demonstrate that there is an enzyme capable of directing this synthesis. By 1959, several investigators, including Samuel Weiss, had independently discovered such a molecule. The enzyme, called **RNA polymerase,** directs the synthesis of the three types of RNA—**messenger, ribosomal,** and **transfer**—that are complementary to DNA templates. RNA polymerase has the identical requirements of DNA polymerase, except that ribonucleoside triphosphates rather than the deoxy forms are used:

$$n(\text{NTP}) + \text{DNA} \xrightarrow[\text{enzyme}]{\text{Mg}^{++}} (\text{NMP})n + \text{DNA} + n(PP_i)$$

The holoenzyme from *E. coli* has been extensively characterized and shown to consist of subunits designated α, β, β', and σ. The active form of the enzyme $\alpha_2\beta\beta'\sigma$ has a molecular weight of approximately 500,000 daltons. The **σ (sigma) subunit** can be removed from the complex without loss of catalytic activity to the remaining **core enzyme.** The sigma component is believed to play a regulatory function and to be involved in recognition of the points along the DNA template where RNA transcription is initiated, called **initiation sites** or **promoters.**

In *E. coli* there are two different RNA polymerases. One is involved in producing the RNA primer during replication (see Chapter 10); the other transcribes mRNA, tRNA, and rRNA. In eukaryotes, separate polymerases are involved in the transcription of these three types of RNA. The nomenclature used in describing these is summarized in Table 18.2.

TABLE 18.2
RNA polymerases in eukaryotes.

Type	Product	Location
I	rRNA	Nucleolus
II	mRNAs	Nucleoplasm
III	5S RNA	Nucleoplasm
	tRNAs	Nucleoplasm

The Promoter Region and Template Binding

The process of transcription occurs on a DNA template and results in the synthesis of a complementary single-stranded RNA molecule. Most evidence suggests that, under cellular conditions, only one of the two strands of the DNA duplex is transcribed. The one transcribed is called the **sense strand,** and its complement is called the **antisense strand.**

Transcription may be divided into four stages: **template binding, chain initiation, chain elongation,** and **chain termination** (Figure 18.1). Template binding involves RNA polymerase and DNA. **Promoter** regions along the sense strand are recognized by the sigma subunit of the holoenzyme. These initiation sites consist of less than 50 deoxyribonucleotide residues and are rich in adenine and thymine. This information supports the idea that the helix is opened up at the point of template binding since A-T base pairs are more easily denatured than G-C pairs.

The promoter regions of DNA are isolated in an ingenious way. DNA representing isolated genes is incubated with copies of the RNA polymerase holoenzyme, but no ribonucleoside triphosphates are added to the reaction mixture. The polymerases bind to the promoter regions, but no transcription occurs. Then, endonucleases are added that cleave DNA except where it is protected by the polymerase. Those protected regions are isolated, and the DNA is separated from the polymerase and sequenced. In this way, the nucleotide sequences of promoter regions from various viruses and bacteria have been ascertained.

Two interesting observations resulted from the determination of the nucleotide sequences of promoter regions. First, all promoters isolated to date from bacteria and viruses have what have been called **Pribnow sequences.** These sequences (named after David Pribnow, who first pointed them out) consist of seven nucleotides taking the general form.

$$3' \ldots \text{–A–T–A–pyrimidine–T–A–pyrimidine} \ldots 5'$$

Although there is some variation found, positions 2 (T) and 6 (A) are invariant. The Pribnow sequence is believed to be essential to the promoter region's function, which is the initial binding of the sigma subunit of the holoenzyme.

A similar sequence has been found in most eukaryotic promoters thus far analyzed. Called the **Goldberg-Hogness box** or **TATA box,** it consists of the six-nucleotide sequence **TATXAX,** where X is usually T or A. Several examples are compared with Pribnow sequences in Table 18.3. Because of the minor variation apparent in all of these, they represent a **consensus**

FIGURE 18.1
Schematic diagram of the
events occurring during
transcription.

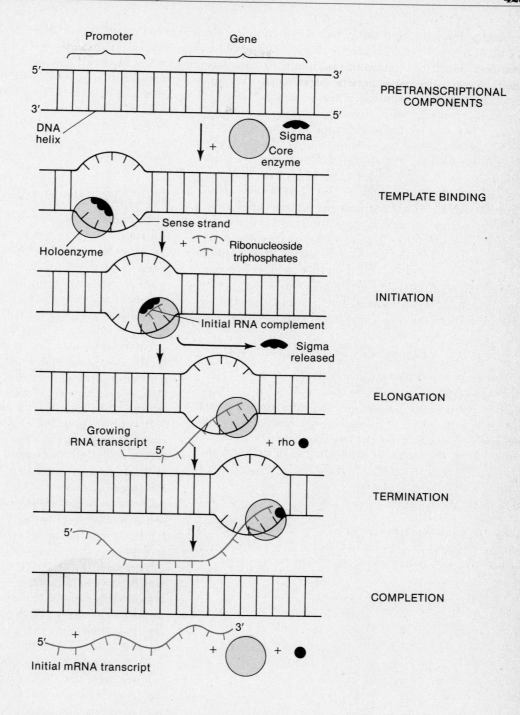

TABLE 18.3
Pribnow and TATA
box sequences.

Type of Sequence	General Group	Organism or Virus	Sequence
Pribnow sequence	Viruses	φX174	· · · TACAGTA · · ·
		SV40	· · · TATAATG · · ·
	Prokaryotes	*E. coli*	· · · TATAATG · · ·
TATA box	Eukaryotes	Mouse	· · · TATAAAG · · ·
		Chicken	· · · TATATAT · · ·

sequence, implying that they serve a common but vital function and have thus been preserved throughout evolution. Since RNA polymerase binds to the promoter region, this interaction represents an obvious point of potential regulation of transcription. There is some evidence that this consensus sequence may be involved. We will return to this topic in Chapter 19.

The second interesting observation is the discovery of **palindromes** in the nucleotide sequences of the complementary DNA strands making up the promoter regions. Palindromes, such as the "**bard saw regal lager was drab**" and "**madam Im adam,**" read identically forward and backward. In the same way the sequence of nucleotide pairs reads identically from right-to-left and from left-to-right along the duplex. Thus, if the sequence of both strands within the palindrome is read in the 3′-to-5′ direction or in the 5′-to-3′ direction, it is identical. For example, the shaded portion of the following sequence is palindromic:

$$3' \cdots \text{G—C—A—T—A—G—C—T—A—T—C—G} \cdots 5'$$
$$5' \cdots \text{C—G—T—A—T—C—G—A—T—A—G—C} \cdots 3'$$

Such a palindrome creates an axis of symmetry that can be split in the middle and have a nonpalindromic sequence of nucleotides (N) inserted:

$$3' \cdots \text{G—C—A—T—A—G—N—N—N—C—T—A—T—C—G} \cdots 5'$$
$$5' \cdots \text{C—G—T—A—T—C—N—N—N—G—A—T—A—G—C} \cdots 3'$$

In theory, a palindrome allows for the formation of the two hairpin loops within the duplex because a complementary sequence exists along each strand:

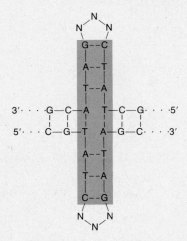

This creates a central structure called a **cruciform** where the junction of four duplexes occurs. While the formation of a hairpin loop produces an obvious recognition site along the helix, and thus might consti-

tute a specific signal for RNA polymerase, there is no direct evidence that such a structure occurs *in vivo.* In fact, such loops are energetically less stable than an uninterrupted duplex of DNA. However, palindromes have also been found in initiating and terminating regions of genes, places where strand separation occurs during transcription. It is unlikely that the conservation of palindromic sequences in specific parts of many genes is due to chance.

The Synthesis of RNA

The residues of the promoter region are themselves not transcribed. Instead, the enzyme moves along the DNA molecule until it encounters the first two nucleotides to be transcribed. In phage and bacterial genes studied, these are found six to seven nucleotides from the Pribnow sequence. The ribonucleotide complements are inserted and linked by a phosphodiester bond, resulting in chain initiation.

Once these first two nucleotides have been linked, **chain elongation** proceeds rapidly in the 5′-to-3′ direction, antiparallel to the DNA template strand. The elongation process produces a temporary DNA/RNA hybrid. After about 10 ribonucleotides have been added to the growing RNA chain, the σ subunit is dissociated from the holoenzyme, and elongation proceeds under the direction of the core enzyme. In *E. coli* this process proceeds at the rate of about 50 nucleotides/second at 37°C. Eventually, the enzyme encounters a **termination signal,** a specific nucleotide sequence, and synthesis is completed in conjunction with the **termination factor, rho (ρ).** In certain cases termination occurs in the absence of the rho factor. In either case, the transcribed RNA molecule is released from the DNA template, and the core enzyme dissociates. Under the direction of RNA polymerase, an RNA molecule is synthesized that is precisely complementary to a DNA sequence representing a gene. Wherever an A, T, C, or G residue existed on the DNA template, a corresponding U, A, G, or C residue has been incorporated into the RNA molecule, respectively. The significance of this synthesis is enormous, for it is the initial step in the process of **information flow** within the cell.

Visualization of Transcription

Electron microscope studies by Oscar Miller, Jr., Barbara Hamkalo, and Charles Thomas have provided striking visual demonstrations of the transcription process. Figure 18.2 shows micrographs and interpretive drawings of the DNA from two organisms, the newt, *Notophthalmus viridescens* and the bacterium *E.*

coli. In both cases, the micrographs capture the process of RNA transcription on DNA templates. In the case of the newt, the segment of DNA is derived from oocytes that are known to produce an enormous amount of rRNA, which then becomes part of the ribosome. To accomplish this synthesis, the genes specific for RNA (**rDNA**) are replicated many times in the oocyte. This process is called **gene amplification.** The micrograph in Figure 18.2(a) shows many of these genes in tandem, each being transcribed simultaneously numerous times. For each rRNA gene, simultaneous transcription results in progressively longer and longer strands of incomplete rRNA molecules as the enzymes move along the DNA strand.

A different picture emerges from the study of *E. coli,* as seen in Figure 18.2(b). Because prokaryotes lack nuclei, cytoplasmic ribosomes are not separated from the bacterial chromosome. As a result, ribosomes are able to attach to partially transcribed mRNA molecules even before transcription is complete. Again, RNA molecules become progressively longer as transcription proceeds, with more and more ribosomes attaching to the longer strands. Visualizing the transcription process confirms what has previously been deduced from biochemical studies.

TRANSLATION: PROTEIN SYNTHESIS

Translation is the biological polymerization of amino acids into polypeptide chains. The process occurs only in association with ribosomes. The central question in translation is, How do triplet ribonucleotides of mRNA direct specific amino acids into their correct position in the polypeptide? This question was answered once transfer RNA (tRNA) was discovered. This class of molecules adapts specific triplet codons in mRNA to their correct amino acids. This adaptor role of tRNA was postulated by Crick in 1957.

In association with a ribosome, mRNA presents a triplet codon that calls for a specific amino acid. Because a specific tRNA molecule contains in its composition three consecutive ribonucleotides complementary to the **codon,** it is thus called the **anticodon** and can base pair with the codon. This tRNA is covalently bonded to the amino acid called for. As this process occurs over and over as mRNA runs through the ribosome, amino acids are polymerized into a polypeptide. The aspect most critical to the fidelity of translation is the codon-anticodon recognition.

In our discussion of translation, we will first consider the structure of the ribosome and transfer RNA. We will then subdivide the translation process into the following four phases and discuss each separately:

tRNA charging, chain initiation, chain elongation, and chain termination.

Ribosomal Structure

Because of its essential role in the expression of genetic information, the ribosome has been extensively analyzed. Bacterial cells contain about 10,000 of these structures, while eukaryotic cells contain many times more. Electron microscopy has revealed that the ribosome is about 250 Å in its largest diameter and consists of a larger and a smaller granule or particle intimately associated with each other. Called **subunits,** these particles can be separated and analyzed biochemically. Although the specific details differ among organisms, particularly between prokaryotes and eukaryotes, all ribosomes share certain characteristics and features:

1 Both subunits consist of one or more molecules of rRNA and an array of **ribosomal proteins.**

2 The large subunit contains two rRNA molecules, the larger of which is always composed of a longer rRNA molecule than that found in the small subunit.

3 The number of proteins is always greater in the large subunit than in the small subunit.

4 The genes coding for the rRNA molecules are redundant, falling into the **moderately repetitive** category of DNA in eukaryotes.

5 The initial transcript of rRNA is larger than the final products incorporated into the subunits. The initial transcript is processed by endonuclease action, presumably in the nucleolus, in the case of eukaryotes. Specific enzymatic digestion leads to the final rRNA components of each subunit.

6 The functional rRNA molecules contain numerous **modified bases,** particularly ones that have been methylated.

With these characteristics and features established, the details of the specific differences between prokaryotic and eukaryotic ribosomes are summarized in Figure 18.3. The subunit and rRNA components are most easily isolated and characterized on the basis of their sedimentation behavior in sucrose gradients.

The two subunits associated with each other constitute a **monosome.** In prokaryotes the monosome is a 70S particle, and in eukaryotes it is approximately 80S. Recall that sedimentation coefficients are not additive. For example, the 70S monosome consists of a 50S and 30S subunit, and the 80S monosome consists of a 60S and 40S subunit. The larger subunit in pro-

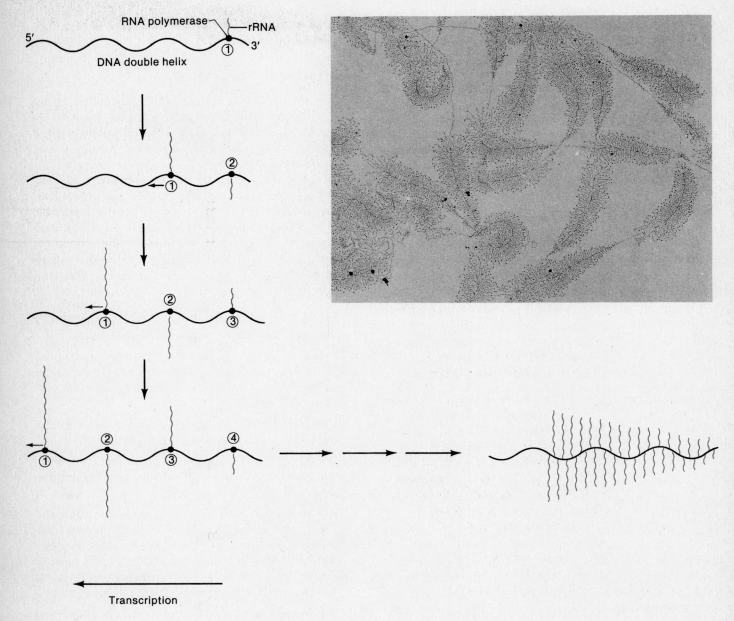

FIGURE 18.2

(a) Direct visualization of simultaneous transcription of multiple copies of the gene coding for rRNA in the oocyte of the newt *Notophthalmus (Triturus) viridescens.* The DNA was derived from one of many small nucleoli present in the oocyte. The genes are separated by noncoding regions called spacer DNA. The diagram illustrates the transcription events leading to what is seen in the electron micrograph. Each circled number represents an independent transcriptive event, initiated on the 3′ end of DNA and proceeding to the 5′ (left) end. (b) Direct visualization of transcription of an unidentified gene of *E. coli* and the simultaneous translation of each mRNA transcript by multiple ribosomes. The diagram illustrates the sequence of transcription and translation events that lead to what is seen in the micrograph. [Part (a) micrograph from Miller and Beatty, 1969, by permission of Alan R. Liss, Inc. Part (b) micrograph from Miller, Hamkalo, and Thomas, 1970. Copyright 1970 by the American Association for the Advancement of Science.]

RNA polymerase
5'
mRNA transcript
5'
DNA helix
3'
①

Ribosome
(a)
②
①

(c)
(b)
(a)
(a)
②
③
①
(a)
(b)

(d)
(c)
(b)
(a)
(b)
(a)
②
③
④
①
(a)
(a)
(b)
(c)

Transcription

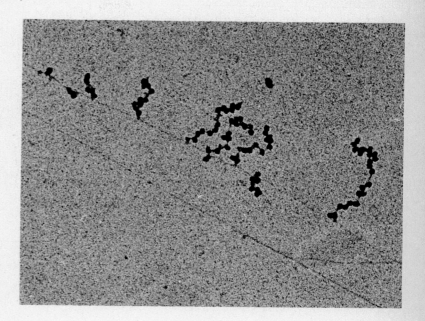

karyotes consists of a 23S RNA molecule, a 5S rRNA molecule, and 34 ribosomal proteins. In the eukaryotic equivalent, a 28S rRNA molecule, a 5S rRNA molecule and many more than 34 proteins are found. In the smaller prokaryotic subunit, a 16S rRNA component and 21 proteins are found. In the eukaryotic equivalent, an 18S rRNA component and many more than 21 proteins are found. The approximate molecular weights and number of nucleotides of these components are shown in Figure 18.3.

Molecular hybridization studies have established the degree of redundancy of the genes coding for the rRNA components. The E. coli genome contains seven copies of a single sequence that codes for all three components—23S, 16S, and 5S. The initial transcript of these genes produces a 30S RNA molecule that is enzymatically cleaved into the above components.

In eukaryotes, many more copies of a sequence encoding the 28S and 18S components are present. In Drosophila, approximately 120 copies per haploid genome transcribe a molecule of about 34S. This is processed to the 28S and 18S rRNA species. In X. laevis, over 500 copies per haploid genome are present. In mammalian cells, the initial transcript is 45S. The rRNA genes are part of the moderately repetitive DNA fraction and are present in clusters at various chro-

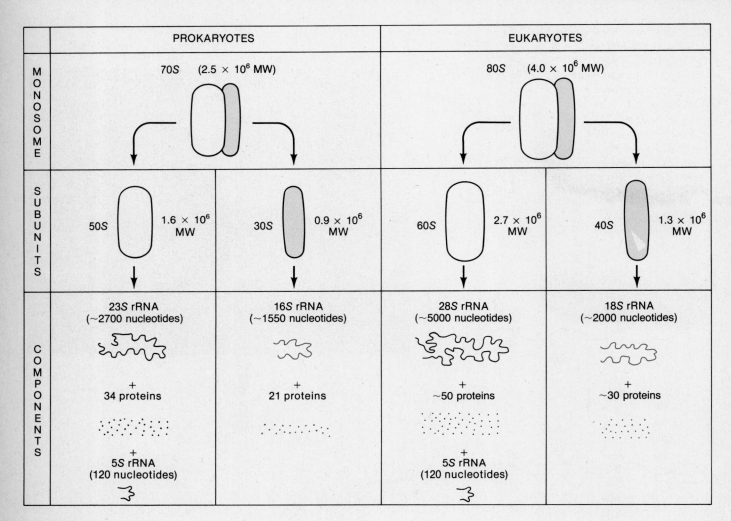

FIGURE 18.3

A comparison of the components in the prokaryotic and eukaryotic ribosomes.

mosomal sites. Each cluster consists of **tandem repeats,** with each unit separated by a noncoding **spacer DNA** sequence [Figure 18.2(b)]. In humans, these gene clusters have been localized on the ends of chromosomes 13, 14, 15, 21, and 22.

The 5S rRNA component of eukaryotes is not part of the larger transcript, as it is in *E. coli.* Instead genes coding for this ribosomal component are distinct and located separately. In humans, a gene cluster encoding them has been located on chromosome 1.

In spite of the detailed knowledge available on the structure and genetic origin of the ribosomal components, a complete understanding of the function of these components has eluded geneticists. This is not surprising, because the ribosome is perhaps the most complex of all cellular organelles. Together in bacteria,

the 3 rRNA components and 55 proteins are precisely organized into a structure with a combined molecular weight of 2.5 million daltons, about 65 percent of which is RNA. The study of assembly mapping of the proteins, particularly by Masayasu Nomura, has provided some insights into the construction of the components. While no single protein is responsible for the binding of other proteins during assembly, groups of specific proteins facilitate the addition of others as the structure is constructed.

Once organized into functional subunits, it is clear that a precise association of ribosomes is essential to the translation process. Some insights to this organization have been gained from studies of the prokaryotic ribosome, the three-dimensional structure of which is illustrated in Figure 18.4.

FIGURE 18.4
Three-dimensional models of the small and large ribosomal subunits and their interaction forming the monosome in *E. coli*. Two views of both subunits are provided. The numerous functional sites, regions, and domains that have been identified on the surface of each subunit are shown. (Drawing courtesy of Ms. Vickie Brewster.)

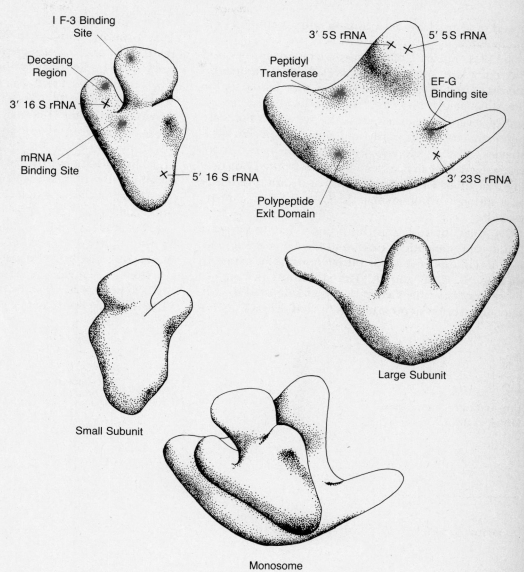

The rRNA molecules are primarily confined to the interior of each ribosomal subunit, while proteins occupy most of the surface. However, studies using immune electron microscopy by James Lake and others show that the 3′ ends of all three rRNAs as well as the 5′ end of the 16*S* rRNA are found on the surfaces of their respective subunits. Certain functional domains have also been identified. The small subunit contains the mRNA binding site, the decoding region that binds the 3′ end of tRNA, and binding sites for the initiation factors. The large subunit contains regions for peptidyltransferase activity, the binding site for one of the elongation factors, and the exit point for the newly formed polypeptide chain. When the two subunits are combined into the functional monosome, each undergoes a conformational change and a groove is created between them where mRNA fits. The role of each of the above-mentioned components will soon become clear as we proceed with our discussion of translation.

tRNA Structure

Because of its small size and stability in the cell, tRNA has been investigated extensively. In fact, it is the best characterized RNA molecule. It is a cylindrical nucleic acid composed of only 75 to 90 nucleotides, displaying a nearly identical structure in bacteria and eukaryotes. In both types of organisms, tRNAs are transcribed as larger precursors, which are cleaved into mature 4*S* tRNA molecules. In *E. coli*, for example, tRNA^tyr is composed of 77 nucleotides, yet its precursor contains 126 nucleotides.

In 1965, Robert Holley and his coworkers reported the complete sequence of tRNA^{ala} (the superscript identifies the specific amino acid that binds to the tRNA) isolated from yeast. Their findings were the result of seven years of research, during which the techniques necessary to purify and separate specific tRNA species were developed. Methods to sequence moderate-sized RNA molecules were also worked out during that time. Their efforts paid handsome dividends, as a number of remarkable discoveries were soon revealed.

Of great interest was the finding that there are a number of nucleotides unique to tRNA. Each is a modification of one of the four nitrogenous bases expected in RNA (G, C, A, and U). The structures of some of these are shown in Figure 18.5. **Inosinic acid (I)** contains the purine **hypoxanthine,** which differs from ad-

enine only at position 6 of the heterocyclic ring. Others, **1-methyl inosinic acid (I^m)**, **1-methyl guanylic acid (G^m)**, and **N,N-dimethyl guanylic acid (G^m)**, vary only in the additional methyl groups. **Pseudouridine (ψ)**, on the other hand, has the uracil moiety attached by its 5-C rather than its normal 3-N to ribose. **Ribothymidylic acid (T)** contains thymine attached to ribose, creating an exception to the normal absence of thymidine in RNA. Finally, **dihydrouridylic acid (U^h)** contains additional hydrogens at the 5 and 6 carbon atoms of the pyrimidine ring.

These modified structures, sometimes referred to as unusual, rare, or odd bases, are created **post-transcriptionally.** That is, the unmodified base is produced during transcription, and then enzymatic reactions catalyze the modifications.

Holley's sequence analysis led him to propose the two-dimensional **cloverleaf model** of transfer RNA. It had been known that tRNA demonstrates a hyper-

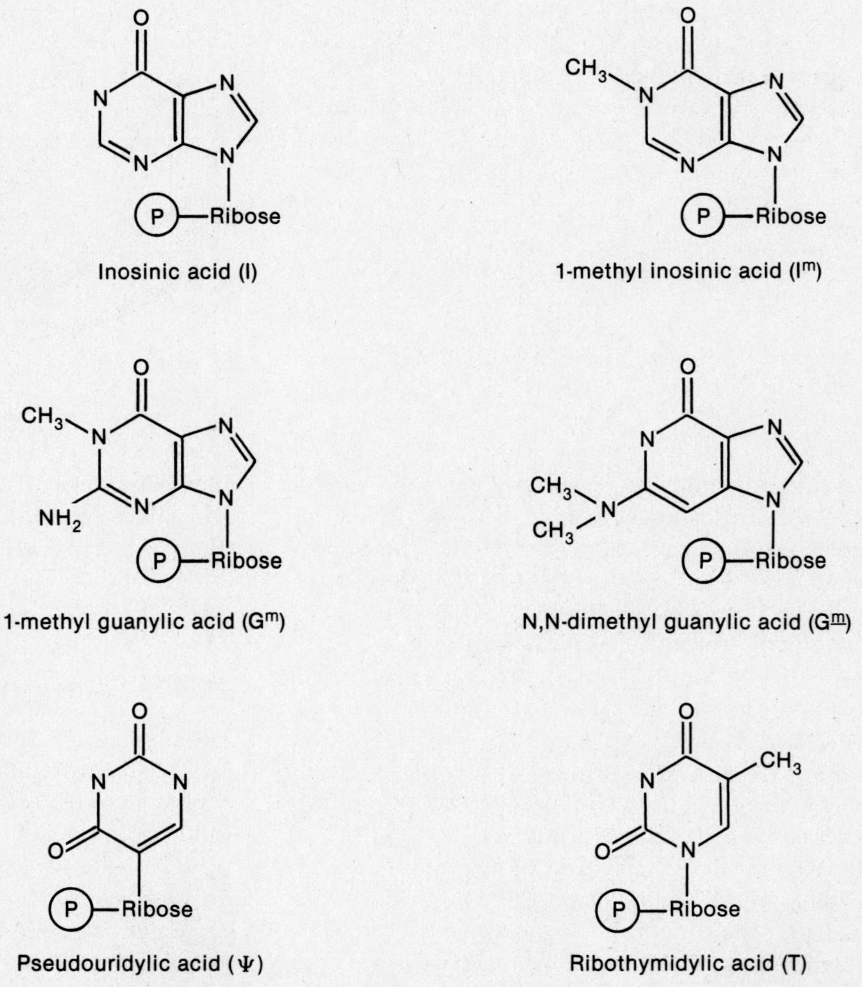

FIGURE 18.5
Unusual nitrogenous bases found in tRNA.

Inosinic acid (I)

1-methyl inosinic acid (I^m)

1-methyl guanylic acid (G^m)

N,N-dimethyl guanylic acid (G^m)

Pseudouridylic acid (ψ)

Ribothymidylic acid (T)

chromic shift when heated, indicative of secondary structure due to base pairing. Holley discovered that he could arrange the linear model in such a way that stretches of base pairing resulted. This arrangement created a series of **paired stems** and **unpaired loops** resembling the shape of a four-leaf clover. Loops consistently contained modified bases, which do not generally form base pairs. Holley's model is shown in Figure 18.6.

Since the triplets GCU, GCC, and GCA specify alanine, Holley looked for the theoretical anticodon of his tRNA^{ala} molecule. He found it in the form of CGI, at the bottom loop of the cloverleaf. Recall from Crick's wobble hypothesis that I is predicted to pair with U, C, or A. Thus, the anticodon loop was established.

As other tRNA species were examined, numerous constant features were observed. First, at the 3′ end, all tRNAs contain the sequence **. . . pCpCpA.** It is to the terminal adenine residue that the amino acid is joined covalently during charging. Interestingly, it appears that the CCA sequence is added post-transcriptionally in some tRNAs. At the 5′ terminus, all tRNAs contain **. . . pG.**

Additionally, the lengths of various stems and loops are very similar. In bacteria, yeast, and animal cells, the **acceptor stem** at the 3′ end consists of seven base pairs, the **T-stem** of five base pairs, and the **D-stem** of three or four base pairs (Figure 18.6). All tRNAs examined also contain an anticodon complementary

to the known amino acid code for which it is specific. These anticodon loops are present in the same position of the cloverleaf as well.

The cloverleaf model was predicted strictly on the basis of nucleotide sequence. Thus, there was great interest in X-ray crystallographic examination of tRNA, which reveals three-dimensional structure. By 1974, Alexander Rich and his coworkers in the United States, and J. Roberts, B. Clark, Aaron Klug, and their colleagues in England had been successful in crystallizing tRNA and performing X-ray crystallography at a resolution of 3 Å. At such resolution, the pattern formed by individual nucleotides is discernible.

As a result of these studies, a complete three-dimensional model is now available (Figure 18.7). The model reveals tRNA to be L-shaped. At one end of the L is the anticodon loop and stem, and at the other end are the acceptor and T-stems. At the corner are the T- and D-loops.

As different tRNAs were so examined, it became apparent that the stem areas were relatively constant. The loop areas, however, showed greater variation. Thus, the areas contributing to the stems provide a rather constant structure for tRNAs. The loops, on the other hand, vary in shape and thus provide specificity to the individual tRNA species. For example, the T-

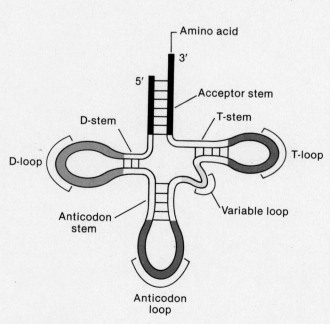

FIGURE 18.6
The two-dimensional cloverleaf model of tRNA.

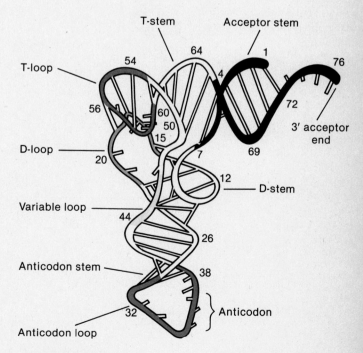

FIGURE 18.7
A three-dimensional model of tRNA. (Redrawn after Rich and Houkim, 1978.)

loop in the initiator tRNA has a different sequence and shape from T-loops of other tRNAs. It has been speculated that the loop shapes may be recognized by specific **aminoacyl synthetases,** the enzymes responsible for adding the amino acid to tRNA. However, this has yet to be shown.

Charging tRNA

Before translation can proceed, the tRNA molecules must be chemically linked to their respective amino acids. This process, called **charging,** occurs under the direction of enzymes called **aminoacyl synthetases.** Because there are 20 different amino acids, there must be at least 20 different tRNA molecules and as many different enzymes. In theory, since there are 61 triplet codes, there could be the same number of specific tRNAs and enzymes. Because of the ability of the third

member of a triplet code to "wobble," however, it is now thought that there are about 30 different tRNAs and corresponding enzymes found in the cytoplasm.

The charging process is outlined in Figure 18.8. In the initial step, the amino acid is converted to an **activated form,** reacting with ATP to form an **aminoacyladenylic acid.** A covalent linkage is formed between the 5′ phosphate group and the amino acid. This molecule remains associated with the enzyme, forming an activated complex which then reacts with a specific tRNA molecule. In this second step, the amino acid is transferred to the appropriate tRNA and bonded covalently. The charged tRNA may participate directly in protein synthesis. Aminoacyl-tRNA synthetases are highly specific enzymes because they recognize only one amino acid and only the corresponding tRNA. This is a crucial point if fidelity of translation is to be maintained.

FIGURE 18.8
(a) Steps involved in charging tRNA. The X denotes that for each amino acid, only a specific tRNA and a specific aminoacyl synthetase enzyme are involved in the charging process. (b) The chemical specificity of the amino acid-AMP interaction. (c) Charged tRNA structure.

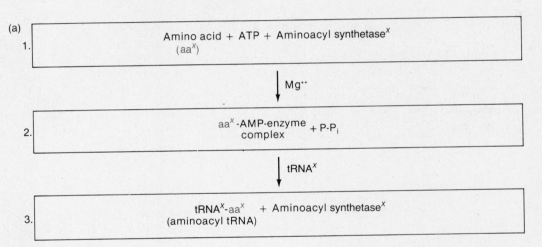

Initiation

Recall that ribosomes serve as a nonspecific workbench for the translation process. Single ribosomes, or monosomes, which are uninvolved in translation, do not exist free in the cytoplasm. Rather, they are dissociated into their large and small subunits. Initiation of translation of *E. coli* involves the ribosomal subunits, an mRNA molecule, a specific initiator tRNA, GTP, Mg^{++}, and at least three protein **initiation factors.** The protein initiator molecules are not part of the ribosome, but are required to enhance the binding affinity of the various translational components (Table 18.4). In prokaryotes, the initiation code of mRNA calls for the modified amino acid formylmethionine.

The overall scheme of initiation events is outlined in Figure 18.9. The small ribosomal subunit binds to an initiation protein, and this complex in turn binds to mRNA. Several other initiation proteins then enhance the binding of charged formylmethionyl-tRNA to the small subunit. This aggregate represents the **initiation complex,** which then combines with the large ribosomal subunit. In this process, a molecule of GTP is hydrolyzed to provide the required energy, and the initiation factors are released.

It is of interest to note that the synthesis of every *E. coli* protein begins with the modified amino acid **N-formylmethionine.** There are two tRNA species that may be charged with methionine. Only one, tRNA^{fmet}, serves as a substrate for the addition of a formyl group at the N-terminus of methionine. It is this charged tRNA that will bind to the small subunit initiation complex when AUG is the initial triplet of the mRNA. The other, tRNA^{met}, is utilized when AUG appears internally in mRNA. It has no affinity for binding to the small ribosomal subunit during the formation of the initiation complex. As we pointed out in Chapter 17, either the formyl group or the entire modified amino acid is subsequently cleaved from the polypeptide chain. In eukaryotes, no formyl group is involved in initiation, but methionine is thought to be the initial amino acid in many, if not all, polypeptides.

Elongation

As illustrated in Figure 18.9, the large ribosomal subunit contains two binding sites for charged tRNA molecules; these are labeled the **P,** or **peptidyl,** and the **A,** or **aminoacyl, sites.** The initiation tRNA binds to the P site, provided the AUG triplet is in the corresponding position of the small subunit. The sequence of the second triplet in mRNA dictates which charged tRNA molecule will become positioned at the A site. Once it is present, **peptidyl transferase** catalyzes the formation of the peptide bond, which links the two amino acids together (Figure 18.10). This enzyme is part of the large subunit of the ribosome. At the same time, the covalent bond between the amino acid and the

TABLE 18.4
Various protein factors involved during translation in *E. coli*.

Process	Factor	Role
Initiation	IF1	Stabilizes 30*S* subunit
	IF2	Binds fmet-tRNA to 30*S*-mRNA complex; Stimulates GTP hydrolysis
	IF3	Binds 30*S* subunit to mRNA; Dissociates monosomes into subunits following termination
Elongation	EF-Tu	Brings aminoacyl-tRNA to the A site
	EF-Ts	Generates active EF-Tu
	EF-G	Stimulates translocation; GTP-dependent
Termination	RF1	Catalyzes release of the polypeptide chain from tRNA and dissociation of the translation complex; Specific for UAA and UAG termination codons
	RF2	Behaves like RF1; Specific for UGA and UAA condons
	RF3	Stimulates RF1 and RF2

FIGURE 18.9
Initiation events in the translation of an mRNA into a polypeptide chain.

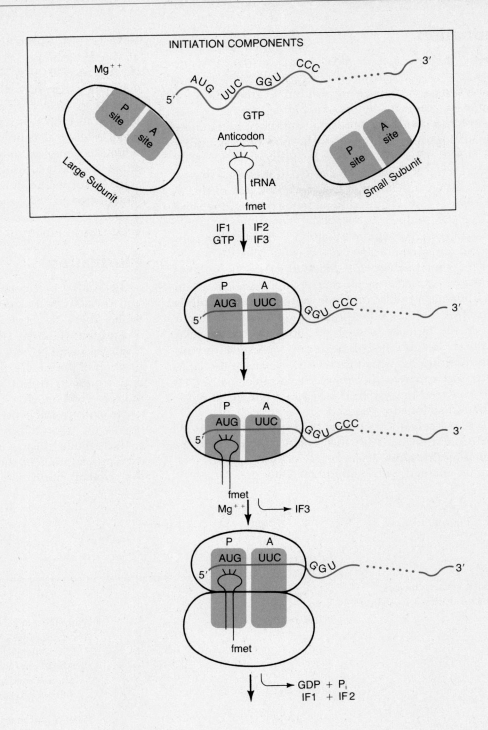

tRNA occupying the P site is hydrolyzed or broken. The product of this reaction is a dipeptide, which is attached to the tRNA at the A site. The step in which the growing polypeptide chain has increased in length by one amino acid is called **elongation.**

Before elongation can be repeated, **translocation** must occur. That tRNA attached to the P site, which is now uncharged, is released from the large subunit. The entire **mRNA–tRNA–aa₂–aa₂** complex now shifts in the direction of the P site by a distance of three nucleotides. This translocation event requires several protein elongation factors as well as the energy derived from hydrolysis of GTP (Table 18.4). The result is that the third triplet of mRNA is now in a position to direct another specific charged tRNA into the A site.

The sequence of elongation is repeated over and over. An additional amino acid is added to the growing polypeptide chain each time the mRNA advances

FIGURE 18.10
Following the formation of the initiation complex, elongation of the polypeptide chain begins under the direction of each successive mRNA triplet. Once the peptide bond is formed between the amino acids in the P and A sites of the large subunit of the ribosome, translocation occurs, bringing the third triplet into the A site. The process is repeated over and over, elongating the growing polypeptide chain by one amino acid at each step.

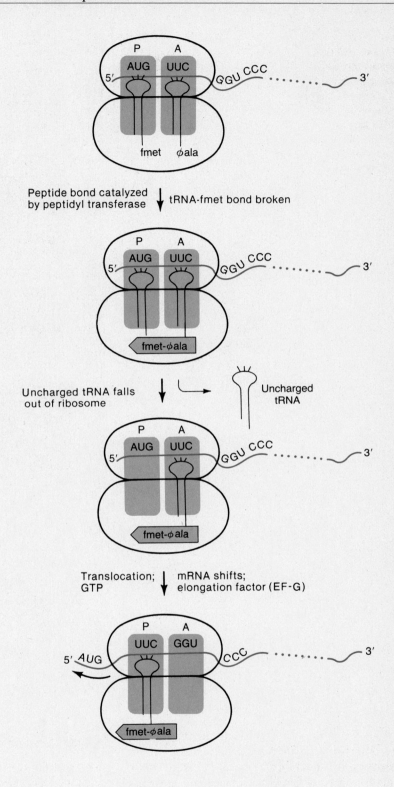

through the ribosome. In *E. coli* this occurs at a rate of about 15 amino acids per second at 37°C. The process of elongation can be likened to a tape moving through a tape recorder. As the tape moves, sound is emitted from the recorder. Likewise, as mRNA moves, a polypeptide is produced by the ribosome.

Termination

The termination of protein synthesis (Figure 18.11) is signaled by one of three triplet codes: UAG, UAA, or UGA. These codons do not specify an amino acid, nor do they direct tRNA into the A site. The finished poly-

FIGURE 18.11
Termination of translation.
The triplet UGA calls for no
amino acid, but is instead
recognized by release factors.
The terminal amino acid
(lysine, in this case) is
cleaved from the tRNA, and
the components of the
translocation complex
dissociate.

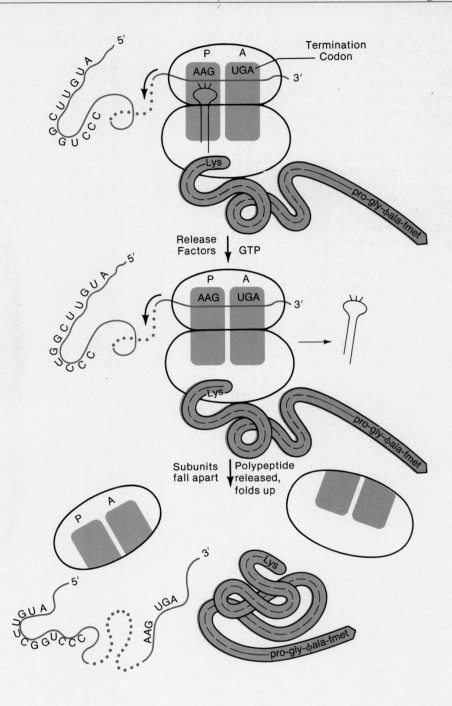

peptide is therefore still attached to the terminal tRNA
at the P site. The termination codon signals the action
of **GTP-dependent release factors** (Table 18.4),
which cleave the polypeptide chain from the terminal
tRNA. Once this cleavage occurs, the tRNA is released
from the ribosome, which then dissociates into its
subunits. If a termination codon should appear in the
middle of an mRNA molecule as a result of mutation,
the same process occurs and the polypeptide chain is
prematurely terminated.

Polyribosomes

As elongation proceeds and the initial portion of
mRNA has passed through the ribosome, the message
is free to associate with another small subunit to form
another initiation complex. This process can be repeated several times with a single mRNA and results
in what are called **polyribosomes** or just **polysomes.**
Polyribosomes can be isolated and analyzed following a gentle lysis of cells. Figures 18.12(a) and (b) illus-

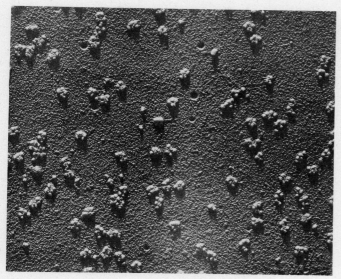

(a)

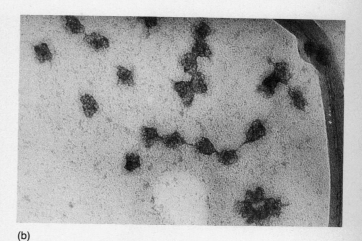

(b)

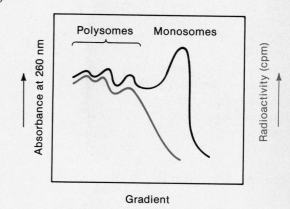

FIGURE 18.12
Polyribosomes (polysomes) visualized under the electron microscope [(a) and (b)] and in a sucrose gradient (c). Those viewed by electron microscopy were derived from rabbit reticulocytes engaged in the translation of hemoglobin mRNA. In (c), ribosome complexes have been analyzed for their absorbance at 260 nm and for radioactivity, as described in the text. (Photomicrographs from Rich et al., 1963. Reproduced by permission of Cold Spring Harbor Laboratory.)

trate these complexes as seen under the electron microscope. Note the presence of mRNA between the individual ribosomes in the micrograph on the right. In Figure 18.12(c), the sedimentation properties of polyribosomes are shown. In this experiment, the cells were exposed briefly to radioactive RNA precursors prior to their lysis. As a result, the mRNA fraction was radioactive. After the ribosome fraction was isolated, it was centrifuged in a sucrose gradient, which was then eluted into successive fractions. Each fraction was then measured for absorbance at 260 nm and for radioactivity (see Figure 8.12). Four absorbance peaks are apparent, each heavier than the one to the right of it. The three heaviest peaks represent polyribosome complexes consisting of four, three, and two ribosomes, respectively. The lightest peak is the single ribosome or monosome fraction. The radioactive mRNA is associated only with the polyribosome peaks, confirming that these complexes are the functional units of protein synthesis.

To complete the analogy with tapes (mRNA) and

tape recorders (ribosomes), in polysome complexes one tape would be played simultaneously, but the transcripts (polypeptides) would all be at different stages of completion.

TRANSCRIPTION AND TRANSLATION IN EUKARYOTES

Much of our knowledge of transcription and translation has been derived from studies of prokaryotes. The general aspects of the mechanics of these processes are believed to be similar in eukaryotes. There are, however, numerous notable differences, several of which have been discussed earlier in this chapter. We will first summarize the differences and then expand on several yet to be mentioned:

1 Transcription in eukaryotes occurs within the nucleus, and for the mRNA to be translated, it must move out into the cytoplasm.

2 Translation occurs on ribosomes that are larger and whose rRNA and proteins are more complex than those present in prokaryotes.

3 Protein factors similar to those in prokaryotes guide initiation, elongation, and termination of translation in eukaryotes. However, there appear to be more factors required during each of these steps.

4 Initiation of eukaryotic translation does not require the amino acid formylmethionine. However, the AUG triplet is essential to the formation of the translational complex and a unique transfer RNA (tRNA$_i^{met}$) is used for its initiation.

5 Most notably, extensive modifications occur to eukaryotic RNA transcripts that eventually serve as mRNAs. The initial transcripts are much larger than those that are eventually translated. Thus, they are called **pre-mRNAs** and are thought to constitute a group of molecules found only in the nucleus—a group referred to generally as **heterogeneous RNA (hnRNA).** Only about 25 percent of hnRNA molecules are converted to mRNA. Those that are have substantial amounts of their ribonucleotide sequence excised, while the remaining segments are spliced back together prior to translation. This phenomenon has given rise to the concept of so-called **split genes** in eukaryotes.

6 Prior to the processing of an mRNA transcript, a cap and a tail are added to the molecule. These modifications are essential to efficient processing and subsequently, to translation.

While we will return to several of these topics again in an ensuing chapter (see Chapter 19), we will conclude our discussion here by elaborating on the modifications that occur to initial RNA transcripts prior to their translation.

Heterogeneous RNA and its Processing

One of the first insights into regions of DNA that do not directly encode proteins has come from the study of RNA. This research has provided detailed knowledge of eukaryotic gene structure. The genetic code, as deciphered and shown in Figure 17.8, is written in the ribonucleotide sequence of mRNA. This information originated, of course, in the sense strand of DNA, where complementary sequences of deoxyribonucleotides exist. In bacteria, the relationship between DNA and RNA appears to be quite direct. The DNA base sequence is transcribed into an mRNA sequence, which is then translated into an amino acid sequence according to the genetic code.

However, in eukaryotes the situation is much more complex than in bacteria. It has been found that many internal base sequences of a gene may never appear in the mature mRNA that is translated. Other modifications occur at the beginning and the end of the mRNA prior to translation. These findings have made it clear that in eukaryotes, complex processing of mRNA occurs before it participates in translation.

By 1970, accumulating evidence showed that eukaryotic mRNA is initially transcribed as a much larger precursor molecule than that which is translated. This notion was based on the observation by James Darnell and his coworkers of **heterogeneous RNA (hnRNA)** in mammalian cells. Heterogeneous RNA is of large but variable size (up to 10^7 daltons), is found only in the nucleus, and is rapidly degraded. Nevertheless, hnRNA was found to contain nucleotide sequences common to the smaller mRNA molecules of the cytoplasm. Thus, it was proposed that the initial transcript of a gene results in a large RNA molecule which must first be processed in the nucleus before it appears in the cytoplasm as a mature mRNA molecule.

A subsequent discovery provided further evidence for this proposal. Both hnRNAs and mRNAs were found to contain at the 3′ end a stretch of up to 200 adenylic acid residues. Such **poly A** fragments are apparently added **post-transcriptionally** to the initial gene transcript. In higher eukaryotes, for example, transcription is terminated, and the 3′ end of the initial transcript is reduced in length close to an AAUAAA sequence and then polyadenylated. Subsequent investigation has shown poly A at the 3′ end of almost all mRNAs studied in a variety of eukaryotic organisms. The exceptions seem to be the products of histone genes and some yeast genes.

As illustrated in Figure 18.13, Darnell proposed that poly A is added to the RNA transcript, which is then processed before its transport to the cytoplasm. It appears that only the segment of nucleotides nearest the 3′ end is retained as mRNA. The majority of the RNA transcript, cleaved from the 3′ poly A fragment, is then rapidly degraded by nucleases. Nonsurviving fragments contain both unique and repetitive sequences. In 1974, Darnell proposed that this process is a means of discriminating which nuclear RNA transcripts survive to be translated.

Still another modification of eukaryotic mRNA has been discovered (Figure 18.14). At the 5′ end of these molecules is found a **7-methyl guanosine (7mG) residue,** or **cap.** This cap is also added post-transcriptionally and is bonded in a unique way. Instead of the traditional 5′-to-3′ phosphodiester bond, 7mG is linked in a 5′-to-5′ configuration to three phosphates of the first nucleotide. It appears that G is first added and then methylated. Additionally, the initial nucleo-

FIGURE 18.13
The conversion of heterogeneous RNA (hnRNA) into mRNA in eukaryotes.

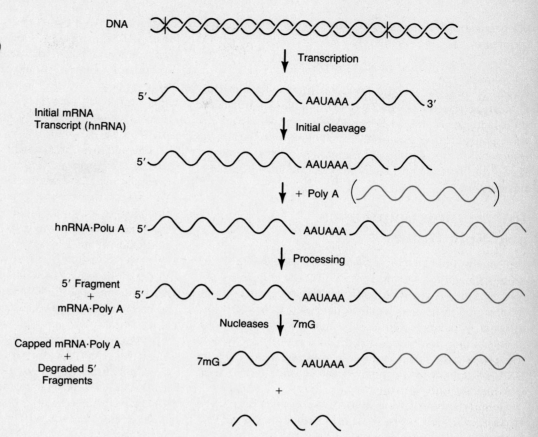

DNA

↓ Transcription

Initial mRNA Transcript (hnRNA)

5′ ——— AAUAAA ——— 3′

↓ Initial cleavage

5′ ——— AAUAAA ———

↓ + Poly A

hnRNA·Polu A   5′ ——— AAUAAA ———

↓ Processing

5′ Fragment
+
mRNA·Poly A

5′ ——— AAUAAA ———

Nucleases ↓ 7mG

Capped mRNA·Poly A
+
Degraded 5′ Fragments

7mG ——— AAUAAA ———

+

FIGURE 18.14
The 7-methyl guanosine cap of eukaryotic mRNA.

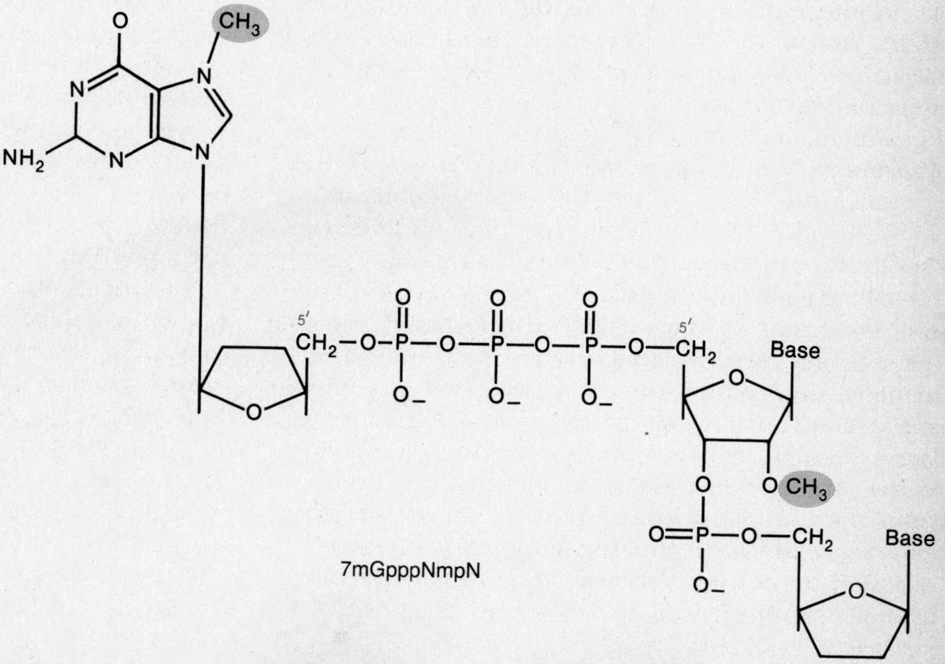

7mGpppNmpN

tides (N^1 and N^2) may also be methylated, yielding the sequence

$$7mGpppN^1mpN^2mpN^3pN^4p \ldots$$

where N^1 and N^2 are the first two nucleosides at the 5' end of the message. The significance of these 5' modifications of eukaryotic mRNAs is not yet clear but they appear to be essential for efficient translation. It may be that they stabilize the message during processing in the nucleus and facilitate the initiation of mRNA translation.

Intervening Sequences and Split Genes

One of the most exciting discoveries in the history of molecular genetics occurred in 1977. At this time, direct evidence was provided by Philip Sharp and others that animal viruses as well as eukaryotic genes contain internal nucleotide sequences that are not expressed in the amino acid sequence of the proteins for which the genes code. That is, certain internal sequences in DNA do not always appear in the mature mRNA that is translated into a protein.

Detailed investigation has revealed numerous examples in which more than one noncontiguous DNA sequence fails to appear in the final mRNA. Such nucleotide segments have been called **intervening sequences,** contained within so-called **split genes.** Those DNA sequences not present in the final mRNA product are also called **introns** ("*int*" for intervening), and those retained and expressed are called **exons** ("*ex*" for expressed).

Intervening sequences were first suggested by studies of the animal viruses **adenovirus 2 (Ad2)** and **SV40.** Viral mRNAs were discovered to contain ribonucleotide sequences derived from widely separated areas of the viral genome. This finding suggested that, in some manner, viral mRNAs were the product of an excision and rejoining process referred to as **splicing.**

Similar discoveries were soon to be made in eukaryotes. Two approaches have been most fruitful. The first involves molecular hybridization of purified, functionally mature mRNAs with DNA containing the gene specifying that message. When hybridization occurs between nucleic acids that are not perfectly complementary, **heteroduplexes** are formed that may be visualized with the electron microscope. As shown in Figure 18.15, if sequences exist in DNA but are absent in the corresponding mRNA, double-stranded DNA regions must loop out so that continuous hybridization may occur. Adjacent to these duplexes is the single-stranded antisense strand of DNA, which has been prevented from reannealing by the mRNA in hybrid-

ization. Such single-stranded structures are called **R-loops.** As Figure 18.15 shows, the absence of any intervening sequences will yield a single R-loop; one intron results in two R-loops; two introns produce three R-loops; and so on. This rather sophisticated technology allows the number of introns and their approximate size and location within the gene to be identified.

The second approach provides more specific information. It involves a comparison of nucleotide sequences of DNA with those of mRNA and amino acid sequences. Such an approach allows the precise identification of all intervening sequences.

So far, a large number of genes from diverse eukaryotes have been shown to contain introns. One of the first so identified was the **beta-globin gene** in mice and rabbits, as studied independently by Philip Leder and Richard Flavell. The mouse gene contains an intron 550 nucleotides long, beginning immediately after the codon specifying the 104th amino acid. In the rabbit (Figure 18.16), there is an intron of 580 base pairs near the codon for the 110th amino acid. Additionally, a second intron of about 120 nucleotides exists earlier in both genes. Similar introns have been found in the beta-globin gene in all mammals examined.

Several genes, notably those coding for histones and interferon, appear to contain no introns. However, intervening sequences have been identified in immunoglobulin genes of the mouse, some tRNA genes in yeast, and rRNA genes in *Drosophila*, among many others. The case of tRNA is interesting. Four different $tRNA^{tyr}$ genes have been sequenced. In each case, an intron of 14 or 15 nucleotide pairs has been found immediately adjacent to the anticodon sequence.

A more extensive set of introns has been located in the **ovalbumin gene** of chickens. The gene has been extensively characterized by Bert O'Malley in the United States and Pierre Chambon in France. As shown in Figure 18.16, the gene contains seven introns. Notice that the majority of the gene's DNA sequence is "silent," being composed of introns. The initial RNA transcript is four times the length of the mature mRNA.

The list of genes containing intervening sequences is growing rapidly. An extreme example of the number of introns in a single gene is that found in one of the chicken genes, *pro-α-2(I) collagen*. One of several genes coding for a subunit of this connective tissue protein, *pro-α-2(I) collagen* contains about 50 introns. The precision with which cutting and splicing occur must be extraordinary if errors are not to be introduced into the mature mRNA. The removal of just one intron and the subsequent splicing is illustrated in Figure 18.17. The loss or addition of a single ribonu-

FIGURE 18.15
R-loop formation in heteroduplexes with zero, one, or two intervening sequences.

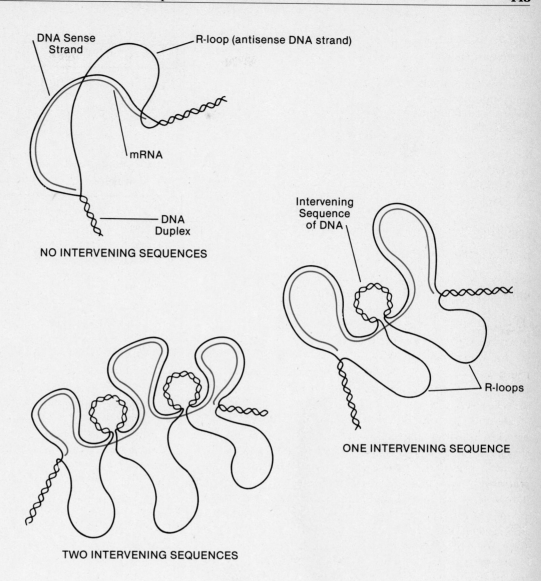

NO INTERVENING SEQUENCES

ONE INTERVENING SEQUENCE

TWO INTERVENING SEQUENCES

cleotide will create a shift in reading and result in missense and/or nonsense triplets.

Splicing Mechanisms

The discovery of split genes represents one of the most exciting genetic findings in recent years. As a result, intensive investigation is now in progress to elucidate the mechanism by which introns of RNA are excised and exons are spliced back together. A great deal of progress has already been made. Interestingly, it appears that somewhat different mechanisms are utilized for each of the three types of RNA as well as for RNAs produced in mitochondria.

The simplest mechanism appears to be used to process **tRNA molecules.** As you will recall, these are small RNAs consisting of approximately 80 nucleotides. Some of them contain a short internal intron of about 15 nucleotides. In such cases, initial folding of the molecule creates a **pre-tRNA** with an extra loop representing the intron. This is enzymatically removed. John Abelson has demonstrated this enzymatic process *in vitro*. The reactions can easily be visualized [Figure 18.8(a)]; a nuclease makes appropriate cuts in the RNA, leaving the ends that are to be joined in juxtaposition to one another. A ligase then links them together.

FIGURE 18.16
Intervening sequences in
various eukaryotic genes.
The numbers indicate
nucleotides present in
various intron and exon
regions.

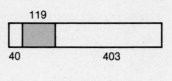

Rat Insulin

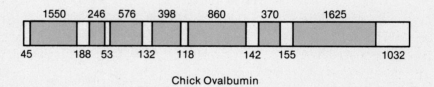

Rabbit β-globin

Chick Ovalbumin

☐ Exons

▨ Introns

FIGURE 18.17
Splicing of the initial RNA
transcript to remove a single
intervening sequence.

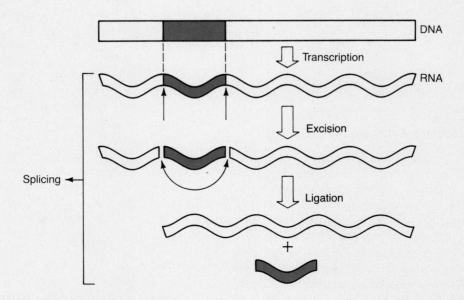

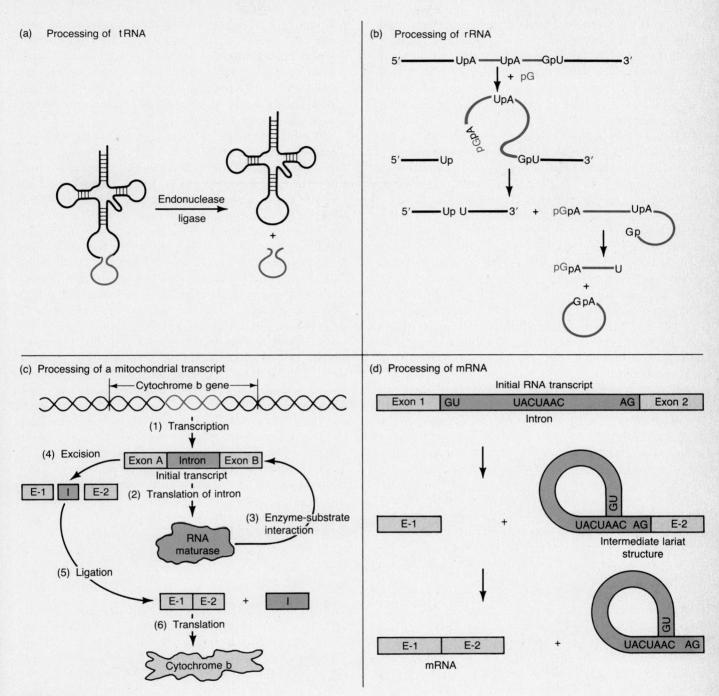

FIGURE 18.18
Various mechanisms utilized to remove introns and ligate exons from the initial RNA transcripts of tRNA, rRNA, mtRNA, and mRNA in eukaryotes. Intron regions are shown in color. Each of the four mechanisms illustrated is discussed in the text.

A second splicing mechanism has been discovered in the processing of **rRNA** by the protozoan, *Tetrahymena thermophila*. We have earlier indicated that 18S and 28S eukaryotic rRNA molecules are derived from a 45S precursor that must be extensively altered. Working with *Tetrahymena*, Thomas Cech and his colleagues have demonstrated that in one of the larger 26S precursors **self-excision** of an intervening sequence occurs. The reaction involves no external catalyst and appears to require only a monovalent and a divalent cation (NH_4^+ and Mg^{++}) as well as guanosine as a cofactor. As illustrated in Figure 18.18(b), the RNA chain is cleaved and a guanosine residue is added to the 5'-linear portion of the intron. Then a second step occurs where the intron is excised and the flanking exons are ligated. The intron, which is 414 nucleotides in length, is converted into a covalently closed circular form. The entire process is mediated by the RNA itself, and no enzymes are required. Thus, the RNA that undergoes this process behaves catalytically; it has been referred to as a **ribozyme.** A similar process has now been demonstrated in the splicing of rRNA introns in yeast and other fungi.

Another interesting splicing mechanism has been discovered for an RNA molecule transcribed on mitochondrial DNA (**mtDNA).** One of the introns of the gene specifying the respiratory molecule cytochrome b encodes a separate enzyme. Once synthesized, this enzyme, called **RNA maturase,** directs the excision and splicing of the initial RNA transcript from which it was derived! Once processed, the resultant RNA is then translated into cytochrome b. This mechanism was proposed originally on genetic grounds, where mutations in an intron inhibited the production of cytochrome b. These steps are illustrated in Figure 18.18(c).

The final mechanism to be discussed is the least well understood and involves eukaryotic **mRNAs.** Introns in mRNA, in comparison to other RNAs thus far discussed, can be much larger—up to 20,000 nucleotides—and they are more plentiful. Thus, their removal requires a much more complex mechanism, which until recently has been very difficult to define. Many clues are now emerging, however. First, the nucleotide sequence around different introns is often similar. Most begin at the 5' end with a GU dinucleotide sequence and terminate at the 3' end with an AG dinucleotide sequence. These, as well as other consensus sequences shared by introns, may attract a molecular complex essential to ligation and splicing. Such a complex has been identified in extracts of yeast as well as mammalian cells. Called a **spliceosome,** it is very large, being 40S in yeast and 60S in mammals. One component of this complex is thought to be a

member of a unique group of RNAs called **small nuclear RNA (snRNA).** One type called **U1** contains sequences complementary to those present at the 5' end of introns. These may serve as guides, causing introns to fold out into space so that their boundaries are placed in proximity to one another, facilitating intron excision and exon ligation.

Intermediate products of processing have now been isolated during *in vitro* studies. An interesting structure called a **lariat** (or closed loop) is formed, as illustrated in Figure 18.18(d). First, the 5' site adjacent to the first exon is cleaved, whereupon the intron folds back on itself, forming a covalent 5'–2' bond such that the lariat is still attached to the second exon. Then, the lariat structure is removed as the two exons are ligated. In many introns the sequence UACUAAC is located between 18 and 37 nucleotides proximal to the 3' end. It seems to be critical to the lariat formation, which is thought to be essential to the formation of the spliceosome.

Even with the above knowledge available, the complexity of the splicing complex renders its analysis most difficult. While we have discussed the removal of one intron in a single RNA transcript, large numbers must be cleaved out and the many exons spliced back together with absolute precision. To ensure such precision an extremely efficient mechanism must be at work.

The finding of "genes in pieces," as split genes have been described, raises many interesting questions and has provided great insights into the organization of eukaryotic genes. Thus, we will return to this topic again in the next chapter.

SUMMARY AND CONCLUSIONS

Transcription and translation—RNA and protein biosynthesis, respectively—are the basis for the expression of genetic information and constitute the central dogma in molecular genetics: DNA $\rightarrow$ RNA $\rightarrow$ protein. All of the various proteins in any cell can therefore be traced back to polynucleotide sequences of DNA known as genes. The complex mechanisms of transcription and translation assure a faithful flow of genetic information. Both processes, like DNA replication, may be subdivided into the stages of initiation, elongation, and termination, and both rely on basepairing affinities between complementary nucleotides.

Transcription, which has been visualized under the electron microscope, is under the control of the enzyme RNA polymerase. One of the strands of the DNA double helix serves as a template for the catalytic synthesis of a complementary mRNA. RNA polymerase is composed of four different subunits. One, the sigma (σ) subunit, is responsible for the binding at specific initiation sites (promoters) along DNA. The remainder of the molecule, called the core enzyme, is responsible for the 5'-to-3' transcription of DNA base sequences. Several different RNA polymerases have been discovered in prokaryotic and eukaryotic cells.

The information transferred to mRNA is coded in triplet sequences of ribonucleotides. Each series of three units specifies one amino acid. Of the 64 possible triplets, 61 code for the 20 amino acids, and 3 are termination codons. One of the 61, AUG, codes for methionine and serves as the initiation codon in prokaryotes.

The synthesis of proteins occurs as the genetic code is deciphered during translation. The process is complex and requires energy. Protein synthesis involves charged tRNA molecules, numerous proteins, ribosomes, and mRNA. Transfer RNA (tRNA) serves as an adaptor molecule. Each tRNA contains an anticodon complementary to a specific triplet codon in mRNA and is charged with the amino acid called for by the triplet. The mRNA moves through the ribosome much like a tape through a tape recorder. As it proceeds, a growing polypeptide chain is produced until, finally, it is released from the ribosome. Prior to termination, the mRNA may initiate transcription with one or more ribosomes, forming polyribosomes.

The processes of transcription and translation in eukaryotes are more complex than in prokaryotes. Of particular note is the presence of nucleotide sequences within genes that are not reflected in the final mRNA subsequent to its translation into a protein. Thus, eukaryotic genes are present in pieces; once an initial RNA transcript is produced, the intervening sequences, or introns, must be excised and these pieces (exons) spliced back together prior to translation. The mechanisms by which such RNA processing occur are varied and have been reviewed. Eukaryotic mRNAs contain the largest and the greatest number of introns compared with tRNA, mtRNA, and rRNA. Their analysis has been the most difficult.

PROBLEMS AND DISCUSSION QUESTIONS

1 Define and differentiate between transcription and translation. Where do these processes fit into the central dogma of molecular genetics?

2 What was the initial evidence for the existence of mRNA?

3 Describe the structure of RNA polymerase. What is the core enzyme? What is the role of the sigma subunit?

4 In a written paragraph, describe the abbreviated chemical reaction shown on page 424 that summarizes RNA polymerase-directed transcription.

5 What differences exist between the "visualization of transcription" studies of *E. coli* and *Notophthalmus?* Why?

6 List all of the molecular constituents present in a functional polyribosome. Including all components, diagram the translation process.

7 How do the three main steps of translation fit the above model? Diagram each step.

8 Contrast the roles of tRNA and mRNA during translation.

9 List all enzymes that participate in the transcription and translation process.

10 Francis Crick proposed the "adaptor hypothesis" for the function of tRNA. Why did he choose that description?

11 What molecule bears the codon? the anticodon?

12 Contrast the roles of the small and large ribosomal subunits from a point prior to initiation to a point following termination of translation.

13 Compare the components of ribosomes in prokaryotes and eukaryotes.

14 What is meant by "fidelity" of transcription and translation? Why is it important, and how is it maintained?

15 The α chain of eukaryotic hemoglobin is composed of 141 amino acids. What is the minimum number of nucleotides in an mRNA coding for this protein chain? Assuming that each nucleotide is 0.34 nm long in the mRNA, how many triplet codes can at one time occupy space in a ribosome that is 20 nm in diameter?

16   Messenger RNA molecules are very difficult to isolate in prokaryotes because they are rather quickly degraded in the cell. Can you suggest a reason why this occurs? Eukaryotic mRNAs are more stable and exist longer in the cell than do prokaryotic mRNAs. Is this an advantage or disadvantage for a pancreatic cell making large quantities of insulin?

17   Summarize the steps involved in charging tRNAs with their appropriate amino acids.

18   In order to carry out its role, each transfer RNA requires at least four specific recognition sites that must be inherent in its tertiary structure. What are they?

19   In 1962, F. Chapeville and others reported an experiment where they isolated radioactive ^{14}C-cysteinyl-tRNA ^{cys} (charged tRNA^{cys} + cysteine). They then removed the sulfur group from the cysteine, creating alanyl-tRNA^{cys}. When alanyl-tRNA^{cys} was added to a synthetic mRNA calling for cysteine but not alanine, a polypeptide chain was synthesized containing alanine. What can you conclude from this experiment?

20   A short RNA molecule was isolated that demonstrated a hyperchromic shift indicating secondary structure. Its sequence was determined to be AGGCGCCGACUCUACU.
   (a) Predict a two-dimensional model for this molecule.
   (b) What DNA sequence would give rise to this RNA molecule through transcription?
   (c) If the molecule were a tRNA fragment containing a CGA anticodon, what would the corresponding codon be?
   (d) If the molecule were part of a message, what amino acid sequence would result from it following translation? (Refer to the code chart in Figure 17.8.)

21   The following represent deoxyribonucleotide sequences derived from the sense strand of DNA:

   Sequence 1   CTTTTTTGCCAT
   Sequence 2   ACATCAATAACT
   Sequence 3   TACAAGGGTTCT

   (a) For each strand, determine the mRNA sequence that would be derived from transcription.
   (b) Using Figure 17.8, determine the amino acid sequence that would result from translation of these mRNAs.
   (c) If we assume that each sequence has been derived from one gene, which represents the initial part? the middle region? the terminal portion?

22   For Sequence 1 in Problem 21, what is the sequence of the antisense strand?

23   For Sequence 2, indicate the anticodons of the tRNAs that would participate in translation.

24   For Sequence 3, draw the structural formula of the polypeptide sequence resulting from translation.

25   Diagram and label the results found in an electron micrograph when the mature beta-globin mRNA is hybridized under appropriate conditions with DNA containing the beta-globin gene.

SELECTED READINGS

ABELSON, J. 1979. RNA processing and the intervening sequence problem. *Ann. Rev. Biochem.* 48: 1035–70.

BARALLE, F. E. 1983. The functional significance of leader and trailer sequences in eukaryotic mRNAs. *Int. Rev. Cytol.* 81: 71–106.

BIELKA, H., ed. 1982. *The eukaryotic ribosome.* New York: Springer-Verlag.

BIRNSTIEL, M., BUSSLINGER, M., and STRUB, K. 1985. Transcription termination and 3′ processing: The end is in site. *Cell* 41: 349–59.

BRENNER, S., JACOB, F., and MESELSON, M. 1961. An unstable intermediate carrying information from genes to ribosomes for protein synthesis. *Nature* 190: 575–80.

BRIMACOMBE, R., and STIEGE, W. 1985. Structure and function of ribosomal RNA. *Biochem. J.* 229: 1–17.

BRODY, E., and ABELSON, J. 1985. The "spliceosome": Yeast pre-messenger RNA associates with the 40S complex in a splicing dependent reaction. *Science* 228: 963–68

CECH, T. R., ZAUG, A. J., and GRABOWSKI, P. J. 1981. *In vitro* splicing of the ribosomal RNA precursor of *Tetrahymena*. Involvement of a guanosine nucleotide in the excision of the intervening sequence. *Cell* 27: 487–96.

CHAMBON, P. 1981. Split genes. *Scient. Amer.* (May) 244: 60–71.

CHAPEVILLE, F., et al. 1962. On the role of soluble ribonucleic acid in coding for amino acids. *Proc. Natl. Acad. Sci.* 48: 1086–93.

COUTELLE, C. 1981. The precursor to animal cell messenger RNA. *Biochem. J.* 197: 1–6.

CRICK, F. 1979. Split genes and RNA splicing. *Science* 204: 264–71.

DANCHIN, A., and SLONIMSKI, P. 1985. Split genes. *Endeavour* 9: 18–27.

DARNELL, J. E. 1978. Implications of RNA: RNA splicing in the evolution of eukaryotic cells. *Science* 202: 1257–60.

_____. 1985. RNA. *Scient. Amer.* (Oct.) 253: 68–87.

_____. 1983. The processing of RNA. *Scient. Amer.* (Oct.) 249: 90–100.

DARNELL, J. E., JELINEK, W. R., and MOLLOY, G. R. 1973. Biogenesis of mRNA: Genetic regulation in mammalian cells. *Science* 181: 1215–21.

DUGAICZK, A., et al. 1978. The natural ovalbumin gene contains seven intervening sequences. *Nature* 274: 328–33.

DYSON, R. D. 1978. *Cell biology—A molecular approach.* 2nd ed. Boston: Allyn and Bacon.

GREENBERG, J. R. 1975. Messenger RNA metabolism of animal cells. *J. Cell Biol.* 64: 269–88.

HALL, B. D., and SPIEGELMAN, S. 1961. Sequence complementarity of T2-DNA and T2-specific RNA. *Proc. Natl. Acad. Sci.* 47: 137–46.

HAMKALO, B. 1985. Visualizing transcription in chromosomes. *Trends in Genet.* 1: 255–60.

HOLLEY, R. W., et al. 1965. Structure of a ribonucleic acid. *Science* 147: 1462–65.

HOOD, L. E., WILSON, J. H., and WOOD, W. B. 1975. *Molecular biology of eukaryotic cells.* Menlo Park, Calif.: W. A. Benjamin.

INGRAM, V. M. 1971. *Biosynthesis of macromolecules.* 2nd ed. Menlo Park, Calif.: W. A. Benjamin.

KINNIBURGH, A. J., and ROSS, J. 1979. Processing of the mouse β-globin mRNA precursor: At least two cleavage-ligation reactions are necessary to excise the larger intervening sequence. *Cell* 17: 915–21.

KISH, V., and PEDERSON, T. 1975. Ribonucleoprotein organization of polyadenylate sequences in HeLa cell heterogeneous nuclear RNA. *J. Mol. Biol.* 95: 227–38.

LAKE, J. A. 1981. The ribosome. *Scient. Amer.* (Aug.) 245: 84–97.

LAZOWSKA, J., JACQ, C., and SLONIMSKI, P. P. 1980. Sequence of introns and flanking exons in wild type and box 3 mutants of cytochrome b reveals an interlaced splicing protein coded by an intron. *Cell* 22: 333–48.

LEHNINGER, A. L. 1982. *Principles of biochemistry.* New York: Worth.

LEWIN, B. 1974. *Gene expression I: Bacterial genomes.* New York: Wiley.

_____. 1985. *Genes.* 2nd ed. New York: Wiley.

LERNER, M. R., and STEITZ, J. A. 1981. Snurps and scyrps. *Cell* 25: 298–300.

MILLER, O. L., and BEATTY, B. R. 1969. Portrait of a gene. *J. Cell Physiol.* 74 (Supplement 1): 225–32.

MILLER, O. L., HAMKALO, B., and THOMAS, C. 1970. Visualization of bacterial genes in action. *Science* 169: 392–95.

NOLLER, H. F. 1984. Structure of ribosomal RNA. *Ann. Rev. Biochem.* 53: 119–62.

NOMURA, M. 1984. The control of ribosome synthesis. *Scient. Amer.* (Jan.) 250: 102–14.

————. 1973. Assembly of bacterial ribosomes. *Science* 179: 864–73.

OHNO, S. 1980. Origin of intervening sequences within mammalian genes and the universal signal for their removal. *Differentiation* 17: 1–15.

PATWARDHAN, S., et al. 1985. Splicing of messenger RNA precursors. *BioEssays* 2: 205–08.

PEDERSON, T. 1981. Messenger RNA biosynthesis and nuclear structure. *Amer. Scient.* 69: 76–84.

REED, R., and MANIATIS, T. 1985. Intron sequences involved in lariat formation during pre-mRNA splicing. *Cell* 41: 95–105.

RICH, A., and HOUKIM, S. 1978. The three-dimensional structure of transfer RNA. *Scient. Amer.* (Jan.) 238: 52–62.

RICH, A., WARNER, J. R., and GOODMAN, H. M. 1963. The structure and function of polyribosomes. *Cold Spr. Harb. Symp.* 28: 269–85.

ROGERS, J., and WALL, R. 1980. A mechanism for RNA splicing. *Proc. Natl. Acad. Sci.* 77: 1877–79.

SHARP, P. A. 1981. Speculations on RNA splicing. *Cell* 23: 643–46.

SHATKIN, A. J. 1985. mRNA cap binding proteins: Essential factors for initiating translation. *Cell* 40: 223–24.

SMITH, J. D. 1972. Genetics of tRNA. *Ann. Rev. Genet.* 6: 235–56.

SOLL, D. 1971. Enzymatic modification of transfer RNA. *Science* 173: 293–99.

STENT, G. S., and CALENDAR, R. 1978. *Molecular genetics—An introductory narrative.* 2nd ed. San Francisco: W. H. Freeman.

STRYER, L. 1981. *Biochemistry.* 2nd ed. San Francisco: W. H. Freeman.

SUDDATH, F. L., et al. 1974. Three-dimensional structure of yeast phenylalanine transfer RNA at 3.0 Å resolution. *Nature* 248: 20–24.

TAYLOR, J. H., ed. 1965. *Selected papers on molecular genetics.* New York: Academic Press.

VOLKIN, E., and ASTRACHAN, L. 1956. Phosphorus incorporation in *E. coli* ribonucleic acids after infection with bacteriophage T2. *Virology* 2: 149–61.

VOLKIN, E., ASTRACHAN, L., and COUNTRYMAN, J. L. 1958. Metabolism of RNA phosphorus in *E. coli* infected with bacteriophage T7. *Virology* 6: 545–55.

WARNER, J., and RICH, A. 1964. The number of soluble RNA molecules on reticulocyte polyribosomes. *Proc. Natl. Acad. Sci.* 51: 1134–41.

WATSON, J. D. 1963. Involvement of RNA in the synthesis of proteins. *Science* 140: 17–26.

WITTMAN, H. G. 1982. Components of bacterial ribosomes. *Ann. Rev. Biochem.* 51: 155–84.

————. 1983. Architecture of prokaryotic ribosomes. *Ann. Rev. Biochem.* 52: 35–65.

WOZNEY, J., et al. 1980. Structure of the pro α2(I) collagen gene. *Nature* 294: 129–35.

ZIFF, E. B. 1980. Transcription and RNA processing by the DNA tumour viruses. *Nature* 287: 491–99.

ZUBAY, G. L., and MARMUR, J. 1973. *Papers in biochemical genetics.* 2nd ed. New York: Holt, Rinehart and Winston.

19

Gene Structure and Organization

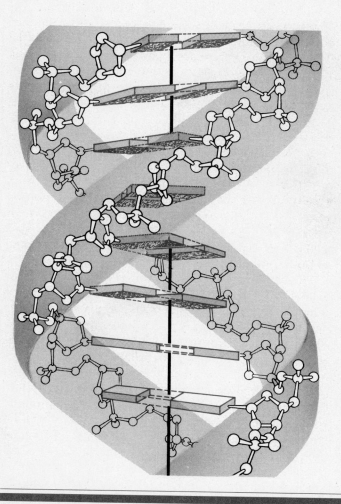

In Chapters 17 and 18, we focused on the chemical language that stores genetic information and the mechanics of its expression. In this chapter, we expand this discussion by describing studies designed to elucidate the precise nature of the gene. Sometimes referred to as **fine structure analysis of the gene,** this topic has been pursued continuously since Watson and Crick presented their model of DNA structure in 1953. At the forefront of genetics, such investigations will continue to be important throughout the 1980s.

Studies performed by Seymour Benzer in the 1950s resulted in the now classical genetic analysis of the *rII* locus in bacteriophage T4. His findings suggested that a gene is an inclusive unit consisting of a linear series of nucleotides, all of which encode information specifying a polypeptide. His work implied that the chromosome consists of a linear series of contiguous genes. To a large extent, these concepts are still considered fairly accurate in bacteriophages and bacteria.

Detailed genetic analysis of individual loci in eukaryotic organisms was also performed for several genes in *Drosophila melanogaster*. However, it was not until the development of recombinant DNA and nucleotide sequencing technologies in the 1970s that we extensively modified our concept of the genes of animal viruses and eukaryotic organisms. These technologies have allowed the detailed biochemical analysis of gene structure.

We learned in Chapter 11 that a large portion of the eukaryotic genome consists of repetitive DNA and that much of this is in the form of noncoding information. In Chapter 18 we established that the majority of areas within eukaryotic genes also consist of noncoding nucleotide sequences. Present in the initial mRNA transcript, these sequences are excised prior to translation, and the coding regions spliced back together. Additionally, many nucleotide sequences preceding and following the region specifying the initial transcript have been shown to be essential to efficient gene expression.

These latter topics represent some of the most exciting and unexpected discoveries made in the field of genetics. For all geneticists, but particularly those trained prior to the 1970s, the findings concerning eukaryotic gene structure serve as a source of wonderment. Perhaps we would have all been better prepared to learn of the much greater complexity of eukaryotic genes compared with those of prokaryotes and viruses if we had paid more attention to the quotation from Lewis Carroll's, *Through the Looking Glass*, which was insightfully included by Richard B. Goldschmidt in his Presidential Address to the IX International Congress of Genetics in 1954:

> "I can't believe that" said Alice. "Can't you?" the Queen said in a pitying tone. "Try again: draw a long breath, and shut your eyes."
>
> Alice laughed. "There's no use trying" she said, "one can't believe impossible things." "I dare say you haven't had much practice" said the Queen. "When I was your age I did it for half-an-hour a day. Why, sometimes I've believed as many as six impossible things before breakfast."

For all of us interested in genetics, it is certain that the next "impossible thing" is just around the corner!

FINE STRUCTURE ANALYSIS OF THE GENE

The experimentation described in the next few sections represents genetic analysis designed to probe the detailed structure of the gene. The findings allow less direct conclusions to be drawn and are more cumbersome to obtain than the more recent biochemical data. Nevertheless, they served as important links while molecular concepts of genetics were being developed. Before we describe these experiments, we will review the concept of **allelism** and **complementation analysis.**

Allelism and Complementation Analysis

When two recessive mutations cause similar phenotypes, they may be suspected to be alleles. If it is possible to map them and they are found at approximately the same chromosome location, the case for allelism is further strengthened. In order to prove this, however, a **complementation test** must be performed. In a diploid organism, a complementation test is conducted by examining the phenotype of an individual carrying both mutations heterozygously. However, there are two configurations by which this can occur. In the *cis* state, both mutations are present on the same chromosome. In the *trans* state, one mutation is on one chromosome and the other mutation is present on its homologue. Since the mutations can either be alleles or not be alleles, four possible conditions are created (Figure 19.1).

When two recessive mutations (m_1 and m_2) are alleles and present heterozygously in the *cis* arrangement (Case A), both are contained within the same gene. As a result, the wild-type phenotype will be expressed. This will occur because the homologue bearing the gene with the wild-type alleles (m_1^+ and m_2^+) will produce the normal gene product. However, if these alleles were present in the *trans* condition (Case B), one mutation would disrupt the function of the gene on one homologue while the other mutation would disrupt the gene on the other homologue. No normal gene product would be made, and a mutant phenotype would be expressed.

Now consider the cases where the two mutations m_1 and m_2 are not alleles of one another and thus affect separate gene products. In both the *cis* and *trans* arrangement, normal products of the two genes can be produced, and the wild-type phenotype will be expressed. In the case of the *cis* arrangement (Case C), both normal products result from the two genes on the same homologue. In the case of the *trans* arrangement (Case D), one normal product is produced from a gene on one homologue, while the other normal product is produced by a gene on the other homologue. In such a situation, **complementation** is said to occur. The genetic information on one chromosome complements that present on another chromosome.

The experimental examination of these possibilities distinguishes between mutations that affect the same gene product and are thus alleles and those that are not. This is called the **cis-trans test.** Barring minor exceptions such as intragenic recombination and intragenic complementation, two mutations that are alleles will always fail to complement one another when present in the *trans* arrangement. As such, they establish a **complementation group,** all the mutations of which affect the same genetic product. On the other hand, mutations that do complement one another in the *trans* arrangement are in separate complementation groups (or genes) and are not alleles.

Complementation testing need not be restricted to diploid organisms. If two viral chromosomes from dif-

FIGURE 19.1

A comparison of two mutations, m_1 and m_2, present heterozygously in the *cis* and *trans* arrangement. The top row indicates the phenotype (mutant or wild type) if the two mutations are in the *cis* arrangement. The bottom row indicates the phenotype if they are in the *trans* arrangement. The outcome is indicated in both cases when m_1 and m_2 are alleles (in a single gene) and when they are not alleles (present in separate genes).

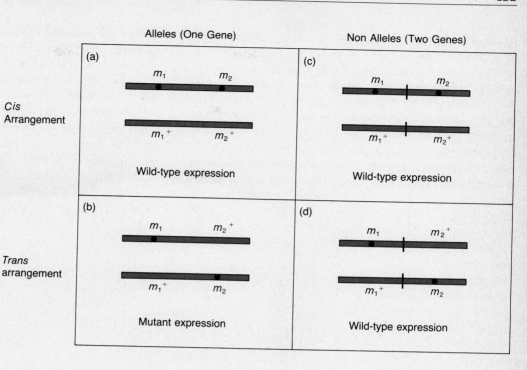

ferent genetic strains simultaneously infect a single bacterium, complementation analysis can be performed. For example, if two viral mutations, each on a separate chromosome, are lethal and alleles of one another, no viral progeny will be produced when tested. If, however, the two lethal mutations represent separate genes and are present on separate chromosomes during simultaneous infection, complementation will occur and normal viral progeny will be produced. We will see an example of this type of testing in the next section (Figures 19.2 and 19.3). Complementation analysis is also an integral part of genetic analysis in somatic cell hybridization studies (Chapter 23).

In general terms, a complementation group is equivalent to a single gene. There are some exceptions, however. A more accurate term, **cistron,** was coined by Benzer. A cistron, determined by the *cis-trans* test, specifies the formation of a single polypeptide chain. Each allele within a cistron is thought to occupy a different mutational site.

The *rII* Locus of Phage T4

A detailed analysis of a single locus, ***rII* in phage T4,** was initiated in 1953 by Seymour Benzer. Mutants at this locus produce distinctive plaques when plated on *E. coli* strain B. Benzer's approach was to isolate a large number of independent *rII* mutants—he even-

tually obtained about 20,000—and to perform recombinational studies in order to produce a genetic map of this locus. While the mapping techniques discussed in Chapter 6 and 14 were of the **intergenic** type, Benzer wished to study **intragenic** recombination; that is, he wanted to establish within the *rII* locus the relative positions of all mutations.

The key to his analysis was that *rII* mutant phages, while capable of infecting and lysing *E. coli* B, could not successfully lyse a second strain, *E. coli* K12 (λ). However, any wild-type phages would lyse both the B and the K12 strains. Benzer realized that he had the potential for a highly sensitive screening system. If he simultaneously infected *E. coli* K12 with two *rII* mutant viruses, extensive lysis would occur only if the two strains complemented one another, resulting in the production of large quantities of wild-type phage. If the mutant phages failed to complement one another, no reproduction would occur. If he then simultaneously infected *E. coli* B with any two *rII* mutants that failed to complement one another, all mutant phages would nevertheless reproduce. If he subsequently plated the mixture of these new phages on *E. coli* K12 and, if there were any wild-type recombinants produced among thousands or millions of *rII* mutant-bearing phage, the recombinants would be the only ones that would grow on *E. coli* K12. Thus, Benzer found that he could easily isolate recombinants occurring as rarely as 1 in 10^8 (or at a rate of 10^{-8}). We will illustrate the application of this screening system as we proceed with our discussion.

First the question of complementation was pursued. If simultaneous infection of *E. coli* K12 by two distinct *rII* mutations occurred, would they complement each other resulting in lysis of the cell (with the consequent production of wild type recombinants)? It was discovered that many mutations did complement one another. A thorough analysis revealed that all *rII*

mutations fell into one of two complementation groups. All mutations in the same group failed to complement one another, but complemented all those in the other group. On the basis of these results, the *rII* locus was divided into two cistrons, A and B.

The comparison of complementation and no complementation is illustrated in Figure 19.2. In the former, the functional product of the A cistron is specified by one mutant DNA molecule, and the functional product of the B cistron is specified by the other mu-

FIGURE 19.2
Comparison of complementation vs. no complementation when two rII mutations are studied together. Complementation occurs when each mutation is in a separate cistron and results in lysis of the cell (with consequent production of wild type recombinants). No complementation occurs if the two mutations are in the same cistron.

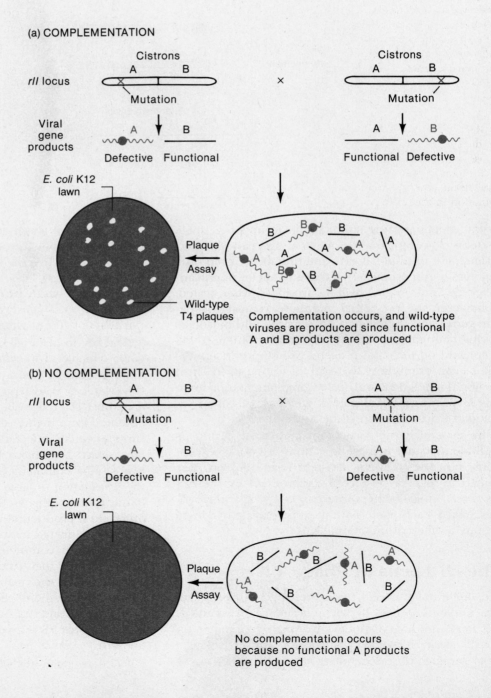

tant DNA molecule. Recall that during simultaneous infection, DNA from both viral strains is injected into the bacterium and replicated, forming a mixed pool of viral chromosomes. Provided that both the A and B functional products can be produced, lysis of *E. coli* K12 will occur.

Of the approximately 20,000 *rII* mutations, roughly half fell into each cistron. Benzer then set about to map the mutations within each one. For example, if two *rIIA* mutants were first allowed to infect *E. coli* B in a liquid culture, and if a recombination event occurred between the mutational sites in the A cistron (Figure 19.3), then rare wild-type progeny viruses would be produced. If samples of the progeny viruses from such an ecperiment were then plated on *E. coli* K12, only the recombinants would lyse the bacteria

and produce plaques. The total number of nonrecombinant progeny viruses could also be determined by plating samples on *E. coli* B. The percentage of recombinants can be determined using these two figures. As in eukaryotic mapping experiments, the frequency of recombination can be taken as an estimate of the distance between the two mutations within the cistron.

For example, if the number of recombinants is equal to 4×10^{-3}/ml, and the number of nonrecombinants is 8×10^{9}/ml, then the frequency of recombination between the two mutants is

$$2\left(\frac{4 \times 10^{-3}}{8 \times 10^{9}}\right) = 2(0.5 \times 10^{-6})$$

$$= 10^{-6}$$

$$= 0.000001$$

FIGURE 19.3
Production of rare wild-type viruses as a result of recombination between two viral strains, each with a mutation in the A cistron.

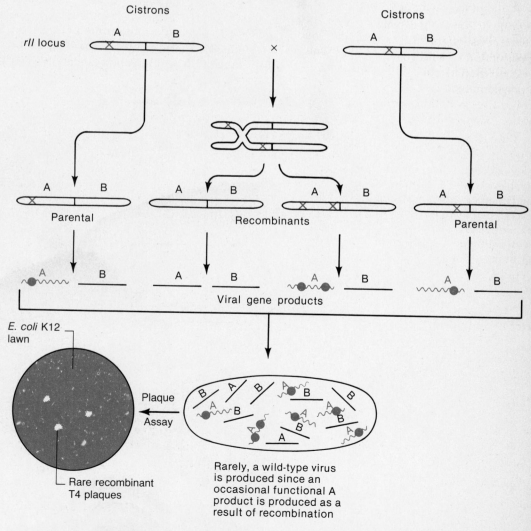

No Complementation; Recombination Occurs

Rarely, a wild-type virus is produced since an occasional functional A product is produced as a result of recombination

Multiplying by 2 is necessary because each recombinant event yields reciprocal products, only one of which—the wild type—is detected.

While the selective system of recombination just described was available to map mutations within each cistron, testing 1000 mutants two at a time in all combinations would have required millions of experiments. Fortunately, Benzer eliminated this step because he discovered that some of the *rII* mutations were in reality **deletions** of small parts of each cistron. That is, the genetic changes giving rise to the *rII* properties were not point mutations involving a single nucleotide change, but instead were due to the loss or deletion of a variable number of nucleotides within the cistron. These deletions could be identified by their failure to revert to wild type, a characteristic of point mutations. Furthermore, it was observed that a deletion, when tested during simultaneous infection with a point mutation located in the deleted part of the cistron, yielded no recombinants. The basis for the failure to do so is illustrated in Figure 19.4.

As shown in Figure 19.5, this second property served as the basis for localization of each mutation. Seven deletions (shown in color) covered variable portions of the A and B regions and were used for initial screening of the point mutations. Each point mutation

was ultimately assigned to an area of the cistron corresponding to one specific deletion. Then, further deletions within each of the seven areas were used to localize or map each *rII* point mutation more specifically. Remember that in each case, a point mutation is localized in the area of a deletion when it fails to give rise to any wild-type recombinants. Once he identified groups of mutations close to one another, Benzer could perform two- and three-point mapping in each small part of the A and B cistrons.

After several years of work, Benzer produced a genetic map of the two cistrons composing the *rII* locus of phage T4 (Figure 19.6). Of the 2000 mutations analyzed, 307 distinct sites within this gene had been mapped in relation to one another. Many mutations fell into the same site. An area containing many mutations was designated as a **hot spot.** Apparently, such an area is more susceptible to mutation than areas where only one or a few mutations are found. Additionally, Benzer found areas within the cistrons where no mutations were localized. He estimated that as many as 200 recombinational units had not been localized by his studies. He coined two terms to describe the units of a cistron. The **muton** is defined as the smallest segment of DNA that can be altered to produce a mutation. Similarly, the **recon** is the small-

FIGURE 19.4
Recombination between one viral strain with a deletion and another with a mutation overlapped by that deletion cannot yield a chromosome with wild-type A and B cistrons.

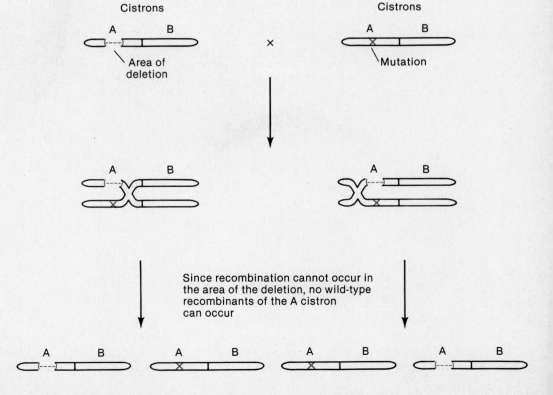

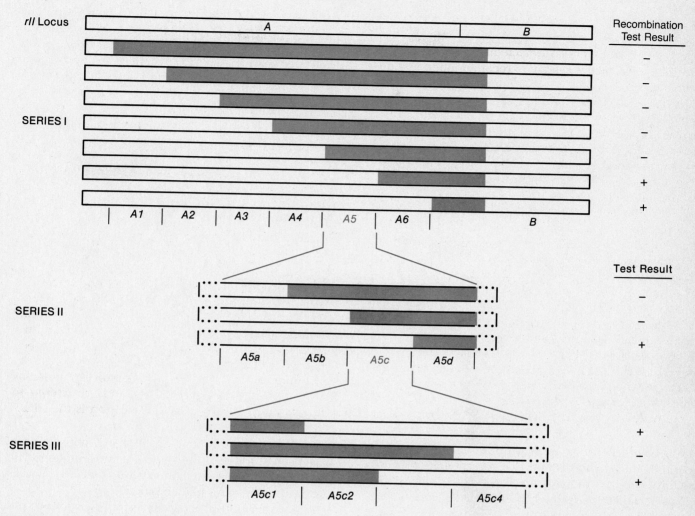

FIGURE 19.5

Three series of overlapping deletions in the *rII* locus used to localize the position of an *r* mutation. For example, if a mutant strain tested against each deletion (shown in color) in Series I for the production of recombinant wild-type progeny shows these results, the mutation must be in segment A5. In Series II, the mutation is further narrowed to segment A5c, and in Series III to segment A5c3. (Redrawn from Benzer, 1961.)

est unit capable of undergoing recombination. Taking into account the probable size of the A and B cistrons, he estimated that they contain less than 10 nucleotides.

The significance of Benzer's work is his application of genetic technologies to the analysis of what had previously been considered an abstract unit—the gene. Benzer demonstrated that a gene is not an indivisible particle but instead consists of mutational and recombinational units. His research was performed shortly after the publication of Watson and Crick's work on DNA and some time before the genetic code was unraveled in the early 1960s. Thus, his analysis is considered one of the classic examples of genetic experimentation.

Eukaryotic Genetic Analysis

Fine structure analysis of eukaryotic genes is much more difficult to perform than that of prokaryotic genes. Screening procedures to detect rare recombinants make such analysis less complex. Nevertheless,

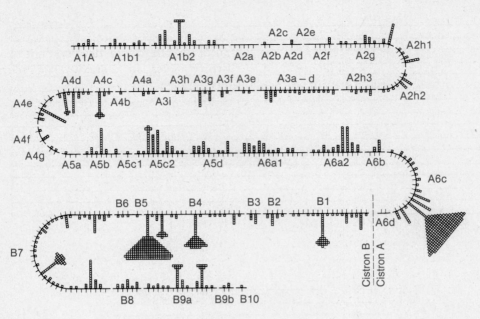

FIGURE 19.6
Map of mutations in the A and B cistrons of the *rII* locus of phage T4. Each square represents an independently isolated mutation. (Redrawn from Benzer, 1961.)

even before Benzer's work, studies lacking such procedures had demonstrated that recombination occurs within regions thought to represent single genes. These studies, performed in *Drosophila*, took advantage of the large number of progeny produced by this organism in relatively short periods of time.

Although many genes in *Drosophila* have been studied, we will focus on three loci that have yielded somewhat different results. The first, the *lozenge* (*lz*) locus, has been thoroughly investigated by Melvin M. Green and K. C. Green. About 20 recessive alleles, all

located at the same position on the X chromosome, have been studied. The most characteristic phenotype produced by alleles in homozygous females involves eye morphology and pigmentation.

In a *cis-trans* test, most alleles fail to complement one another in the *trans* position, as expected. However when various combinations of two alleles present in the *trans* position are tested for recombination, certain pairs recombine at a low but discernible frequency, but others do not. When the results of testing all combinations were sorted out, each allele could be

FIGURE 19.7
Various alleles contained in the subloci of the complex *lozenge* locus of *Drosophila melanogaster*.

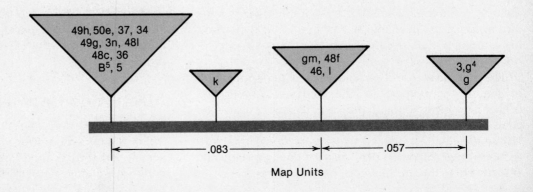

placed in one of four groups. Each would demonstrate recombination with all alleles except those sharing its group (Figure 19.7).

On this basis, the *lozenge* locus was subdivided into four subloci and was deemed a **complex locus.** Recombination, illustrated in Figure 19.8, occurred between 0.03 and 0.09 percent of the time, establishing the relative distances between the subloci. Findings such as these led to the concept of **pseudoalleles,** which test positively for allelism in complementation studies but between which recombination can occur. All those in the same sublocus are considered true alleles, but any two in different subloci are considered pseudoalleles. Edward B. Lewis has suggested that in complex loci, each sublocus is responsible for an essential step in the synthesis of the final gene product and that these steps must occur sequentially in the order established by the subloci along the chromosome. Whether or not this is true must await more detailed biochemical analyses of the DNA making up complex loci.

The second locus to be discussed is *rosy (ry)*, located on chromosome 3 of *Drosophila melanogaster.* Unlike *lozenge*, *rosy* is not considered a complex locus. Detailed analysis of *rosy* mutants is possible because a unique screening technique is available to detect wild-type recombinants. The locus contains the structural gene coding for the enzyme **xanthine dehydrogenase (XDH).** Recessive mutant flies demonstrate a rosy eye color and are missing the pigment isoxanthopterin. Mutant larvae are sensitive to the toxic effects of excess purines, which they cannot metabolize in the absence of XDH activity.

Thus, if a suitable cross has been made involving two different *ry* alleles in the *trans* position—where only recombinants are wild type—the recombinants may be detected by adding excess purines to the medium during larval development. Larvae containing two *ry* alleles fail to survive, but recombinants with at least one ry^+ gene sequence can counteract the lethal effects of the purines because of partial XDH activity. Such a cross is illustrated in Figure 19.9.

Recombinational analysis of many *ry* alleles has established several points. First, the *ry* locus is considerably smaller than the *lz* locus. While *lozenge* consists of 0.140 recombinational unit, *rosy* consists of only 0.005 unit. Second, in this much shorter span, the various alleles are somewhat evenly distributed (Figure 19.10). It has been estimated that the resolution of fine structure analysis in this investigation is nearly comparable to that attained in the *rII* locus. As such, the organization of the *ry* locus appears to be similar to that of each *rII* cistron.

The final locus of interest is that called the *bithorax* gene complex in *Drosophila.* Mutations of this gene complex alter the development of the thorax and abdominal regions of the fly's body. Those affecting the thorax are considered to be part of the *Ultrabithorax* domain of this complex. This domain has been well studied, and genetic testing has established several complementation groups. Mutations in each group display a similar phenotype, affecting a specific compartment or segment of the thorax. For example, the *bithorax (bx)* allele converts the thoracic segment T2 to T3 (there are three such segments). Because the wings originate in T3, a fly homozygous for this mutation has two T3 segments and forms two pairs of wings (see Figure 22.13). The *postbithorax (pbx)* allele behaves in the opposite fashion; segment T3 is converted to T2 and no wings develop. The map of this complex is shown in Figure 19.11.

This complex locus is of great interest in the study of developmental genetics. Thus, we will discuss it again in Chapter 22. Of greatest significance here is the finding that a complex locus occupying much less than one map unit in recombinational studies actually represents more than one functional genetic unit. It appears that each set of mutations is part of a region specifying a distinct product that affects a specific compartment during development. More detailed descriptions of these regions are given in Chapter 22.

THE MOLECULAR ORGANIZATION OF THE EUKARYOTIC GENE

Prior to 1970, there was little evidence to suggest eukaryotic gene structure and organization would turn out to be different from that established in bacteria and their phages. Nevertheless, many geneticists be-

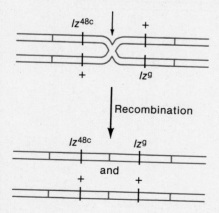

FIGURE 19.8
Intragenic recombination between subloci within the *lozenge* locus.

FIGURE 19.9
Selection for wild-type recombinants from a cross between two *rosy* mutants in *Drosophila melanogaster*.

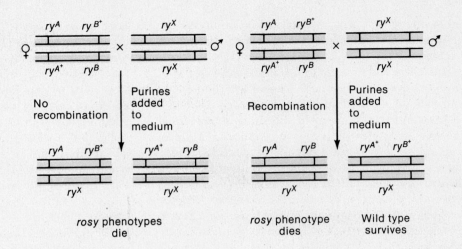

lieved that if the eukaryotic gene could be examined directly, it might well be more complex. Eukaryotes, it was reasoned, have several unique requirements distinguishing them from bacteria and phages. Primarily, cell differentiation, tissue organization, and coordinated development and function depend on genetic expression and its regulation. Do these requirements also depend on different modes of gene structure and organization?

When the technologies of recombinant DNA (Chapter 20) and rapid nucleotide sequencing were devel-

oped, this question was answered. The eukaryotic gene is more complex than we could ever have imagined. In this section, we will review DNA sequencing as well as many findings related to gene structure and organization.

DNA Sequencing

Before turning to some of the findings concerning gene structure and organization, we will discuss briefly the techniques used in DNA sequencing. The

FIGURE 19.10
Approximate map location of numerous *rosy* alleles, as determined by recombinant mapping. (Derived from Gelbart et al., 1976, Fig. 1.)

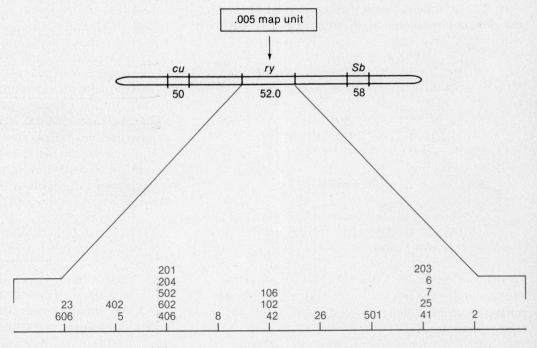

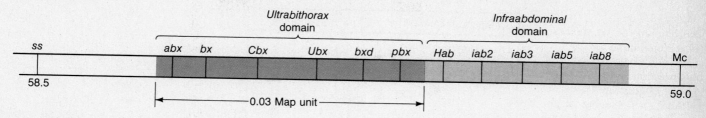

FIGURE 19.11

The *bithorax* complex in *Drosophila* consists of the *Ultrabithorax* and *Infraabdominal* domains. This complex is located on chromosome 3 and is flanked by the unrelated mutations *spineless (ss)* at locus 58.5 and *microcephalus (Mc)* at locus 59.0.

ability to sequence cloned DNA has added immensely to the analysis of gene structure.

Although techniques were available in the 1940s to determine base composition of DNA (see Chapter 9), it was not until the 1960s that those providing direct chemical analysis of nucleotide sequences were successfully developed and used. These methods worked much better for RNA than for DNA, were error-prone, and were extremely time consuming. For example, in 1965 Robert Holley determined the sequence of a tRNA molecule consisting of only 74 nucleotides. This work required about one year of concentrated effort. Because sequencing RNA was more efficiently performed, DNA molecules were often first transcribed *in vitro* and the RNA transcript isolated and sequenced. Thus, DNA sequences were usually determined indirectly.

In the 1970s, more efficient and direct techniques were developed, to allow rapid sequencing of DNA. One of these, conceived in 1977 by Allan Maxam and Walter Gilbert, is most often used in conjunction with recombinant DNA studies. Called the **chemical cleavage method,** it can now be used to determine the sequence of DNA fragments consisting of up to 600 nucleotides in about one week's work.

To explain this procedure, we will follow a theoretical sequence of seven nucleotides:

5′ pACGTCAG

In the first step, the 5′ terminal phosphate is removed and replaced by a radioactive (^{32}P) phosphate group (shown in color below) as a result of polynucleotide kinase action:

5′ pACGTCAG

This step eventually allows the identification of 5′ fragments generated by chemical cleavage as a result of autoradiography.

In the next step, individual DNA samples are subjected to each of four chemical treatments. Each treatment causes cleavage of the molecule at a site specific to one or two nucleotides. For example, treatments have been devised that cleave DNA specifically at the site of either an A, C, G, **or** C and T. In each case, this is accomplished in two steps. First, the base is chemically modified, rendering the nucleotide unstable. Second, cleavage is induced adjacent to the modified base. The specific base is removed in the process. When the treatments are performed, the conditions are adjusted so that each DNA molecule is broken only once, but randomly at any of the numerous sites susceptible to the treatment.

For example, if DNA is treated with dimethyl sulfate, which modifies purines, and is then heated in piperidine, cleavage at G sites occurs. In our theoretical molecule, this will produce

5′ pAC + 5′ TCAG

or

5′ pACGTCA

The results of all four treatments are shown in Figure 19.12.

For each set of treatments, the fragments are then subjected to polyacrylamide gel electrophoresis. Under the conditions of the gel, the different-sized fragments migrate with a velocity inversely proportional to their size. Thus, shorter fragments migrate farthest. So great is the resolution of this procedure that fragments varying by only one nucleotide are distinguished as separate bands, which are identified by subjecting the gel to autoradiography. As a result, only those fragments containing a radioactive 5′ phosphate are detected.

In a sense, then, there is an area on the gel for fragments of every size, up to 200 nucleotides in length. The one migrating farthest is one nucleotide, the sec-

FIGURE 19.12
The rationale of the Maxam-Gilbert technique for sequencing DNA. The short oligonucleotide pACGTCAG is used to illustrate the method. Each specific treatment (A, G, C, or C and T) is shown as a shaded circle.

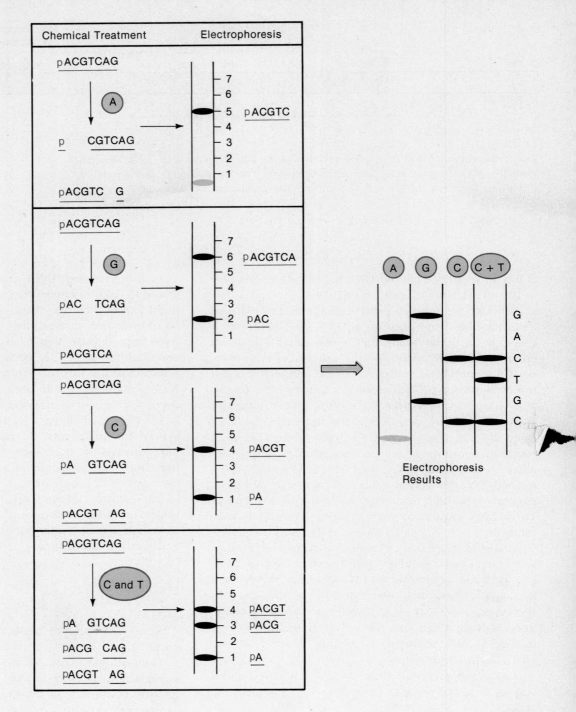

ond farthest is a dinucleotide, and so on. Figure 19.12 shows the gel patterns for each of the four treatments of our theoretical DNA fragment.

The interpretation of the final results is now obvious. The sequence of the nucleotides is read directly from bottom to top while comparing all four gels. At each position, we need only determine which treatment produced a band. In our theoretical example,

reading from bottom to top yields ACGTCAG. Note that the first nucleotide, A, can be ascertained during the "A" treatment by the presence of a faint band representing the radioactive phosphate group. It can also be derived from studies of overlapping fragments.

These and other sequencing techniques may be applied to numerous fragments created by treating DNA with various restriction endonucleases. When the information is pieced together, long stretches of DNA can be analyzed. As we will now see, these and other experimental approaches have been useful in unraveling the structures of eukaryotic genes.

The *C* Value Paradox

The amount of DNA contained in the haploid number of chromosomes of a species is called the *C* **value.** When such values were determined for a large variety of eukaryotic organisms, numerous trends were apparent. The most notable trends are that:

1 Eukaryotes contain substantially more DNA in their genomes than viruses or prokaryotes and exhibit a wide variation between organisms.

2 Evolutionary progression leads to increased amounts of DNA. That is, more advanced forms generally contain more DNA than less advanced forms.

The data from which these trends are inferred are shown in Figure 19.13 (see also Figures 26.19, 26.20, and 26.21).

It might be argued that increasing *C* values are simply the result of a greater need for increased amounts and varieties of gene products in more advanced organisms. However, several observations make this explanation unacceptable. First, the increases are very dramatic. While viruses and bacteria contain 10^4 to 10^6 nucleotide pairs in their single chromosomes, eukaryotes contain 10^7 to 10^{11} nucleotide pairs in each set of chromosomes. Does a eukaryote require 100,000 more genes, as these figures might suggest? Second, closely related organisms with the same degree of complexity in body form, tissue and organ types, and so forth often vary tenfold in DNA content. Third, amphibians and flowering plants, which vary as much as 100-fold within their classes, often contain much more DNA than other advanced eukaryotes. Lastly, there is no real correlation between increased genome size and the morphological complexity of most groups represented in Figure 19.13.

Therefore, it is doubtful that the development of greater complexity during evolution can account for the amount of DNA found in eukaryotic genomes. This conclusion is the basis of what has been called the *C* **value paradox.** Excess DNA is present that does not seem to be essential to the development or survival of eukaryotes. The paradox raises the question, Does all or most of the excess DNA code for proteins? If it does not, as the preceding arguments suggest, are eukaryotic genomes composed mainly of noncoding nucleotide sequences? Is such DNA vital to the organisms carrying it in their genomes, or is it simply excess

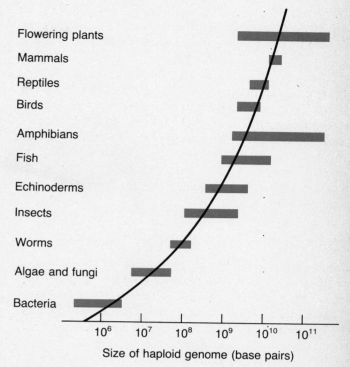

FIGURE 19.13

DNA content of the haploid genome of many representative groups of organisms. While the size of the genome has increased during evolution, a wide range of sizes exist in any single group.

DNA that has somehow accumulated during evolution and has no function?

Recent studies of gene structure are beginning to unravel these mysteries. Some insights have been gained from findings presented earlier in the text. For example, we have already established that noncoding sequences are indeed present in the genomes of higher organisms in the form of **repetitive DNA** (Chapters 9 and 11). Recall that there are two major classes—highly repetitive sequences and moderately or middle repetitive sequences. Can these account for the excess DNA?

To answer this question, consider the human genome. Only about 5 percent of human DNA is highly repetitive. This is in the form of satellite DNA located primarily at the heterochromatic centromeric regions. Middle repetitive DNA comprises only 15 to 30 percent of the human genome. This fraction contains duplicated genes as well as nontranscribable DNA sequences interspersed between unique or single-copy genes. Finally, it is estimated that only 1 to 2 percent of the human genome consists of genes that encode proteins. This leaves over 60 percent of the genome still unaccounted for, much of which is unique sequence DNA.

While the proportion of the genome consisting of repetitive DNA varies between organisms, one feature is common to those studied. Only a small part of the genome codes for proteins. For example, sea urchins contain an estimated 20,000 to 30,000 genes coding for proteins, which occupy less than 10 percent of the genome. In *Drosophila*, only 5 to 10 percent of the genome is occupied by genes coding for proteins. Similar estimates have been made for other organisms. We must conclude that while there is sufficient DNA to code for over one-half million genes in many eukaryotic organisms, the majority of DNA does not serve this function.

Split Genes and Exon Shuffling

One of the first insights into regions of DNA that are neither repetitive nor directly encode proteins has come from the study of RNA. This research has provided detailed knowledge of eukaryotic gene structure.

The genetic code, as deciphered and shown in Figure 17.8, is written in the ribonucleotide sequence of mRNA. This information originated, of course, in the sense strand of DNA, where complementary sequences of deoxyribonucleotides exist. In bacteria, the relationship between DNA and RNA appears to be quite direct. The DNA base sequence is transcribed into an mRNA sequence, which is then translated into an amino acid sequence according to the genetic code.

However, in eukaryotes the situation is much more complex than in bacteria. It has been found that many internal base sequences of genes, called **introns,** never appear in the mature mRNAs that are translated. The remaining areas of each gene, which are reflected in the amino acid sequence of the encoded protein, are called **exons.** While the entire gene is transcribed into a large precursor mRNA, called **heterogeneous RNA (hnRNA),** the introns are excised and the exons spliced back together prior to translation. These findings, which were presented in detail in Chapter 18, have established that eukaryotic genes are exceedingly complex.

Given this information about internal gene sequences, two questions can now be asked. First, do the nucleotide sequences found in introns solve the C value paradox? That is, do introns account for the remainder of the noncoding sequences beyond those of repetitive DNA? The answer is emphatically no! As shown in Table 19.1, the presence of introns may increase by four or five times our estimate of the nucleotide length occupied by single copy genes. However, since this estimate is seldom above 10 percent of the genome and usually much less, a great deal of noncoding DNA is still unaccounted for.

TABLE 19.1

A comparison of mRNAs and the initial transcripts from which they are derived.

Gene	mRNA Size	Minimum Transcript Length	Ratio
Rabbit β-globin	589	1295	2.2
Mouse β^{maj}-globin	620	1382	2.2
Mouse β^{min}-globin	575	1275	2.2
Mouse α-globin	585	850	1.5
Rat insulin I	443	562	1.3
Rat insulin II	443	1061	2.4
Chick ovalbumin	1859	7500	4.0
Chick ovomucoid	883	5600	6.3
Chick lysozyme	620	3700	6.0

Source: From Lewin, 1980, p. 825.

The second question concerns our concept of the gene. Can we any longer provide an accurate definition for it? In his 1980 book *Gene Expression—2, The Eukaryotic Chromosome*, Benjamin Lewin has summarized beautifully this dilemma:

> More than a hundred years of work has led to the concept that the gene is a contiguous region of DNA that is colinear with its protein product. This definition encompasses the complementation test, which formally defines the gene as a unit all of whose parts must be present on one chromosome; and it has found fulfillment in the detailed molecular studies of the past two decades demonstrating that DNA first is transcribed into RNA and then is translated into protein by reading the triplets of the genetic code. Colinearity appeared to be preserved through the stages of gene expression. The recent discovery that eukaryotic genes contain intervening sequences that are not represented in the mRNA from which the protein product is synthesized has caused this concept now to be discarded. . . . This has made it difficult, if not impossible, to arrive at a single satisfactory definition of the gene.

It is clear then that the presence of introns does not entirely resolve the C value paradox and, furthermore, has created a dilemma in arriving at a universal definition of the gene. Nevertheless, studies of introns and exons have expanded considerably our knowledge of eukaryotic gene structure. Such studies further suggest how eukaryotic genes may have arisen during evolution. In 1977, Walter Gilbert suggested that genes in higher organisms may consist of collections of exons present in ancestral genes that are brought together through recombination during the course of evolution. He proposed that exons are modular in the sense that each might encode a functional domain in the structure of a protein. For example, a particular amino acid sequence encoded by a single exon might always create a specific type of fold in a protein. This may impart a unique but characteristic function to all proteins containing such a fold. Serving as the basis of useful parts of a protein, many similar domains might be functional in a variety of proteins. In Gilbert's proposal, many exons could be "mixed and matched" to form unique genes in eukaryotes.

Three observations lend immediate support to this proposal. First, introns and exons are nearly universal in vertebrate organisms, rare in lower eukaryotes such as yeast, and with few exceptions, absent in prokaryotes. This pattern is to be expected if introns and exons are evolutionary products. Second, most exons are fairly small, averaging about 100 to 150 nucleotides and encoding 30 to 50 amino acids. This size is consistent with the production of functional domains in proteins. Third, recombinational events that could lead to **exon shuffling** would be expected to occur within areas of genes represented by introns. If so, many recombinational events at nonspecific regions of introns would ultimately result in random nucleotide sequences and sizes of these intervening sequences. This is, in fact, what is observed. Introns range from 50 to 20,000 bases and exhibit fairly random base sequences.

Since 1977, a vast research effort has been directed toward the analysis of split genes. In 1985, direct evidence in favor of Gilbert's proposal of exon modules was presented. Analysis was made of the human gene encoding the membrane receptor for low density lipoproteins (LDL). This **LDL receptor protein** is essential to the transport of plasma cholesterol into the cell. It mediates **endocytosis** and is expected to have numerous functional domains. These include the capability to bind specifically to the LDL substrates and to interact with proteins at different levels of the membrane during transport across it. In addition, the pro-

tein is modified post-translationally by the addition of an O-linked carbohydrate; a domain must exist that links to this carbohydrate.

Detailed analysis of the gene encoding this protein supports the concept of exon modules and their shuffling during evolution. The gene is quite large—45,000 nucleotides—and contains 18 exons. These represent only slightly less than 2600 nucleotides. These exons are related to the functional domains of the protein *and* appear to have been recruited from other genes during evolution.

Figure 19.14 illustrates these relationships. The first exon encodes a signal sequence that is removed before the LDL receptor becomes part of the membrane. The next five exons represent the domain specifying the binding site for cholesterol. This domain is made up of a 40-amino-acid sequence repeated seven times. This amino acid sequence closely resembles one found in the C9 component of the immunological molecule complement, which is found in human plasma. The next domain consists of a sequence of 400 amino acids bearing a striking homology to the mouse peptide hormone **epidermal growth factor (EGF)**. This region is encoded by eight exons and contains three repetitive sequences of 40 amino acids. A similar sequence is also found in three blood-clotting proteins. The fifteenth exon specifies the domain for the post-translational addition of the O-linked sugar, while the remaining two exons specify regions of the protein that transcend the membrane and anchor the receptor to specific sites called coated pits on the cell surface.

This information strongly implies that the exons encoding the LDL receptor have been preserved through evolution, and that many of them represent functional domains in the product specified by this gene. As such, it has been suggested that this gene is part of a larger **supergene family,** the members of which are related by sequence but not necessarily function. The results of supergene families are mosaic proteins cre-

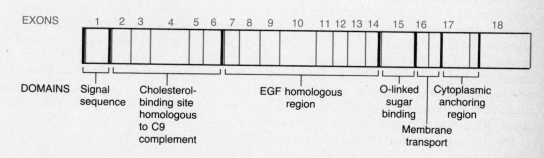

FIGURE 19.14
A comparison of the 18 exons making up the LDL receptor protein and their organization into functional domains. The borders of the exons are shown in color, while the borders of the domains are indicated with black lines.

ated by new combinations of preexisting exons. Further research will undoubtedly uncover other examples of this phenomenon and extend our knowledge of gene structure.

Flanking Regions of Eukaryotic Genes

In the discussion of transcription in Chapter 18, we introduced the concept of the **promoter region.** Found "upstream" from the coding sequence that initiates the 5′ end of mRNA during transcription, the promoter region is responsible for the initial RNA polymerase binding in prokaryotes. A similar but more extensive region, called the **5′ flanking region,** precedes eukaryotic genes.

Three areas of the 5′ flanking region appear to be essential to efficient transcription and have been investigated extensively. The first, introduced in Chapter 18, is called the **Goldberg-Hogness** or **TATA box.** Found at a point about 30 nucleotides upstream from the point where transcription begins, this short region consists of the sequence TATXAX, where X is either T or A. The sequence has been found in most but not all eukaryotic genes examined. While it is not necessarily palindromic, the TATA box is analogous to the Pribnow sequence found associated with prokaryotic gene promoters.

The TATA box has generated much excitement because of its possible involvement in the initiation of transcription. In many instances, both mutations and deletions in this region severely reduce or eliminate the *in vitro* transcription by RNA polymerase II of the associated gene. This is the case in studies of the conalbumin and ovalbumin genes of chickens, the beta-globin gene in rabbits, and some genes in animal viruses. Other evidence suggests that the region is only responsible for fixing the site of initiation of transcription *in vivo.*

The second noncoding region of interest is found farther upstream from the TATA box. In different genes studies, regions anywhere from 50 to 500 nucleotides upstream appear also to modulate transcription. These regions have been located on the basis of the effects on transcription of their deletion. The loss of various noncoding regions appears to drastically reduce *in vivo* transcription. Some noncoding regions, such as those found associated with globin and SV40 genes, are about 50 to 100 nucleotides from the TATA sequence. Others, such as those associated with the sea urchin H2A gene and the *Drosophila* glue protein gene, are 200 to 500 nucleotides upstream. Frequently, the sequence **CCAAT** is part of this noncoding region.

The third noncoding region is represented by elements called **enhancers.** While their location may vary, they often may be found even farther upstream. We will return to a discussion of these elements in Chapter 21.

Recent findings suggest that these regions may be important to the conformation of chromatin structure during gene activity. It has been found that certain regions of genes are hypersensitive to the action of the enzyme DNase I when the gene is transcriptionally active, but are insensitive when the gene is inactive. Nucleotide sequences 100 to 200 units in length found upstream from a variety of genes seem to contain hypersensitive regions. Hypersensitivity is believed to be related to the need for chromatin to achieve a more open or relaxed configuration during the initiation of transcription. Such a configuration is thought to be essential if DNA is to become accessible to the polymerase enzyme and initiation proteins involved in transcription.

The discovery of the TATA box and other flanking sequences has extended our knowledge of gene structure considerably. As shown in Figure 19.15, a more concise view of the typical eukaryotic gene is gradually emerging. This model can be expected to be modified and further extended as research continues.

COMPLEX MULTIGENE FAMILIES

The study of another level of eukaryotic genes has contributed to our understanding of genomic organization. This level consists of the relationship between individual, but functionally related, genes—the so-called **complex multigene families.** For example, the genes coding for the various globin polypeptides or those specifying different subunits of an enzyme with quaternary structure are each considered a family. Those encoding immunoglobulin chains and histone proteins are also gene families. In this section we will see how members of a family are arranged in the genome. In Chapter 26, we will consider the evolutionary origin of gene families.

Individual genes within a family may be widely spread throughout the genome and located on various chromosomes, or they may be clustered together on a single chromosome. In different organisms, a single family may assume both organizations. For example, the genes coding for rRNA in *Drosophila melanogaster* are clustered on the X and Y chromosomes. In humans, this gene family is dispersed on several chromosomes. Usually, individual members of a family contain sufficient nucleotide sequence homology to suggest that they arose through the process of gene duplication and then diverged during evolution. We will examine two gene families, the human beta-like

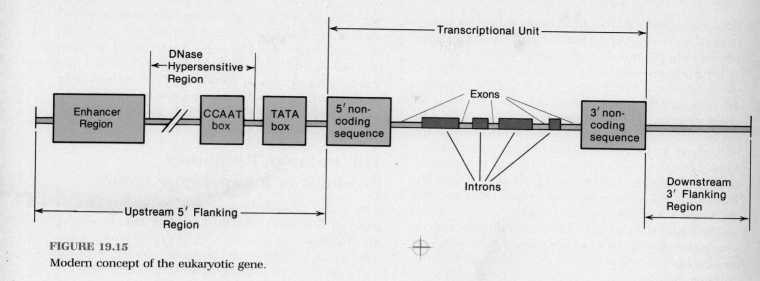

FIGURE 19.15

Modern concept of the eukaryotic gene.

globin and the sea urchin histone genes. In addition, we will discuss the structure of immunoglobulins.

The Beta-Globin and Histone Families

The **human beta-globin gene family** is one of the most extensive clusters to be analyzed. Found on the distal portion of the short arm of chromosome 11, this gene family consists of seven related gene sequences in a chromosome region spanning 60,000 nucleotide pairs (60 kilobases, or 60 kb). Because of their similarity in nucleotide sequence, the seven are said to be **beta-like genes.** These genes have been studied primarily by restriction enzyme analysis and cDNA cloning techniques. As shown in Figure 19.16, a detailed physical map of the entire 60-kb region is now available.

As we learned in Chapter 16, the expression of the β-like genes is coordinately regulated during development (see Figure 16.8). The ε and γ chains appear se-

quentially before birth, whereas the δ and β chains are characteristic of adult hemoglobin. The genes specifying each chain are arranged in the same sequence along the chromosome as they appear during development. Whether this fact is related to the regulation of expresssion, however, is not clear.

Within the cluster, there is a single copy of the ε, δ, and β genes. Two copies of γ, designated $^G\gamma$ and $^A\gamma$, are present. These differ only in the codon specifying amino acid 136, yielding glycine and alanine, respectively. Additionally, a **pseudogene,** $\psi\beta 1$, has been located. Pseudogenes are similar in sequence to other members of a gene cluster, but contain nucleotide substitutions, deletions, and/or additions that prevent their expression.

All five genes (ε, $^A\gamma$, $^G\gamma$, δ, and β) contain two introns found at precisely the same position. The first intron, 125 to 150 base pairs long, occurs between the 30th and 31st codons. The second, 700 to 900 base pairs long, occurs between the 104th and 105th codons. The 5′ flanking region of each gene has also been studied.

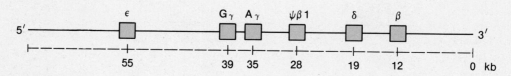

FIGURE 19.16

The human β-like globin gene cluster. The region spans over 60,000 (60 kb) nucleotides. (After Maniatis et al., 1980, Fig. 2. Reproduced, with permission, from the *Annual Review of Genetics*, Vol. 14. © 1980 by Annual Reviews, Inc.)

In all cases, both a TATA box (29 or 30 base pairs upstream) and a CCAAT region (70 to 78 base pairs upstream from the initial codon) have been observed.

Several observations of these genes pertain to the *C* value paradox. Only about 5 percent of the 60-kb region consists of coding sequences. The remaining 95 percent includes the introns, flanking regions, and spacer DNA found between genes. Of this percentage, only about 11 percent consists of introns. Within the spacer regions, multiple copies of a highly repetitive, short DNA sequence have been identified. Characteristic of the human genome and present 300,000 times throughout it, this sequence has also been identified in the alpha-globin gene cluster on chromosome 16. The remainder of the DNA serves no known function and consists of sequences not found elsewhere in the genome. Because it represents the majority of DNA in the cluster, we must conclude that it is either excess nonfunctional DNA *or* DNA whose function has yet to be discovered.

The second gene family considered here is that encoding **histones.** Recall that histones are a small group of five basic proteins functioning as part of the nucleosome characteristic of eukaryotic chromatin. Like the beta-globin family, the genes are clustered together and separated by noncoding spacer DNA. However, several major differences are apparent in the histone cluster. First, the genes encoding these proteins lack introns. Second, the cluster is considerably smaller than the beta-globin cluster, occupying a region consisting of only 5 to 9 kb. The third difference is that in some organisms, each set of five genes is tandemly repeated a variable number of times within the cluster.

As shown in Figure 19.17, the arrangement of genes and their polarity of transcription vary in the sea urchin and *Drosophila*, two organisms where this cluster has been well studied. In sea urchins, this cluster is tandemly repeated 400 to 1200 times, depending on the species. In *Drosophila*, about 100 copies are pres-

ent. However, in mice and humans, members of this gene family appear to be clustered but not present as a tandem array of repeat units. For example, a region of the distal tip of the long arm of the human chromosome 7 contains all members of the histone gene family, but they may be interspersed with other nonhistone genes. Therefore, no single pattern of organization applies to all organisms.

The Immune Response: Products of a Supergene Family

One of the most fascinating yet complex families of genes is that specifying the proteins participating in the **immune response** of vertebrates. These molecules are responsible for the highly specific response mounted when foreign substances called **antigens** enter the blood stream of an organism. The antigen may be an independent molecule, a part of the surface coat of an invading bacterium, virus, or organ transplant, or any other foreign substance. Two types of cells are employed that respond specifically to antigens. **B lymphocytes,** so-named because their origin is bone marrow, react by producing **antibodies** that are secreted into plasma and react with soluble antigens. These antibodies are also called **immunoglobulins. T lymphocytes,** originating and maturing in the thymus, react with antigens present on cell surfaces. They produce a counterpart to antibodies called the **T-cell receptors,** which mediate their specific immune response. T lymphocytes and their receptors are not as well characterized as are B cells and the immunoglobulins they produce. However, it is clear that T lymphocytes play a role in the regulation of the responses of the B lymphocytes besides responding themselves to cell surface antigens. In the latter capacity, T lymphocytes recognize antigens only in conjunction with molecules encoded by the **major histocompatibility complex (MHC).** Therefore, their response is more limited than that of B cells, and they must in some way learn

FIGURE 19.17
The histone gene clusters in *Drosophila* and the sea urchin. The arrows indicate the polarity of transcription.

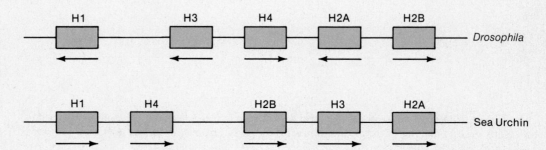

to discriminate between MHC products of "self," to which they do not respond, and "nonself" to which they do.

Although the T-cell receptor has been recently elucidated, we will focus our discussion on immunoglobulins because more is known about them. However, the organization of both classes of immunologically important molecules is comparable, and they are formed in similar ways. Together, the genes encoding each group are multigene families that appear to share common ancestry. As such, together they constitute what is called a supergene family. The components will become clear as we proceed with our discussion of immunoglobulins.

Immunoglobulins

An organism's ability to respond to large numbers of highly specific antigens is tremendous. Mammals, for example, can produce over a million different antibodies, each responding to a different antigen. It is now clear that the basis for this molecular diversity lies in the various amino acid sequences composing the chains of the antibodies. In humans, antibodies, or immunoglobulins (Ig), have been divided into five classes: **IgG, IgA, IgM, IgD,** and **IgE.** The first class, IgG, represents about 80 percent of the antibodies found in blood and is the most extensively characterized.

Each antibody molecule consists of two different polypeptide chains, each present in duplicate. The larger or **heavy chain (H)** consists of 446 amino acids, and the smaller or **light chain (L)** contains 214 amino acids. The first 110 amino acids of the N-terminus of each chain vary in sequence in different IgG molecules and are thus known as the **variable region (V).** The remaining C-terminal amino acid sequence of each chain is invariable and is called the **constant region (C).** The variable regions are responsible for antibody specificity and contain the **antibody combining sites** that bind to the antigens. This general scheme, illustrated in Figure 19.18 was first determined by Rodney Porter and Gerald Edelman in the early 1960s.

FIGURE 19.18
(a) General scheme of the IgG immunoglobulin molecule showing the constant, variable, and hypervariable regions of heavy (H) and light (L) chains. (b) Antibody-binding sites of the variable regions and complexes that may be formed.

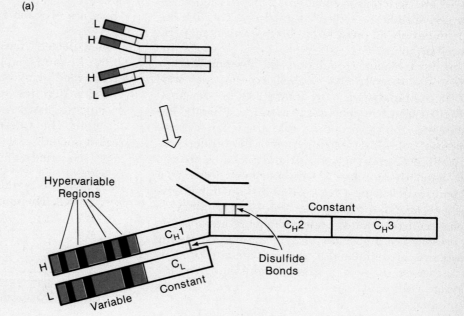

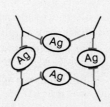

If a foreign organism invades the body, a small part of a surface protein or carbohydrate will serve as an antigen. Once specific antibodies are produced, they combine with the antigenic determinants in a lock-and-key configuration. Since each antibody has two combining sites (Figure 19.18), large complexes are formed that can be engulfed by white blood cells and effectively removed from the body.

Within the variable region are areas that vary more than others. There are three such **hypervariable regions** in light chains and four in heavy chains (Figure 19.18). They are most important to the three-dimensional configuration of the combining site.

The question of major genetic interest is how the vast array of molecular variability demonstrated by antibodies is encoded in the genome. Three hypotheses have been proposed and supported to one degree or another over the past several decades. The first, the **germline theory,** simply suggests that the entire antibody repertoire is encoded by a large number of genes, one for each possible antibody chain. The second, the **somatic mutation theory,** suggests that a limited number of genes exist but that they are hypermutable. During the development of an individual's immune system, large numbers of antibody-producing cells come to contain different genetic sequences, each responsible for a distinct antibody. Through somatic mutation, all possible antibodies are encoded by one cell or another.

The most recent explanation, the **recombination theory,** has received the greatest experimental support. Recombination is now thought to be the most prominent factor in generating antibody diversity. Basically, this theory proposes that there are a large number of genes that encode each of the various portions of the different types of H and L chains. As various antibody-forming B lymphocytes develop, a mechanism involving DNA recombination brings various combinations together so that each lymphocyte comes to encode but one specific type of antibody. In the presence of a specific antigen, the corresponding lymphocyte is stimulated to differentiate and proliferate further. As a result, antibodies containing the appropriate combining sites are produced and interact with the antigen.

To comprehend how the necessary diversity of antibodies can be generated, we must delve more deeply into the structure of immunoglobulins and the genes encoding various regions of their polypeptide components. Recall that there are light and heavy polypeptide chains, each containing variable and constant regions (V_L, C_L, V_H, and C_H). Also, there are five classes of immunoglobulins. As shown in Table 19.2, each class contains one of two types of light chains, λ or κ. Fur-

TABLE 19.2

Categories and components of immunoglobins.

Ig Class	Light Chain	Heavy Chain	Tetramers	
IgA		α	$\kappa_2\alpha_2$	$\lambda_2\alpha_2$
IgD		δ	$\kappa_2\delta_2$	$\lambda_2\delta_2$
IgE	κ or λ	ϵ	$\kappa_2\epsilon_2$	$\lambda_2\epsilon_2$
IgG		γ	$\kappa_2\gamma_2$	$\lambda_2\gamma_2$
IgM		μ	$\kappa_2\mu_2$	$\lambda_2\mu_2$

thermore, each class is characterized by its own specific heavy chain, designated α, γ, δ, ϵ, or μ. Thus, there are ten different types of immunoglobulins that can be formed.

There is even some variation found in the C region of different chains. For example, at least three minor differences in amino acid sequence have been demonstrated in the C_L region of κ chains. Within the C_H region, three different but closely related repeating sequences (C_H1, C_H2, and C_H3) have also been identified. There are probably separate exons encoding each region.

As expected, however, the major variation is found in the V region (Figure 19.19). The V_L region appears to be encoded by two extensive sets of genes, designated V and J. **V genes** specify the N-terminal portion, including the first two hypervariable regions and part of the third. The **J genes** specify the remainder of the V region, including the latter part of the third hypervariable sequence. When an antibody-producing cell is

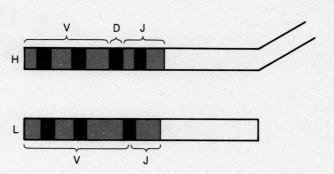

FIGURE 19.19

Schematic diagram of the V and J regions of light chains and the V, D, and J regions of heavy chains.

formed, various combinations of V and J genes are brought close to one another through recombination. If numerous V and J genes can combine at random, an extremely large number of V_L sequences can be encoded.

Does this really happen? In 1976, Susumo Tonegawa applied the techniques of recombinant DNA cloning and nucleotide sequencing to demonstrate that it does. He compared a cloned fragment specifying a λ V_L chain derived from embryonic germline DNA with that derived from DNA of an antibody-producing cell. In the embryonic DNA, the region of embryonic DNA coding for the first 95 amino acids of the V_L region was separated by a 4.5-kb region from that coding for the remaining 13 amino acids (the J region) as well as the C region. However, in the DNA of the antibody-producing cell, the V and J segments were joined contiguously to form a single gene encoding a specific λ V_L chain. This observation strongly supports the hypothesis that immunoglobulin genes are formed through a process of recombination. Similar processes lead to the formation of DNA encoding the κ light chain, which is illustrated in Figure 19.20.

Even more diversity is possible in the V_H region because it is divided into three segments, each thought to be represented by a distinct region (see Figure 19.19). In V_H chains, a V region containing the first two hypervariable regions occurs. Additionally, a D region containing the third hypervariable region and a J region comprising the rest of the V_H segment have been demonstrated. The DNA encoding heavy chains is apparently formed by joining V- and D-coding regions, which are then combined with J-coding regions. The theoretical diagram in Figure 19.21 illustrates the diversity of heavy chains that may be formed in this way.

Based on this information, a general explanation of the origin of antibody diversity is emerging. Certain clusters or gene families encode the various segments of heavy and light chains. Recombination brings together various gene fragments, and the newly formed composite gene encodes a specific chain. Then, various combinations of light and heavy chains can be generated to increase the diversity. Although it is not yet known just how many germline genes exist for each specific fragment, there is likely to be a large number of genes specifying the hypervariable fragments. For example, the number of genes encoding κ chains is known to be between 100 and 300. Furthermore, somatic mutation generates even more genetic diversity.

FIGURE 19.20
The formation of the DNA segments encoding a human κ light chain and the subsequent transcription, mRNA splicing, and translation leading to the final polypeptide chain. In germline DNA, up to 150 different L-V (Leader-Variable) segments are present. These are separated from the J regions by a long noncoding sequence. The J regions are separated from a single C gene by an intervening sequence (intron) that must be spliced out of the initial mRNA transcript. Following translation, the amino acid sequence derived from leader RNA is cleaved off as the mature polypeptide passes across the cell membrane.

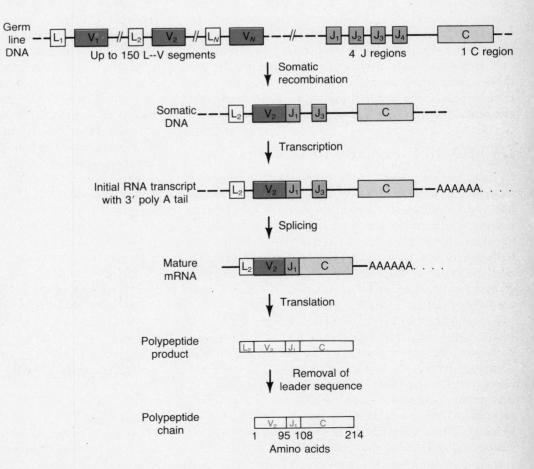

FIGURE 19.21
The theoretical formation of various heavy chain genes during the maturation of an antibody-forming cell.

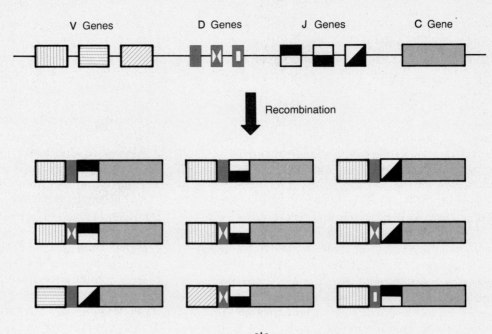

TRANSPOSABLE ELEMENTS IN EUKARYOTES

We conclude this chapter by returning to a discussion of genetic elements that are mobile within the genome. Such units were introduced in Chapter 14 in the description of bacterial **insertion sequences** and the related but more extensive **transposons.**

Ds and *Ac* Elements in Maize

In 1956, about twenty years before the discovery of transposons, the well-known maize geneticist, Barbara McClintock, analyzed the genetic behavior of two mutations, *Dissociation* (*Ds*) and *Activator* (*Ac*). Initially, she determined that *Ds* was located on chromosome 9. Provided that *Ac* was also present in the genome, *Ds* had the effect of inducing breakage at a point on the chromosome adjacent to its location. If breakage occurred in somatic cells during their development, progeny cells often lost part of chromosome 9, causing various phenotypic effects.

Subsequent analysis showed that both the *Ds* and *Ac* genes were sometimes transposed to different chromosomal locations. While *Ds* moved only if *Ac* was also present, *Ac* was capable of autonomous movement. Where *Ds* came to reside determined its genetic effect. Although it might cause breakage, it might instead inhibit gene expression. In cells where

expression was inhibited, *Ds* might move again, releasing this inhibition. In these cases, the *Ds* element is believed to insert into a gene and subsequently to depart from it, causing changes in gene expression. Often, analysis involved an examination of the phenotypes of the kernels of maize resulting from genes expressed in either the endosperm or aleurone layers. Figure 19.22 illustrates the sorts of movements and effects of the *Ds* and *Ac* elements described above. McClintock concluded that the *Ds* and *Ac* genes were **transposable controlling elements.**

Until many years later, nothing comparable to the controlling elements in maize was recognized in other organisms. When the behavior of prophages and plasmids was revealed and insertion sequences and transposons were discovered, many parallels were evident. Transposons and insertion sequences were seen to move into and out of chromosomes, to insert at different positions, and to affect gene expression at the point of insertion. It is assumed that controlling elements in maize behave in a similar, if not identical, manner to bacterial transposons.

Several *Ac* and *Ds* elements have now been isolated and carefully analyzed (Figure 19.23). As a result of this information, the relationship between the two elements has been clarified. The first *Ac* element sequence is 4563 bases long and strikingly similar to the bacterial transposon **Tn3.** This sequence contains two nearly perfect terminal repeats, two coding sequences,

FIGURE 19.22

Two consequences of the influence of the *Activator* (*Ac*) element on the *Dissociation* (*Ds*) element. In (a), *Ds* is transposed to a region adjacent to a theoretical gene *W*. Subsequent chromosomal breakage is induced, the *W*-bearing segment is lost and mutant expression occurs. In (b), *Ds* is transposed to a region within the *W* gene causing immediate mutant expression. *Ds* may also "jump" out of the *W* gene with the accompanying restoration of *W* gene activity and its wild-type expression.

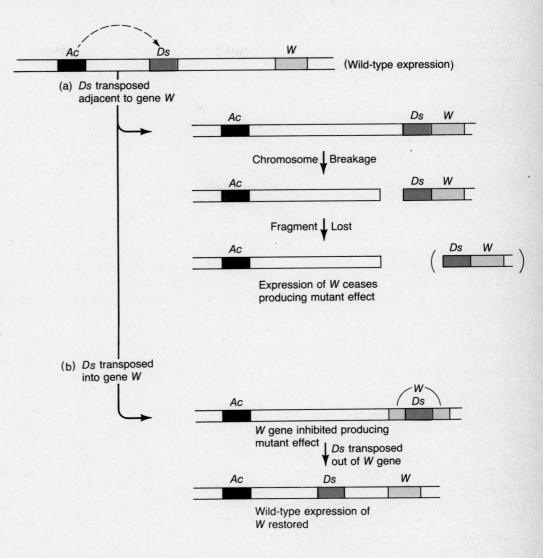

FIGURE 19.23

A comparison of the structure of an *Ac* element with three *Ds* elements, all of which have been isolated and sequenced. The imperfect inverted repeats are designated by a dark black shade, while the transposase gene and gene 2 are shown in different shades of color. Noncoding regions (NC) are unshaded. The dotted lines indicate the area that has been deleted, leading to each successive *Ds* element. As this scheme shows, *Ds-a* appears to be simply an *Ac* element containing a small deletion in the gene encoding the transposase enzyme.

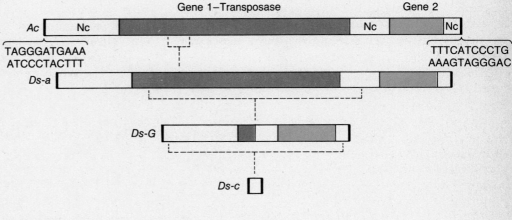

and three noncoding regions. The first *Ds* element studied (*Ds-a*) is nearly identical in structure to *Ac* except for a 194-base segment that has been deleted in the areas of the larger coding sequence (Gene 1 in Figure 19.23). There is some evidence that this gene encodes a **transposase** enzyme, essential to transposition of both the *Ac* and *Ds* elements. The deletion in this gene in the *Ds* element explains its dependence on the *Ac* element for transposition. Several other *Ds* elements have also been sequenced and each reveals an even larger deletion in the region of the Gene 1. In each case, however, the terminal repeats are retained and seem to be essential for transposition, provided that a functional transposase enzyme is supplied by the gene in the *Ac* element.

This information has now clarified at the molecular level the remarkable genetic observations made by McClintock some thirty years ago. As we will see, the terminal sequences and the genetic information encoding a transposase enzyme are universal components of transposons in all organisms studied.

Transposable Elements in *Drosophila*: *Copia* and *Copia*-like Genes

These assumptions have been further strengthened by the discovery of transposable elements in eukaryotic organisms other than maize, notably in yeast, *Drosophila*, and primates, including humans. In 1975, David Hogness and his colleagues David Finnigan, Gerald Rubin, and Michael Young identified a class of genes in *Drosophila melanogaster*, which they designated as *copia*. These genes transcribe "copious" amounts of RNA (thus, their name), which can be isolated in the polyadenylated mRNA fraction. Present up to 30 times in the genome of cells, *copia* genes are

nearly identical in nucleotide sequence. Mapping studies show that they are transposable to different chromosomal locations and are dispersed throughout the genome.

Copia appears to be only one of approximately 30 families of transposable elements, each of which is present 20 to 50 times in the genome of *Drosophila*. Since the discovery of *copia*, many other families have been recognized. These include, among others, the **FB** (**foldback**), **P,** and **I elements,** all of which are referred to as *copia*-like. Together, the many families constitute about 5 percent of the *Drosophila* genome and over half of the middle repetitive DNA of this organism.

Despite the variability in DNA sequence between the members of different families, they share a common structural organization thought to be related to the insertion and excision processes of transposition. Each *copia* gene consists of approximately 5000 base pairs of DNA, including a long family-specific **direct terminal repeat** sequence (**DTR**) of 276 base pairs at each end. Within each repeat is a short **inverted terminal repeat** (**ITR**) of 17 base pairs. These features are illustrated in Figure 19.24. The DTR sequences are found in other transposons in other organisms but are not universal. However, the shorter ITR sequences are considered universal.

Insertion of *copia* and other transposable elements appears to be dependent on ITR sequences and occurs at specific target sites in the genome. This is inferred by the observation that certain regions of chromosomes are more apt to be disposed to transposon-induced changes. Upon insertion, short direct repeats are usually generated at points of insertion. In the case of *copia*, this is 5 nucleotides long. For *Ds* in maize an 8-nucleotide sequence is generated. In general, eukaryotic transposons are strikingly similar to

FIGURE 19.24
Structural organization of a *copia* transposable element in *Drosophila melanogaster*.

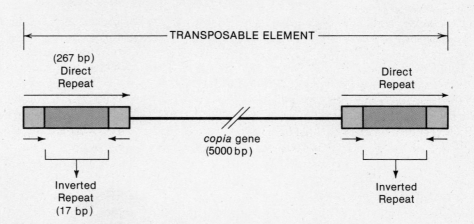

one another and share many features with those in bacteria.

An extensive research effort has been made to determine the role of the *Drosophila* elements. The transcribed RNA from *copia* genes has been isolated, and it can be translated into polypeptides *in vitro*. However, no function has yet been ascribed to these polypeptides. A particularly puzzling feature of *copia*-like elements is that few, if any, are found in the genome of a closely related species, *Drosophila simulans*. Therefore, it is assumed that they serve no essential function and in fact may be superfluous DNA.

Even so, *copia*-like elements demonstrate genetic effects at the point of their insertion in the chromosome. Certain mutations, including ones affecting eye color and segment formation, have been found to be due to insertions within genes. For example, the eye color mutation *white-apricot* (w^a), an allele of the *white* (*w*) mutation, contains a *copia* insertion element. Removing the transposable element restores the wild-type allele. Several unstable mutations in yeast are similarly affected. The insertion of a 10-kb FB into the DNA representing the *white-ivory* (w^i) allele creates the *white-crimson* (w^c) phenotype. Reversion to wild type following the removal of the element frequently occurs.

The P and I elements are associated with a phenomenon in *Drosophila* referred to as **hybrid dysgenesis.** When P or I insertions are contributed paternally to an offspring whose mother lacks them, increased levels of sterility, male recombination, chromosomal rearrangements, and mutations may result. The offspring is said to be "dysgenic" and thus frequently fails to reproduce. The transposability of these elements, which induce most of these problems, is under the control of the maternal cytoplasm in the zygote. One model suggests that P elements encode a transposase enzyme that is active in maternal cytoplasm derived from females lacking them. In such cases, P insertions then induce hybrid dysgenesis. Females already carrying P elements apparently contain cytoplasm that renders the enzyme inactive.

Alu Sequences in Humans

The final class of mobile eukaryotic genetic units that we will discuss is the *Alu* **family** found in mammals. The *Alu* family consists of large numbers of highly repetitive repeat sequences, originally detected using reassociation kinetics. In humans, there are about 300,000 copies found interspersed throughout the genome. These sequences represent most of the highly repetitive DNA in humans and constitute 3 percent of the total genome. The family has received its name because a high percentage of these sequences are cleaved by the restriction endonuclease *Alu I.*

The significant role of the *Alu* family in genetic processes is supported by the following observations:

1 These sequences are found in all primate and rodent DNA.

2 There is conserved within the sequences in mammals a specific 40-base-pair unit.

3 These sequences are reflected in nuclear transcripts generated by RNA polymerase II and III.

These sequences are considered transposable based on several lines of evidence. Most important is the fact that the 300-bp sequences are flanked on either side by direct repeat sequences consisting of 7 to 20 bp. This observation is similar to that of bacterial insertion sequences. These flanking regions are related to the insertion process during transposition. The second line of evidence involves the observation that clustered regions of *Alu* sequences vary in the DNA of normal and diseased individuals and in different tissues of the same individual. Additionally, the sequences have been found extrachromosomally.

Potential mobility of such a large number of short nucleotide sequences is a most intriguing concept. Do these units play a significant role in gene expression and regulation? Have they been important in the evolutionary process of primates? Future research will determine the answers to these questions.

SUMMARY AND CONCLUSIONS

In this chapter, we reviewed various experimental approaches designed to eludicate gene structure and organization. The topic is an exciting one for many reasons. First, it is currently being explored with the most refined technology ever available in this field—recombinant DNA (discussed in Chapter 20) and rapid DNA sequencing techniques. Second, the most recent work, which represents a "direct attack" on the structure of genes, has revealed major differences in the organization of the genetic information in animal viruses and eukaryotes compared with bacterial viruses and bacteria. Third, with each new discovery, new unsolved problems become apparent. When these problems are solved, we are promised an even better understanding of how genetic organization is related to genetic function.

One of the first approaches used was the complementation test, which functionally defines allelism. Two heterozygous mutations present in the *trans* configuration and producing a mutant phenotype are considered to be alleles and belong to the same complementation group. While such a group is generally equivalent to a single gene, a more accurate term, *cistron*, designates the smallest functional genetic unit.

In 1957, Benzer reported the most detailed genetic analysis of a single gene ever performed. He was able to divide the *rII* locus of phage T4 into two complementation groups or cistrons. Intragenic mapping of each cistron was then performed. Relying on deletion mapping and recombinational analysis, he located 2000 mutations at 307 distinct sites within the cistrons. Benzer's work demonstrated that the phage gene is an indivisible particle consisting of mutable units (mutons) and recombinational units (recons).

Similar analyses have been undertaken in eukaryotes, but with much greater difficulty. We discussed studies of the *lozenge, rosy,* and *bithorax* loci in *Drosophila* as examples of these analyses. *Lozenge* is a complex locus divided into subunits. Mutations found in one subunit are pseudoalleles to those located in the other subunits. They test positively in an allelism test, but can be separated by crossing over. Because of the availability of selective methods, the *rosy* gene, which is not a complex locus, has been mapped with a resolution nearly as great as the *rII* locus. The *bithorax* locus is divided into distinct regions, each of which specifies a genetic product affecting a specific segment during development.

Observations of the organization of eukaryotic genes and DNA have pointed out the *C* value paradox. That is, eukaryotic organisms contain much more DNA than necessary to encode the gene products essential to the development and normal functions of an organism. Is this DNA superfluous, or does it serve some yet undiscovered function?

Detailed analysis has revealed numerous categories of noncoding DNA sequences. The initial mRNA transcripts present in the nucleus as heterogeneous RNA are processed into mature mRNA molecules. This is accomplished by excising numerous internal noncoding sequences and splicing together those that encode the gene product. The gene giving rise to such an RNA transcript is said to be "split" and contains noncoding introns and coding exons. Some eukaryotic genes are void of introns, but others contain up to 50 intervening sequences.

The flanking regions of genes, particularly those adjacent to the 5' initiation point, have also been analyzed. Three "upstream" regions appear to be essential to efficient transcription. One, the TATA box, is found about 30 nucleotides from the point of initiation of transcription. The second, hypersensitive to DNase I when active, is located 100 to 500 nucleotides upstream. The third is called an enhancer element.

The final level of organization we have considered has been that of gene families. Usually clustered together, structurally related genes encoding functionally similar products have been discovered. The beta-globin gene family, contained in a 60-kb region of DNA, encodes six genes, including a nonfunctional pseudogene. Only 5 percent of the DNA consists of coding sequences, and the remainder is composed of introns, flanking regions, and spacer DNA. The histone family occupies a smaller DNA region and contains five genes whose polarity varies in different organisms. Unlike the globin family, the histone cluster of genes is sometimes tandemly repeated a variable number of times in different organisms.

We concluded our discussion of gene families by considering those genes involved in the immune response in vertebrates. A description of immunoglobulin structure, the genes that specify the polypeptide chains of these antibodies, and the mechanism for generating the enormous diversity found in this component of the immune system was provided. The genes specifying the various components of the immunoglobulins are perhaps the most extensive sets of gene families in eukaryotes.

Finally, we discussed transposable elements in eukaryotes, first discovered in maize in the late 1940s and now being investigated intensively. Among the most thoroughly studied are the *copia* and *copia*-like genes found in *Drosophila melanogaster*. Together, these constitute about 30 families of genes present 20 to 50 times in each genome. Although each family contains specific nucleotide sequences, they have structural similarities thought to be involved in insertion and excision. These and similar elements discovered in yeast have pronounced genetic effects at their point of insertion. Humans and other primates contain Alu sequences, which are transposable and present some 300,000 times in the genome.

PROBLEMS AND DISCUSSION QUESTIONS

1 Many independent mutants of a haploid organism that cannot synthesize substance X are isolated. In its life cycle, the organism passes through a brief diploid phase as a result of fusion of haploid cells. During this stage, two mutants can be tested for complementation because the mutations are present in the *trans* condition. The following data result when five mutations are tested. Determine the number of complementation groups and which mutations are in each one.

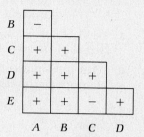

2 Three mutations were discovered with a similar phenotype. In complementation testing, mutants 1 and 2 were shown to be allelic, but neither was allelic to mutant 3. Complete the following table indicating + for positive complementation and − for negative complementation.

Mutants	Condition	
	Cis	Trans
1, 2		
1, 3		
2, 3		

3 In complementation studies of the *rII* locus of phage T4, three groups of three different mutations were tested. For each group, only two combinations were tested. On the basis of each set of data, predict the results of the third experiment.

Group A	Group B	Group C
d × e—lysis	g × h—no lysis	j × k—lysis
d × f—no lysis	g × i—no lysis	j × l—lysis
e × f—?	h × i—?	k × l—?

4 In an analysis of other *rII* mutants, complementation testing yielded the following results:

Mutants	Results (lysis)
1, 2	+
1, 3	+
1, 4	−
1, 5	−

(a) Predict the results of testing 2 and 3, 2 and 4, and 3 and 4 together.

(b) If further testing yielded the following results, what would you conclude about mutant 5?

Mutants	Results
2, 5	−
3, 5	−
4, 5	−

(c) Following mixed infection of mutants 2 and 3 on *E. coli* B, progeny virus were plated in a series of dilutions on both *E. coli* B and K12 with the following results. What is the recombination frequency between the two mutants?

Strain plated	Dilution	Colonies
E. coli B	10^{-5}	2
E. coli K12	10^{-1}	5

(d) Another mutation, 6, was then tested in relation to mutations 1 through 5. In initial testing, mutant 6 complemented mutants 2 and 3. In recombination testing with 1, 4, and 5, mutant 6 yielded recombinants with 1 and 5, but not with 4. What can you conclude about mutation 6?

5 Define *pseudoalleles.* Why was the analysis of the *rosy* locus easier to accomplish than that of the *lozenge* locus in *Drosophila melanogaster?*

6 What is the nucleotide sequence of a DNA fragment yielding the following electrophoresis results?

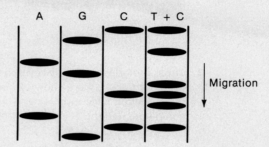

7 What are the differences between the beta-globin and histone gene families?

8 List the various types of distinct noncoding regions found in the eukaryotic genome.

9 What do the following symbols represent in immunoglobulin structure: V_L, C_H, λ, IgG, J, and D? What is the most acceptable theory for the generation of antibody diversity?

10 Figure 19.21 shows nine unique gene sequences formed by recombination. How many other unique sequences can be formed?

11 If embryonic DNA contains 10 V, 30 D, 50 J, and 3 C genes, how many unique DNA sequences can be formed through recombination?

12 If there are 5 V, 10 D, and 20 J genes available to form a heavy-chain gene and 10 V and 100 J genes available to form a light chain, how many unique antibodies can be formed?

SELECTED READINGS

BENZER, S. 1955. Fine structure of a genetic region in bacteriophage. *Proc. Natl. Acad. Sci.* 42: 344–54.

————. 1961. On the topography of the genetic fine structure. *Proc. Natl. Sci.* 47: 403–15.

BOYER, H. W. 1971. DNA restriction and modification mechanisms in bacteria. *Ann. Rev. Microbiol.* 25: 153–76.

BINGHAM, P. M., KIDWELL, M. G., and RUBIN, G. M. 1982. The molecular basis of P-M hybrid dysgenesis: The nature of induced mutations. *Cell* 29: 987–94.

BREGLIANO, J. C., et al. 1980. Hybrid dysgenesis in *Drosophila melanogaster. Science* 207: 606–11.

CALABRETTA, B., et al. 1982. Genome instability in a region of human DNA enriched in Alu repeat sequences. *Nature* 296: 219–25.

CAPRA, J. D., and EDMUNDSON, A. B. 1977. The antibody combining site. *Scient. Amer.* (Jan.) 236: 50–59.

CHAMBON, P. 1981. Split genes. *Scient. Amer.* (May) 244: 60–71.

CHOVNICK, A. 1961. The *garnet* locus in *Drosophila melanogaster.* I. Pseudoallelism. *Genetics* 46: 493–507.

CHOVNICK, A., GELBART, W., McCARRON, M. 1977. Organization of the *rosy* locus in *Drosophila melanogaster. Cell* 11: 1–10.

COLD SPRING HARBOR LABORATORY. 1976. *Origin of lymphocyte diversity. Cold Spr. Harb. Symp.*, Vol. 41.

CRICK, F. 1979. Split genes and RNA splicing. *Science* 204: 264–71.

DARNELL, J. E. 1978. Implications of RNA: RNA splicing in the evolution of eukaryotic cells. *Science* 202: 1257–60.

DORING, H. P. 1985. Plant transposable elements. *Bio. Essays.* 3: 164–66.

DORING, H. P., and STARLINGER, P. 1984. Barbara McClintock's controlling elements: Now at the DNA level. *Cell* 39: 253–59.

DUGAICZK, A., et al. 1978. The natural ovalbumin gene contains seven intervening sequences. *Nature* 274: 328–33.

DULBECCO, R. 1979. Contributions of microbiology to eukaryotic cell biology: New directions for microbiology. *Microbiol. Rev.* 43: 443–52.

EDELMAN, G. M. 1973. Antibody structure and molecular immunology. *Science* 180: 830–40.

FALK, R. 1984. The gene in search of an identity. *Hum. Genet.* 68: 195–204.

FEDEROFF, N. V. 1984. Transposable genetic elements in maize. *Scient. Amer.* (June) 250: 85–98.

FINCHAN, J. R. S., and SASTRY, G. R. K. 1974. Controlling elements in maize. *Ann. Rev. Genet.* 8: 15–50.

FINNEGAN, D. J., RUBIN, G., YOUNG, M., and HOGNESS, M. W. 1978. Repeated gene families in *Drosophila melanogaster. Cold Spr. Harb. Symp.* 42: 1053–64.

FRITSCH, E. F., LAWN, R. M., and MANIATIS, T. 1980. Molecular cloning and characterization of the human β-like globin gene cluster. *Cell* 19: 959–72.

GELBART, W., et al. 1976. Extension of the limits of the XDH structural element in *Drosophila melanogaster. Genetics* 84: 211–32.

GILBERT, W. 1978. Why genes in pieces? *Nature* 271: 501.

———. 1985. Genes in pieces revisited. *Science* 228: 823–24.

GREEN, M. M. 1980. Transposable elements in *Drosophila* and other Diptera. *Ann. Rev. Genet.* 14: 109–20.

GREEN, M. M., and GREEN, K. C. 1949. Crossing-over between alleles at the *lozenge* locus in *Drosophila melanogaster. Proc. Natl. Acad. Sci.* 48: 586–91.

HENTSCHEL, C. C., and BIRNSTIEL, M. L. 1981. The organization and expression of histone gene families. *Cell* 25: 301–13.

HIETER, P. A., et al. 1981. Clustered arrangement of immunoglobulin λ constant region genes in man. *Nature* 294: 536–40.

HOOD, L., CAMPBELL, J. H., and ELGIN, S. C. R. 1975. The organization, expression, and evolution of antibody genes and other multigene families. *Ann. Rev. Genet.* 9: 305–53.

HOOD, L. E., WEISSMAN, I. L., and WOOD, W. B. 1978. *Immunology.* Menlo Park, Calif.: Benjamin-Cummings.

HOOD, L., DRONENBERG, M., and HUNKAPILLER, T. 1985. T cell antigen receptors and the immunoglobulin supergene family. *Cell* 40: 225–229.

HOZUMI, N., and TONEGAWA, S. 1976. Evidence for somatic rearrangement of immunoglobulin genes coding for variable and constant regions. *Proc. Natl. Acad. Sci.* 73: 3628–32.

JERNE, N. K. 1973. The immune system. *Scient. Amer.* (July) 229: 52–60.

———. 1985. The generative grammar of the immune system. *Science* 229: 1057–59.

KARESS, R. E., and RUBIN, G. M. 1984. Analysis of P transposable element functions in *Drosophila. Cell* 38: 135–46.

KEDES, L. H. 1979. Histone genes and histone messengers. *Ann. Rev. Biochem.* 48: 837–70.

KELLER, E. F. 1983. *A feeling for the organism: The life and work of Barbara McClintock.* New York: W. H. Freeman.

KINDT, T. J., and CAPRA, J. D. 1984. *The antibody enigma.* New York: Plenum Press.

KOLATA, G. B. 1981. Genes regulated through chromatin structure. *Science* 214: 775–76.

LEDER, P. 1982. The genetics of antibody diversity. *Scient. Amer.* (May) 246: 102–15.

LEWIN, B. 1980. *Gene expression 2—Eukaryotic chromosomes.* 2nd ed. New York: Wiley.

LEWIS, E. B. 1948. Pseudoallelism and gene evolution. *Cold Spr. Harb. Symp.* 16: 159–74.

MANIATIS, T., et al. 1980. The molecular genetics of human hemoglobins. *Ann. Rev. Genet.* 14: 145–78.

MARX, J. L. 1978a. Antibodies (I): New information about gene structure. *Science* 202: 298–99.

———. 1978b. Antibodies (II): Another look at the diversity problem. *Science* 202: 412–15.

———. 1981. Antibodies: Getting their genes together. *Science* 212: 1015–17.

MAXAM, A., and GILBERT, W. 1977. A new method for sequencing DNA. *Proc. Natl. Acad. Sci.* 74: 560–64.

McCLINTOCK, B. 1956. Controlling elements and the gene. *Cold Spr. Harb. Symp.* 21: 197–216.

MINTY, A., and NEWMARK, P. 1980. Gene regulation: New, old, and remote controls. *Nature* 288: 210–11.

O'HARE, K. 1985. The mechanism and control of P element transposition in *Drosophila melanogaster. Trends in Genet.* 1: 250–54.

ROBERTSON, M. 1981. Genes of lymphocytes I: Diverse means to antibody diversity. *Nature* 290: 625–27.

ROGERS, J., and WALL, R. 1980. A mechanism for RNA splicing. *Proc. Natl. Acad. Sci.* 77: 1877–79.

SANDERS-HAIGH, L. W., ANDERSON, F., and FRANKE, U. 1980. The β-globin gene is on the short arm of chromosome 11. *Nature* 283: 683–86.

SANGER, F. 1981. Determination of nucleotide sequences in DNA. *Science* 214: 1205–10.

SCHMID, C. W., and JELINEK, W. R. 1982. The Alu family of dispersed repetitive sequences. *Science* 216: 1065–70.

SEIDMAN, J. G., et al. 1978. Antibody diversity. *Science* 202: 11–17.

SHAPIRO, J. A. 1984. *Mobile genetic elements.* New York: Academic Press.

SHAPIRO, J. A., and CORDELL, B. 1982. Eukaryotic mobile and repeated genetic elements. *Biol. Cell* 43: 31–54.

SPRADLING, A., and RUBIN, G. M. 1981. *Drosophila* genome organization: Conserved and dynamic aspects. *Ann. Rev. Genet.* 15: 219–64.

———. 1982. Transposition of cloned P elements into *Drosophila* germ line chromosomes. *Science* 218: 341–47.

SÜDHOF, T. C., et al. 1985. The LDL receptor gene: A mosaic of exons shared with different proteins. *Science* 228: 815–22.

TESSMAN, I. 1965. Genetic ultrafine structure in the T4 *rII* region. *Genetics* 51: 63–75.

TONEGAWA, S. 1985. The molecules of the immune system. *Scient. Amer.* (Oct.) 253: 122–31.

VAN den BERG, J., et al. 1978. Comparison of cloned rabbit and mouse β-globin genes showing strong evolutionary divergence of two homologous introns. *Nature* 276: 37–44.

WEIGERT, M., et al.: 1980. The joining of V and J gene segments creates antibody diversity. *Nature* 283: 497–99.

WU, R. 1978. DNA sequence analysis. *Ann. Rev. Biochem.* 47: 607–34.

ZACHAR, Z., and BINGHAM, P. M. 1982. Regulation of *white* locus expression: The structure of mutant alleles at the *white* locus of *Drosophila melanogaster. Cell* 30: 529–41.

20

Recombinant DNA: Technology and Applications

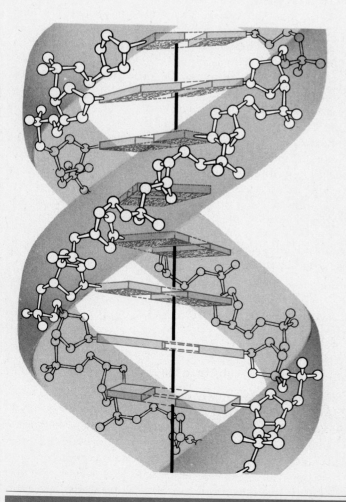

The independent rediscovery of Mendel's work by Hugo de Vries, Carl Correns, and Eric von Tschermak in 1900 marks the beginning of genetics as an organized discipline. During the course of its growth and development, several key discoveries have served as landmarks or turning points, each accelerating the rate at which our knowledge of genetics has grown and opening new fields of investigation.

One of the first such turning points was the concept that chromosomes are the repository for genes. This concept, proposed by Walter Sutton and Theodore Boveri in 1902, was developed by Thomas Morgan and his colleagues using the fruit fly, *Drosophila*. From these studies came our concepts of the gene as a fundamental unit of recombination, mutation, and function, and the technique of using polytene chromosomes to map genes to specific cytological loci.

A second landmark was the discovery by Oswald Avery, Colin MacLeod, and MacLyn McCarty that DNA is the macromolecular carrier of genetic information. This work stimulated the use of viruses and bacteria as organisms for genetic research and led to the Watson-Crick model for the structure of DNA. From this model has come knowledge of the molecular basis for genetic coding, transcription, translation, and gene regulation.

We are now in the beginning stages of another and perhaps the most profound transition in the history of genetics—the development and application of **recombinant DNA technology,** also known as **gene splicing** or **genetic engineering.** This technology, which derives from the seemingly obscure discovery of a class of endonuclease enzymes in bacteria that cleave DNA at specific sites, has potential for generating new knowledge and developing new vaccines and drugs. It has also raised fears about epidemics produced by the accidental release of new pathogenic organisms or government-mandated social control through deliberate manipulation of human genes. In this chapter we will review some of the methods used in recombinant DNA technology to isolate and analyze genes; discuss some of the applications of this technology to agriculture, industry, and medicine; and, finally, explore the possible long-term effects of genetic technology on the individual and society.

RECOMBINANT DNA TECHNOLOGY: AN OVERVIEW

The term **recombinant DNA** refers to the creation of a new association between DNA molecules or parts of DNA molecules that are not found together naturally. Although genetic mechanisms such as crossing over technically produce recombinant DNA molecules, the term is reserved for the artificial creation of unique DNA segments, the parts of which are derived from different biological sources. Popularly called **gene splicing,** the techniques used to create recombinant molecules represent a *tour de force* in the application of enzymology to the study of molecular genetics.

Recombinant DNA technology uses a series of distinctive enzymes in the following steps:

1 Either random or specific segments of eukaryotic DNA are isolated or synthesized *in vitro.*

2 These segments are joined to another DNA molecule, which serves as a **vector.** A vector facilitates the manipulation and recognition of the newly created recombinant molecule.

3 The vector, usually a bacterial plasmid or phage, is taken up by a bacterial cell.

4 Within the bacterial cell, the recombinant DNA molecule is replicated along with endogenous DNA of the host cell during asexual reproduction, a process referred to as **cloning.**

5 The cloned recombinant DNA, produced in abundance in an overnight bacterial culture, can be isolated, purified, and analyzed.

6 Potentially, the cloned DNA may be transcribed, its mRNA translated, and the gene product isolated and studied.

This technology affords many research opportunities, including the characterization of gene structure by the application of DNA sequencing techniques to the cloned DNA and the study of the expression and regulation of eukaryotic genes.

Restriction Enzymes

The cornerstone of recombinant DNA technology is a class of enzymes called **restriction endonucleases.** These enzymes, isolated from bacteria, received their name because they restrict viral infection by degrading the invading nucleic acid. Two classes of restriction endonucleases have been discovered, and more than 80 enzymes of both types are now recognized. Type I restriction enzymes recognize a specific nucleotide sequence and produce a nonspecific double-stranded cut in the DNA molecule close to that sequence. Type II enzymes recognize a specific sequence and produce a double-stranded break precisely within the sequence. In Type II enzymes, the recognition site is **palindromic.** The DNA of bacteria producing a specific enzyme is protected by **modification enzymes** that methylate the nucleotides contained in sequences vulnerable to digestion.

Type II enzymes are used most often. Because of their specificity, some of them produce two DNA fragments with single-stranded complementary tails. For example, the first such enzyme to be discovered was from *E. coli* and was designated **Eco RI.** Its palindromic recognition site and points of cleavage are shown in the following diagram:

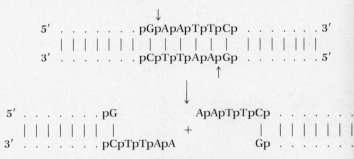

The complementary tails are said to be "sticky" because under hybridization conditions, they will reanneal with each other. If the DNAs from two sources share the same palindromic recognition sites in their DNA sequence, then both will contain complementary single-strand tails if enzymatically treated. When placed together under proper conditions, recombinant DNA molecules can be created (Figure 20.1).

FIGURE 20.1
Construction of a hybrid
plasmid using the *Eco*RI
restriction endonuclease.
Arrows indicate the enzyme
recognition sites.

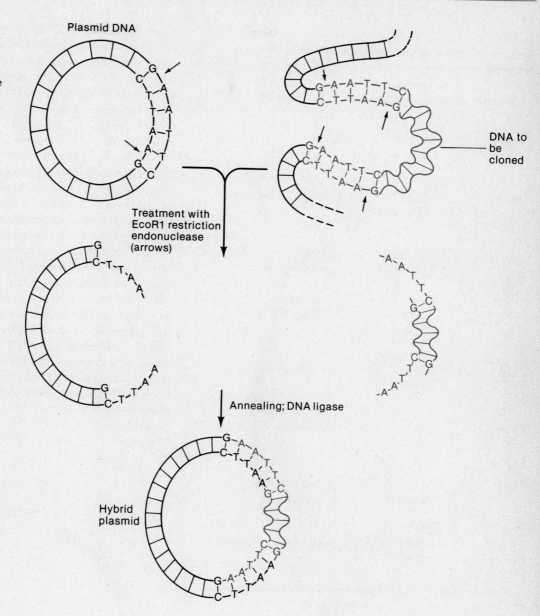

Plasmid DNA

DNA to be cloned

Treatment with
EcoR1 restriction
endonuclease
(arrows)

Annealing; DNA ligase

Hybrid
plasmid

Other restriction enzymes such as *Sma*I, however, produce blunt-end fragments:

$$5' \quad . \quad . \quad . \quad . \quad . \quad . \quad . \quad . \quad \downarrow \quad pCpCpCpGpGpGp \quad . \quad . \quad . \quad . \quad . \quad . \quad 3'$$
$$| \; | \; | \; | \; | \; | \; | \; | \; | \; | \; | \; | \; | \; | \; | \; | \; | \; |$$
$$3' \quad . \quad . \quad . \quad . \quad . \quad . \quad . \quad . \quad pGpGpGpCpCpCp \quad . \quad . \quad . \quad . \quad . \quad . \quad 5'$$
$$\uparrow$$

$$\downarrow$$

$$5' \quad . \quad . \quad . \quad . \quad . \quad pCpCpC \quad \quad GpGpGp \quad . \quad . \quad . \quad . \quad . \quad 3'$$
$$| \; | \; | \; | \; | \; | \; | \; | \; + \; | \; | \; | \; | \; | \; | \; | \; |$$
$$3' \quad . \quad . \quad . \quad . \quad . \quad pGpGpG \quad \quad CpCpCp \quad . \quad . \quad . \quad . \quad . \quad 5'$$

Ligation of DNA fragments with blunt ends is less efficient but can also be used to create recombinant molecules. Commonly, the enzyme **terminal deoxynucleotidyl transferase** may be used to enhance annealing. This enzyme will extend single-stranded ends by adding 50 to 200 adenylic or thymidylic acid residues, depending on which nucleotide is added to the reaction mixture. If a poly A tail is added to vector DNA, and poly dT is added to foreign DNA, more extensive complementary tails are created, and recombinant molecules can be created by ligation. Table 20.1 lists some common restriction enzymes and their rec-

TABLE 20.1
Some restriction enzymes, cleavage pattern and source.

Enzyme	Recognition, Cleavage Sequence	Source
EcoRI	↓ GAATTC CTTAAG ↑	E. coli
HindIII	↓ AAGCTT TTCGAA ↑	Hemophilus influenza
BamHI	↓ GGATCC CCTAAG ↑	Bacillus amyloliquefaciens
TaqI	↓ TCGA AGCT ↑	Thermus aquaticus
AluI	↓ AGCT TCGA ↑	Arthrobacter luteus
HaeIII	GGCC CCGG ↑	Hemophilus aegypticus
BalI	↓ TGGCCA ACCGGT ↑	Brevibacterium albidum
Sau3A	↓ GATC CTAG ↑	Staphylococcus aureus

As described in Chapter 14, plasmids are naturally occurring, extrachromosomal double-stranded DNA molecules that replicate within bacterial cells independent of the host cell chromosome. Many plasmids have been modified or engineered to contain a limited number of restriction sites and specific loci for antibiotic resistance. A plasmid used frequently in DNA research, pBR322 is shown in Figure 20.2.

The bacteriophage most widely used in recombinant DNA work is the phage lambda. Its genes have all been identified and mapped, and the DNA sequence of the entire genome is known. The middle of the lambda chromosome contains a cluster of genes necessary for the lysogenic phase of its life cycle, but not for the lytic phase. As a result, the middle half of the chromosome can be removed and replaced with foreign DNA without affecting the ability of the phage to infect bacterial cells and carry out replication. Over a hundred vectors based on lambda have been derived by removing various segments of the central gene cluster, including the Charon phage, λgt (generalized transducing) phage, and the EMBL phage vectors. Recently, single-stranded filamentous bacteriophage such as M13 have been introduced as vectors. DNA cloned into M13 can be readily sequenced by the

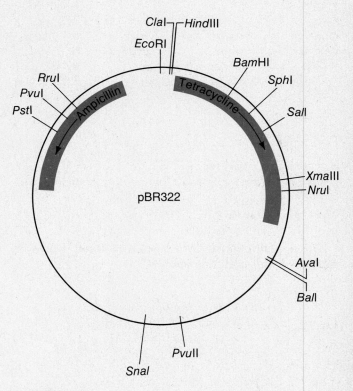

FIGURE 20.2
Restriction map of the plasmid, pBR322. The locations of restriction sites that cleave only once are shown.

ognition sequences, some of which yield cohesive or "sticky" ends, while others generate blunt ends. After its formation, the recombinant molecule must be inserted into a host organism for replication. This is accomplished through the use of vectors.

Vectors

By using a vector or cloning vehicle, the recombinant molecule can gain entry into a host cell and be replicated or cloned. Vectors are carrier molecules into which a DNA fragment can be inserted and cloned, and are capable of independent replication. Three types of vectors are currently in use, **plasmids, bacteriophage,** and **cosmids.**

Sanger method, and can be used as a template for site-specific mutagenesis.

Cosmids are hybrid vectors composed of plasmid sequences containing antibiotic resistance genes and the terminal *cos* sequences of phage lambda, necessary for assembly of chromosomes into phage heads. Cosmids allow the cloning of large segments of DNA, up to 40 kb, while phage vectors can accommodate segments of up to 20 kb; plasmids are usually limited to approximately 10 kb of foreign insert. Other hybrid vectors, composed of eukaryotic viral sequences and bacterial plasmids are known as **shuttle vectors.** These vectors contain markers that are selectable in both prokaryotic and eukaryotic cells, and thus can be shuttled back and forth without recloning.

A variety of microorganisms can be used as replication hosts, including bacteria and yeast. The most commonly used host is a laboratory strain of *E. coli* known as K12. Plasmids are inserted into host cells by the process of transformation. Cells are treated with $CaCl_2$ to permeabilize the membrane and mixed with recombinant plasmids. The bacteria are then plated on a solid medium and form discrete colonies. Since each colony represents the progeny of a single bacterial cell, all members of a colony have the same genetic constitution, and are said to be a **clone.** If cells in the colony contain a recombinant plasmid, multiple copies of the foreign DNA have been replicated or cloned. Similarly, phage containing foreign DNA are used to transfect a lawn of *E. coli.* The resulting plaques each represent the product of a single phage, and each is regarded as a clone. Cosmids attach to and enter *E. coli* like phage, but once inside the cell they replicate like plasmids.

LIBRARY CONSTRUCTION

Since each cloned fragment of DNA is relatively small, many separate clones must be created in order to include even a small portion of the total genome of an organism. Sets of clones derived using a single approach are called **libraries.** As we shall see, these are either genomic, chromosomal, or complementary DNA (cDNA) libraries. The cDNA libraries are more specific than the chromosomal, which are more specific than the genomic libraries.

Genomic Libraries

Often the first step in studying a gene and its flanking sequences is the construction of a **genomic library** or **bank** that contains all of the cloned sequences in the genome of an individual. Since each plasmid or phage chromosome can contain only a few kilobases of for-

eign DNA, a large number of recombinant plasmids or phage chromosomes are necessary to contain the entire genome. Of course, the number of recombinant molecules required to ensure that all the sequences in the genome are represented in the library is dependent on the average size of the cloned inserts and the size of the genome. The number of clones necessary to represent all sequences in a genome is given by the equation

$$N = \ln(1 - P)/\ln(1 - f) \qquad (20.1)$$

where N is the number of required recombinants, P is the probability of recovering a given sequence, and f represents the fraction of the genome present in each clone. If the average clone size in a phage library of the human genome is 17 kb, and the size of the genome is 3.6×10^6 kb, it would be necessary to prepare a library of about 8.1×10^5 phage (where $f = 1.7 \times 10^4/3.0 \times 10^9$) to have a 99 percent ($P = 0.99$) probability that any given human gene is present in at least one phage. Once a genomic library has been constructed, any sequence of interest, from single-copy structural genes through highly repeated satellite sequences, can be recovered for study and analysis from such a library. Libraries for many organisms of genetic interest are now available, including bacteria, yeast, *Drosophila*, and humans.

Chromosome-Specific Libraries

A library of DNA sequences derived from a single chromosome or chromosome region is of great value in the study of chromosome organization and the effects of gene rearrangements on expression. In *Drosophila*, F. Scalenghe and colleagues have successfully cloned the DNA from a small segment of the X chromosome. Using a micromanipulator, they dissected from a single X chromosome a fragment corresponding to a region approximately 50 polytene bands long. The DNA was extracted, cut with restriction endonucleases, and cloned into a bacteriophage lambda vector. This region of the chromosome contains the genes *white*, *zeste*, and *Notch*, and the original site of a transposing element that can translocate a small two- to three-band segment to more than a hundred different chromosomal sites.

Although microdissection has also been used to prepare mammalian chromosomes for cloning, a more convenient technique known as **flow cytometry** has been used by Bryan Young and others to clone the DNA sequences from human chromosomes. In this method, chromosomes isolated from mitotic cells and stained with a fluorescent dye are passed at high speed through laser beams connected to a photometer, which sorts and fractionates the chromosomes by differences in dye binding and light scattering. Purified chromosomes can be sorted at a rate of 100 per second, and as little as 500 nanograms of DNA can be used for cloning. With this technique, cloned libraries for each human chromosome will soon be available.

Libraries constructed using microdissection and flow cytometry can be used to gain access to genetic loci for which there are no convenient selection systems using DNA, mRNA, or proteins. These techniques also provide a means of studying the molecular organization and even the nucleotide sequence in defined chromosomal regions.

cDNA Libraries

A library representing the structural genes active in a specific cell type can be constructed by synthesizing a **cDNA (complementary DNA) molecule** using the enzyme **reverse transcriptase** and the mRNA of the cell as a template. As shown in Figure 20.3, the overall process also involves the use of DNA polymerase I, S1 nuclease, and the transferase enzyme.

This technique is particularly useful where purified mRNA species can be obtained. For example, hemoglobin mRNA is nearly the only mRNA transcript produced at a certain stage of red blood cell development. Similarly, ovalbumin mRNA can be obtained from uterine tissue. In both instances, polysomes are isolated and the mRNA derived from these complexes.

When cDNA is produced, specific genes or gene regions can be inserted into vector DNA and subsequently cloned in bacteria. The technique can also be applied in cases where mRNA can be synthesized directly *in vitro*.

SELECTION OF RECOMBINANT CLONES

Once recombinant DNA fragments have been produced and ligated into a vector, the next task is to select clones that contain a gene or DNA sequence of interest. Often, a mixture of ligation products is present, some of which may be head-to-tail vectors or vectors that have religated upon themselves and do not contain foreign DNA at all. Thus, it is necessary to first eliminate these "empty" vectors and then to isolate the cloned sequence of interest.

Insertional Inactivation

Several techniques are available to select vectors that have indeed incorporated foreign DNA and can thus be a viable source of recombinant DNA clones. In gen-

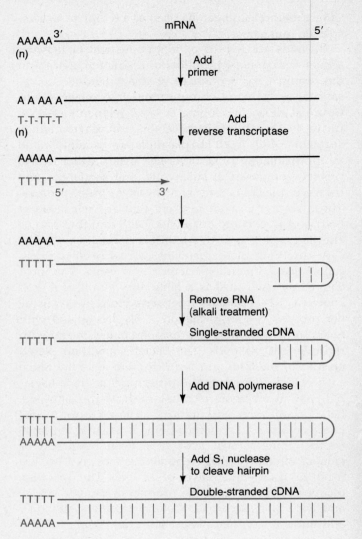

FIGURE 20.3

The production of cDNA from mRNA. Since many eukaryotic mRNAs have a polyadenylated tail of variable length (A_n) at their 3' end, a short oligonucleotide sequence composed of thymidine nucleotides can be annealed. This oligonucleotide acts as a primer for the enzyme reverse transcriptase, which uses the mRNA as a template to synthesize a complementary DNA strand. A characteristic hairpin loop is often formed. The mRNA can be removed by alkaline treatment of the complex, and DNA polymerase is added to synthesize the second DNA strand. S_1 nuclease is used to open the hairpin loop, and the result is a double-stranded cDNA molecule that can be cloned into a suitable vector, or used as a probe in library screening.

eral, these techniques rely on the process of **insertional inactivation** to eliminate religation products that have simply closed back up without incorporating foreign DNA.

In the case of plasmids such as pBR322, this can be accomplished by transferring bacterial cells from a given colony to plates containing ampicillin or tetracycline. If the foreign DNA was inserted at the *Bam*HI site of the plasmid (Figure 20.2), the gene conferring resistance to tetracycline will be inactivated, and colonies will not grow in the presence of tetracycline, but will still grow on plates containing ampicillin.

In the case of phage vectors, many of these have been engineered to produce blue plaques when plated on medium containing a compound called Xgal. The restriction sites for the insertion of foreign DNA occur within the gene responsible for the formation of blue plaques, and consequently, phage carrying insertions produce clear plaques. These methods represent the initial screening step in the isolation of suitable clones.

Colony Hybridization

The next step is to select a cloned sequence of interest from all those in the collection. Many approaches to this problem have been developed, but the most common involves a **gene-specific hybridization probe.**

For screening a plasmid library, the Grunstein-Hogness colony hybridization method is most useful for the detection and recovery of cloned DNAs. Bacteria containing recombinant plasmids are plated out, and the colonies are replica-plated by pressing a nitrocellulose filter onto the plate (Figure 20.4). The bacterial colonies transferred to the filter are then lysed, and the DNA denatured and baked onto the filter. The filter is then hybridized with a solution containing a radioactive probe, which can be an mRNA, cDNA, or a recombinant DNA molecule obtained from another experiment. After any excess probe is washed away, the filter is overlaid with a piece of photographic film. If the RNA or DNA of the probe is complementary to the DNA of one of the lysed colonies, the hybridized colony can be identified as a dark spot on the developed film. That colony can then be recovered from the initial plate and isolated. The procedure can also be used for the selection of recombinant phage plaques.

If a purified protein gene product is available, an alternate method can be used. First, part of the amino acid sequence of the protein must be determined.

FIGURE 20.4
Grunstein-Hogness colony screening method for the detection and recovery of cloned DNAs of interest.

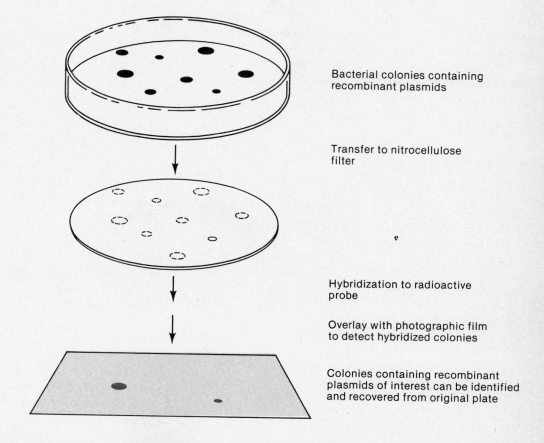

Bacterial colonies containing recombinant plasmids

Transfer to nitrocellulose filter

Hybridization to radioactive probe

Overlay with photographic film to detect hybridized colonies

Colonies containing recombinant plasmids of interest can be identified and recovered from original plate

From this, the possible nucleotide sequences of the gene can be deduced. Because of the degeneracy of the genetic code, several different nucleotide sequences can encode the same amino acid sequence (Figure 20.5). It is not necessary to know the entire amino acid sequence of the protein or to synthesize the entire gene, since a fragment corresponding to 20 nucleotides (6–7 amino acids) is usually sufficient. The synthetic gene fragment is radioactively labeled and used as a probe in library screening. In either technique, each member of the bacterial colony or phage plaque that contains the recombinant DNA of interest may be isolated for study.

CHARACTERIZATION OF CLONED SEQUENCES

The development of recombinant DNA methodology has been instrumental in the study of eukaryotic gene organization and has provided an insight into the mechanisms of gene regulation in both prokaryotic and eukaryotic genomes. The ability to prepare large quantities of a single gene and its flanking sequences allows the use of a wide range of physical and biochemical techniques to study organization, function, and regulation. A comprehensive discussion of these techniques is beyond the scope of this chapter, but we will consider two methods that have widespread applications in recombinant DNA work.

Restriction Mapping

The DNA fragments produced by restriction enzyme digestion of cloned DNA segments can be separated by electrophoresis on agarose or acrylamide gels. Because the length of a DNA fragment is inversely proportional to its migration on the gel, smaller fragments migrate farther than do larger fragments over a given time period. By using molecular weight markers in adjacent lanes, it is possible to determine the molecular weight of the restriction fragments. If different restriction enzymes are utilized alone and in combination, it is possible to construct a map of the order and distance between restriction sites on a cloned DNA segment. This technique is illustrated in Figure 20.6. A **restriction map** produced in this way can be used to order gene sequences along the cloned segment, and the map provides a means of excising specific regions of a clone for further work. The restriction map of phage lambda DNA for the enzymes *Eco*RI and *Bam*HI is shown in Figure 20.6.

Nucleic Acid Blotting

DNA molecules produced by recombinant methods can be used in hybridization reactions to locate specific genes or regions in restriction fragments from entire genomes or fragments produced by cutting other cloned sequences. In this procedure, the DNA to be probed is first cut with a restriction enzyme and the fragments separated by gel electrophoresis. The gel is treated with alkali to denature the DNA, and the DNA on the gel is transferred to a sheet of nitrocellulose, maintaining the same pattern of bands. The denatured DNA is fixed to the nitrocellulose and then hybridized with a radioactive DNA or RNA probe complementary to the DNA sequence of interest. After washing away unbound probe, the hybridized fragments are detected by autoradiography. This technique, developed

FIGURE 20.5

The DNA sequence to be synthesized can be deduced from a partial amino acid sequence of a purified protein. Once synthesized, the DNA sequence can be radioactively labeled and used as a probe to recover the cloned gene by colony screening.

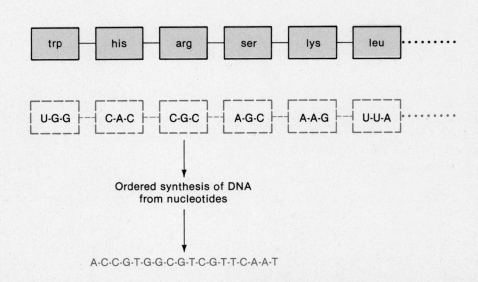

Ordered synthesis of DNA
from nucleotides

A-C-C-G-T-G-G-C-G-T-C-G-T-T-C-A-A-T

FIGURE 20.6
(a) Restriction mapping. A cloned DNA segment (straight line) contains four restriction sites: two for enzyme A, two for enzyme B. Cutting with both enzymes generates 5 fragments of decreasing size from 1 to 5. Vector sequences are shown as a wavy line. (b) Cutting with enzyme A generates fragment 1, shown on the gel in lane A. (For the sake of clarity, vector molecules and cloned fragments attached to vector molecules are not shown.) If the clone is cut first with A, then with B, two smaller fragments, 3 and 4 are generated (lane A then B). If the clone is cut first with B, a single fragment, 2, is generated, shown in lane B. Cutting first with B, then A, generates fragments 3 and 5. (c) In assembling the map, it is clear that fragments 3 and 4 are derived from fragment 1 as shown. Similarly, fragment 2 is composed of fragments 3 and 5. Since fragment 3 is common in both cases, the clone can be mapped by placing fragments 4 and 5 on opposite sides of 3.
(d) Restriction map of phage lambda DNA for the enzymes EcoRI and BamHI.

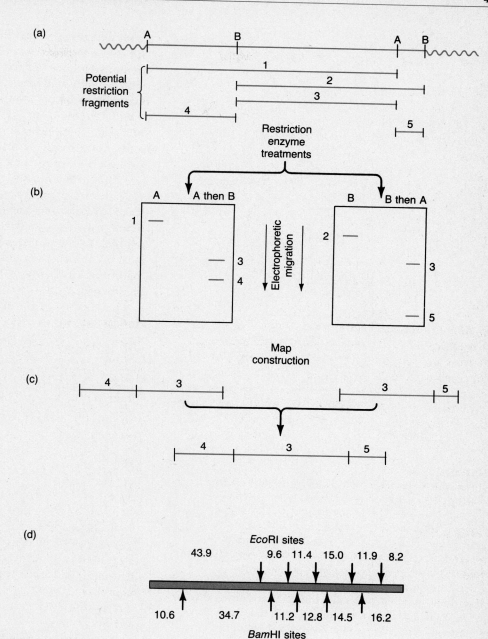

by Edward Southern, has come to be known as a **Southern blot** and is a procedure with many applications in recombinant DNA research (Figure 20.7).

A related technique, developed by James Alwine and colleagues can be used to determine whether a transcript from a cloned DNA sequence is represented in a mixture of cellular RNA molecules. To accomplish this, extracted RNA is fractionated by electrophoresis and the pattern of RNA bands transferred to a sheet of nitrocellulose or DBM paper. The sheet is hybridized to a radioactive cloned DNA probe and detected by autoradiography. This method, known as a **Northern blot,** has been used to study transcriptional patterns of cloned DNA sequences. The array of methods now available makes it possible to identify and characterize almost any gene of interest from prokaryotic or eukaryotic organisms. In many cases, it has been possible to obtain transcription and translation of cloned genes and to isolate biochemical and even commercial quantities of a gene product.

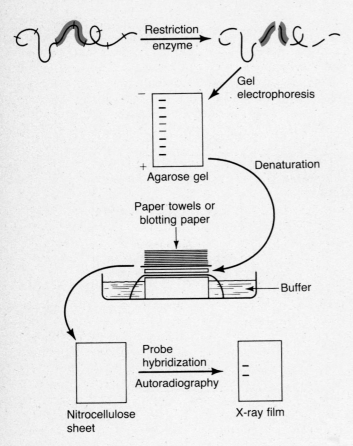

FIGURE 20.7

Southern blotting technique. DNA is first cut with one or more restriction enzymes and the fragments separated by gel electrophoresis. The fragments can be visualized by staining with ethidium bromide. The gel is then placed on a filter paper wick in contact with a buffer solution and covered with a sheet of nitrocellulose. Layers of paper towels or blotting paper are placed on top of the nitrocellulose and held in place with a weight. Capillary action draws the buffer through the gel, transferring the DNA fragments to the nitrocellulose, to which they bind very strongly. The filter is then hybridized with a radioactive probe, washed, and overlaid with a piece of X-ray film for autoradiography. The hybridized fragments show up as bands on the X-ray film.

THE RECOMBINANT DNA DEBATE

Intensive efforts in many laboratories around the world have collectively developed the methodology known as recombinant DNA technology. As scientists developed these techniques, they become aware of the potential hazards associated with their use. As these concerns were articulated, a debate arose within the scientific community about the safety and wisdom of such research. A brief history of the turmoil and ar-

guments surrounding the use of recombinant DNA will demonstrate that scientists do care deeply about the consequences of their work and will point out the controversial effects of genetic technology on individuals and society.

On July 26, 1974, a letter signed by eleven leading molecular biologists appeared in *Science* magazine (and concurrently in *Nature*). They called for a moratorium on certain types of recombinant DNA research. This letter, which is printed on pages 492 and 493, was to spark an ongoing debate on the potential hazards versus the benefits of this research. International in scope, this debate involved scientists, lay people, and politicians. The questions raised required decisions in local communities, academic institutions, the National Institutes of Health (NIH), and subcommittees of the US Senate and the House of Representatives.

The committee's letter proposed a moratorium on the use of viral genes and certain bacterial genes (those specifying antibiotic resistance and toxin production), and urged that experiments with animal genes should proceed only with great caution. It is important to emphasize that this proposed moratorium and all subsequent debate and proposed legislation were predicated only on the *potential* hazards of such research, and that the concerns were first raised by the scientists who themselves were using recombinant DNA in their work. They could cite no case where an accident had released a potentially hazardous organism into the environment.

In the ensuing two years, two major sets of guidelines appeared under the auspices of the NIH. The first was published in 1975. In 1976, a more extensive document was released by the NIH. Both sets of guidelines set rigid controls on all government-supported recombinant DNA research.

After the NIH guidelines were issued, the debate over recombinant DNA research gained rather than lost momentum. Action was initiated in local communities as well as at the federal level. By the spring of 1977, over ten bills regarding the recombinant DNA issue had been introduced in the US Congress.

The majority of scientists, including most who initially called attention to the problem, as well as many who had originally opposed such research, strongly opposed any legislation. Basically, they were opposed to a possible new bureaucratic structure that might needlessly impede the right of free inquiry.

After many hours of testimony and committee reports, the original bills were withdrawn and replaced with less restrictive legislation. However, this legislation did not pass, and most of the bills died in committee. In 1978, revised NIH guidelines lessened the constraints on recombinant DNA research. This relaxation was prompted by the discovery that the K12 laboratory strain of *E. coli* was much safer as a host than previously thought. In addition, no health hazard involving recombinant DNA had been reported. Early in 1982, the NIH moved to downgrade the guidelines even further. No experiments are currently prohibited, although notification of the NIH is required under certain circumstances. What began with a great deal of sound and fury ended with a small notice buried in the pages of the *Federal Register*.

APPLICATIONS OF RECOMBINANT DNA TECHNOLOGY

Although recombinant DNA techniques were originally developed to facilitate basic research into gene organization and regulation of expression, scientists have not been blind to the commercial possibilities of this technology. The basic process used in making recombinant molecules has been patented, and Stanford University is currently earning more than a million dollars a year from licensing agreements. In addition, many scientists have participated in the formation of companies set up to exploit recombinant DNA technology. The first market was primarily medical, but the industry is now turning its attention to plant breeding and agriculture.

Agricultural Applications

Plant and animal breeding has been a human activity for thousands of years. Recent advances in genetic engineering will greatly reduce the time needed to develop new strains and allow geneticists greater freedom in crossing the species barrier for transferring genes. The achievements of the Green Revolution (see Chapter 1) were based on painstaking field work and the development and use of hybrid lines. Although the benefits of this work are undisputed, reliance on hybrid lines carries a major risk—genetic vulnerability.

The genotypes of many crop plants are becoming more and more restricted. For example, when a rapidly spreading leaf blight swept through the US corn belt in 1970, more than 15 percent of the crop was destroyed. The most widely planted hybrid strains proved to be highly susceptible to the fungus. The conventional method for introducing a gene for fungal resistance into the hybrid corn relies on genetic crosses made with other lines known to carry the gene, followed by elaborate field testing to detect the homozygous fungal resistant offspring. This example serves to illustrate that until recently, the plant scientist has been limited to using genes from other strains

or, at most, closely related species to improve crop plants. The development of genetic engineering has made it possible to cross species lines in plant genetics. For example, a cloned gene from the bacterium, *Salmonella*, that confers herbicide tolerance has been transferred to tobacco plants and is being used to transform other crop plants such as cotton, corn, and sunflowers. The ability to use broad spectrum herbicides to control weeds without damage to the crop plants will be of great economic consequence, increasing the available supply of plant products. In the United States, some $500 million in cotton yields, or 15 percent of the crop, is currently lost to weed infestation.

Other advances in agriculture may include increases in the efficiency of photosynthesis and nitrogen fixation. The ability to fix atmospheric nitrogen into biologically useful forms is restricted to certain bacteria and blue-green algae. Some higher plants, such as the legumes (soybeans, clover, and alfalfa), have evolved a symbiotic relationship with nitrogen-fixing bacteria that live in the plant's root nodules. These bacteria use energy produced by the host plant to convert atmospheric nitrogen into forms that can be used by the host plant. In contrast, cereal crops such as corn require heavy applications of nitrogen-containing fertilizers, a process both inefficient and potentially damaging to the environment. The bacterial genes that regulate the fixation of nitrogen (*nif genes*) have been isolated and identified. Techniques of genetic engineering will make possible the transfer of these genes directly to the genome of crop plants, eliminating the need for costly artificial fertilizers.

Industrial and Medical Applications

In industry, the processing of agricultural products and the synthesis of many substances require large quantities of microbe-derived enzymes. Even a small increase of 1 to 2 percent in the efficiency of converting starch to dextrose, for example, can be of enormous financial importance. Consequently, many companies are investing in recombinant DNA research to improve the effectiveness and reduce the cost of commercially useful enzymes. Another possible use of recombinant DNA technology is in the manufacturing of certain bulk chemicals, including methanol, ethanol, and acetone, through the breakdown of cellulose-containing plant waste products by genetically engineered bacteria and yeasts. Obviously, an important application of recombinant DNA technology is the production of medically important products. Most often these substances have direct therapeutic value and are difficult to obtain because they are produced in small

POTENTIAL BIOHAZARDS OF RECOMBINANT DNA MOLECULES

Recent advances in techniques for the isolation and rejoining of segments of DNA now permit construction of biologically active recombinant DNA molecules in vitro. For example, DNA restriction endonucleases, which generate DNA fragments containing cohesive ends especially suitable for rejoining, have been used to create new types of biologically functional bacterial plasmids carrying antibiotic resistance markers[1] and to link *Xenopus laevis* ribosomal DNA to DNA from a bacterial plasmid. This latter recombinant plasmid has been shown to replicate stably in *Escherichia coli* where it synthesizes RNA that is complementary to *X. laevis* ribosomal DNA.[2] Similarly, segments of *Drosophila* chromosomal DNA have been incorporated into both plasmid and bacteriophage DNA's to yield hybrid molecules that can infect and replicate in *E. coli*.[3]

Several groups of scientists are now planning to use this technology to create recombinant DNA's from a variety of other viral, animal, and bacterial sources. Although such experiments are likely to facilitate the solution of important theoretical and practical biological problems, they would also result in the creation of novel types of infectious DNA elements whose biological properties cannot be completely predicted in advance.

There is serious concern that some of these artificial recombinant DNA molecules could prove biologically hazardous. One potential hazard in current experiments derives from the need to use a bacterium like *E. coli* to clone the recombinant DNA molecules and to amplify their number. Strains of *E. coli* commonly reside in the human intestinal tract, and they are capable of exchanging genetic information with other types of bacteria, some of which are pathogenic to man. Thus, new DNA elements introduced into *E. coli* might possibly become widely disseminated among human, bacterial, plant, or animal populations with unpredictable effects.

Concern for these emerging capabilities was raised by scientists attending the 1973 Gordon Research Conference on Nucleic Acids,[4] who requested that the National Academy of Sciences give consideration to these matters. The undersigned members of a committee, acting on behalf of and with the endorsement of the Assembly of Life Sciences of the National Research Council on this matter, propose the following recommendations.

First, and most important, that until the potential hazards of such recombinant DNA molecules have been better evaluated or until adequate methods are developed for preventing their spread, scientists throughout the world join with the members of this committee in voluntarily deferring the following types of experiments.

- *Type 1:* Construction of new, autonomously replicating bacterial plasmids that might result in the introduction of genetic determinants for antibiotic resistance or bacterial toxin formation into bacterial strains that do not at present carry such determinants; or construction of new bacterial plasmids containing combinations of resistance to clinically useful antibiotics unless plasmids containing such combinations of antibiotic resistance determinants already exist in nature.

- *Type 2:* Linkage of all or segments of the DNA's from oncogenic or other animal viruses to autonomously replicating DNA elements such as bacterial plasmids or other viral DNA's. Such recombinant DNA molecules might be more easily disseminated to bacterial populations in humans and other species, and thus possibly increase the incidence of cancer or other diseases.

Second, plans to link fragments of animal DNA's to bacterial plasmid DNA or bacteriophage DNA should be carefully weighed in light of the fact that many types of animal cell DNA's contain sequences common to RNA tumor viruses. Since joining of any foreign DNA to a DNA replication system creates new recombinant DNA molecules whose biological properties cannot be predicted with certainty, such experiments should not be undertaken lightly.

Third, the director of the National Institutes of Health is requested to give immediate consideration to establishing an advisory committee charged with (i) overseeing an experimental program to evaluate the potential biological and ecological hazards of the above types of recombinant DNA molecules; (ii) developing procedures which will minimize the spread of such molecules within human and other populations; and (iii) devising guidelines to be followed by investigators working with potentially hazardous recombinant DNA molecules.

Fourth, an international meeting of involved scientists from all over the world should be convened early in the coming year to review scientific progress in this area and to further discuss appropriate ways to deal with the potential biohazards of recombinant DNA molecules.

The above recommendations are made with the realization (i) that our concern is based on judgments of potential rather than demonstrated risk since there are few available experimental data on the hazards of such DNA molecules and (ii) that adherence to our major recommendations will entail postponement or possibly abandonment of certain types of scientifically worthwhile experiments. Moreover, we are aware of many theoretical and practical difficulties involved in evaluating the human hazards of such recombinant DNA molecules. Nonetheless, our concern for the possible unfortunate consequences of indiscriminate application of these techniques motivates us to urge all scientists working in this area to join us in agreeing not to initiate experiments of types 1 and 2 above until attempts have been made to evaluate the hazards and some resolution of the outstanding questions has been achieved.

PAUL BERG, *Chairman*
DAVID BALTIMORE
HERBERT W. BOYER
STANLEY N. COHEN
RONALD W. DAVIS
DAVID S. HOGNESS
DANIEL NATHANS
RICHARD ROBLIN
JAMES D. WATSON
SHERMAN WEISSMAN
NORTON D. ZINDER

*Committee on Recombinant DNA
Molecules Assembly of Life Sciences,
National Research Council,
National Academy of Sciences,
Washington, D.C. 20418*

[1]S. N. Cohen, A. C. Y. Chang, H. Boyer, R. B. Helling, *Proc. Natl. Acad. Sci. U.S.A.* **70,** 3240 (1973); A. C. Y. Chang and S. N. Cohen, *ibid.,* **71,** 1030 (1974).
[2]J. F. Morrow, S. N. Cohen, A. C. Y. Chang, H. Boyer, H. M. Goodman, R. B. Helling, *ibid.,* in press.
[3]D. S. Hogness, unpublished results; R. W. Davis, unpublished results; H. W. Boyer, unpublished results.
[4]M. Singer and D. Soll, *Science* **181,** 1114 (1973).

SOURCE: From "Letter: Potential Biohazards of Recombinant DNA Molecules" by Paul Berg et al., *Science,* vol. 185, July 26, 1974, p. 303. Copyright 1974 by the American Association for the Advancement of Science.

quantities *in vivo.* However, the production of eukaryotic proteins in bacterial cells has several inherent problems. First, eukaryotic promoter sequences are not recognized by prokaryotic RNA polymerases. Second, many eukaryotic genes contain intervening sequences, and bacterial cells are unable to remove them from the precursor messenger RNA. Third, eukaryotic proteins like insulin are processed from precursors, and bacterial cells are unable to recognize the appropriate processing signals. Several ingenious methods have been employed to solve these problems, including the fusion of eukaryotic genes to prokaryotic promoters, and the use of synthetic genes. A synthetic gene was used to produce somatostatin, a small (14 amino acids) hypothalamic peptide that inhibits the secretion of other hormones, including growth hormone, insulin, and glucagon. Working from the amino acid sequence, a group in Herbert Boyer's laboratory synthesized the DNA fragment for the gene and inserted it near the end of the beta-galactosidase gene of the *lac* operon contained in a plasmid. After transformation into *E. coli,* this totally synthetic gene produced about 10,000 molecules of somatostatin per bacterial cell.

The first commercially distributed pharmaceutical product made using recombinant DNA technology was human insulin, developed by Genentech and marketed by Eli Lilly. At this date, many other eukaryotic gene products are in pilot production or clinical testing, including **factor VIII,** a clotting protein deficient in hemophilia A, **thymosin,** a peptide hormone used in treatment of immune defects, **human growth hormone,** and **TPA (tissue plasminogen activator),** a circulatory system enzyme that is used to dissolve blood clots associated with heart attacks.

Research is proceeding on the use of recombinant DNA to develop vaccines against infectious diseases. For example, at the Rockefeller Foundation, a combination of genetic engineering and protein biochemistry is being used to develop an inexpensive and reliable vaccine against hookworm, a debilitating and potentially fatal parasitic disease that affects 900 million individuals worldwide. It is estimated that recombinant DNA methods can be used to produce vaccines against 14 of the major infectious diseases now common in the United States.

In addition to the manufacture of therapeutic products, recombinant DNA technology is being used for diagnostic purposes. A screening system developed through genetic engineering is being used to test donated blood for antibodies to the HTLV-III virus. Detection of the antibodies indicates that the donor has been exposed to the agent associated with **acquired immunodeficiency syndrome (AIDS).**

Genetic Screening

Human genetic defects occur in 3 to 5 percent of the live births in the United States. Between 200,000 and 300,000 children are born each year with chromosomal aberrations, single gene defects, or congenital malformations. Collectively, these diseases are a leading cause of death in children from the neonatal period through adulthood. The most widely used method for prenatal detection of genetic defects is **amniocentesis.** This is a technique in which a hypodermic needle is inserted through the abdominal wall to withdraw 10 to 20 cc of amniotic fluid (Figure 20.8). The fetal cells contained in the fluid can be collected by centrifugation and cultured to screen for chromosomal disorders, and the fluid itself can be analyzed to detect any of some 75 genetically controlled disorders. Amniocentesis is recommended for pregnant women over the age of 35, or where there is evidence of familial genetic disorders.

Recently, Robert Geever and his colleagues have used recombinant DNA techniques to directly identify the defective gene for sickle-cell anemia in DNA from amniotic cells (Figure 20.9). This genetic defect cannot be otherwise detected prenatally since the beta-globin gene is transcriptionally inactive until the first few days after birth. In sickle-cell anemia, the single nucleotide substitution in the structural gene for beta globin destroys a restriction enzyme cleavage site within the gene, and consequently, longer DNA fragments are produced upon digestion with this enzyme than in DNA from normal individuals. These differences can be detected and quantified using electrophoresis and hybridization techniques. Similar recombinant DNA probes may be used to detect other hemoglobinopathies such as thalassemias, which are defects in the production of either alpha or beta globins.

Gene Therapy: At the Threshold

The technique of *in vitro* **fertilization** has been developed for the treatment of reproductive disorders. In humans, this is a method in which a mature egg is recovered by laparoscopy through a small abdominal incision and transferred to a nutrient medium in a dish. Sperm from the husband or donor is added, and fertilization is monitored. When the embryo has developed to the stage where 16 to 32 cells are present, it is implanted into the uterus of the woman. Louise Brown, born in England on July 26, 1978, is the first human brought to term using this technique. Since then, *in vitro* fertilization has been adopted in other countries, including the United States, as a treatment

FIGURE 20.8

The technique of amniocentesis. The position of the fetus is first determined by ultrasound techniques. Then, a needle is inserted through the abdominal wall to recover fluid and fetal cells for analysis. (From "Prenatal Diagnosis of Genetic Disease" by T. Friedmann. Copyright © 1971 by Scientific American, Inc. All rights reserved.)

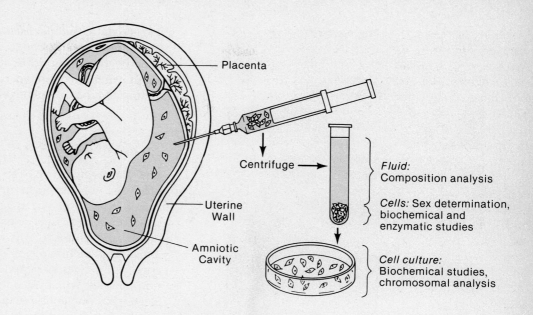

Placenta

Centrifuge →

Fluid: Composition analysis

Cells: Sex determination, biochemical and enzymatic studies

Uterine Wall

Amniotic Cavity

Cell culture: Biochemical studies, chromosomal analysis

(a)

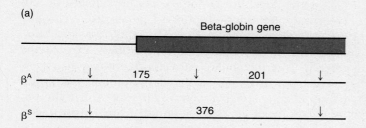

Beta-globin gene

β^A ⟶ ↓ 175 ↓ 201 ↓

β^S ⟶ ↓ 376 ↓

(b)

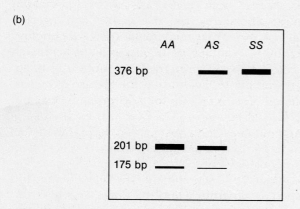

	AA	AS	SS
376 bp		▬	▬
201 bp	▬	▬	
175 bp	▬	▬	

FIGURE 20.9

Direct detection of sickle-cell anemia by restriction enzyme digestion and Southern blotting to a beta-globin probe. (a) The flanking region (thin line) and a portion of the human beta-globin gene (thick line). Sites cut by the enzyme *Dde*I are shown by arrows. Normal beta-globin genes (β^A) produce fragments of 175 and 201 base pairs, while the sickle-cell globin gene (β^S) produces a single fragment (376 bp). (b) Southern blot pattern of normal (*AA*), heterozygous (*AS*), and sickle cell individuals (*SS*).

for cases where blockage of the fallopian tubes prevents the egg from being fertilized.

In vitro fertilization can also be used for gene surgery. Thomas Wagner and a team of researchers have successfully transferred a gene coding for rabbit beta globin into fertilized mouse eggs by microinjection (Figure 20.10). A purified plasmid containing the rabbit beta-globin gene was constructed and grown using recombinant DNA technology. This plasmid was microinjected *in vitro* into the male pronucleus of the fertilized mouse ovum, and the zygote implanted into the uterus. Mice that developed from injected zygotes and their offspring show the presence and expression of the rabbit globin gene, demonstrating that the injected gene becomes part of the cellular genetic information. It is not only expressed, but passed on as a heritable component.

In a similar series of experiments, Richard Palmiter and Ralph Brinster injected a cloned human growth hormone gene into mouse zygotes. These **transgenic** mice produced the human hormone, grew to twice the size of normal mice, and transmitted the gene to their offspring. Because of the success of such gene transfer experiments in animals, it is not too difficult to envision that such gene therapy will be applied to humans.

Under the aegis of the National Institutes of Health, the machinery for reviewing proposals to use gene therapy in the treatment of human diseases is already in place. The guidelines, published in the *Federal Register* for January 22, 1985, recommend a four-stage review, with final approval by the director of the National Institutes of Health. Several genetic diseases are

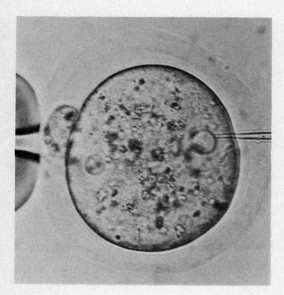

FIGURE 20.10
Microinjection of cloned rabbit beta-globin DNA into a male pronucleus of a mouse zygote. To protect against damage, the fertilized egg is held against a suction pipette by the polar body. (From Wagner et al., 1981, Fig. 1.)

under consideration for treatment by gene therapy, including **Lesch-Nyhan syndrome, adenosine deaminase deficiency,** and **purine nucleoside phosphorylase deficiency.** These candidates have been selected because they are progressive, fatal diseases for which no other therapy is available. Each disease is associated with a single gene defect, and in each case, the normal gene has been successfully cloned. The plan for therapy involves the delivery of the cloned gene to a target somatic tissue *in vitro*, with subsequent reinsertion of the tissue after uptake and switch-on of the transferred gene have been confirmed. The most likely target tissue is bone marrow. However, since Lesch-Nyhan disease is primarily associated with a lack of the enzyme hypoxanthine guanine phosphoribosyl transferase in brain cells, the outcome may be less favorable than in the other diseases affecting stem cells of the immune system in bone marrow.

The use of gene therapy, although in some ways not radically different than other therapies already in use, has raised ethical issues about human experimentation and the possibility of genetically modifying germ cells. Decisions need to be made about the implementation of such treatments and the limitations on their use. To date, the discussion of ethical and social concerns has been thorough and open, and this precedent needs to continue so that intelligent decisions about genetic technology can be made with the participation of a broad spectrum of our society.

SUMMARY AND CONCLUSIONS

The development of recombinant DNA technology promises to transform areas of basic biological research and prenatal diagnosis and to provide the foundation for new classes of drugs, agricultural plants and animals, and even genetically altered humans. This technology is derived in part from the discovery of a class of enzymes known as restriction endonucleases, which recognize and cut DNA molecules at specific nucleotide sequences. Coupled with other developments in microbiology and molecular biology, these methods have allowed researchers to produce abundant copies of DNA sequences ranging from specific genes to entire genomes. The availability of biochemical quantities of purified genes from virtually any organism opens the way to study the structure, organization, regulation, and evolution in systems ranging from metabolic pathways to entire subspecies.

Scientists working on the development and application of gene splicing were the first to realize the potential dangers of this technology and instituted a moratorium on such work in order to evaluate its safety and ethics. After much debate and experimental analysis, there is now general agreement that current laboratory procedures are reasonable and present minimal risk. Now that recombinant DNA technology is becoming a commercial reality, scientists ranging from ecologists and agronomists to molecular biologists are working to ensure that the release of genetically altered organisms into the environment does not have grave consequences. The potential for the use of such methodology for constructing recombinant organisms to be used in biological warfare remains as an issue for consideration.

The impact of recombinant DNA technology on areas such as agriculture and the pharmaceutical and chemical manufacturing industries is only in its beginning stages and is expected to grow rapidly. Even conservative estimates forecast that gene splicing will be a multibillion dollar component of our economy by the year 2000. But it is already clear that the most profound effects of this technology will be on the diagnosis and treatment of genetic diseases in humans. Recombinant probes are now routinely used for prenatal diagnosis at most major medical centers. The vectors necessary to replace defective genes with their normal counterparts are being developed. Gene therapy is in its first stage of development. The problems to be solved are formidable: How can cloned sequences be directed to selected target cells in the body, and once incorporated, how can they be brought under control of the cell's regulatory machinery? With the rapid advances that have been made to date, it would be foolish to suppose that these barriers will remain for very long. Once these problems have been solved, gene therapy will become a common medical procedure rather than a method of last resort. At this point, we will collectively face the question of whether to limit the use of gene therapy to replacing defective genes, or whether to begin tinkering with the genome in the name of improving the human race. It is clear that interesting and challenging times lie ahead.

PROBLEMS AND DISCUSSION QUESTIONS

1 In recombinant DNA studies, what is the role of each of the following: restriction endonucleases, terminal transferase, vectors, calcium chloride, and host cells?

2 Why is poly dT an effective primer for reverse transcriptase?

3 An ampicillin-resistant, tetracycline-resistant plasmid is cleaved with *Eco*RI, which cuts within the ampicillin gene. The cut plasmid is ligated with *Eco*RI-digested *Drosophila* DNA to prepare a genomic library. The mixture is used to transform *E. coli* K12.
   (a) Which antibiotic should be added to the medium to select cells that have incorporated a plasmid?
   (b) What antibiotic resistance pattern should be selected to obtain plasmids containing *Drosophila* inserts?
   (c) How can you explain the presence of colonies that are resistant to both antibiotics?

4 Clones from the previous question are found to have an average length of 5 kb. Given that the *Drosophila* genome is 1.5×10^5 kb long, how many clones would be necessary to give a 99 percent probability that this library contains all genomic sequences?

5 In cloning the large genomes of eukaryotes, it is often advantageous to make a partial digest with a restriction enzyme like *Sau*3A, with a four-base recognition sequence (GATC). For other reasons, many vectors do not contain four-base recognition sequences, but do contain sites like *Bam*HI (GGATCC), where the four internal bases are identical to the *Sau*3A recognition sequence.
   (a) Is it likely that foreign DNA digested with *Sau*3A can be cloned into the *Bam*HI site of a vector? Why?
   (b) What are the drawbacks to this method?

6 The human insulin gene contains a number of introns. In spite of the fact that bacterial cells will not excise introns from mRNA, how can a gene like this be cloned into a bacterial cell and produce insulin?

7 Type II restriction enzymes recognize palindromic sequences in intact DNA molecules and cleave the double-stranded helix at these sites. Inasmuch as the bases are internal in a DNA double helix, how is this recognition accomplished?

8 In a control experiment, a plasmid containing a *Hind*III site within a kanamycin-resistance gene is cut with *Hind*III, religated, and used to transform *E. coli* K12 cells. Kanamycin-resistant colonies are selected, and plasmid DNA from these colonies subjected to electrophoresis. Most of the colonies contain plasmids that produce single bands that migrate at the same rate as the original, intact plasmid. A few colonies, however, produce two bands, one of original size, and one that migrates much higher in the gel. Diagram the origin of this slow band during the religation process.

9 In mice transfected with the rabbit beta-globin gene, the rabbit gene is active in a number of tissues including spleen, brain, and kidney. In addition, some mice suffer from thalassemia, caused by an imbalance in the coordinate production of alpha and beta globins. What problems associated with gene therapy are illustrated by these findings?

10   What facts should you consider in deciding which vector to use in constructing a genomic library of eukaryotic DNA?

11   Although the potential benefits of cloning in higher plants are obvious, the development of this field has lagged behind cloning in bacteria, yeast, and mammalian cells. Can you think of any reason for this?

SELECTED READINGS

ALWINE, J. C., KEMP, D. J., and STARK, G. R. 1977. Method for detection of specific RNAs in agarose gels by transfer to diazobenzyloxymethyl paper and hybridization with DNA probes. *Proc. Natl. Acad. Sci.* 74: 5350–54.

ANDERSON, W. F., and DIACUMAKOS, E. G. 1981. Genetic engineering in mammalian cells. *Scient. Amer.* (July) 245: 106–21.

ANTONARAKIS, S. E., WABER, P. G., KITTUR, S. D., PATEL, A. S., KAZAZIAN, H. H. Jr., MELLIS, M. A., COUNTS, R. B., STAMATOYANNOPOULOS, G., BOWIE, E. J. W., FASS, D. N., PITTMAN, D. D., WUZNEY, J. M., and TOOLE, J. J. 1985. Hemophilia A: Detection of molecular defects and carriers by DNA analysis. *New Engl. J. Med.* 313: 842–48.

BEAUCHAMP, T. L., and CHILDRESS, J. F. 1979. *Principles of biomedical ethics.* New York: Oxford University Press.

BERG, P. 1981. Dissections and reconstructions of genes and chromosomes. *Science* 213: 296–303.

BLANK, R. H. 1981. *The political implications of human genetic technology.* Boulder, Colo.: Westview Press.

BOYER, H. W. 1971. DNA restriction and modification mechanisms in bacteria. *Ann. Rev. Microbiol.* 25: 153–76.

CAVALIERI, L. F. 1981. *The double-edged helix: Science in the real world.* New York: Columbia University Press.

CHANG, J. C., and KAN, Y. W. 1981. Antenatal diagnosis of sickle-cell anemia by direct analysis of the sickle mutation. *Lancet,* pp. 1127–29.

COHEN, S., et al. 1973. Construction of biologically functional bacterial plasmids *in vitro. Proc. Natl. Acad. Sci.* 70: 3240–44.

COLLINS, J., and HOHN, B. 1978. Cosmids: A type of plasmid gene cloning vector that is packageable *in vitro* in bacteriophage heads. *Proc. Natl. Acad. Sci.* 75: 4242–46.

DAVIES, K. E. 1981. The applications of DNA recombinant technology to the analysis of the human genome and human disease. *Hum. Genet.* 58: 351–57.

————. 1982. A comprehensive list of cloned eukaryotic genes. In *Genetic engineering—3,* ed. R. Williamson. New York: Academic Press.

DAVIES, K. E., YOUNG, B. D., ELLES, R. G., HILL, M. E., and WILLIAMSON, R. 1981. Cloning of a representative genomic library of the human X chromosome after sorting by flow cytometry. *Nature* 293: 374–76.

DEMAIN, A. L. 1981. Industrial microbiology. *Science* 214: 987–95.

ELLIS, K. P., and DAVIES, K. E. 1985. An appraisal of the application of recombinant DNA techniques to chromosomal defects. *Biochem. J.* 226: 1–11.

EMERY, A. E. H. 1984. *An introduction to recombinant DNA.* New York: Wiley.

FISHER, E. M. C., CAVANNA, J. S., and BROWN, S. D. M. 1985 Microdissection and microcloning of the mouse X chromosome. *Proc. Natl. Acad. Sci.* 82: 5846–49.

GEEVER, R. F., WILSON, L. B., NALLASETH, F. S., MILNER, P. F., BITTNER, M., and WILSON, J. T. 1981. Direct identification of sickle-cell anemia by blot hybridization. *Proc. Natl. Acad. Sci.* 78: 5081–85.

GLOVER, D. M. 1984. *Gene cloning: The mechanism of DNA manipulation.* London: Chapman and Hall.

GRAF, L. H. 1982. Gene transformation. *Amer. Scient.* 70: 496–505.

GRIESBACH, R. J., KOIVUNIEMI, P. J. and CARLSON, P. S. 1981. Extending the range of plant genetic manipulation. *Bioscience* 31: 754–56.

GROBSTEIN, C. 1979. *A double image of the double helix: The recombinant DNA debate.* San Francisco: W. H. Freeman

HIGGINS, I. J., BEST, D. J., and JONES, J. eds. 1985. *Biotechnology: Principles and applications. Studies in Microbiology,* vol. 3. Palo Alto: Blackwell Scientific.

HOLCENBERG, J. S. 1982. Enzyme therapy: Problems and solutions. *Ann. Rev. Biochem.* 51: 795–812.

HOPWOOD, D. A. 1981. Genetic programming of industrial microorganisms. *Scient. Amer.* (Sept.) 245: 91–102.

JACKSON, D. A., and STICH, S. P., eds. 1979. *The recombinant DNA debate.* Englewood Cliffs, N. J.: Prentice-Hall.

JINKS, J. L., CALIGARI, P. D. S., and INGRAM, N. R. 1981. Gene transfer in *Nicotiana rustica* using irradiated pollen. *Nature* 291: 586–88.

JORDAN, B. R. 1985. Antenatal diagnosis by DNA analysis: Current status, future developments, and a few unanswered questions. *BioEssays* 2: 196–200.

KNORR, D., and SINSKEY, A. J. 1985. Biotechnology in food production and processing. *Science* 229: 1224–29.

KRUMLAUFF, R., JEANPIERRE, M., and YOUNG, B. D. 1982. Construction and characterization of genomic libraries from specific human chromosomes. *Proc. Natl. Acad. Sci.* 79: 2971–75.

MANIATIS, T., FRITSCH, E. F., and SAMBROOK, J. 1982. *Molecular cloning: A laboratory manual.* Cold Spring Harbor: Cold Spring Harbor Laboratory.

ORKIN, S. H. 1982. Genetic diagnosis of the fetus. *Nature* 296: 202–03.

OLD, R. W., and PRIMROSE, S. B. 1985. *Principles of genetic manipulation: An introduction to genetic engineering.* Palo Alto: Blackwell Scientific.

SCALENGHE, F., TURCO, E., EDSTROM, J. E., PIRROTTA, V., and MELLI, M. 1981. Microdissection and cloning of DNA from a specific region of *Drosophila melanogaster* polytene chromosomes. *Chromosoma* 82: 205–16.

SOUTHERN, E. M. 1975. Detection of specific sequences among DNA fragments separated by gel electrophoresis. *J. Mol. Biol.* 98: 503–17.

THOMAS, T. L. and HALL T. C. 1985. Gene transfer and expression in plants: Implications and potential. *BioEssays* 3: 149–53.

TORREY, J. G. 1985. The development of plant biotechnology. *Amer. Scient.* 73: 354–63.

VILLA-KOMAROFF, L., EFSTRADIATIS, A., BROOME, S., LOMEDICA, P., TIZARD, R. NABET, S. P., CHICK, W. L., and GILBERT, W. 1978. A bacterial clone synthesizing proinsulin. *Proc. Natl. Acad. Sci.* 75: 3727–31.

WAGNER, T. E., HOPPE, P. C., JOLLICK, J. D., SCHOLL, D. R., HODINKA, R. L., and GAULT, J. B. 1981. Microinjection of a rabbit beta-globin gene into zygotes and its subsequent expression in adult mice and their offspring. *Proc. Natl. Acad. Sci.* 78: 6376–80.

WATSON, J. D., and KURTZ, D. 1983. *Recombinant DNA. A short course.* New York: W. H. Freeman.

WHITE, R. 1985. DNA sequence ploymorphisms revitalize linkage approaches in human genetics. *Trends in Genet.* 1: 177–80.

WILLIAMSON, B. 1982. Gene therapy. *Nature* 298: 416–18.

21

Genetic Regulation

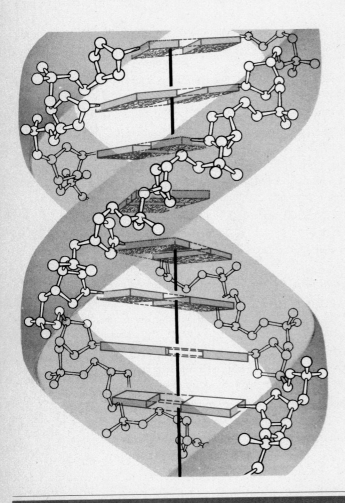

In earlier chapters, we established how DNA is organized into genes, how genes store information, and how this information can be expressed. But, is the information in any gene always expressed, or can genes be turned on and off, depending on the needs of the cell? On theoretical grounds, it is rather easy to argue that the answer to the latter question is yes. Bacterial cells in varying environments may or may not require certain gene products in order to metabolize varying substrates. In mammals, cells of the pancreas do not synthesize hemoglobin and retinal cells do not synthesize insulin. If every cell fully expressed every gene, cellular phenotypes would not vary. Furthermore, organisms would consume phenomenal amounts of energy in the processes of transcription and translation to produce unnecessary gene products.

Thus, it seems clear that some form of **genetic regulation** must exist in organisms. Detailed analysis of proteins in *Escherichia coli* has shown that for 4000 or so polypeptide chains encoded by the genome, there is a vast range of concentration of gene products. Some proteins may be present in as few as 5 to 10 molecules per cell, whereas others, particularly ribosomal proteins as well as those involved in the glycolytic pathway, are present in as many as 100,000 to 500,000 copies per cell. Thus, while a basal level of most gene products exists, it is clear that this level can be increased and subsequently decreased in prokaryotes. In eukaryotic organisms, cells take on distinct functions and are characterized by distinct chemical and structural phenotypes. Thus, fundamental regulatory mechanisms to control the expression of the genetic information must exist in both groups.

In this chapter, we will explore what is known about the regulation of genetic expression. Highly detailed information has been obtained in studies of viruses and bacteria. Whereas regulation can be documented in eukaryotes, the precise mechanisms by which this is accomplished have yet to be determined. We will review a variety of approaches used and types of information obtained to extend our knowledge.

GENETIC REGULATION IN PROKARYOTES: AN OVERVIEW

The regulation of genetic expression has been most extensively studied in prokaryotes, particularly in *E. coli*. What has been found is that highly efficient genetic mechanisms have evolved that turn genes on and off, depending on the cell's metabolic need for the respective gene products. The activity of enzymes may also be regulated once they have been synthesized in the cell, but we will focus primarily on what is known about regulation at the level of gene transcription. It is important to remember that it is the resulting proteins ultimately present or absent that are critical to efficient cell function under varying environmental conditions.

It is not a particularly new concept that microorganisms regulate the synthesis of gene products. As early as 1900 it was shown that when the galactose-glucose-containing disaccharide **lactose** is present in the growth medium of yeast, enzymes specific to lactose metabolism are produced. When this substrate is absent, the enzymes are not manufactured. It was soon shown that bacteria also "adapt" to their chemical environment, producing certain enzymes only when specific substrates are present. Such enzymes were thus referred to as **adaptive.** In contrast, those produced continuously regardless of the chemical makeup of the environment were called **constitutive.**

Since then, the term *adaptive* in descriptive enzymology has been replaced with a more accurate term. We now call them **inducible** enzymes, reflecting the role of the substrate, or **inducer,** in their production.

More recent investigation has revealed other cases where the presence of a specific molecule causes inhibition of genetic expression. This is usually true for molecules that are end products of biosynthetic pathways. For example, an amino acid such as tryptophan can be synthesized by bacterial cells. If an exogenous supply of this amino acid is present in the environment or culture medium, it is energetically inefficient to synthesize the enzymes necessary for tryptophan production. As a result, a mechanism exists whereby tryptophan plays a role in repressing transcription of RNA essential to the translation of the appropriate biosynthetic enzymes. In contrast to the inducible system controlling lactose metabolism, that governing tryptophan is said to be **repressible.**

As we will soon see, instances of regulation as described above, whether inducible or repressible, are examples where **negative control** is exerted. Genetic expression occurs in the absence of any active inhibition. Transcription occurs unless shut off by some form of regulator. This is in contrast to **positive control,** where transcription does not occur unless a regulator molecule directly stimulates RNA production. Thus, genetic regulation can be under positive or negative control; in theory each type of control can govern inducible or repressible systems. Examples discussed in the ensuing sections of this chapter will help distinguish between these possible mechanisms.

LACTOSE METABOLISM IN *E. COLI:* AN INDUCIBLE GENE SYSTEM

The most extensively studied system of gene regulation has involved the metabolism of lactose in *E. coli*. Beginning in 1946 with the studies of Jacques Monod and continuing through the next decade with significant contributions by Joshua Lederberg, François Jacob, and Andre L'woff, genetic and biochemical evidence was amassed. This research provided clear insights into the way in which the genes responsible for lactose metabolism are turned off, or repressed, when lactose is absent but activated or induced in the presence of lactose. In the presence of lactose, the concentration of the enzymes responsible for its metabolism increases rapidly from 5 to 10 molecules to thousands per cell. Thus, the enzymes are **inducible,** and lactose serves as the **inducer.**

Paramount to the understanding of genetic regulation in this system was the discovery of two genes that serve strictly in a regulatory capacity. They do not code for enzymes necessary for lactose metabolism. Three other genes are responsible for the production of enzymes involved in lactose metabolism. Together, the five genes function in an integrated fashion and provide a rapid response to the presence or absence of lactose.

Structural Genes

Genes coding for the primary structure of the enzymes are called **structural genes.** The so-called *lac* z gene specifies the amino acid sequence of the **β-galactosidase enzyme,** which converts lactose to glucose and galactose (Figure 21.1). This conversion is essential if lactose is to serve as the primary energy source in glycolysis. The second gene, *lac* y, specifies the primary structure of **β-galactoside permease,** which facilitates the entry of lactose into the bacterial cell. The third gene, *lac* a, codes for a **transacetylase** enzyme whose physiological role is unknown.

Studies of the genes coding for these three enzymes relied on the isolation of numerous mutations that eliminated the function of one or the other enzyme. Such *lac*⁻ mutants were isolated and studied by Lederberg. Mutant cells fail to produce either active β-galactosidase or permease molecules and so are unable to utilize lactose as an energy source. Mutations also were found in the transacetylase *a* gene. Mapping studies by Lederberg established that all three genes are closely linked or contiguous to one another in the order *z-y-a* (Figure 21.2).

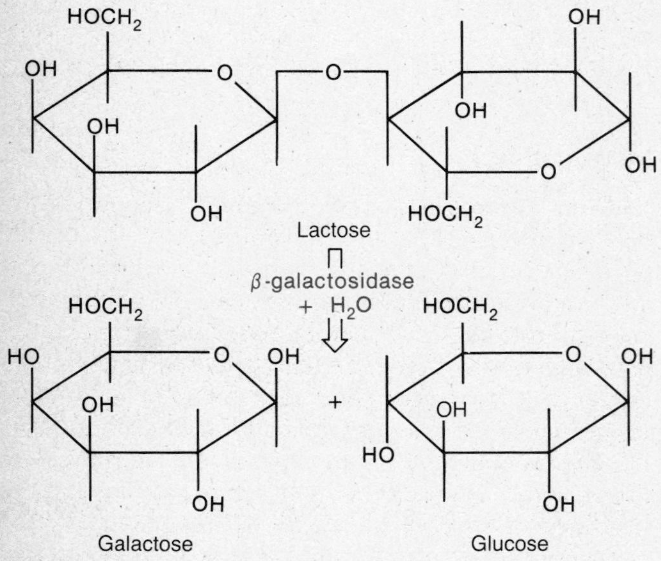

Two other observations are relevant to what became known about the structural genes. First, knowledge of their close linkage led to the discovery that all three genes are transcribed together, resulting in a single **polycistronic message** or **mRNA** (Figure 21.2). Additionally, it has been shown that upon induction by lactose, the rapid appearance of the enzymes results from the *de novo* synthesis or translation of this mRNA. Although this finding might seem obvious or expected, it is in contrast to the proposal that induced enzyme activity may result from the activation of existing but inactive forms of the enzymes.

The Discovery of Regulatory Genes

How, then, can lactose activate structural genes and induce the synthesis of the related enzymes? The discovery and study of **gratuitous inducers** ruled out one possibility. These molecules, which are chemical analogues of lactose, serve as inducers but not as substrates for the enzymatic reaction. One such gratuitous inducer is the sulfur analogue **isopropylthiogalactoside (IPTG),** shown in Figure 21.3. The discovery of such molecules is strong evidence that the primary induction event is not the result of the interaction between the inducer and the enzyme. What, then, is the role of lactose in induction?

The answer to this question required the study of a second class of mutation, the **constitutive mutants.** In this type of mutant, the enzymes are produced regardless of the presence or absence of lactose. These mutations served as the basis of studies that defined the regulatory scheme for lactose metabolism.

Maps of the first type of constitutive mutation, *lac* i⁻, showed that it is located at a site on the DNA close to, but distinct from, the structural genes. A second set of mutations produced identical effects but was found to lie immediately adjacent to the structural genes. This class is designated *lac* o^c. Because the enzymes are continually produced for both types of constitutive mutation, they clearly represent regulatory genes.

The Operon Model: Negative Control

In 1961, Jacob and Monod proposed a scheme of negative control of regulation, which they called the **operon model** (Figure 21.4). In this model, the **operon**

FIGURE 21.1
The catabolic conversion of the disaccharide lactose into its monosaccharide units, galactose and glucose.

FIGURE 21.2

The structural genes of the *lac* operon in *E. coli*. The three genes are transcribed into a single polycistronic mRNA, which is simultaneously translated into the three protein products encoded by the operon.

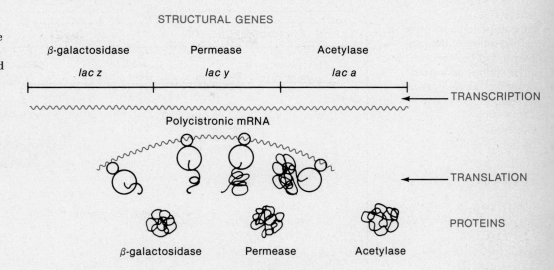

STRUCTURAL GENES

β-galactosidase   Permease   Acetylase

lac z   *lac y*   *lac a*

TRANSCRIPTION

Polycistronic mRNA

TRANSLATION

PROTEINS

β-galactosidase   Permease   Acetylase

---

consists of the structural genes as well as the adjacent region of DNA represented by the *lac o^c* mutation, called the **operator region.** They proposed that the *lac i* gene regulates the transcription of the structural genes by producing a **repressor molecule.** The repressor was hypothesized to be **allosteric.** This concept is applied particularly to proteins that reversibly interact with another smaller molecule, causing a conformational change in three-dimensional shape.

Jacob and Monod suggested that the repressor normally interacts with the DNA sequence of the operator region. When it does so, it inhibits the action of RNA polymerase, effectively repressing the transcription of the structural genes [Figure 21.4(b)]. However, in the presence of lactose, this disaccharide binds to the repressor, causing a conformational change. This change alters the binding site of the repressor capable of interacting with operator DNA [Figure 21.4(c)]. In the

absence of the repressor/operator interaction, RNA polymerase transcribes the structural genes, and the enzymes necessary for lactose metabolism are translated. Since transcription occurs only in the absence of the repressor, negative control is exerted.

The operon model uses these potential molecular interactions to explain the efficient regulation of the structural genes. In the absence of lactose, the enzymes encoded by the genes are not needed and so are repressed. When lactose is present, it indirectly induces the activation of the genes by binding with the repressor. If all lactose is metabolized, none is available to bind to the repressor, which is again free to bind to operator DNA and repress transcription.

Both the *i^-* and *o^c* constitutive mutations interfere with these molecular interactions, allowing continuous transcription of the structural genes. In the case of the *i^-* mutant, the repressor product is altered and cannot bind to the operator region, so the structural genes are always turned on. In the case of the *o^c* mutant, the nucleotide sequence of the operator DNA is altered and will not bind with a normal repressor molecule. The result is the same: Structural genes are always transcribed. Both types of constitutive mutations are illustrated diagrammatically in Figure 21.4(d) and (e).

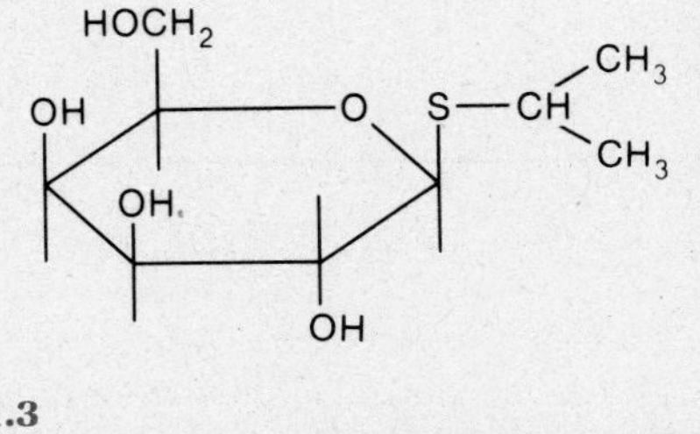

Genetic Proof of the Operon Model

The operon model is a good one because its predictions can be tested to determine its validity. The major assumptions to be tested are (1) the *i* gene produces a diffusible cellular product; (2) the *o* region does not;

FIGURE 21.3

The gratuitous inducer isopropylthiogalactoside (IPTG).

FIGURE 21.4
The components involved in the regulation of the *lac* operon and their interaction under various genotypic conditions, as described in the text.

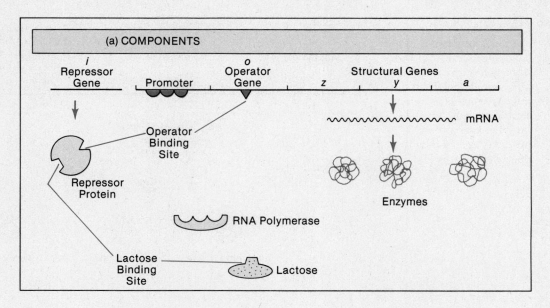

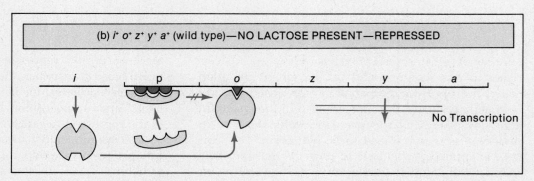

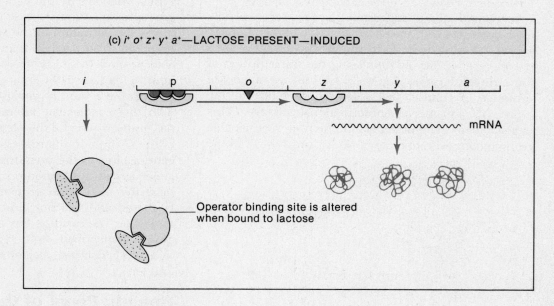

FIGURE 21.4
(continued)

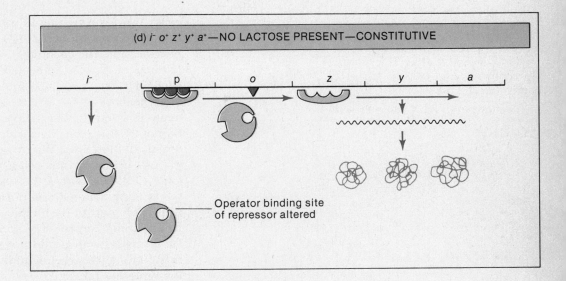

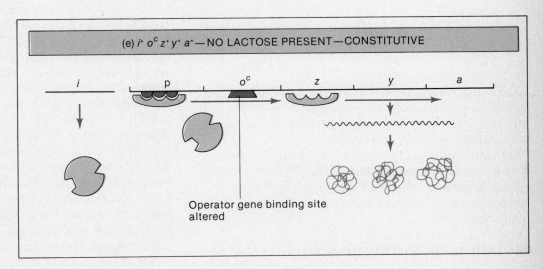

and (3) the *o* region must be adjacent to the structural genes in order to regulate transcription.

The construction of partial diploid bacteria through the use of **sexduction** (see Chapter 14) allows an assessment of these assumptions. For example, it is possible to construct genotypes where an i^+ gene has been introduced into an i^- host, or where an o^+ region is added to an o^c host. The Jacob-Monod operon model predicts that adding an i^+ gene to an i^- cell should restore inducibility, because a normal repressor would again be produced. Adding an o^+ region to an o^c cell should have no effect on constitutive enzyme production, since regulation depends on an o^+ region immediately adjacent to the structural genes.

The results of these experiments are shown in Table 21.1, where *z* represents the structural genes. In both cases just described, the Jacob-Monod model is upheld (part B of Table 21.1). Part C shows the reverse experiments, where either an i^- gene or an o^c region is added to cells of normal inducible genotypes. The model predicts that inducibility will be maintained in these partial diploids, and it is!

Another prediction of the operon model is that certain mutations in the *i* gene should have the opposite effect of i^-. That is, instead of being constitutive by failing to interact with the operator, mutant repressor molecules should be produced, which cannot interact with the inducer, lactose. As a result, the repressor

TABLE 21.1

Genotypes and enzyme activity for the *lac* operon of *E. coli.*

Genotype	Presence of β-galactosidase Activity	
	Lactose Present	Lactose Absent
A. $i^+o^+z^+$	+	−
$i^+o^+z^-$	−	−
$i^-o^+z^+$	+	+
$i^+o^cz^+$	+	+
B. $i^-o^+z^+/F'i^+$	+	−
$i^+o^cz^+/F'o^+$	+	+
C. $i^+o^+z^+/F'i^-$	+	−
$i^+o^+z^+/F'o^c$	+	−
D. $i^so^+z^+$	−	−
$i^so^+z^+/F'i^+$	−	−

would always bind to the operator sequence, and the structural genes would be permanently repressed. If this were the case, sexduction of an i^+ gene would have little or no effect on repression.

In fact, as shown in Table 21.1D, such a mutation, i^s, was discovered. Sexduction of an i^+ gene does not effectively relieve repression of gene activity. These observations again support the operon model for gene regulation.

Isolation of the Repressor

Although the operon theory of Jacob and Monod succeeded in explaining many aspects of genetic regulation in prokaryotes, the nature of the repressor molecule was not known when their landmark paper was published in 1961. While they had assumed that the allosteric repressor was a protein, RNA was also a candidate, since the molecule required binding to DNA. Despite many attempts to isolate and characterize the hypothetical repressor molecule, no direct chemical evidence was immediately forthcoming. Calculations indicated that a single *E. coli* cell contains no more than ten or so copies of the *lac* operon repressor. Direct chemical identification of ten molecules in a population of millions of proteins and RNAs in a single cell presented a tremendous challenge.

In 1966, Walter Gilbert and Benno Müller-Hill reported the isolation of the *lac* repressor in partially purified form. To achieve the isolation, they used a *regulator quantity* (i^q) mutant strain that contains about ten times as much repressor as wild-type *E. coli*

cells. Also instrumental in their success were the use of a gratuitous inducer, IPTG, which binds to the repressor, and the technique of **equilibrium dialysis.** Extracts of i^q cells were placed in a dialysis bag and allowed to attain equilibrium with an external solution of radioactive IPTG, which is small enough to pass freely through the bag. At equilibrium, the concentration of IPTG was higher inside the bag than in the external solution, indicating that an IPTG-binding material was present in the cell extract and that it was too large to diffuse across the wall of the bag. Ultimately the binding material was purified; it was shown to be heat labile and to have other characteristics of a protein as well. Extracts of i^- constitutive cells having no *lac* repressor activity did not exhibit IPTG-binding activity, strongly suggesting that the isolated protein was the repressor molecule.

To demonstrate this further, Gilbert and Müller-Hill grew *E. coli* cells on a medium containing radioactive sulfur and isolated the IPTG-binding protein, which was labeled in its sulfur-containing amino acids. This protein was mixed with DNA from a strain of phage lambda (λ) which carries the *lac* o^+ gene. The DNA sediments at 40*S*, while the IPTG-binding protein sediments at 7*S*. The DNA and protein were mixed and sedimented in a glycerol gradient in an ultracentrifuge. If the radioactive protein binds to the DNA, then it should sediment at a much faster rate, moving in a band with the DNA. This was found to be the case. Further experiments showed that the IPTG-binding or repressor protein binds only to DNA containing the *lac* region. It was also shown not to bind to *lac* DNA containing an operator constitutive (o^c) mutation.

Subsequent study of the repressor protein has revealed that it consists of four identical subunits of 38,000 daltons. Each can bind to one molecule of IPTG. The complete protein has a low binding to DNA in general, but has an affinity about 1000-fold greater for the operator region, which consists of about 30 nucleotide pairs.

The Promoter Region and Catabolite Repression: Positive Control

As we learned in Chapter 18, the nucleotide sequence that binds with RNA polymerase is called the **promoter region.** Within the *lac* operon, the promoter region is located to the left of the operator region and between it and the *i* gene (see Figure 21.4). Certain mutations isolated in the promoter region decrease or increase the effectiveness of polymerase binding. Those that reduce it exhibit a decrease of transcription in the induced or constitutive condition and are called "down" mutants. The "up" mutations increase the transcription rate. Apparently, the i^q mutants used to isolate the repressor were of this type for the *lac i* gene.

Another mode of regulation of the *lac* operon—one which overrides that described thus far—is called **catabolite repression.** In catabolite repression, transcription of the *lac* structural genes is inhibited if glucose is added to the culture medium, even though an inducible or constitutive condition exists. The same type of inhibition occurs in other inducible operons, including those controlling the metabolism of galactose and arabinose. Catabolite repression contributes to the cell's energetic efficiency because glucose can be metabolized in the glycolytic pathway more directly than can lactose, galactose, or arabinose.

Catabolite repression involves a portion of the promoter region called the **CAP site,** a protein called the **catabolite activating protein (CAP),** and a modified nucleotide, **cyclic adenosine monophosphate (cAMP).** These components interact to repress gene activity, as summarized in Figure 21.5.

The promoter contains two regions. The region closest to the operator contains the nucleotide sequence that interacts with the sigma subunit of RNA polymerase. The opposite end, closest to the *i* gene, contains the CAP site. Before RNA polymerase can effectively interact with its promoter region, the CAP site must be bound to CAP. However, this interaction requires that CAP be linked to cAMP.

In the absence of glucose, the enzyme **adenyl cyclase** readily converts ATP to cAMP (Figure 21.5). The availability of cAMP completes the chain of events just described, and, if the *lac* system is induced or constitutive, transcription of the structural genes proceeds.

The presence of glucose, however, drastically reduces the activity of adenyl cyclase, interrupting the events essential to transcription. The concentration of cAMP is also reduced. Under this condition, the CAP-cAMP complex is unavailable to bind to the CAP site, inhibiting transcription.

Unlike regulation by the repressor, the CAP interaction exerts a **positive control.** When bound to cAMP, CAP allows transcription to occur. The repressor, on the other hand, exerts a **negative control.** When it is not bound to lactose, it represses the system. Taken together, both systems involve the direct interaction of DNA with three proteins. The operator binds the repressor, and the promoter binds CAP-cAMP and RNA polymerase. It is interesting to note that all three regions of DNA contain **palindromes** in their nucleotide sequences (see Chapter 18), strengthening the evidence that palindromes are essential to interactions with other molecules.

THE CASE OF THE ARA REGULATOR PROTEIN: POSITIVE AND NEGATIVE CONTROL OF INDUCTION

Before turning to a consideration of repressible systems, we want to make mention of a rather unique inducible operon in *E. coli* that appears to be under both positive and negative control.

In *E. coli*, the metabolism of the sugar arabinose is under the direction of the enzymatic products of the *ara B, A,* and *D* genes. These genes are regulated by the **ara regulatory protein** encoded by the *ara C* locus. In the absence of arabinose, the protein behaves as a repressor, binding to the operator DNA preceding the *ara B A D* genes. Addition of arabinose to the medium results in the release from the operon of the *ara* repressor protein, which binds to arabinose. This newly formed complex then becomes an activator, attaching to a separate initiator site on the operon, stimulating RNA polymerase binding and transcription of the *B, A,* and *D* genes. Thus, the *ara* regulatory protein can alternately behave as a repressor or an activator, depending on the absence or presence of arabinose, respectively. In the former case, negative control is exerted, while in the latter case, positive control occurs. While it is rather easy to visualize negative control, where a repressor blocks the movement of RNA polymerase during transcription, it is not yet clear how the stimulation of transcription accompanying positive control is achieved.

TRYPTOPHAN METABOLISM IN *E. COLI:* A REPRESSIBLE GENE SYSTEM

Although induction had been known for some time, it was not until 1953 that Monod and his coworkers discovered **enzyme repression** in his laboratory at the Pasteur Institute in France. Wild-type bacteria are capable of producing the necessary enzymes essential to the biosynthesis of amino acids and many other molecules. Monod focused his studies on the amino acid tryptophan and the enzyme **tryptophan synthetase.** He discovered that if tryptophan is present in sufficient quantity in the growth medium, the enzymes necessary for its synthesis are **repressed.** Energetically, such enzyme repression is highly economical to the cell, because synthesis is unnecessary in the presence of exogenous tryptophan.

Further investigation has shown that a series of enzymes encoded by five contiguous genes on the *E. coli*

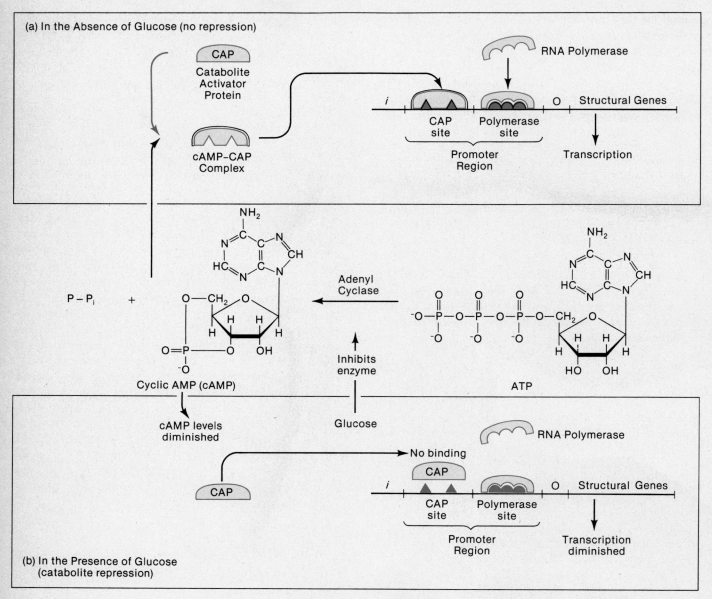

FIGURE 21.5

Catabolite repression. (a) In the absence of glucose, cAMP levels are high. When cAMP binds to CAP, a complex is formed that binds to a promoter region and facilitates transcription. (b) In the presence of glucose, the activity of adenyl cyclase is inhibited and cAMP levels are reduced. CAP, when not bound to cAMP, cannot efficiently bind to the promoter region, and transcription is reduced.

chromosome is involved in tryptophan synthesis. These genes are part of an operon, and in the presence of tryptophan, all genes are coordinately repressed. As a result, none of the enzymes are produced. Because of the great similarity between this repression and the induction of enzymes for lactose metabolism, Jacob and Monod proposed a model of gene regulation analogous to the *lac* system.

To account for repression, they suggested that an inactive repressor is normally made that cannot interact with the operator region of the operon. However, this repressor is also allosteric, interacting as well with tryptophan, if present. The resulting interaction causes a conformational change in the complex, which may now bind to the operator, and transcription is repressed. Thus, when the end product of this anabolic biosynthesis pathway is present, enzyme repression is effected. Since the regulatory complex inhibits transcription of the operon, this repressible system is under negative control.

Evidence for and concerning the *trp* Operon

Support for the concept of a repressible operon was soon forthcoming. Primarily, two distinct categories of constitutive mutations were isolated. The first class, *trp R*⁻, maps at a considerable distance from the structural genes. This locus represents genetic information coding for the repressor. Presumably, the R^- mutation either inhibits the interaction of a mutant repressor with tryptophan or inhibits repressor formation. Thus, no repression ever occurs. Sexduction of an R^+ gene restores repressibility.

The second constitutive mutant is analogous to that of the operator of the lactose operon because it maps immediately adjacent to the structural genes. Furthermore, sexduction of a wild-type operator into cells carrying the mutant does not restore enzyme repression. This is predictable if the mutant operator can no longer interact with the repressor-tryptophan complex.

The entire *trp* operon has now been well defined, as shown in Figure 21.6. Five contiguous structural genes (*trp E, D, C, B,* and *A*) are transcribed as a polycistronic message that directs translation of the polypeptide components. These components catalyze the biosynthesis of tryptophan. As in the *lac* operon, a promotor region (*trp P*), representing the binding site for RNA polymerase, and an operator region (*trp O*), which binds the repressor, have been demonstrated. In the absence of repressor binding, transcription is initiated within the overlapping *trp P–trp O* region and proceeds along a **leader sequence.** Transcription is initiated 162 nucleotides prior to the first structural gene. Within this leader sequence, still another regulatory site has been demonstrated. This component, called an **attenuator,** has been investigated extensively by Charles Yanofsky and his colleagues.

The Attenuator

Yanofsky made two important observations. The first involved constitutive *trp R* mutants in which transcription was expected to proceed regardless of the presence or absence of tryptophan. Instead, he observed that in the presence of tryptophan, transcription was attenuated or somewhat reduced. Since tryptophan cannot bind to the repressor in these mutants, it somehow must be involved in attenuating transcription. What is the role of tryptophan in attenuation?

The second observation involved deletion mutants within the noncoding leader sequence. If the region specifying bases 123 to 150 of the 162 base sequence is deleted, attenuation is eliminated. In fact, in the absence of tryptophan, cells with this deletion synthesize more than five times more of the *trp* operon enzymes than do wild-type cells. This genetic analysis has thus established the location of the regulatory element called the **attenuator.** A similar element has been found in other *E. coli* operons controlling the biosynthesis of six different amino acids (histidine, leucine, isoleucine, threonine, phenylalanine, and valine). Careful analysis has revealed how attenuation occurs. In the presence of tryptophan in wild-type cells, some RNA polymerase may escape the effects of an active repressor complex and initiate transcription of the leader sequence. However, at a point about 140 nucleotides along the sequence, termination of transcription occurs. This point coincides with the attenuator region. Confirmation of this regulatory site occurred when it was shown that starving *E. coli* of tryptophan reduced termination at this position in the leader sequence.

Two major questions are raised by these observations. Why is there a second regulatory element that has the same effect as the repressor? Second, how do high levels of tryptophan lead to termination of transcription? We will consider the initial question first. Attenuation is not as influential as repressor-mediated control. Attenuation reduces synthesis by only an 8- to 10-fold factor, while repressor control mediates transcription reduction by a 70-fold factor. Perhaps attenuation of the *trp* operon provides a fine control under various cellular conditions. In addition, it is important to point out here that in several other operons controlling amino acid synthesis, attenuation has been found to be the only regulatory control mechanism.

The answer to the second question remained an enigma for many years, but in 1977 Yanofsky proposed a model based on many experimental observations. This model has since gained acceptance and further support. The most important observations involved the study of mutations affecting tRNA. When mutations alter the function of either tRNA^trp or the aminoacyl synthetase specific for charging tRNA^trp, attenuation (termination of transcription of the leader sequence) is reduced. Thus, it appeared that attenuation somehow involves charged tRNA^trp rather than the amino acid itself. On the other hand, it was reasoned that the regulatory process might involve translation events that are mediated by this tRNA.

This hypothesis gained support when the nucleotide sequence of the leader region was determined. Yanofsky found that the transcript of the leader region

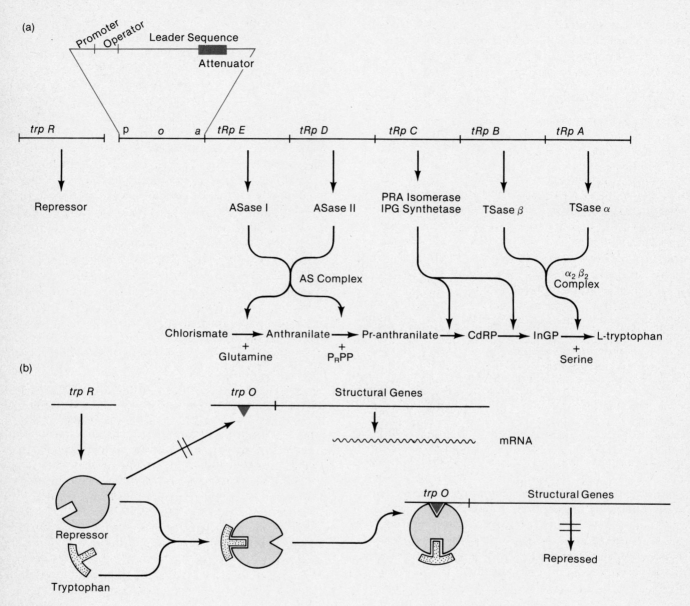

FIGURE 21.6

(a) The tryptophan operon and an abbreviated biosynthetic pathway of tryptophan production under the control of the gene products of the operon. (b) The components involved in the repression of the tryptophan operon. In the absence of this amino acid, the repressor cannot bind to the operator (*trp O*) and transcription occurs. An allosteric interaction between the repressor and tryptophan results in a complex that does bind to the operator and represses transcription. (ASase—anthranilate synthetase; TSase—tryptophan synthetase; P_R—phosphoribosyl; CdRP—carboxyphenyl-amino-deoxyribulose phosphate; InGP—indole-glycerol phosphate)

contained an AUG initiator codon and subsequently two tryptophan codons. Based on this information, he proposed the following model.

Once transcription of the leader sequence begins and the AUG codon is available, ribosomes initiate translation of this newly formed RNA. Recall that transcription and translation can occur simultaneously in bacteria before the RNA transcript is completed (see Figure 18.2). If adequate tryptophan is present in the cell, charged tRNA^{trp} is also present, and translation of this coding region proceeds. If cells are starved of tryptophan, inadequate amounts of tRNA^{trp} are available, and translation "stalls" at the *trp* codons.

Yanofsky has proposed that these two conditions of translation influence the secondary structure of the transcript. If stalling of *translation* occurs because of low levels of tryptophan, *transcription* frequently proceeds through the attenuator region, the operon is fully transcribed, and the enzymes necessary for the biosynthesis of tryptophan are produced.

On the other hand, when no stalling occurs, ribosomal movement is much more rapid and a base-paired hairpin secondary structure is formed downstream that characteristically induces termination of transcription.

The details of the secondary structures of the transcript have now been worked out and described. They satisfy conditions leading to either continued transcription or termination (attenuation). These are illustrated in Figure 21.7. Additionally, various mutations in the leader sequence predicted to alter the secondary structure of the transcript and its impact on attenuation have been isolated. In each case, the predicted result has been upheld. The model, while complex, has significantly extended our knowledge of genetic regulation. Furthermore, the supporting evidence represents an integrated approach involving genetic and biochemical analysis, which is becoming commonplace in molecular genetic research.

GENETIC REGULATION IN PHAGE LAMBDA: LYSOGENY OR LYSIS?

Our understanding of genetic regulation at the transcriptional level has benefited from studies of bacteriophage lambda as well as from studies of operons in bacteria. Lambda DNA contains about 45,000 base pairs, enough bioinformation for about 35 to 40 average-sized genes. After injection of lambda DNA into *E. coli*, either the **lysogenic** or **lytic** pathways may be followed (see Chapter 14). In the first pathway, the phage DNA is integrated into the host genome and is almost totally repressed; in the second, the phage DNA is repressed and viral reproduction ensues.

If the lysogenic pathway is followed, the genes responsible for phage reproduction and lysis are turned off by the **λ repressor protein,** which is produced by one of the virus's own genes, *cI*. If the lytic pathway is followed, the expression of the *cI* gene is repressed by a second protein, **cro,** which is produced by a distinct viral gene of the same name. Both the *cI* repressor (sometimes called the λ repressor) and the cro protein have been isolated and characterized. Their interac-

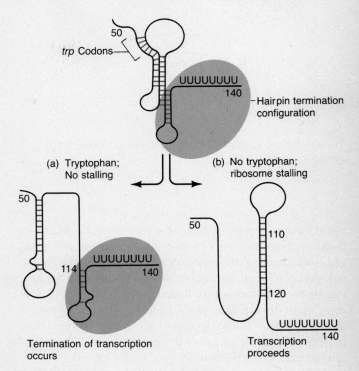

FIGURE 21.7

Diagrammatic representation of the leader sequence of the RNA transcript of the *trp* operon of *E. coli* and its role in attenuation. Shown at the top are the *trp* codons, the potential base-pairing sites, and the hairpin termination configuration in the leader sequence (shown in color shading). In the presence of tryptophan (a), rapid movement of the ribosome across the *trp* codons occurs during translation, and a hairpin configuration forms at the attenuator site. Termination of transcription usually occurs, leading to attenuation. When cells are starved of tryptophan (b), the ribosome stalls at the *trp* codons, no hairpin configuration forms at the attenuation site, and transcription proceeds across it. The model assumes that transcription by RNA polymerase and translation by a ribosome are occurring simultaneously, as is known to happen in bacteria.

tion with λ DNA has also been determined. The various components of the regulatory system are shown in Figure 21.8. The development of this information, which enhances our understanding of gene regulation, is a fascinating story.

The λ repressor was isolated and characterized by Mark Ptashne in 1967. It is a protein consisting of 236 amino acids. The cro protein, isolated more recently, consists of 66 amino acids. When the repressor is produced, it recognizes two different operator regions in the λ DNA, O_L and O_R, found on either side (left or

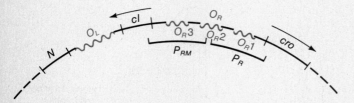

FIGURE 21.8
The regulatory sites controlling lysogeny and lysis in
phage λ.

right) of the *cI* gene. When the repressor is bound to
these regions, two sets of so-called early genes are repressed, causing the remainder of the λ genes to be
turned off as well. As such, the repressor behaves as a
negative control element. Rather surprisingly, it is now
clear that during this same binding, the cI repressor
also stimulates transcription of its own *cI* gene. Thus,
in this role, it behaves as a positive control element
and enhances transcription by a factor of ten.

When the repressor binding to O_L and O_R is absent,
initiation at the respective promoters (P_{RM} and P_R) proceeds, resulting in the production of two proteins, N
and cro. The N protein functions as an antiterminator,
thus allowing completion of transcription of genes essential to reproduction and lysis. The cro protein as
stated earlier, serves as a repressor of *cI* gene transcription. Mutational analysis supports this model.
Mutations in either O_L or O_R prevent repressor binding
and abolish the potential for lysogeny. Cro^- mutations abolish the potential for lysis.

The operator-repressor interactions have also been
worked out in detail. The interaction occurring at O_R
will be used as an illustration. The O_R region consists
of 80 DNA base pairs and is located between the *cI*
gene and the *cro* gene. There are three binding sites
in O_R, each with a distinctive nucleotide sequence.
The three regions are labeled O_R1, O_R2, and O_R3, and
each consists of 17 nucleotides. Overlapping this region are the promoters for the *cI* and *cro* genes, called
P_{RM} and P_R, respectively. Transcription of these genes
occurs in opposite directions, using alternate strands
of the helix to achieve 5′-to-3′ RNA synthesis

The greatest surprise was that both the λ repressor
and cro protein demonstrate a binding affinity to the
three sites of O_R. However, the binding of the two proteins to the three sites varies, depending on the concentration of the proteins. In addition, when either
protein is bound to these sites, the other cannot bind.
The cI protein binds initially to O_R1 and O_R2 coordinately and then to O_R3, but only when present at very
high concentrations. The cro protein behaves in the

opposite fashion. It initially binds O_R3 and subsequently binds O_R1 and O_R2.

Following initial infection, the determination of lysogeny versus lysis appears to depend on which operator region is bound first. If O_R1 and O_R2 are bound,
P_R is inhibited and the *cro* gene is not transcribed.
P_{RM}, on the other hand, is available to RNA polymerase
and the *cI* gene is transcribed, resulting in increased
amounts of the repressor. Lysogeny is now established. However, if O_R3 is bound by the cro protein,
P_{RM} is inhibited and the *cI* gene is repressed. This allows initiation at the P_R site, and transcription essential to lysis proceeds.

While this description explains how the determination is made, it is still not clear why one pathway is
favored over the other. There is some evidence that
when the level of infectivity of viruses to bacteria is
high, lysogeny is preferred, and vice versa. The availability of nutrients in which infected bacteria find
themselves also affects the outcome.

To pursue this question further, Ptashne and others
have found it useful to examine the molecular events
that occur when an induction event transforms the lysogenic pathway to the lytic sequence. Discovered in
1946 by L'woff, treatment of lysogenized *E. coli* by various agents (including UV light and mutagens) induces
a lytic response. While such an event occurs spontaneously about once in a million bacterial cell divisions,
all treated bacteria are induced to enter the lytic cycle.

Induction events have one common property: They
all involve interaction with DNA. In so doing, they
stimulate what has been called the SOS response (see
Chapter 13). During this response, the rec A protein,
normally essential to recombination, is synthesized.
Under inductive conditions, it behaves as a proteolytic
enzyme and plays an essential role in the induction of
lysis.

The λ repressor protein contains two structural domains and may exist in monomer or dimer form. It is
the dimer form that binds strongly to O_R. The rec A
protein cleaves the monomers at a sensitive region between the two domains. Cleavage of monomers prevents their dimerization; as a result, the repressor protein can no longer bind to O_R, stimulating the
expression of the *cro* gene. The cro product, also active in dimer form, then binds to O_R, inhibiting further
transcription of the *cI* gene. Genetic events essential to
viral reproduction then proceed, new viruses are constructed, and lysis of the bacterial cell occurs.

The research leading to the information just described is remarkable. It illustrates the depth of information attainable with regard to what at first may appear as a simple question: How is lysogeny regulated?
Important insights have also been gained regarding

protein/nucleic acid interactions during genetic regulation. Molecular interactions such as these undoubtedly play major roles in many other genetic phenomena.

PHAGE TRANSCRIPTION DURING LYSIS

When a bacteriophage invades its host bacterium and initiates the lytic cycle, it faces a novel problem. How can its genes be transcribed without the presence of RNA polymerase, itself a product of transcription? The strategy that has evolved is an interesting example of parasite-host interaction. You may wish to review the discussion of the phage life cycle and strategies for replication of DNA discussed in Chapter 14.

The study of phage T7, which invades *E. coli*, has provided some answers to this question. This phage has evolved DNA sequences that are recognized as promoter sites by the *E. coli* RNA polymerase. As a result, upon the phage's entry to the host cell, a set of so-called **early genes** is transcribed into mRNAs, which are translated on bacterial ribosomes. One of these gene products is a viral-specific RNA polymerase that recognizes the promoters for the remaining T7 genes. A second early gene product, a protein kinase, inhibits bacterial transcription. The site of inhibitory interaction is presumably the bacterial RNA polymerase.

With the host cell's genetic apparatus secured, the remaining viral genes are expressed. The major products are DNA-replicating enzymes, head and tail components, enzymes involved in DNA packaging, and, finally, a lytic enzyme to break open the host cell.

Studies of the phage **SPO1**, which invades *Bacillus subtilis*, have revealed an even finer control or regulation of transcription during lysis (Figure 21.9). As in T7,

FIGURE 21.9
Initiation and regulation of phage SPO1 genes during lysis.

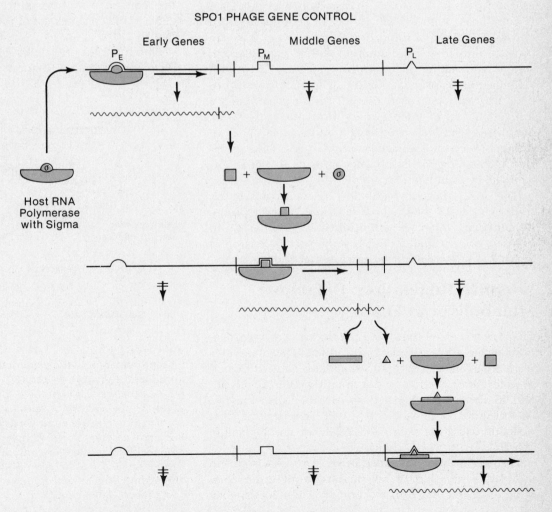

SPO1 PHAGE GENE CONTROL

early genes are read by the host RNA polymerase. One of the early gene products is a protein subunit that associates with the host polymerase. This complex can now interact with the promoter regions of the **middle genes.** This association between the early gene subunit and the bacterial polymerase curbs transcription of the early genes, and the middle genes associated with DNA replication are expressed. Two of the middle genes produce additional polypeptides that can interact with the host polymerase, again altering its specificity such that it directs the expression of the **late genes.** Thus, sequential gene action results, providing a finer mechanism for the regulation of gene expression during the lytic cycle.

REGULATION OF TRANSCRIPTION IN EUKARYOTES

The control of gene expression in eukaryotes is much more complex and less well defined than in prokaryotes and viruses. Each eukaryotic cell contains much greater amounts of genetic information, and this DNA is complexed with a wide variety of proteins to form chromatin. Many chromosomes, rather than one, are present, and they are contained within a nuclear envelope. Translation occurs in the cytoplasm. In multicellular organisms, tissue and cell-specific gene products are restricted, even though each cell contains a full genetic complement. In addition, most eukaryotic organisms are not as amenable to mutagenesis and experimentation as are bacteria and viruses. Thus, progress in defining regulatory mechanisms in higher organisms has not been as rapid.

The study of gene regulation in eukaryotes is the underlying basis for the broader topic of developmental genetics, which we will address in Chapter 22. In this section, we shall focus on some specific examples of what is known about transcriptional control.

Adaptive Regulation: Galactose Metabolism in Yeast

After bacterial operons were characterized, investigations of metabolic regulation in unicellular eukaryotes were intensified. These investigations were designed to determine whether or not operons could be identified in these organisms. If so, do they also exist in multicellular organisms? We will address the first question using the example of galactose metabolism in yeast.

Fermentation of galactose in yeast results from the induced expression of information from three clustered or linked loci, *GAL 1, 7,* and *10.* These loci specify

a **kinase, transferase,** and **epimerase,** respectively. Enzyme activity is induced by galactose, and constitutive mutants have been isolated. Thus, a regulatory gene exists, and mapping studies show the gene not to be linked to the structural genes. Remember that yeasts have numerous distinct chromosomes. Such mutants behave recessively in diploid cells, analogous to the partial diploids for the *i⁻* mutation in *E. coli.*

A rather clear picture of the mechanism of genetic control is now emerging in the case of *GAL 1* and *GAL 10.* Induction is dependent on the action of a protein produced by the expression of a separate locus, *GAL 4.* This protein interacts directly with an **upstream activating sequence (UAS$_G$)** located midway between the *GAL 1* and *10* genes, which are present in an orientation opposite to one another. The GAL 4 protein is an example of a positive control element. However, the action of the GAL 4 protein can be inhibited by the product of still another gene, *GAL 80.* The role of galactose in the induction of these genes is apparently to interfere with the inhibitory action of the GAL 80 protein. When this occurs, the GAL 4 protein exerts positive control over transcription of the *GAL 1* and *10* genes. This complex set of interactions is illustrated in Figure 21.10.

The model is supported by a variety of genetic and biochemical data. Mutations in the *GAL 80* gene be-

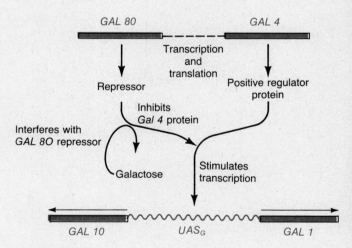

FIGURE 21.10

A representation of the proposed interactions of the products of the *GAL 80* and *GAL 4* genes with galactose during the regulation of the *GAL 1* and *GAL 10* genes in yeast. In the absence of galactose, the *GAL 80* repressor inhibits the positive control exerted by the GAL 4 protein, and transcription is not stimulated. If galactose is added to the medium, it interferes with the action of the *GAL 80* repressor, and transcription is stimulated by the GAL 4 regulatory protein, which interacts with UAS$_G$, the upstream activating sequence.

have as expected. *GAL 80⁻* are constitutive such that galactose is not essential to induction. *GAL 80ˢ* mutations, on the other hand, are not inducible at all. We assume that this superrepressible condition is created by a mutation interfering with the (GAL 80-galactose) interaction essential to the inhibition of the GAL 4 protein.

Of great interest is the nature of the interaction between the GAL 4 protein and the UAS$_G$ region. If UAS$_G$ is experimentally placed in front of another yeast gene, it becomes inducible by galactose. This observation is indicative of the positive control exerted by the GAL 4 protein. It is believed that the protein binds directly to four related 17-base-pair (bp) sites within the UAS$_G$ region. If a consensus sequence of 17 bp is synthesized and artificially placed before another yeast gene, it too becomes regulated by the GAL 4 protein. While the mechanism in yeast is a bit more complex, the similarities between the regulation of the galactose system in yeast and the lactose system in *E. coli* are striking.

The Britten-Davidson Model of Regulation

In 1969, Roy Britten and Eric Davidson proposed a theory to explain gene regulation in the cells of higher organisms. Their model was based on the types of molecular interactions known to occur during the regulation of bacterial and viral genes. However, it took into account the need for more complex gene activation events in higher organisms.

Even though there is currently no direct support for the Britten-Davidson model, it is intriguing. It summarizes many of the observations and assumptions that must be made about regulation in higher organisms. For example, as cells undergo differentiation, it is apparent that previously inactive sets of genes become activated. Such activation is sometimes associated with external signals such as hormonal action or embryonic inductive events. Often, many noncontiguous genes respond to an individual signal.

The essence of the model is the simultaneous regulation of **batteries of genes** during development. Britten and Davidson proposed that **repetitive sequences** (a class of eukaryotic DNA) serve as major control units. When the eukaryotic genome is fractionated and analyzed, much of it is seen to exist as variable numbers of different sizes of repeating nucleotide sequences (see Chapter 11). What is significant is that the distribution of such sequences is not always random with the genome. Rather, many of the shorter sequences (10–20 base pairs) are contiguous to structural genes.

The basic components of the Britten-Davidson model are diagramed in Figure 21.11. As can be seen, a series of batteries of genes is activated by the presence of some signal molecule, such as a hormone. This molecule may or may not require interaction with an existing cytoplasmic molecule. In either case, the hormone or hormone complex interacts with a **sensor gene.** This event activates a contiguous **integrator gene,** which produces an **activator RNA molecule.** It is this molecule that activates genes to produce materials essential to the cell.

Activator RNA (or its protein product) interacts with **receptor genes** (which are comparable to the operator regions in bacteria) to activate transcription. The receptor genes control the transcription of adjacent **producer genes,** which are comparable to the structural genes in bacteria.

FIGURE 21.11
The components of the Britten-Davidson model of gene regulation in eukaryotes.

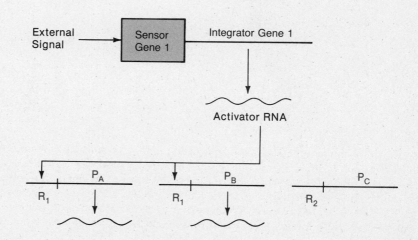

The model proposes that each set of producer genes in a given battery contains a common nucleotide sequence in its adjacent receptor gene site. Thus, a single activator molecule may activate numerous noncontiguous producer genes, called a **battery**. The common sequences of related receptor genes correspond to the enormous numbers of repetitive regions in the genome.

The basic model may be expanded into more complex interrelationships. In Figure 21.12, three different sensor/integrator gene sets are shown in relationship to six receptor/structural gene sets. Such a model is

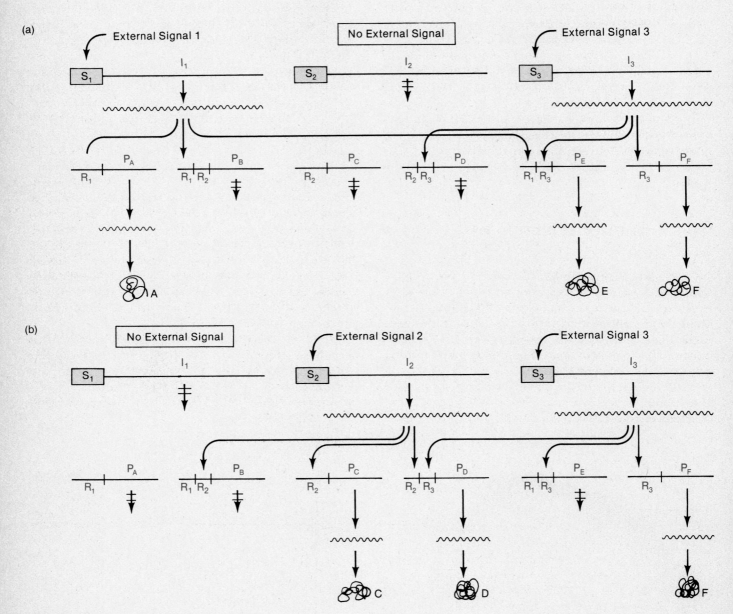

FIGURE 21.12

The Britten-Davidson model of eukaryotic gene regulation for two alternative conditions of stimulation of batteries of genes. In comparison to the simpler model in Figure 21.11, some producer genes are under the control of more than one receptor sequence. In these cases, both sequences must be stimulated in order to activate the adjacent producer genes. In part (a) External Signals 1 and 3 are available, while in part (b) External Signals 2 and 3 are available. Differential gene expression results in these contrasting situations.

complex and, if correct, will be difficult to prove directly. However, it is useful as a starting point in considering regulatory control in higher organisms.

The model predicts that a short repeated nucleotide sequence should be present at the 5′ end of eukaryotic genes that are under coordinate control. There are several instances where this has been found. For example, in yeast, histidine starvation results in the activation of numerous genes controlling the synthesis of histidine, tryptophan, and other amino acids. A short consensus sequence,

$$A_T^A GTGACTC,$$

is found prior to the coding region of the *his 1*, *his 3*, *his 4*, and *trp 5* genes. Deletion mutations within this sequence eliminate inducibility.

In *Drosophila*, members of a set of genes dispersed throughout the genome are induced by heat. Called **heat-shock protein (HSP) genes,** all are preceded by the consensus sequence CTNGAATNTTCTAGA, where N may be any of the four possible nucleotides present in DNA. Finally, in chickens the genes encoding the steroid hormone-induced egg-white proteins, conalbumin, lysozyme, and ovalbumin share a consensus sequence. Located about 140 nucleotides upstream from each gene is the sequence AAAATGGGC. While these observations do not prove the Britten-Davidson model, they support the concept that genes dispersed in the genome and activated by a common stimulus do share short upstream nucleotide sequences. Many researchers believe that nearly identical upstream sequences at many locations would not occur by chance and then be conserved during evolution, were they not integrally involved in gene expression.

Promoters and Enhancers in Eukaryotes

Given the rather clear definition of regulatory elements in prokaryotes, it was expected that similar units would soon be identified in eukaryotes. While the mechanisms of regulation are still far from clear, numerous elements have now been demonstrated, most of which are found upstream of the 5′ coding sequence of the numerous eukaryotic genes studied. In fact, in our discussion of gene function in Chapter 18 and gene structure in Chapter 19, several such units were introduced (see Figure 19.15).

The element closest to the coding region and most universal has been labeled the **TATA box.** Located about 30 bases upstream of the initial point of transcription (−30), and homologous to the Pribnow sequence of prokaryotic promoters, this regulatory element governs initiation of transcription. Various types

of assays have established that the TATA element and GC-rich DNA regions about 15 nucleotides on either side of it represent an area of great significance in the **eukaryotic promoter.** Mutations in TATA severely reduce transcription, while its deletion often alters the initiation point of transcription.

Other areas of the eukaryotic promoter region that have an influence on transcription exist farther upstream. For example, the so-called **CCAAT box,** frequently 40 to 50 nucleotides away from the TATA box (−70 to −80) appears also to modulate transcription. The consensus sequence,

$$GG_C^T CAATCT$$

derived from analysis of many genes, is the basis of the CCAAT box. This region is part of what is more generally referred to as **upstream regulatory sites.** These also include the **DNase-hypersensitive** areas of chromatin associated with genes in the active state but not with the same genes when they are inactive.

While the existence and location of sites such as the CCAAT box are not universal in all genes studied, they are thought to be essential for efficient transcription. In some cases, they control the frequency of initiation. Just how a functional promoter may contain different regions along a DNA molecule separated by 30 or 40 bases is not clear. However, we have a tendency to view DNA as a linear molecule, when in fact, a simple folding would bring the TATA and CCAAT boxes in juxtaposition to one another (Figure 21.13).

While both prokaryotes and eukaryotes thus exhibit genetic promoters of transcription, eukaryotes display a unique element called the **enhancer.** First discovered in the animal virus SV40, this element has now been found to be a part of a variety of gene systems in

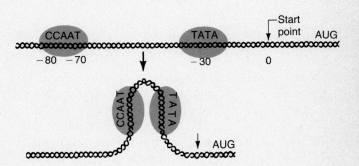

FIGURE 21.13
Folding of DNA, bringing the region of the upstream CCAAT box in close proximity to the TATA box sequence. This is one model of the eukaryotic promoter.

higher organisms. While enhancers stimulate transcription, based on the 100-fold reduction that occurs when they are deleted, they vary significantly from promoters:

1   Their position need not be fixed; they may be upstream, downstream, or even within a gene. Sometimes, they are as far as 3000 base pairs away from the gene they regulate.

2   While they are *cis*-activating, they may have their nucleotide sequence inverted without significant effect.

3   They are not restricted to enhancing transcription of specific genes. If the position of an enhancer in the genome is altered, or if an unrelated gene is placed near an enhancer, stimulation of the expression of the unrelated gene occurs.

Initially discovered independently by Pierre Chambon and George Khoury and their associates, the SV40 enhancer consists of a tandem repeat of a 72-base-pair sequence necessary for the expression of the early genes of the virus. When this repeat sequence is relocated anywhere in the viral genome, these genes are expressed. Removal of the repeats greatly reduces their transcription. Enhancers whose behavior is similar to that of SV40 have been discovered in a number of other viruses that invade and utilize the genetic machinery of eukaryotes.

Whether the cellular genes of eukaryotes also come under the regulation of enhancers was soon made clear when it was shown that immunoglobulin heavy chain genes are affected by such an element. Surprisingly, this enhancer sequence is located downstream from the start point. It is found in the intron between the V-D-J coding region and the C region of the rearranged DNA in lymphocytes. If this intron is deleted experimentally, heavy chain synthesis is diminished severely. It appears that a similar enhancer element may be present as well in one of the introns of the DNA encoding the light chain. Recall that we have previously discussed the structure of DNA encoding immunoglobulins in Chapter 19. Studies also suggest that for cellular genes, these elements may demonstrate tissue specificity. For example, the genetic enhancement by this particular element is specific for cells that produce antibodies.

Since this discovery, other downstream enhancer elements have been found in the human beta-globin gene, the *Xenopus* 5S rRNA genes, and the chicken thymidine kinase gene. The most intriguing question associated with these findings is how control is exerted, especially given the flexibility of location and orientation displayed by enhancers. While there is no clear answer yet, several possibilities have been considered. One is that there are soluble regulatory proteins that interact specifically with enhancer sequences. Perhaps such an interaction alters specific DNA regions that "enhance" the binding of RNA polymerase to the promoter sites. A more specific possibility is that some enhancers function by inducing the formation of **Z-DNA,** which alters the degree of supercoiling. Enhancers characterized thus far do contain a stretch with an alternating sequence of pyrimidine-purine bases, which provides the potential for the formation of Z-DNA. In support of this concept is the observation by Alfred Nordheim and Alexander Rich that the DNA of the SV40 chromosome forms Z-DNA in three regions within and just outside the repeated 72-base-pair enhancer elements. It is of interest to note that DNase-hypersensitive regions of SV40 chromatin lie, among other places, at positions 25 base pairs on either side of each Z-DNA location. However, it is still not known what, if any, functional relationship exists between hypersensitive sites, Z-DNA, and transcriptional activity. We can be certain that future investigations will clarify this issue as well as the way in which enhancer elements function.

Hormonal Regulation of Gene Expression

Higher organisms regulate metabolism and homeostasis primarily through the hormones of the endocrine system. How hormones function has been the subject of extensive investigation. While certain types (notably the polypeptide hormones) bind to the outside of the cell and affect the activity of cytoplasmic enzymes, steroid hormones have been shown to regulate transcription of specific genes.

For example, during insect metamorphosis, the steroid **ecdysone** induces specific changes in the puffing pattern of polytene chromosomes. Recall from Chapter 11 that puffs represent the transcription of specific genes. Another excellent example is the effect of the female sex hormones estrogen and progesterone on uterine tissues in mammals and on the oviduct and liver in birds. Studies of chickens have been particularly informative. The *in vivo* and *in vitro* response of mature oviduct tissue to estrogen is the abundant production of the egg-white protein **ovalbumin.**

The first link between hormonal induction of a specific protein and transcriptional control was the observation that actinomycin D abolishes the response. This antibiotic specifically inhibits DNA-dependent RNA synthesis. A second important observation involved the use of ^{3}H-estrogen. When ^{3}H-estrogen is added to oviduct tissue, radioactivity is first seen to enter the cytoplasm and subsequently the nucleus, where it is found to be bound to chromatin.

Careful study has revealed that estrogen does not enter the nucleus independently, but first binds to a specific **receptor protein** present in the cytoplasm. It is this complex that enters the nucleus, binds to chromatin, and induces the transcription of ovalbumin-specific mRNAs.

Investigation of the cytoplasmic receptor has shown it to be specific to the target tissue of the hormone. It is heat labile, precipitable by ammonium sulfate, and digested by proteolytic enzymes, but not DNase or RNase. These observations confirm its protein nature. Binding studies of the receptor-hormone complex reveal a nonspecific response to DNA, but a specific response to chromatin. Therefore, it was suspected that specific proteins in chromatin are responsible for the nuclear binding leading to the induction of transcription. Various experiments suggest that those involved are nonhistone proteins.

Although a detailed explanation of hormonal regulation of transcription is not yet available, we can see that the action of steroids parallels the inducers in bacterial operons. However, there is no evidence that the precise mechanisms involved are similar. Also, remember that hormone regulation in eukaryotes represents a special case and does not necessarily apply to other examples of gene regulation.

Histone and Nonhistone Proteins

As we discussed earlier in this chapter, eukaryotic cells contain a complete genome; but in any differentiated cell, only part of this genetic information is expressed. One approach in studying regulation has been to analyze the components of chromatin other than DNA as potential gene regulators. Essentially, these components are two general classes of proteins: **histones** and **nonhistones.**

Histones are a group of small proteins (10,000–20,000 daltons) lacking in tryptophan, but enriched in the positively charged amino acids arginine and lysine (see Table 11.2). In the early 1960s, Ru Chih Huang and James Bonner showed that increased amounts of histones in association with DNA inhibited transcription. Maximum inhibition occurred when an equimolar ratio of DNA:histones was used. These results implied that the DNA of a histone-coated gene is turned off and that transcription proceeds once the histone is removed. Still other studies suggested that chemical modification of histone molecules might also be important. The addition and removal of acetate, phosphate, and methyl groups chemically alter histones bound to DNA. It has been suggested that these modifications are related to the expression of genes.

While these observations established a correlation between histones and genetic repression, other findings have caused their importance in specific gene regulation to be questioned. Primarily, there seems to be insufficient heterogeneity among histones to account for the complexity of regulation necessary in eukaryotic organisms. That is, there does not appear to be a sufficient number of different histone molecules to regulate thousands of genes in a specific manner. Histones are separated into five major classes, and even though each class may contain five different related molecules, the total number seems too low. Furthermore, the similarity of histones found in different tissues of an organism indicates a lack of functional diversity.

As we learned in Chapter 11, histones are integral parts of **nucleosomes,** which suggests a yet undefined structural role in chromatin. Although they may also be generalized genetic repressors, it seems doubtful that histones play any role in specific gene regulation.

Recent interest in the **nonhistone chromosomal (NHC) proteins** has stemmed from numerous observations:

1 NHC proteins are much more heterogeneous, both in size and number, than histones (Figure 21.14).

2 NHC proteins are not very stable, and their "turnover" in the cell is much more extensive and rapid than for the more stable histones.

3 While synthesis of histones is synchronized with DNA synthesis in the S phase, synthesis of NHC proteins occurs throughout the cell cycle.

4 Like histones, NHC proteins undergo post-translational modification, including acetylation and phosphorylation.

5 Some NHC proteins exhibit tissue heterogeneity, whereas histones display tissue homogeneity.

Taken together, these observations suggest that NHC proteins play a role in eukaryotic gene regulation.

It is possible to fractionate chromatin into its constituent parts—DNA, histones, and NHC proteins—and then to reconstitute it. These techniques make possible a range of experiments where components with different origins are assembled and transcribed and the resultant gene products analyzed. Such experiments support the role of NHC proteins in gene regulation. For example, if DNA and histones from thymus cells and NHC proteins from bone marrow cells are reconstituted, the RNA transcribed resembles that of bone marrow cells to a greater degree than it does thymus cells. This conclusion was reached in molec-

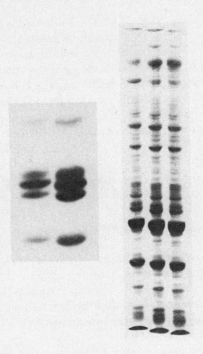

FIGURE 21.14

A comparison of histone and nonhistone chromosomal proteins separated using polyacrylamide gel electrophoresis. Many more bands representing proteins are apparent in the three nonhistone gels shown on the right than in the two histone gels shown on the left. (Courtesy of Gary and Janet Stein.)

ular hybridization experiments between the "reconstituted" DNA product and the RNA derived from marrow and thymus cells. In the converse experiment, where DNA and histones were derived from marrow cells and NHC proteins from thymus cells, the transcribed RNA resembled thymus RNA.

Other experiments have yielded results with a similar pattern, implying that NHC proteins determine the pattern of genes transcribed in different tissues. However, since most evidence has been obtained from *in vitro* studies, there is no solid proof of the regulatory role of NHC proteins. Undoubtedly, further research will show whether such a hypothesis is valid.

PROCESSING TRANSCRIPTS: ANOTHER MODE OF REGULATION IN EUKARYOTES

There are many opportunities where regulation of genetic expression can occur. While the most obvious is at the level of transcription, another theoretical possibility is during post-transcriptional modification of RNA. Changes occur in initial RNA transcripts that are essential to their translation. For example, as discussed in Chapters 18 and 19, most eukaryotic transcripts contain **introns** that must be excised, while the remaining **exons** are precisely spliced back together. This processing step offers several possibilities for regulation, and examples of them have been discovered.

First is the case where a single gene is transcribed, but following processing, two mRNAs, different in their leader sequences, are produced. Since the noncoding leader sequence is believed to be involved somehow in initiation of translation, different efficiencies may characterize the two mRNA products. Thus, variation in splicing specificity may indirectly affect the amount of protein produced. An example of this can be found in the synthesis of **α-amylase** in the rat. Salivary gland cells produce more of this enzyme than liver cells, and, indeed, alternate splicing of identical RNA transcripts in the two tissues results in different leader sequences at the 5′ ends of the respective mRNAs. It is assumed that the two observations are causally related. If so, variation in splicing has led to differential expression of the same primary transcript.

Researchers are especially interested in the possibility that splicing variation might generate different mRNAs responsible for directing the synthesis of different polypeptides. Several instances have now been documented demonstrating that splicing may include or exclude an entire exon. This leads to distinctive mRNAs that specify different polypeptides. This is the case in several systems, including the expression of the **myosin** gene in *Drosophila* and the **α-crystallin** gene in rodents.

Figure 21.15 illustrates still a third example of this phenomenon where the polypeptide products are more clearly distinct from one another. Shown is a diagram of the initial bovine RNA transcript that is processed into one of two **preprotachykinin mRNAs (PPT mRNA).** This mRNA molecule potentially includes the genetic information specifying two neuropeptides called **P** and **K.** These two peptides are members of the family of sensory neurotransmitters referred to as **tachykinins,** and are believed to play different physiological roles. While the P neuropeptide is largely restricted to tissues of the nervous system, the K neuropeptide is found more predominantly in the intestine and thyroid.

The RNA sequences for both neuropeptides are derived from the same gene. However, processing of the initial RNA transcript can occur in two different ways. In one case, exclusion of the K-exon during processing results in the α-PPT mRNA, which upon translation

FIGURE 21.15

Diagrammatic representation of the alternate splicing of the initial RNA transcript of the preprotachykinin gene (PPT). Introns are shown as unshaded and are labeled I. Exons are either numbered or designated by the letters P or K. Inclusion of the P and K exons leads to β-PPT mRNA, which upon translation yields both the P and K tachykinin neuropeptides. When the K exon is excluded, α-PPT mRNA is produced and only the P neuropeptide is synthesized.

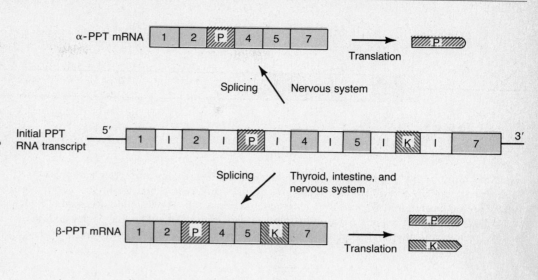

yields neuropeptide P but not K. On the other hand, processing that includes both the P and K exons yields β-PPT mRNA, which upon translation results in the synthesis of both the P and K neuropeptides. Analysis of the relative levels of the two types of RNA has demonstrated striking differences between tissues. In nervous system tissues, α-PPT mRNA predominates by as much as a threefold factor, while β-PPT mRNA is the predominant type in the thyroid and intestine.

While it is not certain what mechanism might lead to alternate splicing in different tissues, the overall effect is clear. The expression of a single gene, leading to two distinct products, is governed by tissue-specific alteration of the initial RNA transcript. Further investigations will very likely uncover still other examples of gene regulation at the post-transcriptional level.

CONCEPTS OF REGULATION IN EUKARYOTES: THE THALASSEMIAS

We conclude this chapter by discussing the **thalassemias,** a diverse group of human anemias, the classes of which are all due to an improper balance of alpha (α) and beta (β) subunits of hemoglobin. Ranging from the complete absence of one or the other chain to an imbalance in the ratio of functional alpha to beta chains, individuals are variously affected. The discussion of this topic is an appropriate way to end this chapter because the thalassemias illustrate several important points concerning the regulation of genetic expression. First, the level of expression of many genes sharing a common function must be precisely coordinated. In prokaryotes, this is most often achieved during the expression of genes in operons where polycistronic mRNAs are produced and translated. In eukaryotes, however, this does not seem to be the case. Genes with related functions are usually dispersed in the genome and mRNAs are monocistronic. Thus genetic regulation must be more precisely coordinated in eukaryotes. Second, there are many steps that occur during eukaryotic gene expression that offer an opportunity for regulation: transcription, post-transcriptional modification of mRNA, transport of mRNA from nucleus to cytoplasm, translation, and post-translational modification. In illustrating many of the above points, we also introduce an important group of human genetic disorders.

As discussed in Chapter 16, the human globin genes are represented by two clusters of coding sequences that are sequentially expressed during prenatal development (see Figure 16.8). The genes for alpha globin are located on chromosome 16, and the beta-globin genes are on chromosome 11. Each gene is part of a series of related coding sequences. Adult hemoglobin is a tetramer composed of two alpha- and two beta-globin polypeptides, each associated with a heme group. Imbalances of either alpha- or beta-globin polypeptides can result in the clinical disease known as thalassemia, characterized by defective or absent synthesis of one or the other of these chains. The use of recombinant DNA technology has elucidated the nature of the molecular defects in these dis-

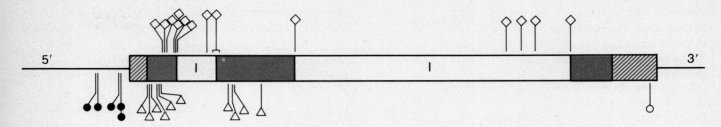

FIGURE 21.16

Location of mutations in the beta-globin gene producing beta thalassemia. The symbols
represent: frameshift and nonsense mutations (△), RNA processing mutations (◇),
transcriptional mutations (●), and RNA cleavage mutations (○). The hatched areas are
noncoding transcribed regions (producing a leader sequence); the shaded areas are coding
regions (exons); and the unshaded areas represent intervening sequences (introns).

orders and provided an insight into the complex nature of gene regulation in higher eukaryotes.

Beta thalassemia is a heterogeneous disease, with
absent (β^o) or defective (β^+) production of the beta-
globin polypeptide. The beta-globin gene contains
three exons and two introns. After transcription and
the addition of a 5′ cap and a 3′ poly A tail, enzymes
within the nucleus remove the introns, and the exons
are ligated together to form the mature mRNA molecule. This molecule is then transported to the cytoplasm for translation. In some forms of β^o thalassemia,
a single base change at an intron/exon junction prevents intron removal, and homozygous individuals
cannot properly splice the mRNA. Consequently, no
functional beta-globin protein is produced. In some
forms of β^+ thalassemia, a mutation creates a second
splice site that acts as an alternative to the one already
present. Random choice of the correct splice site allows some production of mRNA, and thus a reduced
level of beta globin is present.

To date, mutations have been described that affect
transcription, RNA processing and splicing, polyaden-
ylation, and translation into polypeptide chains. In addition, deletion of all or part of the beta-globin gene
has been described. Figure 21.16 shows the location of
30 beta thalassemia mutations that have been analyzed at the molecular level. The distribution of molecular defects indicates that mutations at this locus have
arisen many times in an independent fashion.

In **alpha thalassemia,** there is absent (α^o) or decreased (α^+) synthesis of alpha-globin polypeptides.
As one might expect, alpha thalassemia is also a heterogeneous disease at the molecular level. Each chromosome 16 bears two copies of the alpha-globin gene,
designated as α_1 and α_2, and therefore, since each normal individual has two copies of chromosome 16, four
copies of this gene are present. As in the beta gene,

each gene copy contains three exons and two introns.
Single base changes that result in unstable mRNA,
RNA processing defects, and premature chain termination have been described. In one form of alpha thalassemia, a single base change, from ATG to ACG, alters
the initiation codon and prevents translation of the
mRNA. Thus, the gene is normally transcribed, processed, and polyadenylated, but the mRNA cannot initiate translation, abolishing globin production from
this mutant alpha-globin gene. Although a number of
point mutants have been identified in the alpha-globin
complex, most of the mutations in alpha thalassemia
are the result of deletions within the alpha gene family. This high frequency of deletions may be due to the
high degree of DNA sequence homology that is present within the alpha gene cluster. This homology, also
extending into the 5′ flanking region, provides a large
number of sites for recombination and crossing over,
which may lead to deletions.

The diverse nature of mutations in the thalassemias
should serve to emphasize that gene regulation in eukaryotes can be accomplished in many ways. Mutations that affect the structural integrity of the gene, or
control transcription, RNA processing, polyadenyla-
tion, and translation can reduce or abolish the expression of a gene product. Within a normal cell, these
steps and many others offer an opportunity to regulate the expression of a gene.

SUMMARY AND CONCLUSIONS

If the complete genome is not to be continuously active in transcription throughout the life of every cell of
all species, a system of genetic regulation must exist.
Such a system must be carefully coordinated if energetic efficiency is to be maintained in prokaryotes and
if the specialization associated with differentiation of

eukaryotes is to be accomplished. From all that we know, there is no question that highly refined mechanisms for the regulation of transcription of the genetic information have evolved in both prokaryotes and eukaryotes.

In bacteria, both inducible and repressible operons have been documented and studied. Both involve genes of a regulatory nature in addition to structural genes that code for the enzymes of the system. In the inducible *lac* operon, the regulatory gene produces an allosteric repressor molecule. In the absence of the inducer, lactose, the repressor binds to the operator DNA adjacent to the structural genes, inhibiting transcription. If lactose is present, it binds to the repressor, causing a conformational change and preventing binding of the repressor to the operator. As a result, the structural genes are transcribed. Inducible systems control catabolic processes, producing enzymes to metabolize the substrate only when the inducer is present.

The inducible mode of regulation was proposed in the operon model of Jacob and Monod. Studies of many types of mutations that altered regulation of lactose metabolism served as the basis for the model. Importantly, the operon model allowed for numerous predictions, including gene expression in a variety of partial diploid organisms created by sexduction. In each case, the model was supported. The most critical prediction was the presence of a repressor molecule. Hypothesized strictly on genetic grounds, the molecule has been successfully isolated and shown to be a protein with quaternary structure.

Glucose has also been shown to influence the expression of the *lac* structural genes. Since this simpler monosaccharide can be utilized more efficiently than lactose, another mechanism, called catabolite repression, has evolved that represses the operon when glucose is present. A complex set of molecular interactions involving the CAP protein, adenyl cyclase, cAMP, and the CAP site of the promoter region can either facilitate or inhibit transcription. The promoter, a region of DNA to the left of the operator, is the site of initial binding of the sigma subunit of RNA polymerase.

Repressible systems, on the other hand, control anabolic processes such as tryptophan synthesis. When a high concentration of tryptophan is present, transcription is repressed. An allosteric repressor is made which binds to operator DNA only when it is complexed with the end product of biosynthesis. If tryptophan is not present, the uncomplexed repressor cannot bind to operator DNA. Thus, transcription occurs, resulting in the biosynthesis of this amino acid.

During the study of a constitutive mutation, still another controlling element—the attenuator—was discovered in the *trp* operon. Nucleotide sequences located in a leader sequence of DNA occurring prior to the structural genes apparently lead to termination of transcription. The presence of tryptophan somehow facilitates termination events, causing an inhibition of transcription. When these sequences are deleted, attenuation by tryptophan is relieved.

Throughout the above discussion of regulation, we have distinguished between negative control, where a repressor turns genes off, and positive control, where an activator turns genes on. We have included in our coverage of prokaryotic regulation a discussion of the ara regulator. This protein exerts both positive and negative control over expression of the genes specifying the enzymes that metabolize the sugar arabinose.

Several interesting examples of transcription regulation have been studied in viruses. When bacteria are lysogenized, the viral genome integrates into the bacterial chromosome and is repressed. Repression of transcription in phage lambda has been shown to be due to one of its own gene products (cI), which behaves as a repressor. A second gene product, cro, acts to inhibit the transcription of *cI* and promotes the lytic pathway. The interactions between these proteins and regulatory sites on DNA have now been very carefully worked out.

When bacteriophages are infectious, they must rely on the host RNA polymerase to initiate transcription of their genome. Interestingly, certain viral promoter sequences are recognized by the host polymerase. One of the early gene products in phage T7 is the RNA polymerase that transcribes the remainder of the genome. We described a more complex type of control for the SPO1 phage in *B. subtilis*.

Because they are more complex and contain more genetic material, eukaryotes are assumed to have a more complex system of regulation. The discovery of operon systems in bacteria prompted a search for similar control systems in eukaryotes. In the unicellular yeast, adaptive regulation of galactose metabolism has been documented and regulation has been linked to specific genes and proteins. However, a comprehensive understanding of this system has not yet been attained.

Britten and Davidson have proposed a model that relates repetitive DNA sequences and their transcripts to regulation in eukaryotes. In this model, they proposed that a series of interacting units comprising batteries of genes is controlled by varying signals, including the transcripts of repetitive DNA. The model predicts that short repeated nucleotide sequences should be present at the 5′ end of eukaryotic genes under coordinate control. Several examples in yeast,

Drosophila, and chickens have been discussed, providing support for the model.

We then turned our attention to a consideration of the eukaryotic promoter and a unique regulatory element called the enhancer. While upstream sequences, including the TATA and CCAAT boxes, are part of the promoter, enhancer sequences may be upstream or downstream and even inverted while stimulating transcription. Initially discovered in the animal virus SV40, cellular enhancers have been recognized in genes encoding immunoglobulins, beta globins, and 5S rRNA, among others. It has been speculated that enhancer DNA sequences form Z-DNA, which is related to the induction of transcription.

Another approach to eukaryotic regulation has involved the study of hormones known to activate genes. The estrogen-oviduct system in chickens has been particularly well studied. In this system, the hormone forms a complex with a cytoplasmic receptor protein, which then interacts directly with proteins of chromatin to activate transcription of specific mRNAs.

We also examined the effects of proteins intimately associated with DNA in chromatin. Both the histone and nonhistone proteins have been implicated in gene regulation. However, histones seem to lack heterogeneity and thus specificity in regulating large numbers of genes. Nonhistone proteins are better candidates, but studies have not progressed sufficiently for their acceptance as gene regulators.

In the final sections of the chapter, opportunities for post-transcriptional genetic regulation in eukaryotes were considered. First, alternate patterns of splicing identical RNA transcripts of a single gene were considered and examples cited. Then, the diverse group of inherited anemias referred to as the thalassemias was introduced. The variety of errors that may occur during the expression of alpha- and beta-globin genes indicates the many different instances where genetic regulation can be accomplished.

PROBLEMS AND DISCUSSION QUESTIONS

1 Contrast the need for the enzymes involved in the metabolism of lactose and tryptophan in bacteria in the presence and absence of lactose and tryptophan, respectively.

2 Why is gene regulation assumed to be more complex in a multicellular eukaryote than in a prokaryote? Why is the study of this phenomenon in eukaryotes more difficult?

3 Contrast the role of the repressor in an inducible system and in a repressible system.

4 Describe how the *lac* repressor was isolated. What properties demonstrate it to be a protein? Describe the evidence that it indeed serves as a repressor within the operon scheme.

5 Even though the *z, y,* and *a* structural genes are transcribed as a single polycistronic mRNA, each gene contains the appropriate initiation and termination signals essential for translation. Predict what will happen when a cell growing in the presence of lactose contains a deletion of one nucleotide early in the z gene.

6 For the following *lac* genotypes, predict whether the structural genes (z) are constitutive, permanently repressed, or inducible in the presence of lactose.

Genotype	Constitutive	Repressed	Inducible
$i^+o^+z^+$	—	—	×
$i^-o^+z^+$			
$i^+o^cz^+$			
$i^-o^+z^+/F'i^+$			
$i^+o^cz^+/F'o^+$			
$i^so^+z^+$			
$i^so^+z^+/F'i^+$			

7 For the following genotypes and condition (lactose present or absent), predict whether functional enzymes are made, nonfunctional enzymes are made, or no enzymes are made.

Genotype	Condition	Functional Enzyme Made	Nonfunctional Enzyme Made	No Enzyme Made
$i^+o^+z^+$	No lactose	—	—	×
$i^+o^c z^-$	Lactose			
$i^-o^+z^-$	No lactose			
$i^-o^+z^-$	Lactose			
$i^-o^+z^+/F'i^+$	No lactose			
$i^+o^c z^+/F'o^+$	Lactose			
$i^+o^+z^-/F'i^+o^+z^+$	Lactose			
$i^-o^+z^-/F'i^+o^+z^+$	No lactose			
$i^s o^+z^+/F'o^+$	No Lactose			
$i^+o^c z^-/F'o^+z^+$	Lactose			

8 In a theoretical operon, genes *a*, *b*, *c*, and *d* represent the repressor gene, the promoter sequence, the operator gene, and the structural gene, but not necessarily in that order. This operon is concerned with the metabolism of a theoretic molecule (tm). From the data given below, first decide if the operon is inducible or repressible. Then assign *a*, *b*, *c*, and *d* to the four parts of the operon. (AE = active enzyme; IE = inactive enzyme; NE = no enzyme)

Genotype	tm Present	tm Absent
$a^+b^+c^+d^+$	AE	NE
$a^-b^+c^+d^+$	AE	AE
$a^+b^-c^+d^+$	NE	NE
$a^+b^+c^-d^+$	IE	NE
$a^+b^+c^+d^-$	AE	AE
$a^-b^+c^+d^+/F'a^+b^+c^+d^+$	AE	AE
$a^+b^-c^+d^+/F'^+b^+$	AE	NE
$a^+b^+c^-d^+/F'a^+b^+c^+d^+$	AE + IE	NE
$a^+b^+c^+d^-/F'a^+b^+c^+d^+$	AE	NE

9 If the *cI* gene of phage λ contained a mutation with effects similar to the i^- mutation in *E. coli*, what would be the result?

10 Contrast the characteristics of histones and nonhistones as related to their candidacy as regulators of gene expression.

11 Distinguish between genetic regulation resulting from positive control versus negative control. Cite as many examples of each as you can.

12 Distinguish between the regulatory elements referred to as promoters and enhancers.

13 A bacterial operon is responsible for the production of the biosynthetic enzymes needed to make a theoretical amino acid tisophane (tis). The operon is regulated by a separate gene, *r*. Deletion of the *r* gene causes the loss of synthesis of the enzymes. In the wild-type condition, when tis is present, no enzymes are made. In the absence of tis, the enzymes are made. Mutations in the operator gene (O^-) result in repression regardless of the presence of tis.

Is the operon under positive or negative control? Propose a model for (1) repression of the genes in the presence of tis in wild-type cells; and (2) the O^- mutations.

SELECTED READINGS

ANDERSON, J. E., PTASHNE, M., and HARRISON, S. C. 1985. A phage repressor-operator complex at 7 Å resolution. *Nature* 316: 596–601.

ANTONARAKIS, S., KAZAZIAN, H., and ORKIN, S. 1985. DNA polymorphisms and molecular pathology of the human globin gene clusters. *Hum. Genet.* 69: 1–14.

AZORIN, F., and RICH, A. 1985. Isolation of Z-DNA binding proteins from SV40 minichromosomes: Evidence for binding to the viral control region. *Cell* 41: 365–74.

BECKWITH, J., and ROSSOW, P. 1974. Analysis of genetic regulatory mechanisms. *Ann. Rev. Genet.* 8: 1–13.

BECKWITH, J. R., and ZIPSER, D., eds. 1970. *The lactose operon.* Cold Spring Harbor, N.Y.: Cold Spring Harbor Laboratory.

BERTRAND, K., et al. 1975. New features of the regulation of the tryptophan operon. *Science* 189: 22–26.

BONNER, J. 1965. *The molecular biology of development.* London: Oxford University Press.

BRITTEN, R. J., and DAVIDSON, E. H. 1969. Gene regulation for higher cells: A theory. *Science* 165: 349–57.

CHARNAY, P., et al. 1984. Differences in human α and β-globin gene expression in mouse erythroleukemia cells: The role of intragenic sequences. *Cell* 38: 251–63.

DARNELL, J. E. 1982. Variety in the level of gene control in eukaryotic cells. *Nature* 297: 365–71.

DAVIDSON, E. H., and BRITTEN, R. J. 1973. Organization, transcription and regulation in the animal genome. *Quart. Rev. Biol.* 48: 565–613.

———. 1979. Regulation of gene expression: Possible role of repetitive sequences. *Science* 204: 1052–59.

DAVIDSON, E. H., JACOBS, H. T., and BRITTEN, R. J. 1983. Very short repeats and coordinate induction of genes. *Nature* 301: 468–70.

ENGLESBERG, E., and WILCOX, G. 1974. Regulation: Positive control. *Ann. Rev. Genet.* 8: 219–42.

GEORGIEV, G. P. 1969. Histones and the control of gene action. *Ann. Rev. Genet.* 3: 155–80.

GILBERT, W., and MÜLLER-HILL, B. 1966. Isolation of the *lac* repressor. *Proc. Natl. Acad. Sci.* 56: 1891–98.

———. 1967. The *lac* operator in DNA. *Proc. Natl. Acad. Sci.* 58: 2415–21.

GILBERT, W., and PTASHNE, M. 1970. Genetic repressors. *Scient. Amer.* (June) 222: 36–44.

GILLIES, S. D., et al. 1983. A tissue-specific transcription enhancer element is located in the major intron of a rearranged immunoglobulin heavy chain gene. *Cell* 33: 717–28.

GINIGER, E., VARNUM, S. M., and PTASHNE, M. 1985. Specific DNA binding of GAL 4, a positive regulatory protein of yeast. *Cell* 40: 767–74.

GOLDBERGER, R. F., DEELEY, R. G., and MULLINIX, K. P. 1976. Regulation of gene expression in prokaryotic organisms. *Adv. in Genet.* 18: 1–67.

GROSS, S. R. 1969. Genetic regulatory mechanisms in fungi. *Ann. Rev. Genet.* 3: 395–424.

GUARENTE, L. 1984. Yeast promoters: Positive and negative elements. *Cell* 36: 799.

HERSHEY, A. D., ed. 1971. *The bacteriophage lambda.* Cold Spring Harbor, N.Y.: Cold Spring Harbor Laboratory.

HUANG, R. C., and BONNER, J. 1962. Histone, a suppressor of chromosomal RNA synthesis. *Proc. Natl. Acad. Sci.* 48: 1216–22.

JACOB, F., and MONOD, J. 1961. Genetic regulatory mechanisms in the synthesis of proteins. *J. Mol. Biol.* 3: 318–56.

JOHNSON, A. D., et al. 1981. λ repressor and *cro*—Components of an efficient molecular switch. *Nature* 294: 217–23.

JOHNSTON, H. M., et al. 1980. Model for regulation of the histidine operon of *Salmonella*. *Proc. Natl. Acad. Sci.* 77: 508–12.

KELLER, E. B., and CALVO, J. M. 1979. Alternative secondary structures of leader RNAs and the regulation of the *trp, phe, his, thr,* and *leu* operons. *Proc. Natl. Acad. Sci.* 76: 6186–90.

KHOURY, G., and GRUSS, P. 1983. Enhancer elements. *Cell* 33: 313–14.

KOLTER, R., and YANOFSKY, C. 1982. Attenuation in amino acid biosynthetic operons. *Ann. Rev. Genet.* 16: 113–34.

LEWIN, B. 1974. Interaction of regulator proteins with recognition sequences of DNA. *Cell* 2: 1–7.

————. 1985. *Genes.* 2nd ed. New York: Wiley.

MANIATIS, T., and PTASHNE, M. 1976. A DNA operator-repressor system. *Scient. Amer.* (Jan.) 234: 64–76

MILLER, J. H., and REZNIKOFF, W. S. 1978. *The operon.* Cold Spring Harbor, N.Y.: Cold Spring Harbor Laboratory.

NABESHIMA, Y., et al. 1984. Alternative transcription and two modes of splicing result in two myosin light chains from one gene. *Nature* 308: 333–38.

NAWA, H., KOTANI, H., and NAKANISHI, S. 1984. Tissue-specific generation of two preprotachykinin mRNAs from one gene by alternative RNA splicing. *Nature* 312: 729–34.

NORDHEIM A., and RICH, A. 1983. Negatively supercoiled simian virus 40 DNA contains Z-DNA segments within transcriptional enhancer sequences. *Nature* 303: 674–79.

OCHOA, S. 1979. Regulation of protein synthesis. *CRC Crit. Rev. Biochem.*, November, pp. 7–22.

O'MALLEY, B. W., and BIRNBAUMER, L., eds. 1978. *Receptors and hormone action.* New York: Academic Press.

O'MALLEY, B. W., and MEANS, A. R. 1974. Female steroid hormones and target cell nuclei. *Science* 183: 610–20.

O'MALLEY, B. W., TOWLE, H. C., and SCHWARTZ, R. J. 1977. Regulation of gene expression in eukaryotes. *Ann. Rev. Genet.* 11: 239–76.

PABO, C. O., and SAUER, R. T. 1984. Protein-DNA recognition. *Ann. Rev. Biochem.* 53: 293–321.

PETRUSEK, R. H., DUFFY, J., and GEIDUSCHEK, E. 1976. Control of gene action in phage SPO1 development: Phage-specific modifications of RNA polymerase and a mechanism of positive control. In *RNA polymerase*, eds. R. Losick and M. Chamberlin, pp. 567–85. Cold Spring Harbor, N.Y.: Cold Spring Harbor Laboratory.

PICARD, D., and SCHAFFNER, W. 1984. A lymphocyte-specific enhancer in the mouse immunoglobulin κ gene. *Nature* 307: 80–82.

PLATT, T. 1981. Termination of transcription and its regulation in the tryptophan operon of *E. coli. Cell* 24: 10–23.

PTASHNE, M. 1986. *A genetic switch.* Palo Alto: Blackwell Scientific Publ.

PTASHNE, M. 1967. Isolation of the phage repressor. *Proc. Natl. Acad. Sci.* 57: 306–13.

PTASHNE, M., et al. 1976. Autoregulation and function of a repressor in bacteriophage lambda. *Science* 194: 156–61.

————. 1980. How the λ repressor and *cro* work. *Cell* 19: 1–11.

PTASHNE, M., JOHNSON, A. D., and PABO, C. O. 1982. A genetic switch in a bacterial virus. *Scient. Amer.* (Nov.) 247: 128–40.

REZNIKOFF, W. S. 1972. The operon revisited. *Ann. Rev. Genet.* 6: 133–56.

SERFLING, E., JASIN, M., and SCHAFFNER, W. 1985. Enhancers and eukaryotic gene transcription. *Trends in Genetics* 1: 224–30.

STEIN, G. S., SPELSBERG, T. C., and KLEINSMITH, L. J. 1974. Nonhistone chromosomal proteins and gene regulation. *Science* 183: 817–24.

STEIN, G. S., STEIN, S. S., and KLEINSMITH, L. J. 1975. Chromosomal proteins and gene regulation. *Scient. Amer.* (Feb.) 232: 2, 46–57.

STENT, G. S., and CALENDAR, R. 1978. *Molecular genetics: An introductory narrative.* 2nd ed. San Francisco: W. H. Freeman.

STROYNOWSKI, I., and YANOFSKY, C. 1982. Transcript secondary structures regulate transcription termination at the attenuator of *S. marcescens* tryptophan operon. *Nature* 298: 34–38.

STUDIER, F. W. 1972. Bacteriophage T7. *Science* 176: 367–76.

UMBARGER, H. E. 1978. Amino acid biosynthesis and its regulation. *Ann. Rev. Biochem.* 47: 533–606.

WEISBROD, S. 1982. Active chromatin. *Nature* 297: 289–95.

YANOFSKY, C. 1981. Attenuation in the control of expression of bacterial operons. *Nature* 289: 751–58.

PART FIVE
GENETICS OF ORGANISMS AND POPULATIONS

22

The Role of Genes in Development

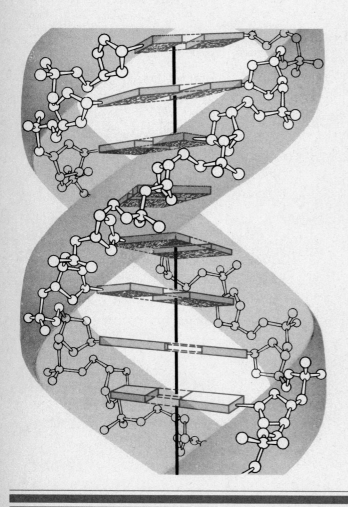

In multicellular plants and animals, a fertilized egg, without further stimulus, begins a cycle of developmental events that ultimately give rise to an adult member of the species from which the egg and sperm were derived. Thousands, millions, or even billions of cells are organized into a cohesive and coordinated unit that we perceive as a living organism. The heterogeneous series of events whereby organisms attain their final adult form is studied by developmental biologists. This area of study is perhaps the most intriguing in biology because comprehension of developmental processes requires knowledge of many biological disciplines. In fact, once development is understood, there may be little left unknown about life processes.

In the past one hundred years, investigations in embryology, genetics, biochemistry, molecular biology, cell physiology, and biophysics have contributed to the study of development. Largely, the findings have pointed out the tremendous complexity of development. Unfortunately, this description of what actually happens does not answer the "why" and "how" of development. Nevertheless, further hypotheses and experimentation will be based on this knowledge and will lead us closer to a comprehensive understanding of development.

In this chapter the primary emphasis will be on the role of genetic information during development. Because genetic information directs cellular function and determines the cellular phenotype, the genetics of development is being actively studied using a wide range of experimental organisms and techniques.

THE USE OF SIMPLE MODELS

Studies of viruses, bacteria, and the less complex eukaryotes have yielded interesting information pertinent to development. These organisms are used because they are more amenable to experimental approaches. In contrast to complex multicellular eukaryotes, these organisms may be cultured under controlled conditions. In addition, their biochemical and developmental responses are less dependent on the interactions of many differentiated cell and tissue types. Thus, before describing more detailed organisms, we shall briefly discuss two model systems that have been studied in some detail. In this discussion we will also illustrate that developmental events in apparently simple systems are in reality complex mechanisms requiring the coordinated action of many gene sets.

Bacterial Sporulation

Certain gram-positive bacilli have developed a mechanism of survival under adverse conditions—the formation of spores. These structures are highly resistant to heat, desiccation, and toxic conditions, and are in-

duced to form in unfavorable environments. Spores are formed within the bacterial cell over a period of several hours (Figure 22.1). When maturation is completed, they possess no metabolic activity; thus, they resemble the seeds of plants. Under favorable conditions, dormancy may be broken and germination may give rise to a vegetative bacterial cell.

Through cytological, biochemical, and molecular investigations, events related to the origin of spores have been elucidated. Following DNA synthesis and chromosome replication, invagination of the cell membrane occurs at one end of the cell, eventually enclosing one of the chromosomes and some of the cytoplasm (Figure 22.1). The partitioning double membrane is called the **spore septum (SPS).** Subsequently, the forming spore becomes independent of the cell membrane of the mother cell and a tough protein layer is laid down, encapsulating the spore.

A number of biochemical events also occur during the sporulation process, including those that are essential and spore-specific; those that are dispensable and spore-specific; and those that are essential in both vegetative and sporulating cells. These biochemical

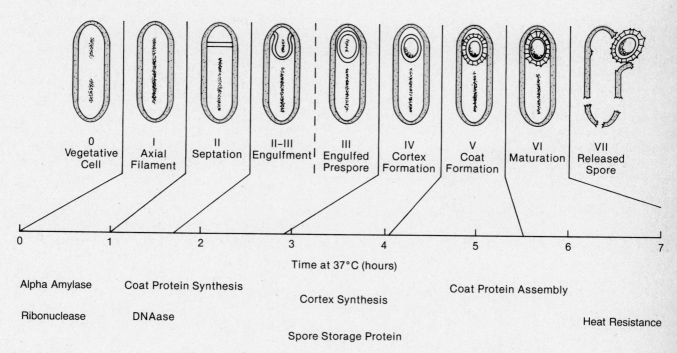

FIGURE 22.1

Morphological stages and some of the biochemical events during sporulation in species of the bacterium *Bacillus.* [With permission from M. Young and J. Mandelstam, 1979, "Early events during bacterial endospore formation," *Adv. Microbiol. Physiol.* 20: 106. Copyright: Academic Press Inc. (London) Ltd.]

events are the result of programmed gene action, under the control of a group of regulatory genes known as the *spoO* loci. Significant to developmental studies are the shifts in genetic activity that take place during sporulation. In the early stages of development at least 20 loci—all of which are spore-specific—are switched on, and transcription of the genes normally used in vegetative growth is shut down. Although the exact sequence of the biochemical events is not yet known, some of the steps are shown in Figure 22.1. Many of the products appearing in a stage-specific fashion, such as spore coat proteins, are used directly in spore production. Others may serve as triggers for the next series of events. If RNA synthesis is inhibited experimentally during sporulation, the entire process is interrupted, thus documenting the dependency of the sporulation process on activity of a different set of genes. As sporulation moves beyond Step II, the stage where two cells are formed, different genes are expressed in the prespore and the mother cell. As differentiation proceeds, the spore retains only a small number of ribosomes, the enzymes necessary for germination, and a copy of the bacterial chromosome.

Recent studies have provided insight into the basis of sporulation. They indicate that a different form of

RNA polymerase is responsible for spore-specific transcription than is used in the same process in vegetative cells. Recall that RNA polymerase is a large enzyme composed of six subunits, including the sigma (σ) factor, which is responsible for the initiation of transcription. Experimental evidence has suggested that during sporulation the vegetative form of sigma is inactivated and a sigma factor specific for genes transcribed during sporulation is utilized.

Thus, a molecular explanation has been proposed for this one type of differentiation, but this apparently simple example of morphogenesis is, in fact, genetically complex. It involves five or more classes of temporally regulated genes, at least six species of sigma factor, and the transcription of 40 to 60 loci. Furthermore, the mechanisms of regulation and interaction of gene products have not yet been fully elucidated.

Slime Mold Development

The cellular slime mold *Dictyostelium discoideum* is a unique organism in studies of development. As shown in Figure 22.2, the slime mold exists in one of two alternating forms during its life cycle. Germinating spores give rise to eukaryotic amoeboid cells called **myxamoebae.** These individual cells feed on bacteria,

FIGURE 22.2
Life cycle of the slime mold *Dictyostelium discoideum.* (Reprinted by permission from Watson, James D., *Molecular biology of the gene,* 3rd ed., Menlo Park, California, The Benjamin/ Cummings Publishing Company, 1976, Fig. 17.7.)

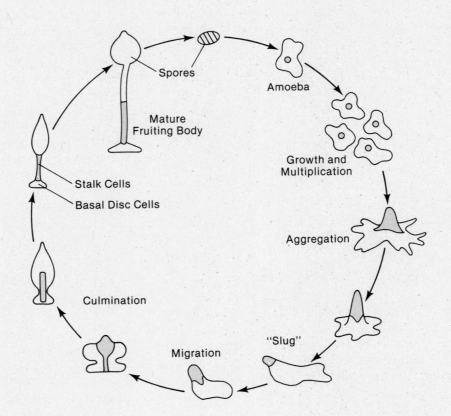

Spores

Amoeba

Mature Fruiting Body

Growth and Multiplication

Stalk Cells

Basal Disc Cells

Aggregation

Culmination

"Slug"

Migration

grow, and multiply in numbers. However, when nutrients become depleted, cyclic AMP is secreted by these cells, which causes them to aggregate and form a multicellular body or slug consisting of approximately 10^5 cells. Two distinct cell types form in the slug, depending on their position in the aggregate. Those at the anterior end become prestalk cells. About 80 percent of the total cells, located at the posterior end, become prespore cells (Figure 22.2). When the spores mature and are released, germination produces myxamoebae and the cycle is repeated.

The slime mold life cycle is an ideal model for investigation of differentiation. Cellular structure and function are completely transformed during the conversion of a myxamoeba cell into a portion of the fruiting body. The proteins present in the individual cells are broken down, and new synthesis of different RNA and protein molecules occurs. In addition, there are qualitative differences in biosynthetic activity between spore-forming cells and stalk cells. As shown in Figure 22.2, the fate of any myxamoeba in the fruiting body is determined by its position and rate of entrance into the intial aggregate. Thus, all cells must have the genetic potential for becoming any part of the fruiting body.

In order to demonstrate genetic control of the developmental events leading to the formation of the fruiting body, biologists have isolated mutations that affect the process in various ways. For example, they have discovered mutants with diverse phenotypic variation in the fruiting body, including effects on upright growth, size, stalk texture, pigmentation, and so on. Such mutations demonstrate that gene products are involved in the phenotypic appearance of the fruiting body.

Still other mutations either block the life cycle at some specific stage or result in only a partial fruiting body. The first class includes strains in which aggregation is prevented or development arrests at the aggregate or the slug stage. Presumably, gene products are essential to the orderly progression of the life cycle. In the second case, one part or another of the fruiting body forms incompletely. A stalk without spores or spores without the stalk are examples.

These mutations demonstrate that a programmed sequence of gene activation is essential for the complete life cycle of *Dictyostelium.* Thus, specific gene products accompany the differentiation of single cells to a multicellular aggregate that gives rise to the fruiting body. However, we do not yet know what causes a cell in the middle of the aggregate to synthesize products specific to the stalk and not to become spore-forming. Because *Dictyostelium* is less complex than higher eukaryotic forms, further study of this organism will help to decipher still other aspects of development.

THE VARIABLE GENE ACTIVITY THEORY

From a genetic perspective, development may be described as the attainment by cells of a differentiated state. **Differentiation** refers to a condition of specialized cellular structure and function. For example, an erythrocyte almost exclusively active in hemoglobin synthesis is differentiated, but a blastula cell is undifferentiated. In order to accomplish this specialization, certain genes are actively transcribed but many other genes are not. Because most eukaryotic organisms are composed of a large number of cell types, differential transcriptive capacities characterize the many functionally diverse cells.

The concept of differential transcription has led to the **variable gene activity hypothesis** of differentiation. This theory was first entertained by Thomas H. Morgan in 1934, later proposed by Edgar and Ellen Stedman as well as Alfred Mirsky in the 1950s, and articulated in modern form by Eric Davidson in *Gene Activity in Early Development.* The theory holds that the forms of differentiation assumed by any specific cell type are qualitatively determined by those genes that are actively transcribed. Its underlying assumption is that each cell contains the entire diploid genome in the nucleus and that only certain gene products characteristic of the cell type are produced. The rest of the genome is actively shut down and not transcribed.

Given our knowledge of molecular biology, the variable gene activity theory is readily acceptable. Every cell is a biochemical entity composed of macromolecules which are either informational, structural, catalytic, or metabolic in nature. In an organized fashion, these molecules constitute the cell. The presence or absence of any molecule in the cell is thus directly or indirectly influenced by gene activity or inactivity. Therefore, a cell is what its active genes direct it to be.

The variable gene activity theory is a very useful model system for experimental design. The remainder of this chapter examines the validity of this theory and its premises and provides examples that offer experimental support. It is important to remember, however, that this type of approach initially tells us what happens. We still do not know why certain genes are active and others inactive and why qualitatively different gene activity occurs between different cells.

Genomic Equivalence

The variable gene activity hypothesis is based on the premise that in multicellular organisms, somatic cells all contain a complete diploid set of genetic information. While biochemical and cytophotometric analysis has shown the quantity of DNA to be equivalent and equal to the diploid content in cells of most species studied, other investigations have approached the question differently. Is it possible to show that differentiated cells are qualitatively equivalent? In other words, is a complete set of genes present in each cell?

One approach is to show that a differentiated cell is **totipotent** and thus capable of giving rise to a complete organism under the proper conditions. In the early part of this century, Hans Spemann demonstrated that a nucleus derived from the 16-cell stage of a newt embryo was capable of supporting total development. By constricting a newly fertilized egg prior to the first cell division, Spemann isolated the zygote nucleus into one half of the cell. The half containing the nucleus proceeded to undergo division and produce 16 cells. At that point, the constriction was loosened and one of the 16 nuclei was allowed to pass back into the nonnucleated cytoplasm on the other side. Both halves subsequently produced complete but separate embryos. Therefore, at the 16-cell stage in this organism, genomic equivalence of the 16 nuclei was demonstrated.

Beginning in the 1950s, more sophisticated experiments were performed by Robert Briggs and Thomas King using the grass frog *Rana pipiens*, and by John Gurdon in the 1960s using an African frog, *Xenopus laevis*. After inactivating or surgically removing the nucleus from an egg, these investigators were able to test the developmental capacity of somatic nuclei by injecting nuclei isolated from cells at various stages of differentiation. Nuclei derived from the blastula stage of development were capable of supporting the development of complete and normal adults when transplanted in this way. In *Rana* (Figure 22.3), transplanted nuclei derived from later stages such as the gastrula and neurula usually allowed only partial development. In *Xenopus*, however, Gurdon's experiments showed that occasionally epithelial gut nuclei of tadpoles were able to support the development of an adult frog. In both cases, it is clear that, normally, progressive restrictions that may prevent the expression of totipotency are placed on differentiating nuclei.

To test whether the nucleus of a highly differentiated adult cell is irreversibly specialized or can support the development of a normal embryo, Gurdon used nuclei from adult frog skin cells in serial transplant experiments (Figure 22.4). In these experiments, a donor nucleus was transplanted into an enucleated egg, and the recipient was allowed to develop for a short time, for example, to the blastula stage. The blastula cells were then dissociated and a nucleus removed from one of them. This nucleus was transplanted into still another enucleated egg and development was allowed to occur. Such serial transfers were repeated a number of times. Subsequently, the blastula was not dissociated, but instead was allowed to continue development as far as it would go. After several such transfers, Gurdon found that 30 percent of the nuclear transplants resulted in the formation of advanced tadpoles (Table 22.1). Because nuclei from specialized epidermal cells can eventually direct the synthesis of gene products such as myosin, he-

FIGURE 22.3
In the process of nuclear transplantation of embryonic frog nuclei, the activated oocyte nucleus is moved out of the egg by a needle (1–3), and a nucleus from another embryo is collected and injected by micropipette (4–6). (From King, 1966, Fig. 1.)

Activation and
Enucleation

Nuclear
Transfer

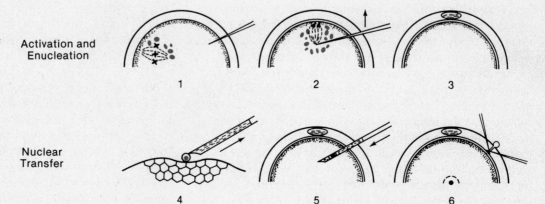

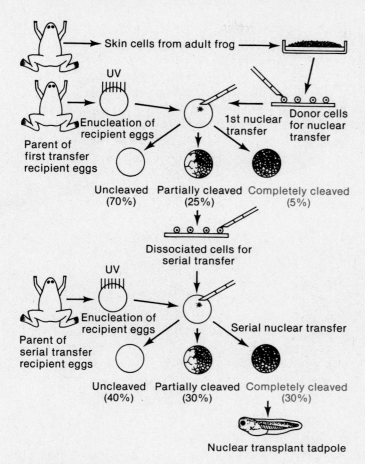

FIGURE 22.4

Serial transplant experiments using adult frog skin cell nuclei. Serial transplantation dramatically increases the percentage (from 5 to 30 percent) of recipient eggs that undergo complete cleavage. (From Gurdon et al., 1975, Fig. 2.)

TABLE 22.1

First transfer vs. serial transplantation in *Xenopus laevis*.

Donor Cells	Percentage Reaching Tadpole Stage with Differentiated Cell Types	
	1st Transfer	Serial Transfer
Intestinal epithelial cells from tadpoles	1.5	7
Cells from adult skin	0.037	8
Blastula or gastrula	36	57

SOURCE: From Gurdon, 1974, p. 24.

moglobin, and crystallin and promote the organization of cells and tissues into a tadpole, we are certain that such cells contain a complete copy of the genome. Under the proper circumstances, genes not expressed in the most specialized cells can become reactivated.

Recently, it has been demonstrated that nuclei from mammalian cells also exhibit totipotency. In 1981, Karl Illmensee and Peter Hoppe collected blastocyst embryos from a genetically marked strain of mice and dissociated them into single cells. Individual nuclei from the dissociated embryo were transplanted into fertilized but enucleated eggs from a different, genetically marked strain (Figure 22.5). The transplant embryos were cultured *in vitro* until they reached the blastocyst stage and then were implanted into the uterus of a foster female to allow the completion of development and birth. Analyses of enzyme patterns, karyotype, and coat color all confirmed that the liveborn mice were exclusively derived from the transplant nuclei. In addition, two of the transplant mice when mated to each other gave rise to normal progeny, all of which exhibited the phenotype of the original nuclear donor.

In plants, the totipotency of terminally differentiated nonmeristematic cells has been demonstrated using the carrot. Fredrick Steward observed that when individual phloem cells are explanted into a liquid culture medium, each cell divides and eventually forms a mass called a callus. Under appropriate conditions, this cell mass will differentiate into a mature plant.

These studies all convincingly argue that differentiated adult cells have not lost any of the genetic information present in the zygote. Instead, the majority of genes in any given cell type are shut off or repressed, but can be reprogrammed to direct normal development. Unfortunately, we do not yet understand the nature of the processes controlling nuclear differentiation during cellular specialization.

EVIDENCE OF DIFFERENTIAL TRANSCRIPTION IN SPECIALIZED CELLS

The entire thrust of the variable gene activity hypothesis of differentiation rests on the concept that differential gene transcription occurs (1) between different cells of an organism at the same time; and (2) within the same cell at different times during development. Although this concept is generally accepted today, we shall look at several lines of evidence that support differential transcription.

FIGURE 22.5
Nuclear transplantation into a fertilized mouse egg. (a) The fertilized egg is held against a suction pipette by the polar body to protect the egg. The injection pipette at left contains a blastocyst nucleus from a different, genetically marked mouse strain. Both the male (short arrow) and female (long arrow) pronuclei are visible. (b) The blastocyst nucleus is injected. (c) The male and female pronuclei are removed by the injection pipette. (d) The injection pipette containing the pronuclei is removed. (From Illmensee and Hoppe, 1981, Fig. 8. Copyright © The MIT Press. Photos provided by K. Illmensee, Laboratory of Cell Differentiation, University of Geneva.)

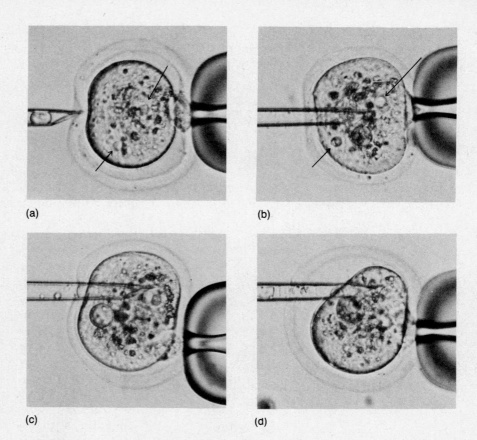

(a)

(b)

(c)

(d)

Chromosome Puffs

As pointed out in Chapter 11, several larval tissues in Dipteran insects contain cells housing giant chromosomes. These polytene chromosomes are composed of many DNA strands in parallel register, are complexed with protein, and show distinctive banding patterns (see Figure 11.6). Each band represents one or several genes. Puffing at particular banded regions is taken as an indication of transcription. Thus, a comprehensive study of puffing patterns during larval development of several tissues within single organisms has been very informative.

What is seen is that salivary gland chromosomes undergo a very specific puffing pattern with time. With great consistency, specific bands show puffing at precise times. As development proceeds, these regions show a regression of puffs, and new puffs are seen associated with different sets of bands [Figure 22.6(b)]. These changing patterns continue until pupation and are interpreted as representing differential synthesis of RNA and thus the transcription of specific but different genes over time.

Of particular interest are the studies performed by Wolfgang Beerman and Ulrich Clever using the fly *Chironomus*. They identified a genetic locus that specifies the production of proteinaceous granules in certain salivary gland cells. In one species, *Ch. pallidivittatus*, a granular secretion is seen. In a second species, *Ch. tentans*, the secretion is clear and lacks the granules. When granules are present (in *Ch. pallidivittatus*), specific puffing occurs on chromosome 4. When no granules are produced (in *Ch. tetans*), this specific puffing is absent.

To provide a firm correlation between the puffing (representing gene transcription) and the granular product, Beerman and Clever produced hybrids between the two species and examined the larval salivary glands. In hybrid cells, the amount of granular secretion is considerably reduced compared with that in *Ch. pallidivittatus*. Since a single salivary chromosome is composed of both members of a homologous pair of chromosomes, only part of the band should puff in the hybrid if it indeed is responsible for the granular product. This is what is observed! This work, summarized in Figure 22.6(a), provides strong support

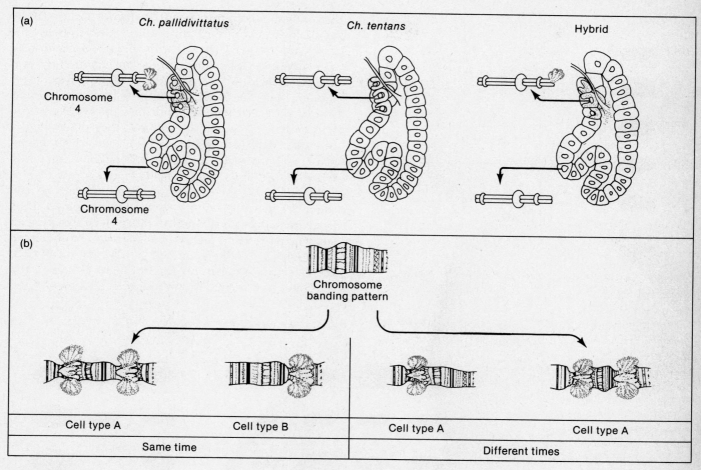

FIGURE 22.6

(a) Puffing patterns in parental species and hybrids of *Chironomus*. Salivary glands of *Ch. pallidivittatus* (left) produce granules. Chromosome 4 from the granule-producing cells has a puff at one end, but chromosome 4 from other cells of the same gland (lower left) has no puff at that location. *Ch. tentans* (center) makes no granules. No puffs at that location are produced in *Ch. tetans*. Hybrids of the two species (right) have a puff only on the chromosome from the *pallidivittatus* parent in the granule-producing cells, and these cells make far fewer granules. (b) Differential puffing patterns are evidence of differential gene activity. (From "Chromosome Puffs" by W. Beerman and U. Clever. Copyright © 1964 by Scientific American, Inc. All rights reserved.)

for the concept of differential transcription of the same gene between two species.

Isozymes

One direct means of demonstrating differential transcription is to show that different gene products are present in the cells of different tissues. One of the most direct approaches has involved studies of **isozymes.** Isozymes are multiple molecular forms of the same enzyme that may be distinguished electrophoretically because of differing net charge. In other words,

an enzyme that catalyzes a single reaction may exist in more than one form. Different forms of the same enzyme are possible when it is composed of separate polypeptide chains or subunits. If the enzyme is composed of a combination of nonidentical subunits, isozymes are observed.

This is best illustrated by examining the isozymic forms of the enzyme **lactate dehydrogenase (LDH).** Clement Markert has shown that this enzyme is a tetrameric molecule composed of four subunits. Each subunit in the molecule may be occupied by either an "A" polypeptide or a "B" polypeptide; thus, there are

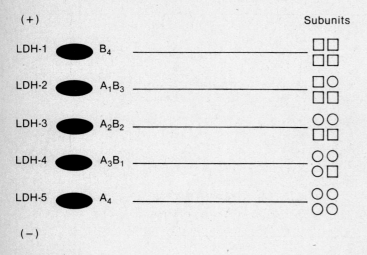

FIGURE 22.7

Zymogram of the multiple forms of the enzyme lactate dehydrogenase (LDH).

five different isozymes possible (Figure 22.7). LDH is ubiquitous among vertebrates and found in many invertebrates as well. This is to be expected because of its catalytic capacity for the interconversion of pyruvic and lactic acid, essential reactions in glycolysis. Even though both the A and B subunits have a molecular weight of 35,000 daltons (and thus the weight of the tetrameric form is 140,000 daltons), they differ slightly in amino acid composition and net electrical charge. Thus, any of the five isozymes will migrate differently when placed in an electric field.

When various tissues of the same species or the same tissues of different species are examined, LDH isozyme patterns are found to differ considerably. The isozyme patterns in Figure 22.8(a) show that either the A subunit (LDH-3,4,5) or the B subunit (LDH-1,2,3) seems to predominate in each of eight tissues of the rat. Since the A and B subunits are presumably the

FIGURE 22.8

(a) Tissue-specific pattern of LDH activity in the rat. (b) Developmental sequence of LDH in the rat heart. (From Clement L. Markert/ Heinrich Ursprung, *Developmental Genetics*, © 1971, pp. 43, 44. Reprinted by permission of Prentice-Hall, Inc., Englewood Cliffs, N.J.)

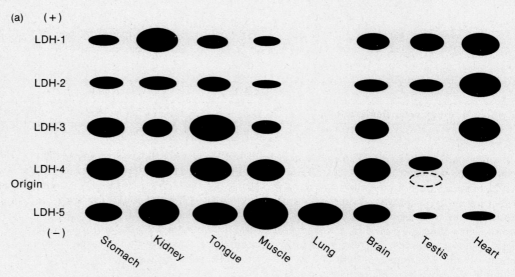

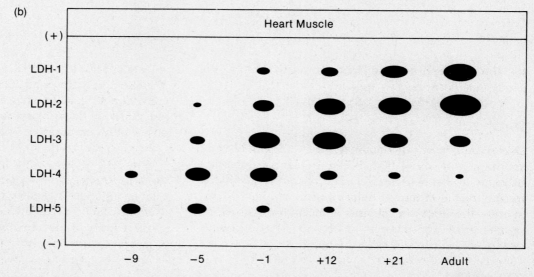

products of separate, distinct genes, these patterns directly reflect differential transcription.

In Figure 22.8(b), LDH isozymes are shown at various stages during the development of the rat heart. Prior to birth (-9, -5 days) LDH-5 predominates, indicative of A gene activity. Just before birth (-1 day) a binomial distribution is seen, indicating that both the A and B genes are equally transcribed. Following birth, there is a gradual shift toward the preponderance of LDH-1 and the B subunit, most notably in the adult. Thus, in the same tissue during development, the A and B genes are differentially transcribed. Similar examples are found in other tissues and between related species.

An even more remarkable variation of LDH is found in warm-blooded animals such as the pigeon. In primary spermatocytes of this species, a third subunit (designated C and clearly distinguishable from A and B) is found. The C-containing tetramer (C_4) is found only in primary spermatocytes and no other cells of the body. Following division and secondary spermatocyte formation, the C gene is turned off. The physiological advantage conferred by one isozymic form or another is not clear. However, isozymes do have slightly different kinetic and substrate specificities, as in the case of the C_4 tetramer.

Human Hemoglobins

The human genome is known to contain at least seven distinct genes coding for different protein chains that function as part of the oxygen-carrying molecule **hemoglobin** (see Figure 16.15). Synthesis of hemoglobin is performed by only one particular cell type of the many composing vascular tissue. Thus, even though many vascular cells and all other nonvascular cells in the body contain the hemoglobin genes in the genetic material, these genes are repressed. This represents another clear case of differential transcription.

The formation of the mature red blood cell is an interesting example of differentiation. Stem cells are continually dividing to produce immature erythroblasts. Neither the stem cell nor the erythroblast contains hemoglobin, but it is during the erythroblast stage that RNA synthesis occurs. The erythroblast gives rise to the reticulocyte, which loses its nucleus. It is this cell that synthesizes hemoglobin and eventually forms the mature erythrocyte or red blood cell. Thus, the hemoglobin mRNA is formed long before its participation in translation of the protein.

Of the seven different hemoglobin polypeptides, only one—the alpha (α) chain—is found throughout fetal and adult development. The two gamma (γ) chains are synthesized only during fetal development, and the beta (β) and delta (δ) chains are made follow-ing birth and throughout adult life. Two chains, epsilon (ϵ) and zeta (ζ), appear almost immediately after fertilization and are present only during the first several weeks of embryonic development. The functional hemoglobin molecule is itself a tetramer composed of two pairs of identical chains. The various combinations and designations are shown in Figure 16.8.

The seven genes responsible for the seven hemoglobin chains show different patterns of transcriptive activity during development. The γ gene is active during fetal development, shuts down shortly after birth, and is never again transcribed. The β and δ genes both become active prior to birth, but the β gene product is much more prominent as a part of HbA than is the δ product as part of HbA$_2$.

Clearly, gene regulation plays an important role during differentiation of even the same cell type—the mature erythrocyte. The variable synthesis of hemoglobin molecules is an important system for the study of the regulation of gene activity during development. Additionally, the relationship of the amino acid compositions of the gene products has served as an important topic in the study of evolutionary genetics. This topic will be pursued further in Chapter 26.

WHAT CAUSES DIFFERENTIAL TRANSCRIPTION DURING DEVELOPMENT?

We have now come to the central issue of developmental biology. In the preceding discussion, we established what is responsible for making cells different from one another (i.e., differential transcription). However, why one cell's fate is different from another's during development is a broader and more critical issue. This question is distinct from that discussed in Chapter 21, where we asked what actually turns any given gene on or off at the cellular level. We consider here the question of how cells become **determined** and eventually acquire one differentiated state or another.

As alluded to earlier, there is no simple answer to this question. However, information derived from the study of many different organisms provides a starting point. We shall examine several of these examples, which together reflect the direction and scope of our knowledge in this area.

The Effect of the Cellular and Extracellular Environment

The egg cytoplasm is complex, heterogeneous, and nonuniform in distribution within the cell. Following fertilization and early cell division, the nuclei of progeny cells will therefore find themselves in different environments as the maternal cytoplasm is distributed into the new cells. One hypothetical model suggests that these different forms of cytoplasm exert variable influences on the genetic material of different cells, causing differential transcription at some point during development. When a specific developmental fate for cells becomes fixed prior to the actual events of differentiation or specialization, the cells have undergone **determination,** an important concept in embryology and developmental biology. Determination is a regulatory event that establishes a specific pattern of gene activity that is characteristic of a given cell type. Although such cells show no evidence of structural or functional specialization, their position in the developing embryo or organism seems to determine the ultimate form of differentiation they will assume. It is as if their fate has been programmed prior to the actual events leading to specialization.

Early gene products further alter the total cytoplasmic content of each cell, producing a still different cellular environment that may in turn lead to the activation of other genes, and so on. In this way, cells embark on different pathways as development proceeds. As the number of cells increases, they influence one another. The total environment acting on the genetic material, therefore, now includes the cell's individual cytoplasm as well as the influence of other cells. Thus, as the developing organism becomes more complex, so does the total environment. In this way, different forms of determination and differentiation occur during development.

We have already discussed several examples that lend credence to the concept of cytoplasmic influence on nuclear activity. For example, Gurdon's nuclear transplantation work with *Xenopus* shows that nuclei from differentiated cell types are capable of reversing their role and serving as zygote nuclei. It may be concluded that the cytoplasmic environment of the enucleated egg exerts a profound influence on nuclear activity.

Similar conclusions may also be drawn from studies of somatic cell hybridization, which will be discussed in detail in Chapter 23. When cultured somatic cells are experimentally fused, two nuclei may exist in a common cytoplasm, forming a heterokaryon. One of the two nuclei usually conforms to the cytoplasm native to the other nucleus so that both nuclei become uniform in their degree of genetic activity.

On a more general environmental level, consider the examples of bacterial sporulation and of the aggregation and differentiation of slime mold myxamoebae discussed earlier in this chapter. In both cases, extracellular conditions trigger specific transcriptional events.

The formation of the body plan of the larva and adult of *Drosophila melanogaster* is an elegant example of the interaction between cytoplasm and nucleus. The body of *Drosophila* is composed of a number of head, thoracic, and abdominal segments. During embryogenesis, a series of nuclear divisions occurs immediately after fertilization. When nine divisions have occurred, the approximately 512 nuclei migrate to the egg's outer surface or cortex, where further divisions take place. The nuclei become enclosed in membranes, forming a single layer of cells (blastoderm) over the embryo surface. Each segment of the adult body is formed from the descendants of cells set aside at the blastoderm stage into discrete structures called **imaginal disks** (Figure 22.9). These disks, formed from

FIGURE 22.9

Imaginal disks of the *Drosophila* larva and the adult structures derived from them.

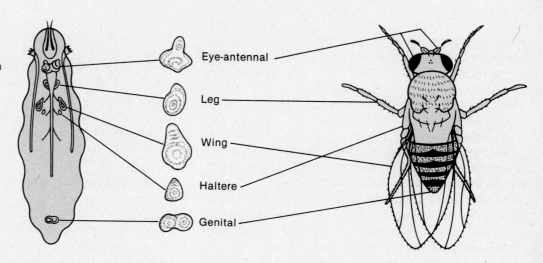

Eye-antennal

Leg

Wing

Haltere

Genital

FIGURE 22.10

Diagram showing the origin of *Drosophila* adult structures from embryonic blastoderm. (From Hotta and Benzer, 1972. Reprinted by permission from *Nature*, Vol. 240, pp. 528–29. Copyright © 1972 Macmillan Journals Limited.)

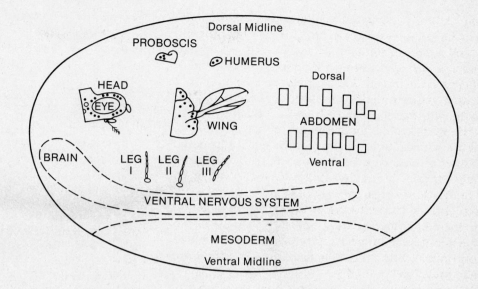

hollow sacs of cells, are determined to form specific parts of the adult body. There are 12 bilaterally paired disks and one genital disk. For example, there are eye-antennal disks, leg disks, wing disks, and so forth. Other cells in the blastoderm will form the internal and external structures of the larva. During the metamorphosis, which occurs during pupation, most of the body parts of the larva histolyze, or break down, and the imaginal disks differentiate into adult structures of the head, thorax, and abdomen.

Studies using a combination of physical and genetic techniques have revealed that the major features of the larval and adult body plans are present by the blastoderm stage of embryonic development in *Drosophila*. The two-dimensional projections of the external adult body parts on the surface of the blastoderm are known as **fate maps** (Figure 22.10).

One of the most striking features of fate maps is that the embryo is organized along an anterior-posterior axis that reflects the body axis of the larva, pupa, and adult (Figure 22.10). The existence of a fate map implies that the developmental fate is governed by the location to which nuclei migrate during blastoderm formation, and that positional information of an unknown nature directs the cleavage nuclei to adopt developmental programs that result in the morphologically distinct segments of the adult body.

At a slightly later stage of development, the precursors of the adult body structures are present as distinct morphological landmarks in the embryo (Figure 22.11). Even at this stage of development, cells in each segment are determined to form either an anterior or a posterior half or **compartment** of the segment. Progressive restrictions continue to occur throughout de-

velopment so that cells in each compartment are committed to forming a limited number of adult structures.

At the present time, very little is known about the nature of spatially distributed information or the mechanisms directing the development of nuclei migrating into a specific region of the cortex. One exception to this is the formation of the germ cells during embryonic development in *Drosophila*. The posterior

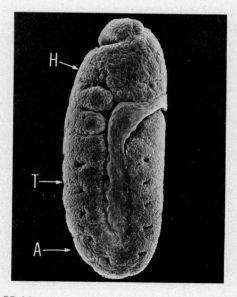

FIGURE 22.11

Scanning electron micrograph of an 8-hour *Drosophila* embryo, showing head (H), thorax (T), and abdominal (A) structures. (Photo by F. R. Turner.)

region of the egg contains cytoplasmic components known as **polar granules,** and nuclei that migrate into this region of the egg form pole cells, which are precursors of the germ cells of the adult. If this region of the egg is destroyed prior to nuclear migration, or if migration is prevented by ligation of the embryo, an adult fly still develops, but no germ cells are present and the fly is sterile. Similarly, the transplantation of polar plasm to anterior regions of the egg induces the formation of pole cells in regions where they are not usually formed. These induced pole cells can form functional germ cells when they are surgically removed and injected into the posterior region of genetically marked embryos. Thus, specific cytoplasm from the posterior pole of an egg propagates a determined state in nuclei that ultimately leads to a specific form of differentiation.

Although we have only rudimentary knowledge of the nature of positional information in the egg and the interaction of nuclei with these components, it is clear that these processes are under genetic control. It has been postulated that in an individual segment or compartment, one or more **selector genes,** cued by positional information, determine a particular developmental pathway for the cells in that compartment. In different compartments and segments, different combinations of selector genes are activated, producing the differentiated structures of the adult body.

Mutations of this selection system, called **homoeotic mutants,** shift the determined state of a group of cells in a compartment or segment to form structures normally seen in another segment. Thus, the mutant *tumorous head* transforms the head into the last abdominal segment and genitalia. The most intensively studied selector gene systems in *Drosophila* are two groups of homoeotic mutants clustered on chromosome 3: the *Antennapedia* complex (*ANT-C*), and the *bithorax* complex (*BX-C*). Work by Tom Kaufmann on *ANT-C* and by E. B. Lewis on *BX-C* suggests that *Antennapedia* controls segmentation of the head and anterior regions of the fly, while *bithorax* acts to regulate the development of the middle and posterior regions of the body. The genes of the *bithorax* complex consist of eight or more genes arranged on the chromosome in an order that corresponds to the anterior-posterior arrangement of body segments in which they are active (Figure 22.12). These genes normally act to direct the correct developmental pathway for thoracic and abdominal segments (Figure 22.13). According to E. B. Lewis, who has studied these genes for more than twenty-five years, none of the genes are active in the second thoracic segment (mesothorax). But, development of the third thoracic segment requires the first gene to be activated, and in each subsequent

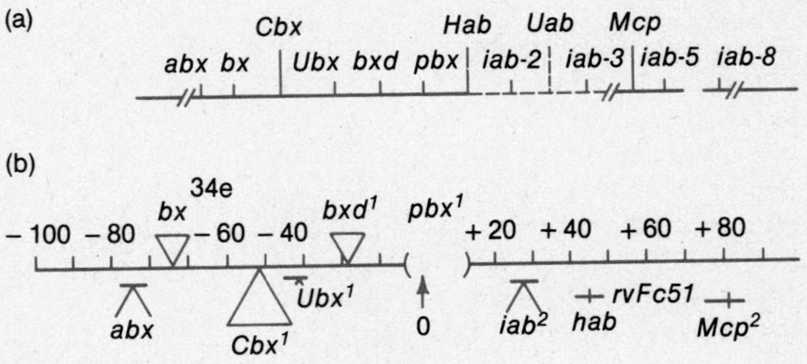

FIGURE 22.12

(a) Map of the *bithorax* locus of *Drosophila* derived from genetic studies. Mutations in the upper row (*Cbx, Hab, Uab, Mcp*) transform anterior segments into posterior ones. Mutations in the lower row cause posterior segments to become more like anterior ones. The order of mutants in the region designated by a dotted line has not been established with certainty. (b) Map derived from studies using recombinant DNA. The zero and arrow indicate the location of the first clone recovered. The numbers along the upper row are kb of DNA. Triangles and inverted V's are insertions; lines are inversions; and *pbx* is a deletion. (From Marx, 1981, pp. 1485–88. Copyright 1981 by the American Association for the Advancement of Science.)

segment one more gene is switched on until, in the last abdominal segment, all the genes in the complex are switched on. In the mutant condition, each of these genes transforms a segment into structures normally formed by other segments, presumably by activating the wrong number of loci in the complex (Figure 22.12).

Using recombinant DNA technology and cloning, several laboratories have undertaken the direct analysis of DNA organization in the *bithorax* complex. The results obtained to date have been somewhat surprising. The recombinant DNA studies have largely confirmed the order of genes derived from genetic studies (Figure 22.12), but they unexpectedly show that the complex is very large—on the order of 200 to 300 kilobases of DNA instead of the 8 to 10 kb predicted from genetic studies. An average gene with no introns is usually thought to be about 1 kb in length, so this region possesses an enormous coding capacity. Only a handful of genes have been identified, however. Another unexpected and important discovery is that most of the mutations in this region are not point mutations involving a change in one or even a few bases, but instead are insertions or deletions covering several thousand base pairs of DNA. The ends of these insertions have characteristics in common with the trans-

FIGURE 22.13

Wild-type fly (left) compared with a fly with two pairs of wings (right) produced by a combination of bithorax mutants. (From Marx, 1981, pp. 1485–88. Photos by E. B. Lewis, Division of Biology, California Institute of Technology, Pasadena, Calif.)

posable elements found in the genomes of other higher organisms.

Not only is there evidence that some DNA regions in the complex may be transposable, but it appears that changing the position of a section of DNA alters its pattern of expression. The pbx^1 mutation is caused by the removal of a 17-kb DNA segment, and Cbx^1 is caused by the insertion of the same segment some 40 kb upstream from its original position. While pbx^1 transforms the rear of the haltere into the rear of the wing, Cbx^1 produces the opposite effect. The conclusion is that the 17-kb piece encodes the information for the rear half of the haltere (or, more generally, the rear half of the third thoracic segment), and the segment in which this information is expressed depends on the location of the DNA within the *bithorax* complex.

Lewis has proposed that genes in the *bithorax* complex and perhaps other genes involved in segmentation, arose from a common ancestral gene by tandem duplication and subsequent divergence of function. In fact, recent evidence has shown that genes of the *BX-C* and *ANT-C* share a highly conserved DNA sequence of about 200 bp called a *homeo box*. Similar sequences have been found in the genomes of other eukaryotes with segmented body plans, including *Xenopus*, chicken, mice, and humans. Homeo boxes from all organisms examined to date are very similar in DNA se-

quence and contain an open reading frame (ORF) that may encode a protein associated with the organization and differentiation of body segments. This suggests that the metameric or segmented body plan may have evolved only once.

All of the steps in determining the segmental arrangement of the body as well as the activation of the correct combination of selector genes depend on gene products produced by the maternal genome and arranged into a positional array in the cortex of the egg. In the second chromosomal mutant *extra sex combs* (*esc*), some of the head and all of the thoracic and abdominal segments develop as posterior abdominal segments (Figure 22.14). The mutation does not affect either the number or the polarity of segments but only their developmental fate, indicating that the esc^+ gene product that is synthesized and stored in the egg by the maternal genome may be required for the correct interpretation of the information gradient in the egg cortex.

These examples represent the striking influence of the cellular and extracellular environment on the fate of cells during development. In many other organisms, determination is not demonstrated as early as it is in *Drosophila*, and it is not as easy to pinpoint and document. This does not imply, however, that determination does not occur. Rather, it seems just a matter of timing during development.

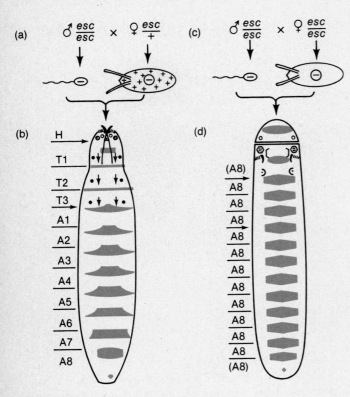

FIGURE 22.14

Action of the *extra sex combs* (*esc*) mutant in *Drosophila*.
(a) Heterozygous females form wild-type *esc* gene product
and store it in the oocyte. (b) Fertilization of *esc⁻* egg by
esc⁻ sperm still produces wild-type larva with normal
segmentation pattern (maternal rescue). (c) Homozygous
esc⁻ females produce defective eggs which, when fertilized
by *esc⁻* sperm, (d) produce a larva in which most of the
segments of the head, thorax, and abdomen are
transformed into the eighth abdominal segment. Maternal
rescue demonstrates that the *esc⁺* gene product is
produced by the maternal genome and is stored in the
oocyte for use in the embryo. (H, head; T1–3, thoracic
segments; A1–8, abdominal segments) Borderlines between
the head and thorax and thorax and abdomen are marked
with arrows. (From Struhl, 1981. Reprinted by permission
from *Nature*, Vol. 293, p. 37. Copyright © 1981 Macmillan
Journals Limited.)

THE STABILITY OF THE DIFFERENTIATED STATE

We will now consider the stability of the differentiated
state. Once a cell has taken on a specific structural
and functional capacity, under normal circumstances
in an adult organism the cell maintains that biological
status. That is, kidney epithelial cells, red blood cells,

muscle cells, and so on are not normally converted
into other cell types. Such observations have led to the
question of whether or not differentiation can be reversed under experimental conditions.

Recall that we have already established that nuclear
differentiation in specialized cell types is not necessarily stable. This conclusion was drawn based on the
nuclear transplantation experiments discussed earlier
in this chapter. As we will see in the following examples, there are certain cases where differentiated cells,
and not just nuclei, may alter their developmental status.

Transdetermination

Ernst Hadorn demonstrated that if imaginal disks are
explanted from a larva and implanted into the abdomen of an adult, these primordia will continue to proliferate by cell division but will not differentiate. The
disk subsequently may be recovered from the adult,
cut in half, and reimplanted—part to another adult
abdomen and part back into a larva (Figure 22.15). The

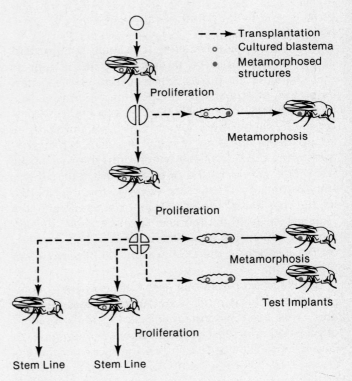

FIGURE 22.15

In vitro culturing of imaginal disk fragments over four
generations. In each generation, a disk fragment is
implanted into a larva to test its developmental potential
and fate. (From Gehring, 1968, p. 139.)

half implanted into a second adult continues to proliferate and can be serially propagated in this way. The part placed into the larva will undergo metamorphosis and differentiate into an adult structure.

Hadorn showed that during the early transfers, the disk primordium always maintained its determined state (i.e., a leg disk always differentiated into leg structures, etc.) following experimental manipulation. After several passages through adult abdomens, atypical structures were occasionally discovered; that is, an antennal disk sometimes produced wing structures, a genital disk leg structures, and so forth. This shift in determination is referred to as **transdetermination.**

The major forms of transdetermination are summarized in Figure 22.16. All disks can be transdetermined into one or another disk type, and certain sequences of transdetermination occur; a disk can give rise only to a limited number of other disks.

Hadorn's experiments showed that developmental programming or determination can be altered by serially "subculturing" disk cells in adult abdomens. This transdetermination involves complete cells, not just the nucleus (as in nuclear transplantation experiments). Thus, it may be concluded that irreversible changes do not occur during preadult development of imaginal disks. Homoeotic mutants show many of the properties of transdetermination; in fact, most of the transdeterminations are also known as homoeotic mutations, emphasizing that these developmental programs are under genetic control.

Regeneration

There are still other examples in which differentiated cells show a lack of developmental stability. Regeneration of amputated limbs occurs in the tadpoles of many frogs and toads and in some larval and adult newts and salamanders. Since replacement of the missing structure often follows the normal developmental process, regeneration has been used to study the stability of the differentiated state as well as normal developmental pathways.

Following limb amputation, a sequence of morphogenetic events follows:

1 The cut end of the stump is covered by a distal migration of epidermal cells, forming a thin, transparent sheet over the wound.

2 Degenerative changes occur in the cells at the cut tip, and phagocytes remove cellular debris.

3 Cells from all or most tissues at the cut tip undergo a dedifferentiation and accumulate under the epidermal covering to form a **blastema,** from which all regenerating tissues will be derived.

The blastema grows by cell division into a cone projecting from the stump and then flattens into a paddle or palette (Figure 22.17). Blastemal mesenchyme cells condense into ridges, forming digits that elongate from the palette. Muscle and nerve differentiation confer movement on the regenerate limb, and growth continues until it reaches full size.

Experiments involving transplantation of ^{3}H-thymidine labeled or triploid blastema tissue onto unlabeled or diploid amputated stumps have demonstrated that the undifferentiated blastema cells (mesenchyme) accumulate from most of the internal tissues of the adjacent stump. Cells from the cartilage, muscle, and

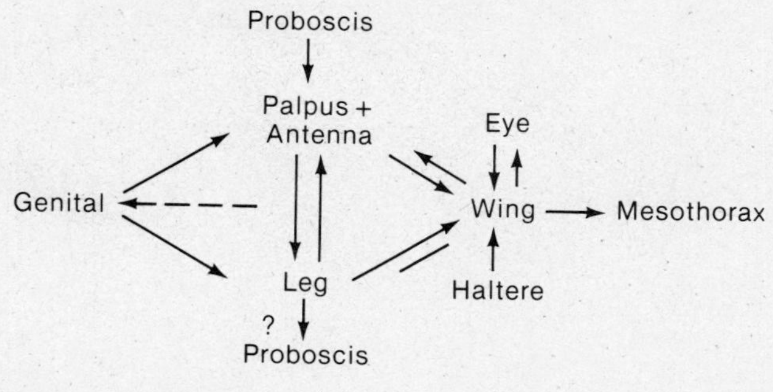

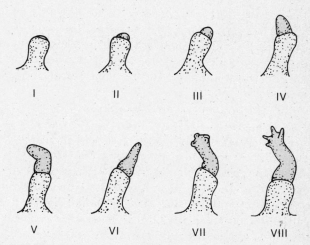

FIGURE 22.16
Observed sequence of transdeterminations from each imaginal disk. (From Gehring, 1968, p. 148.)

FIGURE 22.17
Stages of limb regeneration in the newt *Triturus.* (From Singer, 1952, p. 170.)

connective tissues all contribute to the formation of the blastema. While all of the cells in the blastema have undergone dedifferentiation, some apparently retain their determined state and can redifferentiate only in one direction—a process known as **modulation.** Other cells can transform into several tissue types during differentiation. Cartilage cells, for example, can dedifferentiate into blastemal cells, but undergo modulation and regenerate into cartilage. Muscle cells, on the other hand, can redifferentiate into other cell types, including cartilage. These observations again support the concept that even though cells have attained a highly specialized state where differential transcription is occurring, the potential still exists for inactive genes to be reactivated following cell proliferation.

GENE ACTIVITY DURING AMPHIBIAN DEVELOPMENT

Earlier, we examined the evidence for differential transcription in highly specialized cells. To conclude this chapter, we will discuss a few examples of the role of genes that control specific developmental processes. To assess the nature of a genetic defect in development, two questions are relevant: When is a gene active and where is a gene active? In other words, at what developmental stage does a given gene become active; does this activity continue throughout development; or is it restricted to a discrete period; is the gene active in a specific tissue or organ; or is it active throughout the body? This information may be integrated with that concerning the effect of the cellular and extracellular environment on transcription.

One general approach taken by developmental geneticists is to study mutant genes that interrupt or alter specific developmental processes. By comparing the effects of a mutant gene with normal development, they gain an insight into the role of the nonmutant form (the wild-type allele) of the gene during normal development.

Although developmental mutants have been studied in organisms ranging from *Drosophila* to the mouse, we shall briefly consider several developmental mutants in the Mexican axolotl, *Ambystoma mexicanum.* The axolotl is a salamander that sexually matures during the larval aquatic stage (neoteny) and never metamorphoses to the terrestrial form, thus maintaining gills and a tail fin. This organism has been extensively studied by Rufus Humphrey, Robert Briggs, and their colleagues. More than 30 mutant genes are known to affect oogenesis and embryonic or larval development.

The *o* Gene

The *o* or *ova-deficient* mutation is recessive, and homozygous females (*o/o*) produce eggs that develop normally only through the blastula stage. Abnormalities occur during gastrulation regardless of the genotype of the sperm (Figure 22.18). Thus, the mutant gene exhibits a **maternal effect,** and the main product of the *o* locus is assumed to be produced during oogenesis. This assumption has been confirmed by the injection of cytoplasm from a normal egg into the *o/o* egg, which corrects the defect and allows subsequent gastrulation to occur normally. Furthermore, transplantation of nuclei that have been exposed to the o^+ substance (from normal blastula embryos) into enucleated *o/o* eggs results in normal development, even though no o^+ substance is present in the cytoplasm (Figure 22.19).

These studies demonstrate that the normal allele of the *o* locus produces some substance in the egg that is absolutely essential long after fertilization and multiple divisions have occurred. The mutant effects are first noted as a generalized reduction of the division rate at the mid-to-late blastula stage. Abnormalities during gastrulation are widespread and not specific to one germ layer or another. Characterization of the o^+ substance suggests that it is a protein. Apparently, this protein moves from the cytoplasm to the nucleus and plays an essential role in the stable activation of genes necessary to support gastrulation and normal development.

FIGURE 22.18
Homozygous *o/o* mutant axolotl embryos arrest at gastrulation. (Photo by Ann Janice Brothers.)

FIGURE 22.19
Normal adult axolotls produced by transplanting normal nuclei into enucleated *o/o* eggs. (From Brothers, 1976. Reproduced by permission from *Nature*, Vol. 260, pp. 112–15. Copyright © 1976 Macmillan Journals Limited.)

The *c* Gene

Axolotls homozygous for the *cardiac lethal* (*c/c*) gene show normal development through most of embryogenesis but fail to develop a beating heart. If two embryos, a normal and a *c/c* animal, are joined in parabiosis (where their vascular systems are joined, creating a common circulation), development of the *c/c* embryo proceeds normally except for the heart. This experiment shows that the gene's effect is limited to heart formation. Thus, transcription of the normal allele is specific to heart-forming tissue.

The development of a normal heart is dependent on the interaction of two germ layers. Endodermal cells induce the adjacent mesoderm to form the heart tissue. This is just one example of many known cases of embryonic induction illustrating the importance of cell interaction.

Humphrey tried to determine whether the action of the *c* gene is specific to the inducing endoderm or the heart-forming mesoderm. He performed reciprocal grafts (transplants) of mesoderm between normal and mutant embryos. When mutant mesoderm was grafted into a normal host, the heart developed. However, when normal mesoderm was placed together with mutant endoderm, no heart developed. Therefore, the mutant endoderm is incapable of inducing the formation of the heart. Whatever the gene product of the normal *c* gene, its role is critical to the endodermal cells' normal developmental function.

The *e* Gene

The homozygous recessive mutant *eyeless* (*e*) leads to the complete absence of eyes in adult axolotls. This is an important example because it again illustrates that a single mutation may abolish the formation of an entire organ. This does not, however, imply that only one gene is responsible for eye formation. Rather, production of the eye is likely due to a programmed series of developmental events. The *eyeless* gene is apparently responsible for one of the initial steps.

The normal eye is formed as a result of the inductive influence of chordamesoderm on the overlying ectoderm. When mutant and normal mesoderm were transplanted, it was determined that the ectoderm is the site of action of the *e* gene. That is, mutant mesoderm is capable of inducing the formation of eyes when allowed to interact with normal ectoderm, but not vice versa. Thus, in *eyeless*, the receptor tissue during induction is affected. This is exactly the opposite of the results of studies of *cardiac lethal*.

The *l* Gene

Axolotls homozygous for the *lethal* (*l*) gene develop normally to an advanced embryonic stage, but then demonstrate numerous abnormalities of the head, gills, and limbs. This gene represents a distinct class of mutations called **autonomous cell lethals.** The following effects are noted:

1 There is no maternal effect.
2 Parabiosis and grafting fail to provide any correction of developmental abnormalities.

Thus, whatever the normal *l* gene product might be, it is needed by most cells as they differentiate during a quite advanced period of development.

The mutants described above represent at least three levels of genetic control. Coupled with the study of other mutations in this organism, it is clear that programmed genetic activity that varies from one cell type to another is critical to normal and complete development.

SUMMARY AND CONCLUSIONS

The study of development relies on information drawn from many fields of biology. Although the study of the genetic information's role during development and differentiation is but one approach, it is the most central. Two simple model systems, bacterial sporulation and slime mold development, have been extensively studied. In both systems, environmental conditions induce specific genetic activity which, in turn, leads to altered development.

In higher organisms, developmental genetics has as its foundation the variable gene activity theory. This theory assumes that all somatic cells in an organism contain equivalent genetic information, but that it is differentially expressed. Genomic equivalence has been demonstrated in many studies, including the nuclear transplantation studies in frogs. In these experiments, it has been demonstrated that somatic cells undergoing differentiation retain the entire gene set, which can be reprogrammed to direct the development of the entire organism.

If all cells in any species contain equivalent genetic information, what makes them structurally and functionally distinct is differential gene activity. We have discussed chromosome puffs, isozymes, and human hemoglobin synthesis as examples supporting this concept. Why cells demonstrate differential transcription is not as easy to answer. The total environment of each cell during embryogenesis appears to dictate specific gene activity. The general environment is initiated in the unfertilized egg, and the heterogeneous molecular components of the egg are unequally distributed to newly formed cells following fertilization. In many organisms, the maternal cytoplasm plays a critical role during early development. In *Drosophila*, genetic and molecular studies have confirmed that the egg contains information that specifies the body plan of the larva and adult, and that interaction of embryonic nuclei with the maternal cytoplasm initiates a transcriptional program characteristic of a specific developmental pathway.

During early development, cells become determined; that is, they are progressively restricted in developmental capacity. Determination becomes more and more specific with time and precedes the actual differentiation or specialization of distinctive cell types. In general, the differentiated state of eukaryotic cells is stable. However, certain examples—notably transdetermination in *Drosophila* and regeneration in amphibians—show that this specialized state can be reversed.

In order to illustrate specific gene activity during development, we have examined several mutants in the axolotl. Investigation of abnormalities caused by mutations often provides insights into the function of the normal (nonmutant) gene during development.

To unravel the precise mechanisms that lead to determination and differentiation is a formidable task. Thousands of cells all develop simultaneously and interact as an organism is forming. Furthermore, specific findings from one related group of organisms may not apply directly to other unrelated organisms. As a result, developmental genetics encompasses a large number of investigations that do not fit easily into a unifying model system. All such studies are valuable and, in the future, will help us to understand the intriguing question of how we and all living organisms came to exist.

PROBLEMS AND DISCUSSION QUESTIONS

1 Carefully distinguish between the terms *differentiation* and *determination*. Which phenomenon occurs initially during development?

2 The *Drosophila* mutant *spineless aristapedia* (ss^a) results in the formation of a miniature tarsal structure (normally part of the leg) on the end of the antenna. Such a mutation is referred to as *homoeotic*. From your knowledge of imaginal disks and Hadorn's transdetermination studies, what insight is provided by ss^a concerning the role of genes during determination?

3 How many isozymic forms are possible for a molecule that is a dimer and may contain any combination of three nonidentical polypeptide subunits?

4 How does the concept of isozymes relate to the theory of gene duplication discussed in Chapter 12? How would you test for the molecular homology between two genes coding for two polypeptide subunits, as in LDH?

5 In the sea urchin, early development up to gastrulation may occur even in the presence of actinomycin D, which inhibits RNA synthesis. However, if actinomycin D is removed at the end of blastula formation, gastrulation does not proceed. In fact, if actinomycin D is only present between the 6th and 11th hours of development, gastrulation (normally occurring at the 15th hour) is arrested. What conclusions can be drawn concerning the role of gene transcription between hours 6 and 15?

6 In this chapter, we have pointed out several examples of a specific molecule being synthesized by only one specific differentiated cell type (e.g., skin cells and keratin; erythrocytes and hemoglobin). From your past training in biology, what similar examples can you cite?

7 In the slime mold, the enzyme UDPG-pyrophosphorylase is essential to carbohydrate metabolism necessary for formation of the fruiting body. At a specific time during fruiting-body formation, mRNA specific to the enzyme is produced, translation of the enzyme ensues, and transcription ends. This is an example of specific gene activity during development. After transcription of UDPG-pyrophosphorylase mRNA has stopped, the cells of the fruiting body may be dissociated and allowed to reaggregate. Again, at the same stage in the newly formed fruiting body, transcription and translation begin on schedule. What conclusions can you draw concerning the activation of the UDPG-pyrophosphorylase gene during development?

8 Nuclei from almost any source may be injected into frog oocytes. Studies have shown that these nuclei remain inactive in transcription and translation. How can such an experimental system be useful in developmental genetic studies?

9 The concept of epigenesis indicates that an organism develops by forming cells that acquire new structures and functions, which become greater in number and complexity as development proceeds. This theory is in contrast to the preformationist doctrine that miniature adult entities are contained in the egg that must merely unfold and grow to give rise to a mature organism. What sorts of isolated evidence presented in this chapter might have led to the preformation doctrine? Why is the epigenetic theory held as correct today?

SELECTED READINGS

ARMSTRONG, J. B. 1985. The axolotl mutants. *Dev. Genet.* 6: 1–26.

BEACHY, P. A., HELFAND, S. L., and HOGNESS, D. S. 1985. Segmental distribution of bithorax complex proteins during *Drosophila* development. *Nature* 313: 545–51.

BEERMAN, W., and CLEVER, U. 1964. Chromosome puffs. *Scient. Amer.* (April) 210: 50–58.

BENDER, W., AKAM, M., KARCH, F., BEACHY, P., PEIFER, M., SPIERER, P., LEWIS, E. B., and HOGNESS, D. 1983. Molecular genetics of the *bithorax* complex in *Drosophila melanogaster*. *Science* 221: 23–29.

BISHOP, J. M. 1985. Trends in oncogenes. *Trends in Genet.* 1: 245–49.

BRIGGS, R. 1973. Developmental genetics of the axolotl. In *Genetic mechanisms of development*, ed. F. H. Ruddle, pp. 169–200. New York: Academic Press.

BRIGGS, R., and KING, T. 1952. Transplantation of living nuclei from blastula cells into enucleated frog eggs. *Proc. Natl. Acad. Sci.* 38: 455–63.

BROTHERS, A. J. 1976. Stable nuclear activation dependent upon a protein synthesized during oogenesis. *Nature* 260: 112–15.

BROWER, D. L. 1985. The sequential compartmentalization of *Drosophila* segments revisited. *Cell* 41: 361–64.

DAVIDSON, E. H. 1976. *Gene activity in early development.* 2nd ed. New York: Academic Press.

DuBOW, M. S., ed. 1981. *Bacteriophage assembly.* New York: Alan R. Liss.

EBERT, J. 1965. *Interacting systems in development.* New York: Holt, Rinehart and Winston.

FULTON, C., and KLEIN, A. 1976. *Explorations in developmental biology.* Cambridge, Mass.: Harvard University Press.

GEHRING, W. 1968. The stability of the differentiated state in cultures of imaginal disks in *Drosophila*. In *The stability of the differentiated state*, ed. H. Unsprung. New York: Springer-Verlag.

GODFREY, S., and SUSSMAN, M. 1982. The genetics of development in *Dictyostelium discoideum*. *Ann. Rev. Genet.* 16: 385–404.

GURDON, J. B. 1968. Transplanted nuclei and cell differentiation. *Scient. Amer.* (Dec.) 219: 24–35.

————. 1974. *The control of gene expression in animal development.* Cambridge, Mass.: Harvard University Press.

GURDON, J. B., LASKEY, R. A., and REEVES, O. R. 1975. The developmental capacity of nuclei transplanted from keratinized skin cells of adult frogs. *J. Embryol. Exp. Morphol.* 34: 93–112.

HADORN, E. 1968. Transdetermination in cells. *Scient. Amer.* (Nov.) 219: 110–20.

HOTTA, Y., and BENZER, S. 1973. Mapping of behavior in *Drosophila* mosaics. In *Genetic mechanisms of development*, ed. F. H. Ruddle, pp. 129–67. New York: Academic Press.

ILLMENSEE, K., and HOPPE, P. 1981. Nuclear transplantation in *Mus musculus*: Developmental potential of nuclei from preimplantation embryos. *Cell* 23: 9–18.

KING, T. J. 1966. Nuclear transplantation in amphibia. *Methods in cell physiology*, ed. D. Prescott, vol. 2. New York: Academic Press.

LEIGHTON, T., and LOOMIS, W. F., eds. 1980. *The molecular genetics of development.* New York: Academic Press.

LEVINE, M., RUBIN, G., and TJIAN, R. 1984. Human DNA sequences homologous to a protein coding region conserved between homoeotic genes of *Drosophila*. *Cell* 38: 667–73.

LEWIS, E. B. 1976. A gene complex controlling segmentation in *Drosophila*. *Nature* 276: 565–70.

LOSICK, R., and PERO, J. 1981. Cascades of sigma factors. *Cell* 25: 582–84.

LOSICK, R. 1973. The question of gene regulation in sporulating bacteria. In *Genetic mechanisms of development*, ed. F. H. Ruddle, pp. 15–28. New York: Academic Press.

MARX, J. L., 1981. Genes that control development. *Science* 213: 1485–88.

McGINNIS, W., HART, C. P., GEHRING, W. J., and RUDDLE, F. H. 1984. Molecular cloning and chromosome mapping of a mouse DNA sequence homologous to homoeotic genes of *Drosophila*. *Cell* 38: 675–80.

MARKERT, C. 1975. *Isozymes: Genetics and evolution.* New York: Academic Press.

REINERT, J., and HOLTZER, H., eds. 1975. *The cell cycle and cell differentiation.* New York: Springer-Verlag.

ROY, A. K., and CLARK, J. H. 1980. *Gene regulation by steroid hormones.* New York: Springer-Verlag.

RUDDLE, F. H., HART, C. P., and McGINNIS, W. 1985. Structural and functional aspects of the mammalian homeo-box sequences. *Trends in Genetics* 1: 46–50.

SINGER, M. 1952. Influence of the nerve in regeneration of the amphibian extremity. *Quart. Rev. Biol.* 27: 169–200.

STRUHL, G. 1981. A gene product required for correct initiation of segmental determination in *Drosophila*. *Nature* 293: 36–41.

STRUHL, G., and BROWER, D. 1982. Early role of the esc^+ gene product in the determination of segments in *Drosophila*. *Cell* 31: 285–92.

SUBTELNY, S., and GREEN, P. B., eds. 1982. *Developmental order: Its origin and regulation.* New York: Alan R. Liss.

SUSSMAN, M., ed. 1972. *Molecular genetics and developmental biology.* Englewood Cliffs, N.J.: Prentice-Hall.

————. 1973. *Developmental biology, its cellular and molecular foundations.* Englewood Cliffs, N.J.: Prentice-Hall.

WATSON, J. D. 1976. *Molecular biology of the gene.* 3rd ed. Menlo Park, Calif.: W. A. Benjamin.

WEBER, R. 1975. *Biochemistry of animal development: Molecular aspects of animal development.* Vol. 3. New York: Academic Press.

YOUNGMAN, P., et al. 1985. New ways to study developmental genes in spore-forming bacteria. *Science* 228: 285–91.

23

Somatic Cell Genetics

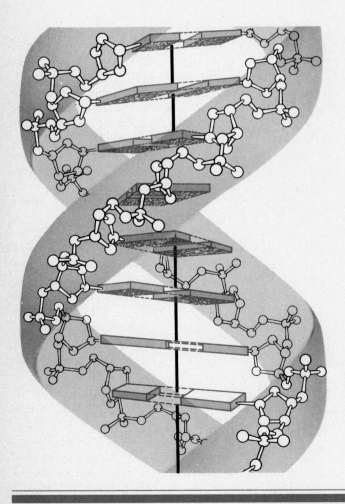

Although *in vitro* studies of eukaryotic cells have the obvious disadvantage of dealing with less than the intact organism, such investigations have been extremely important in genetics. In many instances, problems can be approached with *in vitro* studies that would otherwise be impossible to study using whole organisms. In proper nutrient media, dissociated tissue from embryos or adults yields cells that grow and divide. These cells adhere to the surface of the culture vessel and divide until a monolayer of cells forms. The inhibition of further growth is related to intercellular contact and is called **contact inhibition,** a property of nonmalignant cells. Contact-inhibited cells may be removed from the culture vessel, diluted in number, and added to fresh medium. When this subculture process is performed, growth and division are reinitiated. Subculturing may be performed many times, yielding large numbers of generations of cells in a relatively short time. Eventually, however, these cell strains derived from normal diploid tissue fail to divide further after a transfer, and the culture is lost. On the other hand, if the original tissue is tumorous, subculturing may perpetuate the cell line almost indefinitely. These permanent cell lines are almost always abnormal in their karyotype compared with the diploid genome. Most often, such cells show a wide range of aneuploidy, and specific cell lines can only be characterized by an average number of chromosomes.

In 1960, Georges Barski was working with two permanent mouse cell lines that differed in their morphology and karyotype. Barski was hoping to observe transformation between the two mouse cell lines similar to that known to occur in bacteria. What he discovered when growing the mouse cell lines together was quite unexpected. From this mixed culture, he observed the formation of a new cell type that had apparently been formed by the fusion of the two different cells. The number of chromosomes in the new cell nearly equaled the sum of the chromosomes of the individual cell lines. Analysis of the karyotype of the new cell type confirmed the hybrid nature of the newly created cells. The nuclei contained chromosomes morphologically similar to both of the original cell lines. Barski's findings were quickly reproduced in the laboratory of Boris Ephrussi, confirming the observation that mammalian cells could indeed undergo fusion and that consolidation of the genetic information into a single nucleus could occur.

In the relatively short time since these observations, this procedure, called **somatic cell hybridization,** has been used extensively in many areas of genetics. As the techniques of somatic cell hybridization have developed, it has become possible to achieve ever higher degrees of resolution in the transfer of genes between cells. An entire genome can be transferred through cell fusion, and microcells can transfer a smaller number of chromosomes. Chromosome-mediated gene transfer involves genes on a single linkage group, and DNA-mediated gene transfer can be used to incorporate gene-sized DNA fragments into cells. In this chapter, we will survey the techniques of creating and isolating somatic cell hybrids and discuss how these unique cells are used in the field of genetics.

THE PRODUCTION, ISOLATION, AND CHARACTERISTICS OF SOMATIC CELL HYBRIDS

Spontaneous Cell Fusion

When two genetically distinct cell lines are grown together in culture, cell fusion is a spontaneous but rare event. Following the interaction between the cell membranes and subsequent fusion of the two cytoplasms, a hybrid cell is created. This new cell, which contains two nuclei, is called a **heterokaryon.** If nuclear fusion then occurs, the heterokaryon is converted to a synkaryon. Heterokaryons remain viable for a short period of time, but unless nuclear fusion occurs, such cells do not replicate and thus die off. Therefore, the only viable hybrid cells have a single

fused nucleus containing genetic information approximately equivalent to the sum of the two parent cell lines. These somatic cell hybrids will undergo a continuous cell cycle of replication and division almost indefinitely.

When examined cytologically, hybrid cells may be identified by their increased size and substantially larger number of chromosomes. Fusion between cells of the two distinct lines is usually confirmed by the presence in the hybrid nucleus of unique marker chromosomes from both lines. Cell fusion may involve more than two cells, in which case the heterokaryon contains three or more nuclei; and following nuclear fusion, the total number of chromosomes is greatly increased.

Agents that Promote Cell Fusion

In 1962, the use of the hemagglutinating virus of Japan (HVJ) was shown to increase the rate of cell fusion to 100 to 1000 times the spontaneous rate. This virus, a member of the parainfluenza group of Myxoviruses and called **Sendäi virus,** had originally been shown by Y. Okada to produce cell fusion of Ehrlich ascites cells (derived from mice) when added to the cultures. An attenuated form of the virus produced by ultraviolet light treatment was used so that infection did not occur. When added to mixed cell cultures, the virus adsorbs to cell surfaces of adjacent cells, causing them to adhere to one another. This action of the virus promotes cytoplasmic bridge formation and cytoplasmic fusion. Heterokaryons containing two or more nuclei are formed, which may be converted to viable hybrid cells following nuclear fusion.

The direct application of Sendäi virus to hybridization studies was first performed by Henry Harris in 1965 when he successfully fused cells derived from two different species. The two cell lines were of human and mouse origin. The human line was of the **HeLa cell** variety originally isolated in 1951 from a cervical carcinoma from a black female, Henrietta Lacks. The mouse cells were the Ehrlich ascites cells.

Although the human-mouse hybrid cell remained a heterokaryon and never formed a viable hybrid cell line, many subsequent interspecific cell fusions were successful in producing viable hybrids. A wide variety of hybrid cell lines were permanently established for study using cell lines derived from mice, rats, chickens, hamsters, humans, and monkeys.

Other agents that chemically modify the cell surface have been used to promote the formation of hybrid cells, including lysolethicin and polyethylene glycol (PEG). PEG is easily obtainable, inexpensive, and highly active in inducing cell fusion. George Pontecorvo first showed that PEG induces the formation of viable somatic cell hybrids, and Richard Davidson and his colleagues have refined the technique into a simple and effective procedure that can increase the frequency of hybridization to approximately 4 percent of the cells in culture.

With the use of Sendäi virus and PEG, viable hybrids may now be produced from any two permanent cell lines (Figure 23.1). By producing interspecific hybrids, geneticists have overcome the incompatibility of sexual reproduction between two species and expanded the experimental subjects available for genetic study.

FIGURE 23.1
(a) Metaphase spread of chromosomes from a mouse cell line. (b) Metaphase spread from a Syrian hamster cell line. (c) The metaphase chromosomes from a hybrid cell line created by the fusion of mouse and hamster cells.

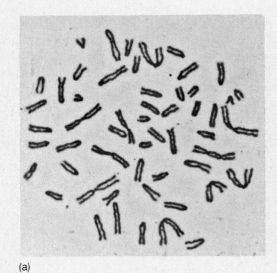

(a)

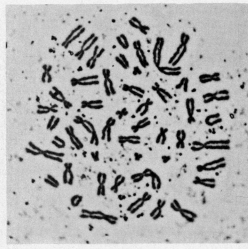

(b)

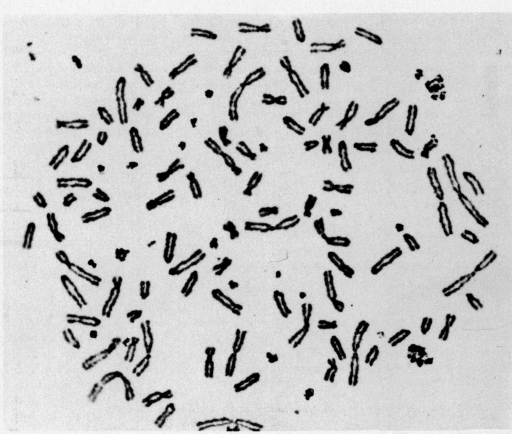

(c)

Isolation Procedures

In mixed cultures, hybrid cells are produced along with much larger numbers of parental cells. Before the hybrid cells can serve as the basis for experimentation, they must be isolated and grown as a pure line. In some of the earlier studies, it was found that the hybrid cells grew faster and divided more often when the mixed cultures were grown at a lowered temperature. As a result, the desired hybrid cells would overgrow the parental lines and could be easily isolated, cytologically identified, and cloned. Often, however, a differential growth rate was not attained at a lower temperature, and a more efficient technique was sought.

The first sophisticated isolation technique was used by John Littlefield in 1964. This selection system was based on biochemical information concerning the synthesis of nucleotides essential to the replication of DNA by the cultured cells. As shown in Figure 23.2, nucleotides may be synthesized *de novo* (i.e., from the beginning; anew) by the metabolic conversion of carbohydrate and amino acid precursors present in the culture medium. Although this is the major pathway, nucleotides may also be constructed as a result of an independent "salvage" pathway from nitrogenous bases and nucleosides. This salvage pathway is dependent upon several enzymes including **thymidine kinase (TK)** and **hypoxanthine guanine phosphori-**

bosyl transferase (HGPRT). So long as one of the two pathways is active, nucleotide synthesis is maintained, DNA synthesis occurs, and the cell cycle may be repeated over and over.

Littlefield's technique was ingeniously simple. He selected two cell lines, neither of which could utilize the salvage pathway. One cell line contained a mutation resulting in loss of thymidine kinase activity (TK^-), and the other cell line carried a mutation resulting in the lack of HGPRT activity ($HGPRT^-$). Thus, both parental cell lines, by necessity, could only synthesize nucleotides *de novo*. After the two cell lines were initially mixed in culture, the chemical antagonist **aminopterin** was added to the medium. This molecule completely inhibits the use of the *de novo* major synthetic pathway. As a result, neither of the parental cell lines could synthesize nucleotides, and their growth and division were arrested. However, any hybrid cell that formed should contain at least one normal gene for the TK and one for the HGPRT enzymes (Figure 23.2). Essentially, the following "cross" has been made as a result of the hybridization:

$$(TK^-; HGPRT^+) \times (TK^+; HGPRT^-)$$
$$\text{Parental A} \qquad \text{Parental B}$$
$$\downarrow$$
$$(TK^+; TK^-; HGPRT^+; HGPRT^-)$$
$$\text{Hybrid}$$

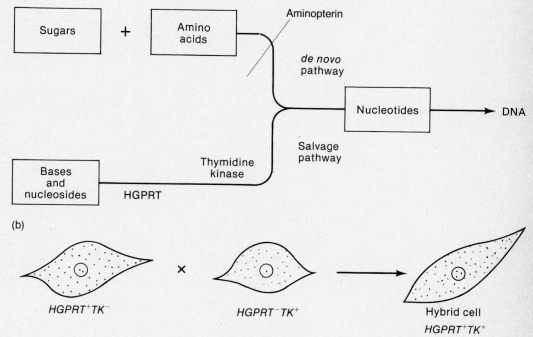

FIGURE 23.2

(a) Metabolic pathways for nucleotide synthesis by *de novo* and salvage pathways. Cells lacking either thymidine kinase (TK) or hypoxanthine guanine phosphoribosyl transferase (HGPRT) cannot use the salvage pathway and must use the *de novo* pathway. (b) Aminopterin blocks the *de novo* pathway when added to the medium, and only cells that can utilize the salvage pathway (hybrid cells) can grow. (c) The formation of hybrid $HGPRT^+TK^+$ cells by fusion of $HGPRT^+TK^-$ and $HGPRT^-TK^+$ cell lines.

The newly formed hybrid cells can utilize the salvage pathway to synthesize nucleotides and thus can replicate DNA and continue through the cell cycle. Therefore, hybrid cells are the only survivors under these conditions. They may be isolated and cloned. Since it is possible to select cells deficient for either TK or HGPRT from many cell lines, this selection technique has been important in the initial isolation of somatic cell hybrids.

The small number of well-characterized genetic markers in mammalian cells, however, has been a limiting factor in the use of mutant cell lines in cell hybridization. To rectify this, Theodore Puck developed a method for the selection of nutritionally deficient (auxotrophic) mammalian cell lines that can be employed in somatic cell hybridization. In this method, Chinese hamster ovary (CHO-K1) cells are exposed to a chemical mutagen and transferred to a minimal medium deficient in many nutrients. In this medium, only the normal, nonmutant cells will grow and divide; the mutant cells will survive but not grow. Dividing cells are killed by exposure to **bromodeoxyuridine (BrdU),** a nucleobase analogue, and near-visible ultraviolet light. An enriched medium lacking BrdU is added, and the nutritionally deficient mutant cells are grown in colonies (Figure 23.3). To discover the nature of the biochemical block, subcultures of the deficient cells are grown in minimal medium supplemented with single nutrients. Over 55 such mutants have been recovered. Many of these have been analyzed biochemically and the defect shown to involve the loss or reduced activity of a single enzyme. These mutants can be used to form somatic cell hybrids with cells from many mammalian species, greatly facilitating the assignment of genes to chromosomes.

Characterization of Cell Hybrids

The degree of usefulness of hybrid cells in genetic studies is related to the demonstration of genetic activity in those cells. Of initial concern is whether or not the information derived from both of the parental cell lines is active in the hybrid cell. Studies of several enzymes taken from hybrid cells have shown that both parental genomes are transcribed and lead to "hybrid" proteins.

As we learned in Chapter 22, many functional enzymes are **isozymes** composed of subunits that may

FIGURE 23.3
A method for the derivation of nutritional mutants (auxotrophs) from a mammalian cell line. Mutant cells are unable to grow in the minimal medium. Growing (nonmutant) cells are killed by BrdU and UV light, and the surviving mutant cells can be grown in enriched medium and recovered as colonies. (After Kao and Puck, 1974, p. 24.)

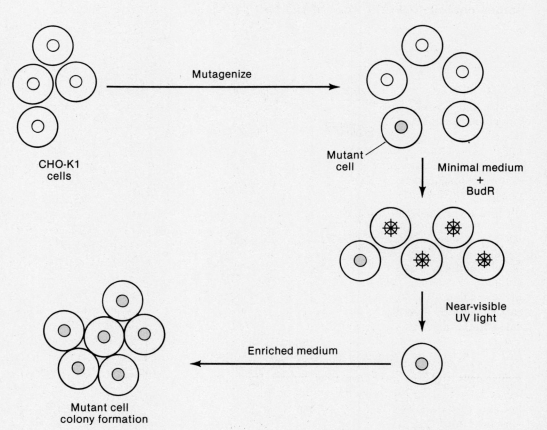

be distinguished electrophoretically. Figure 23.4 shows three banding patterns resulting from electrophoresis of the enzyme **malate dehydrogenase.** Cultured cell lines of hamster or mouse origin produce a single but characteristic form, and the mouse and hamster enzymes can be resolved by electrophoresis. When hybrid cells are used as the source of the enzyme, the hybrid nature of malate dehydrogenase is apparent. Two bands characteristic of the mouse and the hamster enzyme are revealed, and, in addition, a third band with an intermediate migration rate can be seen. Since the hamster and mouse forms of the enzyme are each composed of two identical subunits that are species-specific, the intermediate band of the hybrid represents the combination of one mouse subunit and one hamster subunit. The examination of this and other enzymes has confirmed an important rule in somatic cell hybrids: **Both parental genomes are genetically active in a variety of interspecific somatic cell hybrids.** If each parental line synthesizes a particular protein before fusion, then both genomes in the hybrid will continue to produce the protein, barring chromosome loss.

Karyotypic analysis of successive generations of newly formed hybrids has revealed a second important characterization of such cells: **While the initial hybrid nucleus may contain a number of chromosomes equivalent to the sum of the two parental cell lines, a variable number of chromosomes is lost with many succeeding generations of the new cell line.** When intraspecific hybrid cell lines are carefully studied, a slow random loss of about 10 to 20 percent of the chromosomes occurs. In interspecific hybrids formed from organisms that are not too distantly related, a similar loss of chromosomes occurs, but it is preferential toward one of the two parental lines. For example, hybrids formed between mouse and rat lines tend to lose rat chromosomes. Hamster-mouse hybrids tend to lose mouse chromosomes during successive generations. Following the loss of 10 to 20 percent of primarily mouse chromosomes, the nucleus stabilizes with respect to the number of remaining chromosomes over successive generations.

A similar but more pronounced phenomenon occurs in hybrids formed from distantly related organisms. Chromosome loss is preferential, but much more rapid and extensive. For example, human-mouse hybrids retain almost all mouse chromosomes, but lose human chromosomes until the hybrid nucleus contains only one to three human chromosomes. Which human chromosomes remain is the result of a random process. Therefore, when numerous hybrid clones are derived from the same parental cell lines, each new hybrid cell line randomly retains different combinations of a relatively few chromosomes characteristic of one of the parental lines.

This finding is extremely significant, because loss of functional activity of any kind in a hybrid cell line may be correlated with the loss of specific chromosomes. As we shall see, these circumstances have provided a new approach to mapping human chromosomes.

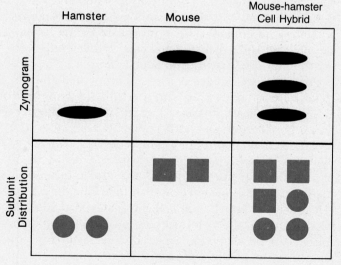

FIGURE 23.4
Zymogram of malate dehydrogenase from hamster cells, mouse cells, and a mouse-hamster cell hybrid. The hybrid cells contain not only both parental forms of the enzyme, but also a third form derived from the subunits of the parental forms.

THE APPLICATION OF SOMATIC CELL HYBRIDIZATION TO GENETICS

As we discussed earlier, two distinct cytological models are available for cell hybridization investigations: heterokaryons and synkaryons. In heterokaryons, a common cytoplasm contains two or more nuclei which may be closely or distantly related and in a similar or dissimilar state of genetic activity or stage in the cell cycle. Even though heterokaryons survive at most only two to three weeks, many basic problems may be investigated during this interval. On the other hand, synkaryons contain a single nucleus with genetic information from different origins and may be serially propagated indefinitely. Since chromosomes are lost, a correlation between retained function and the remaining chromosomes may be made.

Both model systems have already been exploited experimentally with great success. We shall briefly consider several major findings from both models.

Heterokaryons and Genetic Regulation

Several investigations have demonstrated the cytoplasm's importance in the regulation of genetic activity. The initial fusion of two cells in different stages of the cell cycle yields results similar to David Prescott and Lester Goldstein's nuclear transplantation experiments with *Amoeba* (see Chapter 2). If a cell in the G_1 stage is fused with a cell in the S phase, the G_1 nucleus is induced to synthesize DNA prematurely. On the other hand, the combination of a G_1 stage or S phase nucleus with a mitotic nucleus sometimes results in the premature condensation of the G_1 or S chromatin. This condition is characteristic of the G_2 nucleus just prior to its initiation of mitosis. One can conclude that some component from the mitotic cell has entered the G_1 or S nucleus following cell fusion and affected the state of the chromatin.

More specific information has been obtained by Henry Harris in his studies of fusion of cells with various capacities for RNA and DNA synthesis. For example, rabbit macrophage white blood cells normally synthesize RNA but not DNA. The hen erythrocyte (which, unlike that of mammals, retains its nucleus) is highly differentiated and synthesizes almost no RNA or DNA. The human-derived HeLa cell, on the other hand, synthesizes both RNA and DNA. When the various combinations of these three cells are used two at a time as the source of heterokaryon formation and studied, several general observations can be made:

1 If either cell normally synthesizes RNA prior to fusion, both nuclei of the heterokaryon synthesize RNA.

2 If either cell normally synthesizes DNA prior to fusion, both nuclei of the heterokaryon synthesize DNA.

3 If neither cell normally synthesizes RNA or DNA prior to fusion, neither of the nuclei synthesizes RNA or DNA, respectively, following heterokaryon formation.

Since the three cells above are from quite diverse sources (rabbit, chicken, and human), a very significant conclusion may be drawn. The signals initiating either RNA or DNA synthesis in a previously inactive nucleus are shared by the three organisms and are not at all species-specific. As we shall see, the signals given appear to originate in the cytoplasm, and they are not

"false" signals resulting in some form of random genetic activity.

When hen erythrocytes are used as a source for heterokaryon formation, the cells, prior to fusion, contain a small nucleus with highly condensed chromatin. Using Sendäi virus on these erythrocytes causes hemolysis, thus rupturing the cell membrane and dispersing the cytoplasmic contents. As shown in Figure 23.5, what remains is the intact nucleus and parts of the cell membrane, together referred to as an **erythrocyte ghost (EG).** These EGs attach to whatever cells are being utilized in the fusion study and are ultimately engulfed into the second cell's cytoplasm. Of importance here is that the heterokaryon that is formed does not contain hen erythrocyte cytoplasm, but only cytoplasm of the other cell type used in the experiment. Thus, any initiating signals for reactivation of DNA and RNA synthesis or morphological change of the hen nucleus must originate from the foreign cytoplasm.

We have already seen that when HeLa cells are fused with hen erythrocytes, synthesis of both RNA and DNA is detected. Even more striking are the morphological changes that may be observed in the hen nucleus following heterokaryon formation. Not only does the erythrocyte nucleus proceed to enlarge its volume 20 to 30 times, but the highly condensed chromatin, characteristic of genetic inactivity, becomes greatly dispersed (Figure 23.5). Presumably, this dispersion makes more of the DNA accessible to transcription and is related to the increasing amount of RNA that is synthesized as the nucleus enlarges. Further studies by Harris have shown that hen-specific antigens and enzymes are synthesized in the heterokaryon, and HeLa cell proteins can be detected entering the hen nucleus prior to gene action.

From this investigation it is apparent that the nucleus from a highly specialized cell, upon entering a foreign cytoplasm that is actively metabolizing, reverses its differentiated and genetically inert state. The original genetic control of the intact erythrocyte is lost when its cytoplasm is lost, and the regulation of genetic activity in the nucleus is strongly influenced by the new cytoplasmic environment following fusion. Interestingly, although hen-specific gene products can be detected in the heterokaryon, no hemoglobin synthesis can be detected. Since this was the primary gene product of the erythrocyte before fusion, its absence is conspicuous. However, other studies, particularly of synkaryons, have also shown that the so-called household products necessary for everyday functions (metabolism, for example) are usually expressed. The "luxury" products that characterize specialized cell function, such as hemoglobin in erythrocytes, are almost always lost following fusion.

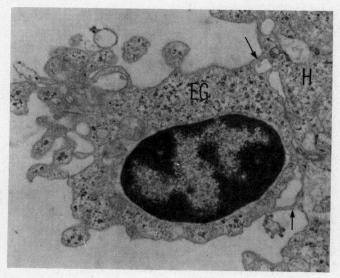

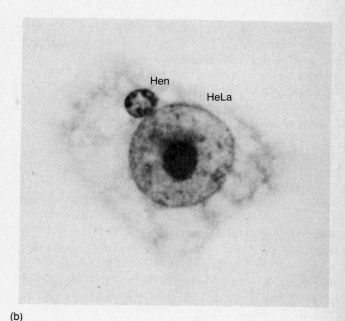

(a)

(b)

(c)

FIGURE 23.5

Stages in heterokaryon formation between HeLa cells and hen erythrocytes. (a) Cytoplasmic bridges between hen erythrocyte ghost (EG) and HeLa cell (H). (b) Heterokaryon contains a large HeLa cell nucleus and a small hen erythrocyte nucleus. (c) Enlargement and dispersion of chromatin in hen nucleus following incorporation into HeLa cytoplasm. (From Harris, 1974.)

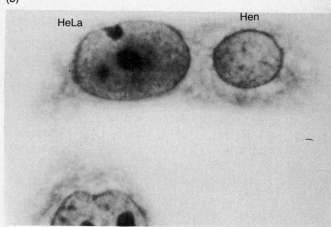

Heterokaryons and Genetic Complementation

When mutation disrupts a particular functional capacity of an organism, the loss of function may be due to any one of a number of causes. A particular macromolecule, essential to a particular function, may be present in a cell as the result of a series of metabolic steps controlled by many enzymes, themselves controlled by many structural genes. In addition, there may be other regulatory genes that directly influence the expression of the structural genes. The study of **genetic complementation** (see Chapter 19) is critical to understanding the complexity of the genetic control of any particular function. For example, earlier in this chapter we described Littlefield's selection technique. The hybrid cell was capable of using the salvage pathway of nucleotide synthesis even though both parental cells could not. When fused, the parental cells complement one another, leading to a normal function. From this observation, we may conclude that *at least* two genes and two gene products are involved in the salvage synthesis of nucleotides.

In a similar way, heterokaryon formation has served as the basis for complementation studies of other functions. For example, normal cultured fibroblasts are capable of UV-induced repair synthesis of DNA. Individuals with the autosomal recessive disorder, xeroderma pigmentosum, however, are unable to repair damage caused by UV light. These patients show a high rate of sunlight-induced skin cancers. As shown in Figure 23.6, heterokaryons produced by fusion of cells from different individuals with this defect may show complementation (XP4/XP12), or may not (XP4/XP16). Analysis of cells from many different patients has shown the existence of seven complementation groups. The ability to repair DNA is restored when cells from any group are fused with cells from any of the other groups, suggesting that at least seven different genes are involved in the excision repair of DNA. In the former case, the defects in XP4 and XP12 patients are known to be in different genes, since complementation occurs between them, while XP4 and XP16 have been identified as different mutations of the same gene because heterokaryons derived from these patients do not show complementation.

Still another example of genetic heterogeneity substantiated by heterokaryon analysis is an autosomal recessive disorder in humans, **maple syrup urine dis-ease (MSUD).** Affected individuals are unable to degrade the branch-chain amino acids leucine, isoleucine, and valine, which results in severe keto acidosis. Convulsions, mental retardation, and a maple syrup odor of the urine are the clinical symptoms. The phenotype varies in unrelated individuals from a severe, early onset to a mild, later onset. The enzyme that decarboxylates the branch-chain keto-amino acids, **BCKA-decarboxylase,** is not expressed in fibroblasts derived from affected persons. A positive complementation test occurs in heterokaryons formed from different MSUD fibroblast sources when enzyme activity is restored. Of four cell lines studied by Linda Lyons and her colleagues, one was shown to complement each of the other three. When the latter three were tested, none of them complemented the other two lines. Therefore, MSUD is a disease with at least two complementation groups—perhaps representing two separate genes. This also indicates that the enzyme, BCKA-decarboxylase is a multimer, made up of at least two different polypeptide subunits.

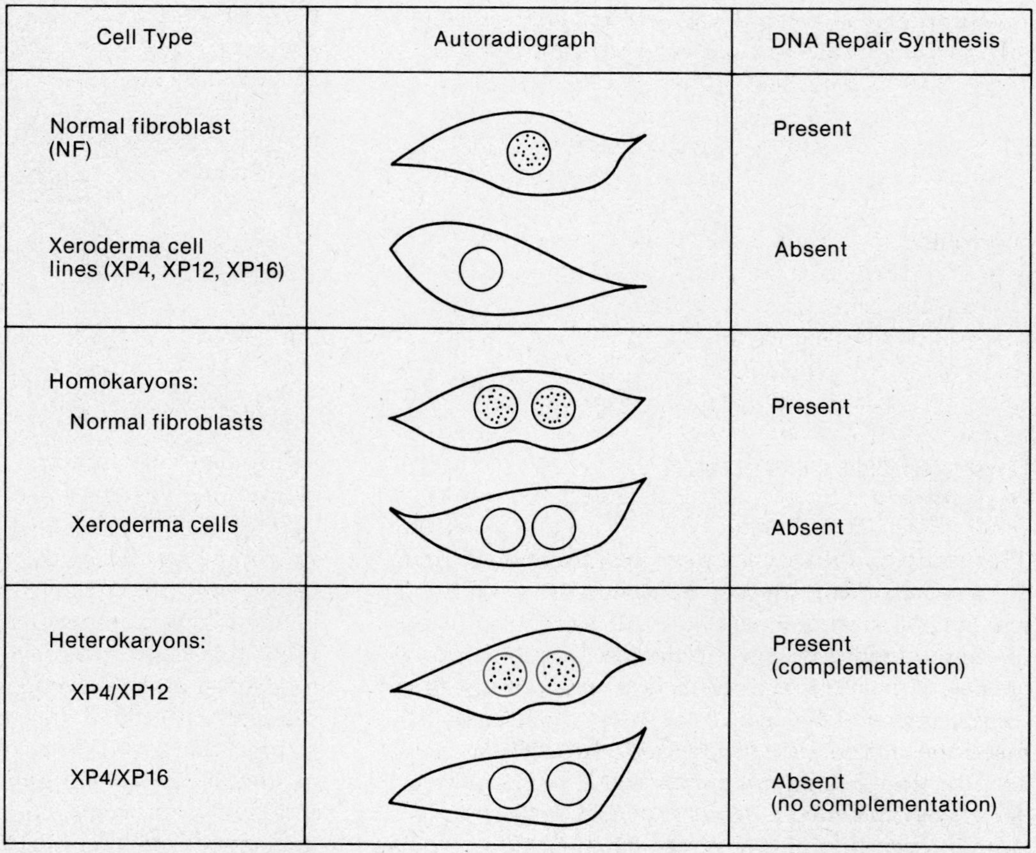

FIGURE 23.6
Complementation analysis in xeroderma pigmentosum. Cells active in DNA repair incorporate ³H-thymidine, which is detected in an autoradiogram as silver grains (dots) over the nuclei. Nuclei without silver grains are defective in DNA repair. Heterokaryons that show complementation and, thus, function in DNA repair, have silver grains over the nuclei (XP4/XP12).

Cell Type	Autoradiograph	DNA Repair Synthesis
Normal fibroblast (NF)		Present
Xeroderma cell lines (XP4, XP12, XP16)		Absent
Homokaryons: Normal fibroblasts		Present
Xeroderma cells		Absent
Heterokaryons: XP4/XP12		Present (complementation)
XP4/XP16		Absent (no complementation)

Although these examples are simple ones, the technique described is very valuable and may be applied to more complex cases of genetic interaction. Loss of **hexosaminidase** activity in humans is related to lysosomal defects and leads to both **Tay-Sachs disease** and to **Sandhoff disease.** There are two forms and activities of the enzyme (A and B). In Tay-Sachs disease, the A form is absent; and in Sandhoff disease, both forms are missing. Complementation does occur in fibroblast heterokaryons derived from individuals with one or the other disease. One hypothesis is that two genes located on separate chromosomes are involved. A mutation at each locus results in a separate defect, as indicated by the complementation study.

Synkaryons and Chromosome Linkage of Human Genes

Synkaryons represent true hybrid cells where the genetic material from the two cell lines is contained within a single nucleus. Since these hybrid cells may be cloned and cultured almost indefinitely, large amounts of material for biochemical analysis are available. As a result, many known gene products may be detected in such cells. These advantages, combined with the observed preferential loss of human chromosomes in certain hybrids, have allowed geneticists to assign human genes to linkage groups and to estimate the relative proximity of genes determined to be linked to a single chromosome.

In the most simple cases, the approach to this mapping exercise is straightforward. If a hybrid cell contains only one or a few human chromosomes, and a specific human gene product is detected, the gene controlling that product is linked to one of the remaining chromosomes. If only one chromosome remains, direct linkage is determined. If more than one chromosome remains, further study may reveal the subsequent simultaneous loss of one more chromosome and of the gene product. Again, the gene is determined to reside on a specific chromosome. In the same way, when two or more gene products are always present or always lost together, the respective genes appear to be linked to the same chromosome.

In practice, the actual experimentation and analysis are more complex. Mary C. Weiss and Howard Green, who first assigned linkage of a human gene by this method, studied the gene responsible for thymidine kinase (TK) in human-mouse hybrids. They first selected mouse cells that had lost TK activity (TK^-) and fused these deficient cells with human cells that were TK^+. In the presence of aminopterin, the only hybrid cells that can survive must contain the human TK^+ gene. Since the remaining mouse and human chromosomes can be distinguished, it is possible to determine which human chromosomes remain.

After many generations, most of the human chromosomes were lost. It was ultimately determined that one of the E group (17–18) carried the TK locus in the human genome. This conclusion was further substantiated by selecting against cells that synthesize TK. By adding the analogue bromodeoxyuridine to the culture, such cells are killed. None of the cells that survived the analogue addition were found to retain the E chromosome later shown to be human chromosome 17. Thus, the structural locus for the enzyme thymidine kinase can be assigned to human chromosome 17.

A more elegant method of assigning genes to human chromosomes has been made possible by the development of **hybrid clone panels.** These are hybrid cell lines that have been selected so that the pattern of retention or loss of each human chromosome is unique. In Figure 23.7, note that each chromosome has a unique pattern of representation. Only three cloned lines are needed to localize a gene to one of the eight human chromosomes defined by this panel. A combination of only five selected clones is needed to define a unique pattern of retention or loss for each of the 24 different human chromosomes; and if a human gene product can be detected in the hybrid cells, the gene responsible can be assigned to a chromosome in a single experiment. In the example shown in Figure 23.7, suppose that the human form of **acid phosphatase-1 (Acp-1)** is found in cell lines A and B but not in C. This pattern is consistent with the representation of chromosome 2 in the panel, and the gene for this enzyme can be provisionally assigned to chromosome 2.

A third approach to mapping human genes employs auxotrophic cell lines in which the nature of the enzyme defect is known. Theodore Puck and his col-

Human Chromosomes

Hybrid Clones	1	2	3	4	5	6	7	8
A	+	+	+	+	−	−	−	−
B	+	+	−	−	+	+	−	−
C	+	−	+	−	+	−	+	−

FIGURE 23.7
A hybrid clone panel. Each cell line (A, B, and C) contains a unique combination of human chromosomes. [(+): chromosome present; (−): chromosome absent]

leagues have created a collection of mutant hamster cell lines that require specific supplements for growth. In one such cell line designated ade⁻C, the medium must be supplemented with purines, because one of the early enzymes in the purine biosynthetic pathway, **glycineamide ribonucleotide synthetase (GARS)** is missing. To locate the structural gene for this enzyme in the human genome, the ade⁻C cell line is fused with human lymphocytes and plated on medium lacking adenine. Since human lymphocytes are nondividing, they will not grow, and since the medium is deficient in adenine, the parental hamster ade⁻C cells will not grow. Only hybrid cells containing the human chromosome that corrects this defect will grow. All such hybrids retain one small human chromosome, and one subline retains a single human chromosome (Figure 23.8). On the basis of its size and banding characteristics, it is identified as human chromosome 21. Biochemical assays indicate that the hybrid cells with this chromosome contain the enzyme GARS, but the ade⁻C cell line does not. On the basis of this evidence, the gene for GARS has been assigned to chromosome 21.

By using this general procedure, over 300 human genes have been investigated. Particularly through the efforts of Frank Ruddle, Thomas Shows, and their colleagues, these genes have either been confirmed or tentatively assigned to specific chromosomes. Of particular importance in this work has been the application of **chromosome banding techniques** (see Chapter 11). Recall that without such techniques, similar-sized chromosomes with identical centromere placement cannot be distinguished from one another.

Synkaryons and Human Gene Mapping

The techniques described in the previous section have enabled geneticists to assign specific human genes to specific chromosomes. By extending this approach to translocations and deletions, it is possible to assign specific loci to particular regions of the chromosome carrying the gene. When the data of many experiments involving many genes are accumulated and compiled, the result is a series of **chromosome maps.**

One experiment involved a human cell line that had been shown to have the long arm of the X chromosome translocated to chromosome 14. Since three genes—*PGK* (phosphoglycerate kinase), *G6PDH* (glucose-6-phosphate dehydrogenase), and *HGPRT* (hypoxanthine guanine phosphoribosyl transferase)—were all known to be on the X chromosome, it was possible to determine which ones were involved in the translocation.

The human cell line was hybridized with mouse cells, and the hybrid cell lines were selected for the presence of the human *HGPRT*⁺ gene. All surviving hybrids were then assayed for the presence of the human G6PDH and PGK gene products. Invariably, both gene products were present along with the *HGPRT* locus. This observation suggests that all three loci remained together in each surviving hybrid, and that they must all be located on *either* the translocated long arm of the X chromosome *or* on the short arm of the X chromosome. This conclusion is valid since there is no reason why both the remaining short arm of the X chromosome *and* chromosome 14 (containing

FIGURE 23.8

(a) Karyotype of an auxotrophic hamster cell line requiring adenine for growth. (b) The hybrid cell line that grows on minimal medium contains chromosome 21 as its only human chromosome, suggesting that the gene which corrects this defect is located on human chromosome 21. (Photos by M. Cummings.)

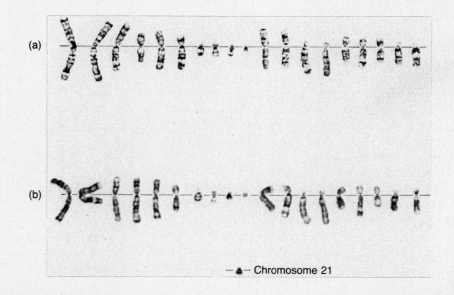

(a)

(b)

—♠— Chromosome 21

the long arm of the X) would *always* be present in every hybrid clone. Karyotypic analysis of the hybrid cells showed which of the alternatives was correct. Invariably, the X/14 translocation complex was present in surviving cells. Subsequently, translocations with break points at different locations along the long arm of the X were used to assign these three loci to specific regions, as shown in Figure 23.9.

A second experiment involved the *TK* locus on chromosome 17. When human-mouse hybrids were selected for the presence of the human TK gene product (thymidine kinase), occasional cell lines were shown to maintain human *TK* activity but to have lost chromosome 17! Careful analysis showed that part of the long arm of the human chromosome 17 had been translocated to one of the mouse chromosomes. By the use of chromosome banding analysis, it was established which part of chromosome 17 had been translocated and thus where the *TK* locus resides.

When translocations of specific segments of human chromosomes occur and a particular gene has been "mapped," it is then possible to look for the presence of other gene products that also correlate with the translocation. In this way, the proximity of genes may be established. In other cases, specific deletions may occur and be correlated with loss of function.

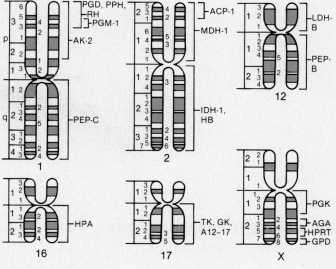

FIGURE 23.9

Regional gene assignments made by translocational analysis.

Chromosome-Mediated Gene Transfer

In addition to the transfer of entire chromosome sets through cell fusion, it is possible to transfer a small number of genes belonging to a single linkage group using preparations of metaphase chromosomes. O. Wesley McBride and Harvey Ozer prepared metaphase chromosomes from *HGPRT*⁺ Chinese hamster cells by inhibiting the cell cycle at metaphase with colchicine, followed by disruption of the cells and ultracentrifugation to separate the metaphase chromosomes from cell debris. This purified chromosome preparation was incubated with *HGPRT*⁻ mouse cells, and *HGPRT*⁺ cells were selected by growth in HAT medium. It was shown that the active HGPRT enzyme in these cells is the hamster form of the enzyme, thus providing a new system for transferring subgenomic amounts of genetic information from one cell to another.

Cells transformed with metaphase chromosome preparations are of two types: unstable transformants that rapidly lose genetic markers, and stable transformants that carry fragments of donor chromosomes integrated into the host genome (Figure 23.10). The analysis of these stable transformants can also be used to map genes to specific chromosome regions. Purified metaphase chromosomes from human HeLa cells incubated with mouse *TK*⁻ cells produce *TK*⁺ cells under HAT selection. Figure 23.10 shows the results of several experiments in which fragments of human chromosome 17 (lightly stained) became fused with host mouse chromosomes (darkly staining). The expression of the human forms of thymidine kinase (*TK*), galactokinase (*GALK*), and procollagen I (*ProCol I*) are also shown. By comparing the banding pattern of the human chromosome fragments with expression of human genes, it can be established that all three genes are on the long arm of the chromosome, with *GALK* closest to the centromere.

DNA-Mediated Gene Transfer

Although the transfer of genetic information by foreign DNA is well known in microbial systems (Chapter 14), the successful application of this method to higher eukaryotic cells was slow in developing and took place in three stages. DNA-mediated gene transfer in mammalian cells was first reported by Elizabeth Szybalska and Wclaw Szybalski in 1962, using DNA from *HGPRT*⁺ cells to transform *HGPRT*⁻ cells followed by selection for *HGPRT*⁺ cells in HAT medium. However, because the transfer was from one human cell line to another, it was difficult to prove that the gene expressed in the

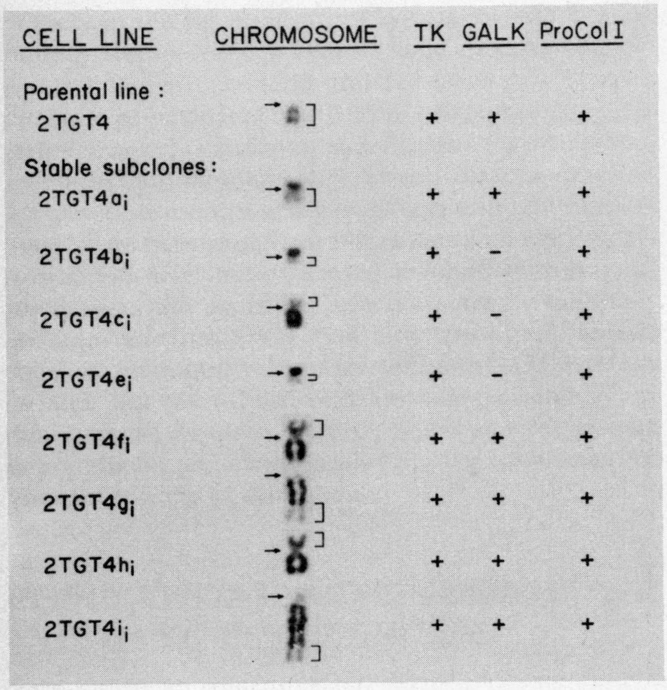

CELL LINE	CHROMOSOME	TK	GALK	ProCol I
Parental line :				
2TGT4		+	+	+
Stable subclones :				
2TGT4a₁		+	+	+
2TGT4b₁		+	−	+
2TGT4c₁		+	−	+
2TGT4e₁		+	−	+
2TGT4f₁		+	+	+
2TGT4g₁		+	+	+
2TGT4h₁		+	+	+
2TGT4i₁		+	+	+

FIGURE 23.10
Chromosome-mediated transfers in which fragments of
transferred human chromosome 17 (lightly staining) have
become associated with the darkly staining mouse
chromosomes. Expression of human chromosome 17 genes
is shown at right: (+) expression of genes, (−) no
expression. Brackets indicate the length of the human
fragment; arrows mark the centromeres.

transformed cells was that of the donor DNA, and not
the result of a mutation causing reexpression in the
host cells. Nonetheless, this experiment provided the
impetus for more work on DNA-mediated gene transfer in mammalian cells.

The second step in the development of transformation in higher eukaryotes was the use of herpes simplex virus (HSV) as a source of transfecting DNA. This
virus carries a gene for the enzyme thymidine kinase
(TK), which is immunologically different from the TK
produced by mammalian cells. William Munson and
his colleagues treated TK^- mouse cells with HSV that
was incapable of causing lytic infection. TK^+ cells were
selected in HAT medium, and the active TK enzyme in
these cells was shown to be viral, not mammalian.
This was concrete proof that mammalian cells are capable of taking up exogenous DNA and can express
genes encoded on incorporated DNA sequences.

At this stage, DNA-mediated gene transfer was used
in the isolation of several genes, but application was
limited to selectable markers. To overcome this difficulty, Michael Wigler, Richard Axel, and their colleagues demonstrated the feasibility of cotransferring
cells with two physically unlinked genes, one of which
is a selectable marker. They incubated mouse TK^-
cells with purified HSV TK^+ DNA and other DNA, such
as a cloned rabbit beta-globin gene, and obtained a
high proportion (6/8) of TK^+ transformed cells that
contained a cotransferred rabbit beta-globin gene. This
work demonstrated that two physically unlinked
genes can be transferred into the same cell with a high
frequency. While the details of the mechanism are unclear, subsequent work has demonstrated that following entry into the cell, the unrelated DNA sequences
become ligated together and can become integrated
into a random chromosomal site in the recipient cell.
In addition, it has been shown that these integrated
DNA sequences can be expressed in the host cells.

The coupling of somatic cell genetics with **recombinant DNA techniques** has increased the degree of
resolution in mapping genes to specific chromosome
regions compared with conventional mapping using
deletions or translocations. In particular, the beta-globin gene family has been localized to a 4500-kb region
on the short arm of chromosome 11 (see Chapter 19).
A panel of hybrid clones containing various deletions
of chromosome 11 (Figure 23.11) essentially divides the
chromosome into six regions. To localize the beta-globin gene, DNA from each of these cell lines was cut
with restriction enzymes, the DNA fragments separated by electrophoresis and transferred to nitrocellulose paper by the **Southern technique.** Then, radioactive cloned DNA fragments of the beta-globin gene
isolated by recombinant DNA techniques were hybridized to the immobilized cellular DNA to test for the
presence or absence of the globin gene. Table 23.1
summarizes the results of these experiments and indicates that the gene is present in the DNA from J1–
23, J1–10, and J1–11, but not from J1–7 or J1–9. Thus,
the beta-globin gene is located between the break-
points for J1–9 and J1–10. Cytogenetic examination of
the difference between these partially deleted chromosomes reveals a small, 4500-kb segment (Region III
in Figure 23.11) within which the beta-globin gene is
contained.

Restriction Fragment Length Polymorphisms (RFLPS)

Although an increasing number of human genes have
been localized to chromosomal sites, in the majority
of human genetic diseases the primary defect is un-

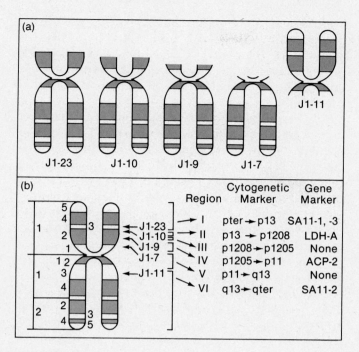

FIGURE 23.11

(a) Diagram of terminal deletions of chromosome 11 in five hybrid cell lines. (b) Human chromosome 11. The arrows indicate the break points for the deletions shown above. The breaks divide the chromosome into six regions, each of which is characterized by cytogenetic and genetic markers. SA11-1, -2, and -3 are cell surface antigens; LDH-A is lactate dehydrogenase; and ACP-2 is acid phosphatase. (From Gusella et al., 1979, p. 5240.)

known, and without a marker the gene cannot be mapped. For example, cystic fibrosis is an autosomal recessive condition afflicting one in 2000 individuals. To date, intensive efforts have failed to identify the mutant gene product(s) responsible for this disease, and, as a result, the chromosomal locus of this defect remains unknown.

The development of recombinant DNA technology has provided a new way to study the inheritance of allelic variants in humans that makes it possible to map genes without the need to identify variants or defective protein products. This method uses **restriction endonucleases,** enzymes that make double-stranded cuts in DNA molecules at recognition sites consisting of specific nucleotide sequences (see Table 20.1). Single base changes in the recognition sequence can alter the pattern of cuts made in DNA, giving rise to a detectable variation that is inherited in a Mendelian codominant fashion (Figure 23.12). These allelic variants or polymorphisms are called **restriction fragment length polymorphisms** or **RFLPs** for short. Such variations can be used to map the chromosomal location of genes for which no mutant gene product has been identified.

The use of RFLP markers for mapping requires the selection of cloned DNA sequences that recognize variations in DNA cutting sites, and the mapping of these sequences to specific human chromosomes by using a panel of somatic cell hybrids. Over 200 such RFLP markers have now been identified, and at least one marker is available for each human chromosome. As a final step, one must identify large families in which both the RFLP marker and a specific genetic disease are inherited. This is necessary in order to demonstrate linkage of the marker with the genetic disease.

TABLE 23.1

Demonstration of relationships between amount of chromosome 11 lost through terminal deletions in various mutant hybrid clones and presence of human beta-globin gene.

Clone	Terminal Deletion	Presence of Specific Human Markers					Assay for Human Hbβ Gene
		SA11-1	SA11-3	LDH$_A$	ACP$_2$	SA11-2	
J1	None	+	+	+	+	+	+
J1-23	p13→pter	−	−	+	+	+	+
J1-10	p1208→pter	−	−	−	+	+	+
J1-9	p1205→pter	−	−	−	+	+	+
J1-7	p11→pter	−	−	−	−	+	−
J-11	q13→qter	+	+	+	+	−	+

SOURCE: From Gusella et al., 1979.

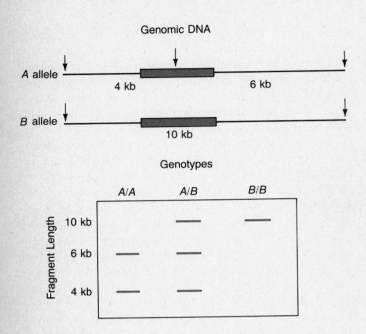

FIGURE 23.12
Restriction fragment length polymorphisms. The lines represent genomic DNA, and the hatched areas correspond to a single copy sequence that has been cloned from this region of the genome. Restriction endonuclease sites are indicated by arrows. If the central cutting site is present (as in *A*), fragments of 4 kb and 6 kb are generated. If this site is missing (as in *B*), a single 10-kb fragment is produced. The absence of the site could be the result of a single base change in the nucleotide sequence of the recognition site. The alleles carried by any individual can be detected by digestion of genomic DNA (obtained from leucocytes or skin fibroblasts) with a given restriction enzyme, separation of the DNA fragments by agarose gel electrophoresis, transfer of DNA to a filter, and hybridization with the radioactive cloned sequence. The location of the hybridized fragments is detected by autoradiography. The patterns for the three possible genotypes are shown in the box.

This approach has been used by James Gusella and his colleagues to map the gene responsible for Huntington's disease (HD), a degenerative and fatal disease of the nervous system that is inherited as an autosomal dominant (see Chapter 24 for a description of HD). These researchers assembled cloned DNA markers representing each of the human autosomes. To map the *HD* locus, they used a large, multigenerational family in which *HD* was segregating. In the family pedigree (Figure 23.13), an RFLP marker allele called G8 and *HD* show very tight linkage. Previous work had shown that this RFLP marker was located on the short arm of chromosome 4, indicating that the *HD* gene is

also located on chromosome 4. Future work with flanking RFLP markers will help define the precise location of the *HD* gene and, in combination with other recombinant DNA techniques, the resulting information can be used to clone the *HD* gene and isolate its gene product.

This modification of the usual method of mapping human genes is a significant breakthrough and can be applied to many other genetic diseases where the primary defect is unknown. Along with the other somatic cell hybrid techniques described above, this method will revolutionize mapping of the human genome and promises to produce a human gene map approaching the level of resolution achieved for other eukaryotic organisms such as *Drosophila* and the mouse.

At this writing, more than 400 genes have been individually assigned to specific autosomes, and approximately half of these have been provisionally or absolutely assigned to a particular chromosome region. Another 250 or so X-linked genes have not been included in this calculation. Altogether, it has been estimated that about 2 percent of the genes in the human genome have been identified. Table 23.2 lists some of the various human genes that have been assigned to a specific chromosome and a particular region, if known. Note that the chromosome assignment has been confirmed positively (C); provisionally assigned (P); or inconsistent in assignment (I).

Although the total number of loci mapped is still small, these maps represent remarkable progress when we consider that our own species is not amenable to conventional mapping procedures.

Synkaryons: Other Applications

Somatic cell hybrids have been used to investigate other broad, unsolved questions in genetics. First, we still do not know the means by which specific genes are activated in a eukaryotic organism's specialized cells and why they are inactive in most other cells (see Chapter 22). Second, the possible genetic basis of malignancy is a question of great significance. We will address each of these basic problems briefly.

When somatic cell hybrids are formed between a cell line derived from highly differentiated cells that synthesize a specialized product and a cell line that does not make that product, what is the fate of the gene product in the hybrid line? The most usual occurrence is that the gene product under investigation is no longer produced; that is, it is **extinguished.** For example, Syrian hamster melanoma cells make large amounts of melanin under the direction of the enzyme **dopa oxidase (tyrosinase),** which catalyzes the multistep conversion of tyrosine to melanin. When

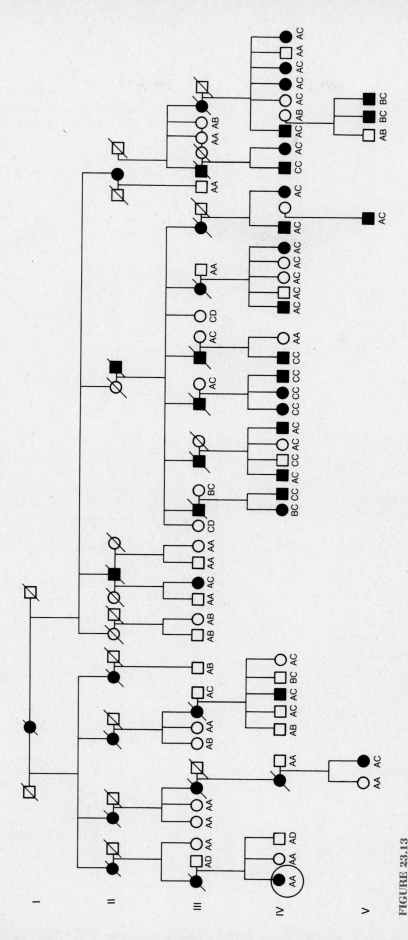

FIGURE 23.13

Analysis of Huntington's disease (HD) pedigree. The RFLP marker, G8 (mapped to chromosome 4), recognizes two polymorphic sites, giving rise to four codominant alleles, *A*, *B*, *C*, and *D*. The pedigree shows the coinheritance of the *C* allele with HD (filled symbols), leading to the conclusion that the structural gene for HD is on chromosome 4. Statistical analysis indicates that the odds are greater than 1 trillion to 1 in favor of such linkage. A single-crossover event, indicating recombination between the G8 marker and the *HD* locus is circled in generation *IV*. Deceased individuals are indicated by a diagonal slash. (From Gusella et al., 1984, with permission.)

TABLE 23.2
A partial list of mapped human genes.

Marker Symbol	Marker Name	Location of Gene Determining Marker Phenotype	
		Chromosome	Region
A12M1	Adenovirus-12 chromosome modification site 1C	1 (P)	q42–q43
ABO	ABO blood group	9	q34
ACP1	Acid phosphatase-1	2	p23 or p25
ACP2	Acid phosphatase-2	11	p12–cen
ADA	Adenosine deaminase	20	q13–ter
AG(HBAC)	ALPHA GLOBIN GENE CLUSTER	16	p12–pter
ALB	Albumin	4	q11–q13
AMY	alpha amylase (salivary)	1	p21
AT3	Antithrombin III	1 (P)	q23.1–23.9
C2	Complement component-2	6 (P)	p 21.3
CAH	Congenital adrenal hyperplasia	6	p21
CBD	Color blindness (deutan)	X	q28
CBP	Color blindness (protan)	X	q28
CC	Congenital cataract	—	—
CHE2	Cholinesterase (serum)-2	16 (P)	cen–q22
COL1A1	Collagen I alpha-1 chain	7	q210–q220
F7	Clotting factor VII	8 (I)	q34
F7E	Clotting factor VII, expression	8	—
FY	Duffy blood group	1	q12–q21
G1	Immunoglobin γ^1 chain marker	6 (I), 7 (I), 14 (I), X (I)	—
GALK	Galactokinase	17	q21–q22
GALT	Galactose-1-phosphate uridyltransferase	9	p13–p21
GLA	alpha-galactosidase	X	q22–q24
GLB1	beta-galactosidase-1	3	p21–cen
G6PD	Glucose-6-phosphate	X	q28
H	HISTONE GENE FAMILY	7	q32–q36
HBB	Hemoglobin beta chain	11	p15
HBD	Hemoglobin delta chain	11	p15
HBG1	Hemoglobin gamma ala as AA 136	11	p15
HBG2	Hemoglobin gamma gly as AA 136	11	p15
HD	Huntington's disease	4	—
HEMA(F8C)	Classic hemophilia (A)	X	q28
HEXA	Hexosaminidase-A (α subunit)	15 (C)	q22–q2501
HEXB	Hexosaminidase-B (β subunit)	5	q13
HLA-A,B,C	Human leucocyte antigens	6	p21.3
HLAD/DR	Human leucocyte antigen, D related	6	p21.3
HPRT	Hypoxanthine phosphoribosyl transferase	X	q26–q28
IF1	Interferon-1	2 (P)	p32–qter
IF2	Interferon-2	5 (P)	p
IFF	Interferon-fibroblast type	9	p24–p13
IFL(IFA)	INTERFERON, LEUCOCYTE GENE CLUSTER	9	p13–qter
IGH	IMMUNOGLOBULIN, HEAVY CHAIN GENE CLUSTER	14	q32.3
IGK	IMMUNOGLOBULIN, KAPPA LIGHT CHAIN GENE CLUSTER	2	p

TABLE 23.2
(cont.)

INS	Insulin	11	p15–p15.5
JK	Kidd blood group	2	q
LARS	Leucyl-tRNA synthetase	5(P)	—
MN	MN blood group	4	q28–q31
NPS1	Nail-patella syndrome type 1	9	q3
OA1	Ocular albinism-1(Nettleship-Falls Type)	X	p22
PAH(PKU)	Phenylalanine hydroxylase	12	—
PG	Pepsinogen (isozyme-5)	6 (I)	—
PGM1	Phosphoglucomutase-1	1	p34–p33
PGM2	Phosphoglucomutase-2	4	p14–q12
PGM3	Phosphoglucomutase-3	6	q12–qter
PI	alpha-1-antitrypsin	14	q
PKU	Phenylketonuria = PAH, q.v.	1	
RH	Rhesus blood group	1	p32–p36.11
RNR	Ribosomal RNA	13, 14, 15, 21, 22	p12
RN5S	5S RNA	1 (C)	q42–q43
TK1	Thymidine kinase (soluble)	17	q21–q22
WARS	Tryptophanyl-tRNA synthetase	14	q21–qter
XG	Xg blood group	X	—

SOURCE: After McKusick, 1984.

these cells are fused with mouse fibroblasts that do not produce melanin, the hybrids lose the capacity for pigment production. Through appropriate experimentation with suitable controls, the data are consistent with the hypothesis that the fibroblast genome is responsible for the production of a diffusible substance that exerts negative control on the hamster genes essential for pigment synthesis.

In a second example, a particular human chromosome has been correlated with the regulation of a kidney-specific enzyme from mouse cells. The enzyme esterase-2 (ES-2) was initially extinguished in human-mouse hybrids, but then reactivated as human chromosomes were lost. The loss of chromosome 10 was correlated with reactivation. Experiments such as these provide excellent opportunities to pursue problems with a genetic base.

Investigation of the **genetic basis of malignancy** using somatic cell hybridization techniques has yielded conflicting results. When malignant cells are fused with nonmalignant cells, it is possible to test hybrid cells for their tumor-inducing capacity. In some studies, malignancy has resulted and is therefore dominant to nonmalignancy in hybrid cells. In other cases, the nonmalignancy phenotype was shown to be dominant and to suppress tumor induction in hybrid cells. This result suggests that a negative genetic control mechanism is operational in the hybrid cells. Under this assumption, attempts to localize this control mechanism on one or more chromosomes may be made. In theory, this may be accomplished by correlating the presence of specific chromosomes with the regulatory action and subsequently the loss of regulation with the loss of chromosomes. Although such experiments have been conducted and tentative findings are available, no general conclusions have been drawn. Such a technique may be particularly valuable in basic investigations of cancer.

Hybridomas

The technique of somatic cell hybridization has also been used to produce cloned hybrid cells involved in antibody production. Cesar Milstein pioneered this method, which employs the fusion of myeloma cells, a tumor line that grows indefinitely in tissue culture, with B lymphocytes from spleen. The B lymphocytes have the capacity to synthesize a single antibody, but will not grow *in vitro.* Hybrids between these cells, known as **hybridomas,** have the ability to grow in culture and synthesize large amounts of a single antibody. After hybrids are formed, they are diluted and subcultured to produce clones derived from single cells. These clones can be screened to identify those producing an antibody of interest. Such antibodies, derived from a single cloned cell, are known as **monoclonal antibodies.** This powerful technology has many applications in research and medicine. For example, since many forms of cancer possess unique cellular surface antigens, monoclonal antibodies to these antigens can be used to identify tumor cells in a tissue or organ and may serve as vehicles for delivering antitumor agents directly to cancerous cells.

SUMMARY AND CONCLUSIONS

The discovery that somatic cells cultured *in vitro* can fuse or hybridize with other cells from the same or different species led to the development of somatic cell hybridization techniques. Heterokaryons, in which two nuclei share a common cytoplasm, and synkaryons, in which the nuclei subsequently fuse, have been used in many ways in the study of eukaryotic genetics.

Hybrid cells may be formed spontaneously in a mixture of two cell lines, but are most efficiently obtained using agents such as Sendäi virus or chemicals that promote cell fusion. Selection of hybrid cells involves using mutations in the two cell lines so that only hybrid cells grow as a result of complementation. Auxotrophic cell lines have also been employed in developing selection systems. Complementation studies, where wild-type genes from one genome compensate for mutant genes of a second genome, and vice versa, can provide insights into the genetic basis of disease. If complementation is demonstrated by heterokaryons produced from cells derived from different individuals with the same disorder, then more than one genetic factor can produce the disorder. Such studies have established this to be the case in the human disorders xeroderma pigmentosum and maple syrup urine disease.

The most significant application of somatic cell hybridization techniques in humans is their use in assigning genes to linkage groups and mapping their location on specific chromosomes. Certain interspecific hybrids preferentially lose human chromosomes randomly. The correlation between the expression of a human gene product and the invariable presence of a specific human chromosome determines linkage of the gene. By using translocations and deletions of specific regions and chromosome banding techniques, geneticists have determined the location of many human genes. Over 260 genes have been assigned to specific autosomes, and half of these have been assigned to a particular region of the chromosome. About 100 other genes have been located on the X chromosome.

Other applications of somatic cell hybridization include the investigation of eukaryotic gene regulation and the genetic basis of malignancy. In the future, we can expect that this technique will be applied in many ways in the study of eukaryotic genetics.

PROBLEMS AND DISCUSSION QUESTIONS

1 What is the major difference between a heterokaryon and a synkaryon?

2 What experimental genetic approaches using synkaryons cannot be used with heterokaryons?

3 A guanine analogue, 8-azaguanine, causes cell death when it is incorporated into DNA. In the Littlefield selection technique, *HGPRT*$^-$ cells are initially selected on the basis of being resistant to 8-azaguanine. How do these observations relate to the pathways by which nucleotides are synthesized?

4 Based on the information in Problem 3, what is the fate of a mixed population of parental cells that is placed in a medium containing both 8-azaguanine *and* aminopterin?

5 In Chapter 20, the enzyme LDH was shown to be a tetrameric molecule made up of either 4 A subunits, 4 B subunits, or a mixture of A and B subunits in various combinations. The isozymic forms of LDH were examined electrophoretically in two different cell lines from two species. One species contained only LDH-1 (A_0B_4), while the other species showed only LDH-5 (A_4B_0). Predict the pattern of isozymes of LDH that would be revealed in an interspecific hybrid derived from these two cell lines.

6 What is the evidence from the study of heterokaryons which suggests that cytoplasmic information plays an important role in regulating the genetic activity or inactivity of the nucleus of a cell?

7 Compare and contrast the data and conclusions drawn concerning regulation of the cell cycle from heterokaryon studies discussed in this chapter with the nuclear transplantation studies of Prescott and Goldstein using *Amoeba* (see Chapter 6).

8 The data below represent three theoretical studies of genetic complementation. A, B, C, and D represent four separate cell lines, each defective in their ability to synthesize a particular amino acid. Each set of data represents all possible heterokaryon formations. "+" indicates that amino acid synthesis was restored, and "−" represents no restoration. For each case, indicate the number of complementation groups and which cell lines are in the same complementation group.

(a)

	A	B	C	D
A		−	+	−
B			+	−
C				+

(b)

	A	B	C	D
A		+	+	−
B			−	+
C				+

(c)

	A	B	C	D
A		+	+	+
B			+	+
C				−

9 Describe the rationale of assigning the linkage of genes to human chromosomes using somatic cell hybridization.

10 The theoretical data shown in the following table represent the activity of five human enzymes as well as the human chromosomes remaining in five clones of human-mouse hybrid cells. Based on this information, draw an appropriate conclusion for a linkage assignment of each gene specifying each enzyme.

		HYBRID CLONES				
		A	B	C	D	E
Enzyme	I	+	−	−	+	−
	II	−	+	+	−	−
	III	+	−	−	+	−
	IV	+	+	+	+	+
	V	−	−	−	−	−
Chromosomes	2	−	−	+	−	+
	14	−	+	+	−	−
	22	+	+	+	+	+
	X	+	−	−	+	−

11 Discuss how translocations and chromosome banding techniques are used in somatic cell hybrids to assign specific genes to specific human chromosomal regions.

12 How might *in situ* hybridization techniques (see Chapter 10) be used to confirm somatic cell hybridization studies that have assigned linkage groups for genes coding for rRNA?

SELECTED READINGS

BARSKI, G., SORIEUL, S., and CORNEFERT, F. 1961. Hybrid cell types in combined cultures of two different mammalian cell strains. *J. Natl. Cancer Inst.* 26: 1269–91.

CARLSON, P. S. 1973. Somatic cell genetics of higher plants. In *Genetic mechanisms of development*, ed. F. H. Ruddle, pp. 329–55. New York: Academic Press.

CASKEY, C. T., and ROBBINS, D. C., eds. 1982. *Somatic cell genetics.* New York: Plenum Press.

Cell Fusion. 1984. CIBA Foundation Symposium, Number 103. London: Pitman.

CROCE, C. M., and KOPROWSKI, H. 1978. The genetics of human cancer. *Scient. Amer.* (Feb.) 238: 117–25.

DAVIDSON, R. L. 1973a. Control of the differentiated state in somatic cell hybrids. In *Genetic mechanisms of development*, ed. F. H. Ruddle, pp. 295–328. New York: Academic Press.

————. 1973b. *Somatic cell hybridization: Studies on genetics and development.* Reading, Mass.: Addison-Wesley.

————. 1974. Genetic expression in somatic cell hybrids. In *Annual review of genetics*, Vol. 8, eds. H. L. Roman, et al., pp. 195–218. Palo Alto, Calif.: Annual Reviews.

————. 1984. *Somatic cell genetics.* Benchmark Papers in Genetics, Vol. 14. Stroudsburg, Pa.: Hutchinson Ross.

DAVIES, K. E. 1985. New hypervariable DNA markers for mapping human genetic disease. *Trends in Genetics* 1: 97–98.

EPHRUSSI, B. 1972. *Hybridization of somatic cells.* Princeton, N.J.: Princeton University Press.

EPHRUSSI, B., and WEISS, M. C. 1969. Hybrid somatic cells. *Scient. Amer.* (April) 220: 26–35.

GORDON, S. 1973. Regulation of differentiated phenotypes in heterokaryons. In *Genetic mechanisms of development*, ed. F. H. Ruddle, pp. 269–94. New York: Academic Press.

GUSELLA, J., VARSANYI-BREINER, A., KAO, F. T., JONES, C., PUCK, T. T., KEYS, C., ORKIN, S., and HOUSMAN, D. 1979. Precise localization of human beta-globin gene complex on chromosome 11. *Proc. Natl Acad. Sci.* 76: 5239–43.

GUSELLA, J., TANZI, R. E., ANDERSON, M. A., HOBBS, W., GIBBONS, K., RASCHTCHIAN, R., GILLIAM, T. C., WALLACE, M. R., WEXLER, N. S., and CONNEALLY, P. M. 1984. DNA markers for nervous system diseases. *Science* 225: 1320–26.

HARRIS, H. 1970. *Cell fusion.* Cambridge, Mass.: Harvard University Press.

————. 1974. *Nucleus and cytoplasm.* 3rd ed. Oxford, England: Clarendon Press.

HOOD, L. E., WILSON, J. H., and WOOD, W. B. 1975. *Molecular biology of eukaryotic cells: A problems approach.* Menlo Park, Calif.: W. A. Benjamin.

KAO, F. T., and PUCK, T. 1974. Induction and isolation of auxotrophic mutants in mammalian cells. In *Methods in cell biology*, ed. D. Prescott, vol. 8, pp. 23–39. New York: Academic Press.

KEATS, B. 1982. Genetic mapping: Chromosome 6–22. *Amer. J. Hum. Genet.* 34: 730–34.

LEWIN, B. 1974. *Gene expression: Eukaryotic chromosomes.* Vol. 2. New York: Wiley.

LITTLEFIELD, J. W. 1964. Selection of hybrids from matings of fibroblasts *in vitro* and their presumed recombinants. *Science* 145: 709–10.

McBRIDE, O. W., and OZER, H. 1973. Transfer of genetic information by purified metaphase chromosomes. *Proc. Natl. Acad. Sci.* 70: 1258–62.

McKUSICK, V. A. 1984. The human gene map 15 November 1983. *Clin. Genet.* 25: 90–123.

McKUSICK, V., and RUDDLE, F. H. 1977. The status of the gene map of the human chromosomes. *Science* 196: 390–405.

MOORE, E. E., JONES, C., KAO, F. T., and OATES, D. 1977. Synteny between glycinamide ribonucleotide synthetase and superoxide dismutase (soluble). *Amer. J. Hum. Genet.* 29: 386–96.

MUNYON, W., KRAISELBURD, E., DAVIS, D., and MANN, J. 1971. Transfer of thymidine kinase to thymidine kinaseless L cells by infection with ultraviolet-irradiated herpes simplex virus. *J. Virol.* 7: 813–20.

PELLICER, A., ROBINS D., WOLD, B., SWEET, R., JACKSON, J., LOWY, I., ROBERTS, J. M., SIM, G.-K., SILVERSTEIN, S., and AXEL, R. 1980. Altering genotype and phenotype by DNA-mediated gene transfer. *Science* 209: 1414–22.

PUCK, T., and KAO, F. T. 1982. Somatic cell genetics and its application to medicine. *Ann. Rev. Genet.* 16: 225–72.

ROBINS, D. M., RIPLEY, S., HENDERSON, A. S., and AXEL, R. 1981. Transferring DNA integrates into the host chromosome. *Cell* 23: 29–39.

RUDDLE, F. H. 1981. A new era in mammalian gene mapping: Somatic cell genetics and recombinant methodologies. *Nature* 294: 115–20.

RUDDLE, F. H., and CREAGAN, R. P. 1975. Parasexual approaches to the genetics of man. In *Annual Review of Genetics*, ed. H. L. Roman, et al., pp. 407–86. Palo Alto, Calif.: Annual Reviews.

RUDDLE, F. H., and KUCHERLAPATI, R. S. 1974. Hybrid cells and human genes. *Scient. Amer.* (July) 231: 36–49.

SEEGMILLER, J. E., and BOSS, G. R. 1982. Genetic defects in human purine and pyrimidine metabolism. *Ann. Rev. Genet.* 16: 297–328.

SHOWS, T. B., and McALPINE, P. J. 1979. The 1979 catalog of human genes and chromosome assignments. *Cytogenet. Cell Genet.* 25: 117–29.

SHOWS, T. B., SAKAGUCHI, A. Y., and NAYLOR, S. L. 1982. Mapping the human genome, cloned genes, DNA polymorphisms, and inherited disease. *Adv. in Hum. Genet.* 12: 341–452.

SIDEBOTTOM, E. 1974. Heterokaryons and their uses in studies of nuclear function. In *The cell nucleus*, vol. 1, ed. H. Busch, pp. 441–70. New York: Academic Press.

SIXTH INTERNATIONAL WORKSHOP ON HUMAN GENE MAPPING. 1982. *Cytogenet. Cell Genet.* 32: 1–240.

SZYBALSKA, E. H., and SZYBALSKI, W. 1962. Genetics of human cell lines. IV. DNA-mediated heritable transformation of a biochemical trait, *Proc. Natl. Acad. Sci.* 48: 2026–34.

WESTERVELD, A., et al. 1973a. *Mammalian cell hybridization: I.* New York: MSS Information Corp.

————. 1973b. *Mammalian cell hybridization: II.* New York: MSS Information Corp.

ZEUTHEN, J. 1975. Heterokaryons in the analysis of genes and gene regulation. *Humangenetik* 27: 275–301.

24

Genes and Behavior

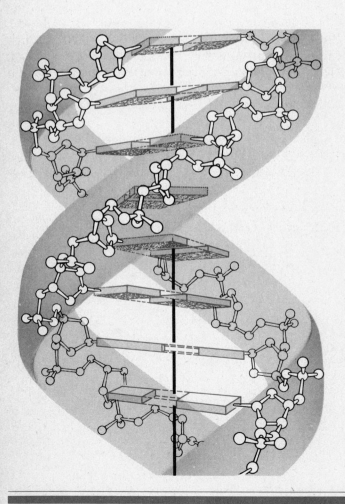

Behavior is defined generally as reaction to stimuli or environment. In broad terms, every action, reaction, and response represents a type of behavior. Animals run, remain still, or counterattack in the presence of a predator; birds build complex and distinctive nests; fruit flies execute intricate courtship rituals; plants bend toward light; and humans reflexively avoid painful stimuli as well as "behave" in a variety of ways as guided by their intellect, emotions, and culture.

Even though clearcut cases of genetic influence on behavior were known in the early 1900s, the study of behavior was of greater interest to psychologists, who were concerned with learning and conditioning. While some traits were recognized as innate or instinctive, behavior that could be modified by prior experience received the most attention. Such traits or patterns of behavior were thought to reflect the previous environmental setting to the exclusion of the organism's genotype. This philosophy served as the basis of the **behaviorist school.**

Such thinking provided a somewhat distorted view of the nature of behavioral patterns. It is logical that a genotype can be expressed within a series of environmental levels (e.g., cell, tissue, organ, organism, population, surrounding environment) and that a behavioral pattern must rely on the expression of the individual genotype for its execution. Nevertheless, the so-called **nature-nurture controversy** flourished well into the 1950s. By that time it became clear that while certain behavioral patterns, particularly in less advanced animals, seemed to be innate, others were the result of environmental modifications limited by genetic influences. The latter condition is particularly true in more advanced organisms with more complicated nervous systems.

Since about 1950, studies of the genetic component of behavioral patterns have intensified, and support for the importance of genetics in understanding behavior has increased. The prevailing view is that all behavior patterns are influenced both genetically and environmentally. The genotype provides the physical basis and/or mental ability essential to execute the behavior and further determines the limitations of environmental influences.

Behavior genetics has blossomed into a distinct specialty within the larger field of genetics as more and more behaviors have been found to be under genetic control. This chapter provides an overview of the role of genes in behavior. More extensive treatment can be found in the list of selected readings at the end of the chapter. There seems to be no question that this topic will be one of the most exciting areas of genetics in the years to come.

THE STUDY OF BEHAVIOR GENETICS: METHODOLOGY

In 1965, David Merrell summarized the three major approaches used in studying behavior genetics. The first involves the determination of behavioral differences between genetic strains of the same species or between closely related species. If such closely related organisms exist in similar environments and their survival needs are identical, any observed behavioral differences may be correlated with genetic differences. In the second approach, a modified behavioral trait is selected from a heterozygous population. If such a modified trait can be established in a new genetic strain, the positive influence of the genotype is established. The first two approaches identify behavioral patterns as being under the control of genes. Genetic investigation through controlled breeding experiments can then be performed in an attempt to establish inheritance patterns. In most cases where these approaches have been used successfully, it has only been shown that the inheritance is not due to simple Mendelian patterns. Instead, behavioral traits have been attributed to polygenic inheritance (i.e., many genes controlling one trait).

The third approach, infrequent in 1965 but used extensively today, is to study the effects of single genes on behavior. Within inbred strains of organisms with relatively constant genotypes, spontaneous mutants with behavioral deviations may arise. Even more productive is the induction and isolation through selection of mutations altering behavior. While a single complex behavioral trait may be controlled by many genes, disruption of the trait by a single mutation may occur and be analyzed. This approach is the most attractive because it offers the most objective information concerning the role of genes in behavior. Nevertheless, as we proceed through this chapter, all three approaches will be illustrated.

COMPARATIVE APPROACHES IN STUDYING BEHAVIOR GENETICS

Before turning to specific examples of genetic influence on behavior, we shall present several case studies to illustrate two of the approaches used in behavior genetics. In the first approach, behavior is compared in closely related strains. Alcohol preference in mice, open-field behavior in mice, and mate selection in baboons are case studies representative of this method. The second approach involves selection within heterozygous populations for a modification of behavior. To illustrate this mode of investigation, we will describe studies of learning in rats and geotaxis in *Drosophila*.

Alcohol Preference in Mice

Many studies of **alcohol preference** in mice have been reported. Different inbred strains have been compared for preference or aversion to ethanol, with all studies indicating genetic control of this response.

For example, D. A. Rogers and Gerald E. McClearn compared four strains of inbred mice over a period of three weeks. Each strain was presented with seven vessels containing either pure water or alcohol varying in strength from 2.5 to 15.0 percent. Daily consumption was measured. Table 24.1 shows the proportion of absolute alcohol to total liquid consumed on a weekly basis.

Examination of these data shows clearly that the C57BL and C3H/2 strains exhibit a preference for alcohol, while BALB/c and A/3 demonstrate an aversion toward it. Since the environment in raising the mice over many generations has been constant, the preference differences are attributed to the genotypes of each strain. Presumably, these strains contain identical sets of genes and vary only by the fixation of different alleles at these loci.

In other similar studies, crosses between strains differing in alcohol preference have been performed. The results have not established any simple mode of in-

TABLE 24.1
Alcohol consumption in mice.

Strain	Week	Proportion of Absolute Alcohol to Total Liquids	$\bar{x}$
C57BL	1	0.085	
	2	0.093	0.094
	3	0.104	
C3H/2	1	0.065	
	2	0.066	0.069
	3	0.075	
BALB/c	1	0.024	
	2	0.019	0.020
	3	0.018	
A/3	1	0.021	
	2	0.016	0.017
	3	0.015	

SOURCE: From Rogers and McClearn, 1962. Reprinted by permission from *Quarterly Journal of Studies on Alcohol*, Vol. 23, pp. 26–33, 1962. Copyright by Journal of Studies on Alcohol, Inc. New Brunswick, NJ 08903.

heritance, however. It appears that two or more genes may be involved.

It has been suggested that alcohol preference may be related to enzyme activity of the liver. While some differences between strains have been noted in the levels of alcohol dehydrogenase (ADH) and acetaldehyde dehydrogenase (ACH), the results are still inconclusive. Both enzymes are involved in the metabolism of alcohol.

Open-Field Behavior in Mice

First used in 1934, an **open-field test** was designed to study exploratory and emotional behavior in mice. When placed in a new environment, mice normally explore the surroundings, but they are a bit cautious or "nervous" about the new setting. The latter response is evidenced by their elevated rate of defecation and urination. To study these behaviors in the laboratory, an enclosed, brightly illuminated box was devised with the floor marked into squares. Exploration is measured by counting the number of movements into different sections, and emotion is measured by counting the number of defecations. These measurements thus attempt to quantify the two behavioral patterns.

As in the study of alcohol preference, different inbred strains of mice have been shown to vary significantly in their response to the open-field setting. Of greatest interest is the experimentation of John C. DeFries and his associates in 1966. DeFries concentrated his work on two strains. BALB/cJ and C57BL/6j. The BALB strain is homozygous for a coat color allele, *c*, and is albino, while the C57 strain has normal pigmentation (*CC*). BALB demonstrates low exploratory activity and is very emotional, while C57 is field-active and nonemotional.

DeFries proceeded to cross the two strains and then to interbreed each generation, creating F_2, F_3, F_4, etc., generations. Each generation beyond the F_1 contained albino and nonalbino mice, and these were tested. Regardless of generation, pigmented mice behaved as strain C57, while albino mice behaved as BALB. The general conclusion may be drawn that the *c* allele behaves **pleiotropically,** affecting both coat color and behavior.

Detailed heritability and variance analysis has been performed to assess the degree of input of the *c* gene to these behavioral patterns. Such analysis has shown that this locus accounts for 12 percent of the additive genetic variance in open-field activity and 26 percent of defecation-related emotion. Thus, these behaviors are under polygenic control, presumably in an additive way.

What might be the relation between albinism and behavior? DeFries sought an answer by testing albinos and nonalbinos under both white and red light. Red light provides little visual stimulation to mice. The behavioral differences between mice with the two types of coat pigmentation disappeared under red light. Thus, albino open-field responses are visually mediated. This is not surprising, since albino mice are photophobic, and lack of pigmentation in albinos extends to the iris as well as to the coat.

Baboon Mate Selection

Baboons are large, quadrupedal terrestrial monkeys found in Africa (Figure 24.1). Because they are primates and exhibit a variety of social behavioral patterns, they have been investigated rather thoroughly. They are classified into one genus, *Papio*, and several species. *Papio anubis* is a savanna-dwelling group, while *Papio hamadryas* is a desert dweller. Their ranges sometimes overlap.

P. hamadryas has a social system that includes harems. Each male is associated permanently with one or more adult females. *P. anubis* displays the more common social structure of baboons, in which permanent associations are not formed. Instead, matings are promiscuous but affected by dominance ranking of males.

FIGURE 24.1
A male and female Gelada baboon. (Photo © Zoological Society of San Diego.)

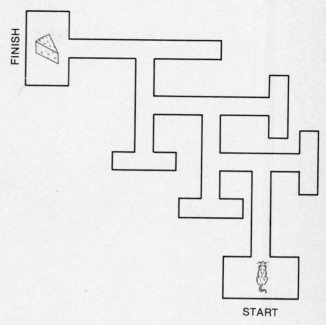

FIGURE 24.2
A multiple T-maze used in learning studies with rodents.

In the overlapping range, which is an arid region resembling that of the hamadryas more than that of the anubis, hamadryas males have been observed to kidnap juvenile anubis females. These females learned to cooperate with the herding behavior of males in harem life. Hybrid males resulting from such matings were not so efficient in herding females as the pure hamadryas. Since both hamadryas and hybrid males appear to have the same opportunity to learn the herding behavior, it has been concluded that herding and harem formation are at least partly genetically determined.

Selection: Maze Learning in Rats

The second major approach in behavior genetics—selection for behavior modification from a heterozygous population—leads to the production of inbred lines with differences in behavior. The end result is similar to the previous approach of initially comparing inbred lines. The classical illustration of the second approach involves measurement of **maze learning** in rats.

The first experiment of this kind was reported by E. C. Tolman in 1924. He began with 82 white rats of heterozygous ancestry and measured their ability to "learn" to obtain food at the end of a multiple T-maze (Figure 24.2) by recording the number of errors and trials. When first exposed to the maze, a rat explores all alleys and eventually arrives at the end, to be rewarded with food. In succeeding trials, fewer and

fewer mistakes are made as the rat learns the correct route. Eventually, a hungry rat may proceed to the food with no errors.

From the initial 82 rats, nine pairs each of the "brightest" and "dullest" rats were selected and mated to produce two lines. In each generation, selection was continued. Even in the first generation, Tolman demonstrated that he could select and breed rats whose offspring performed more efficiently in the maze. Subsequently, his approach was pursued by others, notably R. C. Tryon, who in 1942 published results of 18 generations of selection.

As shown in Figure 24.3, two lines, one clearly superior and one clearly inferior in maze learning, were established. Each point on the graph represents the mean ($\bar{x}$) number of errors normalized at each generation. There is some variation around the mean, but by the eighth generation there was no overlap between lines even in this variation. That is, the dullest of the bright rats were then superior to the brightest of the dull rats.

In other studies, bright and dull rats were tested for other related traits with varying results. Brights were found to be better in solving hunger-motivation problems but inferior in escape-from-water tests. Brights were also found to be more emotional in open-field experiments. Thus, it was concluded that selection of

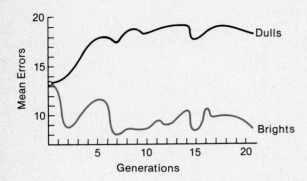

FIGURE 24.3
Selection for the ability and inability of rats to learn to
negotiate a maze. (Redrawn from Tryon, 1942.)

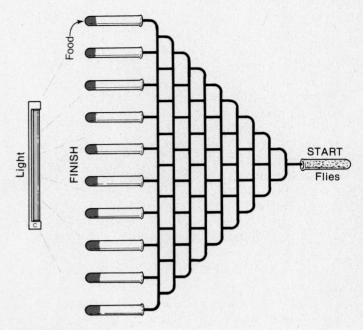

FIGURE 24.4
Schematic drawing of a maze used to study geotaxis in
Drosophila.

genetic strains superior in certain traits is possible,
but care must be taken not to generalize such studies
to overall intelligence, which is composed of many
learning parameters. Tryon seems to have selected for
specific capacities rather than for general intelligence.

Geotaxis in *Drosophila*

Taxis is a general term referring to the movement of a
free-living organism toward the source of an external
stimulation. The response may be positive or negative,
and the sources of stimulation may include chemicals,
gravity, light, and so on. The investigation of **geotaxis**
(response to gravity) in *Drosophila* illustrates this general behavior response as well as the selection technique. In addition, this investigation used a novel approach that determines the genetic influences of
specific chromosomes on geotaxis.

Jerry Hirsch and his colleagues designed a mass
screening device that allows about 200 flies to be
tested per trial, as shown in Figure 24.4. The maze is
placed vertically, a fluorescent light illuminates the
side opposite the entry point, and flies are added.
Those that continue to turn up at each junction will
finally arrive at the top; those that always turn down
will arrive at the bottom; and those making both "up"
and "down" decisions along the way will ultimately
reside somewhere in between. It was found that flies
could be selected for both positive and negative geotropism, establishing the genetic influence on this behavioral response.

As shown in Figure 24.5, mean scores may vary
from +4.0 to −6.0, corresponding to the number of T-
junctions the fly encounters going up or down. These
data show that the extreme of negative geotropism is
stronger than the extreme of positive geotropism. As
the two lines were compared over many generations,

clearcut but fluctuating differences were observed.
Such results indicate the additive effects of polygenic
inheritance.

Hirsch and his colleagues then turned to an analysis of the importance of genes located on different
chromosomes to geotaxis. They were able to differen-

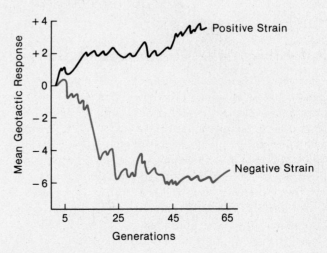

FIGURE 24.5
Selection for positive and negative geotaxis in *Drosophila*
over many generations. (Redrawn from Hirsch, 1967, Figure
12.3.)

tiate between genes located on chromosomes 2 and 3 and the X in a rather ingenious way. They executed a mating design that produced flies that were either heterozygous or homozygous for the chromosome (of a positively or negatively geotaxic fly) to be assayed.

Figure 24.6 shows how this is accomplished. From either a selected or unselected line, a male is crossed to a "tester" female whose chromosomes are each suitably marked with a dominant mutation. Each marked chromosome carries an inversion to suppress the recovery of any crossover products. As shown, one of the F_1 females, now heterozygous for each chromosome, is back-crossed to a male from the original line. The resulting female offspring contain all combinations of chromosomes. The dominant mutations make it possible to recognize which chromosomes are present. Thus, by subjecting these flies to the geotaxic maze, it is possible to assess the influence of genes on any given chromosome on the behavioral response.

Many flies of each type were tested, yielding the results shown in Table 24.2. When an unselected line was tested, the X chromosome and chromosome 2 contributed to a positive response, and chromosome 3 contributed to a negative response, in comparison with the tester chromosomes.

In comparing the positive geotaxic line with the unselected line, we see little effect of chromosome 2, but

TABLE 24.2

Chromosomal effects on geotaxis in *Drosophila*.

| | Chromosome | | |
Line	X	2	3
Positive geotaxis	+1.39*	+1.81	+0.12
Unselected	+1.03	+1.74	−0.29
Negative geotaxis	+0.47	+0.33	−1.08

*Each unit represents one level up or down in the maze.
SOURCE: After Hirsch, 1967, Table 12.1.

a positive effect of the X and 3. Comparison of the negative geotaxic line with the unselected line shows a reduction in positive effect by the X and 2 and an enhancement of the negative response by 3. Thus, it may be concluded that geotaxis is under polygenic control and that the responsible genes are distributed on all three major chromosomes in *Drosophila*.

SINGLE-GENE EFFECTS ON BEHAVIOR

By far the most definitive information on the genetic influence on behavior has come from the study of the

FIGURE 24.6
In this mating scheme in *Drosophila*, the effect of genes located on specific chromosomes that contribute to geotaxis can be assessed. The progeny produced by back-crossing the F_1 female contain all combinations of chromosomes. Examination of the phenotype of these flies makes it possible to determine which chromosomes from the selected strain are present. Subsequent testing for geotaxis is then performed (*B–Bar* eyes; *Cy–Curly* wing; *Sb–Stubble* bristles).

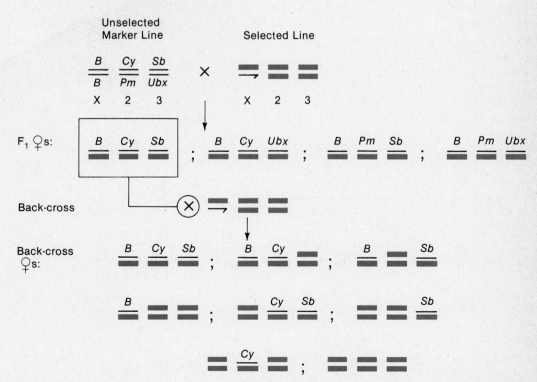

effects of single genes on behavior. In this more recent approach, either spontaneous or, more often, induced mutations are analyzed in order to infer general principles about how normal behavior is created and regulated.

There are obvious advantages to this approach. By and large, in these laboratory studies the environmental effect on the behavioral response is minimized or eliminated. Thus, the genetic influence is more straightforward and easier to define than in studies where the environment is a major factor. As a result, it is theoretically possible to dissect a behavioral pattern into its components.

In this section we shall look at a number of single-gene influences. However, because of the large amount of information resulting from such studies, we shall necessarily be selective in our discussion.

Nest-Cleaning Behavior in Honeybees

Honeybee nests are frequently infected with *Bacillus larvae*, the agent causing American foulbrood disease. The disease may be counteracted by what is called **hygienic behavior** by worker bees. The cells of infected combs containing afflicted larvae are opened, and the diseased organisms are removed from the hive. Hy-

gienic hives are resistant to infection, while hives containing strains that do not display removal behavior are susceptible to the disease.

In 1964, Walter Rothenbuhler published results of his cross between a hygienic (Brown) line with a nonhygienic (Van Scoy) line. This work strongly favors the hypothesis that two recessive independently assorting genes (*u* and *r*) or a gene complex are responsible for hygienic behavior.

The F₁ hybrids were all nonhygienic. However, when F₁ drones were back-crossed to hygienic queens, four phenotypes were produced in roughly equal proportions, as shown in Figure 24.7. While one group was hygienic and one group nonhygienic, the other two groups were most interesting. One could uncap cells but not remove infected larvae. The fourth group, which at first appeared nonhygienic, was shown to be able to remove larvae if the cells were artificially uncapped. Thus, they were not able to uncap.

It appears that one gene pair (*u/u*) or a linked complex of genes determines uncapping behavior, and a second gene pair (*r/r*) or complex determines removal ability. We have begun with this example to illustrate the way a genetic study has allowed components of a more complex behavior to be dissected. Nest cleaning in honeybees is also one of the most striking examples of the far-reaching effects of genes on behavioral responses.

FIGURE 24.7
Results of a honeybee cross between hygienic diploid females and nonhygienic haploid males.

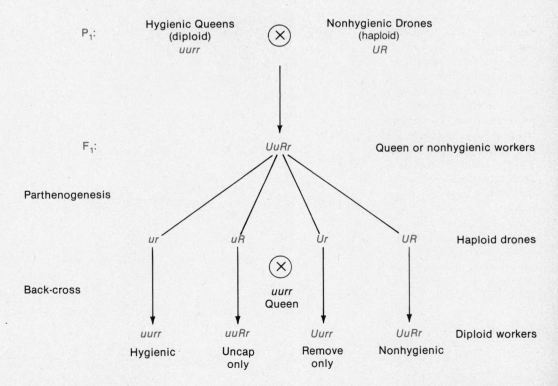

Taxes in Bacteria

Behavioral responses exist even in single-celled organisms such as bacteria. These responses are in the form of taxes (plural of *taxis*) and are mediated by flagella or cilia. Bacteria demonstrate **chemotaxis** and are attracted to or avoid a variety of stimuli. These reponses have now been carefully analyzed, and mutants have been isolated that disrupt normal behavior.

Motile bacteria such as *E. coli* and *Salmonella* are capable of monitoring gradients of chemicals and moving along the gradient by flagellar action. Cells exhibit smooth swimming by counterclockwise rotation of flagella, which act much like propellers, driving the bacteria through the medium. However, bacteria also "tumble" by reversing the rotation of flagella. Tumbling serves to change the direction of swimming. In the absence of stimuli, smooth swimming is interrupted by brief random periods of tumbling. In the presence of an attractant, tumbling activity is suppressed. Avoidance, however, is accomplished by increasing the frequency of tumbling.

In 1969, Julius Adler showed that *E. coli* contain large numbers of chemosensitive membrane proteins that underlie the response to changes in chemical composition in their growth medium. Over a dozen chemoreceptors for attractants (galactose, glucose, ribose, serine, etc.) have been described, and numerous repellents (fatty acids, alcohols, aromatic molecules) have been identified. Some chemoreceptor proteins can be removed from the membrane, but others are tightly bound.

Adler and his colleagues have used mutagens and isolated dozens of true-breeding mutations that disrupt chemotaxis in *E. coli*. Analysis of these mutants has shown that various levels of reception of stimuli exist, creating an integrated pathway leading to a response. This general pathway is shown in Figure 24.8. One class of mutation, the **specifically nonchemotactic mutants,** eliminates the response to only one particular chemical. In the second class, the so-called **multiply nonchemotactic mutants,** responsiveness to groups of chemicals is interrupted. Genetic analysis has placed these in three complementation groups: *tsr, tar,* and *trg.* In the third class, the **generally nonchemotactic mutants,** the response loss is even more general. A fourth class of mutants is called **nonmotile.** The defect in this class lies in flagellar function, which is normal in the first three classes. While the specific basis of each class of mutation is not yet clear, this genetic analysis shows that there are various components of a more complex behavioral response.

Because of the chemical nature of the receptors, Adler has been able to study the behavioral response

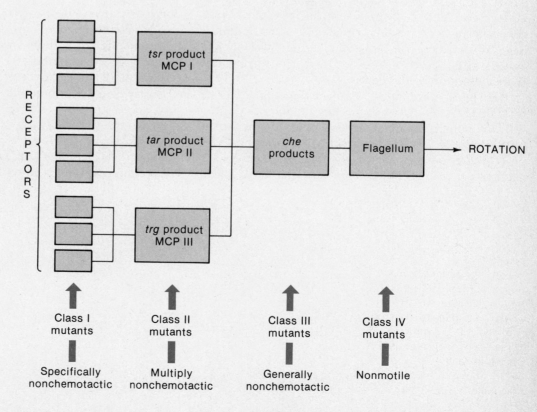

FIGURE 24.8
Integrated pathway of the general chemotactic response in *E. coli.* Mutations that interrupt this pathway at various points cause characteristic forms of the loss of the response. (From Springer, Goy, and Adler, 1979. Reproduced with permission from *Nature,* Vol. 280, pp. 279–84. Copyright © 1979 Macmillan Journals Limited.)

at the molecular level. He and his colleagues have proposed that information flows along an integrated pathway leading to a reponse as a result of a **methylation reaction.** This theory is based on several observations. First, the extent of methylation of some membrane proteins is altered by chemicals leading to the taxis. Second, the methylation of these proteins is eliminated in certain generally nonchemotactic mutants. The methylation response reappears if these mutants revert to wild type.

The isolation of *che R* mutations has led to further support of this theory. These mutations have the receptor proteins present but are unable to methylate them. Such mutants either lose the ability to adapt to chemical changes of the environment or do so only after extremely prolonged periods of time. Finally, if methionine (the amino acid serving as the methyl donor in the reaction) is depleted from the medium, response time in nonmutant cells is also prolonged.

Adler has designated the **methyl-accepting chemosensitive proteins** as MCPs and is now beginning to identify them by genetic and biochemical means. This work represents one of the premier efforts to link a behavioral response to its underlying molecular basis.

Avoidance Reactions in *Paramecium*

A great deal is known about the neurophysiology of ciliary action in *Paramecium*. This organism swims in a more coordinated fashion than a bacterium. Upon encountering a physical barrier or noxious stimulus, it reverses the direction in which all cilia are beating, causing it to back up. A reversal again occurs, and it swims off in a new direction.

The close coordination of ciliary direction suggests that there is an electrical phenomenon associated with the outer membrane that is similar to a nerve impulse. Indeed, this has been shown to be true. A depolarization passes electrically throughout the cell, leading to a calcium influx. The increased Ca^{++} inside

the cell is thought to initiate the reversal of direction of beating cilia.

Through the isolation and study of mutants unable to execute this avoidance reaction, genetic factors controlling the response have been identified. The *Pawn* mutant, named after its chess counterpart, cannot swim backward under normal conditions. However, if the membrane is disrupted with detergent and the calcium ion concentration is increased sufficiently, backward swimming is observed. Thus, the ciliary apparatus is completely functional. The pattern shown in Figure 24.9 confirms that the mutation has eliminated electrical potential changes in the membrane and thus affects the electrical conductance essential to reversal.

A second mutation, *Paranoic*, is also aptly named. Its electrical discharge patterns are compared with those of *Pawn* and wild type in Figure 24.9. A *Paranoic* mutant shows random reversals of ciliary direction and backing up in the presence of increased sodium ions, even in the absence of barriers. One might describe it as a paramecium that doesn't know if it's coming or going. While little is known about what part of the membrane is affected, it is known that the membrane becomes more permeable to many ions, including calcium. This undoubtedly explains the extended periods of ciliary reversal.

Behavior Genetics of a Nematode

In 1968, in one of the boldest attempts to define the total genetic influence on the behavior of a single organism, Sidney Brenner began an investigation of the nematode *Caenorhabditis elegans*. Brenner had previously made valuable contributions in the field of molecular genetics, including studies of DNA replication, F factors, mRNA, and the genetic code. When he turned to the study of behavior, he hoped that it would be possible to dissect genetically the nervous system of *C. elegans* using techniques previously applied successfully to other organisms.

FIGURE 24.9
Recordings of electrical potential from wild-type and mutant *Paramecium* in response to barium. (From Satow and Kung, 1974. Reproduced with permission from *Nature*, Vol. 247, pp. 69–71. Copyright © 1974 Macmillan Journals Limited.)

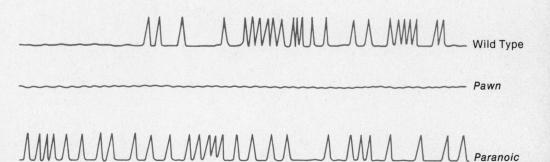

He chose this nematode because it was possible to determine the complete structure of the nervous system. Adult worms are about 1 mm long and are composed of only 600 cells, about 300 of which are neurons. As a result, it is possible to cut serial sections of an embedded organism and reconstruct the entire organism and its nervous system in a three-dimensional model. Brenner then hoped to induce large numbers of behavioral mutations and to correlate aberrant behavior with structural and biochemical alterations in the nervous system. Because of the vast scope of this research endeavor, work is still underway. Some progress has been made, particularly in isolating large numbers of mutations. Three types of behavioral mutants have been characterized. The worms are positively chemotactic to a variety of stimuli (cyclic AMP and GMP; anions such as Cl^-, Br^-, and I^-; and cations such as Na^+, Li^+, K^+, and Mg^{++}). As shown in Figure 24.10, positive attraction can be tracked in gradients on agar plates. The study of mutants has shown that sensory receptors in the head alone mediate the orientation reponses to attractants.

A second class of behavior studied involves **thermotaxis.** Cryophilic mutants move toward cooler temperatures, and thermophilic mutants move toward warmer temperatures. However, this behavior has not yet been correlated with the responsible component of the nervous system.

The third class of behavior involves generalized movement on the surface of an agar plate. Of 300 induced mutations, 77 affected the movement of the animal. While wild-type worms move with a smooth, sinuous pattern, mutants are either **uncoordinated** (*unc*) or **rollers** (*rol*). Those that are uncoordinated vary from the display of partial paralysis to small aberrations of movement, including twitching. Rollers move by rotating along their long axis, creating circular tracks on an agar surface. Of these mutants, some have been correlated with defects in the dorsal or ventral nerve cord or in the body musculature.

Linkage mapping has also begun. The vast number of mutants are distributed on six linkage groups, corresponding to the haploid number of chromosomes characteristic of this organism. Numerous *unc* mutants are found on each of the six chromosomes, indicating extensive genetic control of nervous system development.

Brenner's work is being extended in many laboratories throughout the world. In fact, an international congress now meets regularly to discuss work on *C. elegans*. Research now includes genetic analysis of development, and nongenetic problems are also being

FIGURE 24.10

Chemotactic response to ammonium chloride (NH_4Cl) of wild-type and mutant *Caenorhabditis*. (From Ward, 1973.)

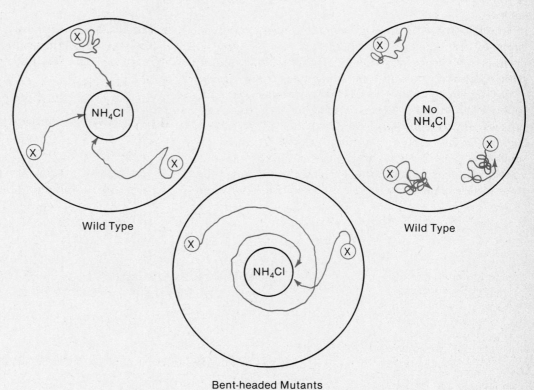

Wild Type

No NH_4Cl

Wild Type

Bent-headed Mutants

approached. This organism promises to rank with *Drosophila* in the amount of information acquired about it. It is hoped that study of *C. elegans* will unlock the mysteries of how genes control the structure of the nervous system.

Genes and Mating Behavior in *Drosophila*

We have already discussed geotaxis in *Drosophila* as an example of the selection methodology, but much more information on the behavior genetics of this organism is available. This is not at all surprising because of our extensive knowledge of its genetics and the ease with which it can be manipulated experimentally. Advances made in the 1970s, particularly by Seymour Benzer, represent significant strides in the field of behavior genetics.

As early as 1915, Alfred Sturtevant observed that the sex-linked recessive gene *yellow* affects mating preference, besides conferring the more obvious pigmentation difference. Recall that such a situation is an example of **pleiotropic** gene expression (see Chapter 4). Sturtevant found that both wild-type and *yellow* females, when given the choice of wild-type or *yellow* males, prefer to mate with wild type. Wild-type and *yellow* males prefer to mate with *yellow* females. These conclusions were based on quantitative measurements of success in mating between all combinations of *yellow* and wild-type (grey-bodied) males and females.

In 1956, Margaret Bastock extended these observations by investigating which, if any, component of courtship behavior was affected by the *yellow* mutant gene. Courtship in wild-type *Drosophila* is a complex ritual. The male first undergoes **orientation,** where he follows the female, perhaps circles her, and then orients usually at right angles and taps her on the abdomen. Once he has her attention, male wing display or **vibration** occurs. The wing closest to the female is raised and vibrates rapidly for several seconds. He then moves behind her, and contact is made between the male proboscis and female genitalia, described as "licking." Following this phase, if she has signaled acceptance by remaining in place, he mounts her and copulation occurs.

Bastock compared wild-type and *yellow* males for courtship rituals. What she observed was that *yellow* males prolong orientation but spend much less time in the vibrating and licking phases. Courtship patterns displayed by wild-type and *yellow* males are represented in Figure 24.11.

It appears that the *yellow* mutation has disrupted the intricate sequences of male courtship. Differences in the finer aspects of courtship have also been noted between related species of *Drosophila*. In these instances, the behavioral differences are thought to serve as possible isolating mechanisms during evolution.

The Genetic Dissection of Behavior in *Drosophila*

In 1967, Seymour Benzer and his colleagues initiated a comprehensive study of behavior genetics in *Drosophila*. Benzer's approach is an excellent example of the use of genetic techniques to "dissect" a complex biological phenomenon into its simpler components. Furthermore, the use of such techniques allows geneticists to study the underlying basis of the phenomenon—in this case, behavior. Benzer's goals in this research illustrate the "genetic dissection" approach:

1 To discern the genetic components of behavioral responses by the isolation of mutations that disrupt normal behavior.

2 To identify the mutant genes by chromosome localization and mapping.

3 To determine the actual site within the organism at which the gene expression influences the behavioral response.

4 To learn, if possible, how the particular gene expression influences behavior.

FIGURE 24.11
Courtship patterns of wild-type and *yellow* males in *Drosophila*. The duration of the behavioral patterns of licking, vibration, and orientation are recorded over time. (After Bastock, 1967.)

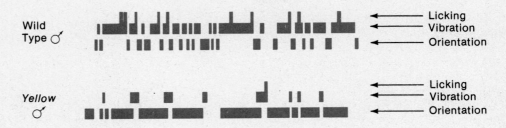

All four steps are illustrated in a discussion of **phototaxis,** one of the first behaviors studied by Benzer. Normal flies are positively phototactic; that is, they move toward a light source. Mutations are induced by feeding male flies sugar water containing **EMS** (ethylmethanesulfonate, a potent mutagen) and mating them to attached-X virgin females. As shown in Figure 24.12, the F_1 males receive their X chromosome from their fathers. Because they are hemizygous, any sex-linked recessive mutations are expressed.

The F_1 males must be tested for their response to light, and those that are not attracted to it are isolated. Benzer found *runner* mutants, which move quickly to and from light; *negatively phototactic* mutants, which move away from light; and *nonphototactic* mutants, which show no preference for light or darkness. He established that the behavior changes were due to mutations by mating these F_1 males to attached-X virgin females. Male progeny of this cross also showed the abnormal phototactic responses, confirming them as products of sex-linked recessive mutations.

We shall now delve into further work concerning just one group, the *nonphototactic* mutants. Such flies behave in the light as normal flies do in the dark. They can walk normally, but respond as if they were blind. Benzer and Yoshiki Hotta tested the electrical activity at the surface of mutant eyes in response to a flash of light. The pattern of electrical activity was recorded as an **electroretinogram.** Various types of abnormal responses were detected, and in none of the mutants was a normal pattern evident. When these mutations were mapped, they were not all allelic; instead, they were shown to occupy several loci on the X chromosome. Thus, it can be concluded that several gene products contribute to the formation of the behavioral response to light.

Where, within the fly, must adequate gene expression occur to yield a normal pattern? In an ingenious approach aimed at answering this question, Benzer turned to the use of mosaics. In mosaic flies, some tissues are mutant and others are wild type. If it can be ascertained which part must be mutant in order to yield the abnormal behavior, the **primary focus** of the genetic alteration can be determined.

To facilitate the production of mosaic flies, Benzer used a strain that has one of its X chromosomes in an unstable ring shape. When present in a zygote undergoing cell division, the ring-X is frequently lost by nondisjunction. If the zygote is female and has two X chromosomes (one normal and one ring-X), this will result in two cells—one with a single X (normal X) and one with two X chromosomes. The former cell goes on to produce male tissue (XO) and expresses all alleles on the remaining X, while the latter produces female tissue and does not express heterozygous recessive X-linked genes. Such an occurrence is illustrated in Figure 24.13. One can see that loss of the ring-X will produce mosaic flies with male and female parts that express or do not express sex-linked recessive genes, respectively.

In the embryo, when and where the ring-X is lost determines the pattern of mosaicism. The loss usually occurs early in development, before the cells migrate to the surface of the blastula to form the **blastoderm.** As shown in Figure 24.14, depending on the orientation of the spindle when the loss occurs, different types of mosaics will be created. If the stable X chromosome contains the behavior mutation and an obvious mutant gene (*yellow,* for example), the pattern of mosaicism will be readily apparent. For example, the fly may consist of a mutant head on a wild-type body, a normal head on a mutant body, one normal and one mutant eye on a normal or mutant body, and so on.

When such mosaics for *nonphototactic* mutants were studied, it was found that the **focus** of the genetic defect was in the eye itself. In mosaics where every part of the fly except the eye was normal, abnormal behavior was still detected. Even if only one eye was mutant, a modified abnormal behavior was evident. Instead of crawling straight up toward light as the normal fly does, the single mutant-eyed fly crawls upward to light in a spiral pattern. In the dark, such a fly will move in a straight line. Thus, mosaic studies have established that the focus of gene expression is

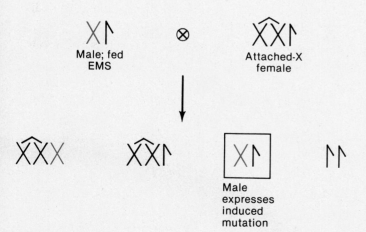

FIGURE 24.12

Genetic cross in *Drosophila* that facilitates the recovery of X-linked induced mutations. The female parent contains two X chromosomes that are attached, in addition to a Y chromosome. In a cross between this female and a normal male that has been fed the mutagen ethylmethanesulfonate, all surviving males receive their X chromosome from their father and express all mutations induced on that chromosome.

FIGURE 24.13

Production of a mosaic fruit fly as a result of fertilization by a gamete carrying an unstable ring-X chromosome (shown in color). If this chromosome is lost in one of the two cells following the first mitotic division, the body of the fly will consist of one part which is male (XO) and the other part which is female (XX). The male side will express all mutations contained on the X chromosome.

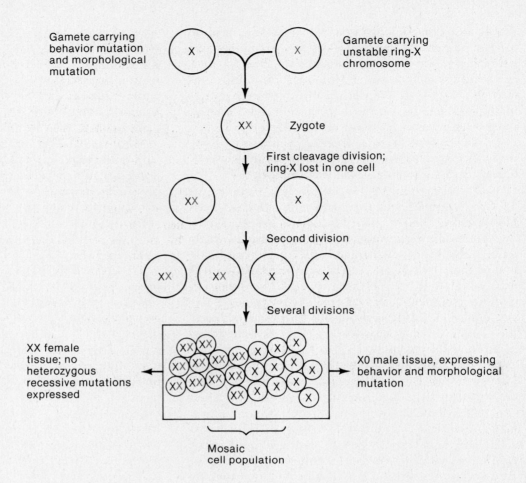

Gamete carrying behavior mutation and morphological mutation

Gamete carrying unstable ring-X chromosome

XX Zygote

First cleavage division; ring-X lost in one cell

Second division

Several divisions

XX female tissue; no heterozygous recessive mutations expressed

X0 male tissue, expressing behavior and morphological mutation

Mosaic cell population

FIGURE 24.14

The effect of spindle orientation on the production of mosaic flies as shown in Figure 24.13. (From Benzer, 1972. Reproduced with permission from *Nature*, Vol. 240, pp. 527–35. Copyright © 1972 Macmillan Journals Limited.)

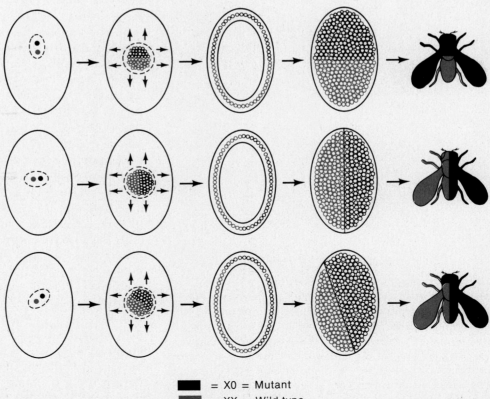

■ = X0 = Mutant

■ = XX = Wild type

in the eye itself. Additionally, the abnormal behavior is due to altered electrical conductance of the cells of the eye.

Benzer and other workers in the field have identified a large number of genes affecting behavior in *Drosophila*. As shown in Table 24.3, mutants have been isolated that affect locomotion, response to stress, circadian rhythm, sexual behavior, visual behavior, and even learning. The mutations have received very descriptive and often humorous names.

Many mutations have been analyzed with the mosaic technique in order to localize the focus of gene expression. While it was easy to predict that the focus of the *nonphototactic* mutant would be in the eye, other mutants are not so predictable. For example, the

focus of mutants affecting circadian rhythms has been located in the head, presumably in the brain. The *wings up* mutant might have a defect in the wings, articulation with the thorax, the thorax musculature, or the nervous system. Mosaic studies have pinpointed the indirect flight muscles of the thorax as the focus. Cytological studies have confirmed this finding, showing a complete lack of myofibrils in these muscles. Temperature-sensitive *paralytic* mutants are paralyzed at a raised temperature (29°C) but recover rapidly if the temperature is lowered. Mosaic studies have revealed that both the brain and thoracic ganglia represent the focus causing this abnormal behavior.

The *drop-dead* mutant is most interesting. Such mutant flies appear completely normal for the first few

TABLE 24.3
Some behavioral mutants of *Drosophila*.

Class	Name	Characteristics
Locomotor	*sluggish*	Moves slowly
	hyperkinetic	High consumption of O_2; shaking of legs; early death
	wings up	Wings perpendicular to body
	flightless	Does not fly well, although wings are well developed
	uncoordinated	Lacks coordinated movements
	nonclimbing	Fails to climb
Response to stress	*easily shocked*	Mechanical shock induces "coma"
	stoned	Stagger induced by mechanical shock
	shaker	Vibrates all legs while etherized
	freaked out	Grotesque, random gyrations under the influence of ether
	paralyzed	Collapses above a critical temperature
	parched	Dies quickly in low humidity conditions
	tko	Epilepticlike response
	comatose	Paralyzed by cool temperature
	out-cold	Similar to comatose
Circadian rhythm	*period^o*	Eclosion at any time; locomotor activity spread randomly over the day
	period^s	19-hour cycle rather than 24-hour cycle
	period^l	28-hour cycle rather than 24-hour cycle
Sexual	*savoir-faire*	Males unsuccessful in courtship
	fruity	Males pursue each other
	stuck	Male is often unable to withdraw after copulation
	coitus interruptus	Males disengage in about half the normal time
Visual	*nonphototactic*	Blind
	negatively phototactic	Moves away from light
Learning	*dunce*	Fails to learn conditioned response

days of adult life. Then they begin to stagger, fall over, and die. The cause of death could be related to any vital function anywhere in the body. However, mosaic study revealed that the defect is in the head. Most flies with mutant heads and normal bodies drop dead, while most with normal heads and mutant bodies do not. As shown in Figure 24.15, the brain of the *drop-dead* mutant appears to be full of holes. Examination of the brain of mutants before death shows them to be normal.

The mosaic technique has also been used to determine which regions of the brain are associated with sex-specific aspects of courtship and mating behavior. Jeffrey Hall and his associates have shown that mosaics with male cells in the most dorsal region of the brain, the protocerebrum, exhibit the initial stages of male courtship toward females. Later stages of male sexual behavior including wing vibrations and attempted copulations require male cells in the thoracic ganglion. Similar studies of female-specific sexual behavior have shown that the ability of a mosaic to induce courtship by a male depends on female cells in the posterior thorax or abdominal region. A region of the brain within the protocerebrum must be female for receptivity to copulation. Anatomical studies have confirmed that there are fine structural differences in the brains of male and female *Drosophila*, indicating that some forms of behavior may be dependent on the development and maturation of specific parts of the nervous system.

Single-Gene Effects in Mice

Surely the oldest recorded behavior mutant is the *waltzer* mutation in the mouse, recorded in 80 B.C. in China and described as a mouse "found dancing with its tail in its mouth." Waltzing mice run in a tight circle and demonstrate both horizontal and vertical head shaking as well as hyperirritability. Some waltzers are also deaf.

Genetic crosses between mutant and normal house mice (*Mus musculus*) reveal a simple recessive inheritance pattern characteristic of Mendel's monohybrid matings. Investigation of the inner ear of mutants has revealed degeneration of both the cochlea and semicircular canals, accounting for the deafness and circling behavior. This is an example of a mutation causing a structural anomaly which, in turn, alters behavior.

Many other single-gene behavior effects are known in the mouse, an organism particularly well characterized genetically. Of over 300 mutants discovered representing about 250 loci, over 90 are neurological in nature and alter behavior. These are classified into three groups of syndromes, as shown in Table 24.4. The neurological basis of many of these has now been determined.

The incoordination mutants *quaking* and *jumping* are due to faulty myelination of nervous tissue. *Quaking* is an autosomal recessive mutation, and *jumping* is inherited as a sex-linked recessive. The former cannot

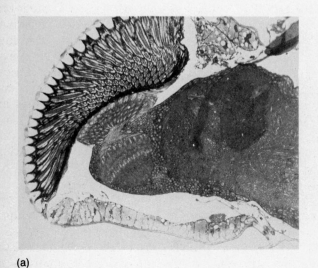

(a)

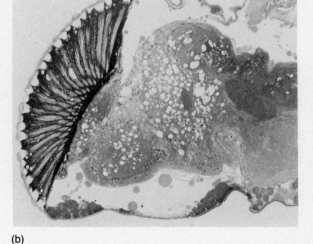

(b)

FIGURE 24.15
Photomicrographs of sections cut through the brains of (a) wild-type and (b) *drop-dead* mutant of *Drosophila melanogaster*. (From Benzer, 1973, p. 34. Photos courtesy of Seymour Benzer.)

TABLE 24.4
Inherited neurological defects in mice.

Group	Name
Waltzer-shaker	*Waltzer, shaker, pirouette, jerker, fidget, twirler, zig-zag*
Convulsive	*Trembler, tottering, spastic*
Incoordination	*Jumping, quaking, reeler, staggerer, leaner, Purkinje cell degeneration, cerebral degeneration*

synthesize adequate myelin, while the latter demonstrates a degenerative process involving myelin. *Cerebral degeneration* is a mutant exhibiting progressive deterioration of behavior. As its name implies, part of the brain deteriorates.

The study of such mutants clearly establishes the genetic basis of normal development. Additionally, and more germane to this chapter, abnormal development is linked directly to altered behavior by these findings.

DROSOPHILA BEHAVIOR GENETICS

Learning in *Drosophila*

To study the genetics of behavior such as learning, it would be advantageous to use an organism like *Drosophila*, in which methods of genetic analysis are highly advanced. However, the first question is: Can *Drosophila* learn? Recent work from a number of laboratories indicates that organisms such as *Drosophila* are, in fact, capable of learning. Using a simple apparatus, flies are presented with a pair of olfactory cues, one of which is associated with an electrical shock. Flies quickly learn to avoid the odor associated with the shock. That this response is learned is indicated by a number of factors: first, performance is associated with the pairing of a stimulus/response with a reinforcer; second, the response is reversible. Flies can be trained to select an odor that they previously avoided, and flies exhibit short-term memory for the training they have received.

Other evidence suggests that more innate forms of behavior, such as courtship, are also associated with learning. Richard Siegel and his colleagues have found that following unsuccessful courtship of sexually unreceptive females, males show a reduction of courtship behavior for about 3 hours, even in the presence of sexually receptive females. Memory-deficient mutants, such as *amnesiac*, resume active courtship in about 1 hour, presumably because the previous experience has been forgotten.

The demonstration that *Drosophila* can learn opens the way to selecting mutants that are defective in learning and memory. To accomplish this, males from an inbred wild-type strain are mutagenized and mated to females from the same strain. Their progeny are recovered and mated to produce populations of flies, each of which carries a mutagenized X chromosome. Mutants that affect learning are selected by testing the population for response in the olfactory/shock apparatus. A number of learning-deficient mutants including *dunce, turnip, rutabaga,* and *cabbage* have been recovered. In addition, a memory-deficient mutant, *amnesiac*, that learns normally but forgets four times faster than normal has been recovered. Each of these mutations represents a single gene defect that affects a specific form of behavior. Because of the method used to recover them, all the mutants found so far are X-linked genes. Presumably similar genes controlling behavior are also located on autosomes.

Molecular Biology of Behavior

Since the result of many mutations is the alteration or abolition of a single protein, biochemical study of the learning mutants described above can provide a link between behavior and molecular biology. The *dunce* mutation is the first for which this link has been established. The locus for *dunce* has been shown to encode the structural gene for the enzyme cyclic AMP phosphodiesterase. It appears that *rutabaga* is a mutation in the structural gene for adenylate cyclase and that the *turnip* mutation is in the gene for a GTP-binding protein associated with adenylate cyclase. The unexpected clustering of these independently derived mutations in the biochemical pathway of the adenyl cyclase system suggests a role for cyclic nucleotides in learning. This conclusion is consistent with work in the sea hare, *Aplysia*, indicating that short-term memory is associated with an increase in cyclic nucleotides within neurons. Further work will be necessary to clarify the role of such molecules in behavior.

Because behavior is controlled and modified by cells of the nervous system, another approach to understanding behavior at the molecular level centers on the nerve cell. Barry Ganetzky and his colleagues have identified and characterized a number of mutations in *Drosophila* that affect the generation and propagation of nerve impulses. Using recombinant DNA techniques, they have successfully cloned several of the genes responsible for these mutations. These cloned

DNA sequences can be used to study fundamental questions about the function of the nervous system and the molecular basis of memory and learning. Certainly, work undertaken in this field in the next few years will be exciting and important.

HUMAN BEHAVIOR GENETICS

The genetic input to behavior in humans is much more difficult to characterize than that of other organisms. Not only are humans unavailable as experimental subjects in genetic investigations, but the types of responses considered to be interesting behavior are extremely difficult to study. The most popular forms of studied behavior all include some aspects of **intelligence, language, personality,** or **emotion.** There are two problems in examining such traits. First, all are difficult to define objectively and to measure quantitatively. Second, they are the traits most affected by the environment. In each case, while there is undoubtedly a genetic basis, it is a complex one. Furthermore, the environment is extremely important in shaping, limiting, or facilitating the final phenotype for each trait.

The study of human behavior genetics has also been hampered by two other factors. Many studies of human behavior have been performed by psychologists without adequate input from the biologist or geneticist. Second, traits involving intelligence, personality, and emotion have the greatest social and political significance. As such, these traits are more likely to be the subject of sensationalism when reported to the lay public. Because their study comes closest to infringing upon individual liberties such as the right to privacy, these traits are the basis of the most controversial investigations.

In lamenting the gulf between psychology and genetics in explaining human behavior genetics, C. C. Darlington in 1963 wrote, "Human behavior has thus become a happy hunting ground for literary amateurs. And the reason is that psychology and genetics, whose business it is to explain behavior, have failed to face the task together." Since 1963, great progress has been made in bridging this gap.

Mental Disorders with a Clearcut Genetic Basis

Many genetic disorders in humans result in some behavioral abnormality. One of the most prominent examples is **Huntington's disease** (sometimes referred to as Huntington's chorea). Inherited as an autosomal dominant disorder, it affects the nervous system, including the brain. Onset of the disease occurs usually in one's thirties with a loss of motor function and coordination. Degeneration of the nervous system is progressive, and personality changes occur. The affected individual soon is unable to care for himself. Since onset is usually after a family has been started, all children of an affected person must live with the knowledge that they face a 50 percent probability of developing the disorder. No early detection method or cure is now available, but recombinant DNA technology (see Chapter 20) may provide a method of diagnosis.

The **Lesch-Nyhan syndrome** is inherited as a sex-linked recessive disorder. Onset is within the first year, and the disease is most often fatal early in childhood. The disorder is of a metabolic nature, involving purine biosynthesis. Mental and physical retardation occurs, and these individuals demonstrate uncontrolled self-mutilation. They also strike out at individuals attempting to care for them.

Other metabolic disorders are also known to affect mental function. For example, **Tay-Sachs disease,** an autosomal recessive disorder also referred to as amaurotic idiocy, involves severe mental retardation among other phenotypic characteristics. The disease is apparent soon after birth and is fatal. **Porphyria,** under the control of an autosomal dominant allele, usually has a much later onset and is marked by recurring periods of dementia. The autosomal recessive disease **phenylketonuria,** unless detected and treated early, results in mental retardation. All of these disorders alter the normal biochemistry of the affected individuals and are inherited in a Mendelian fashion.

Chromosome abnormalities also produce syndromes with behavioral components. **Down syndrome** (trisomy 21) results in mental retardation. While there is a wide range of variability, the mean IQ of affected individuals is estimated to be 25 to 50. The onset of walking and talking is often delayed until four to five years of age. Both Klinefelter syndrome (XXY) and Turner syndrome (XO) may also result in diminished mental capacity.

Human Behavior Traits with Less-Defined Genetic Bases

Other forms or aspects of human behavior, notably **schizophrenia, manic-depressive illness,** and **general intelligence,** have been the subject of extensive study. Investigations have sought to relate the development of the mental disorder or the display of intelligence to the closeness of family relationships or twin studies. In all three cases, it has been concluded that a genetic component influences the trait, but that environment also plays a substantial role.

A discussion of schizophrenia may serve to illustrate the methodology used. This mental disorder is characterized by withdrawn, bizarre, and sometimes delusional behavior. It is clearly a family disease, with relatives of schizophrenics having a much higher incidence of this disorder than the general population. Furthermore, the closer the relationship to the index case or proband, the greater is the probability of the disorder occurring.

Such observations, taken alone, could argue equally for a genetic cause, environmental causes, or a combination of both for the development of schizophrenia. The use of twin studies, as first devised by Galton in 1883, has improved the accuracy of our understanding of this problem. The concordance of schizophrenia in monozygotic and dizygotic twins has been the subject of many studies (see Chapter 7).

In almost every investigation, concordance has been higher in monozygotic twins than in dizygotic twins reared together. Although these results suggest that a genetic component exists, they do not reveal the precise genetic basis of schizophrenia. Simple monohybrid and dihybrid inheritance as well as multiple gene control have been proposed for schizophrenia. However, it seems unlikely that only one or two loci are involved, nor is it likely that the control is strictly quantitative, as in polygenic inheritance.

In schizophrenia, manic-depressive illness, and intelligence alike, it is most sound to conclude that each individual is endowed with a genetic predisposition for normal or abnormal behavior and that environmental factors can serve to alter the final phenotype. An individual may or may not be predisposed to the mental disorders. In combination with a mentally healthy or unhealthy environment, the disease may or may not develop. With regard to intelligence, genetic factors may provide an upper and lower potential range, but an individual's environment may modify the development of intelligence.

There is a long-standing controversy about genetic differences in intelligence between races. While IQ testing has established intelligence differences in populations of different races, there is currently no strong evidence to support the conclusion that this is due to the genetic component of intelligence. The environment undoubtedly has a profound effect on the type of intelligence measured by the various forms of IQ tests. It seems likely that the genetic component of intelligence does not differ any more significantly between races than it does between individuals within the same race.

SUMMARY AND CONCLUSIONS

While early studies focused on the influence of learning and the environment on behavior, it has now become clear that the genotype also plays an important part in determining reactions to stimuli or situations. As a result, behavior genetics has emerged as an important specialty within the field of genetics. Because both the genotype and environment have an impact on behavioral traits, it is particularly difficult to provide adequate controls and to derive purely objective data from investigations in this area.

Three major approaches have been used in behavior genetics. First, geneticists have studied behavioral differences between closely related organisms in similar environments whose survival needs appear to be identical. In such cases, a genetic component is believed to be responsible for the behavior pattern. The studies of alcohol preference and open-field behavior in mice and mate selection in baboons illustrate this approach. Each of these behaviors is strongly influenced by the genotype.

The second approach is to select in the laboratory for a modified behavioral trait and demonstrate that is is inherited. Studies of maze learning in rats and geotaxis in *Drosophila* have ascribed a genetic component to these behaviors. In the case of maze learning, both bright and dull lines of rats have been established. In *Drosophila*, strains that are either positively or negatively geotaxic have been obtained, and researchers have determined the relative contributions of genes on specific chromosomes to this behavior.

The third approach, which is a more direct appraisal of the role of genetics in behavior, is to study the effects of single genes on behavioral patterns. By isolating mutations causing deviations from normal behavior, the role of the corresponding wild-type allele in the respective response is established. In this chapter we have examined numerous examples of this approach in a variety of organisms. These include hygienic behavior in honeybees; responses to chemical stimuli in bacteria and paramecia; chemotaxis, thermotaxis, and general movement in nematodes; and a variety of behaviors in *Drosophila*. In the latter set of studies, a unique method involving mosaic flies allows the determination of the body region affected by the mutant gene response. The single gene effect is also demonstrated in various neurological mutations in mice where brain function is impaired.

The study of the genetic influence on any aspect of human behavior is difficult to approach because of the important contribution of the environment to the development of these traits. Abnormal behavior such as schizophrenia often is seen to run in families. Twin studies have confirmed that a genetic predisposition, which can be modified by the environment, does play a role in the development of this disorder. General intelligence is also the product of genetic and environmental influences.

PROBLEMS AND DISCUSSION QUESTIONS

1 Contrast the methodologies used in studying behavior genetics. What are the advantages of each method with respect to the type of information gained?

2 Contrast the advantages of using *Drosophila* versus *Caenorhabditis* for studying behavior genetics.

3 Theoretical data concerning the genetic effect of *Drosophila* chromosomes on geotaxis are shown below. What conclusions can you draw? See Table 24.2 for comparative data.

	Chromosome		
	X	2	3
Positive geotaxis	+0.2	+0.1	+3.0
Unselected	+0.1	−0.2	+1.0
Negative geotaxis	−0.1	−2.6	+0.1

4 If a haploid honeybee drone of the genotype *uR* is mated to a queen of genotype *UuRR*, what ratio of behavioral phenotypes will be observed in the hive in the offspring with respect to American foulbrood disease? For *uR* × *UURr*? For *Ur* × *UURr*?

5 Assume that you discovered a fruit fly that walked with a limp and was continually off balance as it moved. Describe how you would determine if this behavior was due to an injury (induced in the environment) or was an inherited trait. Assuming that it is inherited, what are the various possibilities for the focus of gene expression causing the imbalance? Describe how you would locate the focus experimentally if it were X-linked.

6 In humans, the chemical phenylthiocarbamide (PTC) is either tasted or not. When the offspring of various combinations of taster and nontaster parents are examined, the following data are obtained:

PARENTS	Both tasters	Both tasters	Both tasters	One taster One nontaster	One taster One nontaster	Both nontasters
OFFSPRING	All tasters	1/2 tasters 1/2 nontasters	3/4 tasters 1/4 nontasters	All tasters	1/2 tasters 1/2 nontasters	All nontasters

Based on these data, how is PTC tasting behavior inherited?

7 Discuss why the study of human behavior genetics has lagged behind that of other organisms.

8 J. P. Scott and J. L. Fuller studied 50 traits in five pure breeds of dog. Almost all traits varied significantly in the five breeds, but very few bred true in crosses. What can you conclude with respect to the genetic control of these behavioral traits?

SELECTED READINGS

BASTOCK, M. 1956. A gene which changes a behavior pattern. *Evolution* 10: 421–39.

————. 1967. *Courtship—An ethological study.* Chicago: Aldine.

BYERS, D., DAVID, R. L., and KIGER, J. A. Jr. 1981. Defect in the cyclic AMP phosphodiesterase due to the *dunce* mutation of learning in *Drosophila. Nature* 289: 79–81.

BENZER, S. 1973. Genetic dissection of behavior. *Scient. Amer.* (Dec.) 229: 24–37.

BODMER, W. F., and CAVALLI-SFORZA, L. L. 1976. *Genetics, evolution and man.* San Francisco: W. H. Freeman.

BRENNER, S. 1974. The genetics of *Caenorhabditis elegans*. *Genetics* 77: 71–94.

EHRMAN, L., and PARSONS, P. A. 1981. *Behavior genetics and evolution*. New York: McGraw-Hill.

FARBER, S. L. 1980. *Identical twins reared apart*. New York: Basic Books.

FULLER, J. L. 1960. Behavior genetics. *Ann. Rev. Psychol.* 11: 41–63.

FULLER, J. C., and THOMPSON, W. R. 1978. *Foundations of behavior genetics*. St. Louis: C.V. Mosby.

GAILEY, D. A., HALL, J. C., and SIEGEL, R. W. 1985. Reduced reproductive success for a conditioning mutant in experimental populations of *Drosophila melanogaster*. *Genetics* 111: 795–804.

GANETZKY, B., and WU, C.-F. 1985. Genes and membrane excitability in *Drosophila*. *Trends in Neurosciences* (in press).

GOODMAN, R. M., ed. 1970. *Genetic disorders of man*. Boston: Little, Brown.

GOTTESMAN, I. I., and SHIELDS, J. 1972. *Schizophrenia and genetics: A twin study vantage point*. New York: Academic Press.

HALL, J. C., GREENSPAN, R. J., and HARRIS, W. A. 1982. *Genetic neurobiology*. Cambridge, Mass.: The MIT Press.

HARRIS, W. A. 1985. Genetics and development of the nervous system. *J. Neurogenet.* 2: 179–96.

HAY, D. A. 1985. *Essentials of behaviour genetics*. Palo Alto: Blackwell Scientific Publ.

HESTON, L. L. 1970. The genetics of schizophrenia and schizoid disease. *Science* 167: 249–56.

HIRSCH, J., ed. 1967. *Behavior-genetic analysis*. New York: McGraw-Hill.

HOTTA, Y., and BENZER, S. 1972. Mapping behavior of *Drosophila* mosaics. *Nature* 240: 527-35.

KAPLAN, A. R. 1976. *Human behavior genetics*. Springfield, Ill.: Charles C. Thomas.

KIDD, K. K., and CAVALLI-SFORZA, L. L. 1973. An analysis of the genetics of schizophrenia. *Social Biol.* 20: 254–65.

KUNG, G., CHANG, S. Y., SATOW, Y., VanHOUTEN, J., and HANSMA, H. 1975. Genetic dissection of behavior in *Paramecium*. *Science* 188: 898–904.

KUNG, C., and NAITOH, Y. 1973. Calcium-induced ciliary reversal in the extracted models of *pawn*, a behavioral mutant of *Paramecium*. *Science* 179: 195–96.

LEWONTIN, R. C. 1975. Genetic aspects of intelligence. *Ann. Rev. Genet.* 9: 387–405.

LINDZEY, G., LOEHLIN, J., MANOSEVITZ, M., and THIESSEN, D. 1971. Behavioral genetics. *Ann. Rev. Psychol.* 22: 39–94.

LINDZEY, G., and THIESSEN, D. D. 1970. *Contributions to behavior-genetic analysis: The mouse as a prototype*. New York: Appleton-Century-Crofts.

LIVINGSTONE, M. S. 1985. Genetic dissection of *Drosophila* adenylate cyclase. *Proc. Natl. Acad. Sci.* 82: 5795–99.

McCLEARN, G. E. 1970. Behavioral genetics. *Ann. Rev. Genet.* 4: 437–68.

MERRELL, D. J. 1965. Methodology in behavior genetics. *J. Hered.* 56: 263–66.

PLOMIN, R., DeFRIES, J. C., and McCLEARN, G. E. 1980. *Behavioral genetics*. 2nd ed. San Francisco: W. H. Freeman.

QUINN, W. B., and GREENSPAN, R. J. 1984. Learning and courtship in *Drosophila*: Two stories with mutants. *Ann. Rev. Neurosci.* 7: 67–93.

QUINN, W. G., and GOULD, J. L. 1979. Nerves and genes. *Nature* 278: 19–23.

RIDDLE, D. L. 1978. The genetics of development and behavior in *Caenorhabditis elegans*. *J. Nematol.* 10: 1–15.

ROGERS, D. A., and McCLEARN, G. E. 1962. Mouse strain differences in preference for various concentrations of alcohol. *Quart. J. Stud. Alc.* 23: 26–33.

ROTHENBUHLER, W. C., KULINCEVIC, J. M., and KERR, W. E. 1968. Bee genetics. *Ann. Rev. Genet.* 2: 413–38.

SATOW, Y., and KUNG, C. 1974. Genetic dissection of active electrogenesis in *Paramecium aurelia. Nature* 247: 69–71.

SCOTT, J. P., and FULLER, J. L., eds. 1974. *Dog behavior: The genetic basis.* Chicago: The University of Chicago Press.

SIEGEL, R. W., HALL, J. C., GAILEY, D. A., and KYRIACOU, C. P. 1984. Genetic elements of courtship in *Drosophila:* Mosaics and learning mutants. *Behav. Genet.* 14: 383–410

SPRINGER, M. S., GOY, M. F., and ADLER, J. 1979. Protein methylation in behavioral control mechanisms and in signal transduction. *Nature* 208: 279–84.

STANSBURY, J. B., WYNGAARDEN, J. B., and FREDERICKSON, D. S., eds. 1972. *The metabolic basis of inherited disease.* 3rd ed. New York: McGraw-Hill.

TULLY, T. 1984. *Drosophila* learning: Behavior and biochemistry. *Behav. Genet.* 14: 527–57.

TRYON, R. C. 1942. Individual differences. In *Comparative psychology,* 2nd ed., ed. F. A. Moss. Englewood Cliffs, N. J.: Prentice-Hall.

VALE, J. R. 1980. *Genes, environment, and behavior: An interactionist approach.* New York: Harper & Row.

WARD, S. 1973. Chemotaxis by the nematode, *Caenorhabditis elegans:* Identification of attractants and analysis of the response by use of mutants. *Proc. Natl. Acad. Sci.* 70: 817–21.

25

The Genetics of Populations

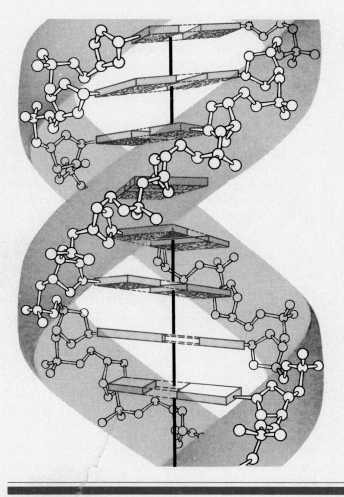

When Charles Darwin published *The Origin of Species* in 1859 following his briefer statement, coauthored with Alfred Russell Wallace on the likely mechanism of natural selection, the foundation for the modern interpretation of evolution was established. Although organisms are capable of reproducing in a geometric fashion, Darwin observed that this growth potential of species is not realized. Instead, population numbers remain relatively constant in nature. He deduced that some form of competitive struggle for survival must therefore occur. He also observed that there is a natural variation between individuals within a species; on this observation, he based his theory of natural selection: ". . . that any being, if it vary however slightly in any manner profitable to itself . . . will have a better chance of surviving." Darwin included in his concept of survival ". . . not only the life of the individual, but success in leaving progeny."

While Mendel was familiar with Darwin's work, Darwin was unaware of any underlying mechanism to account for the morphological variation he had observed. However, with the development of the concept of genes and alleles, the genetic basis of inherited variation was established.

As others pursued the study of evolution, it became apparent that the population rather than the individual was the functional unit in this process. In order to study the role of genetics in the process of evolution, therefore, it was necessary to consider gene frequencies in populations rather than offspring from individual matings. Thus arose the discipline of **population genetics.**

Early in the twentieth century various workers, including Gudny Yule, William Castle, Godfrey Hardy, and Wilhelm Weinberg, formulated ideas that became the basic principles of this field of investigation. In the earlier years of population genetics, the emphasis was on theory and mathematical models used to describe the genetic structure of a population. Only in the past thirty years have experimentalists and field workers attempted to test these theories, extending their comprehension and exploring new phenomena. Gene frequencies and the forces that alter these frequencies, such as mutation, migration, selection, and random genetic drift, have been and are being examined. In this chapter, we shall consider some general aspects of population genetics and also discuss other areas of genetics relating to evolution.

POPULATIONS, GENE POOLS, AND GENE FREQUENCIES

Members of a species are often distributed over a wide geographic range. A **population,** however, is a local group belonging to a single species, within which mating is actually or potentially occurring. Within the population, the set of genetic information carried by all interbreeding members is called the **gene pool.** For a given gene, this pool includes all the alleles of that gene that are present in the population. In population genetics the focus is on groups rather than on individuals, and on the measurement of gene (allelic) and genotype frequencies from generation to generation rather than on the distribution of genotypes resulting from a single mating. The term **gene frequency** will be used throughout the chapter and will represent the frequency of alleles in contrast to genotype frequencies. Gametes produced by one generation represent withdrawals from the gene pool to form the zygotes of the next generation. This new generation has a reconstituted gene pool that may differ from that of the preceding generation.

Obviously, then, populations are dynamic; they may grow and expand or diminish and contract through changes in birth or death rates, by migration, or by merging with other populations.

One approach used to study a population's genetic structure is to measure the frequency of a given gene controlling a known trait. This is possible once the mode of inheritance and the number of different alleles present in the population have been established. Gene frequencies cannot always be determined directly because in many cases only phenotypes, and not genotypes, can be observed. However, if alleles expressed in a codominant fashion are considered, phenotypes are equivalent to genotypes. Such is the case with the autosomally inherited **MN blood group** in humans.

In this case, the gene L on chromosome 2 has two alleles, L^M and L^N (often referred to as M and N, respectively).* Each controls the production of a distinct antigen on the surface of red blood cells. Thus, an individual may be type M ($L^M L^M$), N ($L^N L^N$), or MN ($L^M L^N$).

The genotypes, phenotypes, and immunological reactions of the MN blood group are shown in Table 25.1. Because they are codominant, the frequency of M and N in a population can be determined simply by counting the number of alleles for each phenotype. For example, consider a population of 100 individuals of which 36 are MM, 48 are MN, and 16 are NN. The 36 MM individuals represent 72 M alleles, and the 48 MN heterozygotes represent 48 additional M alleles. There are 72 + 48 or 120 M alleles out of a total of 200 alleles in the population (100 individuals, each with two alleles at this locus). Therefore, the frequency of the M allele in this population is 0.6 (120/200 = 60% = 0.6). The frequency of the N allele can be calculated in a similar fashion (80/200 = 40%) and is 0.4. Table 25.2 illustrates two methods for computing the frequency of M and N alleles in a hypothetical population of 100 individuals, and Table 25.3 lists the frequencies of M and N alleles actually measured in several human populations.

THE HARDY-WEINBERG LAW

In the example of the MN blood group in humans, M and N are codominant. If, on the other hand, one allele had been recessive, the heterozygotes would have been phenotypically identical to the homozygous dominant individuals, and the frequency of the alleles

TABLE 25.1
MN blood groups.

Genotype	Blood Type	Reaction with Antibodies	
		Anti-M	Anti-N
$L^M L^M$	M	+	−
$L^M L^N$	MN	+	+
$L^N L^N$	N	−	+

*L stands for Karl Landsteiner, the geneticist for whom the locus is named.

TABLE 25.2
Methods of determining allele frequencies for codominant alleles.

A. Counting Alleles

Genotype/Phenotype	MM	MN	NN	Total
No. of individuals	36	48	16	100
No. of M alleles	72	48	0	120
No. of N alleles	0	48	32	80
Total no. of alleles	72	96	32	200

Frequency of M in population: $\frac{120}{200} = 0.6 = 60\%$

Frequency of N in population: $\frac{80}{200} = 0.4 = 40\%$

B. From Genotypes

Genotype/Phenotype	MM	MN	NN	Total
No. of individuals	36	48	16	100
Genotype frequency	36/100 = 0.36	48/100 = 0.48	16/100 = 0.16	1.00

Frequency of M in population: $36 + (1/2)48 = 0.36 + 0.24 = 0.60 = 60\%$
Frequency of N in population: $16 + (1/2)48 = 0.16 + 0.24 = 0.40 = 40\%$

could not have been directly determined. However, a mathematical model developed independently by the British mathematician Godfrey H. Hardy and a German physician, Wilhelm Weinberg, can be used to calculate allele frequencies in this case. The **Hardy-Weinberg law (HWL)** is one of the fundamental concepts in population genetics. As is the case with most mathematical models, certain assumptions must be made. In the Hardy-Weinberg law, the following conditions are presumed:

1 The population is infinitely large, or large enough that sampling error is negligible.

2 Mating within the population occurs at random.

3 There is no selective advantage for any genotype; that is, all genotypes produced by random mating are equally viable and fertile.

4 There is an absence of other factors such as mutation, migration, and random genetic drift.

To illustrate the application of the Hardy-Weinberg law, suppose that in a given population, a gene has two alleles, A and a. The frequency of the dominant allele A in both eggs and sperm is represented by p, and the frequency of the recessive allele in gametes is

represented by q. Because the sum of p and q represents 100 percent of the alleles for that gene in the population, then $p + q = 1$. A Punnett square can be used to represent the random combination of gametes containing these alleles and the resulting phenotypes:

Sperm

	$A(p)$	$a(q)$
$A(p)$	AA (p^2)	Aa (pq)
$a(q)$	Aa (pq)	aa (q^2)

Eggs

In the random combination of gametes in the population, the probability that sperm and egg both contain the A allele is $p \times p = p^2$. Similarly, the chance that gametes will carry unlike alleles is $(p \times q) + (p \times q) = 2pq$, and the chance that a homozygous recessive

TABLE 25.3
Frequencies of M and N alleles in various populations.

Population	Genotype Frequency (%)			Allele Frequency	
	MM	MN	NN	M	N
U.S. whites	29.16	49.38	21.26	0.540	0.460
U.S. blacks	28.42	49.64	21.94	0.532	0.468
U.S. Indians	60.00	35.12	4.88	0.776	0.224
Eskimos (Greenland)	83.48	15.64	0.88	0.913	0.087
Ainus (Japan)	17.86	50.20	31.94	0.430	0.570
Aborigines (Australia)	3.00	29.60	67.40	0.178	0.822

individual will result is $q \times q = q^2$. Note that these terms also describe an important aspect of the HWL: Allele frequencies determine genotype frequencies. In other words, while the value p^2 is the probability that both gametes will carry the A allele, it is also a measure of the frequency of AA homozygotes. In a similar way, $2pq$ describes the frequency of Aa heterozygotes, and q^2 is a measure of the frequency of homozygous recessive (aa) zygotes. Thus, the distribution of genotypes produced by random mating in the next generation can be expressed as

$$p^2 + 2pq + q^2 = 1 \qquad (25.1)$$

Let us consider a population in which 70 percent of the alleles for a given gene are A, and 30 percent are a. Thus, $p = 0.7$ and $q = 0.3$, and $p(0.7) + q(0.3) = 1$. The distribution of genotypes produced by random mating is shown in the following Punnett square.

Sperm

	$A(p = 0.7)$	$a(q = 0.3)$
A $(p = 0.7)$	AA p^2 (0.49)	Aa pq (0.21)
Eggs		
a $(q = 0.3)$	Aa pq (0.21)	aa q^2 (0.09)

In this new generation, 49 percent (p^2) of the individuals will be homozygous dominant, 42 percent ($2pq$) will be heterozygous, and 9 percent (q^2) will be ho-

mozygous recessive. The frequency of the A allele in the new generation is

$$p^2 + \tfrac{1}{2}\,2pq \qquad (25.2)$$
$$0.49 + \tfrac{1}{2}\,(0.42)$$
$$0.49 + 0.21 = 0.70$$

For a, the frequency is

$$q^2 + \tfrac{1}{2}\,2pq \qquad (25.3)$$
$$0.09 + \tfrac{1}{2}\,(0.42)$$
$$0.09 + 0.21 = 0.30$$

Since $p + q = 1$, the value for a could have been calculated as

$$q = 1 - p \qquad (25.4)$$
$$q = 1 - 0.70 = 0.30$$

The values of A and a in the new generation are the same as in the previous generation.

A population in which the frequency of a given gene remains constant from generation to generation is said to be in state of **genetic equilibrium** for that gene. In this case, since the frequencies of A and a remain constant, we can presume that the conditions listed for Hardy-Weinberg equilibrium hold true in this population.

Several important points are relevant to the preceding example. First, while the hypothetical genes we have considered were at equilibrium, not all genes in a population are. This is particularly true when the assumptions made in the Hardy-Weinberg law do not hold. Second, the examples illustrate why dominant traits do not tend to increase in frequency as new generations are produced. Finally, the examples demonstrate that genetic equilibrium maintains a state of **genetic variability** in a population. Once fixed in a population, allele frequencies remain unchanged during equilibrium—a factor important to the evolutionary process.

Sex-Linked Genes

In considering genotype and gene frequencies for autosomal recessive traits using the Hardy-Weinberg equation, we assumed that the frequency of A is the same in both sperm and eggs. What about sex-linked genes? Since human females have two X chromosomes, one maternal and the other paternal in origin, they carry two copies of all genes on the X chromosome. Males, on the other hand, receive an X chromosome only from the mother. Genes on the X chromosome are therefore distributed unequally in the population, with females carrying two-thirds and males one-third of the total number. Can the Hardy-Weinberg law be applied to calculate allele frequencies and genotype frequencies in such circumstances?

It is easy to determine frequencies for X-linked genes in males. Because they have only one copy of all genes on this chromosome, the phenotype of both dominant and recessive genes is usually expressed. Therefore, the frequency of an X-linked gene is the same as the phenotypic frequency. In western Europe, a form of sex-linked color blindness occurs with a frequency of 8 percent in males ($q = 0.08$). Since females have two doses of all genes on the X chromosome, color-blind females behave as an autosomal recessive subpopulation, and the genotype and frequencies can be calculated by using the standard Hardy-Weinberg equation. For example, since color blindness in males has a frequency of 0.08, the expected frequency in females is q^2, or 0.0064. This means that 800 out of 10,000 males surveyed would be expected to be color-blind, but only 64 of 10,000 females would show this trait. Expected values of the frequency of X-linked traits in males and females are compared in Table 25.4.

TABLE 25.4
Expected relative frequency of X-linked traits in males and females.

Frequency of Males with Trait	Expected Frequency in Females
9/10	81/100
5/10	25/100
1/10	1/100
1/100	1/10,000
1/1000	1/1,000,000
1/10,000	1/100,000,000

If the frequency of an X-linked allele differs between males and females, then the population is not in equilibrium. In contrast to autosomal recessive genes, equilibrium will not be reached in a single generation, but will be approached over a series of succeeding generations. Because males inherit their X chromosome maternally, the gene frequency in females will determine the frequency in some of the next generation. Daughters will inherit both a maternal and a paternal X, and their gene frequency is the average of that found in the parents. Even though the population's overall gene frequency remains constant, there is an oscillation in gene frequencies for the two sexes in each generation, with the differences being halved in each succeeding generation until an equilibrium is reached. This concept is illustrated in Figure 25.1 for the simple case where an allele has an initial frequency of 1.0 in females and 0.0 in males.

Multiple Alleles

In addition to autosomal recessive and sex-linked genes, it is common to find several alleles of a single locus in a population. The ABO blood group in humans is such an example. The single locus I (isoagglutinin) has three alleles (I^A, I^B, and I^O), yielding six possible genotypic combinations ($I^A I^A$, $I^B I^B$, $I^O I^O$, $I^A I^B$, $I^A I^O$, $I^B I^O$). Recall that in this case types A and B are codominant and both of these are dominant to type O. The result is that homozygous AA and heterozygous AO individuals are phenotypically identical, as are BB and BO individuals, so we can distinguish only four phenotypic combinations.

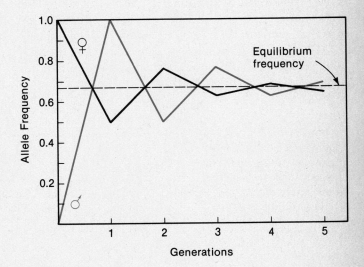

FIGURE 25.1
Approach to equilibrium for an X-linked trait with an initial frequency of 1.0 in females and 0.0 in males.

By adding another term to the Hardy-Weinberg equation, we can calculate both genotype and gene frequencies for the situation involving three alleles. In an equilibrium population, the frequency of the three genes can be described by

$$p(A) + q(B) + r(O) = 1 \qquad (25.5)$$

and the distribution of genotypes will be given by

$$(p + q + r)^2 \qquad (25.6)$$

In our hypothetical population, the genotypes *AA*, *AB*, *AO*, *BB*, *BO*, and *OO* will be found in the ratio

$$p^2(AA) + 2\,pq(AB) + 2pr(AO) + q^2(BB)$$
$$+ 2qr(BO) + r^2(OO) = 1 \qquad (25.7)$$

Knowing the frequencies of *A*, *B*, and *O* for a population, we can then calculate both the genotypic and phenotypic frequencies for all combinations of these three genes. In an Armenian population, the frequency of *A* (*p*) is 0.38; *B* (*q*) is 0.11; and *O* (*r*) is 0.51. By using the formula $(p + q + r)^2 = 1$, we can calculate the genotypic and phenotypic frequencies for this population as shown in Table 25.5, combining phenotypically identical genotypes to obtain the phenotypic frequencies.

Heterozygote Frequency

One of the practical applications of the Hardy-Weinberg law, mentioned briefly in the discussion of multiple alleles, is the calculation of heterozygote frequency in a population. When a deleterious recessive trait is investigated, the frequency of the recessive phenotype usually can be determined by counting such individuals in a sample of the population. This information and the Hardy-Weinberg equation are used to calculate the gene and genotype frequencies.

Albinism, an autosomal recessive trait, has an incidence of about 1/10,000 (0.0001) in some populations.

Albinos are easily distinguished from the population at large by a lack of pigment in skin, hair, and iris. Since this is a recessive trait, albino individuals must be homozygous. Their frequency in a population is represented by q^2, provided that mating has been at random and all Hardy-Weinberg conditions have been met in the previous generation.

The frequency of the recessive allele therefore is

$$\sqrt{q^2} = \sqrt{0.0001} \qquad (25.8)$$
$$q = 0.01, \text{ or } 1/100$$

Since $p + q = 1$, then the frequency of *p* is

$$p = 1 - q \qquad (25.9)$$
$$= 1 - 0.01$$
$$= 0.99, \text{ or } 99/100$$

In the Hardy-Weinberg equation, the frequency of heterozygotes is given as $2pq$. Therefore, we have

$$\text{Heterozygote frequency} = 2pq \qquad (25.10)$$
$$= 2[(0.99)(0.01)]$$
$$= 0.02, \text{ or } 2\%, \text{ or } 1/50$$

Thus, heterozygotes for albinism are rather common in the population (2%), even though the incidence of homozygous recessives is only 1/10,000.

In general, the frequencies of all three genotypes can be calculated once the frequency of either allele is known. The relationship between genotype frequency and gene frequency is shown in Figure 25.2. It is important to note how fast heterozygotes increase in a population as the values of *p* and *q* move away from zero. This observation confirms the conclusion that when a trait like albinism is rare, the majority of those carrying the allele are heterozygotes. In some cases, where the frequencies of *p* and *q* are between 0.33 and 0.67, heterozygotes actually constitute the major class in the population.

TABLE 25.5

Calculating genotypic and phenotypic frequencies for multiple alleles where the frequency of *A*(*p*) is 0.38, *B*(*q*) is 0.11, and *O*(*r*) is 0.51

Genotype	Genotypic Frequency		Phenotype	Phenotypic Frequency
AA	p^2	$= (0.38)^2 = 0.14$	A	0.53
AO	$2pr$	$= 2(0.38 \times 0.51) = 0.39$		
BB	q^2	$= (0.11)^2 = 0.01$	B	0.12
BO	$2qr$	$= 2(0.11 \times 0.51) = 0.11$		
AB	$2pq$	$= 2(0.38 \times 0.11) = 0.082$	AB	0.08
OO	r^2	$= (0.51)^2 = 0.26$	O	0.26

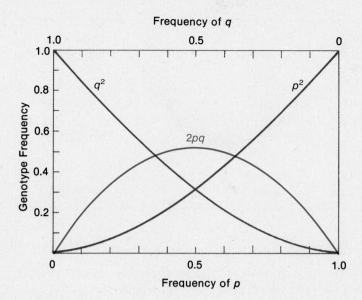

Frequency of q

Frequency of p

FIGURE 25.2
Relationship between genotype frequency and gene frequencies derived from the Hardy-Weinberg equation.

Demonstrating Equilibrium

The Hardy-Weinberg law can also be used to determine whether a given population is in equilibrium or is evolving. In order to do this, one must be able to distinguish heterozygotes phenotypically. If this is possible, then it must be ascertained whether the existing population fits a $p^2 + 2pq + q^2 = 1$ relationship. If so, the population has probably reached equilibrium. If not, some factor, presumably either selection, mutation, migration, or some other force, is causing gene frequencies to shift with each successive generation.

The MN blood group in one human population, the Australian aborigines, is a good example of this application of the Hardy-Weinberg law. From the values of allele frequencies in Table 25.3 (0.178 for M and 0.822 for N), the expected frequencies of blood types M, MN, and N can be calculated to determine if the population is in equilibrium:

Expected frequency of type M $= p^2$ $\qquad$ (25.11)

$$= (0.178)^2 = 0.032 = 3.2\%$$

Expected frequency of type MN $= 2pq$

$$= 2(0.178)(0.822) = 0.292 = 29.2\%$$

Expected frequency of type N $= q^2$

$$= (0.822)^2 = 0.676 = 67.6\%$$

These **expected frequencies** are nearly identical to the **observed frequencies** shown in Table 25.3, confirming that the population is in equilibrium. If there were a question as to whether the observed frequencies varied significantly from the expected frequencies, a chi-square analysis could be performed (see Chapter 3). This calculation could ascertain whether the values agree statistically.

In these calculations, the values are so close that we shall accept them as being in agreement. In addition to demonstrating that equilibrium has been reached, we may conclude that all prerequisites of the HWL are being met: mating between individuals of these genotypes is random in this population; there is no selective advantage to any of the three genotypes; the population is fairly large; there is no net change of gene frequency as a result of mutation; and there is insignificant migration between this and any other population. If any of the above conditions are not maintained, then at least they are balancing their effects so that the population remains in equilibrium.

On the other hand, if the Hardy-Weinberg test demonstrates that the population is not in equilibrium, one or more of these necessary conditions are not being met. Using the values listed for M and N frequencies in Table 25.3, we can construct a hypothetical mixed population of 500 Australian aborigines and 500 American Indians. The frequencies for M and N for this hypothetical population are shown in Table 25.6. In such a mixed population, the frequency of M (p) would be

$$0.315 + 1/2(0.324) = 0.477 \qquad (25.12)$$

and that of N (q) would be

$$0.361 + 1/2(0.324) = 0.523 \qquad (25.13)$$

If these gene frequencies had been derived by random mating, then the proportions of phenotypes we should expect are MM (p^2), 22.8 percent; MN $(2pq)$, 49.8 percent; and NN (q^2), 27.4 percent. Since the expected genotype frequencies do not fit those observed (and a chi-square test would confirm this), we would conclude that the population is in a state of nonequilibrium brought about, obviously, by a lack of random mating in this hypothetical population.

What would happen, however, if such a mixed population were to be established on an island with no previous inhabitants and mating did occur at random? How long would it take for equilibrium to be established? Applying the Hardy-Weinberg equation, we can predict that after one generation the observed and expected genotype frequencies would converge and the population would be in equilibrium. Using the p

TABLE 25.6

Frequencies of M and N alleles and genotypes in natural and artificial populations.

Population	Genotype Frequencies			Allele Frequencies	
	MM	MN	NN	p	q
Australian aborigines	0.03	0.296	0.674	0.178	0.822
American Indians	0.60	0.351	0.049	0.776	0.224
Mixed population (500 + 500):					
Observed	0.315	0.324	0.361	0.477	0.523
Expected	0.228	0.498	0.274	$\longleftarrow\qquad\longrightarrow$	

and q values just calculated, and remembering that the gene frequencies represent those of the total gene pool of the first generation mixed population, we can construct the following Punnett square:

♀ \ ♂	$M(p = 0.477)$	$N(q = 0.523)$
M $(p = 0.477)$	MM p^2 0.228	MN pq 0.249
N $(q = 0.523)$	MN pq 0.249	NN q^2 0.274

We can see that after one generation, the observed genotype frequencies of 22.8 percent for MM (p^2); 49.8 percent for MN ($2pq$); and 27.4 percent for NN (q^2) are those expected for an equilibrium population. The gene frequencies of $M = 0.477$ and $N = 0.523$ are also in equilibrium.

FACTORS THAT ALTER GENE FREQUENCIES

The Hardy-Weinberg law allows us to examine the relationship between gene frequencies and genotype frequencies in static, nonevolving populations in which our initial assumptions about random mating, absence of selection and mutation, and equal viability and fecundity hold true. Obviously, it is difficult—if not impossible—to find natural populations in which these conditions are met. Populations in nature are dynamic, and changes in size and structure are part of their life cycles. This does not mean that the Hardy-Weinberg relationship is invalid; rather, it provides an opportunity to study those forces that introduce and maintain genetic diversity in populations. In this section, we will discuss factors that prevent populations from reaching equilibrium and the relative contribution of these factors to evolutionary change.

Mutation

Within a population, the gene pool is reshuffled each generation to produce new combinations in the genotypes of the offspring. Because the number of possible combinations is so large, the members of a population alive at any given time represent only a small fraction of all possible genotypes. Although an enormous genetic reserve is present in the gene pool, genetic variability is produced by Mendelian assortment and recombination, and these processes do not produce any new alleles. **Mutation** alone acts to create new alleles and is a force in increasing genetic variation. But in the absence of other forces, mainly selection, mutation is a negligible force in changing gene frequencies. Details of the chemical and molecular aspects of mutation have been presented in Chapter 13.

For our purposes here, we need only consider that mutational events occur at random—that is, without regard for any possible benefit or disadvantage to the organism. We will discuss only the generation of mutant alleles and in a later section will consider the spread and distribution of new alleles through the population.

To know whether mutation is a significant force in changing gene frequencies, we must measure the rate at which mutations are produced. Since most mutations are recessive, it is difficult to observe mutation rates directly in diploid organisms. Indirect methods using probability and statistics or large-scale screening programs must be employed. For certain dominant mutations, however, a direct method of measurement can be used. To ensure accuracy, several conditions must be met:

1 The trait must produce a distinctive phenotype that can be distinguished from similar traits produced by recessive alleles.

2 The trait must be fully expressed or completely penetrant so that mutant individuals can be identified.

3 The trait must never be produced by nongenetic agents such as drugs or chemicals.

In humans, a dominant form of dwarfism known as **achondroplasia** fulfills the requirements for the measurement of mutation rates. Individuals with this skeletal disorder have an enlarged skull and short arms and legs and can be diagnosed by radiologic examination at birth. In a survey of almost 250,000 births, the mutation rate (μ) for achondroplasia has been calculated as

$$1.4 \times 10^{-5} \pm 0.5 \times 10^{-5} \qquad (25.14)$$

Knowing the rate of mutation, we can then employ the Hardy-Weinberg law to calculate the change in frequency for each generation. If initially only the normal d allele exists (i.e., all individuals are homozygous dd), the frequency of d (p_0) is 1.0 and the frequency of D (dwarf) is $q_0 = 0$. If the rate of mutation from d to D is μ, then in each successive generation,

$$\text{Frequency of } D = q_1 = p_0\mu \qquad (25.15)$$

The new frequency for d, which will be lowered by the rate of mutation, can be expressed as

$$\text{Frequency of } d = p_1 = 1 - p_0\mu \qquad (25.16)$$

Initially, the rate of addition of the mutant allele is relatively high; however, it becomes more gradual with each passing generation.

As a more general example, let us assume that the mutation rate for genes in the human genome is 1.0×10^{-5}. With such a low mutation rate, changes in gene frequency brought about by mutation alone are very small. In fact, Figure 25.3 shows that if A is the only allele of the a locus in the population ($p = 1$), with a mutation rate of 1.0×10^{-5} for A to a, it would require about 70,000 generations to reduce the frequency of A to 0.5. Even if the rate of mutation were to increase through exposure to higher levels of radioactivity or chemical mutagens, the impact of mutation on gene frequencies would be extremely weak. This example once again emphasizes that mutation is a major force in generating genetic variability, but by itself has an insignificant role in changing gene frequencies.

Migration

Frequently, a species of plant or animal becomes divided into subpopulations that to some extent are geographically separated. Differences in mutation rate and selective pressures can establish different allele frequencies in the subpopulations. **Migration** occurs when individuals move between these populations.

Consider a single pair of alleles, A (p) and a (q), where

$\quad p$ = frequency of A in existing population
$\quad p_m$ = frequency of A in immigrants
$\quad \Delta p$ = change in one generation
$\quad m$ = migration rate (proportion of migrant genes entering the population per generation)

The change in the frequency of A in one generation can be expressed as

$$\Delta p = m(p_m - p) \qquad (25.17)$$

For example, assume that $p = 0.4$ and $p_m = 0.6$. Ten percent of the parents giving rise to the next generation are immigrants ($m = 0.1$). Then, the change in the frequency of A in one generation is

$$\begin{aligned} \Delta p &= m(p_m - p) \\ &= 0.1(0.6 - 0.4) \\ &= 0.1(0.2) \\ &= 0.02 \end{aligned}$$

In the next generation, the frequency of A (p_1) will increase as follows:

$$\begin{aligned} p_1 &= p + \Delta p \qquad (25.18) \\ &= 0.40 + 0.02 \\ &= 0.42 \end{aligned}$$

If either m is large and/or if p is much larger or smaller than p_m, then a rather large change in the frequency of A will occur in a single generation. All other factors being equal, an equilibrium will be attained

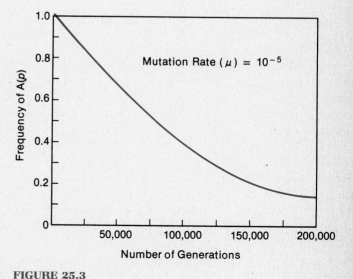

FIGURE 25.3

Rate of replacement of an allele by mutation alone, assuming an average mutation rate of 1.0×10^{-5}.

when $p = p_m$. These calculations reveal that the change in gene frequency attributable to migration is proportional to the differences in frequency between the donor and recipient populations. Since the migration coefficient can have a wide range of values, the effect of migration can substantially alter gene frequencies in populations. However, migration can be somewhat difficult to quantify unless all individuals in two separated populations can be monitored.

Migration can also be regarded as the flow of genes between two populations that were once but are no longer geographically isolated. For example, West African populations were the source of most of the blacks brought to the United States as slaves. In Africa, the frequency of the Duffy blood group allele Fy^0 is almost 100 percent. Among Europeans, the source of US Caucasians, the frequency is close to zero percent, and almost all individuals are Fy^a or Fy^b. By measuring the frequency of Fy^a or Fy^b among US blacks, we can estimate the amount of Caucasian gene flow into the black population. Figure 25.4 shows the frequency of the Fy^a allele among blacks in several regions of the country. Using the average frequency and assuming an equal rate of gene flow in each generation, we calculate the migration of the Fy^a allele at 5 percent per generation.

Selection

Mutations and migration introduce new alleles into populations. **Natural selection,** on the other hand, is the principal force that shifts gene frequencies within populations and is one of the most important factors in evolutionary change.

Darwin's main contribution to the study of evolution was his recognition of selection as the mechanism that leads to the divergence and eventual separation of populations into distinct species. In any population at a given time, there are individuals with different genotypes. Because of these inherent genetic differences, some of the progeny produced will be better adapted to the environment than others, leading to the differential survival and reproduction of some genotypes over others. Natural selection is thus the differential reproduction of genotypes. Therefore, gene frequencies will change over time. Natural selection, then, challenges the Hardy-Weinberg assumption that all genotypes have equal viability and fertility.

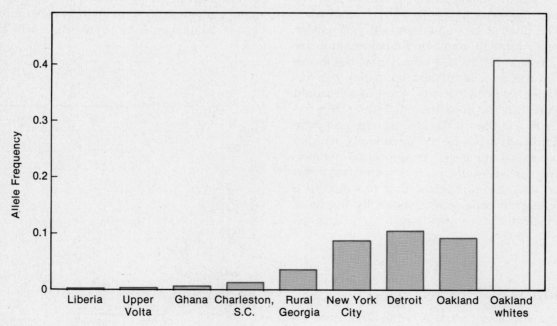

FIGURE 25.4

Distribution of the Fy^a allele in African and US black populations. (From T. E. Reed, "Caucasian Genes in American Negroes," *Science* 165: 762–768, Table 4, 22 August 1969. Copyright 1969 by the American Association for the Advancement of Science.)

When a particular genotype/phenotype confers an advantage to organisms in competition with others harboring an alternate combination, selection occurs. The relative strength of selection varies with the amount of advantage provided. The probability that a particular phenotype will survive and leave offspring is a measure of its **fitness.** Thus, fitness refers to a total reproductive potential or efficiency. The concept of fitness is usually expressed in relative terms by comparing a particular genotypic/phenotypic combination with one regarded as optimal. Fitness is a relative concept because as environmental conditions change, so does the advantage conferred by a particular phenotype.

Mathematically, the difference between the fitness of a given genotype and another regarded as optimal is called the **selection coefficient (s).** For a phenotype conferred by the genotype *aa*, when only 99 of every 100 organisms successfully reproduce, $s = 0.01$. If the *aa* genotype is a homozygous lethal, then $s = 1.0$, and the *a* allele is propagated only in the heterozygous carrier state. Starting with any original frequencies of A (p_0) and a (q_0) in a population, when $s = 1.0$, the effect of selection on any successive generation can be calculated using the formula

$$q_n = \frac{q_0}{1 + nq_0} \qquad (25.19)$$

where n is the number of generations elapsing since p_0 and q_0.

Figure 25.5 demonstrates the change in allele frequencies when A $(p_0) = a$ $(q_0) = 0.5$. Initially, because

FIGURE 25.5
Changes in allele frequency under selection ($s = 1.0$).

Generation	p	q	p^2	2pq	q^2
0	0.50	0.50	0.25	0.50	0.25
1	0.67	0.33	0.44	0.44	0.12
2	0.75	0.25	0.56	0.38	0.06
3	0.80	0.20	0.64	0.32	0.04
4	0.83	0.17	0.69	0.28	0.03
5	0.86	0.14	0.73	0.25	0.02
10	0.91	0.09	0.84	0.15	0.01
20	0.95	0.05	0.91	0.09	< 0.01
40	0.98	0.02	0.95	0.05	< 0.01
70	0.99	0.01	0.98	0.02	< 0.01
100	0.99	0.01	0.98	0.02	< 0.01

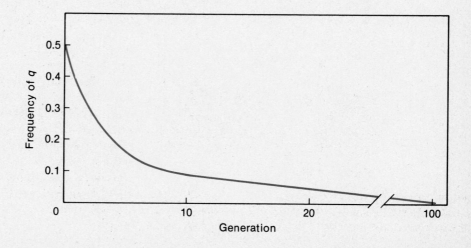

of the high percentage of *aa* genotypes, the frequency of the *a* allele is reduced rapidly. The frequency of *a* is halved (0.25) in only two generations. By the sixth generation, the frequency of *a* is again reduced twofold (0.12). By now, however, the majority of *a* alleles are carried heterozygously (heterozygotes are not selected against), and subsequent reductions in frequency occur very slowly in successive generations. For this reason, it is practically impossible to eliminate a recessive gene totally from the population as long as heterozygotes continue to mate.

When *s* is less than 1.0, it is possible to calculate the effects of selection on each successive generation with any *s*, p_0, and q_0 values using the formula

$$q_1 = \frac{q_0 - sq_0^2}{1 - sq_0^2} \qquad (25.20)$$

Figure 25.6 demonstrates a variety of initial conditions and the change of the frequency of *q* (*a*) through a large number of generations. If q_0 is initially 0.5 and *s* = 0.5, five generations must elapse before *q* is halved. Then, another nine generations must elapse to cut *q* in half again. Figure 25.6 shows the relative trends as q_0 and *s* change.

There has been much study of the action of selection on both laboratory and natural populations. A classic example of selection in natural populations is that of the peppered moth, *Biston betularia*, in England. Before 1850, 99 percent of the moth population was light colored, allowing the moths to fit well into their surroundings. As toxic gases produced by industry killed the lichens and mosses growing on trees and buildings, and soot deposits darkened the landscape, the light-colored moths became easy targets for their predators. The rare forms of dark-colored moths suddenly gained a great adaptive advantage because of their natural camouflage. A rapid shift in frequency of this phenotype occurred, probably in 50 generations or less. Today, in industrial areas, 90 percent of these moths are the "dark" phenotype.

Laboratory experiments have demonstrated that the dark form is due to a single dominant allele, *C*. Figure 25.7 demonstrates the protective advantage from predators provided by the dark phenotype. Following the adoption of laws to restrict environmental pollution in 1964, the frequency of nonmelanic forms of the moth has started to increase around the Manchester area, showing the close relationship between the environment and selection.

Because selection acts on the organism's genotypic/phenotypic combination, it also acts on polygenic or quantitative traits—those that are controlled by a number of genes and may be susceptible to environmental influences. Such quantitative traits, including adult body height and weight in humans, often demonstrate a continuously varying distribution resembling a bell-shaped curve. Selection for such traits can be classified as (1) directional, (2) stabilizing, or (3) disruptive.

In **directional selection,** which is important to plant and animal breeders, desirable traits, often representing phenotypic extremes, are selected for. In fact, if the trait is polygenic, the most extreme phenotypes that the genotype can express will appear in the population only after prolonged selection. An example is the selection for oil content in domestic corn kernels. In upward selection, C. M. Woodworth and others were able to raise the oil content of corn more than threefold, from about 4 percent to just over 15 percent in 50 generations, with no sign of a plateau being reached (Figure 25.8). Similarly, downward selection reduced the content to about 1 percent. Directional selection favors an extreme phenotype and tends to produce genetic uniformity in the population. In nature, directional selection can occur when one of the phenotypic extremes becomes selected for or against, usually as a result of changes in the environment.

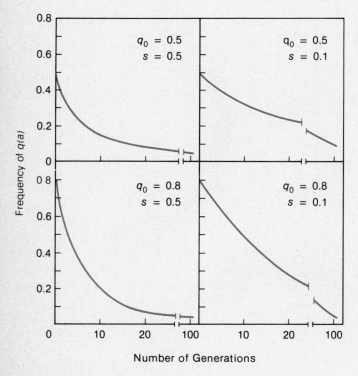

FIGURE 25.6
Changes in allele frequencies with different selection coefficients and different initial gene frequencies.

Stabilizing selection, on the other hand, tends to favor intermediate types, with both extreme phenotypes being selected against. One of the clearest demonstrations of stabilizing selection is the data of M. N. Karn and Sheldon Penrose on human birth weight and survival. Figure 25.9 shows the distribution of birth weight for 13,730 children born over an 11-year period and the percentage of mortality at four weeks of age. Infant mortality increases dramatically on either side of the optimal birth weight of 7.5 pounds. At the genetic level, stabilizing selection represents a situation where a population is adapted to its environment.

Disruptive selection is selection against intermediates and for both phenotypic extremes. It may be viewed as the opposite of stabilizing selection. John Thoday applied disruptive selection to progeny of *Drosophila* strains with high and low bristle number. Matings were carried out using females from one strain and males from the other. Progeny were selected for high and low bristle number, and matings were carried out for a number of generations. Despite the fact that opportunities for gene flow between the two lines through random mating were experimentally provided at each generation, the two lines diverged rapidly. In natural populations, such a situation may exist for a population in a heterogeneous environment. Gene flow may occur between two niches in which selection is favoring opposite extremes.

The types and effects of selection are summarized in Figure 25.10.

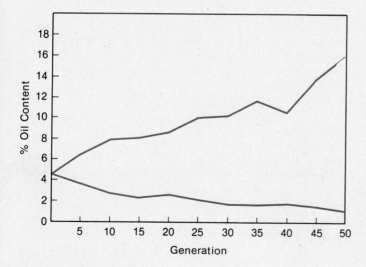

FIGURE 25.8
Long-term selection to alter the oil content of corn kernels. (Reprinted from *Agronomy Journal*, Volume 44, Feb. 1952, No. 2, p. 61, by permission of the American Society of Agronomy.)

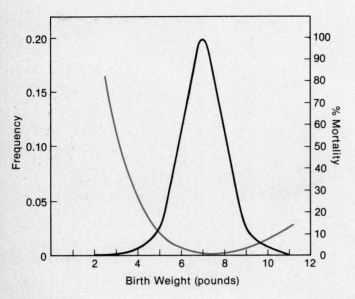

FIGURE 25.9
Relationship between birth weight and mortality in humans. Optimal birth weight (7.5 lb) is associated with lowest mortality. (Data from Karn and Penrose, 1951.)

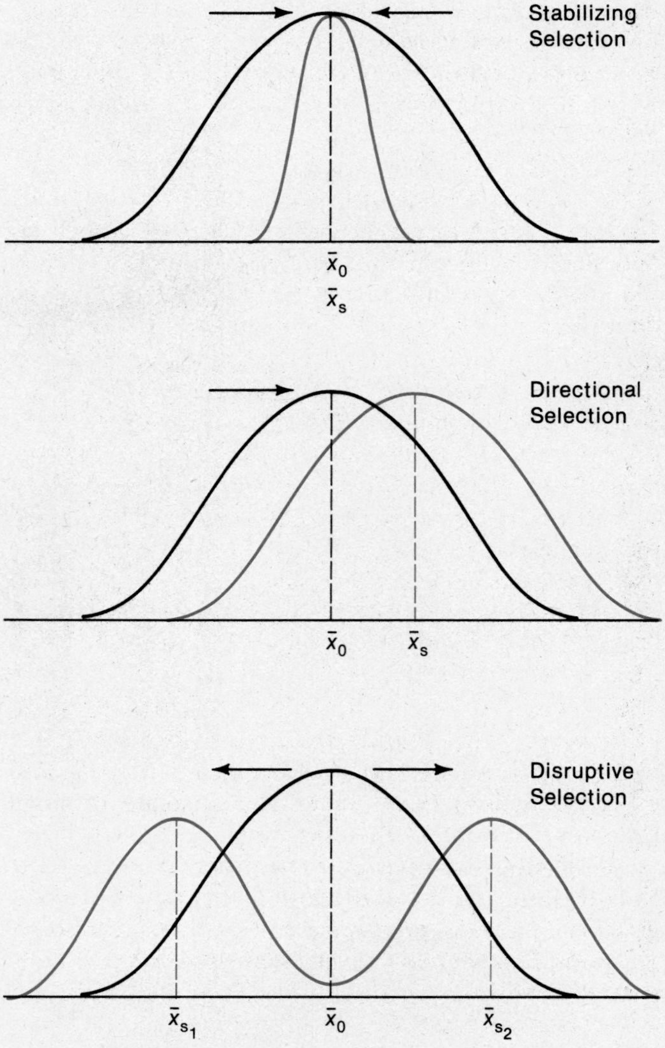

Genetic Drift

One condition essential to realizing theoretical genetic ratios (i.e., 3:1, 1:2:1, 9:3:3:1) is a fairly large sample size. This condition is equally important to the study of population genetics as gene and genotype frequencies are examined or predicted. For example, if a population consisted of 1000 heterozygotes (Aa) and they randomly mated, the next generation would consist of approximately 25 percent AA, 50 percent Aa, and 25 percent aa genotypes. Provided the initial population was large, there would be at most only minor fluctuations from this ideal ratio, and the frequency of A would remain about equal to a (0.50).

However, if a population formed from only one set of heterozygous parents and they produced only two offspring, the probability of the occurrence of genotypes and gene frequencies in the subsequent generation can be predicted as summarized in Table 25.7. As shown, in 10 of 16 times such a cross is made, the new allele frequencies would be altered dramatically. In 2 of 16 times, either the A or a allele would be eliminated in a single generation.

The conditions used to generate these calculations are extreme, but they illustrate why large interbreeding populations are essential to the Hardy-Weinberg equilibrium. In small populations, large random fluctuations in gene frequencies are possible strictly on the basis of chance deviation. The degree of fluctua-

Stabilizing Selection

Directional Selection

Disruptive Selection

FIGURE 25.10
Comparison of the impact of stabilizing, directional, and disruptive selection. In each case, $\bar{x}_0$ is the mean of an original population, and $\bar{x}_s$ is the mean of the population following selection.

TABLE 25.7
All possible pairs of offspring produced by two heterozygous parents ($Aa \times Aa$).

Possible Genotypes of Two Offspring	Probability	Allele Frequency	
		A	a
AA and AA	$(1/4)(1/4) = 1/16$	1.00	0.00
aa and aa	$(1/4)(1/4) = 1/16$	0.00	1.00
AA and aa	$2(1/4)(1/4) = 2/16$	0.50	0.50
Aa and Aa	$(2/4)(2/4) = 4/16$	0.50	0.50
AA and Aa	$2(2/4)(1/4) = 4/16$	0.75	0.25
aa and Aa	$2(1/4)(2/4) = 4/16$	0.25	0.75

tion expected will increase as the population size decreases. Such changes illustrate the concept of **genetic drift.** In the extreme case, genetic drift may lead to the chance fixation of one allele to the exclusion of another allele.

Using laboratory populations of *Drosophila melanogaster,* Warwick Kerr and Sewall Wright set up over 100 lines, with four males and four females as the parents for each line. Within each line, the frequency of the sex-linked bristle mutant *forked* (*f*) and its wild-type allele (*f*$^+$) was 0.5. For each generation, four males and four females were chosen at random to be parents of the next generation. After 16 generations, fixation had occurred in 70 lines—29 in which only *forked* was present and 41 in which the wild-type allele had become fixed. The rest of the lines were still segregating or had been lost. If fixation had occurred randomly, then an equal number of lines should have become fixed for each allele. In fact, the experimental results do not differ statistically from the expected ratio of 35:35, demonstrating that alleles can spread through a population and eliminate other alleles by chance alone.

How are small populations created in nature? In one instance, a large group of organisms may be split by some natural event, creating an isolate of small numbers. A natural disaster such as an epidemic might also occur, leaving a small number of survivors to constitute the breeding population. Third, a small group might emigrate from the larger population, as founders, to a new environment, such as a volcanically created island.

Gene frequencies in certain human isolates best support the factor of drift as an evolutionary force in natural populations. The Pingelap atoll in the western Pacific Ocean (lat. 6° N, long. 160° E) has in the past been devastated by typhoons and famine, and in 1775 there were only 30 survivors. Today there are fewer than 2000 inhabitants, and their ancestry can be traced to the original founders of the population. Approximately 5 percent of the current population is affected by the recessively determined disorder **achromatopsia,** which causes ocular disturbances and a form of color blindness. This disorder is extremely rare in the human population as a whole. However, the responsible allele has by chance become fixed at a relatively high frequency in the Pingelap population. From genealogical reconstruction, it was found that one of the original survivors was a chief who was heterozygous for the condition. If we assume that he was the only carrier in the founding population, the initial gene frequency was 1/60, or 0.016. Since some 5 percent of the current population is affected, they represent homozygous recessives, and the frequency of the gene in the population is calculated as 0.23.

Another example of genetic drift involves the Dunkers, a small religious isolate living in Pennsylvania who emigrated from the German Rhineland. Since their religious beliefs do not permit marriage with outsiders, the population has grown only through intermarriage. When the frequencies of the ABO and MN blood group alleles are compared, significant differences are found among the Dunkers, the German population, and the US population. The frequency of blood group A in the Dunkers is about 60 percent, in contrast to 45 percent in the US and German populations. The I^B allele is nearly absent in the Dunkers. Type M blood is found in about 45 percent of the Dunkers, compared with about 30 percent in the US and German populations. Since there is no evidence for a selective advantage of these alleles, it is apparent that their observed frequencies are the result of chance fixation in a relatively small isolated population.

Inbreeding and Heterosis

One of the assumptions of the Hardy-Weinberg equilibrium is random mating within the population. This is an ideal condition and as such does not always occur in nature. Within a widespread population, individuals tend to mate with those in physical proximity rather than with those separated by some distance. The result is the restriction of gene flow within that population.

Nonrandom mating occurs in many species, including humans. One form of nonrandom mating is called **assortative.** In humans, pair bonds may be established by religious practices, physical characteristics, professional interests, and so forth on a nonrandom basis. In nature, phenotypic similarity plays the same role.

Another form of nonrandom mating is **inbreeding,** where mating occurs between relatives. Inbreeding results in increased homozygosity. To illustrate this concept, we shall consider the most extreme form of inbreeding, **self-fertilization.**

Figure 25.11 shows the consequences of four generations of self-fertilization starting with a single individual heterozygous for one pair of alleles. By the fourth generation, only about 6 percent of the individuals are still heterozygous. Note that the frequencies of *A* and *a* still remain at 50 percent.

In humans, inbreeding through cousin marriages (called **consanguineous marriages**) is related to population size, mobility, and social customs governing such marriages. To determine the probability that two alleles at the same locus in an individual are derived from a common ancestor, Sewall Wright devised the **coefficient of inbreeding.** Expressed as *F*, the coefficient is also used to express the proportion of

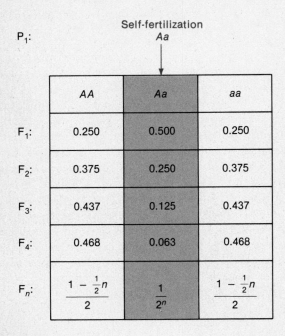

FIGURE 25.11
Reduction in heterozygote frequency brought about by self-fertilization. After n generations, the frequencies of the genotypes can be calculated according to the formulas in the bottom row.

probability of the presence of a is shown. The pedigrees then culminate in the final probability, F, that the lethal allele will become homozygous in the offspring of first or second cousins. Since every population carries as its genetic burden a number of heterozygous recessive alleles which when homozygous are lethal or deleterious, mortality will rise in matings between relatives.

Thus, from an evolutionary standpoint, if a population is split into smaller subpopulations, the effects of inbreeding will be evident. **Inbreeding depression,** an expression of reduced vigor in a population caused by inbreeding, may occur; and because an increase in homozygosity results, genetic variability will decrease. Both favorable and unfavorable alleles will become more frequently homozygous. From the standpoint of Darwinian fitness, the adaptive nature of the population will decrease as well.

This information has been applied in the domestication of plant and animal species. First, an inbreeding program is initiated. As homozygosity increases, some organisms will fix favorable alleles and others unfavorable alleles. By selecting the more viable and vigorous plants or animals, desirable traits can be increased.

If members of two favorable inbred lines are mated, hybrid offspring are often more vigorous in desirable traits than is either of the parental lines. This phenomenon has been called **hybrid vigor** or **heterosis.** Such an approach was used in the breeding programs established for corn. Not only were crop yields increased

loci that are homozygous in an individual. Shown in Figure 25.12 are pedigrees of first- and second-cousin marriages. A lethal allele, a, is assumed to be present heterozygously in one parent. For each relative, the

FIGURE 25.12
Changes in the coefficient of inbreeding (F) brought about by consanguineous marriages. Numbers in color indicate the probability of carrying the recessive allele a.

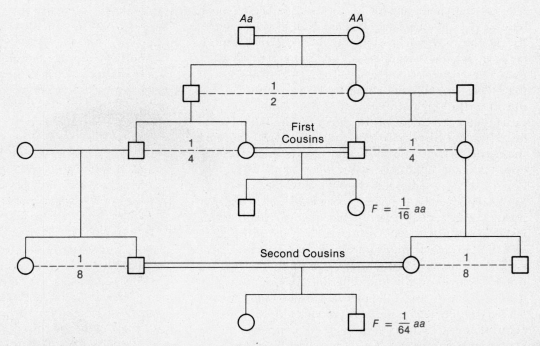

tremendously, but disease resistance was also incorporated into existing strains. Unfortunately, the hybrid vigor extends only through the first generation. Many hybrid lines are sterile, and those that are fertile show subsequent declines in yield. Consequently, the hybrids must be regenerated each time by crossing the inbred parental lines.

Heterosis has been explained in two ways. The first hypothesis, the **dominance theory,** incorporates the obvious reversal of inbreeding depression, which inevitably must occur in out-crossing. Those deleterious alleles present homozygously will be masked by the more favorable dominant allele in hybrids. Thus, such masking is thought to cause the hybrid vigor.

The second theory, **overdominance,** holds that in many cases the heterozygote is superior to either homozygote. Thus, the cumulative effect of heterozygosity at many loci accounts for the hybrid vigor. Most likely, such vigor results from a combination of phenomena explained by both theories.

SUMMARY AND CONCLUSIONS

When Darwin proposed his theory of evolution and natural selection, he had no knowledge of the underlying basis of variation within groups of organisms. When Mendel's work was rediscovered in the early 1900s, genes were established as the basis for this variation. It was apparent to those investigating evolution that the population is the effective unit of evolution. Therefore, they began to study the frequencies of genes and genotypes in populations rather than in individual matings. From these ideas emerged the discipline of population genetics.

The Hardy-Weinberg law established the conditions for genetic equilibrium within a population. These conditions include a large, randomly breeding population occurring in the absence of selection, mutation, and migration. If any of these conditions are not met, gene or genotype frequencies will change from generation to generation.

The Hardy-Weinberg formula is also useful in demonstrating whether or not a population is in equilibrium and in calculating the degree of heterozygosity within a population for any particular locus.

Because changing gene frequencies provide genetic variation within a population, conditions leading to change are precisely those necessary for evolution. Thus, the mathematical derivation of the Hardy-Weinberg equilibrium has served to assess the forces of evolution: selection, mutation, migration, genetic drift, and inbreeding. In this chapter, we have considered each of these topics and the mathematical analysis of their effects on gene frequencies in a population.

Mutation and migration introduce new alleles into a population, but the retention or loss of these alleles depends on the degree of fitness conferred and the action of selection.

Natural selection is the most powerful force altering a population's genetic constitution. In small populations, however, inbreeding and genetic drift may be important factors. In Chapter 26 we will examine how these forces are related to the divergence of populations and the formation of species.

PROBLEMS AND DISCUSSION QUESTIONS

1 The ability to taste the compound PTC is controlled by a dominant allele T, while individuals homozygous for the recessive allele t are unable to taste this compound. In a genetics class of 125 students, 88 were able to taste PTC, 37 could not. Calculate the frequency of the T and t alleles in this population, and the frequency of the genotypes.

2 Calculate the frequencies of the AA, Aa, and aa genotypes after one generation if the initial population consists of 0.2 AA, 0.6 Aa, and 0.2 aa genotypes. What frequencies will occur after a second generation?

3 Consider rare disorders in a population caused by an autosomal recessive mutation. From the following frequencies of the disorder in the population, calculate the percentage of heterozygous carriers:
   (a) 0.0064
   (b) 0.000081
   (c) 0.09
   (d) 0.01
   (e) 0.10

4 What must be assumed in order to validate the answers in Problem 3?

5   In a population where only the total number of individuals with the dominant phenotype is known, how can you calculate the percentage of carriers and homozygous recessives?

6   For the following two sets of data, determine whether they represent populations that are in equilibrium. Use chi-square analysis if necessary.
   (a) MN blood groups: *MN*, 60%; *MN*, 35.1%; *NN*, 4.9%.
   (b) Sickle-cell hemoglobin: *AA*, 75.6%; *AS*, 24.2%; *SS*, 0.2%.

7   If 4 percent of the population in equilibrium expresses a recessive trait, what is the probability that the offspring of two individuals who do not express the trait will express it?

8   Consider the gene frequencies $p = 0.7$ and $q = 0.3$ where selection occurs against the homozygous recessive genotype. What will be the gene frequencies after one generation if
   (a) $s = 1.0$
   (b) $s = 0.5$
   (c) $s = 0.1$
   (d) $s = 0.01$?

9   If initial gene frequencies are $p = 0.5$ and $q = 0.5$, and $s = 1.0$, what will be the gene frequencies after 1, 5, 10, 25, 100, and 1000 generations?

10   Assume that the initial frequency of the *A* allele is 100 percent and that *A* mutates to *a* at a rate of 10^{-6}. What will be the frequency of *A* and *a* after 10, 100, 500, and 1000 generations?

11   Determine the frequency of allele *A* after one generation under the following conditions of migration:
   (a) $p = 0.6; p_m = 0.1; m = 0.2$
   (b) $p = 0.2; p_m = 0.7; m = 0.3$
   (c) $p = 0.1; p_m = 0.2; m = 0.1$

12   Assume that a recessive autosomal disorder occurs in one of 10,000 individuals (0.0001) in the general population. Assume that in this population about 2 percent (0.02) of the individuals are carriers for the disorder. Estimate the probability of this disorder occurring in the offspring of a marriage between first cousins and between second cousins. Compare these probabilities to the population at large.

13   What is the basis of inbreeding depression?

14   Does inbreeding cause an increase in frequency of recessive alleles within a population?

15   Describe how inbreeding may be used in the domestication of plants and animals. Discuss the theories underlying these techniques.

SELECTED READINGS

BODMER, W. F., and CAVALLI-SFORZA, L. L. 1976. *Genetics, evolution and man.* San Francisco: W. H. Freeman.

CAVALLI-SFORZA, L. L., and BODMER, W. F. 1971. *The genetics of human populations.* San Francisco: W. H. Freeman.

COOK, L. M., ASKEW, R. R., and BISHOP, J. A. 1970. Increasing frequency of the typical form of the peppered moth in Manchester. *Nature* 227: 1155.

CROW, J. F., and KIMURA, M. 1970. *An introduction to population genetic theory.* New York: Harper and Row.

DOBZHANSKY, T. 1955. A review of some fundamental concepts and problems of population genetics. *Cold Spr. Harb. Symp.* 20: 1–15.

FELDMAN, M. W. and CHRISTIANSEN, F. B. 1985. *Population genetics.* Palo Alto: Blackwell Scientific Publ.

FISHER, R. A. 1930. *The genetical theory of natural selection.* Oxford, England: Clarendon Press. (Reprinted in 1958 by Dover Press.)

HARTL, D. L. 1981. *A primer of population genetics.* Sunderland, Mass.: Sinauer Associates.

HEDRICK, P. W. 1983. *Genetics of populations.* Boston: Science Books International.

JONES, J. S. 1981. How different are human races? *Nature* 293: 188–90.

KARN, M. N., and PENROSE, L. S. 1951. Birth weight and gestation time in relation to maternal age, parity and infant survival. *Ann. Eugen.* 16: 147–64.

KERR, W. E., and WRIGHT, S. 1954. Experimental studies of the distribution of gene frequencies in very small populations of *Drosophila melanogaster.* I. *Forked. Evolution* 8: 172–77.

KETTLEWELL, H. B. D. 1961. The phenomenon of industrial melanism in *Lepidoptera. Ann. Rev. Entomol.* 6: 245–62.

————. 1973. *The evolution of melanism: The study of a recurring necessity, with special reference to industrial melanism in the Lepidoptera.* London: Oxford University Press.

KIMURA, M., and OHTA, T. 1971. *Theoretical aspects of population genetics.* Princeton, N.J.: Princeton University Press.

LEWONTIN, R. C. 1973. Population genetics. *Ann. Rev. Genet.* 7: 1–18.

————. 1974. *The genetic basis of evolutionary change.* New York: Columbia University Press.

LI, C. C. 1977. *First course in population genetics.* Pacific Groves, Calif.: Boxwood Press.

METTLER, L. E., and GREGG, T. 1968. *Population genetics and evolution.* Englewood Cliffs, N.J.: Prentice-Hall.

NEEL, J. V., and ROTHMAN, E. 1981. Is there a difference among human populations in the rate with which mutation produces electrophoretic variants? *Proc. Natl. Acad. Sci.* 78: 3108–12.

PROVINE, W. 1971. *The origins of theoretical population genetics.* Chicago: The University of Chicago Press.

REED, T. E. 1969. Caucasian genes in American Negroes. *Science* 165: 762–68.

SPIESS, E. B. 1977. *Genes in populations.* New York: Wiley.

WALLACE, B. 1981. *Basic population genetics.* New York: Columbia University Press.

WOODWORTH, C. M., LENG, E. R., and JUGENHEIMER, R. W. 1952. Fifty generations of selection for protein and oil in corn. *Agron. J.* 44: 60–66.

26

Genetics and Evolution

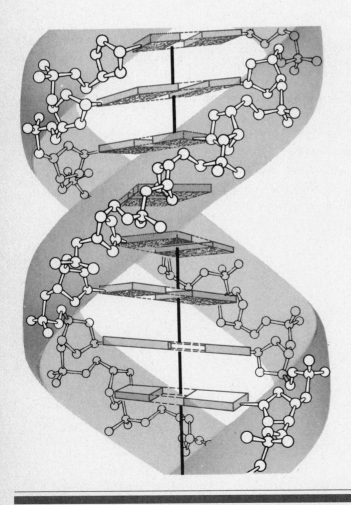

In Chapter 25, we described populations in terms of allele frequencies and genetic equilibria and outlined the forces that alter the genetic constitution of populations. Mutation, migration, selection, and drift are some of the forces that individually and collectively alter gene frequencies and lead to evolutionary divergence and species formation. This process depends not only upon genetic divergence but also upon environmental or ecological diversity. If a population is spread over a geographic range encompassing a number of subenvironments or **niches,** the subpopulations occupying these niches will adapt and become genetically differentiated. This process may lead to the formation of **races** (ecotypes) or subspecies. Because the formation and maintenance of these races depend ultimately on the interaction of the genotype and environment, races are dynamic entities that may remain in existence, become extinct, merge with the parental population, or continue to separate from the parental population until they become reproductively isolated and form new species.

It is difficult to define the exact moment of transition between a subspecies and species. A **species** can be defined as one or more groups of interbreeding or potentially interbreeding organisms that are reproductively isolated in nature from all other organisms. The process of **speciation** depends on changing gene frequencies and involves the division of a homogeneous population (gene pool) into two or more reproductively isolated subunits (separate gene pools). Changes in morphology, physiology, and adaptation to an ecological niche may also occur, but are not necessary components of the speciation event. Speciation can take place gradually over a long time period or within a generation or two. Throughout this process, the degree of genetic diversity is the key factor. Individuals of a species are members of a common gene pool that distinguishes them from members of other species. Therefore, it is clear that genetic divergence should be the starting point for a discussion of speciation and evolution.

In this chapter, we will focus on three central topics: first, the extent of genetic diversity in closely related organisms and its potential contribution to speciation; second, analysis of the evolution of diverse organisms at the molecular level; and third, the role of mutation in evolution, a matter of current controversy.

GENETIC DIVERSITY AND SPECIATION

If a population of organisms constituting a species is well suited to inhabit and reproduce successfully in its environment, how much genetic diversity does it possess? We might assume that members of a population are fit because the most favorable allele at each locus has become fixed homozygously. Certainly, an examination of most populations of plants and animals reveals them to be phenotypically similar, if not identical. However, current evidence supports the opinion that a high percentage of heterozygosity is maintained within the genome of diploid individuals in a population. This built-in diversity is concealed, so to speak, because it is not necessarily apparent in the phenotype. It is believed that such diversity better adapts a population to the inevitable changes in the environment and better assures the survival of the species.

The detection of this concealed genetic variation is not a simple task. Nevertheless, such investigation has been successful using several techniques.

Inbreeding Depression

One method of measuring genetic diversity is to determine what fraction of the alleles carried by individuals of a population are deleterious. Deleterious alleles in a population can be detected by monitoring increasing homozygosity resulting from inbreeding. As a result of inbreeding, recessive alleles previously present in the heterozygous condition become homozygous, and if these alleles are deleterious, a reduction in viability called **inbreeding depression** is evident in successive generations.

Theodosius Dobzhansky and his colleagues have succeeded in making groups of alleles and sometimes entire chromosomes of *Drosophila* species completely homozygous. When this is accomplished, viability is depressed. As shown in Table 26.1, such studies reveal lethal, semilethal, subvital, and male and female sterility alleles to be prominent in natural populations of different species. In *Drosophila pseudoobscura*, almost

TABLE 26.1

Percentage of homozygous chromosomes tested that reveal recessive modifiers of viability and fertility.

Drosophila Species	Chromosome	Lethals and Sublethals	Subvitals	Supervitals	Female Sterility	Male Sterility
D. pseudoobscura	2	33.0	62.6	<0.1	10.6	8.3
(Mather, Calif.)	3	25.0	58.7	<0.1	13.6	10.5
	4	22.7	51.8	<0.1	4.3	11.8
D. persimilis	2	25.5	49.8	0.2	18.3	13.2
(Mather, Calif.)	3	22.7	61.7	2.1	14.3	15.7
	4	28.1	70.7	0.3	18.3	8.4
D. prosaltans	2	32.6	33.4	<0.1	9.2	11.0
(Brazil)	3	9.5	14.5	3.0	6.6	4.2
D. willistoni	2	38.8	57.5	<0.1	40.5	64.8
(Venezuela, British Guiana, and Trinidad)	3	34.7	47.1	1.0	40.5	66.7

SOURCE: From Spiess, 1977. p. 288.

100 percent of chromosomes 2 and 3 from a sample of the population cause a depression in viability when made homozygous. In *Drosophila willistoni*, almost 70 percent of chromosomes 2 and 3 contain male sterile alleles when made homozygous.

Although these and other studies reveal the substantial genetic burden carried by organisms, we discuss them here as examples of genetic variation maintained in natural populations. If detrimental alleles are maintained heterozygously, certainly alleles that are effectively neutral or potentially advantageous must also be retained in the heterozygous condition. As recessive alleles, they represent concealed genetic variation within the "normal" heterozygous genotypes characterizing the population.

Protein Polymorphisms

The use of Mendelian genetics to estimate genetic diversity is both laborious and slow. Gel electrophoresis (see Appendix A), on the other hand, is a simpler technique that can be used to separate protein molecules on the basis of differences in electrical charge. If variation in a structural gene results in the substitution of a charged amino acid such as glutamic acid for an uncharged amino acid such as glycine, the net charge on the protein will be altered. This change in charge can be detected by electrophoresis. In the mid-1960s, John Hubby and Richard Lewontin introduced the use of gel electrophoresis to measure protein variation in natural populations of *Drosophila*. Figure 26.1 shows the separation of the enzyme phosphoglucomutase from *Drosophila pseudoobscura*. Since then, genetic variation has been studied using gel electrophoresis in a wide range of organisms.

When two alleles from one locus produce slightly different electrophoretic forms, yet perform identical functions, the resultant proteins are designated as **allozymes.** As shown in Table 26.2, a surprisingly large percentage of loci examined from diverse species produce allozymes. Of the populations shown in Table 26.2, approximately 40 loci per species were examined,

and about 30 percent of the loci were polymorphic (i.e., they revealed allozymes). An approximate average of 10 percent heterozygosity per diploid genome was revealed.

These values apply only to genetic variation detectable by altered protein migration in an electric field. Electrophoresis is thought to detect only about 30 percent of the actual variation that is due to amino acid substitutions. Richard Lewontin has thus estimated that about two-thirds of all loci are polymorphic in a population, and that in any individual within that group about one-third of the loci exhibit genetic variation in the form of heterozygosity.

There is some controversy over the significance of genetic variation as detected by electrophoresis. Some workers feel that alleles producing allozymes are effectively neutral and therefore do not play any role in evolution. We shall address this argument later in this chapter.

Chromosomal Polymorphisms

While the chromosome number is commonly regarded as invariant for a given species, the arrangement of chromosomal material is often polymorphic as a result of chromosomal inversions and translocations. These chromosomal aberrations usually have little direct effect on the phenotype because gene content is rearranged but not altered.

Inversions are produced by two breaks in a chromosome followed by rejoining of the broken fragment rotated 180° (see Chapter 12). The presence of an inversion does not imply that any new genes are present, nor does the presence of a variety of specific inversions prove that gene frequencies vary between any two arrangements. However, we have learned in Chapter 12 that inversion heterozygotes only rarely produce viable crossover products involving genes contained within the inversion. Therefore, an inversion suppresses the recovery of crossover gametes. In so doing, inversions tend to preserve specific allele arrangements along the inverted stretch of the chromosome involved. The net effect of inversions in populations, therefore, is a reduction in available, realized genetic variability. Thus, if specific inversions are present in populations well adapted to specific environments, it can be hypothesized that there is a causal relationship. It can be postulated that the inversion has preserved a favorable gene arrangement, which at least partially accounts for the fitness of the population in a specific habitat. In other words, linkage becomes a possible criterion for selection. The role of inversions in natural populations has been intensively studied in *Drosophila* and will be discussed later in this section.

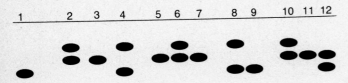

FIGURE 26.1
Phosphoglucomutase isozymes in *D. pseudoobscura*. Each column represents the isozyme pattern from a single fly.

TABLE 26.2
Heterozygosity at the molecular level.

Species	Number of Populations Studied	Number of Loci Examined	Loci per Population	Heterozygosity per Locus
Homo sapiens (humans)	1	71	28	6.7
Mus musculus (house mouse)	4	41	29	9.1
Drosophila pseudoobscura (fruit fly)	10	24	43	12.8
Limulus polyphemus (horseshoe crab)	4	25	25	6.1

SOURCE: From Lewontin, 1974, p. 117.

Translocations, in which chromosome segments move to other, nonhomologous chromosomes, are also found in populations. In simple translocations, a segment of one chromosome becomes joined to a nonhomologous chromosome. In reciprocal translocations, chromosome parts are exchanged between two nonhomologous chromosomes. Translocations also reduce realized genetic variability in populations because translocation heterozygotes have reduced frequencies of crossing over and are usually subfertile because they produce some proportion of abnormal gametes. Although somewhat rare in animal populations, translocation heterozygotes are found in many plants. In a population of the flowering plant *Clarkia elegans*, Harlan Lewis found that 13 percent of the individuals carried a translocation. In the extreme case, the angiosperm *Rhoeo discolor* has all of its chromosomes in translocation heterozygotes.

DNA Sequence Polymorphisms

Obviously, the most direct way to estimate genetic variation is by examination of the actual nucleotide sequence diversity between individuals of a population. With the development of techniques for cloning and sequencing DNA, nucleotide sequence variations have been cataloged for an increasing number of gene systems. Using restriction endonucleases to detect polymorphisms (see Chapter 23 for the methodology), Alec Jeffrey surveyed 60 unrelated individuals to estimate the total number of DNA sequence variants in humans. His results show that within the genes of the beta-globin cluster, 1 in 100 base pairs shows polymorphic variation. If this region is representative of the genome, this indicates that at least 3×10^7 nucleotide variants per genome are possible. Data from other organisms such as *Drosophila*, the rat, and the mouse have produced similar estimates of nucleotide diversity, indicating that there is an enormous reservoir of genetic variability within a population, and that at the level of DNA, most and perhaps all genes exhibit diversity from individual to individual.

The Adaptive Norm

A population of interbreeding organisms is adapted to its surrounding environment. Adaptive phenotypes are thought to be a consequence of the array of genotypes possessed by the individuals constituting the group. These genotypes are present as a result of the evolutionary history of the population.

Dobzhansky has called the array of genotypes present in the population the **adaptive norm.** Ideally, each individual should possess a genotype and phenotype best suited to the immediate environment. However, the preceding discussion on genetic variability in natural populations suggests that there is a wide variation in genotypes in many populations. Therefore, the adaptive norm tends to be one of balanced heterozygosity. Because numerous recessive alleles are concealed and do not alter the phenotype, it follows that a variety of genotypes can be well adapted to the same environment. As Dobzhansky has pointed out, the variety making up the adptive norm constitutes a co-adapted set of genotypes that must be able to meet the environmental diversity and stresses over space and time.

Alleles that at one point in time do not seem to be of major significance to individual fitness may be of great value to the population in future generations. Under changing environmental conditions, previously insignificant alleles may become essential to the maintenance of fitness. The concept of **preadaptation** describes the so-called hidden or concealed genetic variation as a storehouse of genetic information available to enhance survival under new environmental conditions.

Demonstration of an adaptive norm comprised of extensive heterozygosity provides strong support for the fundamental theorem set down by Ronald Fisher in 1930. Fisher's theorem relates genetic variability to a population's evolutionary fitness as determined by natural selection. The greater the genetic variability upon which selection for fitness may act, the greater is the improvement in fitness. In mathematical terms, this theorem of natural selection states that the rate of increase in fitness equals the variance in fitness present at any time in a population.

Speciation

In the classical sense, the process of splitting a genetically homogeneous population into two or more populations that undergo genetic differentiation and eventual reproductive isolation is called **speciation.** According to Ernst Mayr, species originate in two predominant ways. In the first mode, often called **phyletic speciation,** species A over a long period of time becomes transformed into species B. In the second method, one species gives rise to one or more derived species, bringing about multiplication of species. This second process can occur over a long period of time or abruptly in a generation or two. Table 26.3 summarizes the principal methods of speciation.

The most developed classical model of speciation is **geographic** or **allopatric speciation,** first proposed by Moritz Wagner in 1868. According to Wagner, physical isolation of populations by geographic features such as lakes, rivers, or mountains that act as barriers to gene flow is the first step toward species formation. In a second step, these isolated populations undergo independent evolution and may continue to diverge to produce two distinct species.

Twentieth-century workers such as Mayr and Dobzhansky have refined and updated this model but have retained the general features of Wagner's hypothesis. In the refined model, the first step requires that populations become separated; that is, gene flow must be interrupted. The absence or interruption of gene flow is a prerequisite for the development of genetic differences by adaptation to local conditions. Genetic diversity brought about by natural selection, or by random genetic drift, will be reflected in the presence of new alleles, changes in allele frequency, or the presence of new chromosomal arrangements. Eventually, a point will be reached when the populations have

TABLE 26.3
Modes of speciation.

I. Transformation of Species (phyletic speciation)   1. Autogenous speciation
II. Reduction in Number of Species (fusion of two species)

III. Multiplication of Species (true species)
  A. Instant speciation (through individuals)
    1. Genetic
      (a) Single mutation in asexual species
      (b) Macrogenesis
    2. Cytological
      (a) Chromosomal mutation (translocations, etc.)
      (b) Autopolyploidy
      (c) Amphidiploidy
  B. Gradual speciation (through populations)
    1. Sympatric speciation
    2. Semigeographic speciation
    3. Geographic speciation (allopatric)

SOURCE: From Mayr, 1963, Table 15.1.

enough differences that they can be identified as distinct races or semispecies. This process may continue uninterrupted until two or more species are present.

If at any time during the process of genetic divergence the conditions halting gene flow between the populations are removed, two outcomes are possible: (1) the two populations may fuse into a single gene pool, because hybridization does not reduce fertility or viability; or (2) the gene pools of the populations may have diverged to the point where isolating mechanisms may have arisen.

The various biological and behavioral properties of organisms acting to prevent or reduce interbreeding are called **reproductive isolating mechanisms.** These mechanisms are classified in Table 26.4. For example, genetic divergence may have reached the stage where the viability or fertility of hybrids is reduced. Hybrid zygotes may be formed, but all or most may be inviable. Alternately, the hybrids may be viable but have reduced fertility or be sterile. In another possibility, the hybrids themselves may be fertile, but their progeny may have lowered viability or fertility. These mechanisms, called **postzygotic,** are a byproduct of genetic divergence. Such isolating mechanisms waste gametes and zygotes and lower the fitness of hybrid survivors. Selection will therefore favor the spread of

TABLE 26.4

Reproductive isolating mechanisms.

Prezygotic Mechanisms (prevent fertilization and zygote formation)

1. Geographic or ecological: The populations live in the same regions but occupy different habitats.

2. Seasonal or temporal: The populations live in the same regions but are sexually mature at different times.

3. Behavioral (only in animals): The populations are isolated by different and incompatible behavior before mating.

4. Mechanical: Cross-fertilization is prevented or restricted by differences in reproductive structures (genitalia in animals, flowers in plants).

5. Physiological: Gametes fail to survive in alien reproductive tracts.

Postzygotic Mechanisms (fertilization takes place and hybrid zygotes are formed, but these are nonviable or give rise to weak or sterile hybrids)

1. Hybrid nonviability or weakness.

2. Developmental hybrid sterility: Hybrids are sterile because gonads develop abnormally or meiosis breaks down before completion.

3. Segregational hybrid sterility: Hybrids are sterile because of abnormal segregation to the gametes of whole chromosomes, chromosome segments, or combinations of genes.

4. F_2 breakdown: F_1 hybrids are normal, vigorous, and fertile, but F_2 contains many weak or sterile individuals.

SOURCE: From G. Ledyard Stebbins, *Processes of Organic Evolution*, 3rd edition, © 1977, p. 143. Reprinted by permission of Prentice-Hall, Inc., Englewood Cliffs, N.J.

alleles that will reduce the formation of hybrids, leading to the development of **prezygotic isolating mechanisms.** Speciation and the development of isolating mechanisms may gradually occur in the absence of selection, as when populations remain permanently isolated on two or more islands. Selection accelerates speciation in those cases where some reproductive isolation has resulted from genetic diversity, but not all isolating mechanisms are represented in each speciation event. Usually at least two isolating mechanisms are developed by natural selection drawing on the genetic variability present in the evolving populations.

Formation of Races and Species

Although the theory of geographic speciation is well developed, it has been more difficult to observe directly the processes involved. Diversification of isolated populations occurs gradually over thousands or hundreds of thousands of years. In addition, the geographic changes that paralleled this divergence may be complex and completely unknown. In most cases, therefore, the formation of species is a historical event, and biologists studying this process must rely on the present-day distribution of races, subspecies, and sibling species to reconstruct the stages of the evolutionary process.

To study speciation, therefore, biologists must first find examples in nature where all or most of the stages of race formation and speciation can be documented. The intensive studies carried out on natural populations of *Drosophila* provide good examples of the stages involved in **geographic** or **allopatric** speciation.

To illustrate the first step, the formation of races, we shall consider the case of *Drosophila pseudoobscura* as studied by Dobzhansky and his colleagues. This species is found over a wide range of environmental habitats, including the western and southwestern United States. Although the flies throughout this range are morphologically similar, Dobzhansky discovered that populations from different locations varied substantially in the arrangement of genes on chromosome 3. He found a variety of different inversions of this chromosome that could be detected by loop formations in the salivary chromosomes. Each particular inversion sequence was named after the locale in which it was first detected (i.e., AR = Arrowhead, British Columbia; CH = Chiricahua Mountains, etc.) and was compared with one standard sequence, designated ST.

For example, Figure 26.2 shows a comparison of three arrangements detected in populations found at three different elevations in the Sierra Nevada in California's Yosemite region. The ST arrangement is most common at low elevations but declines in frequency as elevation increases. At 8000 feet, AR is most common and ST least common. The CH arrangement gradually increases with elevation. It seems likely that the gradual change of inversion frequencies is the result of natural selection and thus parallels the gradual environmental changes occurring at ascending elevations. Populations along this gradient or **cline** can be designated as **ecotypes,** or racially distinct groups.

Dobzhansky also found that if populations were collected at only one site throughout the year, inversion frequencies also changed. During the different seasons cyclic variation occurred, as shown in Figure

FIGURE 26.2
Inversions in chromosome 3
of *D. pseudoobscura* found at
different elevations in the
Sierra Nevada near Yosemite,
California. (From
Dobzhansky, 1948, Fig. 1.)

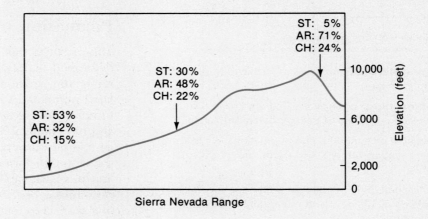

26.3. Such variation was observed to be repeated over
a period of several years. ST was always observed to
decline during the spring, with a concomitant increase in CH during the same period.

To test the hypothesis that this cyclic change is a
response to natural selection, Dobzhansky devised the
following laboratory experiment. He constructed a
large population cage in which samples could be periodically removed and studied. He began with a population of a known inversion frequency, 88 percent CH
and 12 percent ST. He reared them at 25°C and sam-
pled them over a one-year period. As shown in Figure
26.4, the frequency of ST increased gradually until it
was present at a level of 70 percent. At this point, an
equilibrium between ST and CH was reached. When
the same experiment was performed at 16°C, no
change in inversion frequency occurred. It can be concluded that the equilibrium reached at 25°C was in
response to the elevated temperature, the only variable in the experiment.

The results of Dobzhansky's experiment are strong
evidence that a balanced condition of the two inversions and their respective gene arrangements is superior to either inversion by itself. The equilibrium attained presumably represents the greatest degree of
fitness in the population under varying laboratory
conditions. This interpretation of the experiment suggests that natural selection is the driving force toward
equilibrium. If so, then we may conclude that the existence of various inversions is representative of genetic variation.

In a more extensive study, Dobzhansky went on to
sample populations over a much broader geographic
range. Twenty-two different gene arrangements were
found in populations from 12 locations. In Figure 26.5,
the frequencies of five inversions are shown according
to geographic location. The differences are largely
quantitative, with most populations differing only in
relative frequencies of inversions. However, in some
cases, there are qualitative differences. For example,

FIGURE 26.3
Changes in the ST and CH
arrangements in *D.
pseudoobscura* throughout
the year. (From Dobzhansky,
1966.)

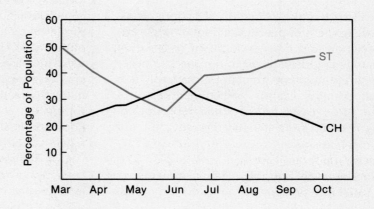

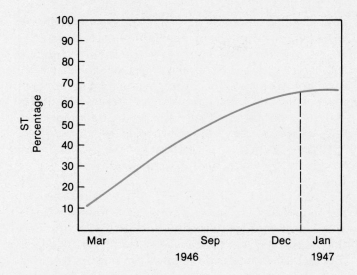

FIGURE 26.4
Increase in the ST arrangement in *D. pseudoobscura* in population cages under laboratory conditions. (From Dobzhansky, 1947, Fig. 5.)

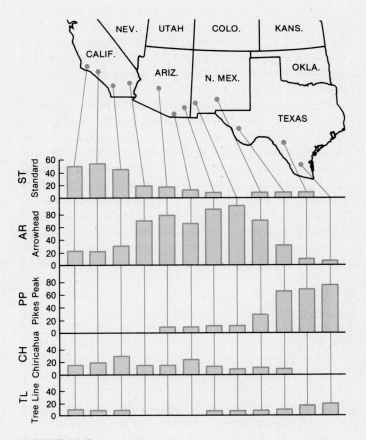

FIGURE 26.5
Frequencies of five chromosomal inversions in *D. pseudoobscura* in different geographic regions. (From Dobzhansky, 1947, Fig. 1.)

PP is absent in all four California locales where ST is most predominant. In the Texas locations, ST is absent or present in very low frequency and PP is predominant. Some general trends are also apparent. ST increases in frequency from east to west, while just the opposite is true of PP. AR is least common in the east-west extremes, yet predominant in Arizona and New Mexico.

Because these locations represent varied environments and the inversions preserve different gene combinations, we can conclude that numerous races of *D. pseudoobscura* have been formed. Not only do the studies of *D. pseudoobscura* illustrate the principles of the initial step in speciation, but they strongly support the concept that the adaptive norm consists of balanced genotypes within a population.

The development of races must be followed by reproductive isolation, the second step in speciation. We might then ask whether the evolution of *D. pseudoobscura* has gone beyond the formation of races. Dobzhansky investigated this question by examining the chromosome structure of other related **sibling species.** Sibling species have become reproductively isolated but remain very similar morphologically.

One sibling species, *Drosophila persimilis*, has provided very interesting information. *D. persimilis*, like *pseudoobscura*, contains five pairs of chromosomes and inversions within chromosome 3. In fact, all five chromosomes have identical arrangements of salivary gland chromosome bands except for a region on chromosome 2 and a small segment of the X chromosome. When Dobzhansky examined the chromosome 3 inversions carefully, 11 such arrangements were found. One, ST, is also found in *D. pseudoobscura*. Based on the common arrangement, it was possible to construct a phylogenetic sequence for all arrangements found in both species. The phylogeny was based on the similarity of any two inversions. If a fly is heterozygous for different inversions (each member of the pair of chromosome 3 contains a different inversion), only one inversion loop may be necessary to achieve homologous pairing. If this criterion is met, the populations bearing the inversions are on a direct line; that is, one can give rise directly to the other. However, while the sequence of arrangements constitutes a phylogeny, we cannot be absolutely sure of its direction.

As shown in Figure 26.6, ST is shared by both species. Only one hypothetical arrangement is necessary to complete the continuity of the tree. It appears that an ancestral population with the ST arrangement gave rise to many different inversions. Some were incorporated into races as members of the *pseudoobscura* species, and others gave rise to the *persimilis* species.

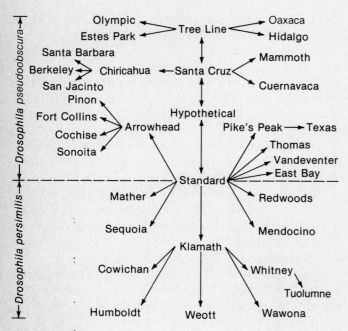

FIGURE 26.6
Inversion phylogeny for arrangements of chromosome 3 in *D. pseudoobscura* and *D. persimilis*. The ST arrangement is shared by both species. (From Anderson et al., 1975, Fig. 2.)

Today, even when the geographic distributions of these sibling species overlap, several isolating mechanisms keep them from interbreeding. They are isolated by prezygotic mechanisms such as habitat selection, with *persimilis* preferring higher elevations and cooler temperatures. Differences in courtship rituals allow females to distinguish between males of the two species and choose only males of their own species for mating. Even the time of day when courtship and mating occur is different in the species, with *persimilis* tending to court in the morning and *pseudoobscura* more active in the evening. Postzygotic mechanisms also maintain reproductive isolation in these species. When cross-fertilized in the laboratory, the species produce sterile F_1 hybrid males, with male sterility being associated with interactions between the X chromosome and chromosome 2. Back-crosses between F_1 females and parental males exhibit hybrid breakdown through lowered viability of the offspring.

Quantum Speciation

At the heart of the classical theory of speciation is the concept of **gradualism.** According to this concept, speciation is a microevolutionary event resulting from the accumulation of many minute gene differences over a long period of time under the influence of nat-

ural selection. Recently, however, more **stochastic** or **catastrophic models of speciation** have been proposed. These models emphasize the role of chromosome rearrangements as isolating mechanisms, with resultant speciation occurring within a few generations. They represent evolutionary events that occur suddenly and intermittently and are therefore referred to as **quantum speciation.** We will briefly discuss two examples and then consider a mechanism that has long been known to exist in angiosperms—speciation through polyploidy.

In his study of the evolution of Hawaiian *Drosophila*, Hampton Carson has proposed a **founder-flush speciation theory.** According to this theory, populations and even species can be started by a single individual. The founder principle is not new, as this idea was advanced much earlier by Mayr. In Carson's proposal, however, gradualism is not a necessary component of speciation, and more importantly, reproductive isolation *precedes* adaptation rather than being a consequence of genetic diversification. In other words, Carson has changed the order of steps in the classical model of speciation. According to the founder-flush theory, a single fertilized female can colonize an isolated territory previously unoccupied by members of this species. If conditions are favorable, the population founded by this individual will undergo a **flush,** or rapid expansion. After several generations, it is likely that the population growth will outstrip the environment's capacity, causing a **population crash.** The crash, a completely random event, causes the death or dispersal of almost the entire population. A single survivor, or at most a few survivors, may rebuild the population, which eventually undergoes several flush-crash cycles before coming to equilibrium with the environment. This cycle is diagramed in Figure 26.7. Through the genetic revolution it has undergone, the colony will, in all probability, have acquired adaptations making it unable to interbreed with the parental population.

According to Carson's theory, genetic changes are brought about in two ways. First, the original founding of the population by one or a very few individuals can establish allele frequencies different from those in the ancestral population. Second and most important, selection is relaxed during a population flush. If descendants of the founder can invade a new niche, they expand their numbers in a flush. Carson proposes that normally some blocks of genes on chromosomes remain tightly linked or "closed" to recombination because of some selective advantage conferred by their configuration. Any new genotypes produced by recombination within these closed areas have reduced fitness. In fact, the advantage conferred by balanced

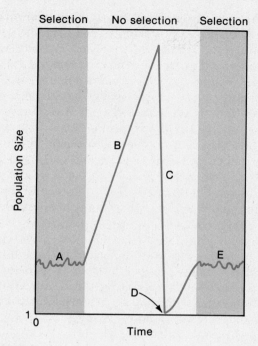

FIGURE 26.7

The flush-crash cycle. Small population descended from a single founder (A) undergoes a population flush (B), followed by a crash (C). A single survivor in the form of a fertilized female (D) builds a new population (E). [From H. L. Carson, "The Genetics of Speciation at the Diploid Level," *American Naturalist* 109: 83–92 (1975), Fig. 2. Reprinted by permission of The University of Chicago Press.]

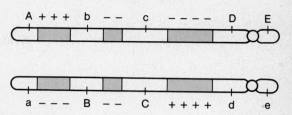

FIGURE 26.8

Model for open and closed regions of chromosomes. Products of crossing over anywhere within the open system (unshaded) result in offspring with high fitness, while crossovers in the closed regions (shaded) produce zygotes with low fitness. The letters represent genes or polygenes, and the pluses and minuses represent internally balanced gene complexes. Recombination within these blocks produces unfit gene combinations. [From H. L. Carson, The Genetics of Speciation at the Diploid Level," *American Naturalist* 109: 83–92 (1975), Fig. 1. Reprinted by permission of The University of Chicago Press.]

polymorphisms for inversion heterozygotes may derive from the protection of such closed gene complexes. A diagram of open and closed chromosomal regions is shown in Figure 26.8.

During the flush period, genotypes produced by recombination in closed regions of the genome may survive under the relaxed conditions of selection. After a crash, the survivor's reshuffled genome is acted upon by selection to produce a new combination of open and closed gene groups adapted to the environment. Several such cycles can produce enough genetic differences so that crosses between the progenitor and colony populations produce hybrids with lowered fitness commonly displayed by interspecific hybrids. Carson views the cycles of disorganization and reorganization of the genomes as the essence of speciation rather than as the gradual genetic divergence of isolated populations over long periods of time.

The evolution of *Drosophila* species in the Hawaiian Islands appears to have followed such a founder-flush cycle. Geologic evidence indicates that the northwesternmost islands are the oldest, and that the south-

eastern island of Hawaii is the youngest at about 700,000 years old. The relationships between species of *Drosophila* can be traced by mapping the location and frequency of inversions in the banded polytene chromosomes of larval salivary glands. One group, the *planitibia* complex, has three species with the same basic set of chromosome inversions: *D. planitibia*, *D. heteroneura*, and *D. silvestris*. *D. planitibia* is found on the older island of Maui, and the other two are found on Hawaii. From this evidence, it is postulated that an immigrant fertilized female belonging to an ancestral stock on Maui, chromosomally related to the present-day *D. planitibia*, crossed the Alenuihaha Channel to Hawaii. Subsequent flush-crash cycles led to the reconstruction of the colony's genotype, giving rise to the two species found on Hawaii—*D. heteroneura* and *D. silvestris* (Figure 26.9). An alternate theory proposes that one of the species on Hawaii could have arisen as the result of a second colonization from Maui. Similar evidence indicates that two other groups on Maui have given rise to a total of five species on Hawaii.

In laboratory tests of this theory, Jeffrey Powell has found that 15 generations after several flush-crash cycles, laboratory strains of *Drosophila* species showed significant premating (behavioral) isolation from other strains. Subsequent testing several months later showed that these differences were not transitory. These experimental results are important in establishing that the first stages of speciation can occur rapidly under certain circumstances.

The second model of quantum speciation that we will discuss, Michael White's **statispatric speciation,**

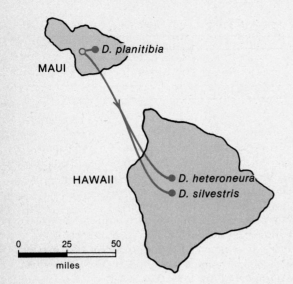

FIGURE 26.9
Proposed pathway of colonization of Hawaii by members of the *D. planitibia* species complex. Open circles represent a population ancestral to the three present-day species. (From H. L. Carson, "Chromosome Tracers and Origin of Species," *Science* 168: 1414–18, Fig. 3, 19 June 1970. Copyright 1970 by the American Association for the Advancement of Science.)

was originally derived to account for the evolution of flightless grasshoppers in Australia. In this model, a chromosomal aberration such as a translocation arises by chance in a small population. If heterozygotes for the translocation have a slightly reduced fitness, perhaps caused by abnormal meiosis, selection will favor either homokaryotype (two copies of the translocation or two normal chromosomes). The translocation homokaryotype, with a chromosome number different from the original, may spread and partially displace the ancestral population. Semispecies would then be reproductively isolated because hybrids would have an unbalanced chromosome complement and a lowered fitness. Thus, in White's as well as Carson's model, reproductive isolation precedes the development of genetic variability. The statispatric model has also been applied to the development of closely related species of mole rats differing in chromosome number, the *Spalax ehrenbergi* complex.

The final example illustrating quantum speciation involves polyploidy in plants. The formation of species by polyploidy in animals is rare, but has been an important factor in the evolution of plants. It is estimated that one-half of all flowering plants have evolved by polyploidy. One such form of polyploidy is **allopolyploidy** (see Chapter 12), which is produced by dou-

bling the chromosome number in an interspecific hybrid. If two species of related plants have the genetic constitution *SS* and *TT*, where *S* and *T* represent the haploid set of chromosome in each species, then the F_1 hybrid would have the chromosome constitution *ST*. Normally such a plant would be sterile because there are few or no homologous chromosome pairs, and aberrations would arise during meiosis. If the hybrid undergoes a spontaneous doubling of chromosome number, however, a tetraploid *SSTT* would be produced. This might occur in somatic tissue, giving rise to a partially tetraploid plant that would produce some tetraploid flowers. Alternately, aberrant meiotic events may produce *ST* gametes, which when fertilized would yield *SSTT* zygotes. The *SSTT* plants would be fertile because they would possess homologous chromosomes producing viable *ST* gametes. This new, true-breeding tetraploid would have a combination of characters derived from the parental species, and would be reproductively isolated from them because F_1 hybrids would be triploids and consequently sterile. The tobacco plant *Nicotiana tabacum* ($2n = 48$) is the result of the doubling of the chromosome number in the hybrid between *N. otophora* ($2n = 24$) and *N. silvestris* ($2n = 24$).

MOLECULAR EVOLUTION

The diversity of life forms inhabiting the earth is overwhelming. Nearly two million species of plants and animals have been described, and surely there are many yet to be classified. Nevertheless, all organisms share the same molecular features: they are composed principally of carbon, hydrogen, nitrogen, and oxygen atoms; they use nucleic acids to store and transfer chemical information; and proteins are, for them, the indispensable products of the stored genetic information.

The recent development of techniques to analyze and sequence proteins and nucleic acids has allowed biologists to determine relatedness of organisms and to construct phylogenetic sequences. In this section we shall review some of the findings in this area of evolutionary study and examine the genetic variability demonstrated at the molecular level within populations.

Amino Acid Sequence Phylogeny

Evolutionary relatedness or divergence can be measured by comparing the amino acid sequences of proteins common to various organisms. Proteins are generally large molecules. In order to determine the protein's primary structure (or amino acid sequence), the amino acid chain is first enzymatically digested into smaller polypeptide fragments. These are separated, and the amino acid sequence of each fragment is determined by chemical means. The experiment is repeated using a second enzyme that cleaves the molecule at different sites and yields a second set of sequenced fragments. From the overlap of sequences of the two sets of fragments, it is possible to determine their order in the protein chain, thus establishing the complete amino acid sequence. This technique is illustrated in Figure 26.10. Two sets of sequenced fragments are derived from the digestion by two enzymes in separate experiments. The fragments are then put together correctly into a single sequence based on the overlap between the two sets. Finally, the points of initial cleavage are shown for both enzymes.

Having described the technique, we will now describe some of the findings based on amino acid homology between identical molecules in different organisms. Rather than presenting the highly detailed lists of sequences of molecules, we shall concentrate on those findings pertaining to evolution. The interested reader may examine the precise sequences of the molecules by consulting the references listed at the end of this chapter.

The first protein to be sequenced was **insulin,** composed of only 51 amino acids. In the early 1950s, insulin was examined in a variety of mammals, including cattle, pigs, horses, sperm whales, and sheep. With the exception of a stretch of three amino acids, the protein was shown to contain an identical sequence in each of these mammals.

FIGURE 26.10
Summary of amino acid sequencing technique. The numbers 1 through 7 are used to represent amino acids.

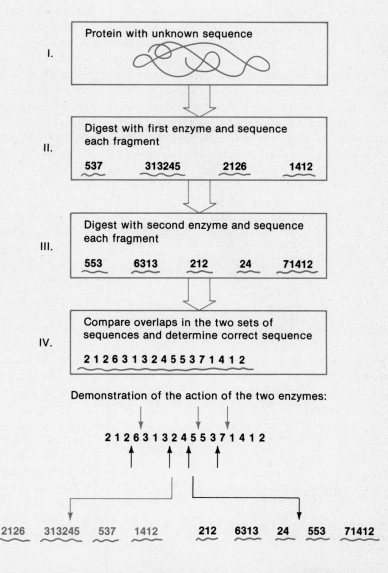

Cytochrome c is another commonly investigated protein. It is a respiratory pigment found in the mitochondria of plants and animals. The molecule consists of 104 amino acids in many vertebrates and a slightly higher number in most other organisms. Cytochrome c has changed very slowly during evolution. For example, the amino acid sequence in humans and chimpanzees is identical; between humans and rhesus monkeys only one amino acid is different. This is remarkable considering that lines leading to humans and monkeys diverged from a common ancestral form approximately 20 million years ago.

Table 26.5 shows the number of amino acid differences between a variety of organisms, using human cytochrome c as a standard. Even as distantly removed from humans as yeasts are, only 38 amino acids are different. In a comparison of a large number of organisms, more than 15 percent of the sequences remain unchanged.

In addition to the number of amino acid differences, it is possible to assess the number of nucleotide changes that must have occurred during evolution. This assessment requires knowledge of the genetic code and is a more refined analysis. For example, more than one nucleotide change may have been necessary to establish any given amino acid change found between two organisms. Thus, over evolutionary time, two or more independent mutations may have been essential in order to produce the observed change. When all necessary nucleotide changes for all amino acid differences are totaled, the **minimal mutational distance** between any two species is established. Table 26.5 shows such an analysis of the genes coding for cytochrome c in ten organisms. One can see that, as expected, these values are larger than the corresponding number of amino acids separating humans from the other nine organisms.

Minimal mutational distance provides an estimation of evolutionary divergence between organisms. Furthermore, on the basis of present-day sequence data, the order of derivation of one species from another and from common ancestors can be deduced. Thus, it is possible to construct **divergence dendrograms** or **phylogenetic trees** based on the amino acid sequences of a single protein from diverse organisms. This is perhaps the most fascinating application of this information to evolutionary study.

A molecule such as cytochrome c is not analyzed easily. Sequences of over 100 amino acids from numerous organisms must be studied, and each amino acid difference must be correlated with potential triplet code changes. Properly programmed, a computer can handle these computations.

The underlying assumption of this analysis is that all present-day sequences from different species represent gene products that diverged from common ancestral sequences at various points in evolutionary time. By determining minimal mutational distances between all species under analysis and in which specific amino acids have changed, the computer program can establish the most likely relationships between the species. It can also determine the point in these relationships at which a now extinct ancestral sequence must have existed in order to lead to evolutionary divergence.

This information may be summarized in the form of a phylogenetic tree. The phylogenetic tree shown in Figure 26.11 is based on the sequences used to derive the data of Table 26.5. Each open circle represents an extinct ancestral sequence. The number along each line represents the minimal mutational distance between any two points. The tree is plotted so that the ordinate represents proportional amounts of distance. If a constant mutation rate is assumed, the ordinate represents a relative estimate of geologic time.

When phylogenetic trees are constructed in this way, they agree remarkably well with trees constructed using more conventional approaches such as morphological and paleontological evidence. Once a number of proteins from a variety of organisms are sequenced and analyzed simultaneously, even more accurate trees may be constructed.

When closely related species are to be examined in this way, proteins that have evolved more rapidly than

TABLE 26.5

A comparison of the number of amino acid differences and the minimal mutational distance in cytochrome c.

Organism	Number of Amino Acid Differences	Minimal Mutational Distance
Human	0	0
Chimpanzee	0	0
Rhesus monkey	1	1
Rabbit	9	12
Pig	10	13
Dog	10	13
Horse	12	17
Penguin	11	18
Moth	24	36
Yeast	38	56

SOURCE: From W. M. Fitch and E. Margoliash, "Construction of Phylogenetic Trees," *Science* 155: 279–84, 20 January 1967. Copyright 1967 by the American Association for the Advancement of Science.

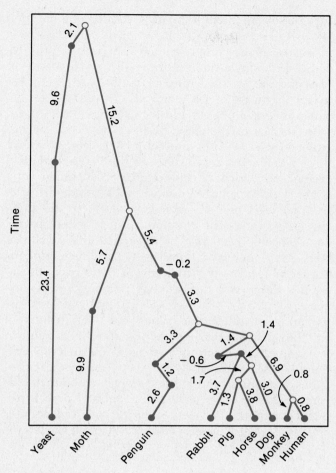

FIGURE 26.11
Phylogenetic sequence constructed by comparison of homologies in cytochrome c amino acid sequences. (From W. M. Fitch and E. Margoliash, "Construction of Phylogenetic Trees," *Science* 155: 279–84, 20 January 1967. Copyright 1967 by the American Association for the Advancement of Science.)

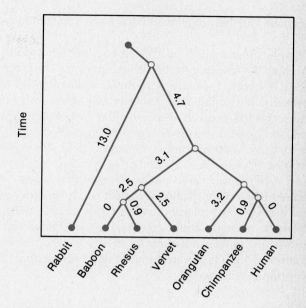

FIGURE 26.12
Phylogenetic sequence of carbonic anhydrase amino acid substitutions. (From Tashian and Carter, 1976.)

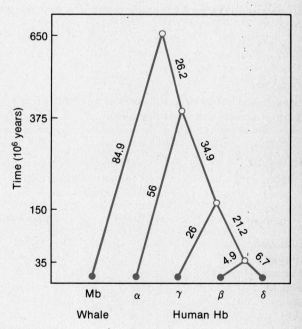

FIGURE 26.13
Phylogenetic sequence of myoglobin and hemoglobin proteins. These genes arose by duplication and subsequently diverged in amino acid sequence. (From W. M. Fitch and E. Margoliash, "Construction of Phylogenetic Trees," *Science* 155: 279–84, Figure 3, 20 January 1967. Copyright 1967 by the American Association for the Advancement of Science.)

cytochrome c are more useful. The 115-amino-acid protein carbonic anhydrase is used to analyze more accurately the relationship between humans and several other primates (Figure 26.12).

This phylogenetic approach may also be applied to related molecules that have arisen during evolution through gene duplication. (The importance of duplication in producing genetic variation has been discussed in Chapter 12.) The various hemoglobin chains and myoglobin have been analyzed in this way. As in the comparison of species, each protein chain may be studied using comparative sequences. As shown in Figure 26.13, the oxygen-carrying myoglobin molecule and all hemoglobin chains are proposed to have arisen from a common ancestral sequence. Furthermore, all four hemoglobin chains have a common origin.

Single chains such as the alpha and beta globins can also be compared between species. The differences between humans and their close relatives are noteworthy. Human and chimpanzee alpha and beta chains are identical in sequence, but the gorilla differs from the human in one amino acid in each chain. Table 26.6 shows the number of amino acid differences between humans and a number of species.

All in all, amino acid sequence data have been extremely useful in evolutionary studies. These independent analyses complement other types of evolutionary evidence. Perhaps their greatest value is that these studies are performed directly at the level of genetic variation, which is the underlying resource supporting evolutionary change.

Nucleotide Sequence Phylogeny

The molecular hybridization of DNA from different sources has been a valuable technique in evolutionary studies. Although it has been discussed in Chapter 9 and is described in Appendix A, we shall briefly review it here.

Hybridization of nucleic acids is based on nucleotide sequence homology. Molecules are heated until they dissociate (melt) into single strands and are then allowed to reanneal by slow cooling of a given mixture. In reannealing, complementary sequences join together in a stable double-stranded form at the lower temperature. Molecular hybridization will occur between mixtures of DNA strands (DNA:DNA) or between mixtures of DNA and RNA strands (DNA:RNA).

In studies of sequence diversity between species, radioactive DNA from one species is prepared and those sequences present once in the haploid genome (single copy fraction) are isolated. This single copy tracer is dissociated into individual strands and reassociated with melted single copy DNA of the same species (homologous reaction) and of other species (heterologous reaction). The difference in thermal stability (ΔT_m) between homologous and heterologous duplex molecules is a measure of the nucleotide sequence divergence between the two species. For example, a ΔT_m of 1°C corresponds to 1 percent mismatching in nucleotide sequences. The results of experiments with two species of sea urchin are shown in Figure 26.14. From differences in thermal stability, there is a sequence divergence between *Strongylocentrotus purpuratus* and *S. franciscanus* of about 19 percent. A similar experiment shows about 7 percent sequence diversity between *S. purpuratus* and another species, *S. drobachiensis*. From the data on nucleotide

---

TABLE 26.6

A comparison of the alpha and beta hemoglobin chains between humans and other organisms.

Organism	Amino Acid Differences Between Humans and Various Organisms	
	α Chains	β Chains
Chimpanzee	0	0
Gorilla	1	1
Macaque	5	10
Mouse	19	31
Sheep	26	33
Pig	20	28
Horse	22	30
Rabbit	28	16
Chicken	45	—
Kangaroo	—	54
Carp	93	—
Lamprey	113	—

SOURCE: Data from various sources.

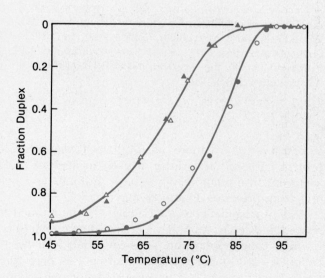

FIGURE 26.14

Measurement of nucleotide sequence diversity in the sea urchin. The thermal stability of *S. purpuratus* single copy DNA reassociated with *S. purpuratus* (circles) and *S. franciscanus* (triangles) DNA is measured. Lowered stability with *franciscanus* DNA is a measurement of nucleotide sequence divergence. (From Hall et al., 1980, Fig. 1.)

diversity and what evidence is available in the fossil record, it has been demonstrated that *S. purpuratus* diverged from *S. franciscanus* some 15 to 20 million years ago.

Where the nucleotide sequence divergence can be compared with other indicators of genetic variability, such as protein polymorphisms and chromosomal rearrangements, the evidence indicates that nucleotide sequence diversity is a more sensitive indicator of evolutionary divergence than amino acid replacements. *Drosophila heteroneura* and *D. silvestris*, found only on the island of Hawaii, are thought to have diverged only about 300,000 years ago, but it is difficult to demonstrate significant differences between these species in chromosomal inversion patterns or protein polymorphisms (Figure 26.15). Figure 26.16 shows the results obtained when labeled single copy DNA from *D. silvestris* is hybridized with itself and with DNA from *D. heteroneura* and *D. picticornis*. (*D. picticornis* is another member of the *planitibia* group and is found only on Kauai, a much older island.) The sequence diversity between the two species from Hawaii is about 0.55 percent, but *D. silvestris* and *D. picticornis* show a 2.1 percent difference in nucleotide sequences. Thus, nucleotide sequence diversity may precede the development of protein or chromosomal polymorphisms.

The fact that *D. heteroneura* and *D. sylvestris* share identical chromosome arrangements, have almost no detectable protein differences, are 98 percent homologous in DNA sequence, and yet are classified as separate species may seem paradoxical. However, the two species are clearly separated from each other by different and incompatible courtship and mating behaviors (a prezygotic isolating mechanism), by morphology, and by pigmentation of the body and wings (Figure 26.17). The available evidence suggests that these differences are controlled by a relatively small number of genes. In a genetic analysis of the differences in head shape and pigmentation, F. C. Val has estimated that only 15 to 19 loci are responsible for the morphological differences between these species,

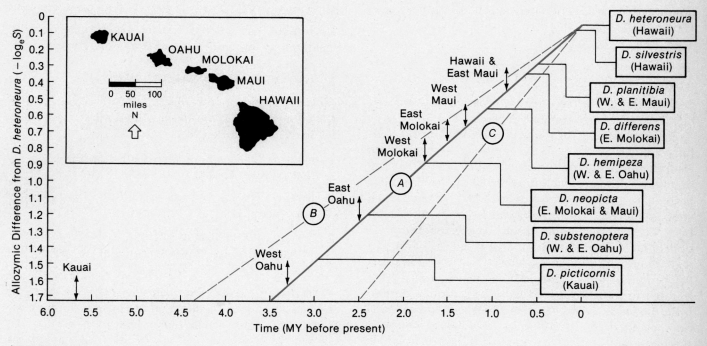

FIGURE 26.15

Detectable allozyme differences and proposed time of species origin in some Hawaiian *Drosophila* (line *A*). Line *B* assumes a slower accumulation of genetic differences and *C* a faster accumulation. In any case, genetic differences based on enzymes in the most recently evolved species (*planitibia, silvestris, heteroneura*) are difficult to quantify. (From Carson, 1976. Reprinted by permission from *Nature*, Vol. 259, pp. 395–96. Copyright © 1976 Macmillan Journals Limited.)

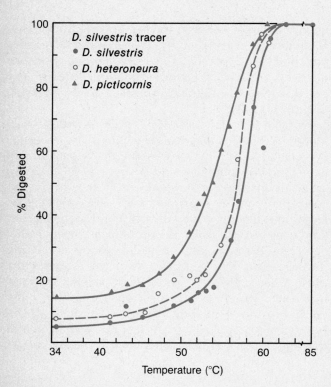

FIGURE 26.16

Nucleotide sequence diversity in the *planitibia* species complex. The degree of shift to the left by the heterologous hybrids is an indication of relatedness. (From Hunt et al., 1981, p. 363.)

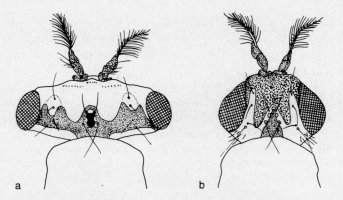

FIGURE 26.17

Head morphology and pigmentation patterns in (a) *D. heteroneura* and (b) *D. sylvestris*.

demonstrating that the process of speciation need only involve a small number of genes.

Problems similar to those encountered in measuring genetic differences in the Hawaiian *Drosophila* are also encountered in studies of higher primates. The hominoid primates include the chimpanzee, gorilla, orangutan, gibbon, and human. Past chromosomal studies and data on protein and isozyme differences have failed to accurately resolve the taxonomic relationships among the chimpanzee, gorilla, and humans, because they are so closely related. Using hybridization of single copy DNA sequences from a large number of individuals and using calibrations derived from the fossil record, Charles Sibley and Jon Ahlquist have clarified the evolutionary branching pattern in the hominoids and have estimated the times at which divergence occurred (Figure 26.18). According to their data, humans and chimpanzees are more closely related than either is to the gorilla. The measure of relatedness is called the $\Delta T_{50}H$ and is related to the T_m discussed earlier. In man and chimpanzee, the $\Delta T_{50}H$ is 1.9, somewhat less than the value of 2.1 to 2.3 for the

distance between the gorilla line and the chimpanzee/human line. Their calculation of divergence from the chimpanzee line is 8 to 10 million years (MY) for the gorilla, 6.3 to 7.7 MY for humans, and 2.4 to 3.0 MY for the pygmy chimpanzee. The high degree of nucleotide homology, similarity in chromosome patterns, and protein homology are reminiscent of the situation in *D. heteroneura* and *D. sylvestris* and lend support to the notion that distinct species are not necessarily separated by large numbers of gene differences.

Evolution of Genome Size

An early study on the relationship between cellular DNA content and evolution by Alfred Mirsky and Hans Ris in 1951 concluded that

1 Nuclear DNA content increased from the lower to higher invertebrates.

2 Closely related organisms usually have similar amounts of DNA.

3 The development of terrestrial vertebrates has been accompanied by reductions in nuclear DNA content.

DNA content measurements from a wide range of species are now available, and organisms can be classified into four somewhat overlapping groups, as shown in Figure 26.19. The lowest DNA content in free-living organisms is found in bacteria (Class 1), with a range between 0.003 and 0.01 picograms (pg) per cell. Fungi (Class 2) average less than 0.1 pg. Class 3 includes most animals and plants, including sea urchins, reptiles, birds, and mammals. Class 4 is composed of organisms whose genomes are larger than 10

FIGURE 26.18
Phylogenetic sequence of the hominoid primates and Old World monkeys (Cercopithecoidae). The numbers at the branch points are $\Delta T_{50}H$ measurements. The evolutionary branch points are dated by reference to the fossil record and rates of nucleotide divergence.

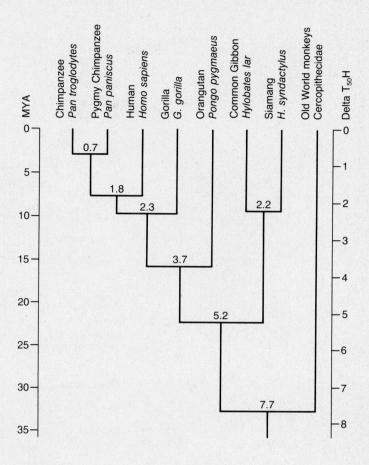

pg, including salamanders, some fish, and many plant species. The clustering of haploid DNA content in higher organisms and the clustering of related organisms is shown in Figure 26.20.

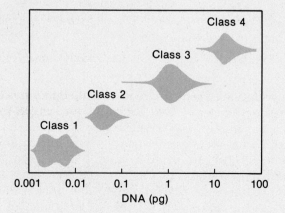

FIGURE 26.19
Classes or organisms grouped according to their DNA content. Class 1, bacteria; Class 2, fungi; Class 3, most animals and some plants; Class 4, many plants, salamanders, and some fish. (From Hinegardner, 1976, Fig. 4.)

Although the relationship between DNA content and evolution is somewhat complex, several general statements can be made. A large increase in DNA content has occurred during evolution from bacteria to higher plants. Few higher eukaryotes have genomes smaller than 0.1 pg, suggesting that a minimum DNA content is necessary to support this level of organization. Interestingly, while there is no necessary correlation between DNA content and chromosome number, there is a direct relationship between DNA content and nuclear and cellular size. In fact, a rough estimate of DNA content can be made from measurements of nuclear size.

Several explanations have been offered for the trend toward increases in DNA content per haploid genome during evolution, including the development of control system redundancy, generation of repetitive DNA sequences, and slower cell division and development. Unfortunately, we are not sure whether increases in DNA content are the raw material for selection or are produced as a byproduct. An examination of genome size in closely related organisms suggests that changes in DNA content have occurred mainly by duplication of small segments of DNA rather than by saltational increases associated with polyploidy. Figure 26.21

FIGURE 26.20
Haploid genome size clusters
in higher organisms. (From
Britten and Davidson, 1971,
Fig. 1.)

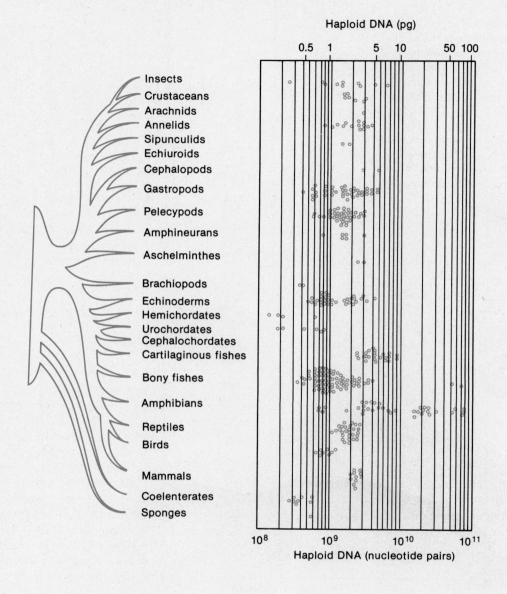

shows the distribution of DNA content in teleost fishes. If changes had arisen through polyploidy, a distribution into discrete classes reflecting multiples of a basic genome size would be expected. However, the differences in DNA content are seen to be the result of small incremental increases. Such studies confirm the role of gene duplication in evolution.

Evolution of Gene Families by Duplication

By examining functionally distinct but sequence-related genes that arose by duplication, we can study the evolution of gene function and the development of regulatory systems. The origin of the globin genes from an ancestral oxygen-carrying molecule has already been mentioned. Cloning and nucleotide sequencing have produced a great deal of information about the evolution of this gene family. Figure 26.22 summarizes what is known about the organization of the human alpha- and beta-globin gene families. Several features of these gene clusters are worth noting:

1   Both families consist of several genes packed together over a relatively short distance, all of which are oriented in the same direction with 5′ ends to the left and 3′ ends to the right.

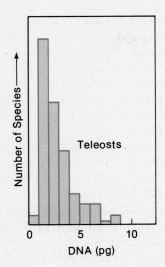

FIGURE 26.21
Haploid genome sizes in teleost fish. (From Bachman et al., 1972, p. 419.)

2 In addition to sequences that code for all known globin polypeptides, several related gene sequences called **pseudogenes** are present.

3 The genes in each family are arranged in the order in which they are expressed in development.

In the beta-globin cluster, the embryonic epsilon form is first, followed by the fetal gamma genes, and the

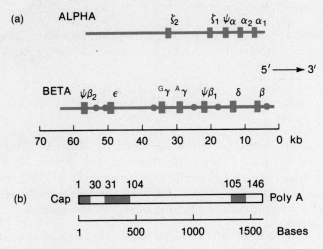

FIGURE 26.22
(a) Organization of the alpha- and beta-globin genes in humans. The genes are represented as rectangles, and repeated DNA sequences within the beta-globin family are shown as circles. (b) The coding (solid) and noncoding regions of the $^G\gamma$-globin gene. The top numbers are amino acid residues. (From Smithies et al., 1981, Fig. 1.)

adult beta and delta genes. It is not known whether this order is related to mechanisms regulating gene expression or is a reflection of evolutionary history.

The complete DNA sequences for all these genes are now available. A phylogenetic tree based on these sequences is shown in Figure 26.23. The original duplication that gave rise to the alpha and beta gene clusters occurred some 500 million years ago. The adult sequences diverged from the prenatal forms about 200 MY ago, sometime during the evolution of reptiles, which are ancestral to mammals. Separate embryonic and fetal sequences diverged at the beginning of the mammalian radiation some 100 MY ago, and this event may be linked to the development of placental mammals with their unique problems of oxygen transport. The adult beta and delta genes diverged some 40 MY ago, before the separation of higher primates into two lines—one leading to the New World monkeys and the other to the Old World monkeys, including great apes and humans.

Comparison of the nucleotide sequences in the coding and noncoding regions of the genes indicates that these regions have continued to diverge by base substitutions as well as by duplication and deletion of short regions. An unexpected outcome of this analysis has been the detection of pseudogenes in each gene family. **Pseudogenes** are stretches of DNA that have some sequence homology with protein-coding regions but are themselves inactive. Such sequences have most probably arisen as a result of increased crossing over between homologous chromosome regions. These extra copies—free from selection because they have no phenotypic effect—can accumulate mutations that would cause the gene to become inactive, but re-

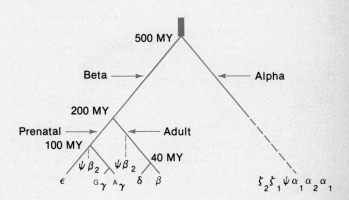

FIGURE 26.23
Phylogenetic sequence of globin genes based on nucleotide sequence of coding regions. Dotted line indicates relationships not yet clarified. (From Smithies et al., 1981, Fig. 2.)

tain many of the features associated with active genes such as promoter sites, splicing sequences, and poly-A addition sites. In the human globin gene families, pseudogenes separate the embryonic and adult alpha genes, and the prenatal and adult beta genes. The significance and function of such pseudogenes are matters of intense speculation. Oliver Smithies has proposed that because of their location in human globin families, pseudogenes modulate the expression of the protein-producing alpha and beta genes. Philip Leder and his colleagues, on the other hand, have shown that pseudogenes of the alpha-globin complex in the mouse are located on different chromosomes than the active genes. Dispersed single pseudogenes (called **orphons**) derived from tandemly repeated families, including histone and ribosomal genes, have recently been found in a wide range of organisms. Drawing on these results, Leder suggests that the genomes of higher eukaryotes are mosaics of dispersed pseudogenes, which serve as a reservoir of sequences that can evolve new function. This interpretation implies that gene duplication leading to the formation and dispersal of pseudogenes has been a critical factor in the evolution of higher organisms.

THE ROLE OF MUTATION IN EVOLUTION

Selectionists versus Neutralists

When we discussed the concept of the adaptive norm in an earlier section, it was pointed out that there is a controversy over the importance of extensive genetic variation within a population. This controversy centers particularly on the variation detected at the molecular level. On the one hand, the **classical hypothesis** proposes that natural selection favors the fixation of the most favorable alleles at each locus. Homozygosity therefore should be the rule rather than the exception. On the other hand, the **balance hypothesis** favors the maintenance through natural selection of a high degree of heterozygosity in populations.

The classical hypothesis is supported by the observation that induced mutations almost invariably lower fitness. Therefore, if most mutations are detrimental, the accumulation of genetic variability will result in a substantial genetic burden to the fitness of a population. According to the classical theory, this detrimental variation will be purged by natural selection.

In contrast, Dobzhansky and his colleagues have provided the initial support for the balance hypothesis. Having observed a high degree of heterozygosity in

adapted *Drosophila* populations, they postulated that populations displaying genetic polymorphism will have added fitness. The observation of protein polymorphisms by Lewontin in *Drosophila* and by Henry Harris in humans, and the evidence for numerous amino acid substitutions in single proteins throughout evolution, attest to a high degree of heterozygosity, particularly at the molecular level, in many diploid and polyploid populations. These findings do not confirm the balance hypothesis, however. Instead, they lend support to it and argue against the classical theory.

However, the existence of a high degree of genetic variation does not prove that it is the basis of evolutionary fitness. A third theory, the **neutralist hypothesis** proposed by James Crow and Motoo Kimura, argues that mutations leading to amino acid substitutions are rarely favorable. They are more often detrimental, but most often neutral or genetically equivalent to the allele that is replaced. Those that are favorable or detrimental are either preserved or removed from the population, respectively, by natural selection. However, neutral genetic changes will not be affected by selection and will instead become randomly fixed in the population. Their frequency will be determined by the rate of mutation and the principles of random genetic drift.

The neutralist interpretation, as most recently articulated by Kimura, is based on several observations and theoretical considerations. These ideas focus on several important aspects of genetic variation as an evolutionary phenomenon. The neutralist theory is based on the following points:

1 The rate of amino acid substitution in a given protein is about the same in organisms with diverse lineages.

2 The types of substitutions do not seem to demonstrate any particular pattern, but are instead random.

With regard to these observations, the neutralists believe that if selection favored the amino acid changes even slightly, the rate would vary in different organisms exposed to different selective pressures. The substitutions would therefore not be random.

3 The overall rate of mutation leading to amino acid substitutions is relatively high (it is about equal to the substitution of one nucleotide per genome every two years in mammals) and has remained relatively constant for a long period of time.

For example, when nucleotide substitution as estimated from amino acid changes in seven proteins from 16 pairs of mammals is estimated over the last 150 MY, the points fall close to a straight line. A constant rate of change is indicated. The neutralists argue that if natural selection were acting on all such mutations, the rate of fixation would not be constant in fluctuating environments over time.

4 The number of amino acid substitutions is higher in the less important parts of molecules (those parts not critical to tertiary structure and the active site of proteins) and in molecules less critical to the organism.

This argument centers on **functional constraint.** That is, those regions that determine whether the molecule functions or not are less able to tolerate change. The regions of lesser importance—those determining only how well the molecule functions—can tolerate greater amounts of change. It is argued that if such changes in less constrained regions of proteins are neutral, the rate of amino acid substitution in these regions will be substantially higher than in constrained regions. The same argument is applicable to total proteins of lesser importance to organisms. Examination of less-constrained regions of proteins and less-critical proteins does show a higher amino acid substitution rate. This is interpreted by the neutralists as support for their contentions.

5 The rate of amino acid substitution is much too high to be accounted for by selection.

If all heterozygosity within a population represented alleles with even a slight advantage, the neutralists argue that the **cost of selection** necessary to fix and maintain them in populations would be astronomically high. The cost of selection is the genetic death of those organisms that must be replaced by those bearing the more favorable change. Mathematical calculations can be made to support this theory.

The neutralist theory runs counter to the theory of natural selection as set down by Darwin and as expanded by modern knowledge of genetics. In the 1960s there was general agreement that all biological characteristics could be interpreted as arising by adaptive evolution through natural selection. In this sense, most mutations that have survived through evolution should be at least slightly adaptive in homozygotes, heterozygotes, or both.

Those who oppose the neutralist theory have been called **selectionists.** They point out examples where enzyme or protein polymorphism is associated with adaptation to certain environmental conditions. The well-known advantage of sickle-cell anemia carriers in malarial regions is such an example.

Selectionists also stress that enzyme polymorphism may often appear to offer no advantage, but exists in such a frequency that it is impossible to explain as a random occurrence. Thus, even though no currently available analytical technique can detect any physiological difference, it cannot be proved that some slight advantage associated with any given amino acid substitution does not exist.

On many specific points of contention, selectionists are able to offer persuasive theoretical arguments and calculations. For example, although the nucleotide substitution rate appears on the average to be constant, localized variation is sometimes as much as 2.5 times higher than expected by chance.

Even though this controversy is highly theoretical and esoteric, it is important that we not lose sight of several factors. For example, the neutralists do not discount natural selection as a guiding force in evolution. Rather, they suggest that some features of organisms' genotypes are nonadaptive, fluctuate randomly, and may have been fixed by genetic drift. On the other hand, the selectionists certainly do not deny that genetic drift is an important factor in establishing gene frequencies. Rather, they recognize genetic drift as an important evolutionary force.

Finally, it should be pointed out that the two theories are not mutually exclusive. It is difficult to argue against the notion that some genetic variation must be neutral. The difference between the two theories is in the degree of neutrality that exists. While current data are insufficient to resolve the problem one way or the other, one important point has emerged from the arguments. In considering natural selection, there are clearly two levels that must be examined: the phenotypic level, including all morphological and physiological characteristics imparted by the genotype, and the molecular level, represented by the precise nucleotide and amino acid sequences of DNA and proteins. There is no question about selection acting at the phenotypic level. The extent to which selection occurs at the molecular level, however, is in question.

Recent Trends

One of the major premises of the classical theory of speciation is that mutations (along with recombination) are the source of all variation. In this context, new alleles are inherited in Mendelian fashion and may gradually increase in frequency if they confer greater fitness in accord with the principles of population genetics. Speciation involves a large number of allelic substitutions sequentially incorporated into the gene pool. We have already seen that the concept of gradualism is being challenged by chromosomal theories of quantum speciation. Work in molecular biology, particularly the discovery of intervening sequences, transposable elements, and pseudogenes, is causing fundamental changes in our concepts of genome structure, stability, and mutation.

The genome of higher organisms has been regarded as a largely static entity, with low rates of mutation and chromosome aberration. We now know that the genome is highly dynamic and in constant flux, with families of repeated sequences being created, fixed, deleted, and turned over. Mutation has been thought of in terms of nucleotide substitution or, at most, small duplications or deletions. However, nucleotide sequence analysis of the *bithorax* gene complex in *Drosophila* has shown that most mutations in these genes involve the insertion or deletion of pieces of DNA several thousand base pairs in length. Insertion of transposable elements into genes produces mutant phenotypes, and their removal produces reversion to wild type. In addition, since many transposable elements contain start and stop signals for transcription, the dispersal and mobility of such elements in the genome can create new patterns of gene expression in a single generation.

The evolutionary role of regulatory mutations in development is also receiving increased attention. Many workers now believe that mutations that change the timing of gene expression in embryogenesis can have a significant impact on the rates of speciation. Such mutations would not change the frequency of protein polymorphisms in a population, but rather the order or duration of gene expression during embryonic or preadult development, producing instant morphological variation. Such a mutant has already been described in the *bithorax* region of *Drosophila*. By moving a gene some 40 kb within the complex, the mutant changes the gene's time of expression during development. This gene, *Cbx*, produces a fly with a greatly altered phenotype.

Finally, pseudogenes are seen as possible reservoirs of inactive genes that can be rapidly mobilized to perform new functions. In sum, the classical theory of speciation appears ready to undergo a new and dramatic synthesis, and it is likely that this synthesis will arise from within molecular biology.

SUMMARY AND CONCLUSIONS

The process of evolution depends on the formation of species. Speciation, which depends on genetic variation and forces controlling its distribution (e.g., natural selection), is initiated when a population becomes separated into smaller, reproductively semi-isolated breeding groups. As these semi-isolates are exposed to the forces of evolution, each acquires its own unique set of genetic variations. This partitioning of a population's gene pool results in a better fitness in the new environment for each group and the formation of races. As evolution proceeds, races may become reproductively isolated from one another, in which case they represent separate species. The formation of races and speciation are illustrated by several studies of *Drosophila*.

Substantial genetic variation also exists at the molecular level, particularly at the level of nucleotide sequence. Two approaches have been especially fruitful in evolutionary study at the molecular level: (1) comparison of complementary sequences present in the DNA of different organisms; and (2) comparison of amino acid substitutions in proteins common to a variety of organisms. Such comparisons provide not only a measure of evolutionary relatedness, but *further* allow the assessment of genetic variation during evolution. Perhaps of greatest interest is the fact that phylogenetic relationships of diverse organisms may be predicted on the basis of molecular changes.

When the importance of variation at the molecular level is considered, the central question is whether or not all nucleotide and amino acid changes preserved through evolution in DNA and proteins are adaptive to the population. Some workers believe that the majority of these changes are neutral or genetically equivalent. Others adhere to the theory of adaptive evolution through natural selection. The arguments are, by and large, theoretical, but not mutually exclusive. In all likelihood, some mutations are indeed neutral, while many may be adaptive and increase fitness in the population.

Evolution, a unifying concept in biology, cannot be fully understood or appreciated without at least a preliminary understanding of population genetics and molecular biology.

PROBLEMS AND DISCUSSION QUESTIONS

1 Discuss the rationale behind the statement that inversions in chromosome 3 of *Drosophila pseudoobscura* represent genetic variation.

2 Describe the process of race formation. What is the role of natural selection?

3 Contrast the classical and balance hypotheses as they relate to the adaptive norm.

4 What types of nucleotide substitutions will not be detected by electrophoretic studies of a gene's protein product?

5 In a sequencing experiment (using the numbers 1 through 6 to represent amino acids), the following two sets of peptide fragments were obtained in independent experiments with different proteolytic enzymes:

Proteins	Fragments		
Enzyme I	624	24635	135
Enzyme II	13	6356	524	24

Determine the sequence of fragments and amino acids in the protein.

6 Shown below are two homologous lengths of the alpha and beta chains of human hemoglobin. Consult the genetic code dictionary (Figure 17.8) and determine how many amino acid substitutions may have occurred as a result of a single nucleotide substitution. For any that cannot occur as the result of a single change, determine the minimal mutational distance.

Alpha:	Ala	Val	Ala	His	Val	Asp	Asp	Met	Pro
Beta:	Gly	Leu	Ala	His	Leu	Asp	Asn	Leu	Lys

7 Determine the minimal mutational distances between the following amino acid sequences of cytochrome c from various organisms. Compare the distance between humans and each organism.

Human:	Lys	Glu	Glu	Arg	Ala	Asp
Horse:	Lys	Thr	Glu	Arg	Glu	Asp
Pig:	Lys	Gly	Glu	Arg	Glu	Asp
Dog:	Thr	Gly	Glu	Arg	Glu	Asp
Chicken:	Lys	Ser	Glu	Arg	Val	Asp
Bullfrog:	Lys	Gly	Glu	Arg	Glu	Asp
Fungus:	Ala	Lys	Asp	Arg	Asn	Asp

8 The genetic difference between *D. heteroneura* and *D. sylvestris* as measured by nucleotide diversity is about 1.8 percent. The difference between chimpanzee (*Pan troglodytes*) and humans (*Homo sapiens*) is about the same, yet the latter species are classified in different genera. In your opinion, is this valid? If so, why; if not, why not?

9 As an extension of the previous question, consider the following: In sorting out the complex taxonomic relationships among birds, species with $\Delta T_{50}H$ values of 0.4 are placed in the same genus, even by traditional taxonomy based on morphology. Using the data in Figure 26.18, construct a phylogeny that obeys this rule, using any of the appropriate genus names (*Pongo*, *Pan*, *Homo*), or constructing new ones.

10 The use of nucleotide sequence data to measure genetic variability is complicated by the fact that the genes of higher eukaryotes are complex in organization and contain 5' and 3' flanking regions as well as introns. Slightom and colleagues have compared the nucleotide sequence of two cloned alleles of the gamma-globin gene from a single individual and found a variation of 1 percent. Those differences include 13 substitutions of one nucleotide for another, and three short DNA segments that have been inserted in one allele or deleted in the other. None of the changes take place in the exons (coding regions) of the gene. Why do you think this is so, and should it change the concept of genetic variation?

11 Discuss the arguments supporting the neutral mutation theory. What counterarguments are proposed by the selectionsists?

12 Of what value to our understanding of genetic variation and evolution is the debate concerning the neutral mutation theory?

SELECTED READINGS

ANDERSON, W., et al. 1975. Genetics of natural populations, XLII. Three decades of genetic change in *Drosophila pseudoobscura. Evolution* 29: 24–36.

AYALA, F. J. 1976. *Molecular evolution.* Sunderland, Mass.: Sinauer Associates.

————. 1984. Molecular polymorphism: How much is there, and why is there so much? *Dev. Genet.* 4: 379–91.

BACHMAN, K. O. B., et al. 1972. Nuclear DNA amounts in vertebrates. In *Evolution of genetic systems,* ed. H. B. Smith. Brookhaven Symposium in Biology, vol. 23.

BRITTEN, R. J., and DAVIDSON, E. H. 1971. Repetitive and non-repetitive DNA sequences and a speculation on the origin of evolutionary novelty. *Quart. Rev. Biol.* 46: 111–38.

CARSON, H. 1970. Chromosome tracers of the origin of species. *Science* 168: 1414–18.

————. 1975. The genetics of speciation at the diploid level. *Amer. Natur.* 109: 83–92.

DAYHOFF, M. O. 1969. Computer analysis of protein evolution. *Scient. Amer.* (July) 221: 86–95.

DOBZHANSKY, T. 1947. Adaptive changes induced by natural selection in wild populations of *Drosophila. Evolution* 1: 1–16.

————. 1948. Genetics of natural populations, XVI. Altitudinal and seasonal changes produced by natural selection in certain populations of *Drosophila pseudoobscura* and *Drosophila persimilis. Genetics* 33: 158–76.

————. 1955. *Genetics of the evolutionary process.* New York: Columbia University Press.

DOBZHANSKY, T., et al. 1966. Genetics of natural populations, XXXVIII. Continuity and change in populations of *Drosophila pseudoobscura* in Western United States. *Evolution* 20: 418–27.

DOBZHANSKY, T., AYALA, F., STEBBINS, G., and VALENTINE, J. 1977. *Evolution.* San Francisco: W. H. Freeman.

DOVER, G. 1982. Molecular drive: A cohesive mode of species evolution. *Nature* 299: 11–17.

DOVER, G. A., and FLAVELL, R. B., eds. 1982. *Genome evolution.* New York: Academic Press.

EFSTRATIADIS, A., et al. 1980. The structure and evolution of the human beta-globin gene family. *Cell* 21: 653–68.

FITCH, W. M. 1973. Aspects of molecular evolution. *Ann. Rev. Genet.* 7: 343–80.

FITCH, W. M., and MARGOLIASH, E. 1967. Construction of phylogenetic trees. *Science* 155: 279–84.

————. 1970. The usefulness of amino acid and nucleotide sequences in evolutionary studies. *Evol. Biol.* 4: 67–109.

GOULD, S. J. 1982. Darwinism and the expansion of evolutionary theory. *Science* 216: 380–87.

GRANT, V. 1977. *Organismic evolution.* San Francisco: W. H. Freeman.

HALL, T., GRULA, J., DAVIDSON, E. H., and BRITTEN, R. J. 1980. Evolution of sea urchin non-repetitive DNA. *J. Mol. Evol.* 16: 95–110.

HINEGARDNER, R. 1976. Evolution of genome size. In *Molecular evolution,* ed. F. J. Ayala, pp. 179–99. Sunderland, Mass.: Sinauer Associates.

HUNT, J., et al. 1981. Evolution distance in Hawaiian *Drosophila. J. Mol. Evol.* 17: 361–67.

JEFFREY, A. 1979. DNA sequence variation in the $^G\gamma$-, $^A\gamma$-, δ- and β-globin genes of man. *Cell* 18: 1–10.

JENKINS, N., COPELAND, N. G., TAYLOR, B. A., and LEE, B. K. 1981. Dilute(d) coat colour mutation of DBA.2 J mice is associated with the site of integration of an ecotropic MuLV genome. *Nature* 293: 370–74.

JUKES, T. H. 1966. *Molecules and evolution.* New York: Columbia University Press.

KIMURA, M. 1979a. Model of effectively neutral mutations in which selective constraint is incorporated. *Proc. Natl. Acad. Sci.* 76: 3440–44.

————. 1979b. The neutral theory of molecular evolution. *Scient. Amer.* (Nov.) 241: 98–126.

KIMURA, M., and OHTA, T. 1971. Protein polymorphism as a phase of molecular evolution, *Nature* 299: 467–69.

KING, M. C., and WILSON, A. C. 1975. Evolution at two levels: Molecular similarities and biological differences between humans and chimpanzees. *Science* 188: 107–16.

KOHNE, D. E., CHISCON, J. A., and HOYER, B. H. 1972. Evolution of primate DNA sequences. *J. Hum. Evol.* 1: 627–44.

LEDER, A., SWAN, D., RUDDLE, F. H., D'EUSTACHIO, P., and LEDER, P. 1981. Dispersion of alpha-like globin genes of the mouse to three different chromosomes. *Nature* 293: 196–200.

LEWONTIN, R. C. 1974. *The genetic basis of evolutionary change.* New York: Columbia University Press.

LEWONTIN, R. C., and HUBBY, J. L. 1966. A molecular approach to the study of genic heterozygosity in natural populations. II. Amount of variation and degree of heterozygosity in natural populations of *Drosophila pseudoobscura. Genetics* 54: 595–609.

LEWONTIN, R. C., et al., eds. 1981. *Dobzhansky's genetics of natural populations, I-XLII.* New York: Columbia U. Press.

MAYR, E. 1963. *Animal species and evolution.* Cambridge, Mass.: Harvard University Press.

PATTERSON, C. 1978. *Evolution.* Ithaca, N.Y.: Cornell University Press.

POWELL, J. 1978. The founder-flush speciation theory: An experimental approach. *Evolution* 32: 465–74.

SIBLEY, C., and AHLQUIST, J. 1984. The phylogeny of the hominoid primates, as indicated by DNA-DNA hybridization. *J. Mol. Evol.* 20: 2–15.

SMITH, J. M., ed. 1982. *Evolution now: A century after Darwin.* San Francisco: W. H. Freeman.

SMITHIES, O., BLECHL, A. E., SHEN, S., SLIGHTOM, J. L., and VANIN, E. F. 1981. Co-evolution and control of globin genes. In *Levels of genetic control in development*, eds. S. Subtelny and U. Abbot, pp. 185–200. New York: Alan R. Liss.

SPIESS, E. B. 1977. *Genes in populations.* New York: Wiley.

STEBBINS, G. L. 1977. *Processes of organic evolution.* 3rd ed. Englewood Cliffs, N. J.: Prentice-Hall.

STEBBINS, G. L., and AYALA, F. J. 1981. Is a new evolutionary synthesis necessary? *Science* 213: 967–71.

TASHIAN, R. E., and CARTER, N. D. 1976. Biochemical genetics of carbonic anhydrase. In *Advances in human genetics*, eds. H. Harris and K. Hirschhorn, pp. 1–56. New York: Plenum Press.

TEMPLETON, A. R. 1985. Phylogeny of the hominoid primates: A statistical analysis of the DNA-RNA hybridization data. *Mol. Biol. Evol.* 2: 420–33.

VAL, F. C. 1977. Genetic analysis of the morphological differences between two interfertile species of Hawaiian *Drosophila. Evolution* 31: 611–29.

WHITE, M. J. D. 1977. *Modes of speciation.* San Francisco: W. H. Freeman.

YUNIS, J. J., and PRAKASH, O. 1982. The origin of man: A chromosomal pictorial legacy. *Science* 215: 1525–30.

ZUCKERKANDL, E. 1965. The evolution of hemoglobin. *Scient. Amer.* (May) 212: 110–18.

APPENDIX A
Experimental Methods

In addition to the techniques of genetic analysis, physical and chemical techniques for the separation and analysis of macromolecular components of the cell nucleus and cytoplasm have been instrumental in advancing our understanding of genetics at the molecular level. Some of these, such as DNA sequencing (Chapter 19) and the formation of somatic cell hybrids (Chapter 23), are explained in the text; and others, including nucleic acid hybridization and density gradient centrifugation, are mentioned (Chapter 9) but not discussed in detail. In this appendix, we will describe the background and theoretical basis of some techniques that have been important in molecular genetics.

ISOTOPES

Isotopes are forms of an element that have the same number of protons and electrons but differ in the number of neutrons contained in the atomic nucleus. For example, the most common form of carbon has an atomic number of 6 (the number of protons in the nucleus) and an atomic weight of 12 (the sum of the protons and neutrons in the nucleus). In a very small percentage of carbon atoms, a seventh neutron is present, producing an atom with an atomic weight of 13. This is an example of a so-called **heavy isotope.** Since the number of protons and electrons, and thus the net charge, has not changed, the atom has the same chemical properties as **carbon-12** (^{12}C) and differs only in mass. **Carbon-13** (^{13}C) is thus a stable, heavy isotope of carbon.

Although the addition of neutrons does not alter the chemical properties of an atom, it can produce instabilities in the atomic nucleus. If another neutron is added to a carbon-13 atom, the isotope **carbon-14** (^{14}C) results. However, the presence of eight neutrons and six protons is an unstable condition, and the atom undergoes a nuclear reaction in which radiation is emitted during the transition to a more stable condition. Therefore, carbon-14 is a **radioactive isotope** of carbon.

The type of radiation emitted and the rate at which these nuclear events take place are characteristic of the element. Table A.1 lists types of radioactivity. The rate at which a radioactive isotope emits radiation is expressed as its **half-life,** which is the time required for a given amount of a radioactive substance to lose one-half of its radioactivity. Table A.2 lists some of the isotopes available for use in research.

DETECTION OF ISOTOPES

The choice of which isotope to use as a tracer in biological experiments depends on a combination of its physical and chemical properties, which enable the investigator to quantitate the amount of radioactivity or measure the ratio of heavy to light isotopes. For the detection of heavy isotopes, two methods are commonly employed: **mass spectrometry** and **equilibrium density gradient centrifugation** (discussed in the following section). Although the use of heavy isotopes has been more restricted than that of radioisotopes, they have been instrumental in several basic advances in molecular biology, e.g., demonstrating the semiconservative nature of DNA replication and the existence of messenger RNA (mRNA).

TABLE A.1
Properties of ionizing radiation.

Type	Relative Penetration	Relative Ionization	Range in Biological Tissue
Alpha particle (2 protons + 2 neutrons)	1	10,000	Microns
Beta particle (electron)	100	100	Microns–mm
Gamma ray	>1000	<1	∞

TABLE A.2

Some isotopes used in research.

Element	Isotope	Half-life	Radiation
H	^{2}H	—	Stable
	^{3}H	12.3 years	β
C	^{13}C	—	Stable
	^{14}C	5700 years	β
N	^{15}N	—	Stable
O	^{18}O	—	Stable
P	^{32}P	14 days	β
S	^{35}S	87 days	β
K	^{40}K	1.2×10^9 years	β, gamma
Fe	^{59}Fe	45 days	β, gamma
I	^{125}I	60 days	β, gamma
	^{131}I	8 days	β, gamma

There are a number of methods to detect radioisotopes, the foremost being **liquid scintillation spectrometry,** which provides quantitative information about the amount of radioactive isotope present in a sample, and **autoradiography,** which is used to demonstrate the cytological distribution and localization of radioactively labeled molecules.

In recording radioactivity by liquid scintillation counting, a small sample of the material to be counted is solubilized and immersed in a solution containing a **phosphor,** an organic compound that emits a flash of light after it absorbs energy released by decay of the radioactive compound. The counting chamber of the liquid scintillation spectrometer is equipped with very sensitive photomultiplier tubes that record the light flashes emitted by the phosphor. The data are recorded as counts of radioactivity per unit time and are displayed on a printout or can be fed into a computer for storage.

In autoradiography, a gel, chromatogram, or plant or animal part is placed against a sheet of photographic film. Radioactive decay from the incorporated isotope behaves just as light energy does and reduces silver grains in the emulsion. After exposure, the sheet of film is developed and fixed, revealing a deposit of silver grains corresponding to the location of the radioactive substance (Figure A.1).

Alternately, to record the subcellular localization of an incorporated labeled isotope, cells or chromosomal preparations that have been incubated with radioactively labeled compounds are affixed to microscope slides and covered with a thin layer of liquid photographic emulsion. After they are exposed, the slides are processed to develop and fix the reduced silver grains in the emulsion. After staining, microscopic examination reveals the location and extent of labeled isotope incorporation (Figure A.2).

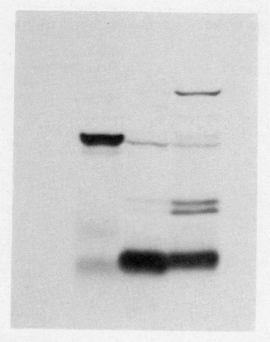

FIGURE A.1

Autoradiogram of radioactive bacterial proteins synthesized in a minicell system. (Photo by Ronald E. Law.)

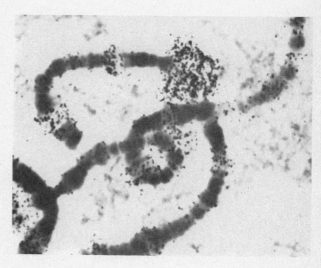

FIGURE A.2

Autoradiogram of RNA synthesis in salivary glands of *Drosophila* larva. Silver grains are deposited over sites of RNA synthesis at chromosome puffs. (Photo by Ralph M. Sinibaldi.)

CENTRIFUGATION TECHNIQUES

The centrifugation of biological macromolecules is widely employed to provide information about their physical characteristics (e.g., size, shape, density, and molecular weight) and to purify and concentrate cells, organelles, and their molecular components.

Differential centrifugation is commonly used to separate materials such as cell homogenates according to size. Initially, the homogenate is distributed uniformly in the centrifuge tube. After a period of centrifugation, the pellet obtained is enriched for the largest and most dense particles, such as nuclei, in the homogenate. After each step, the supernatant can be recentrifuged at higher speeds to pellet the next heavier component. A typical fractionation scheme for cell homogenates is shown in Figure A.3. Further fractionation using density gradient techniques can be used to purify any of the fractions obtained by differential centrifugation.

Rate zonal centrifugation is used to separate particles on the basis of differences in their sedimentation rates. It may be used to separate mixtures of macromolecules such as proteins or nucleic acids and cellular organelles such as mitochondria. In this technique, which employs a medium of increasing density, the rate at which particles sediment depends on size, shape, density, and the frictional resistance of the solvent.

In addition to the preparation and purification of macromolecules and cellular components, rate zonal centrifugation can be used to determine the **sedimen-** **tation coefficients** and **molecular weights** of biological macromolecules. If a purified molecule such as a protein is spun in a centrifugal field, the molecule will eventually sediment toward the bottom at a constant velocity. At this point, the molecular weight (M) can be calculated as

$$M = f \times v/\omega^2 r$$

where f is the frictional coefficient of the solvent system (which has been calculated from other measurements) and $v/\omega^2 r$ is the rate of sedimentation per unit applied centrifugal field. The latter value is given the symbol S, or sedimentation coefficient. The S value for most proteins is between 1×10^{-13} sec and 2×10^{-11} sec. A sedimentation coefficient of 1×10^{-13} is defined as one **Svedberg unit (S);** this unit is named for The Svedberg, a pioneer in the field of centrifugation. Thus, a protein with a sedimentation value of 2×10^{-11} sec would have a value of 200S. Figure A.4 shows the S values of selected molecules and particles.

Isopycnic centrifugation, or **equilibrium density gradient centrifugation,** is one of the most widely used techniques in genetics and molecular biology. In this technique, the solvent varies in density from one end of the tube to the other. A mixture of molecules layered on top and centrifuged through this gradient will migrate toward the bottom of the tube until each particle reaches its isopycnic point—that is, the place in the gradient where the density of the solvent equals the buoyant density of the particle. When each molecular species in the mixture migrates to its own char-

FIGURE A.3
Fractionation scheme for cell homogenates.

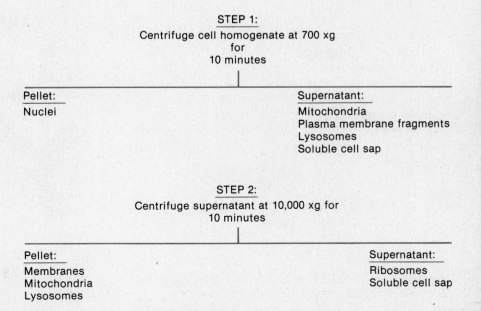

STEP 1:
Centrifuge cell homogenate at 700 xg
for
10 minutes

Pellet:
Nuclei

Supernatant:
Mitochondria
Plasma membrane fragments
Lysosomes
Soluble cell sap

STEP 2:
Centrifuge supernatant at 10,000 xg for
10 minutes

Pellet:
Membranes
Mitochondria
Lysosomes

Supernatant:
Ribosomes
Soluble cell sap

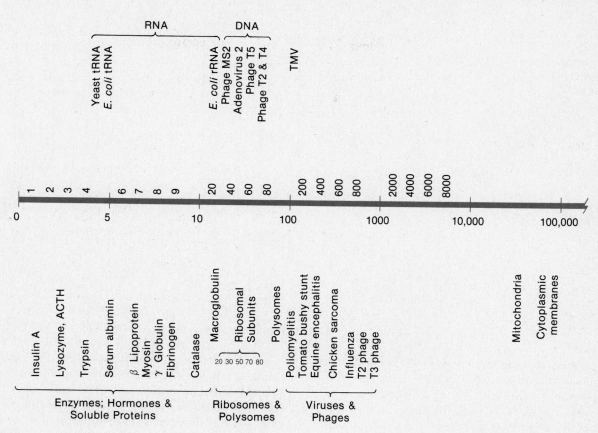

FIGURE A.4

S values of some common biological molecules and particles. (From O. M. Griffith, Beckman Instruments.)

acteristic isopycnic point, it no longer moves and is at equilibrium no matter how much longer the centrifugal field is applied. After separation by this method, components may be recovered by puncturing the bottom of the tube and collecting fractions. Two methods of forming density gradients are commonly employed (Figure A.5), one using preformed gradients of sucrose, or soluble salts of heavy metals such as cesium chloride or cesium sulfate. In the second method, the gradient is formed by the action of the centrifugal field on the salt solution.

RENATURATION AND HYBRIDIZATION OF NUCLEIC ACIDS

The ability of separated complementary strands of nucleic acids to unite and form stable, double-stranded molecules has been used to measure the relatedness of nucleic acids from different parts of the cell, different organs, and even different species. If both strands are DNA, the process is known as **renaturation** or **reassociation.** If one strand is RNA and the other DNA, it is known as **hybridization.** Renaturation and hybridization involve two steps: (1) a rate-limiting step, in which collision or nucleation between two homologous strands initiates base pairing, and (2) a rapid pairing of complementary bases, or "zippering" of the strands, to form a double-stranded molecule. For DNA renaturation, the formation of double-stranded molecules can be assayed at any time during an experiment. A sample is passed over a column of **hydroxyapatite,** which selectively binds double-stranded DNA but allows single-stranded molecules to pass through. The double-stranded molecules can be released from the column by raising the temperature or salt concentration.

FIGURE A.5
Two methods of forming
density gradients. (From O.
M. Griffith, Beckman
Instruments.)

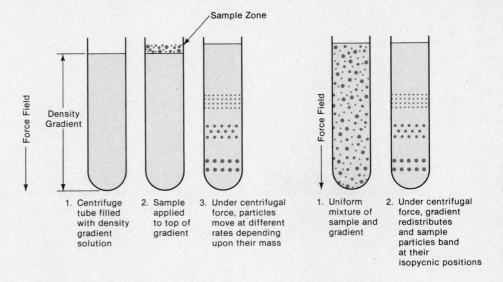

The process of renaturation follows second-order kinetics according to the equation

$$\frac{C}{C_0} = \frac{1}{1 + kC_0t}$$

where C is the single-stranded DNA concentration at time t, C_0 is the total DNA concentration, and k is a second-order rate constant. It is usually convenient to express the data from renaturation experiments as the fraction of single-stranded DNA at any time t versus the product of total DNA concentration and time, as shown in Figure A.6.

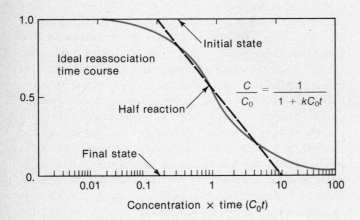

FIGURE A.6
Idealized C_0t curve. (From Britten and Kohne, 1968, Fig. 2. Copyright 1968 by the American Association for the Advancement of Science.)

In the process of renaturation, the DNA is initially single-stranded and is renatured to double-stranded structures in the final state. The time necessary to reassociate half of the DNA in a sample at a given concentration should be proportional to the number of different pieces of DNA present. Consequently, the half reassociation time should be proportional to the DNA content of the genome, with smaller genomes having shorter half-renaturation rates. Figure A.7 confirms this expectation and shows increases in C_0t values as the size of the genome increases. This proportional relationship between C_0t and genome size is valid only in cases where repetitive DNA sequences are absent from the DNA being studied. DNA from calf thymus (and many other eukaryotic sources) exhibits a complex pattern of reassociation, indicating that bovine DNA contains some sequences (in this case, 40 percent of the total DNA) that reassociate rapidly and others that reassociate more slowly. The rapidly reassociating fraction must therefore contain sequences that are present in many copies. The more slowly renaturing DNA, however, contains sequences present in only one copy per genome. The *E. coli* DNA shown in Figure A.8 renatures with a pattern close to that of an ideal second-order reaction, indicating the absence of repeated DNA sequences.

In the case of hybridization between DNA and RNA, two approaches can be used—either the RNA or the DNA can be in excess. RNA-excess hybridization is usually preferred because most double-stranded molecules that are formed are RNA:DNA hybrids. Since DNA is present in low concentration, and since single-stranded RNA molecules lack complementary RNA

FIGURE A.7
Changes in C_0t value as genome size increases. (After Britten and Kohne, 1968, Fig. 2. Copyright 1968 by the American Association for the Advancement of Science.)

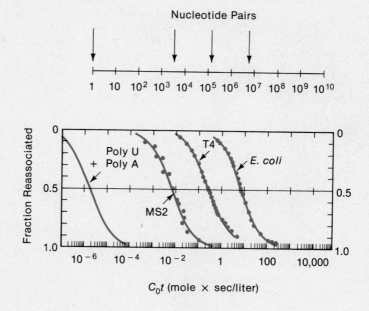

strands in the mixture, the number of RNA:RNA and DNA:DNA hybrids is negligible.

The extent of hybridization can be followed by using hydroxyapatite columns or by using radioactively labeled RNA or DNA. In practice, one of the components of the hybridization reaction is usually immobilized on a substrate such as nitrocellulose paper. For example, DNA may be sheared to a uniform size, denatured to single strands, immobilized on nitrocellulose filters, and hybridized with an excess of labeled RNA. After hybridization, the unbound RNA is removed by washing, and the hybrids assayed by liquid scintillation counting. DNA:RNA hybridization can also be performed using cytological preparations, a technique called **in situ hybridization.** DNA, which is a part of an intact chromosome preparation fixed to a slide, can be denatured and hybridized to radioactive RNA. Hybrid formation is detected using autoradiography (Figure A.9).

ELECTROPHORESIS

Electrophoresis is a technique that measures the rate of migration of charged molecules in a liquid-con-

FIGURE A.8
Renaturation curve for *E. coli* DNA, containing no repetitive DNA sequences, and for calf thymus DNA, containing several families of repetitive DNA. (From Britten and Kohne, 1968, Fig. 3. Copyright 1968 by the American Association for the Advancement of Science.)

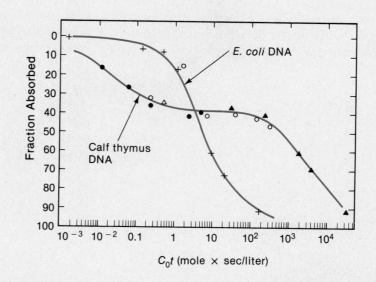

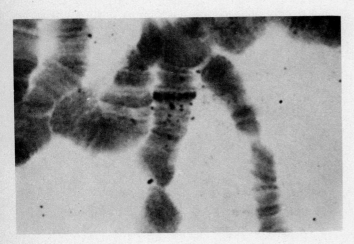

FIGURE A.9
Light micrograph of *in situ* hybridization of radioactive 5*S* RNA to a single band of a polytene chromosome in the salivary gland of a *Drosophila* larva. (Photo by Ralph M. Sinibaldi.)

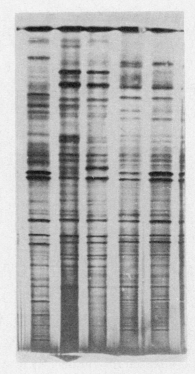

FIGURE A.10
Coomassie blue-stained protein slab gel. (Photo by Suzanne McCutcheon.)

taining medium when an electrical field is applied to the liquid. Negatively charged molecules (anions) will migrate toward the positive electrode (anode) and positively charged molecules (cations) will migrate toward the negative electrode (cathode). Several factors affect the rate of migration, including the strength of the electrical field and the molecular sieving action of the medium (paper, starch, or gel) in which migration takes place. Since proteins and nucleic acids are electrically charged, electrophoresis has been used extensively to provide information about the size, conformation, and net charge of these macromolecules. On a larger scale, electrophoresis provides a method of fractionation that can be used to isolate individual components in mixtures of proteins or nucleic acids.

More recently, the analytical separation of proteins has been enhanced with the development of two-dimensional electrophoresis techniques, in which separation in the first dimension is by net charge, and in the second dimension by size and molecular weight.

In practice, electrophoresis employs a buffer system; a medium (paper, cellulose acetate, starch gel, agarose gel, or polyacrylamide gel); and a source of direct current. Samples are applied, and current is passed through the system for an appropriate time. Following migration of the molecules, the gel or paper may be treated with selective stains to reveal the location of the separated components (Figure A.10).

APPENDIX B
Human Genetic Disorders

Interest in human genetics has been enhanced by the development of new techniques for mapping human genes, and the prospect of genetic therapy or gene surgery to repair mutant genes. The use of somatic cell hybrids, restriction fragment length polymorphisms, and marker chromosomes has accelerated the rate at which genes are being added to the human map. The catalog of human genetic traits, *Mendelian Traits in Man* by Victor McKusick, listed 1487 traits in the first edition in 1966. The sixth edition in 1983 listed some 3368 traits, and as of December 1, 1984, 3675 such loci had been identified.

The application of recombinant DNA technology to human genetics has resulted in the isolation of cloned DNA sequences that completely or partially encode over two hundred human genes. The availability of automated equipment for synthesis of polypeptides and DNA segments will undoubtedly increase the rate at which cloned human genes become available. With cloned probes in hand, geneticists are now developing methods that will be used to modify the human genome. Genetically engineered retroviruses are being developed that will serve as vehicles to deliver normal genes to cells carrying defective alleles that result in genetic disease.

With these developments in mind, this appendix lists a cross section of human genetic diseases, many of which are common in the human population, some of which have been used as models for gene therapy, and others may be used in gene surgery in the near future. Information is provided on the mode of inheritance, phenotypic characteristics, and where known, the biochemical or molecular basis of the disease. The final section of the appendix is a genetic map of loci associated with heritable diseases in humans and should serve as a useful introduction to human genetic disorders (see Figure B.1).

Albinism. A family of genetic diseases in which the affected individual exhibits a lack of pigmentation in the eye and/or the skin. The focus of this defect is the pigment cell or melanocyte. Those diseases that involve both the eye and skin are known as ocular-cutaneous albinism (OCA), while those affecting only the eye are known as ocular albinism (OA). Ten distinct forms of OCA have been identified, nine are inherited as autosomal recessive traits, and one as an autosomal dominant. All types of OCA are characterized by a reduction or absence of melanin pigment in the eyes, skin, and hair, accompanied by a reduction in visual acuity and rapid, involuntary eye movements. OA occurs as an X-linked recessive and an autosomal recessive disease, with normal pigmentation of the skin and hair, but absent or reduced pigmentation in the eyes. The most characteristic form of OCA is the classical autosomal recessive, tyrosine-positive type of albinism. It has a frequency of 1 in 37,000 in the US white population, and a frequency of 1 in 15,000 in the US black population.

Ataxia telangectasia. A recessive autosomal trait that affects the nervous system, immune system, and the cardiovascular system. The affected child often appears normal during the first year of life, and repeated respiratory infections and thyroid hypoplasia are among the earliest features of the disease. Neurological symptoms include tremor, wide-based gait, and lack of coordination. Small vascular lesions (telangectasias) appear on sun-exposed areas of the head and neck. Patients with this disease tend to develop malignancies of the lymphatic system, including leukemia. Death usually occurs before adulthood. The nature of the basic defect remains unknown, but patients with this disease are unusually sensitive to X-rays, and X-ray treatment of malignancies in these patients can be fatal.

Cri-du-chat syndrome (5p monosomy). A chromosomal abnormality caused by deletion of all or part of the short arm of chromosome 5. The most distinctive characteristic of this syndrome is the infant's cry, which is identical to that of a mewing kitten. In addition, there are characteristic craniofacial abnormalities, which include a small head, moonlike face, and receding lower jaw. Affected individuals are severely retarded. Fatality is low, and many individuals survive well into adulthood. In most cases, the deletion appears *de novo* in the affected individual.

Cystic fibrosis. An autosomal recessive trait that affects function of exocrine secretory glands. Homozygous individuals develop obstructive lung disease with susceptibility to infection, abnormal pancreatic function resulting in improper absorption of proteins and fats from the intestine, and excrete abnormally high levels of sodium and chloride in sweat. Early death is characteristic, although there is an unexplained increase in the survival of male patients. Despite intensive research efforts, the basic metabolic defect remains unknown. The clinical symptoms point to a defect in the mechanism of secretion or in the exocrine secretory glands. Reliable methods for prenatal detection and heterozygote screening are not yet available. The disease is common among US Caucasians, with a frequency of 1 in 2000 live births, but is very rare among African and US blacks.

Down syndrome (trisomy 21). Aneuploid condition caused by trisomy for chromosome 21 (47,21 +). Individuals with Down syndrome are usually of short stature, with a broad skull, round face, epicanthic fold, and mental retardation. About 1 in 6 children with this condition die within the first year, usually of congenital heart defects. The frequency of leukemia is 10 to 14 times higher in these individuals than in the population at large. The average life expectancy is about 30 years. Down syndrome occurs with a frequency of 1 in 700 births and is directly related to maternal age. Forty percent of Down syndrome patients are born to women over 40 years old, although these women produce only about 4 percent of all births.

Edwards syndrome (trisomy 18). Condition associated with trisomy for chromosome 18 (47,18 +). Characterized by congenital malformations affecting many organ systems. Low set, faunlike ears with pointed pinnae are characteristic. The hands are usually tightly closed in fists, with the index finger folded to overlap the third digit, and the fifth overlapping the fourth digit. Cardiac malformations include septal defects. Mean survival rate is 2 to 3 months for males and 10 months for females.

Fabry's disease. An X-linked recessive disease caused by inactivity of the enzyme alpha galactosidase A. Affected individuals are unable to properly metabolize glycosphingolipids, and these compounds are deposited as crystals in the walls of blood vessels, in the heart, kidneys, and eyes. Patients suffer from episodes of intense, burning pain in the extremities, with progressive cardiac and renal complications. Onset is during childhood or adolescence, and death usually occurs in the fourth or fifth decade of life. The disease is rare, with an estimated incidence of 1 in 40,000, and can be detected by prenatal diagnosis. Because the molecular nature of the disease is well known, and the activity of the enzyme is normally limited to a small number of tissues, this disease has been used as a model for gene therapy. Several modes of therapy have been employed, including injection of purified enzyme microencapsulated in liposomes, and organ transplants of liver and kidneys from normal individuals.

Familial hypercholesterolemia. An autosomal dominant disease resulting from one of several defects in cell surface receptors that regulate the metabolism of low density lipoproteins (LDL). At least three mutant alleles have been identified: a nonfunctional receptor that is unable to bind LDL; a receptor that has greatly reduced binding activity; and a third type that binds LDL normally, but is unable to transport LDL into the cell, where it is metabolized in lysosomes. The result in each of these defects is elevated levels of LDL-derived cholesterol in the blood serum, and increased deposition of cholesterol in the arteries, leading to premature coronary heart disease. Homozygous individuals are more severely affected than heterozygotes, and death from myocardial infarction commonly occurs before age 20. The frequency of heterozygotes in the European, American, and Japanese populations is 1 in 500. Because the condition is inherited as a dominant trait, it is one of the most common genetic diseases in these populations and is a leading cause of heart disease.

Galactosemia. An autosomal recessive trait associated with a defect in the enzyme galactose-1-phosphate uridyl transferase. Affected individuals are normal at birth but symptoms appear within a few days of milk feeding. The clinical features include dehydration, jaundice, cataract formation, liver enlargement, and mental retardation. In severe cases, death may occur in infancy, but those with mild clinical symptoms may be undiagnosed for a number of years. Accumulation of galactose-1-phosphate in the tissues may be responsible for the damage to the liver, brain, and kidneys. If dietary treatment is instituted early, brain damage can be prevented. Treatment before three months of age can cause regression of liver and kidney damage. Dietary treatment consists of avoiding milk and milk-related products. A prevalence of 1 in 57,000 is estimated.

Hemophilia (classical hemophilia A). An X-linked recessive trait associated with a defect in antihemophilic globulin (Factor VIII). Females are usually heterozygous carriers who transmit the disease to their sons. Treatment of this condition involves measures to control bleeding and transfusions to replace the deficient factor. Factor VIII is degraded upon storage and can be replaced only with fresh blood or a cryoprecipitate from plasma.

Huntington's disease (Huntington's chorea). A progressive neurological disease inherited as an autosomal dominant. This is a particularly insidious disease because the age of onset is usually between 30 and 40 years. Patients develop both psychological and physical symptoms, with progressive dementia and involuntary movements and gestures. The disease progresses slowly and death usually occurs 5 to 15 years after onset. The molecular basis for this disease is unknown. Prevalence is estimated at 4 to 7 in 100,000.

Klinefelter syndrome (XXY). A chromosomal abnormality caused by the presence of two X chromosomes in addition to one Y chromosome (47,XXY). Individuals with this condition are phenotypically male, but often show a tendency toward female secondary sexual characteristics, such as enlarged breasts. Diagnosis is usually at puberty when testicular atrophy begins. Individuals are aspermatozoic and sterile. Intellectual development is in the normal range, and survival is also within the normal range. Chromosomal nondisjunction is the cause of this condition, which occurs with a frequency of 2 in 1000 male births. In a small percentage of cases, more complex karyotypes such as 48,XXXY are seen. In general, higher numbers of sex chromosomes are associated with more severe hypogonadism and the development of mental retardation.

Lesch-Nyhan syndrome. Inherited as an X-linked recessive, this condition is associated with a deficiency of the enzyme hypoxanthine-guanine phosphoribosyltransferase (HPRT). This defect produces abnormalities in purine metabolism that result in the development of the characteristic involuntary movements, mental retardation, and self-mutilation. Joint involvement, gout, and renal damage are often observed in older children. Treatment with allopurinol has been used to control renal damage and gout, but the disease is progressive and fatal. Affects 1 in 10,000 males.

Maple syrup urine disease (MSUD). An autosomal recessive condition in which affected individuals are unable to metabolize the branch-chain amino acids leucine, isoleucine, and valine. These amino acids and their keto acids accumulate in the plasma and urine. The name of the disease is derived from the characteristic odor of the urine and sweat. Undetected and untreated, the disease begins in the first month after birth, leading to lethargy, withdrawal from feeding, rigidity, and progressive decline to death. The disease results from a deficiency of branch-chain keto-acid decarboxylase activity. The mechanism of brain involvement is unknown. Dietary therapy is difficult, but is successful if instituted soon after birth. There is no indication that dietary restrictions can ever be lifted, as

in PKU, so the treatment is burdensome to the patient, as the diet consists of gelatin supplemented with essential amino acids other than the branch-chain acids. A frequency of 1 in 200,000 is estimated.

Marfan syndrome. A pleiotropic disorder of connective tissue that is inherited as an autosomal dominant. Affected individuals have abnormalities of the eye, skeletal system, and the cardiovascular system. Most patients with this disease have myopia, elongated eyeballs, and upward displacement of the lens of the eye. Skeletal abnormalities include long narrow extremities (spider fingers), excessive height, and irregular body proportions. Aortic aneurysms and mitral valve problems are common. The nature of the defect is unknown. The only detectable biochemical abnormality is an increased urinary excretion of hydroxyproline, suggesting that the disorder affects a structural protein in the connective tissue such as elastin or collagen. A frequency of 4 to 6 in 100,000 is projected for this disorder.

Muscular dystrophy (Duchenne type). Muscular dystrophy is a group of muscle diseases that produce weakness and progressive wasting and are inherited as dominant, recessive, autosomal, and sex-linked diseases. Duchenne muscular dystrophy is an X-linked recessive disease of childhood with an onset between 1 and 6 years. Muscle weakness and atrophy progress until the patient is confined to a wheelchair by the age of 12. Three-fourths of the cases die before age 20. Death is usually attributed to respiratory infections or cardiac failure. The nature of the underlying defect is unknown.

Patau syndrome (trisomy 13). An aneuploid condition caused by trisomy for chromosome 13 (47,13+). The phenotype is somewhat variable, but includes mental retardation, cleft lip and/or palate, deafness, cardiac malformations, and posterior protrusions of the heel. Most children with this condition die within the first 3 months of life. This condition, caused by chromosome nondisjunction, occurs with a frequency of 1 in 5000 births.

Phenylketonuria. First described in 1934, this autosomal recessive disease spurred interest in heritable metabolic defects. The major form of the disease is caused by a lack of the enzyme phenylalanine hydroxylase, which converts phenylalanine to tyrosine. This enzyme deficit causes severe mental retardation. Increased concentrations of phenylalanine or its alternate metabolic products are responsible for the irreversible brain damage, but the mechanism of action is unknown. Brain damage is preventable if the patient is put on a diet low in phenylalanine soon after birth. In most cases, dietary restrictions can be withdrawn around the age of 6 or 7 years. A frequency of 1 in 12,000 is estimated.

Porphyria, acute intermittent (AIP). Inherited as an autosomal dominant disorder, AIP is the result of reduction by about 50 percent of the enzyme porphobilinogen (PBG) deaminase (previously called uroporphyrinogen synthetase). The enzyme is involved in the synthesis of heme. The attacks, involving acute episodes of a variety of neuropathic symptoms, rarely occur before puberty and are often precipitated by porphyrinogenic drugs such as barbiturates and sulfonamides. Abdominal pain, paresthesias, and paralysis occur as well as seizures, psychotic episodes, and hypertension. Many latent AIP subjects never experience acute attacks but characteristically excrete excess porphyrin precursors in urine. The dominant nature of the enzyme disorder is unusual because 50 percent activity levels are seldom rate-limiting. A high prevalence (1/1000) is observed in northern Sweden but a much lower rate of occurrence (1/50,000) is predicted for all other ethnic groups.

Retinoblastoma. An embryonic neoplasm of retinal origin, inherited as an autosomal dominant disorder. It is almost always manifested in childhood and most often is bilateral. Early detection and surgical removal of the eye alleviates complications that are normally lethal. Most unilateral, sporadic cases are not hereditary and are assumed to be of somatic rather than germ-cell origin. The dominantly inherited form of the disease demonstrates incomplete penetrance, estimated at about 80 percent. In some cases, a deletion in the q arm of chromosome 13 has been associated with the disorder. An occurrence of 1 in 23,000 live births is estimated.

Tay-Sachs disease. An autosomal recessive condition, usually fatal in the second or third year of life. Development is usually normal until the eighth month after birth, when cerebral functions begin to degenerate, producing the characteristic paralysis, epilepsy, enlargement of the head, and blindness. The disease is associated with a defect in the enzyme hexosaminidase A. This enzyme is composed of alpha subunits (coded by a gene on chromosome 15) and beta subunits (coded by a gene on chromosome 5). Tay-Sachs disease is associated with a defect in the alpha subunits, while a related disease, Sandhoff's, is associated with a defect in the beta subunit. The enzyme defect leads to the accumulation of gangliosides in the neuronal cytoplasm in both the central and peripheral

nervous system. The frequency of this disease is much higher in Ashkenazi Jews of eastern European descent. In this population, 1 in 30 to 1 in 60 individuals are carriers.

Thalassemias. A heterogeneous and ubiquitous group of hereditary anemias characterized by defective alpha- or beta-globin synthesis. The normal balance of synthesis of the two types of chains is impaired, resulting in a failure to produce normal hemoglobin. In alpha thalassemia, defects range from mild to complete suppression of synthesis of the alpha chain. The most common cause of the alpha form of the disorder is the presence of internal deletions affecting one or both of the loci encoding the alpha chain. Other lesions have included a nonsense mutation causing premature termination of the alpha chain and an intron mutation affecting mRNA processing. The beta thalassemia phenotypes are also heterogeneous, but seldom caused by deletion of the single beta-globin locus. More often, failure to adequately process the primary transcript to the mature beta-globin mRNA is at fault. The disorder may be detected using prenatal diagnostic techniques.

Turner syndrome (45,X). This disorder was the first chromosomal anomaly involving the sex chromosomes to be described. Affected individuals are females, having only one X chromosome. The phenotypic characters include small stature, swelling of the hands and feet as infants, and an excess of skin at the nape of the neck. Gonadal atrophy in the form of streak gonads is characteristic. Overall, the symptoms are not usually severe, and the survival rate is not different from the normal population. This condition, arising from chromosomal nondisjunction, occurs with a frequency of 1 in 3000 female births.

Wilms tumor. A chromosomal abnormality associated with a deletion within the short arm of chromosome 11 (band p13). This interstitial deletion results in aniridia (absence of the iris), mental retardation, microcephaly, and malignant tumors of the kidney. Tumor growth usually occurs in infants and young children, and leads to death.

XXX syndrome. The 47,XXX female karyotype was first described in 1959 and has been shown to occur in about 1 in 1200 births. In most cases, the phenotype is completely normal, including fertility. There is a slight disturbance in intellectual development in two-thirds of the cases, but otherwise the addition of an extra chromosome has little or no effect on the individual. Normality is attributed to the inactivation of two X chromosomes (as Barr bodies) in somatic cells.

XYY syndrome. This chromosomal condition was first described in 1961, and subsequent population surveys revealed that 7 XYY males were detected in a group of 197 mentally retarded males institutionalized for violent or criminal tendencies. On this basis, it was suggested that the presence of an extra Y chromosome might predispose individuals toward aggressive behavior. A number of surveys afterward have determined that the frequency of XYY individuals in the general population is about 0.11, while up to 2 percent of the individuals institutionalized as mentally ill and dangerous are XYY. On the other hand, XYY is a common condition, occurring with a frequency of 1 in 1000 male births, and only a small percentage of these individuals (about 3.5 percent) are ever institutionalized. A wide range of phenotypes is present in this condition, but only two, tallness and subnormal intelligence, are constant.

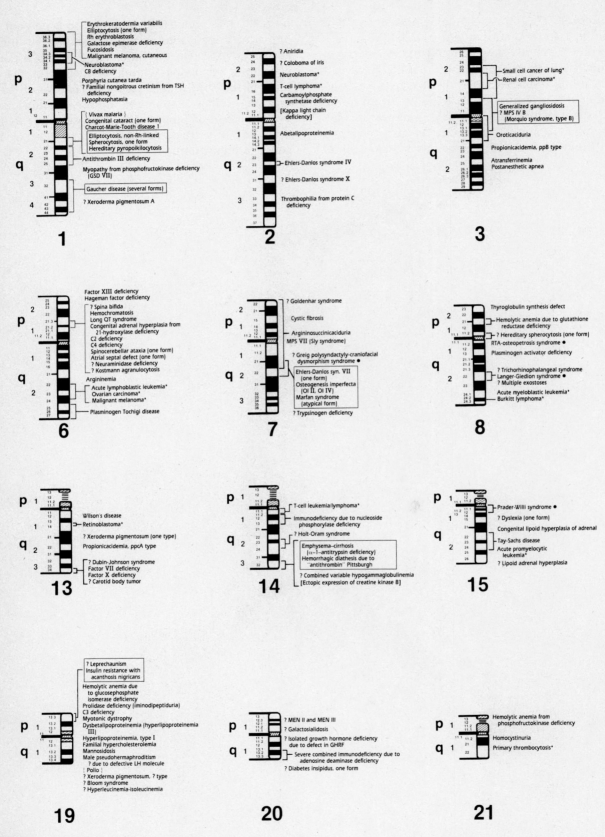

FIGURE B.1
Map of human genetic diseases.

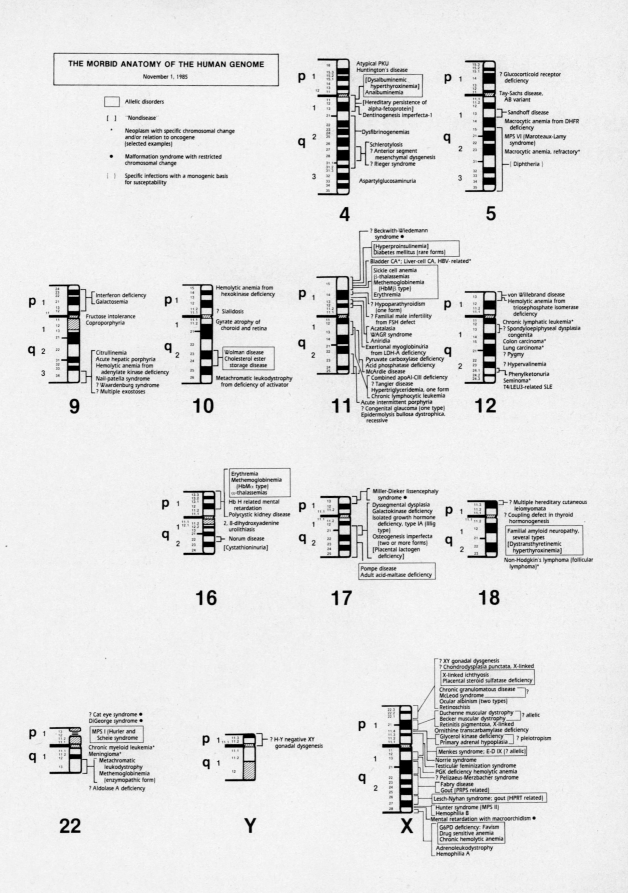

THE MORBID ANATOMY OF THE HUMAN GENOME

November 1, 1985

Allelic disorders

[ ]   "Nondisease"

*   Neoplasm with specific chromosomal change
       and/or relation to oncogene
       (selected examples)

●   Malformation syndrome with restricted
       chromosomal change

{ }   Specific infections with a monogenic basis
       for susceptability

4

Atypical PKU
Huntington's disease

[Dysalbuminemic
hyperthyroxinemia]
Analbuminemia

[Hereditary persistence of
alpha-fetoprotein]
Dentinogenesis imperfecta-1

Dysfibrinogenemias

Schlerotylosis
? Anterior segment
mesenchymal dysgenesis
? Rieger syndrome

Aspartylglucosaminuria

5

? Glucocorticoid receptor
deficiency

Tay-Sachs disease,
AB variant

Sandhoff disease

Macrocytic anemia from DHFR
deficiency

MPS VI (Maroteaux-Lamy
syndrome)

Macrocytic anemia, refractory*

{ Diphtheria }

9

Interferon deficiency
Galactosemia

Fructose intolerance
Coproporphyria

Citrullinemia
Acute hepatic porphyria
Hemolytic anemia from
adenylate kinase deficiency
Nail-patella syndrome
? Waardenburg syndrome
? Multiple exostoses

10

Hemolytic anemia from
hexokinase deficiency

? Sialidosis

Gyrate atrophy of
choroid and retina

Wolman disease
Cholesterol ester
storage disease

Metachromatic leukodystrophy
from deficiency of activator

11

? Beckwith-Wiedemann
syndrome ●

[Hyperproinsulinemia]
Diabetes mellitus (rare forms)

Bladder CA*; Liver-cell CA, HBV- related*

Sickle cell anemia
β-thalassemias
Methemoglobinemia
(HbMβ type)
Erythremia

? Hypoparathyroidism
(one form)
? Familial male infertility
from FSH defect
Acatalasia
WAGR syndrome
Aniridia
Exertional myoglobinuria
from LDH-A deficiency
Pyruvate carboxylase deficiency
Acid phosphatase deficiency
McArdle disease
Combined apoAI-CIII deficiency
? Tangier disease
Hypertriglyceridemia, one form
Chronic lymphocytic leukemia
Acute intermittent porphyria
? Congenital glaucoma (one type)
Epidermolysis bullosa dystrophica,
recessive

12

von Willebrand disease
Hemolytic anemia from
triosephosphate isomerase
deficiency
Chronic lymphatic leukemia*
? Spondyloepiphyseal dysplasia
congenita
Colon carcinoma*
Lung carcinoma*
? Pygmy

? Hypervalinemia

Phenylketonuria
Seminoma*
T4/LEU3-related SLE

16

Erythremia
Methemoglobinemia
(HbMα type)
α-thalassemias

Hb H related mental
retardation
Polycystic kidney disease

2, 8-dihydroxyadenine
urolithiasis

Norum disease
[Cystathioninuria]

17

Miller-Dieker lissencephaly
syndrome ●

Dyssegmental dysplasia
Galactokinase deficiency
Isolated growth hormone
deficiency, type IA (Illig
type)
Osteogenesis imperfecta
(two or more forms)
[Placental lactogen
deficiency]

Pompe disease
Adult acid-maltase deficiency

18

? Multiple hereditary cutaneous
leiomyomata
? Coupling defect in thyroid
hormonogenesis

Familial amyloid neuropathy,
several types
[Dystransthyretinemic
hyperthyroxinemia]

Non-Hodgkin's lymphoma (follicular
lymphoma)*

22

? Cat eye syndrome ●
DiGeorge syndrome ●

MPS I (Hurler and
Scheie syndrome

Chronic myeloid leukemia*
Meningioma*

Metachromatic
leukodystrophy
Methemoglobinemia
(enzymopathic form)
? Aldolase A deficiency

Y

? H-Y negative XY
gonadal dysgenesis

X

? XY gonadal dysgenesis
? Chondrodysplasia punctata, X-linked

X-linked ichthyosis
Placental steroid sulfatase deficiency

Chronic granulomatous disease
McLeod syndrome   ?
Ocular albinism (two types)
Retinoschisis
Duchenne muscular dystrophy
Becker muscular dystrophy   ? allelic
Retinitis pigmentosa, X-linked
Ornithine transcarbamylase deficiency
Glycerol kinase deficiency   ? pleiotropism
Primary adrenal hypoplasia

Menkes syndrome; E-D IX (? allelic)

Norrie syndrome
Testicular feminization syndrome
PGK deficiency hemolytic anemia
? Pelizaeus-Merzbacher syndrome
Fabry disease
Gout (PRPS related)

Lesch-Nyhan syndrome; gout (HPRT related)

Hunter syndrome (MPS II)
Hemophilia B
Mental retardation with macroorchidism ●

G6PD deficiency: Favism
Drug sensitive anemia
Chronic hemolytic anemia

Adrenoleukodystrophy
Hemophilia A

655

APPENDIX C
Answers to Selected Problems

CHAPTER 2

3 The genetic loci along the length of the chromosome.

5 In mitosis, 16 chromosomes containing 32 chromatids are visible at the end of prophase. 16 chromosomes move to each pole during anaphase.

13 (a) 8 tetrads
(b) 8 dyads
(c) 8 monads

16 As shown in Figure 2.8, an alternate alignment at metaphase II would have resulted in a new combination of maternally and paternally derived chromosomal segments in each spermatid. Had the initial alignment at metaphase I been different than shown (one other alignment is possible), two other arrangements would be possible at metaphase II. Therefore, three other unique combinations of four spermatids are possible in addition to the one shown.

17 (a) 7
(b) 5
(c) trisomy

18 Prophase I and metaphase I.

19 (a) pachynema
(b) zygonema
(c) leptonema
(d) diplonema

20 Meiosis results in the production of haploid microspores and megaspores, which themselves develop into haploid gametes. Following fertilization between male and female gametes, the diploid zygote develops into the prominent sporophyte.

CHAPTER 3

1 (a) P_1: $WW \times ww$
F_1: $Ww \times Ww$
F_2: 1/4 WW; 1/2 Ww; 1/4 ww
(b) All white
(c) Cross 1: $WW \times WW$ or $WW \times Ww$
Cross 2: $Ww \times Ww$

2 (a) $Aa \times Aa$
(b) $AA \times aa$
(c) $Aa \times aa$

3 Dominance/recessiveness; unit factors in pairs; segregation.

6 The checkered pattern is dominant to the plain pattern. Let P = checkered and p = plain.
(a) $PP \times PP$ yields all PP
(b) $PP \times pp$ yields all Pp
(c) $pp \times pp$ yields all pp
(d) $PP \times pp$ yields all Pp
(e) $Pp \times pp$ yields 1/2 Pp and 1/2 pp
(f) $Pp \times Pp$ yields 1/4 PP, 1/2 Pp, and 1/4 pp
(g) $PP \times Pp$ yields 1/2 PP and 1/2 Pp

7 P_1: $WWGG \times wwgg$
F_1: $WwGg$
F_2: 9/16 round, yellow (W–G–)
3/16 round, green (W–gg)
3/16 wrinkled, yellow (wwG–)
1/16 wrinkled, green ($wwgg$)

8 (a) $WwGG \times WwGG$ or $WwGG \times WwGg$
(b) $wwGg \times WwGg$
(c) $WwGg \times WwGg$
(d) $WwGg \times wwgg$

9 (d) is a test cross.

10 Independent assortment

14 (a) F_1: All grey, long ($EeVv$)
F_2: 9/16 grey, long (E–V–); 3/16 grey, vestigial (E–vv); 3/16 ebony, long (eeV–); 1/16 ebony, vestigial ($eevv$)
(b) Same answer as (a).
(c) F_1: All grey, long ($EEVv$)
F_2: 3/4 grey, long (EEV–); 1/4 grey, vestigial ($EEvv$)

15 (a) 4: AB, Ab, aB, ab
(b) 2: AB, aB
(c) 8: ABC, ABc, AbC, Abc, aBC, aBc, abC, abc
(d) 2: ABc, aBc
(e) 4: ABc, Abc, aBc, abc
(f) 32: $ABCDE$, $ABCDe$, $ABCdE$, $ABCde$
$ABcDE$, $AbcDe$, $AbCDE$, $AbCDe$
$aBCDE$, $aBCDe$, $abCDE$, $abCDe$
$aBcDE$, $aBcDe$, $aBCdE$, $aBCde$
$AbcDE$, $AbcDe$, $ABcdE$, $ABcde$
$abcDE$, $abcDe$, $aBcdE$, $aBcde$
$abCdE$, $abCde$, $AbcdE$, $Abcde$
$AbCdE$, $AbCde$, $abcdE$, $abcde$

16 (a) *Genotypic Ratio:*

1/16 $AABBCC$	2/16 $AaBBCC$	1/16 $aaBBCC$
1/16 $AABBCc$	2/16 $AaBBCc$	1/16 $aaBBCc$
1/16 $AABbCC$	2/16 $AaBbCC$	1/16 $aaBbCC$
1/16 $AABbCc$	2/16 $AaBbCc$	1/16 $aaBbCc$

Phenotypic Ratio:
12/16 A–B–C–
4/16 aaB–C–

(b) *Genotypic Ratio:*

1/8 *AaBBCC*   1/8 *aaBBCC*

2/8 *AaBBCc*   2/8 *aaBBCc*

1/8 *AaBBcc*   1/8 *aaBBcc*

Phenotypic Ratio:

3/8 *A–BBC–*   3/8 *aaBBC–*

1/8 *A–BBcc*   1/8 *aaBBcc*

(c) *Genotypic Ratio:*   *Phenotypic Ratio:*

1/64 *AABBCC*   27/64 *ABC*

2/64 *AABBCc*   9/64 *ABc*

1/64 *AABBcc*   9/64 *AbC*

2/64 *AABbCC*   9/64 *aBC*

4/64 *AABbCc*   3/64 *Abc*

2/64 *AABbcc*   3/64 *aBc*

1/64 *AAbbCC*   3/64 *abC*

2/64 *AAbbCc*   1/64 *abc*

1/64 *AAbbcc*

2/64 *AaBBCC*

4/64 *AaBBCc*

2/64 *AaBBcc*

4/64 *AaBbCC*

8/64 *AaBbCc*

4/64 *AaBbcc*

2/64 *AabbCC*

4/64 *AabbCc*

2/64 *Aabbcc*

1/64 *aaBBCC*

2/64 *aaBBCc*

1/64 *aaBBcc*

2/64 *aaBbCC*

4/64 *aaBbCc*

2/64 *aaBbcc*

1/64 *aabbCC*

2/64 *aabbCc*

1/64 *aabbcc*

17 P_1: *GG* × *gg*

F_1: All *Gg*

F_2: 1/4 *GG* (breed true); 2/4 *Gg* (produce 3/4 yellow: 1/4 green); 1/4 *gg* (breed true)

18 In 94 cases, the black parent was homozygous for the dominant black allele. In 6 cases, the black parent was heterozygous. The F_2 and F_3 generations of the 6 cases will produce the identical results of the F_1 generation.

19 1/4 *WwGg* (round, yellow); 1/4 *Wwgg* (round, green); 1/4 *wwGg* (wrinkled, yellow); 1/4 *wwgg* (wrinkled, green)

20 (a) $\chi^2 = 0.07; p = 0.80$

(b) $\chi^2 = 0.38; p = 0.55$

21 (a) For 9:3:3:1, $\chi^2 = 0.51; p = 0.90$.

(b) For 3 smooth: 1 wrinkled, $\chi^2 = 0.35; p = 0.63$.

(c) For 3 yellow: 1 green, $\chi^2 = 0.01; p = 0.85$.

22 For the 3:1 ratio, $\chi^2 = 33.33$. For the 1:1 ratio, $\chi^2 = 25.00$. In both cases, the probability of obtaining the data with the observed deviation strictly by chance is much less than 0.0001. Therefore, both null hypotheses are rejected.

23 The pedigree fits an autosomal recessive mode of inheritance. I-1 is *aa*; I-2 is *Aa*; I-3 is *Aa*; and I-4 is *Aa*.

24 Myopia is inherited as an autosomal recessive disorder according to this pedigree.

25 If the "minor" anemia went undetected and such individuals were considered phenotypically normal, then the major anemia would be classified as a recessive disorder. However, on the basis of red blood cell and hemoglobin analysis, the heterozygote is intermediate in phenotypic expression. Therefore, dominance is incomplete and neither condition is considered recessive.

CHAPTER 4

1 Incomplete dominance; roan is *Aa*.

AA × *AA* → *AA*; *aa* × *aa* → *aa*; *AA* × *aa* → *Aa*

Aa × *Aa* → 1/4 *AA*; 1/2 *Aa*; 1/4 *aa*

3 2/3 platinum; 1/3 silver. The *P* allele behaves as a recessive lethal, but as a dominant in its influence on coat color.

4 The F_1 ratio appears to be 1:1; the cross between selected F_1 mice yields a 2:1 ratio. The short mutation is dominant, but homozygous lethal. Let *S* = short and *s* = normal (long). The crosses are thus:

Ss × *ss* → 1/2 *Ss* : 1/2 *ss*

short   long   short   long

Ss × *Ss* → 1/4 *SS* : 2/4 *Ss* : 1/4 *ss*

short   short   lethal   short   long

5 (10) (35) (20) (30) or 210,000 different haplotypes can be formed. $(210,000)^2$ different HLA genotypes are possible, or $(2.1 \times 10^5)^2 = 4.41 \times 10^{10}$ genotypes.

6 The I^A and I^B alleles are dominant to the recessive I^O allele. The I^A and I^B alleles are codominant to each other.

7 Theoretically, x^{20}, where x is the number of amino acids in the protein chain. The term is taken to the 20th power because there are 20 different amino acids that can be substituted at each position of the protein. The theoretical limit for the number of alleles for a protein with 100 amino acids is 100^{20}.

8 Blood type AB would exclude the male in question. If he were type A, B, or O, he could contribute an I^O allele, and could be the father. However, countless males are of these blood types, so no proof is provided.

9 The secretor allele behaves dominantly to the nonsecretor allele.

10 (a) $c^h a \times c^h a \longrightarrow c^a c^a$ 1/2 $c^a c^a$ albino

$\xrightarrow{\text{x}}$

$Cc^{ch} \times c^a c^a \longrightarrow c^{ch} c^a$ 1/2 $c^{ch} c^a$ chinchilla

(b) $c^a c^a \times c^{ch} c^a \longrightarrow c^a c^a$ 1/2 Cc^a full color

$\xrightarrow{\text{x}}$

$C- \times c^a c^a \longrightarrow Cc^a$ 1/2 $c^a c^a$ albino

(c) $c^{ch}c^h \times c^a c^a \longrightarrow c^h c^a$ 1/4 $c^h c^h$ Himalayan
                 x $\longrightarrow$ 1/2 $c^h c^a$ Himalayan
     $Cc^h \times c^a c^a \longrightarrow c^h c^a$ 1/4 $c^a c^a$ albino

11 (a) $c^k c^a \times c^d c^a$ = 1/2 sepia: 1/4 cream: 1/4 albino
    (b) $c^k c^a \times c^d c^d$ = 1/2 sepia: 1/2 cream (or $c^k c^a \times c^d c^a$)
    (c) $c^k c^k \times c^d c^d$ = all sepia (or $c^k c^k \times c^d c^a$)
    (d) $c^k c^a \times c^d c^d$ = 1/2 sepia: 1/2 cream

12 3/16 red, straight; 1/16 red, curled; 6/16 pink, straight; 2/16 pink, curled; 3/16 white, straight; 1/16 white, curled

13 As in Problem 12, pink is the heterozygote, and white and red are the two homozygotes. Personate is dominant to peloric. F$_1$ (a) × F$_1$ (b) cross: 3/16 red, personate; 1/16 red, peloric; 6/16 pink, personate; 2/16 pink, peloric; 3/16 white, personate; 1/16 white, peloric.

14 (a) F$_1$: All wild type
        F$_2$: 9/16 wild; 3/16 brown; 3/16 scarlet; 1/16 white
    (b) F$_1$: All wild type
        F$_2$: 3/4 wild; 1/4 scarlet
    (c) F$_1$: All brown
        F$_2$: 3/4 brown; 1/4 white

15 (a) F$_1$: All grey (*CcAa*)
        F$_2$: 9/16 grey (*C–A–*); 3/16 black (*C–aa*); 4/16 albino (*ccA–* and *ccaa*)
    (b) (1) *CcAA*   (2) *CCAa*   (3) *CcAa*

16 Two genes. One product is an enzyme catalyzing the conversion of the precursor to cyanidin, and the other converts cyanidin to a purple pigment. *A–B–* yields purple; *A–bb* yields red; *aaB–* yields colorless; *aabb* yields colorless. The *A* gene converts the precursor to cyanidin, while the *B* gene converts cyanidin to purple.

17 (a) All grey
    (b) All grey
    (c) 16/32 albino; 9/32 grey; 3/32 yellow; 3/32 black; 1/32 cream
    (d) 9/16 grey; 3/16 black; 4/16 albino
    (e) 3/8 grey; 1/8 yellow; 4/8 albino

18 (a) *AaBbCC* × *AaBbCC* or *AaBbCC* × *AaBbCc*
    (b) *AABbCc* × *AABbCc* or *AABbCc* × *AaBbCc*
    (c) *AaBbCc* × *AaBbCc*
    (d) *aaBbCc* × *aabbCc*
    (e) *aaBbCc* × *aaBbcc*

19 The 37:17:11 ratio can be reduced to a 9:4:3 modified dihybrid ratio, where *A–B–* yields white, *A–bb* and *aabb* yields brown, and *aaB–* yields black. In the initial cross between *AABB* (white) and *aabb* (brown), the F$_1$ generation is *AaBb* (white), while the F$_2$ yields the predicted 9:4:3 ratio of white: brown: black.

20 Since neither parent is white and they are of different colors, one must be brown and one must be black. The genotypes *AAbb* (brown) and *aaBB* (black) will produce all white F$_1$ dogs (*AaBb*) when involved in such a cross.

21 In a complex set of gene interactions, any combination of *A* or *a* alleles and *bb* yields black. Any time *aa* and at least one *B* allele are present, jet is produced. Any time *AA* and at least one *B* allele are present, red occurs. The

remaining genotypes (*AaBB* and *AaBb*) yield sooty. In summary,

      – –bb: black   *aaB–: jet*
      AAB–: red   *AaB–:* sooty

22 This is an excellent example of recessive epistasis in humans. In the presence of the *hh* genotype, the expression of the I^A, I^B, and I^O alleles is masked.

23 3/16 type A: 6/16 type AB: 3/16 type B: 4/16 type O. All of those which type as O actually lack the H substance.

24 1–C; 2–d; 3–b, 4–e, 5–a. No!

25 (a) Three, based on the F$_2$ ratios which fall into 64s.
    (b) Two gene pairs control eye color while one controls croaking. This is determined by analyzing the ratios of eye color separately from croaking.
    (c) blue = *A–B–*   rib-it utterer = *R–*
        purple = *A–bb*   knee-deep mutterer = *rr*
        green = *aaB–* and *aabb*
        P$_1$ genotypes: *AAbbRR* × *aaBBrr*
        F$_1$ genotype: *AaBbRr*
    (d) blue-eyed, rib-it utterer: *A–B–R*
        green-eyed, rib-it utterer: *aa– –R–*
        blue-eyed, knee-deep mutterer: *A–B–rr*
        purple-eyed, rib-it utterer: *A–bbR–*
        green-eyed, knee-deep mutterer: *aa– –rr*
        purple-eyed, knee-deep mutterer: *A–bbrr*
    (e) F$_1$ all blue-eyed, rib-it utterers
        F$_2$: 9/16 blue-eyed, rib-it utterer
           3/16 blue-eyed, knee-deep mutterer
           3/16 purple-eyed, rib-it utterer
           1/16 purple-eyed, knee-deep mutterer
    (f) The results differed because true-breeding green-eyed frogs can be of either *aaBB*, *aabb*, or *aaBb* genotypes. Each mating with a true-breeding purple-eyed frog will yield different results.
    (g) Both were *AabbRr*.

CHAPTER 5

2 (a) F$_1$: 1/2 wild ♀; 1/2 wild ♂
        F$_2$: 1/4 wild ♀; 1/4 reduced ♀; 1/2 wild ♂
    (b) F$_1$: 1/2 wild ♀; 1/2 reduced ♂
        F$_2$: 1/4 wild ♀; 1/4 reduced ♀; 1/4 wild ♂; 1/4 reduced ♂
    (c) No

5 3/8 wild ♀; 1/8 wild ♂ (transformed); 4/8 singed ♂

7 The Y chromosome is essential for initial differentiation of testicular tissue. However, hormones subsequently produced by the testes are responsible for differentiation of the remainder of the male reproductive tract.

9 As a result of crossing over between the X and Y chromosomes during male meiosis, the X chromosome normally passed on to daughters would receive male-determining genes.

11 Klinefelter, one; Turner, none; 47,XYY, none; 47,XXX, two; 48,XXXX, three.

12 Because the mosaic coat pattern is due to the expression of sex-linked heterozygous alleles according to the Lyon hypothesis.

14 The male hormones of the male twin are produced earlier than the female hormones of the female twin. As a result, sexual differentiation in the female is abnormal because of the common placenta.

15 In *Drosophila*, the condition results in a normal, fertile female. In humans, the condition produces a male with Klinefelter syndrome.

16 Because X̂XX and YY flies are inviable, all females are X̂XY and are white eyed. All XY males receive their X from their father and Y from their mother and exhibit the miniature-wing phenotype.

17 If nondisjunction occurred and the parthenogenic egg retained two X chromosomes.

18 3/4 will be ♂ (*Xa, Xb,* and *XaXa*); 1/4 will be ♀ (*XaXb*).

20 1/4 black ♀; 1/4 tortoise-shell ♀; 1/4 black ♂; 1/4 yellow ♂. There is no chance of a tortoise-shell male since mammalian males cannot be heterozygous for a sex-linked gene.

21 (a) 1/4 (b) 1/2 (c) 1/4 (d) 0

22 If the women is $I^A I^A$, then 1/4 A $Xg^{(a+)}$♀; 1/4 AB $Xg^{(a+)}$♀; 1/4 A $Xg^{(a-)}$♂; 1/4 AB $Xg^{(a-)}$♂. If the women is $I^A I^O$, all females will be $Xg^{(a+)}$, with 1/2A, 1/4 B, and 1/4 AB. All males will be $Xg^{(a-)}$ with 1/2 A, 1/4 B, and 1/4 AB.

23 (a) F_1: 1/2 scalloped ♂; 1/2 normal ♀
   F_2: 1/4 scalloped ♀; 1/4 scalloped ♂; 1/4 normal ♀; 1/4 normal ♂
   (b) Assuming the female is homozygous, the F_1 consists of all normal males and females. In the F_2, 1/2 normal ♀; 1/4 normal ♂; 1/4 scalloped ♂.

24 Assuming the P_1 female is homozygous,
   F_1: normal-winged, normal-bodied females; scalloped-winged, normal-bodied males
   F_2: 3/16 normal, normal ♀; 3/16 normal, normal ♂; 1/16 normal, ebony ♀; 1/16 normal, ebony ♂; 3/16 scalloped, normal ♀; 3/16 scalloped, normal ♂; 1/16 scalloped, ebony ♀; 1/16 scalloped, ebony ♂.

25 F_1: 1/2 red, normal ♀; 1/2 red, normal ♂
   F_2: 6/16 red, normal ♀; 2/16 normal, dumpy ♀; 3/16 red, normal ♂; 3/16 white, normal ♂; 1/16 red, dumpy ♂; 1/16 white, dumpy ♂

26 F_1: All wild-type ♂ and ♀
   F_2: 8/16 wild-type ♀; 5/16 wild-type ♂; 3/16 vermilion ♂

27 (a) F_1: 1/2 wild ♀; 1/2 vermilion ♂
   F_2: 3/16 vermilion ♀; 1/16 white ♀; 3/16 wild ♂; 1/16 brown ♀; 3/16 vermilion ♂; 1/16 white ♂; 3/16 wild ♂; 1/16 brown ♂
   (b) F_1: All wild ♂ and ♀
   F_2: 6/16 wild ♀; 2/16 brown ♀; 3/16 vermilion ♂; 1/16 white ♂; 3/16 wild ♂; 1/16 brown ♂
   (c) F_1: 1/2 wild ♀; 1/2 vermilion ♂; F_2: 3/16 wild ♀: 1/16 brown ♀: 3/16 vermilion ♀: 1/16 white ♀: 3/16 wild ♂: 1/16 brown ♂: 3/16 vermilion ♂: 1/16 white ♂.

28 Color is determined by two alleles of a gene located on an autosome. The expression of the trait is sex influenced such that the heterozygote is mahogany if male and red if female. If *RR* = red and *rr* = mahogany, then the P_1 genotypes of the initial cross are *RR* ♀♀ × *rr* ♂♂. In the F_1, all cattle are *Rr*, and the males are mahogany and the females are red. In the F_2, the genotypic ratio of 1/4 *RR* : 2/4 *Rr* : 1/4 *rr* will yield the observed result when interpreted as described above. The reciprocal cross works exactly the same way.

29 F_1: all hen-feathered males and females
   F_2: 4/8 hen-feathered females
   3/8 hen-feathered males
   1/8 cock-feathered males.

30 In females, all three genotypes result in hen-feathering; in males, however, only *H* – results in this phenotype. The genotype *hh* produces cock-feathering in males. The cross described involves two heterozygous parents (*Hh*)

CHAPTER 6

3 Because each crossover event occurs randomly along the length of the chromosome. The farther apart two loci are, the more frequently a random event will occur between them.

4 Because a single exchange can involve only two of the four chromatids in a tetrad, the other two nonsister chromatids remain as noncrossover products.

5 Because the probability of two independent events occurring simultaneously is equal to the product of the probability of each individual event.

7 One parent must be heterozygous for all genes to be mapped. Second, the phenotypes of the offspring must reveal the genotypes of the gametes formed by the crossover parent that gave rise to these progeny.

8 *dp–cl–ap*

9 (a) 1/4 express the *A* and *B* phenotypes; 1/4 express the *A* and *b* phenotypes; 1/4 express the *a* and *B* phenotypes; 1/4 express the *a* and *b* phenotypes.
   (b) The same ratio of phenotypes as in (a).
   (c) If the *A* and *B* alleles are coupled (*AB/ab*), then 1/2 will express *A* and *B* and 1/2 will express *a* and *b*. If they are not coupled (*Ab/aB*), then 1/2 will express *A* and *b* and 1/2 will express *a* and *B*.
   (d) Assuming that *A* and *B* are coupled as shown in (c), the expected phenotypes are 45% *AB*; 5% *Ab*; 45% *ab*; and 5% *aB*.

10 The unknown plant's genotype is *RY/ry*, and its phenotype is colored, green. The two loci are 10 map units apart.

11 Regardless of the percentage of offspring produced as a result of crossover vs. noncrossover gametes, both groups yield progeny in a $2 + : 1\ ca : 1\ e$ ratio. It is impossible to determine from the data the map distance between the e and ca loci.

12 The unknown organism contained the P and Z alleles on one homologue and the p and z alleles on the other. It expressed the P and Z phenotype. The two loci are 14 map units apart.

13 *Female A:* 3 and 4 are NCO; 7 and 8 are SCO between the d and b loci; 1 and 2 are SCO between the b and c loci; 5 and 6 are DCO. These occur in the relative frequency such that 3, 4 > 1, 2 > 7, 8 > 5, 6.

Female B: 7 and 8 are NCO; 5 and 6 are SCO between the d and b loci; 3 and 4 are SCO between the b and c loci; 1 and 2 are DCO. These occur in the relative frequency such that 7, 8 > 3, 4 > 5, 6 > 1, 2

14 (a) $P_1\ \dfrac{sc\ v\ s}{sc\ v\ s} \times \dfrac{+\ +\ +}{\longrightarrow}$.

$F_1\ \dfrac{sc\ \ v\ \ s}{+\ +\ +} \times \dfrac{sc\ v\ s}{\longrightarrow}$.

(b) $sc\overset{33}{\rule{1.2cm}{0.4pt}}v\overset{10}{\rule{1.2cm}{0.4pt}}s$

(c) Fewer; 33 are expected; $c = 0.727$; positive interference.

15 (a) $\dfrac{y\ w\ +}{+\ +\ ct} \times \dfrac{y\ w\ +}{\longrightarrow}$

(b) $\overset{0.0\quad 1.5\qquad\qquad 20.0}{\underset{y\quad\ w\qquad\qquad ct}{\rule{5cm}{0.4pt}}}$

(c) Yes; $(0.015)(0.185)(1000) = 2.775$

(d) No; the phenotype of the females would not always reflect their genotypes.

16 (b) The F_1 heterozygote is $\dfrac{D\ +\ +}{+\ p\ e}$, with *pink* in the middle. The interlocus distances are 3.0 for D to p and 18.5 for p to e.

17 (a) The homozygote was the male parent.

(b) The sequence is h–fz–eg. The distance between h and fz is 14 map units and between fz and eg is 6 map units.

(c) Negative interference; $C = 2.380$; $I = (-1.380)$

18 Since Sb is dominant and lethal when homozygous, *curled* male $\left(\dfrac{+\ cu}{+\ cu}\right)$ would be the suitable choice.

19 (a) $\dfrac{+\ b\ c}{a\ +\ +}$

(b) The order is correct as shown *(a–b–c)*. The distance between a and b is 7 map units and between b and c is 2 map units.

(c) The double-crossover phenotypes $(+\ +\ c$ and $a\ b\ +)$ are missing. They are expected to occur in about 1 of 1000 offspring $(0.07 \times 0.02 = 0.0014)$, and by chance, none were produced.

20 (a) Provided that *short* is linked to *black* on chromosome 2 and that no crossing over occurs in the

heterozygous males, the observed result will occur. The cross is

$$\frac{b\ sh}{+\ +};\frac{p}{+}\ \text{♂} \times \frac{b\ sh}{b\ sh};\frac{p}{p}$$

(b) Since the female is now heterozygous, crossing over will occur, resulting in the additional phenotypes. Each crossover chromatid from chromosome 2 will independently assort with the noncrossover chromatids of chromosome 3. Since 15 percent crossover gametes are recovered, the distance between the b and sh loci is 15 map units.

21 10 map units

22 The a and b genes are not linked on the same chromosome. Additionally, each locus must be very close to the centromere of its respective chromosome, because no exchange has been observed.

23 The genes represented by the a and b alleles are independently assorting (Cross 1) and are therefore on separate chromosomes. Cross 2 shows that the genes represented by b and c are linked. They are 12 map units apart. The genes represented by a and c cannot be linked and will produce approximately equal P and NP tetrads. The percentage of all tetrads showing the P and NP arrangement will depend on the distances between the two genes and their respective centromeres because these distances correlate with the percentage of T tetrads that will be formed.

25 (a) P = 1, 2; NP = 3; T = 4, 5, and 6

(b) Category 2 will arise if a SCO occurs between one of the genes and the centromere.

(c) $\dfrac{6 + 1/2\ (22)}{71} = 0.239 = 24$ map units

26 Because the "twin" spots arise by a single exchange and the "singed" spots arise as the result of a double exchange during mitosis. If *tan* and *forked* were studied, the "tan" spot would be highest, the "twin" spot would be less frequent, and the "forked" spot would be least frequent. The centromere of the *Drosophila* X chromosome is located at about locus 66.

27 No, in both cases.

29
$a\ b\ c$ —168	$a\ b\ +$ —10
$+\ +\ +$ —168	$+\ +\ c$ —10
$a\ +\ +$ — 20	$a\ +\ c$ — 2
$+\ b\ c$ — 20	$+\ b\ +$ — 2 Total = 400

Map distances: a to b = 11 map units; b to c = 6 map units.

30 Because it would be impossible to distinguish which progeny phenotypes represent crossovers and which represent noncrossovers. The heterozygous female should be test-crossed to an abc/abc male.

CHAPTER 7

1 (a) 40 cm

(b) F_1: All $AaBb$, 30 cm

F_2: 1/16 *AABB*, 40 cm; 4/16 $\begin{bmatrix} AABb \\ AaBB \end{bmatrix}$ 35 cm;

4/16 $\begin{bmatrix} aaBb \\ Aabb \end{bmatrix}$, 25 cm; 6/16 $\begin{bmatrix} AAbb \\ AaBb \\ aaBB \end{bmatrix}$, 30 cm;

1/16 *aabb*, 20 cm

2 Only one possible genotype will have only two additive alleles: *AabbCc*. It will occur in 1/8 of all progeny.

3 (a) Polygenic inheritance
   (b) 4 gene pairs
   (c) Each additive allele of the eight possible contributes 3 cm.
   (d) P_1: *AABBccdd* × *aabbCCDD*
       F_1: *AaBbCcDd*
   (e) Any plant with two additive alleles will be 18 cm tall (e.g., *AAbbccdd*, *AaBbcc*, *aaBBccdd*, etc.) Any plant with seven additive alleles will be 33 cm tall. There are only four possibilities: *AABBCCDd*, *AABBCcDD*, *AABbCCDD*, and *AaBBCCDD*.

8 Determination of height has a substantial genetic component since MZ twins, whether reared together or apart, are much more similar than DZ twins and sibs reared together. Determination of weight is under much weaker genetic control than height. Both height and weight are also influenced by the environment (diet).

9 F_1: All wild-type ♂ and ♀
   F_2 8/16 wild-type ♀; 5/16 wild-type ♂; 3/16 vermilion ♂

10 If four gene pairs control height such that each of the eight potential additive alleles adds 2 in. to the base height of 4 in., then the following parental genotypes are consistent with the data shown:
   (a) *aabbccdd* (4″) × *AABBCCDD* (20″)
   (b) *AAbbccdd* (8″) × *AABBCCDD* (20″)
   (c) *AABBccdd* (12″) × *AABBCCDD* (20″)
   (d) *AABBCCdd* (16″) × *AABBCCDD* (20″)
   Other genotypes will fit equally well for the shorter parents in (b), (c), and (d). For example, in (b), *aaBBccdd*, *aabbCCdd*, and *aabbccDD* will all yield offspring that are 14″ tall.

11 (a) $\overline{X} = 140$ cm
   (b) $s^2 = 374.176$
   (c) $s = 19.343$
   (d) $S_{\overline{X}} = 0.6892$

12 (a) Because the F_2 ratios are in sixteenths, 2-locus inheritance is indicated, with 2 alleles at each locus.
   (b) Yes.
   (c) White has lack of color, is homozygous recessive. Therefore, recessive alleles produce no color, dominants each produce an equal amount of color, and dosage of the dominent alleles determines color.
   (d)

1 dark red :	4 medium dark red :	6 medium red :	4 light red :	1 white
AABB	*AaBB*	*AaBb*	*Aabb*	*aabb*
	AABb	*AAbb*	*aaBb*	
		aaBB		

13 (a) Yes. Individuals of moderate height are partially heterozygous. Since dominant alleles contribute more units of height than recessive alleles, offspring can be significantly taller or shorter. For example:

$R/r\ S/s\ t/t\ u/u \times r/r\ s/s\ T/t\ U/u$
↓
$R/r\ S/s\ T/t\ U/u$ tall offspring
or
$r/r\ s/s\ t/t\ u/u$ short offspring

   (b) No. Minimum height is represented by the homozygous recessive parent. Matings will not place any additional dominant alleles in the offspring, so they can never be taller than tall parent.

$r/r\ s/s\ t/t\ u/u \times R/r\ S/s\ T/t\ U/u$
↓
tallest child: $R/r\ S/s\ T/t\ U/u$ same height as tall parent

14 Heritability is estimated as $H^2 = \dfrac{V_G}{V_P}$,
   where $V_P = V_E + V_G$.
   Because the parental strains are true breeding, we can assume that they are genetically homogeneous, and that variance in corolla length is due to environmental factors. The average V_E for the P_1 is therefore 3.50. If we further assume that the F_1 is genetically homogeneous, then the overall V_E is $= \dfrac{3.50 + 4.743}{2} = 4.1215$

   If we now define the phenotypic variance V_P as the variance shown by the F_2 ($V_P = 47.708$), then the genetic variance becomes

$$V_G = V_P - V_E$$
$$V_G = 47.708 - 4.1215$$
$$= 43.5865$$

   In this case then,

$$H^2 = \frac{43.5865}{47.708} = 0.913$$

   About 91% of the variance in corolla length is contributed by genetic factors.

CHAPTER 8

5 How does the transforming DNA enter the cell? Does it enter preferentially at any particular time of the host's cell cycle? In what form and size does it enter? How does it recombine with the host DNA?

7 Technically,^{32}P will label RNA as well as DNA. However, since mature T2 phages lack RNA, DNA must be the genetic material of the virus that enters the bacterium.

CHAPTER 9

3 Guanine: 2-amino, 6-oxy purine
   Cystosine: 2-oxy, 4-amino pyrimidine
   Thymine: 2,4-dioxy, 5-methyl pyrimidine
   Uracil: 2,4-dioxy pyrimidine

7 They might have concluded that adenine:guanine and thymine:cytosine base pairs form between the two strands. Since A and G are purines and T and C are pyrimidines, the diameter between strands would not be constant, as observed by Franklin and Wilkins.

11 The nitrogenous base is responsible for UV absorption.

21 Fraction A contains a mixture of types of DNA sequences. One group, which reanneals more quickly than *E. coli* DNA, is highly repetitive and possibly short. Another group reanneals more slowly and appears to be moderately repetitive. The third group consists of sequences that are single copies and of greater complexity than the *E. coli* single copy sequences. Fraction B consists of all single copy, nonrepetitive DNA sequences of greater complexity than those in either fraction A or *E. coli*.

22 The curve would be similar to that produced by fraction A DNA in Problem 21.

23 (1) Shown.

(2) N at 8 position instead of 9 position in purine ring of adenine.

(3) Thymine lacks a methyl group at its C-5 position.

(4) In the adenylic acid residue, an extra hydroxyl group is present at the C'-2 position of the sugar.

(5) In the thymidylic acid residue, the C'-5 of d-ribose has an extra hydroxyl group.

(6) In the adenylic acid residue, the bond from d-ribose is joined to the N-3 position of the purine instead of the N-9 position.

24 (a) Several models could be proposed. An obvious one is a triple helix with A, T, and H forming hydrogen bonds with one another and C, G, and X forming hydrogen bonds with one another. The backbone of each chain would be similar to the Watson-Crick model except that erythrose would replace ribose.

(b) See answer in (a).

(c) Both must be purines or pyrimidines to maintain a constant diameter. In fact, both are purines.

25 Since T4, MS-2, and *E. coli* do not contain repetitive sequences, a direct proportionality between $C_0t_{1/2}$ and X (complexity) exists. Thus, MS-2 DNA consists of 200 base pairs, and *E. coli* DNA consists of 2×10^6 base pairs.

CHAPTER 10

3 The conservative mode can be ruled out after the first round; the dispersive mode can be ruled out after the second round.

5 Semiconservative replication would be very difficult, if not impossible, for a triple helix. Conservative replication, while difficult, might work better.

7 Vast quantities of *E. coli* were necessary in order to obtain small amounts of the highly purified enzyme because purification methods were not as refined as those subsequently developed.

22 (a) No, nearest neighbor data reflects proportionality, not absolute amount.

(b) Yes, since nearest neighbor frequencies are dependent on a similar frequency of the bases.

(c) Not necessarily. A group of 10 unique sequences, each 20 nucleotides long, could be put together in a number of ways and still yield similar nearest neighbor data.

23

(a)

Base	Nearest Neighbor
G $\longrightarrow$	C, C, T
C $\longrightarrow$	T, A
T $\longrightarrow$	T, A, A, G
A $\longrightarrow$	C, A, G

(b)

Base	Nearest Neighbor
G $\longrightarrow$	C, T, A
C $\longrightarrow$	A, T, G
T $\longrightarrow$	T, A, G
A $\longrightarrow$	C, A, T

24 (a) $T \rightarrow T = A \rightarrow A$

$C \rightarrow C = G \rightarrow G$

(b) If parallel: for $C \rightarrow T =$ then $G \rightarrow A$

for $C \rightarrow A =$ then $G \rightarrow T$

If antiparallel: for $C \rightarrow T = A \rightarrow G$

for $C \rightarrow A = T \rightarrow G$

(c) $G \rightarrow A = T \rightarrow C$

$G \rightarrow T = A \rightarrow C$

$T \rightarrow C = G \rightarrow A$

$T \rightarrow G = C \rightarrow A$

Each expectation is realized, except where $T \rightarrow C$ should equal $G \rightarrow A$.

25 They must be of identical polarity.

CHAPTER 11

11 (a) 5×10^6 nucleosomes

(b) 10^6 solenoids

(c) 4.5×10^7 histone molecules

(d) 6.8×10^6 nm

12 Yes; using the value of π as 3.14 and the formulas πr^2 for the area of a circle and $4/3 \pi r^3$ for the volume of a sphere, the total volume of the chromosome is approximately 1.57×10^8 Å^3, while the viral head has a volume of 2.68×10^8 Å^3.

CHAPTER 12

8 The hybrid was an alloploid with one member of each chromosome pair from each species. Treatment with colchicine to double the chromosome number may create a fertile plant.

12 Haploid, 9; triploid, 27; tetraploid, 36; trisomic, 19; monosomic, 17

16 No; they suppress *the recovery* of crossover products when the crossover occurs within the inverted region.

18 Crossover products between *d* and *e* are the only ones recovered. Genes *a*, *b*, and *c* are part of an inversion on one of the homologues. The distance between *d* and *e* can be mapped. They are 10 map units apart.

23 Since she received her single *X* chromosome from her father, her mother contributed the gamete devoid of sex chromosomes.

24 If the chromosome in question is called *a*, the offspring would appear in the ratio of 1/2 *aa*: 1/4 *aaa*: 1/4 *a*.

26 *P. kewensis* is an allotetraploid that arose from a cross between *P. floribunda* and *P. verticillata*. The initial hybrid was sterile and contained 18 chromosomes which then duplicated in a few reproductive cells, giving rise to fertile offspring containing 36 chromosomes.

27 The varieties are examples of autopolyploidy. Each contains an even ploidy value, e.g., 2, 4, 6, 8, and 10*n*. The variety with 27 chromosomes is triploid and produces unbalanced gametes.

28 The second crossover may negate the possible deleterious effects of the first crossover, which would have led to duplications and deletions as well as dicentric and acentric chromosomes if the inversion were paracentric.

29

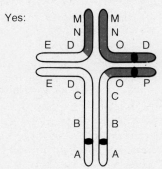

Yes. However, because of the acentric and dicentric chromosomes, gametes with unbalanced genetic information will result following segregation during gamete formation.

30 No, at most, only one-fourth of the gametes will be normal because an acentric and a dicentric chromosome have been created.

CHAPTER 13

9. (a) Sex-linked necessive lethal mutations.
   (b) It is critical to the screening of F_2 cultures lacking male offspring.
   (c) To insure that both the *Bar* and lethal alleles remain together on the same homologue.
   (d) An inversion, which suppresses the recovery of crossover products.

10 The induced mutation rate is probably less than 0.002. However, because of the small sample size, one cannot be certain. The experiment should be repeated and the data pooled.

11 In the F_1, select single *Cy L* males and back-cross them to *Cy L/Pm* females. In the F_2, cross flies that are *Cy L* with one another in single-pair matings. In the F_3, any cultures lacking wild-type flies were derived from parents bearing an induced recessive lethal on chromosome 2. Any induced morphological mutation will be expressed in flies that are not *Cy L*.

15 X-rays are derived from the portion of the electromagnetic spectrum with a shorter and more energetic wavelength.

18 Because of the dangers inherent in X-ray exposure.

24 Unscheduled DNA synthesis represents excision repair. The data are consistent with the hypothesis that the cell lines fall into three complementation groups. In each case, all members of a group fail to complement one another but complement all other cell lines.
Group I: XP1, XP2, XP3; *Group II:* XP4;
Group III: XP5, XP6, XP7
All members of a complementation group are involved in the production of a single, essential gene product.

25 Since each individual is the product of two gametes, the average number of new mutations in each individual is:
$$2(5 \times 10^{-5})(10^5) = 10 \text{ new mutations.}$$
Thus, the current human population contains 43 billion (4.3×10^{10}) newly arisen mutations.

26 The frequency of mutant cells following one round of replication is
$$\frac{18 \times 10^1}{6 \times 10^7} = 3 \times 10^{-6}$$
The rate of mutation per cell replication must take into account the fact that the total number of cells replicating was only one-half the total assayed, since each replicating cell produces two progeny cells. Thus, the mutation rate per cell replication is 1.5×10^{-6}.

CHAPTER 14

11 A circular chromosome of sequence TCHROMBAK can be derived from the data. Note that recombination of various sequences runs in either direction.

12 Because the F factor is always the last part of the chromosome to enter the recipient.

15 The data are consistent with the fact that genes *a* and *b* are not linked (or not close enough to each other to be included in a single transformation event).

16 The *a* and *b* genes are linked. The *c* gene is not linked to either of them. As a result, single transformation events can produce a^+b^+c, a^+bc, ab^+c, and abc^+. The $a^+b^+c^+$, a^+bc^+, and ab^+c^+ genotypes all require two simultaneous events. Consistent with our conclusion, these all occur at much lower frequencies.

20 In this theoretical mapping experiment, abc^+ and a^+b^+c appear as double exchanges in the lowest frequency. Therefore, the c gene is located between the a and b genes. The distance between a and c is calculated as

$$(740 + 670 + 90 + 110)/10{,}000 = 16.1\%$$

The distance between c and b is

$$(160 + 140 + 90 + 110)/10{,}000 = 5\%$$

The percentages represent map units.

Interference is negative. A double exchange produces the abc^+ and a^+b^+c genotypes, which are expected to occur about 80 times in each 10,000 viruses. 200 are observed.

23 Expt. 1: $e \times f \rightarrow +$
Expt. 2: $h \times i \rightarrow -$
Overall, d and f are in the same gene, and e, g, h, and i are located together in a separate gene.

24 (a) Culture 1: total number of bacteria in the irradiated culture;
Culture 2: total number of bacteria with leu^- to leu^+ mutations present in the irradiated culture;
Culture 3: total number of bacteria in the control culture;
Culture 4: total number of bacteria with leu^- to leu^+ mutations present in the control culture.
The values in cultures 1 and 3 should be very similar if they were set up and cultured identically. In culture 1, a density of 24×10^9 cells/ml was observed, while in culture 3, a density of 12×10^9 cells/ml was observed.

(b) Induced rate: $\dfrac{12 \times 10^2}{24 \times 10^9} = 0.5 \times 10^{-7}$

Spontaneous rate: $\dfrac{3 \times 10^1}{12 \times 10^9} = 0.25 \times 10^{-8}$

CHAPTER 15

2 In both cases, only dextrally coiled F_1 snails would result.

3 In the original cross, all progeny cells would be resistant to streptomycin. In the reciprocal cross, one-half of the progeny would be sensitive and one-half resistant, with the phenotypes dependent on the nuclear genotype.

4 (a) Green leaves
(b) White leaves
(c) Green and white leaves
(d) Green leaves
In each case, the phenotype is derived from the ovule that contributes the chloroplast to the progeny plants.

5 *Segregational petites* are inherited as a recessive gene.

6 The diploid zygote would be normal because it carries a normal nuclear allele derived from the *neutral petite* parent. This zygote produces 1/2 normal: 1/2 petite ascospores.

9 Remember that a *Paramecium* may have a K allele but not contain kappa particles. Such a strain would be sensitive. Assuming that cytological exchange always occurs during conjugation, the Mendelian ratios produced may be generated by the following genotypes:

$$KK \times Kk$$
$$KK \times kk$$
$$Kk \times Kk$$

10 (a) The protein synthesis capability in both chloroplasts and mitochondria is likely to be affected. As a result, aerobic respiration and photosynthesis will be inhibited.

(b) If the mutation is in cpDNA, then only the mt^+ strains contribute functional chloroplasts to progeny cells.

11 The maternal parent was Dd and the paternal parent was dd.

12 The egg, but not the sperm, of a fly must impart a factor that allows the development of the male or female gonad of each offspring. A homozygous female can arise only if her mother was heterozygous *(Gs/gs)* and her father contributed a *gs* allele. She is fertile, but none of her eggs contain the normal factor necessary for gonad development. Thus, her offspring are viable but sterile. Homozygous recessive males can have grandchildren.

CHAPTER 16

4 The conversion of substance w to x is controlled by gene product **b**; of substance x to y by gene product **c**; of substance y to z by gene product **a**.

5 In the cross $v ♀ \times cn ♂$,
F_1: 1/2 wild-type ♀: 1/2 bright red ♂
F_2: 5/16 bright red ♀; 3/16 wild-type ♀; 5/16 bright red ♂; 1/16 wild-type ♂
In the cross $cn ♀ \times v ♂$,
F1: All wild-type ♂ and ♀
F2: 6/16 wild-type ♀; 2/16 bright red ♀; 5/16 bright red ♂; 3/16 wild-type ♂

6 (a) Two independently assorting autosomal recessive genes are involved. The P_1 generation is $AAbb \times aaBB$. The F_1 is $AaBb$, and in the F_2, the following genotypic ratios occur:

$$
\begin{array}{ll}
A-B- & 9/16 \\
A-bb & 3/16 \\
a\,a\,B- & 3/16 \\
a\,a\,b\,b & 1/16
\end{array}
$$

Only those with at least one dominant A and one dominant *B* allele are purple (9/16). All other genotypes (7/16) are white.

(b) The same general explanation as in (a) accounts for inheritance, except that

    A — B —   purple
    A — bb   pink
    a a B —   white
    a a b b   white

7   Precursor $\xrightarrow{trp\text{-}8}$ Anthranilic Acid $\xrightarrow{trp\text{-}2}$ IGP $\xrightarrow{trp\text{-}3}$ Indole $\xrightarrow{trp\text{-}1}$ Trp

8   Precursor $\xrightarrow{thi\text{-}2}$ Pyrimidine $\rightharpoondown$
Precursor $\xrightarrow{thi\text{-}1}$ Thiazole $\dashv$ $\xrightarrow{thi\text{-}3}$ Thiamine

14 The amino acid difference in the hemoglobin of the two individuals involves no change in charge. Thus, the net charge is not different in the two molecules. Fingerprinting, which involves chromatography (in addition to electrophoresis) may detect an amino acid difference on the basis of peptide migration in the solvent used.

17 If the mutations are mapped within the gene, and if colinearity holds, each successive mutant should result in a slightly longer polypeptide chain following termination.

CHAPTER 17

1 If only a single central amino acid is considered, then the first and third amino acid can be any of the 20 possibilities. Therefore, 20^2 or 400 different tripeptides can be created. Since any of 20 amino acids can serve as the central one, a total of $20(20^2)$ or 20^3 different tripeptides can be formed using all amino acids.

2 No

3 No. Either a triplet or sextuplet code would allow restoration of the reading frame.

4 (a) GGG = $(3/4)^3$ = 27/64; each 2G:1C (GGC, GCG, and CGG) = $(3/4)^2(1/4)$ = 9/64; each 1G:2C (GGC, CGC, and CCG) = $(3/4)(1/4)^2$ = 3/64; CCC = $(1/4)^3$ = 1/64

  (b) Glycine: GGG and one 2G:1C
Alanine: one 2G:1C and one 1G:2C
Arginine: one 2G:1C and one 1G:2C
Proline: one 1G:2C and CCC

  (c) Glycine: GGG and GGC
Proline: CCC and CCG
Arginine: CGC and CGG, or GCC and GCG
Alanine: CGC and CGG, or GCC and GCG

5 3C: proline
2C:1A: proline, histidine, threonine
1C:2A: threonine, glutamine, asparagine
3A: lysine

7 With a repeating dinucleotide sequence, initiation can occur at one nucleotide or the other on any given synthetic mRNA (e.g., ACACACAC . . . or CACACACAC. . .). In either case, the same polypeptide, repeating two amino acids, will occur. However, with a repeating trinucleotide, initiation can occur at one of three points (e.g., ACTACTACT . . . , CTACTACTA . . . , or TACTAC-TAC . . .). Three different polypeptides, each containing a single but different amino acid, will result.

8 For the repeating UUAC sequence, two of the triplets (UUA and CUU) code for leucine. For the repeating UAUC sequence, all four possible triplets code for a different amino acid.

9 ACA

10 AGA: arginine; GAG and GAA: glutamic acid; AAG: lysine

11 *In vivo*, the enzyme degrades RNA.

15 The most surprising is that CUA, which normally codes for leucine, specifies threonine (normally coded by ACA, ACU, ACC, and ACG). The other changes involve triplets related by two of the three nucleotides.

16 You should find that almost all of them can be related by a single nucleotide change.

17
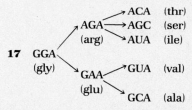

18 Poly U and Poly A are complementary strands, forming a double-helical structure.

19 mRNA must be single-stranded.

20 Sequence 1: met-pro-asp-tyr-ser
Sequence 2: met-pro-asp
Sequence 2 contains a frameshift mutation, creating UAA (a termination codon) as the fourth triplet.

21 For the codon GGA, four amino acid substitutions can result: arginine (AGA and CGA), glutamic acid (GAA), alanine (GCA), and valine (GUA). Additionally, a change to UGA results in a termination codon. For the codon GGU, six amino acid substitutions can occur: serine (AGU), arginine (CGU), cysteine (UGU), aspartic acid (GAU), alanine (GCU), and valine (GUU).

22 (a) Two initiation points (AUG triplets) exist in the mRNA sequence giving rise to:
met-his-thr-tyr-glu-thr-leu-gly . . .
met-arg-pro-leu-gly . . .

  (b) The mutation has produced a termination codon directly after the second initiation triplet. This causes an amino acid substitution in the first polypeptide chain, but immediately terminates the synthesis of the second polypeptide chain.

CHAPTER 18

15 426 nucleotides, including a termination codon. Each triplet must occupy 3(0.34) or about 1 nm; about 20 triplets could potentially occupy a single ribosome.

19 The recognition of the corrected charged tRNA involves the tRNA portion of this molecule and not the amino acid portion.

20 (a)

```
        G — C       C — C
      /       \   /       \
   A — G       G           G
   |   |       |           |
   U — C       C           A
      \       /   \       /
        A — U       U — C
```

(b) TCCGCGGCTGAGATGA

(c) GCU

(d) arg-arg-arg-leu-tyr

21 (a) Sequence 1: GAAAAAACGGUA
Sequence 2: UGUAGUUAUUGA
Sequence 3: AUGUUCCCAAGA

(b) Sequence 1: glu-lys-thr-val
Sequence 2: cys-ser-tyr-nonsense
Sequence 3: met-phe-pro-arg

(c) Initial: sequence 3
Middle: sequence 1
Terminal: sequence 2

22 Sequence 1: GAAAAAACGGTA

23 Sequence 2: ACAUCAAUAACU

CHAPTER 19

1 There are three complementation groups. *A* and *B* are in one; *C* and *E* are in another; *D* is in the third group.

2

Mutants	Cis	Trans
1, 2	+	−
1, 3	+	+
2, 3	+	+

3 Group A: $e \times f \rightarrow$ lysis
Group B: $h \times i \rightarrow$ no lysis
Group C: $k \times l \rightarrow$ no lysis

4 (a) 2 and 3: −
2 and 4: +
3 and 4: +

(b) Since 5 does not complement any of the other mutations, it is either a double mutation in both cistrons or a deletion that overlaps both cistrons.

(c) $2(2.5 \times 10^{-4}) = 5.0 \times 10^{-4}$

(d) It is a deletion in the cistron containing mutants 1 and 4 and overlaps the region containing mutant 4 but not the region containing mutant 1.

6 5′–GCATCTGATGC–3′

10 3^3 or 27 variations can be formed. Thus, 18 are not shown.

11 (10)(30)(50)(3) = 45,000

12 $(5)(10)(20) \times (10)(100) = (1000)(1000)$
$$= (10^3)(10^3)$$
$$= (10^6)$$

CHAPTER 20

1 Restriction endonucleases cleave double-stranded DNA helices at specific nucleotide sequences. Terminal transferase aids in reannealing DNA fragments that are to be cloned with plasmids by extending the length of single-stranded "sticky" tails. A vector is the DNA strand that will accept the fragment that is under study. Calcium promotes bacterial cells to take up the recombined plasmids. Host cells are the bacteria in which the recombined plasmids will undergo replication.

2 Because mRNA to be copied usually has a poly dA tail to pair with the poly dT.

3 (a) Cells that have incorporated a plasmid should be resistant to tetracycline.

(b) Growth on tetracycline but no growth on ampicillin.

(c) Colonies resistant to both antibiotics contain plasmids that religated without incorporating *Drosophila* DNA.

4 $N = \ln (1 - P)/\ln (1 - f)$
$N = \ln (1 - 0.99)/\ln (1 - [5 \times 10^3/1.5 \times 10^8])$
$N = \ln (1 - 0.99)/\ln (1 - 3.3 \times 10^{-5})$
$N = \ln (0.01)/\ln (9.99 \times 10^{-1})$
$N = 1.3955 \times 10^5$

5 (a) Yes. The degree of homology present between *Sau*3A cohesive ends and *Bam*HI ends will allow reannealing to occur, and ligation by DNA ligase.

(b) The *Bam*HI site is often destroyed by insertion of an incorrect base in the unpaired region between the vector and the insert. As a result, the cloned insert cannot be recovered intact.

6 First, a full length cDNA copy of the insulin gene is prepared from the mRNA. This will not contain introns. This cDNA is cloned through the use of linkers into a vector containing a prokaryotic promoter, and transformed into *E. coli* K12.

7 Although the bases are internal to the DNA molecule, the side groups attached to the bases extend into the major groove of the helix and serve as the basis of recognition.

8 The slow band is formed by the religation of two plasmids in opposite orientation.

9 The problem is one of regulating expression of the transferred gene. In the case of the mouse, the rabbit globin gene is expressed in inappropriate tissues, and in improper amounts in the target tissue. Both considerations are of major importance in gene therapy.

10 The size of the genome to be cloned and the cloning capacity of the vector. Generally, it is better to clone large genomes in high capacity vectors so that fewer clones need to be screened to select a gene of interest.

11 There are very few suitable vectors for cloning in plants.

CHAPTER 21

5 The remainder of the z gene and all of the y and a genes will be out of frame, creating missense. Likely, one of the out-of-frame triplets will be a nonsense codon causing premature termination of the garbled polypeptide during translation.

6

Genotype	Constitutive	Repressed	Inducible
$i^+o^+z^+$			×
$i^-o^+z^+$	×		
$i^+o^cz^+$	×		
$i^-o^+z^+/F'i^+$			×
$i^+o^cz^+/F'o^+$	×		
$i^so^+z^+$		×	
$i^so^+z^+/F'i^+$		×	

7

Genotype	Condition	Functional Enzyme Made	Nonfunctional Enzyme Made	No Enzyme Made
$i^+o^+z^+$	No lactose			×
$i^+o^cz^-$	Lactose		×	
$i^-o^+z^-$	No lactose		×	
$i^-o^+z^-$	Lactose		×	
$i^-o^+z^+/F'i^+$	No lactose			×
$i^+o^cz^+/F'o^+$	Lactose	×		
$i^+o^+z^-/F'i^+o^+z^+$	Lactose	×	×	
$i^-o^+z^-/F'i^+o^+z^+$	No lactose			×
$i^so^+z^+/F'o^+$	No lactose			×
$i^+o^cz^-/F'o^+z^+$	Lactose	×	×	

8 The operon is inducible. The data are consistent with a being the operator (a^- is constitutive and not corrected by a^+), b being the promoter (b^- is permanently repressed unless an entire operon is sexduced into the merozygote), c being the structural gene (c^- yields inactive enzymes), and d being the repressor (d^- is constitutive and corrected by d^+).

11 Positive control occurs when genes are normally inactive and an activator stimulates transcription. Negative control occurs when genes are normally transcribed unless repressed by a regulator molecule. Positive control is seen in catabolite repression of the *lac* operon, in the *ara* operon, and in the regulation of galactose metabolism in yeast. Negative control is seen in the induction of *lac* operon, the repression of the *trp* operon, and the repression of the *ara* operon in *E. coli*.

12 Promoters are *cis*-activating elements found immediately upstream of the genes which they control. They are the sites of initial RNA polymerase binding and are found in prokaryotes as well as eukaryotes. Enhancers, unique to eukaryotes, stimulate transcription but may be upstream or downstream of the genes which they regulate.

13 The operon is under positive control. In the presence of tis, the activator produced by the r gene is bound to tis and inhibited from stimulating the operon. The O^- mutations prevent stimulation by the activator, causing repression of the genes.

CHAPTER 22

1 Undifferentiated cells become determined preceding differentiation into specialized form.

3 Six; for A, B, and C polypeptides, AA, BB, CC, AB, AC, and BC

5 Gene products essential for gastrulation are synthesized between hours 6 and 15 during development. While

blastula cells appear to be undifferentiated, biochemical events occurring well in advance of gastrulation are critical to subsequent development.

CHAPTER 23

3 $HGPRT^-$ cells cannot phosphorylate the analogue and thus cannot incorporate it into DNA. This prevents cell death.

4 Aminopterin inhibits the *de novo* pathway for nucleotide synthesis and 8-azaguanine kills all cells that are not $HGPRT^-$.

5 Both the *A* and *B* genes are present and may be transcribed in the hybrid. If both are transcribed equally, all possible tetramers will be assembled.

8 (a) Two complementation groups: ABD and C
   (b) Two complementation groups: AD and BC
   (c) Three complementation groups: A, B, and CD

10 Enzyme I: X chromosome; enzyme II: chromosome 14; enzyme III: X chromosome; enzyme IV: chromosome 22; enzyme V: not on any of these chromosomes

CHAPTER 24

3 Genes on chromosome 2 influence negative geotropism, and genes on chromosome 3 have an impact on positive geotropism.

4 (a) $uR \times UuRR \rightarrow$ 1/2 *uuRR* (uncap): 1/2 *UuRR* (nonhygienic)
   (b) $Ur \times UURr \rightarrow$ 1/2 *UURr* (nonhygienic): 1/2 UUrr (remove if uncapped)
   (c) $uR \times Uurr \rightarrow$ 1/2 *UuRr* (nonhygienic): 1/2 *uuRr* (uncap)

6 If PTC tasting is inherited as a dominant allele such that *TT* or *Tt* individuals show the trait, various genotypic combinations can be devised consistent with the data in this problem.

8 The traits are not under the control of single loci.

CHAPTER 25

1 Allele frequency:
   homozygous recessives $= q^2 = 37/125 = 0.296$
   frequency of recessive allele $= q = 0.544$
   $p = 1 - q$
     $= 1 - 0.544$
     $= 0.456$
   Genotype frequency:
   $p^2 + 2pq + q^2 = 1$
   $TT = p^2 = (0.456)^2 = 0.208 = 20.8\%$
   $Tt = 2pq = 2(0.456 \times 0.544) = 0.496 = 49.6\%$
   $tt = q^2 = (0.544)^2 = 0.296 = 29.6\%$.

To test for equilibrium, a chi-square test must be done comparing the observed frequencies with the frequencies expected for a single locus, two-allele, dominant/recessive system.

2 After one generation, $AA = 0.25$, $Aa = 0.50$, and $aa = 0.25$. In the absence of evolutionary forces, the same genotypic frequencies will occur in subsequent generations.

3 (a) 14.72%   (b) 1.78%   (c) 42%
   (d) 18%   (e) 44.2%

4 The population is in equilibrium.

5 From the information provided, the frequency of *aa* (q^2), and therefore the frequency of *a(q)* can be determined. Then, *p* can be determined and the frequency of *Aa* (*2pq*) can be calculated.

6 (a) The population is in equilibrium.
   (b) The population is not in equilibrium.

7 1/4

8 (a) For $s = 1.0$, $p_1 = 0.77$ and $q_1 = 0.23$
   (b) For $s = 0.5$, $p_1 = 0.733$ and $q_1 = 0.267$.
   (c) For $s = 0.1$, $p_1 = 0.707$ and $q_1 = 0.293$.
   (d) For $s = 0.01$, $p_1 = 0.7006$ and $q_1 = 0.2994$.

9 After 1 generation, $q = 0.330$.
   After 5 generations, $q = 0.143$.
   After 10 generations, $q = 0.083$.
   After 25 generations, $q = 0.037$.
   After 100 generations, $q = 0.010$.
   After 1000 generations, $q = 0.001$.

10 0.9999

11 (a) 0.5   (b) 0.35   (c) 0.11

12 For first cousins, there is a $2(0.02) = 0.04 = 1/25$ probability that one of the two common grandparents is a carrier. If one is a carrier, the offspring of a first-cousin marriage have a 1/16 probability of being homozygous for the disorder. The overall probability is $(1/25)(1/16) = 1/400$. For second-cousin offspring, $p = (1/25)(1/64) = 1/1600$. For the population at large, $p = 0.0001$ or 1/10,000.

CHAPTER 26

4 (a) Those that do not change the code.
   (b) Those that convert a positively charged amino acid to one of the same charge.
   (c) Those that convert a negatively charged amino acid to one of the same charge.
   (d) Those that convert one neutral amino acid to another neutral amino acid.

5 13524635624

7 The minimal mutational distance between humans and each organism shown is horse, 3; pig, 2; dog, 3; chicken, 3; bullfrog, 2; and fungus, 6.

GLOSSARY

A-DNA An alternate form of the right-handed double-helical structure of DNA in which the helix is more tightly coiled, with 11 base pairs per full turn of the helix. In this form, the bases in the helix are displaced laterally and tilted in relation to the longitudinal axis. It is not yet clear whether this form has biological significance.

abortive transduction An event in which transducing DNA fails to be incorporated into the recipient chromosome. (See *transduction*.)

acentric chromosome Chromosome or chromosome fragment with no centromere.

acrocentric chromosome Chromosome with the centromere located very close to one end. Human chromosomes 13, 14, 15, 21, and 22 are acrocentric.

active immunity Immunity gained by direct exposure to antigens followed by antibody production.

active site That portion of a protein, usually an enzyme, whose structural integrity is required for function (e.g., the substrate binding site of an enzyme).

adaptation A heritable component of the phenotype which confers an advantage in survival and reproductive success. The process by which organisms adapt to the current environmental conditions.

additive genes See *polygenic inheritance*.

albinism A condition caused by the lack of melanin production in the iris, hair, and skin. In humans, most often inherited as an autosomal recessive.

aleurone layer In seeds, the outer layer of the endosperm.

alkaptonuria An autosomal recessive condition in humans caused by the lack of an enzyme, homogentisic acid oxidase. Urine of homozygous individuals turns dark upon standing due to oxidation of excreted homogentisic acid. The cartilage of homozygous adults blackens from deposition of a pigment derived from homogentisic acid. Such individuals often develop arthritic conditions.

allele One of the possible mutational states of a gene, distinguished from other alleles by phenotypic effects.

allele frequency Measurement of the proportion of individuals in a population carrying a particular allele.

allelic exclusion In plasma cell heterozygous for an immunoglobulin gene, the selective action of only one allele.

allopatric speciation Process of speciation associated with geographic isolation.

allopolyploid Polyploid condition formed by the union of two or more distinct chromosome sets with a subsequent doubling of chromosome number.

allosteric effect Conformational change in the active site of a protein brought about by interaction with an effector molecule.

allotetraploid Diploid for two genomes derived from different species.

allozyme An allelic form of a protein that can be distinguished from other forms by electrophoresis.

alpha fetoprotein (AFP) A 70-kd glycoprotein synthesized in embryonic development by the yolk sac. High levels of this protein in the amniotic fluid are associated with nerural tube defects such as spina bifida. Lower than normal levels may be associated with Down syndrome.

***Alu* sequence** An interspersed DNA sequence of approximately 300 bp found in the genome of primates that is cleaved by the restriction enzyme *Alu* I. *Alu* sequences are composed of a head to tail dimer, with the first monomer approximately 140 bp and the second approximately 170 bp. In humans, they are dispersed throughout the genome and are present in 300,000 to 600,000 copies, constituting some 3 to 6 percent of the genome. See *SINES*.

amber codon The codon UAG, which does not code for an amino acid but for chain termination.

Ames test A procedure devised by Bruce Ames to test the carcinogenic properties of chemicals by their ability to induce mutations in the bacterium *Salmonella*.

amino acid Any of the subunit building blocks that are covalently linked to form proteins.

aminoacyl tRNA Covalently linked combination of an amino acid and a tRNA molecule.

amniocentesis A procedure used to test for fetal defects in which fluid and fetal cells are withdrawn from the amniotic layer surrounding the fetus.

amphidiploid See *allotetraploid*.

anabolism The metabolic synthesis of complex molecules from less complex precursors.

analogue A chemical compound structurally similar to another, but differing by a single functional group (e.g., 5-bromodeoxyuridine is an analogue of thymidine).

anaphase Stage of cell division in which chromosomes begin moving to opposite poles of the cell.

aneuploidy A condition in which the chromosome number is not an exact multiple of the haploid set.

angstrom Unit of length equal to 10^{-10} meter. Abbreviated Å.

antibody Protein (immunoglobulin) produced in response to an antigenic stimulus with the capacity to bind specifically to the antigen.

anticodon The nucleotide triplet in a tRNA molecule which is complementary to, and binds to, the codon triplet in a mRNA molecule.

antigen A molecule, often a cell surface protein, that is capable of eliciting the formation of antibodies.

antiparallel Describing molecules in parallel alignment, but running in opposite directions. Most commonly used to describe the opposite orientations of the two strands of a DNA molecule.

apoenzyme The protein portion of an enzyme that requires a cofactor or prosthetic group to be functional.

ascospore A meiotic spore produced in certain fungi.

ascus In fungi, the sac enclosing the four or eight ascospores.

asexual reproduction Production of offspring in the absence of any sexual process.

assortative mating Nonrandom mating between males and females of a species. Selection of mates with the same genotype is positive; selection of mates with opposite genotypes is negative.

ATP Adenosine triphosphate.

attached-X chromosome Two conjoined X chromosomes that share a single centromere.

attenuator A nucleotide sequence between the promoter and the structural gene of some operons that can act to regulate the transit of RNA polymerase and thus control transcription of the structural gene.

autogamy A process of self-fertilization resulting in homozygosis.

autoimmune disease The production of antibodies that results from an immune response to one's own molecules, cells, or tissues. Such a response results from the inability of the immune system to distinguish self from nonself. Diseases such as arthritis, scleroderma, systemic lupus erythematosus, and perhaps diabetes are considered to be autoimmune diseases.

autopolyploidy Polyploid condition resulting from the replication of one diploid set of chromosomes.

autoradiography Production of a photographic image by radioactive decay. Used to localize radioactively labeled compounds within cells and tissues.

autosomes Chromosomes other than the sex chromosomes. In humans, there are 22 pairs of autosomes.

auxotroph A mutant microorganism or cell line which requires a substance for growth that can be synthesized by wild-type strains.

B-DNA See *double helix*.

back-cross A cross involving an F_1 heterozygote and one of the P_1 parents (or an organism with a genotype identical to one of the parents).

bacteriophage A virus that infects bacteria (synonym is *phage*).

bacteriostatic A compound that inhibits the growth of bacteria, but does not kill them.

balanced lethals Recessive, nonallelic lethal genes, each carried on different homologous chromosomes. When organisms carrying balanced lethal genes are interbred, only organisms with genotypes identical to the parents (heterozygotes) survive.

balanced polymorphism Genetic polymorphism maintained in a population by natural selection.

Barr body Densely staining nuclear mass seen in the somatic nuclei of mammalian females. Discovered by Murray Barr, this body is thought to represent an inactivated X chromosome.

base analogue See *analogue*.

biotechnology Commercial and/or industrial processes that utilize biological organisms or products.

bivalents Synapsed homologous chromosomes in the first prophase of meiosis.

BrdU (5-bromodeoxyuridine) A mutagenically active analogue of thymidine in which the methyl group at the 5′ position in thymine is replaced by bromine.

buoyant density A property of particles (and molecules) that depends upon their actual density, as determined by partial specific volume and degree of hydration. Provides the basis for density gradient separation of molecules or particles.

CAAT box A highly conserved DNA sequence found about 75 base pairs 5′ to the site of transcription in eukaryotic genes.

canonical sequence See *consensus sequence*.

cAMP Cyclic adenosine monophosphate. An important regulatory molecule in both prokaryotic and eukaryotic organisms.

CAP Catabolite activator protein. A protein that binds cAMP and regulates the activation of inducible operons.

carcinogen A physical or chemical agent that causes cancer.

carrier An individual heterozygous for a recessive trait.

cassette model First proposed to explain mating type interconversion in yeast, this model proposes that both genes for mating types, *a* and *alpha* are present as silent or unexpressed genes in transposable DNA segments (cassettes) that are activated (played) by transposition to the mating type locus.

catabolism A metabolic reaction in which complex molecules are broken down into simpler forms, often accompanied by the release of energy.

catabolite activator protein See *CAP*.

catabolite repression The selective inactivation of an operon by a metabolic product of the enzymes encoded by the operon.

cDNA DNA synthesized from an RNA template by the enzyme reverse transcriptase.

cell cycle Sum of the phases of growth of an individual cell type; divided into G_1 (gap 1), S (DNA synthesis), G_2 (gap 2), and M (mitosis).

cell-free extract A preparation of the soluble fraction of cells, made by lysing cells and removing the particulate matter, such as nuclei, membranes, and organelles. Often used to carry out the synthesis of proteins by the addition of specific, exogenous mRNA molecules.

centimeter A unit of length equal to 10^{-2} meter. Abbreviated cm.

centimorgan A unit of distance between genes on chromosomes. One centimorgan represents a value of 1 percent crossing over between two genes.

central dogma The concept that information flow progresses from DNA to RNA to proteins. Although exceptions are known, this idea is central to an understanding of gene function.

centric fusion See *Robertsonian translocation*.

centriole A cytoplasmic organelle composed of nine groups of microtubules, generally arranged in triplets. Centrioles function in the generation of cilia and flagella and serve as foci for the spindles in cell division.

centromere Specialized region of a chromosome to which the spindle fibers attach during cell division. Location of the centromere determines the shape of the chromosome during the anaphase portion of cell division. Also known as the primary constriction.

centrosome Region of the cytoplasm containing the centriole.

character An observable phenotypic attribute of an organism.

charon phage A group of genetically modified lambda phage designed to be used as vectors for cloning foreign DNA. Named after the ferryman in Greek mythology who carried the souls of the dead across the River Styx.

chemotaxis Negative or positive response to a chemical gradient.

chiasma (pl., **chiasmata**) The crossed strands of nonsister chromatids seen in diplotene of the first meiotic division. Regarded as the cytological evidence for exchange of chromosomal material, or crossing over.

chi-square (χ^2) analysis Statistical test to determine if an observed set of data fits a theoretical expectation.

chloroplast A cytoplasmic self-replicating organelle containing chlorophyll. The site of photosynthesis.

chromatid One of the longitudinal subunits of a replicated chromosome, joined to its sister chromatid at the centromere.

chromatin Term used to describe the complex of DNA, RNA, histones, and nonhistone proteins that make up chromosomes.

chromatography Technique for the separation of a mixture of solubilized molecules by their differential migration over a substrate.

chromocenter An aggregation of centromeres and heterochromatic elements of polytene chromosomes.

chromomere A coiled, beadlike region of a chromosome most easily visible during cell division. The aligned chromomeres of polytene chromosomes are responsible for their distinctive banding pattern.

chromosomal aberration Any change resulting in the duplication, deletion, or rearrangement of chromosomal material.

chromosomal mutation See *chromosomal aberration*.

chromosomal polymorphism Alternate structures or arrangements of a chromosome that are carried by members of a population.

chromosome In prokaryotes, an intact DNA molecule containing the genome; in eukaryotes, a DNA molecule complexed with RNA and proteins to form a threadlike structure containing genetic information arranged in a linear sequence.

chromosome banding Technique for the differential staining of mitotic or meiotic chromosomes to produce a characteristic banding pattern or selective staining of certain chromosomal regions such as centromeres, the nucleolus organizer regions, and GC- or AT-rich regions. Not to be confused with the banding pattern present in unstained polytene chromosomes, which is produced by the alignment of chromomeres.

chromosome map A diagram showing the location of genes on chromosomes.

chromosome puff A localized uncoiling and swelling in a polytene chromosome, usually regarded as a sign of active transcription.

***cis* configuration** The arrangement of two mutant sites within a gene on the same homologue, such as

$$\frac{a^1 \ a^2}{+ \ +}$$

Contrasts with a *trans* arrangement, where the mutant alleles are located on opposite homologues.

***cis* dominance** The ability of a gene to affect the expression of other genes adjacent to it on the chromosome.

***cis-trans* test** A genetic test to determine whether two mutations are located within the same cistron.

cistron That portion of a DNA molecule that codes for a single polypeptide chain; defined by a genetic test as a region within which two mutations cannot complement each other.

cline A gradient of genotype or phenotype distributed over a geographic range.

clonal selection Theory of the immune system that proposes that antibody diversity precedes exposure to the antigen, and that the antigen functions to select the cells containing its specific antibody to undergo proliferation.

clone Genetically identical cells or organisms all derived from a single ancestor by asexual or parasexual methods. For example, a DNA segment that has been enzymatically inserted into a plasmid or chromosome of a phage or a bacterium and replicated to form many copies.

cloned library A collection of cloned DNA molecules representing all or part of an individual's genome.

code See *genetic code*.

codominance Condition in which the phenotypic effects of a gene's alleles are fully and simultaneously expressed in the heterozygote.

codon A triplet of bases in a DNA or RNA molecule which specifies or encodes the information for a single amino acid.

coefficient of coincidence A ratio of the observed number of double-crossovers divided by the expected number of such crossovers.

coefficient of selection A measurement of the reproductive disadvantage of a given genotype in a population. If for genotype *aa*, only 99 of 100 individuals reproduce, then the selection coefficient (*s*) is 0.1.

colchicine An alkaloid compound that inhibits spindle formation during cell division. Used in the preparation of karyotypes to collect a large population of cells inhibited at the metaphase stage of mitosis.

colicin A bacteriocidal protein produced by certain strains of *E. coli* and other closely related bacterial species.

colinearity The linear relationship between the nucleotide sequence in a gene (or the RNA transcribed from it) and the order of amino acids in the polypeptide chain specified by the gene.

competence In bacteria, the transient state or condition during which the cell can bind and internalize exogenous DNA molecules, making transformation possible.

complementarity Chemical affinity between nitrogenous bases as a result of hydrogen bonding. Responsible for the base pairing between the strands of the DNA double helix.

complementation test A genetic test to determine whether two mutations occur within the same gene. If two mutations are introduced into a cell simultaneously and produce a wild-type phenotype (i.e., they complement each other), they are often nonallelic. If a mutant phenotype is produced, the mutations are noncomplementing and are often allelic.

complete linkage A condition in which two genes are located so close to each other that no recombination occurs between them.

complexity The total number of nucleotides or nucleotide pairs in a population of nucleic acid molecules as determined by reassociation kinetics.

complex locus A gene within which a set of functionally related pseudoalleles can be identified by recombinational analysis (e.g., the *bithorax* locus in *Drosophila*).

concatemer A chain or linear series of subunits linked together. The process of forming a concatemer is called concatenation (e.g., multiple units of a phage genome produced during replication).

concordance Pairs or groups of individuals identical in their phenotype. In twin studies, a condition in which both twins exhibit or fail to exhibit a trait under investigation.

conditional mutation A mutation that expresses a wild-type phenotype under certain (permissive) conditions and a mutant phenotype under other (restrictive) conditions.

conjugation Temporary fusion of two single-celled organisms for the sexual exchange of genetic material.

consanguine Related by a common ancestor within the previous few generations.

consensus sequence A nucleotide sequence most often found in a defined segment of DNA.

cosmid A vector designed to allow cloning of large segments of foreign DNA. Cosmids are hybrids composed of the cos sites of lambda inserted into a plasmid. In cloning, the recombinant DNA molecules are packaged into phage protein coats, and after infection of bacterial cells, the recombinant molecule replicates and can be maintained as a plasmid.

coupling conformation See *cis configuration*.

covalent bond A nonionic chemical bond formed by the sharing of electrons.

cri-du-chat syndrome A clinical syndrome in humans produced by a deletion of a portion of the short arm of chromosome 5. Afflicted infants have a distinctive cry which sounds like that of a cat.

crossing over The exchange of chromosomal material (parts of chromosomal arms) between homologous chromosomes by breakage and reunion. The exchange of material between nonsister chromatid during meiosis is the basis of genetic recombination.

cross-reacting material (CRM) Nonfunctional form of an enzyme, produced by a mutant gene, which is recognized by antibodies made against the normal enzyme.

C-terminal amino acid The terminal amino acid in a peptide chain which carries a free carboxyl group.

cytogenetics A branch of biology in which the techniques of both cytology and genetics are used to study heredity.

cytokinesis The division or separation of the cytoplasm during mitosis or meiosis.

cytological map A diagram showing the location of genes at particular chromosomal sites.

cytoplasmic inheritance Non-Mendelian form of inheritance involving genetic information transmitted by self-replicating cytoplasmic organelles such as mitochondria, chloroplasts, etc.

dalton A unit of mass equal to that of the hydrogen atom, which is 1.67×10^{-24} gram. A unit used in designating molecular weights.

Darwinian fitness See *fitness*.

deficiency (deletion) A chromosomal mutation involving the loss or deletion of chromosomal material.

degenerate code Term used to describe the genetic code, in which a given amino acid may be represented by more than one codon.

deletion See *deficiency*.

deme A local interbreeding population.

denatured DNA DNA molecules that have been separated into single strands.

de novo Newly arising; synthesized from less complex precursors rather than having been produced by modification of an existing molecule.

density gradient centrifugation A method of separating macromolecular mixtures by the use of centrifugal force and solvents of varying density. In sedimentation velocity centrifugation, macromolecules are separated by the velocity of sedimentation through a preformed gradient such as sucrose. In density gradient equilibrium centrifugation, macromolecules in a cesium salt solution are centrifuged until the cesium solution establishes a gradient under the influence of the centrifugal field, and the macromolecules sediment until the density of the solvent equals their own.

dermatoglyphics The study of the surface ridges of the skin, especially of the hands and feet.

deoxyribonucleic acid (DNA) A macromolecule usually consisting of antiparallel polynucleotide chains held together by hydrogen bonds, in which the sugar residues are deoxyribose. The primary carrier of genetic information.

determination A regulatory event that establishes a specific pattern of gene activity and developmental fate for a given cell.

diakinesis The final stage of meiotic prophase I in which the chromosomes become tightly coiled and compacted and move toward the periphery of the nucleus.

dicentric chromosome A chromosome having two centromeres.

differentiation The process of complex changes by which cells and tissues attain their adult structure and functional capacity.

dihybrid cross A genetic cross involving two characters in which the parents possess different forms of each character (e.g., tall, round × short, wrinkled peas).

diploid A condition in which each chromosome exists in pairs; having two of each chromosome.

diplotene A stage of meiotic prophase I immediately after pachytene. In diplotene, one pair of sister chromatids begins separating from the other, and chiasmata become visible. These overlaps move laterally toward the ends of the chromatids (terminalization).

directional selection A selective force that changes the frequency of an allele in a given direction, either toward fixation or toward elimination.

discontinuous variation Phenotypic data that fall into two or more distinct classes that do not overlap.

discordance In twin studies, a situation where one twin shows a trait but the other does not.

disjunction The separation of chromosomes at the anaphase stage of cell division.

disruptive selection Simultaneous selection for phenotypic extremes in a population, usually resulting in the production of two discontinuous strains.

dizygotic twins Twins produced from separate fertilization events; two ova fertilized independently. Also known as fraternal twins.

DNA See *deoxyribonucleic acid.*

DNA gyrase One of the DNA topoisomerases that functions during DNA replication to reduce molecular tension caused by supercoiling. DNA gyrase produces, then seals, double-stranded breaks.

DNA ligase An enzyme that forms a covalent bond between the 5′ end of one polynucleotide chain and the 3′ end of another polynucleotide chain. Also called polynucleotide-joining enzyme.

DNA polymerase An enzyme that catalyzes the synthesis of DNA from deoxyribonucleotides and a template DNA molecule.

DNase Deoxyribonucleosidase, an enzyme that degrades or breaks down DNA into fragments or constitutive nucleotides.

dominance The expression of a trait in the heterozygous condition.

dosage compensation A genetic mechanism that regulates the levels of gene products at certain autosomal loci; this results in homozygous dominants and heterozygotes having the same amount of a gene product. In mammals, random inactivation of one X chromosome in females leads to equal levels of X chromosome-coded gene products in males and females.

double-crossover Two separate events of chromosome breakage and exchange occurring within the same tetrad.

double helix The model for DNA structure proposed by James Watson and Francis Crick, involving two antiparallel, hydrogen-bonded polynucleotide chains wound into a right-handed helical configuration, with 10 base pairs per full turn of the double helix. Often called B-DNA.

duplication A chromosomal aberration in which a segment of the chromosome is repeated.

dyad The products of tetrad separation or disjunction at the first meiotic prophase. Consists of two sister chromatids joined at the centromere.

effector molecule Small, biologically active molecule that acts to regulate the activity of a protein by binding to a specific receptor site on the protein.

electrophoresis A technique used to separate a mixture of molecules by their differential migration through a stationary phase in an electrical field.

endogenote The segment of the chromosome in a partially diploid bacterial cell (merozygote) which is homologous to the chromosome transmitted by the donor cell.

endomitosis Chromosomal replication that is not accompanied by either nuclear or cytoplasmic division.

endonuclease An enzyme that hydrolyzes internal phosphodiester bonds in a polynucleotide chain or nucleic acid molecule.

endoplasmic reticulum A membraneous organelle system in the cytoplasm of eukaryotic cells. The outer surface of the membranes may be ribosome-studded (rough ER) or smooth ER.

endopolyploidy The increase in chromosome sets that results from endomitotic replication within somatic nuclei.

endosymbiont theory The proposal that self-replicating cellular organelles such as mitochondria and chloroplasts were originally free-living organisms that entered into a symbiotic relationship with nucleated cells.

enhancer Originally, a 72-bp sequence in the genome of the virus, SV40, that increases the transcriptional activity of nearby structural genes. Similar sequences that enhance transcription have been identified in the genomes of eukaryotic cells. Enhancers can act over a distance of thousands of base pairs and can be located 5′ or 3′ to the gene they affect, and thus are different from promoters.

environment The complex of geographic, climatic, and biotic factors within which an organism lives.

enzyme A protein or complex of proteins that catalyzes a specific biochemical reaction.

episome A circular genetic element in bacterial cells that can replicate independently of the bacterial chromosome or integrate and replicate as part of the chromosome.

epitope That portion of a macromolecule or cell that acts to elicit an antibody response; an antigenic determinant. A complex molecule or cell can contain several such sites.

epistasis Nonreciprocal interaction between genes such that one gene interferes with or prevents the expression of another gene. For example, in *Drosophila*, the recessive gene *eyeless*, when homozygous, prevents the expression of eye color genes.

equatorial plate See *metaphase plate*.

euchromatin Chromatin or chromosomal regions that are lightly staining and are relatively uncoiled during the interphase portion of the cell cycle. The region of the chromosomes thought to contain most of the structural genes.

eukaryotes Those organisms having true nuclei and membranous organelles and whose cells demonstrate mitosis and meiosis.

euploid Polyploid with a chromosome number that is an exact multiple of a basic chromosome set.

evolution The origin of plants and animals from preexisting types. Descent with modifications.

excision repair Repair of DNA lesions by removal of a polynucleotide segment and its replacement with a newly synthesized, corrected segment.

exogenote In merozygotes, the segment of the bacterial chromosome contributed by the donor cell.

exon (extron) The DNA segment(s) of a gene that are transcribed and translated into protein.

exonuclease An enzyme that breaks down nucleic acid molecules by breaking the phosphodiester bonds at the $3'$ or $5'$ terminal nucleotides.

expression vector Plasmids or phage carrying promoter regions designed to cause expression of cloned DNA sequences.

expressivity The degree or range in which a phenotype for a given trait is expressed.

extranuclear inheritance Transmission of traits by genetic information contained in cytoplasmic organelles such as mitochondria and chloroplasts.

F^+ **cell** A bacterial cell having a fertility (F) factor. Acts as a donor in bacterial conjugation.

F^- **cell** A bacterial cell that does not contain a fertility (F) factor. Acts as a recipient in bacterial conjugation.

F factor An episome in bacterial cells that confers the ability to act as a donor in conjugation.

F' **factor** A fertility (F) factor that contains a portion of the bacterial chromosome.

F_1 **generation** First filial generation; the progeny resulting from the first cross in a series.

F_2 **generation** Second filial generation; the progeny resulting from a cross of the F_1 generation.

facultative heterochromatin Chromatin that may alternate in form between euchromatic and heterochromatic. The Y chromosome of many species contains facultative heterochromatin.

familial trait A trait transmitted through and expressed by members of a family.

fate map A diagram or "map" of an embryo showing the location of cells whose developmental fate is known.

fertility (F) factor See *F factor*.

filial generations See F_1, F_2 *generations*.

fingerprint The pattern of ridges and whorls on the tip of a finger. The pattern obtained by enzymatically cleaving a protein or nucleic acid and subjecting the digest to two-dimensional chromatography or electrophoresis.

fitness A measure of the relative survival and reproductive success of a given individual or genotype.

fixation In population genetics, a condition in which all members of a population are homozygous for a given allele.

fluctuation test A statistical test developed by Salvadore Luria and Max Delbruck to determine whether bacterial mutations arise spontaneously or are produced in response to selective agents.

fMet See *formylmethionine*.

formylmethionine (fMet) A molecule derived from the amino acid methionine by attachment of a formyl group to its terminal amino group. This is the first amino acid inserted in all bacterial polypeptides. Also known as N-formyl methionine.

founder effect A form of genetic drift. The establishment of a population by a small number of individuals whose genotypes carry only a fraction of the different kinds of alleles in the parental population.

fragile site A heritable gap or nonstaining region of a chromosome that can be induced to generate chromosome breaks.

frameshift mutation A mutational event leading to the insertion of one or more base pairs in a gene, shifting the codon reading frame in all codons following the mutational site.

fraternal twins See *dizygotic twins*.

gamete A specialized reproductive cell with a haploid number of chromosomes.

gene The fundamental physical unit of heredity whose existence can be confirmed by allelic variants and which occupies a specific chromosomal locus. A DNA sequence coding for a single polypeptide.

gene amplification The process by which gene sequences are selected for differential replication either extrachromosomally or intrachromosomally.

gene duplication An event in replication leading to the production of a tandem repeat of a gene sequence.

gene flow The gradual exchange of genes between two populations, brought about by the dispersal of gametes or the migration of individuals.

gene frequency The percentage of alleles of a given type in a population.

gene interaction Production of novel phenotypes by the interaction of alleles of different genes.

gene mutation See *point mutation*.

gene pool The total of all genes possessed by reproductive members of a population.

generalized transduction The transduction of any gene in the bacterial genome by a phage.

genetic burden Average number of recessive lethal genes carried in the heterozygous condition by an individual in a population. Also called genetic load.

genetic code The nucleotide triplets that code for the 20 amino acids or for chain initiation or termination.

genetic counseling Analysis of risk for genetic defects in a family and the presentation of options available to avoid or ameliorate possible risks.

genetic drift Random variation in gene frequency from generation to generation. Most often observed in small populations.

genetic engineering The technique of altering the genetic constitution of cells or individuals by the selective removal, insertion, or modification of individual genes or gene sets.

genetic equilibrium Maintenance of allele frequencies at the same value in successive generations. A condition in which allele frequencies are neither increasing nor decreasing.

genetic fine structure Intragenic recombinational analysis that provides mapping information at the level of individual nucleotides.

genetic load See *genetic burden.*

genetic polymorphism The stable coexistence of two or more discontinuous genotypes in a population. When the frequencies of two alleles are carried to an equilibrium, the condition is called balanced polymorphism.

genetics The branch of biology that deals with heredity and the expression of inherited traits.

genome The array of genes carried by an individual.

genotype The specific allelic or genetic constitution of an organism; often, the allelic composition of one or a limited number of genes under investigation.

graft versus host disease (GVHD) In transplants, reaction by immunologically competent cells of the donor against the antigens present on the cells of the host. In human bone marrow transplants, often a fatal condition.

gynandromorph An individual composed of cells with both male and female genotypes.

gyrase One of a class of enzymes known as topoisomerases. Gyrase converts closed circular DNA to a negatively supercoiled form prior to replication, transcription, or recombination.

haploid A cell or organism having a single set of unpaired chromosomes. The gametic chromosome number.

haplotype The set of alleles from closely linked loci carried by an individual and usually inherited as a unit.

Hardy-Weinberg law The principle that both gene and genotype frequencies will remain in equilibrium in an infinitely large population in the absence of mutation, migration, selection, and nonrandom mating.

heat shock A transient response following exposure of cells or organisms to elevated temperatures. The response involves activation of a small number of loci, inactivation of previously active loci, and selective translation of heat shock mRNA. Appears to be a nearly universal phenomenon observed in organisms ranging from bacteria to humans.

helicase An enzyme that participates in DNA replication by unwinding the double helix near the replication fork.

hemizygous Conditions where a gene is present in a single dose. Usually applied to genes on the X chromosome in heterogametic males.

hemoglobin (Hb) An iron-containing, conjugated respiratory protein occurring chiefly in the red blood cells of vertebrates.

hemophilia A sex-linked trait in humans associated with defective blood-clotting mechanisms.

heredity Transmission of traits from one generation to another.

heritability A measure of the degree to which observed phenotypic differences for a trait are genetic.

heterochromatin The heavily staining, late replicating regions of chromosomes that are condensed in interphase. Thought to be devoid of structural genes.

heteroduplex A double-stranded nucleic acid molecule in which each polynucleotide chain has a different origin. These structures may be produced as intermediates in a recombinational event, or by the *in vitro* reannealing of single-stranded, complementary molecules.

heterogametic sex The sex that produces gametes containing unlike sex chromosomes.

heterogenote A bacterial merozygote in which the donor (exogenote) chromosome segment carries different alleles than does the chromosome of the recipient (endogenote). A heterozygous merozygote.

heterogeneous nuclear RNA (hnRNA) The collection of RNA transcripts in the nucleus, representing precursors and processing intermediates to rRNA, mRNA, and tRNA. Also represents RNA transcripts that will not be transported to the cytoplasm, such as snRNA (small nuclear RNAs).

heterokaryon A somatic cell containing nuclei from two different sources.

heterosis The superiority of a heterozygote over either homozygote for a given trait.

heterozygote An individual with different alleles at one or more loci. Such individuals will produce unlike gametes and therefore will not breed true.

Hfr A strain of bacteria exhibiting a high frequency of recombination. These stains have a chromosomally integrated F factor that is able to mobilize and transfer all or part of the chromosome to a recipient F⁻ cell.

histocompatibility antigens See *HLA.*

histones Proteins complexed with DNA in the nucleus. They are rich in the basic amino acids arginine and lysine and function in the coiling of DNA to form nucleosomes.

HLA Cell surface proteins, produced by histocompatibility loci, which are involved in the acceptance or rejection of tissue and organ grafts and transplants.

hnRNA See *heterogeneous nuclear RNA.*

Hogness box A short nucleotide sequence 20 to 30 bp 5′ to the initiation site of eukaryotic genes to which RNA polymerase II binds. The consensus sequence is TATAAAA. Also known as a TATA box or Goldberg-Hogness box.

holandric A trait transmitted from males to males. In humans, genes on the Y chromosome are holandric.

homoeotic mutation A mutation that causes a tissue normally determined to form a specific organ or body part to alter its differentiation and form another structure. Alternately spelled: homeotic.

homogametic sex The sex that produces gametes that do not differ with respect to sex chromosome content; in mammals, the female is homogametic.

homogeneously staining regions (hsr) Segments of mammalian chromosomes that stain lightly with Giemsa following exposure of cells to a selective agent. These regions arise in conjunction with gene amplification and are regarded as the structural locus for the amplified gene.

homogenote A bacterial merozygote in which the donor (exogenote) chromosome carries the same alleles as the chromosome of the recipient (endogenote). A homozygous merozygote.

homologous chromosomes Chromosomes that synapse or pair during meiosis. Chromosomes that are identical with respect to their genetic loci and centromere placement.

homozygote An individual with identical alleles at one or more loci. Such individuals will produce identical gametes and will therefore breed true.

hybrid An individual produced by crossing two parents of different genotypes.

hybridoma A somatic cell hybrid produced by the fusion of an antibody-producing cell and a cancer cell, specifically, a myeloma. The cancer cell contributes the ability to divide indefinitely, and the antibody cell confers the ability to synthesize large amounts of a single antibody.

hybrid vigor See *heterosis*.

hydrogen bond An electrostatic attraction between a hydrogen atom bonded to a strongly electronegative atom such as oxygen or nitrogen and another atom that is electronegative or contains an unshared electron pair.

identical twins See *monozygotic twins*.

Ig See *immunoglobulin*.

imaginal disk Discrete groups of cells set aside during embryogenesis in holometabolous insects which are determined to form the external body parts of the adult.

immunoglobulin The class of serum proteins having the properties of antibodies.

inborn error of metabolism A biochemical disorder that is genetically controlled; usually an enzyme defect that produces a clinical syndrome.

inbreeding Mating between closely related organisms.

inbreeding depression A loss or reduction in fitness that usually accompanies inbreeding of organisms.

incomplete dominance Expression of heterozygous phenotype which is distinct from, and often intermediate to, that of either parent.

incomplete linkage Occasional separation of two genes on the same chromosome by a recombinational event.

independent assortment The independent behavior of each pair of homologous chromosomes during their segregation in meiosis I. The random distribution of genes on different chromosomes into gametes.

inducer An effector molecule that activates transcription.

inducible enzyme system An enzyme system under the control of a regulatory molecule, or inducer, which acts to block a repressor and allow transcription.

insertion sequence See *IS element*.

in situ hybridization A technique for the cytological localization of DNA sequences complementary to a given nucleic acid or polynucleotide.

intercalating agent A compound that inserts between bases in a DNA molecule, disrupting the alignment and pairing of bases in the complementary strands (e.g., acridine dyes).

interference A measure of the degree to which one crossover affects the incidence of another crossover in an adjacent region of the same chromatid. Positive interference increases the chances of another crossover; negative interference reduces the probability of a second crossover event.

interferon One of a family of proteins that act to inhibit viral replication in higher organisms. Some interferons may have anticancer properties.

interphase That portion of the cell cycle between divisions.

intervening sequence See *intron*.

intron A portion of DNA between coding regions in a gene which is transcribed, but which does not appear in the mRNA product.

inversion A chromosomal aberration in which the order of a chromosomal segment has been reversed.

inversion loop The chromosomal configuration resulting from the synapsis of homologous chromosomes, one of which carries an inversion.

in vitro Literally, in glass; outside the living organism; occurring in an artificial environment.

in vivo Literally, in the living; occurring within the living body of an organism.

IS element A mobile DNA segment that is transposable to any of a number of sites in the genome.

isoagglutinogen An antigenic factor or substance present on the surface of cells that is capable of inducing the formation of an antibody.

isochromosome An aberrant chromosome with two identical arms and homologous loci.

isolating mechanism Any barrier to the exchange of genes between different populations of a group of organisms. In general, isolation can be classified as spatial, environmental, or reproductive.

isotopes Forms of a chemical element that have the same number of protons and electrons, but differ in the number of neutrons contained in the atomic nucleus. Unstable isotopes undergo a transition to a more stable form with the release of radioactivity.

isozyme Any of two or more distinct forms of an enzyme

that have identical or nearly identical chemical properties, but differ in some property such as net electrical charge, pH optima, number and type of subunits, or substrate concentration.

kappa particles DNA-containing cytoplasmic particles found in certain strains of *Paramecium aurelia*. When these self-reproducing particles are transferred into the growth medium, they release a toxin, paramecin, which kills other sensitive strains. A nuclear gene, *K*, is responsible for the maintenance of kappa particles in the cytoplasm.

karyokinesis The process of nuclear division.

karyotype The chromosome complement of a cell or an individual. Often used to refer to the arrangement of metaphase chromosomes in a sequence according to length and position of the centromere.

kilobase A unit of length consisting of 1000 nucleotides. Abbreviated kb.

Klenow fragment A part of bacterial DNA polymerase that lacks exonuclease activity, but retains polymerase activity. It is produced by enzymatic digestion of the intact enzyme.

Klinefelter syndrome A genetic disease in human males caused by the presence of an extra X chromosome. Klinefelter males are XXY instead of XY. This syndrome is associated with enlarged breasts, small testes, sterility, and, occasionally, mental retardation.

lagging strand In DNA replication, the strand synthesized in a discontinuous fashion, 5′ to 3′ away from the replication fork. Each short piece of DNA synthesized in this fashion is called an Okazaki fragment.

lampbrush chromosomes Meiotic chromosomes characterized by extended lateral loops, which reach maximum extension during diplotene. Although most intensively studied in amphibians, these structures occur in meiotic cells of organisms ranging from insects through humans.

leader sequence That portion of an mRNA molecule from the 5′ end to the beginning codon; may contain regulatory or ribosome binding sites

leading strand During DNA replication, the strand synthesized continuously 5′ to 3′ toward the replication fork.

lectins Carbohydrate-binding proteins found in the seeds of leguminous plants such as soybeans. Although their physiological role in plants is unclear, they are useful as probes for cell surface carbohydrates and glycoproteins in animal cells. Proteins with similar properties have also been isolated from organisms such as snails and eels.

leptotene The initial stage of meiotic prophase I, during which the chromosomes become visible and are often arranged in a bouquet configuration, with one or both ends of the chromosomes gathered at one spot on the inner nuclear membrane.

lethal gene A gene whose expression results in death.

LINES Long interspersed repetitive sequences found in the genomes of higher organisms, such as the 6-kb KpnI sequences found in primate genomes.

linkage Condition in which two or more nonallelic genes tend to be inherited together. Linked genes have their loci along the same chromosome, do not assort independently, but can be separated by crossing over.

linkage group A group of genes that have their loci on the same chromosome.

locus The site or place on a chromosome where a particular gene is located.

long period interspersion Pattern of genome organization in which long stretches of single copy DNA are interspersed with long segments of repetitive DNA. This pattern of genome organization is found in *Drosophila* and the honeybee.

long terminal repeat (LTR) Sequence of several hundred base pairs found at the ends of retroviral DNAs.

Lutheran blood group One of a number of blood group systems inherited independently of the ABO, MN, and Rh systems. Alleles of this group determine the presence or absence of antigens on the surface of red blood cells. Gene is on human chromosome 19.

Lyon hypothesis The idea proposed by Mary Lyon that random inactivation of one X chromosome in the somatic cells of mammalian females is responsible for dosage compensation and mosaicism.

lysis The disintegration of a cell brought about by the rupture of its membrane.

lysogenic bacterium A bacterial cell carrying a temperate bacteriophage integrated into its chromosome.

lysogeny The process by which the DNA of an infecting phage becomes repressed and integrated into the chromosome of the bacterial cell it infects.

lytic phase The condition in which a temperate bacteriophage loses its integrated status in the host chromosome (becomes induced), replicates, and lyses the bacterial cell.

major histocompatibility loci See *MHC*.

map unit A measure of the genetic distance between two genes, corresponding to a recombination frequency of 1 percent. See *centimorgan*.

maternal effect Phenotypic effects on the offspring produced by the maternal genome. Factors transmitted through the egg cytoplasm which produce a phenotypic effect in the progeny.

maternal inheritance A type of inheritance usually controlled by cytoplasmic genetic factors (such as mitochondria) and transmitted through the female parent to all offspring.

mean The arithmetic average.

median The value in a group of numbers below and above which there is an equal number of data points or measurements.

meiosis The process in gametogenesis or sporogenesis during which one replication of the chromosomes is followed by two nuclear divisions to produce four haploid cells.

melting profile See T_m.

merozygote A partially diploid bacterial cell containing, in addition to its own chromosome, a chromosome fragment introduced into the cell by transformation, transduction, or conjugation.

messenger RNA See *mRNA*.

metabolism The sum of chemical changes in living organisms by which energy is generated and used.

metacentric chromosome A chromosome with a centrally located centromere, producing chromosome arms of equal lengths.

metafemale In *Drosophila*, a poorly developed female of low viability in which the ratio of X chromosomes to sets of autosomes exceeds 1.0. Previously called a superfemale.

metamale In *Drosophila*, a poorly developed male of low viability in which the ratio of X chromosomes to sets of autosomes is less than 0.5. Previously called a supermale.

metaphase The stage of cell division in which the condensed chromosomes lie in a central plane between the two poles of the cell, and in which the chromosomes become attached to the spindle fibers.

metaphase plate The arrangement of mitotic or meiotic chromosomes at the equator of the cell during metaphase.

MHC Major histocompatibility loci. In humans, the HLA complex; and in mice, the H2 complex.

micrometer A unit of length equal to 1×10^{-6} meter. Previously called a micron. Abbreviated μm.

micron See *micrometer*.

migration coefficient An expression of the proportion of migrant genes entering the population per generation.

millimeter A unit of length equal to 1×10^{-3} meter. Abbreviated mm.

minimal medium A medium containing only those nutrients that will support the growth and reproduction of wild-type strains of an organism.

missense mutation A mutation that alters a codon to that of another amino acid, causing an altered translation product to be made.

mitochondrion Found in the cells of eukaryotes, a cytoplasmic, self-reproducing organelle that is the site of ATP synthesis.

mitogen A substance that stimulates mitosis in nondividing cells; e.g., phytohemagglutinin.

mitosis A form of cell division resulting in the production of two cells, each with the same chromosome and genetic complement as the parent cell.

mode In a set of data, the value occurring in the greatest frequency.

monohybrid cross A genetic cross between two individuals involving only one character (e.g., $AA \times aa$).

monosomic An aneuploid condition in which one member of a chromosome pair is missing; having a chromosome number of $2n - 1$.

monozygotic twins Twins produced from a single fertilization event; the first division of the zygote produces two cells, each of which develops into an embryo. Also known as identical twins.

mRNA An RNA molecule transcribed from DNA and translated into the amino acid sequence of a polypeptide.

mtDNA Mitochondrial DNA.

multiple alleles Three or more alleles of the same gene.

multiple-factor inheritance See *polygenic inheritance*.

multiple infection Simultaneous infection of a bacterial cell by more than one bacteriophage, often of different genotypes.

mu phage A phage group in which the genetic material behaves like an insertion sequence, capable of insertion, excision, transposition, inactivation of host genes and induction of chromosomal rearrangements.

mutagen Any agent that causes an increase in the rate of mutation.

mutant A cell or organism carrying an altered or mutant gene.

mutation The process which produces an alteration in DNA or chromosome structure; the source of most alleles.

mutation rate The frequency with which mutations take place at a given locus or in a population.

muton The smallest unit of mutation in a gene, corresponding to a single base change.

nanometer A unit of length equal to 1×10^{-9} meter. Abbreviated nm.

natural selection Differential reproduction of some members of a species resulting from variable fitness conferred by genotypic differences.

nearest-neighbor analysis A molecular technique used to determine the frequency with which nucleotides are adjacent to each other in polynucleotide chains.

neutral mutation A mutation with no immediate adaptive significance or phenotypic effect.

noncrossover gamete A gamete which contains no chromosomes that have undergone genetic recombination.

nondisjunction An accident of cell division in which the homologous chromosomes (in meiosis) or the sister chromatids (in mitosis) fail to separate and migrate to opposite poles; responsible for defects such as monosomy and trisomy.

nonsense mutation A mutation that alters a codon to one which encodes no amino acid; i.e., UAG (amber codon), UAA (ochre codon), or UGA (opal codon). Leads to premature termination during the translation of mRNA.

NOR See *nucleolus organizer region*.

normal distribution A probability function that approximates the distribution of random variables. The normal curve, also known as a Gaussian or bell-shaped curve, is the graphic display of the normal distribution.

np See *nucleotide pair*.

N-terminal amino acid The terminal amino acid in a peptide chain that carries a free amino group.

nu body See *nucleosome*.

nuclease An enzyme that breaks bonds in nucleic acid molecules.

nucleoid The DNA-containing region within the cytoplasm in prokaryotic cells.

nucleolus organizer region (NOR) A chromosomal region

containing the genes for rRNA; most often found in physical association with the nucleolus.

nucleolus A nuclear organelle that is the site of ribosome biosynthesis; usually associated with or formed in association with the NOR.

nucleoside A purine or pyrimidine base covalently linked to a ribose or deoxyribose sugar molecule.

nucleosome A complex of four histone molecules, each present in duplicate, wrapped by two turns of a DNA molecule. One of the basic units of eukaryotic chromosome structure. Also known as a nu body.

nucleotide A nucleoside covalently linked to a phosphate group. Nucleotides are the basic building blocks of nucleic acids. The nucleotides commonly found in DNA are deoxyadenylic acid, deoxycytidylic acid, deoxyguanylic acid, and deoxythymidylic acid. The nucleotides in RNA are adenylic acid, cytidylic acid, guanylic acid, and uridylic acid.

nucleotide pair The pair of nucleotides (A and T, or G and C) in opposite strands of the DNA molecule that are hydrogen-bonded to each other.

nucleus The membrane-bounded cytoplasmic organelle of eukaryotic cells that contains the chromosomes and nucleolus.

nullisomic Describes an individual with a chromosomal aberration in which both members of a chromosome pair are missing.

ochre codon A codon that does not code for the insertion of an amino acid into a polypeptide chain, but signals chain termination. The ochre codon is UAA.

Okazaki fragment The small, discontinuous strands of DNA produced during DNA synthesis.

operator region A region of a DNA molecule that interacts with a specific repressor protein to control the expression of an adjacent gene or gene set.

operon A genetic unit that consists of one or more structural genes (that code for polypeptides) and an adjacent operator gene that controls the transcriptional activity of the structural gene or genes.

orphon Single copies of tandemly repeated genes found dispersed in the genome. For example, histone genes are members of a multigene family present in several hundred copies clustered in a tandem array. A single copy of a histone gene found elsewhere in the genome is said to have lost its family and is regarded as an orphon.

overlapping code A genetic code first proposed by George Gamow in which any given nucleotide is shared by two adjacent codons.

pachytene The stage in meiotic prophase I when the synapsed homologous chromosomes split longitudinally (except at the centromere), producing a group of four chromatids called a tetrad.

palindrome A word, number, verse, or sentence that reads the same backward or forward (e.g., *able was I ere I saw elba*). In nucleic acids, a sequence in which the base pairs read the same on complementary strands ($5' \rightarrow 3'$). For example: $5'GAATTC3'$, $3'CTTAAG5'$. These often occur as sites for restriction endonuclease recognition and cutting.

paracentric inversion A chromosomal inversion that does not include the centromere.

parasexual Condition describing recombination of genes from different individuals which does not involve meiosis, gamete formation, or zygote production. The formation of somatic cell hybrids is an example.

parental gamete See *noncrossover gamete*.

parthenogenesis Development of an egg without fertilization.

partial diploids See *merozygote*.

partial dominance See *incomplete dominance*.

passive immunity A form of immunity produced by receipt of antibodies synthesized by another individual.

patroclinous inheritance A form of genetic transmission in which the offspring have the phenotype of the father.

pedigree In human genetics, a diagram showing the ancestral relationships and transmission of genetic traits over several generations in a family.

penetrance The frequency (expressed as a percentage) with which individuals of a given genotype manifest at least some degree of a specific mutant phenotype associated with a trait.

peptide bond The covalent bond between the amino group of one amino acid and the carboxyl group of another amino acid.

pericentric inversion A chromosomal inversion that involves both arms of the chromosome and thus involves the centromere.

permissive condition Environmental conditions under which a conditional mutation (such as a temperature-sensitive mutant) expresses the wild-type phenotype.

phage See *bacteriophage*.

phenocopy An environmentally induced phenotype (nonheritable) which closely resembles the phenotype produced by a known gene.

phenotype The observable properties of an organism that are genetically controlled.

phenylketonuria (PKU) A hereditary condition in humans associated with the inability to metabolize the amino acid phenylalanine. The most common form is caused by the lack of the liver enzyme phenylalanine hydroxylase.

phosphodiester bond In nucleic acids, the covalent bond between a phosphate group and adjacent nucleotides, extending from the 5' carbon of one pentose (ribose or deoxyribose) to the 3' carbon of the pentose in the neighboring nucleotide. Phosphodiester bonds form the backbone of nucleic acid molecules.

photoreactivation repair Light-induced repair of damage caused by exposure to ultraviolet light. Associated with an intracellular enzyme system.

pilus A filamentlike projection from the surface of a bacterial cell. Often associated with cells possessing F factors.

plaque A clear area on an otherwise opaque bacterial lawn caused by the growth and reproduction of phages.

plasmid An extrachromosomal, circular DNA molecule (often carrying genetic information) that replicates independently of the host chromosome.

platysome Term originally used in electron and X-ray diffraction studies to describe the flattened appearance of the DNA in the nucleosome core.

pleiotropy Condition in which a single mutation simultaneously affects several characters.

ploidy Term referring to the basic chromosome set or to mutiples of that set.

point mutation A mutation that can be mapped to a single locus. At the molecular level, a mutation that results in the substitution of one nucleotide for another.

polar body A cell produced at either the first or second meiotic division in females which contains almost no cytoplasm as a result of an unequal cytokinesis.

polycistronic mRNA A messenger RNA molecule that encodes the amino acid sequence of two or more polypeptide chains in adjacent structural genes.

polygenic inheritance The transmission of a phenotypic trait whose expression depends on the additive effect of a number of genes.

polymerases The enzymes that catalyze the formation of DNA and RNA from deoxynucleotides and ribonucleotides, respectively.

polymorphism The existence of two or more discontinuous, segregating phenotypes in a population.

polypeptide A molecule made up of amino acids joined by covalent peptide bonds. This term is used to denote the amino acid chain before it assumes its functional three-dimensional configuration.

polyploid A cell or individual having more than two sets of chromosomes.

polyribosome See *polysome*.

polysome A structure composed of two or more ribosomes associated with mRNA, engaged in translation. Formerly called polyribosome.

polytene chromosome A chromosome that has undergone several rounds of DNA replication without separation of the replicated chromosomes, forming a giant, thick chromosome with aligned chromomeres producing a characteristic banding pattern.

population A local group of individuals belonging to the same species, which are actually or potentially interbreeding.

position effect Change in expression of a gene associated with a change in the gene's location within the genome.

postzygotic isolation mechanism Factors that prevent or reduce inbreeding by acting after fertilization to produce nonviable, sterile hybrids or hybrids of lowered fitness.

preadaptive mutation A mutational event which later becomes of adaptive significance.

prezygotic isolation mechanism Factors that reduce inbreeding by preventing courtship, mating, or fertilization.

Pribnow box A 6-bp sequence 5′ to the beginning of transcription in prokaryotic genes, to which the sigma subunit of RNA polymerase binds. The consensus sequence for this box is TATAAT.

primary protein structure Refers to the sequence of amino acids in a polypeptide chain.

primary sex ratio Ratio of males to females at fertilization.

primer In nucleic acids, a short length of RNA or single-stranded DNA which is necessary for the functioning of polymerases.

prion An infectious pathogenic agent devoid of nucleic acid and composed mainly of a protein, PrP, with a molecular weight of 27,000 to 30,000 daltons. Prions are known to cause scrapie, a degenerative neurological disease in sheep, and are thought to cause similar diseases in humans, such as kuru and Creutzfeldt-Jakob disease.

probability Ratio of the frequency of a given event to the frequency of all possible events.

proband See *propositus*.

progeny The offspring produced from a mating.

prokaryotes Organisms lacking nuclear membranes, meiosis, and mitosis. Bacteria and blue-green algae are examples of prokaryotic organisms.

promotor site Region having a regulatory function and to which RNA polymerase binds prior to the initiation of transcription.

prophage A phage genome integrated into a bacterial chromosome. Bacterial cells carrying prophage are said to be lysogenic.

propositus (female, **proposita**) An individual in whom a genetically determined trait of interest is first detected. Also known as a proband.

protein A molecule composed of one or more polypeptides, each composed of amino acids covalently linked together.

protoplast A bacterial or plant cell with the cell wall removed. Sometimes called a spheroplast.

prototroph A strain (usually microorganisms) that is capable of growth on a defined, minimal medium. Wild-type strains are usually regarded as prototrophs.

pseudoalleles Genes that behave as alleles to one another by complementation, but that can be separated from one another by recombination.

pseudodominance The expression of a recessive allele on one homologue caused by the deletion of the dominant allele on the other homologue.

pseudogene A nonfunctional gene with sequence homology to a known structural gene present elsewhere in the genome. They differ from their functional relatives by insertions or deletions and by the presence of flanking direct repeat sequences of 10 to 20 nucleotides.

puff See *chromosome puff*.

quantitative inheritance See *polygenic inheritance*.

quantum speciation Formation of a new species within a single or a few generations by a combination of selection and drift.

quaternary protein structure Types and modes of inter-action between two or more polypeptide chains within a protein molecule.

race A phenotypically or geographically distinct subgroup within a species.

rad A unit of absorbed dose of radiation with an energy equal to 100 ergs per gram of irradiated tissue.

radioactive isotope One of the forms of an element, differ-ing in atomic weight and possessing an unstable nucleus that emits ionizing radiation.

random mating Mating between individuals without re-gard to genotype.

reading frame Linear sequence of codons (groups of three nucleotides) in a nucleic acid.

reannealing Formation of double-stranded DNA mole-cules from dissociated single strands.

recessive Term describing an allele that is not expressed in the heterozygous condition.

reciprocal cross A paired cross in which the genotype of the female in the first cross is present as the genotype of the male in the second cross, and vice versa.

reciprocal translocation A chromosomal aberration in which nonhomologous chromosomes exchange parts.

recombinant DNA A DNA molecule formed by the joining of two heterologous molecules. Usually applied to DNA molecules produced by *in vitro* ligation of DNA from two different organisms.

recombinant gamete A gamete containing a new combi-nation of genes produced by crossing over during meiosis.

recombination The process that leads to the formation of new gene combinations on chromosomes.

recon A term coined by Seymour Benzer to denote the smallest genetic units between which recombination can occur.

redundant genes Gene sequences present in more than one copy per haploid genome (e.g., ribosomal genes).

regulatory site A DNA sequence that is involved in the control of expression of other genes, usually involving an interaction with another molecule.

rem Radiation equivalent in man; the dosage of radiation that will cause the same biological effect as one roentgen of X-rays.

renaturation The process by which a denatured protein or nucleic acid returns to its normal three-dimensional structure.

repetitive DNA sequences DNA sequences present in many copies in the haploid genome.

replicating form (RF) Double-stranded nucleic acid mol-ecules present as an intermediate during the reproduc-tion of certain viruses.

replication The process of DNA synthesis.

replication fork The Y-shaped region of a chromosome associated with the site of replication.

replicon A chromosomal region or free genetic element containing the DNA sequences necessary for the initiation of DNA replication.

repressible enzyme system An enzyme or group of en-zymes whose synthesis is regulated by the intracellular concentration of certain metabolites.

repressor A protein that binds to a regulatory sequence adjacent to a gene and blocks transcription of the gene.

reproductive isolation Absence of interbreeding between populations, subspecies, or species. Reproductive isola-tion can be brought about by extrinsic factors, such as behavior, and intrinsic barriers, such as hybrid invia-bility.

resistance transfer factor (RTF) A component of R plas-mids that confers the ability for cell-to-cell transfer of the R plasmid by conjugation.

resolution In an optical system, the shortest distance be-tween two points or lines at which they can be perceived to be two points or lines.

restriction endonuclease Nuclease that recognizes spe-cific nucleotide sequences in a DNA molecule, and cleaves or nicks the DNA at that site. Derived from a va-riety of microorganisms, those enzymes that cleave both strands of the DNA are used in the construction of re-combinant DNA molecules.

restrictive condition Environmental conditions under which a conditional mutation (such as a temperature-sensitive mutant) expresses the mutant phenotype.

restrictive transduction See *specialized transduction*.

retrovirus Viruses with RNA as genetic material that uti-lize the enzyme reverse transcriptase during their life cy-cle.

reverse transcriptase A polymerase that uses RNA as a template to transcribe a single-stranded DNA molecule as a product.

reversion A mutation that restores the wild-type pheno-type.

R factor (R plasmid) Bacterial plasmids that carry anti-biotic resistance genes. Most R plasmids have two com-ponents: an r-determinant, which carries the antibiotic resistance genes, and the resistance transfer factor (RTF).

Rh factor An antigenic system first described in the rhe-sus monkey. Recessive *r/r* individuals produce no Rh an-tigens and are Rh negative, while *R/R* and *R/r* individuals have Rh antigens on the surface of their red blood cells and are classified as Rh positive.

ribonucleic acid A nucleic acid characterized by the sugar ribose and the pyrimidine uracil, usually a single-stranded polynucleotide. Several forms are recognized, including ribosomal RNA, messenger RNA, transfer RNA, and heterogeneous nuclear RNA.

ribosomal RNA See *rRNA*.

ribosome A ribonucleoprotein organelle consisting of two subunits, each containing RNA and protein. Ribosomes are the site of translation of mRNA codons into the amino acid sequence of a polypeptide chain.

RNA See *ribonucleic acid*.

RNA polymerase An enzyme that catalyzes the formation

of an RNA polynucleotide strand using the base sequence of a DNA molecule as a template.

RNase A class of enzymes that hydrolyze RNA molecules.

Robertsonian translocation A form of chromosomal aberration that involves the fusion of long arms of acrocentric chromosomes at the centromere.

roentgen A unit of measure of the amount of radiation corresponding to the generation of 2.083×10^9 ion pairs in one cubic centimeter of air at 0°C at an atmospheric pressure of 760 mm of mercury. Abbreviated R.

rRNA The RNA molecules that are the structural components of the ribosomal subunits. In prokaryotes, these are the 16*S*, 23*S*, and 5*S* molecules; and in eukaryotes, they are the 18*S*, 28*S*, and 5*S* molcules.

RTF See *resistance transfer factor.*

satellite DNA DNA that forms a minor band when genomic DNA is centrifuged in a cesium salt gradient. This DNA usually consists of short sequences repeated many times in the genome.

SCE See *sister chromatid exchange.*

secondary protein structure The alpha helical or pleated-sheet form of a protein molecule brought about by the formation of hydrogen bonds between amino acids.

secondary sex ratio The ratio of males to females at birth.

secretor An individual having soluble forms of the blood group antigens A and/or B present in saliva and other body fluids. This condition is caused by a dominant, autosomal gene unlinked to the *ABO* locus (*I* locus).

sedimentation coefficient See *Svedberg coefficient unit.*

segregation The separation of homologous chromosomes into different gametes during meiosis.

selection The force that brings about changes in the frequency of alleles and genotypes in populations through differential reproduction.

selection coefficient (*s*) A quantitative measure of the relative fitness of one genotype compared with another.

selfing In plant genetics, the fertilization of ovules of a plant by pollen produced by the same plant. Reproduction by self-fertilization.

semiconservative replication A model of DNA replication in which a double-stranded molecule replicates in such a way that the daughter molecules are composed of one parental (old) and one newly synthesized strand.

semisterility A condition in which a proportion of all zygotes are inviable.

sex chromosome A chromosome, such as the X or Y in humans, which is involved in sex determination.

sexduction Transmission of chromosomal genes from a donor bacterium to a recipient cell by the F factor.

sex-influenced inheritance Phenotypic expression that is conditioned by the sex of the individual. A heterozygote may express one phenotype in one sex and the alternate phenotype in the other sex.

sex-limited inheritance A trait that is expressed in only one sex even though the trait may not be X-linked.

sex-linkage The pattern of inheritance resulting from genes located on the X chromosome.

sex ratio See *primary* and *secondary sex ratio.*

sexual reproduction Reproduction through the fusion of gametes, which are the haploid products of meiosis.

Shine-Dalgarno sequence The nucleotides AGGAGG present in the leader sequence of prokaryotic genes that serves as a ribosome binding site. The 16*S* RNA of the small ribosomal subunit contains a complementary sequence to which the mRNA binds.

short period interspersion Pattern of genome organization in which stretches of single copy DNA (about 1000 bp) are interspersed with short segments of repetitive DNA (300 bp). This pattern is found in *Xenopus*, humans, and the majority of organisms examined to date.

shotgun experiment The cloning of random fragments of genomic DNA into a vehicle such as a plasmid or phage, usually to produce a bank or library of clones from which clones of specific interest will be selected.

sickle-cell anemia A genetic disease in humans caused by an autosomal recessive gene, usually fatal in the homozygous condition. Caused by an alteration in the amino acid sequence of the beta chain of globin.

sickle-cell trait The phenotype exhibited by individuals heterozygous for the sickle-cell gene.

sigma factor A polypeptide subunit of the RNA polymerase which recognizes the binding site for the initiation of transcription.

SINES Short interspersed repetitive sequences found in the genomes of higher organisms, such as the 300-bp *Alu* sequence.

sister chromatid exchange (SCE) A crossing over event which can occur in meiotic and mitotic cells; involves the reciprocal exchange of chromosomal material between sister chromatids (joined by a common centromere). Such exchanges can be detected cytologically after BrdU incorporation into the replicating chromosomes.

small nuclear RNA (snRNA) Species of RNA molecules ranging in size from 90 to 400 nucleotides. The abundant snRNAs are present in 1×10^4 to 1×10^6 copies per cell. snRNAs are associated with proteins and form RNP particles known as snRNPs or snurps. Six uridine-rich snRNAs known as U1–U6 are located in the nucleoplasm, and the complete nucleotide sequence of these is known. snRNAs have been implicated in the processing of pre-mRNA and may have a range of cleavage and ligation functions.

snurps See *snRNA.*

solenoid structure A level of eukaryotic chromosome structure generated by the supercoiling of nucleosomes.

somatic cell genetics The use of cultured somatic cells to investigate genetic phenomena by parasexual techniques.

somatic cells All cells other than the germ cells or gametes in an organism.

somatic mutation A mutational event occurring in a somatic cell. In other words, such mutations are not heritable.

somatic pairing The pairing of homologous chromosomes in somatic cells.

SOS response   The induction of enzymes to repair damaged DNA in *E. coli*. The response involves activation of an enzyme that cleaves a repressor, activating a series of genes involved in DNA repair.

spacer DNA   DNA sequences found between genes, usually repetitive DNA segments.

specialized transduction   Genetic transfer of only specific host genes by transducing phages.

speciation   The process by which new species of plants and animals arise.

species   A group of actually or potentially interbreeding individuals that is reproductively isolated from other such groups.

spheroplast   See *protoplast*.

spindle fibers   Cytoplasmic fibrils formed during cell division which are involved with the separation of chromatids at anaphase and their movement toward opposite poles in the cell.

spontaneous mutation   A mutation that is not induced by a mutagenic agent.

spore   A unicellular body or cell encased in a protective coat that is produced by some bacteria, plants, and invertebrates; is capable of survival in unfavorable environmental conditions; and can give rise to a new individual upon germination. In plants, spores are the haploid products of meiosis.

stabilizing selection   Preferential reproduction of those individuals having genotypes close to the mean for the population. A selective elimination of genotypes at both extremes.

standard deviation   A quantitative measure of the amount of variation in a sample of measurements from a population.

standard error   A quantitative measure of the amount of variation in a sample of measurements from a population.

sterility   The condition of being unable to reproduce; free from contaminating microorganisms.

strain   A group with common ancestry which has physiological or morphological characteristics of interest for genetic study or domestication.

structural gene   A gene that encodes the amino acid sequence of a polypeptide chain.

sublethal gene   A mutation causing lowered viability, with death before maturity in less than 50 percent of the individuals carrying the gene.

submetacentric chromosome   A chromosome with the centromere placed so that one arm of the chromosome is slightly longer than the other.

subspecies   A morphologically or geographically distinct interbreeding population of a species.

supercoiling   A form of DNA structure in which the helix is coiled upon itself. Such structures can exist in stable forms only when the ends of the DNA are not free, as in a covalently closed circular DNA molecule.

superfemale   See *metafemale*.

supermale   See *metamale*.

suppressor mutation   A mutation that acts to restore (completely or partially) the function lost by a previous mutation at another site.

Svedberg coefficient unit   A unit of measure for the rate at which particles (molecules) sediment in a centrifugal field. This unit is a function of several physico-chemical properties, including size and shape. A sedimentation value of 1×10^{-13} sec is defined as one Svedberg coefficient (S) unit.

symbiont   An organism coexisting in a mutually beneficial relationship with another organism.

sympatric speciation   Process of speciation involving populations that inhabit, at least in part, the same geographic range.

synapsis   The pairing of homologous chromosomes at meiosis.

synaptonemal complex (SC)   An organelle consisting of a tripartite nucleoprotein ribbon that forms between the paired homologous chromosomes in the pachytene stage of the first meiotic division.

syndrome   A group of signs or symptoms that occur together and characterize a disease or abnormality.

synkaryon   The nucleus of a zygote that results from the fusion of two gametic nuclei. Also used in somatic cell genetics to describe the product of nuclear fusion.

syntenic test   In somatic cell genetics, a method for determining whether or not two genes are on the same chromosome.

T_m The temperature at which a population of double-stranded nucleic acid molecules is half-dissociated into single strands. This is taken to be the melting temperature for that species of nucleic acid.

target theory   In radiation biology, a theory which states that damage and death from radiation is caused by the inactivation of specific targets within the organism.

tautomeric shift   A reversible isomerization in a molecule brought about by a shift in the localization of a hydrogen atom. In nucleic acids, tautomeric shifts in the bases of nucleotides can cause changes in other bases at replication and are a source of mutations.

telocentric chromosome   A chromosome in which the centromere is located at the end of the chromosome.

telomere   The terminal chromomere of a chromosome.

telophase   The stage of cell division in which the daughter chromosomes reach the opposite poles of the cell and reform nuclei. Telophase ends with the completion of cytokinesis.

temperate phage   A bacteriophage that can become a prophage and confer lysogeny upon the host bacterial cell.

temperature-sensitive mutation   A conditional mutation that produces a mutant phenotype at one temperature range and a wild-type phenotype at another temperature range.

template   The single-stranded DNA or RNA molecule that specifies the nucleotide sequence of a strand synthesized by a polymerase molecule.

teratocarcinoma   Embryonal tumors that arise in the yolk sac or gonads and are able to undergo differentiation into a wide variety of cell types. These tumors are used to investigate the regulatory mechanisms underlying development.

terminalization The movement of chiasmata toward the ends of chromosomes during the diplotene stage of the first meiotic division.

tertiary protein structure The three-dimensional structure of a polypeptide chain brought about by folding upon itself.

test cross A cross between an individual whose genotype at one or more loci may be unknown and an individual who is homozygous recessive for the genes in question.

tetrad The four chromatids that make up paired homologues in the prophase of the first meiotic division. The four haploid cells produced by a single meiotic division.

tetrad analysis Method for the analysis of gene linkage and recombination using the four haploid cells produced in a single meiotic division.

tetranucleotide theory An early theory of DNA structure which proposed that the molecule was composed of repeating units, each consisting of the four nucleotides adenosine, thymidine, cytosine, and guanine.

tetraparental mouse A mouse produced from an embryo that was derived by the fusion of two separate blastulas.

thymine dimer A pair of adjacent thymine bases in a single polynucleotide strand between which chemical bonds have formed. This lesion, usually the result of damage caused by exposure to ultraviolet light, inhibits DNA replication unless repaired by the appropriate enzyme system.

topoisomerase A class of enzymes that convert DNA from one topological form to another. During DNA replication, these enzymes facilitate the unwinding of the double-helical structure of DNA.

totipotent The ability of a cell or embryo part to give rise to all adult structures. This capacity is usually progressively restricted during development.

trailer sequence A transcribed but nontranslated region of a gene or its mRNA that follows the termination signal.

trait Any detectable phenotypic variation of a particular inherited character.

***trans* configuration** The arrangement of two mutant sites on opposite homologues, such as

$$\frac{a^1 \; +}{+ \; a^2}$$

Contrasts with a *cis* arrangement, where they are located on the same homologue.

transcription Transfer of genetic information from DNA by the synthesis of an RNA molecule copied from a DNA template.

transdetermination Change in developmental fate of a cell or group of cells.

transduction Virally mediated bacterial recombination.

transfer RNA See *tRNA*.

transformation Heritable change in a cell or an organism brought about by exogenous DNA.

transition A mutational event in which one purine is replaced by another, or one pyrimidine is replaced by another.

translation The derivation of the amino acid sequence of a polypeptide from the base sequence of an mRNA molecule in association with a ribosome.

translocation A chromosomal mutation associated with the transfer of a chromosomal segment from one chromosome to another. Also used to denote the movement of mRNA through the ribosome during translation.

transposable element A defined length of DNA that translocates to other sites in the genome, essentially independent of sequence homology. Usually such elements are flanked by short, inverted repeats of 20 to 40 base pairs at each end. Insertion into a structural gene can produce a mutant phenotype. Insertion and excision of transposable elements depends on two enzymes, transposase and resolvase. Such elements have been identified in both prokaryotes and eukaryotes.

transversion A mutational event in which a purine is replaced by a pyrimidine, or a pyrimidine is replaced by a purine.

triploidy The condition in which a cell or organism possesses three haploid sets of chromosomes.

trisomy The condition in which a cell or organism possesses two copies of each chromosome, except for one, which is present in three copies. The general form for trisomy is therefore $2n + 1$.

tritium (^{3}H) A radioactive isotope of hydrogen, with a half-life of 12.46 years.

tRNA Transfer RNA; a small ribonucleic acid molecule that contains a three-base segment (anticodon) that recognizes a codon in mRNA, a binding site for a specific amino acid, and recognition sites for interaction with the ribosome and the enzyme that links it to its specific amino acid.

Turner syndrome A genetic condition in human females caused by a 45,X genotype (XO). Such individuals are phenotypically female but are sterile because of undeveloped ovaries.

unequal crossing over A crossover between two improperly aligned homologues, producing one homologue with three copies of a region and the other with one copy of that region.

unique DNA DNA sequences that are present only once per genome. Single copy DNA.

unwinding proteins Nuclear proteins that act during DNA replication to destabilize and unwind the DNA helix ahead of the replicating fork.

up promoter A promoter sequence, often mutant, that increases the rate of transcription initiation. Also known as strong promoter.

variable region Portion of an immunoglobulin molecule that exhibits many amino acid sequence differences between antibodies of differing specificities.

variance A statistical measure of the variation of values from a central value, calculated as the square of the standard deviation.

variegation Patches of differing phenotypes, such as color, in a tissue.

vector In recombinant DNA, an agent such as a phage or plasmid into which a foreign DNA segment will be inserted.

viability The measure of the number of individuals in a given phenotypic class that survive, relative to another class (usually wild type).

virulent phage A bacteriophage that infects and lyses the host bacterial cell.

W, Z chromosomes Sex chromosomes in species where the female is the heterogametic sex (WZ).

wild type The most commonly observed phenotype or genotype, designated as the norm or standard.

wobble hypothesis An idea proposed by Francis Crick which states that the third base in an anticodon can align in several ways to allow it to recognize more than one base in the codons of mRNA.

X inactivation In mammalian females, the random cessation of transcriptional activity of one X chromosome. This event, which occurs early in development, is a mechanism of dosage compensation. Molecular basis of inactivation is unknown, but loci on the tip of the short arm of the X can escape inactivation. See *Barr body, Lyon hypothesis.*

X linkage See *sex linkage.*

X-ray crystallography A technique to determine the three-dimensional structure of molecules through diffraction patterns produced by X-ray scattering by crystals of the molecule under study.

Y chromosome Sex chromosome in species where the male is heterogametic (XY).

Y linkage Mode of inheritance shown by genes located on the Y chromosome.

Z-DNA An alternate structure of DNA in which the two antiparallel polynucleotide chains form a left-handed double helix. Z-DNA has been shown to be present along with B-DNA in chromosomes and may have a role in regulation of gene expression.

zein Principal storage protein of corn endosperm, consisting of two major proteins, with molecular weights of 19,000 and 21,000 daltons.

zygote The diploid cell produced by the fusion of haploid gametic nuclei.

zygotene A stage of meiotic prophase I in which the homologous chromosomes synapse and pair along their entire length, forming bivalents. The synaptonemal complex forms at this stage.

NAME INDEX

SUBJECT INDEX